APPLIED BIOMECHANICS

SECOND EDITION

CONCEPTS AND CONNECTIONS

John McLester, PhD
Professor
Department of Exercise Science and Sport Management
WellStar College of Health and Human Services
Kennesaw State University
Kennesaw, GA

Peter St. Pierre, PhD
Associate Professor
Department of Health Promotion and Physical Education
WellStar College of Health and Human Services
Kennesaw State University
Kennesaw, GA

JONES & BARTLETT
LEARNING

World Headquarters
Jones & Bartlett Learning
5 Wall Street
Burlington, MA 01803
978-443-5000
info@jblearning.com
www.jblearning.com

Jones & Bartlett Learning books and products are available through most bookstores and online booksellers. To contact Jones & Bartlett Learning directly, call 800-832-0034, fax 978-443-8000, or visit our website, www.jblearning.com.

Substantial discounts on bulk quantities of Jones & Bartlett Learning publications are available to corporations, professional associations, and other qualified organizations. For details and specific discount information, contact the special sales department at Jones & Bartlett Learning via the above contact information or send an email to specialsales@jblearning.com.

17016-0

Production Credits
VP, Product Management: Amanda Martin
Director of Product Management: Cathy L. Esperti
Product Manager: Sean Fabery
Product Assistant: Andrew LaBelle
Project Specialist: Nora Menzi
Project Specialist: Rachel DiMaggio
Digital Products Specialist: Angela Dooley
Director of Marketing: Andrea DeFronzo
VP, Manufacturing and Inventory Control: Therese Connell
Composition and Project Management: Exela Technologies
Cover Design: Kristin E. Parker
Text Design: Michael O'Donnell
Rights & Media Specialist: John Rusk
Media Development Editor: Troy Liston
Cover Image (Title Page, Chapter Opener): © technotr/Getty Images
Printing and Binding: LSC Communications
Cover Printing: LSC Communications

Library of Congress Cataloging-in-Publication Data
Names: McLester, John, author. | St. Pierre, Peter, author.
Title: Applied biomechanics : concepts and connections / John McLester and Peter St. Pierre.
Description: 2. | Burlington, MA : Jones & Bartlett Learning, [2020] | Includes bibliographical references and index.
Identifiers: LCCN 2018042621 | ISBN 9781284170047 (pbk. : alk. paper)
Subjects: | MESH: Biomechanical Phenomena | Movement–physiology | Sports–physiology
Classification: LCC QH513 | NLM WE 103 | DDC 612.76–dc23 LC record available at https://lccn.loc.gov/2018042621

6048

Printed in the United States of America
23 22 21 20 19 10 9 8 7 6 5 4 3 2 1

DEDICATION

To my wife Cherilyn, beautiful, loving, intelligent, and strong
To my daughter Reagan, just like her mother
To my mother Carolyn, an angel on Earth
To my father Richard, one of the last true men I ever knew
To my sisters Richie, Tanya, and Mytesa, you shaped my life in ways that you will never know

—John McLester

To Martha and John Zocchi, mentors and role models
To Kimberly Fletcher, my muse and rock

—Peter St. Pierre

BRIEF TABLE OF CONTENTS

TABLE OF CONTENTS

PREFACE

In writing the first edition of *Applied Biomechanics: Concepts and Connections*, we tried to keep in mind that all-too-important question, "When will I ever use this again?" As two professors at the beginning of our careers, we have been on both sides of the classroom in dealing with this question. As teachers observing our own students, we see the same quizzical looks we gave when we were students. An exercise scientist and a teacher educator shared a common problem: How do we make biomechanics content more relevant for our undergraduate students across multiple disciplines? Through several long conversations, the essence of this text evolved.

The goal of this text is to address the unfortunate fact that much of education is compartmentalized. All too often, individual courses within a curriculum are presented with little regard for other related courses required of students both within and outside a specific program. We believe that understanding the interconnectedness of *all* information is fundamental to the ability to think critically and to synthesize information. Education can be thought of as trying to put links together to form a chain. One individual link is rigid and not especially useful by itself, but each added link allows for greater and greater mobility. Therefore, the goal of this text is to make relevant connections between the physics of human movement and their application to related topics of study.

OBJECTIVES

This text has two major objectives: (1) to provide a clear understanding of the topics in the field of biomechanics and (2) to relate those topics to other fields of study. To meet our objectives, we have constructed each chapter with a Concepts section and a Connections section. The Concepts are the core, or "nuts and bolts," of understanding the mechanics of movement. The Connections are designed to show how the Concepts are used in the many diverse areas within the movement sciences. The text was written to meet the needs of students who aspire to any field of human movement or performance, making the concepts of biomechanics relevant no matter their ultimate choice of profession.

THE STRUCTURE OF THE TEXT

The order of topics introduced is one of the most difficult decisions in constructing a text because there simply is no way to determine what is best for everyone. However, after teaching biomechanics for several years, I discovered a sequence of teaching the course that students seem to follow with relative ease. The reader will find that some aspects of the order are traditional and that other sequencing choices are novel. Overall, each chapter builds on elements of all previous chapters. However, teachers will find that each chapter can stand alone, with only a minimal amount of preface required, and some can be omitted altogether. This design enables great flexibility in adapting the course to meet the specific needs of a particular curriculum. The rationale for the chosen sequence of chapters follows.

Chapter 1 is designed to introduce students to both the text and the discipline. Students will learn the special features of the text and the ways in which its design will enhance their understanding of biomechanics and its connection to other fields. In addition, students will be introduced to terminology used in the various disciplines that are discussed throughout the text—for example, adapted motion, biomechanics, exercise physiology, kinesiology, kinetics, kinematics, motor control, motor development, motor learning, and pedagogy. By providing a clear definition of each discipline and its related terminology, this chapter enhances students' understanding of the connections between the disciplines.

Chapter 2 gives students the tools necessary to describe the motion of a system, the location of the system within the environment, and the type of motion exhibited. In this chapter, students begin to look at motion in an entirely different way. They have to pay attention to aspects of the motion that they perhaps have ignored to this point. The placement of this chapter is important because it provides examples that lead students to classify several movements, but it engages them in a relatively low level of movement analysis. Higher-level movement analysis is covered immediately in the next chapter.

Chapter 3 introduces students to various qualitative and quantitative methods of studying and analyzing motion. Qualitative methods are introduced before quantitative ones so students can better understand how to "look at" or "see" movement before analyzing it technically. Within the quantitative methods, graphical methods are presented before trigonometric methods so students can once again practice the skill of "visualizing" forces before learning the trigonometric methods on which the graphical methods are based.

Chapter 4 further elucidates the concept of force and introduces the various forces both encountered by, and acting within, the system. Students should gain some understanding of the implications of exposure to these forces. The forces covered in this chapter are explained in greater detail in later chapters to further student knowledge of force application. Students are also introduced to Newton's laws for the first time. More detailed coverage of Newtonian laws comes in later chapters to show application and interrelationship.

In Chapter 5, attention is focused specifically upon linear motion. For example, students will gain a deeper understanding of velocity and acceleration, and of the influence of gravity on the system. The concepts of kinematics are covered before kinetics, because kinematic equations and terminology are helpful in explaining kinetic concepts to the student. Newtonian laws are described in greater detail, especially the relationships among force, mass, and acceleration. Concepts of linear motion are discussed in terms of their relationship to Newton's laws and energy transfer.

Chapter 6 is dedicated to angular motion of the system. This chapter is critical to students' understanding of biomechanics because angular motion is present during any movement of the musculoskeletal system and during almost all sports activities. The concept of torque is introduced, and Newtonian laws are discussed in relationship to angular motion. Topics in this chapter are developed in a format similar to that of Chapter 5. So, once again, the concepts of kinematics are covered first to enhance students' understanding of kinetics. The concepts in this chapter are discussed as analogs of topics in the previous chapter on linear motion. In this way, the student will realize there is a theme to physical laws and mathematical derivations.

Many important concepts such as force, gravity, torque, and center of gravity are introduced in Chapters 1 through 6. In Chapter 7, these concepts are applied to situations in which stability and balance are of the utmost importance. Balance and stability are important concepts in sports situations as well as in activities of daily living. Understanding stability requires a base knowledge of both linear and rotary concepts, and stability is therefore covered after Chapters 5 and 6.

With prerequisite information covered in previous chapters (e.g., linear versus angular kinematics, torque, and stability), the student is introduced to machines in Chapter 8. The basic properties of machines are introduced, and levers, wheel-axle arrangements, and pulley systems are then covered in detail, as well as the musculoskeletal configurations that act as machines within the human body.

By Chapter 9, students will have an in-depth understanding of linear motion, angular motion, equilibrium, and machines. With these concepts in mind, it is easier to comprehend concepts that apply when the system moves through fluids. For example, rotary motion through a fluid causes a curvilinear path. One must first understand rotary motion before understanding the behavior of objects as they move through fluids. Drag and lift are more fully elucidated in the context of knowledge that students have gained since these topics were introduced in Chapter 4.

In Chapter 10, projectiles are introduced. Projectiles are subject to linear, rotary, and fluid forces. Therefore, their flight can only be fully understood at this point in the text. Students will gain an understanding of the mechanics of projecting an object for vertical and horizontal distance as well as for accuracy. This chapter begins with a review of previously covered concepts that relate to projectiles and then elucidates the mechanisms by which these factors affect projection.

Chapter 11 is designed to help students understand biomechanics more deeply by applying previously learned concepts to the musculoskeletal system. Muscle physiology and contraction are covered in detail. In addition, the biomechanical implications of muscle location, shape, and design are covered, as well as how muscles work together to produce movement and reduce injury. The chapter applies many concepts previously covered in the text to help students realize that the human body is the result of many interrelated biomechanical principles.

Chapter 12 includes a detailed analysis of skill in one sport—golf—using concepts from throughout the text; a second analysis of recent research relates to the sport of soccer and connects with concepts in this text. A final analysis is presented at this point because students now know more specifically the aspects of movement on which they should focus. In addition, the chapter serves as a review of all previous concepts and shows the integrated nature of the field of biomechanics. Though any sport could have been used for this chapter, golf was chosen because elements of every chapter must be considered in this sport.

NEW TO THE *SECOND EDITION*

Notable updates for the *Second Edition* include the following:

- A Workbook has been created for this edition to allow students more practice with, and application of, the material outside of the classroom.
- Focus on Research sections now feature more recent research studies. The studies in this edition span a wide range of topics, many of which are discussed on a daily basis in our field. There is also research on areas that might be unexpected, such as the wearing of high heels. Also included is research on injuries that are of concern in sport such as ACL tears, football uniform materials and potential traumatic brain injury, and Tommy John surgery. The latest research on prosthetics is included, as well as barefoot running.
- The concluding chapter now includes an analysis of soccer in addition to the detailed analysis of golf that was included in the previous edition.
- Learning Objectives have been added at the beginning of each chapter.
- References and suggested readings have been updated throughout the text.
- The art package for this edition has been greatly updated and improved.

OPTIMUM USE OF THIS TEXT

The importance of understanding the interrelated nature of the discipline of biomechanics cannot be overstated. Therefore, the underlying theme of this text is to demonstrate biomechanical connections to other fields of study. Only through an enhanced understanding of the connections between disciplines can students fully appreciate biomechanics. Ideally, that appreciation will motivate students of movement to learn more and enable them to fully comprehend how useful this deeper understanding can be.

The title of this text contains several key words other than *Biomechanics: Applied, Concepts,* and *Connections. Applied* is the first word, and application is one of the major goals of this text. However, students should realize that, in any course, some content may not be directly useful but will be helpful for understanding other course content. Therefore, throughout this text each Concept is not only demonstrated through examples of application but also connected to other concepts both in this text and in other disciplines.

In fact, those Connections are one of the most important features of this text. Each biomechanical concept is introduced and subsequently demonstrated through applied examples and connections to previous concepts. At the end of each chapter is a Connections section in which the concepts from the chapter are connected to the disciplines of exercise physiology, injury science, motor behavior, pedagogy, and adapted movement. These connections provide a foundation of knowledge for teaching and applying the concepts to promote enhanced performance and are demonstrated through practical examples that relate the biomechanical concepts to motor skills used in fundamental movements, sports, dance, and recreation.

Understanding the biomechanics of the human movement in isolation is possible, but for students in any movement-related field it is more relevant when combined with other areas that use common principles to help students, athletes, and those who may be injured to learn or relearn sport and life skills. The human body is a collection of systems that must work together to perform motor skills. We hope this text will help students understand that the study of human movement must take into account all of these systems and the mechanisms by which they work together.

FEATURES

The educational elements of this text have evolved since its initial conception. We began with the premise of Concepts and Connections. Over the course of writing this text, many important features were developed in response to excellent suggestions from reviewers and the fantastic team at Jones & Bartlett Learning. The following are the resulting pedagogical features of this text.

Learning Objectives at the beginning of each chapter focus students on key concepts and the material they will learn.

The **Concepts** section of each chapter is written to provide a clear understanding of the basic topics of the field of biomechanics. Within each Concepts section, Newtonian physics have been included and applied in as many situations as possible. We have also attempted to relate each individual topic to the other topics in that Concepts section. Further, each Concepts section of a given chapter refers to elements of other chapters to enhance students' overall understanding of the field of biomechanics.

A **Connections** section is included at the end of each chapter with the goal of enhancing students' understanding of the ways in which other movement-related disciplines apply the Concepts of that chapter. Not every discipline could be included in every Connections section, so we attempted to provide a wide range of connections throughout the text. Connections are made most directly to the following fields: exercise physiology, motor behavior (motor control, motor development, and motor learning), ergonomics, injury sciences (physical therapy and sports medicine), pedagogy, adapted motion, and sport science.

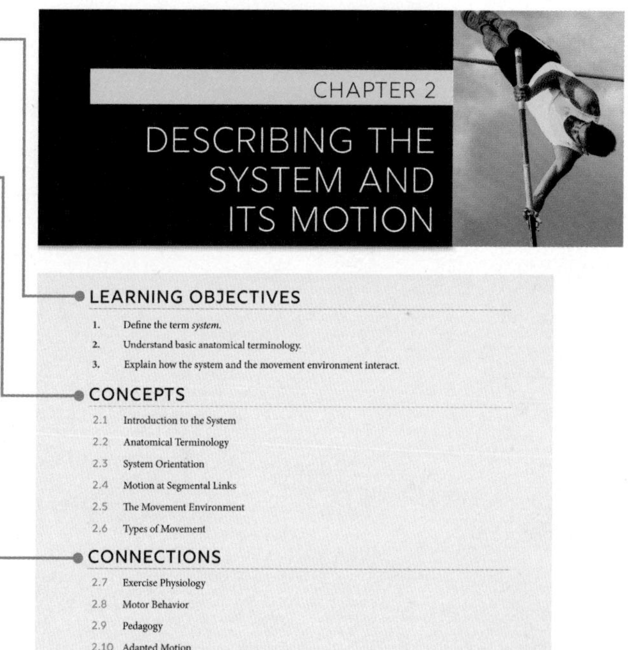

CHAPTER 2
DESCRIBING THE SYSTEM AND ITS MOTION
LEARNING OBJECTIVES
1. Define the term *system*.
2. Understand basic anatomical terminology.
3. Explain how the system and the movement environment interact.
CONCEPTS
2.1 Introduction to the System
2.2 Anatomical Terminology
2.3 System Orientation
2.4 Motion at Segmental Links
2.5 The Movement Environment
2.6 Types of Movement
CONNECTIONS
2.7 Exercise Physiology
2.8 Motor Behavior
2.9 Pedagogy
2.10 Adapted Motion

2. **mesomorphic**, being muscular, strong, and possessing weight relatively proportional to height, which is also called *athletic*; and
3. **endomorphic**, which is rounder and relatively heavy for height.

A person does not have to fit perfectly into any given category but can possess varying degrees and combinations of traits from all three categories. As previously mentioned, a quantitative (points-based) system exists for classifying the degree to which a person possesses qualities of each category. However, most people tend to use this system in a more qualitative manner by visually categorizing a person. Even qualitatively, it can be assumed that most athletes are going to have a significant mesomorphic component blended with one of the other categories, depending on the sport. For example, an offensive lineman in football could be considered to have qualities of both the mesomorphic and endomorphic categories. In contrast, a combination of the ectomorphic and mesomorphic categories better describes a wrestler in a light weight class. In which category do you think a sumo wrestler would fall?

Rather than classifying or describing the entire body, biomechanical analysis is often more productive if we look at the anthropometrics of various individual parts of the body in relation to each other. Learning to pay close attention to individual body segments is important because of the variations in angular kinematics produced by varying limb proportions.

Body Segment Proportions

The ratio of waist to hip circumferences is another anthropometric measure that is often used to assess health risk. A person with a larger ratio (more "apple shaped") is at higher risk for disease than a person with a lower ratio ("pear shaped"). Even though the **waist-to-hip ratio** is associated with various disease states, the student of biomechanics should be aware of the difference in location of the whole body center of gravity that accompanies the change in distribution of weight from either above or below the waist. Consider the injured soccer player. What is the likely difference in waist-to-hip ratio for a female versus a male soccer player? Is this piece of information relevant in terms of the injury?

Many comparisons of system body-segment parameters are associated with various aspects of performance. One such comparison is that of limb proportion in the **crural index (CI)**:

$$CI = \frac{\text{length of the tibia}}{\text{length of the femur}} \times 100 \qquad (2.3)$$

As you can see from the equation, an animal with a large CI possesses a long distal segment (segment farther from the body) in proportion to the proximal segment (segment closer to the body) (Figure 2.3). As you will discover in later chapters, this quality provides the animal with an advantage in terms of running and jumping.

Crural index is just one example of segmental comparison. You can find other advantageous trends within sports such as long arms in successful baseball pitchers and high torso-to-leg-length ratios in swimmers (i.e., two people of the same height do not necessarily have the same leg length). As previously mentioned, all of these anthropometric variables have some biomechanical influence on a person's ability to succeed within a given sport.

Having looked at some ways of evaluating the system itself and its characteristics, we can turn to the movement situation and the environment in which the system is moving. In following sections of this chapter, we will learn how to clearly define the movement space. Here we also begin

Mesomorphic Somatotype described as being muscular, strong, and possessing weight relatively proportional to height.

Endomorphic Somatotype described as being rounder and relatively heavy for height.

Waist-to-hip ratio Ratio of waist circumference to hip circumference often associated with disease risk.

Crural index Ratio of the length of the tibia to the length of the femur.

Key Terms appear in boldface type the first time they are mentioned. Their definitions also appear in the margins near the relevant discussion, making it easy for students to review.

injury or muscular imbalance at one segment or link in the chain can force other links to adapt their motion. These **compensatory movements** are adaptations at normal kinetic chain links as a result of abnormal motion at another link so that a task can still be accomplished. Awareness of compensatory movement is important for two reasons: (1) the task will now probably require more energy than before because it is no longer being carried out with use of the most efficient movement pattern (i.e., the pattern that minimizes muscle force requirements); and (2) changes in force-loading patterns on the compensating links can eventually lead to musculoskeletal abnormalities in other areas of the chain. In other words, compensatory motion arising from injury or muscular imbalance not only makes activities more difficult but also will possibly result in other injuries.

Compensatory movements Adaptations at normal kinetic chain links as a result of abnormal motion at another link.

FOCUS ON RESEARCH

As we proceed through the textbook, we hope it will become clear that the concepts in biomechanics are surprisingly related to many aspects of our daily lives. For example, millions of women wear high heels on a daily basis probably without the term *biomechanics* ever coming to mind. However, the relationship between wearing high heels, muscle activity, and dynamics of human walking was examined in a study by Simonsen et al. (2012).

First, we must explain several pieces of research equipment that are frequently used to study many aspects of biomechanics. One common tool for studying biomechanics is three-dimensional motion capture (3D MOCAP for short). The 3D MOCAP process records the motion of the system of interest using multiple cameras (often with high frame rates), each having a different view. Analyzing the videos via special software allows re-creation of the motion in a virtual 3-D Cartesian coordinate system and then the calculation of many static and dynamic parameters. Next, force platforms or force plates are instruments often recessed into the floor that are capable of measuring the forces exerted by a system moving on or across them. The force plates can measure the forces in all three directions of the Cartesian coordinate system discussed earlier. Simultaneous use of MOCAP with the force plates allows for calculation of many aspects of statics and dynamics that will be discussed throughout the book. Finally, electromyography (EMG) is the process of using a special instrument with electrodes to record the electrical signals generated by muscle fibers when activated. The signals can then be analyzed to evaluate muscle activation level, abnormalities, recruitment parameters, and so on.

Simonsen et al. (2012) used all of the just cited equipment to examine muscle activity and walking dynamics while wearing high heels (9 cm). Dynamics of the hip, knee, and ankle were measured using MOCAP along with force plates. EMG was used to measure activity of the following muscles of the leg: soleus, gastrocnemius medialis, tibialis anterior, vastus medialis, vastus lateralis, rectus femoris, biceps femoris, and semimembranosus. The researchers found that in the first half of the stance phase of walking on high heels, knee extensor moment peak was doubled. It was also found that the knee joint was flexed significantly more in the first half of the stance phase. A significant increase in knee and hip joint abductor moments in the frontal plane was also observed. Those differences, along with increases in several EMG parameters, led the authors to speculate about walking on high heels. The researchers conclude that the large increased knee joint extensor moment could be caused by the observed increased knee joint flexion during the stance phase along with increased EMG in the quadriceps muscle. Along with the significant increase in knee joint

Focus on Research modules within each chapter introduce students to research in the field of biomechanics. Research topics were chosen not only for their applicability to the Concepts and Connections within that chapter but also for the novelty of the experiment adding to the existing body of knowledge. Students should be aware that experimental research in the field of biomechanics requires skill on the part of the researcher and complex technological instrumentation. Therefore, each experiment presented in the Focus on Research sections has been greatly summarized to enhance understanding by a wide audience of readers.

horizontal component (also called *parallel*). Later, we will be more specific about the terminology of vector components.

We should mention before we begin that a vector may possess only one of these components. For example, if force is applied perfectly perpendicular to the ground (orthogonal to the *x*-axis), only a vertical component exists (i.e., all of the force is in the positive *y* direction). So when resolving a vector into its components, remember that the magnitude of one of the components may be equal to zero.

SAMPLE PROBLEM

Depending on your point of view, there are two ways to achieve our goal of resolving a vector into its vertical and horizontal components. Again, we will use examples of both to fully demonstrate the process. We begin with resolution using *vector parallelograms*. In other words, we are starting with one vector and would like to draw a parallelogram in which it fits corner to corner. Remember that the answer will always be correct if we follow some simple steps.

The first step is to draw the given vector to the established scale, being sure that it has the correct direction and orientation. Before proceeding to step two and while looking at the vector, remember the goal: We want to draw a parallelogram around our vector.

The corners of our parallelogram are defined by the tip and tail of the given vector (Figure 3.17). Also recall that the components for which we are solving (vertical and horizontal) are perpendicular to each other, so the parallelogram we are forming is always a square or rectangle. So our second step is to draw the sides of our parallelogram that represent the horizontal component. Do not be afraid to draw the lines longer than you know the final components will be (it actually helps).

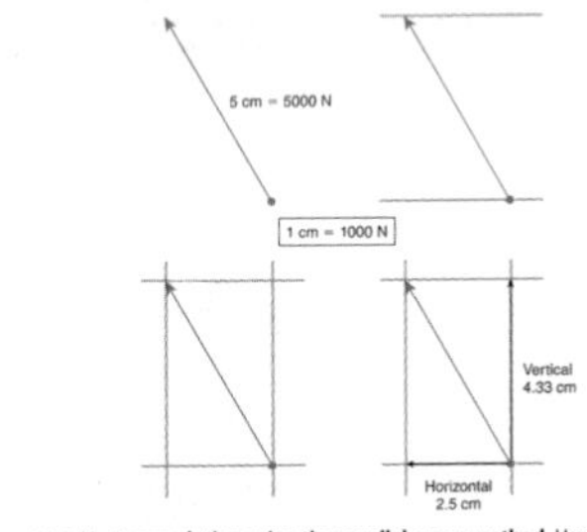

Figure 3.17 **Vector resolution using the parallelogram method.** Use your hands to define the tips of the methods.

Throughout each chapter the student is introduced to the mathematical equations that will enhance understanding of the overall concepts. Some equations are relatively straightforward and require only a brief explanation and problem-solving sample. In other cases, students are "walked through" **Sample Problems** that may be more complex or especially important to understanding the material in that chapter.

In addition, **Review Questions** and **Practice Problems** are included at the end of each chapter to reinforce student comprehension of the material and basic problem-solving skills. Answers to odd-numbered Practice Problems are included as part of the appendices.

Qualitative analysis of motion describes how the human body "looks" on visual inspection as it performs skills, including its position in space, the position of body parts relative to each other, and in some cases the position of *segments* of body parts in relation to each other. Two general approaches to qualitative analysis are the composite approach and the component approach. The former views the whole body as a system that progresses through stages or phases as it refines movement patterns; the latter approach uses the same phase and stage method but, rather than looking at the whole body as a global system, breaks it down into component sections, with each section progressing through more refined steps toward mature movement patterns.

Quantitative methods of motion analysis arise from the need to further elucidate mechanisms underlying observed events or a requirement for highly precise numerical data. Although advanced software programs are often used to analyze motion data, the theoretical base for quantitative motion analysis can be understood through graphical and trigonometric methods of vector analysis.

REVIEW QUESTIONS

1. When can quantitative measures provide valuable information for physical education teachers?
2. What are "critical features" of a skill?
3. What is the most common qualitative assessment tool used by teachers?
4. What types of qualitative assessment tools are used by both physical therapists and athletic trainers?
5. In what units are mass and weight measured? Why are they not the same units?
6. Where are some places that you could go where your weight would change, but your mass would stay the same?
7. Name the four properties of a force and give brief definitions of each.
8. a. What is the difference between a scalar and a vector?
   b. Which one is mass?
   c. Which one is weight?
9. a. What does the length of a vector represent?
   b. The arrow point?
   c. The tail?
10. a. A muscle force can be resolved into which two components?
    b. What function does each of these perform?
11. Explain the Q angle, its normal values, and the consequences for an abnormally large value.

PRACTICE PROBLEMS

1. What are the values of the two muscle force vector components in the following problems? Interpret the answers to this question. 1 cm = 100 N.

CHAPTER 1

DISCIPLINES

NAME

SECTION DATE

1.1 List some other courses within your major to which you think biomechanics might apply. In what way do you think they are related?

1.2 List three occupations in which you are potentially interested and how biomechanics might be used in that occupation.

New to this edition, a perforated **Workbook** has been included at the end of the text for additional student practice and application.

SUPPLEMENTS

In writing *Applied Biomechanics: Concepts and Connections, Second Edition*, we have provided the following resources for instructors to use in conjunction with the text.

- Test Bank, containing more than 300 multiple choice, true/false, short answer, and essay questions
- Slides in PowerPoint format, featuring more than 450 slides
- Image Bank, collecting photographs and illustrations that appear in the text
- Answer Key, documenting solutions to the in-text Review Questions, Practice Problems, and Workbook

ACKNOWLEDGMENTS

This text has been a part of our lives for several years, so I would be remiss if I did not first express by deepest love and gratitude to my wife, Cherilyn, and my daughter, Reagan. I also wish to acknowledge the many mentors who guided me and shaped my career so profoundly: Dr. John Hammett, Dr. Phillip Bishop, Dr. Mark Richardson, Dr. Joe Smith, and Dr. Keith Tennant. I would also like to express my gratitude to two fantastic department chairs who understood the time commitment involved with writing a text and provided the encouragement to follow through: Dr. Thad Crews and Dr. Mitchell Collins. Two true friends and collaborators who were patient enough to allow many manuscripts to sit idle during the writing of this work and must also be acknowledged: Dr. Matt Green and Dr. Scott Lyons. Finally, I would like to thank Dr. Peter St. Pierre, a friend and colleague who trusted me enough to join in the writing of this text. Thanks to all of you!

—John McLester

A work like this is a major undertaking that requires support and motivation. I will always be thankful for the guidance of my mentors at the University of New Hampshire and the University of Georgia. I would also like to thank Mark Smith and Tina Hall, two colleagues who continue to inspire my pursuit of excellence in teaching. A special thank you is extended to Dr. John McLester, who invited me to join him in this project.

We would like to give a special thanks to Andrew LaBelle and the team at Jones & Bartlett Learning.

—Peter St. Pierre

ABOUT THE AUTHORS

John McLester received his doctorate at the University of Alabama specializing in kinesiology under the mentorship of Dr. Phillip Bishop. Dr. McLester is currently a Professor in the Department of Exercise Science and Sport Management at Kennesaw State University, Georgia. He taught and performed research at the University of West Georgia from 2000 to 2002 and at Western Kentucky University from 2002 to 2005. Dr. McLester's research interests include physiological and biomechanical relationships.

Peter St. Pierre received his doctorate at the University of Georgia, specializing in teacher education. Dr. St. Pierre is currently an Associate Professor in the Department of Health Promotion and Physical Education at Kennesaw State University, Georgia. Dr. St. Pierre is responsible for the preparation of future health and physical education teachers. Dr. St. Pierre's research interests include expertise in teaching and coaching.

REVIEWERS

We would like to acknowledge the following scholars. Your criticisms, ideas, and suggestions were invaluable in creating and refining this text. We sincerely thank you.

Harish Chander, PhD
Assistant Professor
Department of Kinesiology
College of Education
Mississippi State University
Starkville, MS

Alfred Finch, PhD, HFI, FISBS, DCT
Professor
Department of Kinesiology, Recreation, and Sport
College of Health and Human Services
Indiana State University
Terre Haute, IN

David H. Fukuda, PhD
Assistant Professor and Division Head
Division of Kinesiology
College of Health Professions and Sciences
University of Central Florida
Orlando, FL

Mark Geil, PhD
Professor and Chair
Department of Kinesiology and Health
Center for Research on Atypical Development and Learning
Georgia State University
Atlanta, GA

Chad Smith, PhD
Assistant Professor
Department of Kinesiology
College of Science
Coastal Carolina University
Conway, SC

CHAPTER 1

BIOMECHANICS AND RELATED MOVEMENT DISCIPLINES

© technotr/Getty Images

LEARNING OBJECTIVES

1. Understand the discipline of biomechanics.
2. Understand the relationship of biomechanics to related movement disciplines.

1.1 BENEFITS OF A COMPREHENSIVE UNDERSTANDING OF BIOMECHANICS

Have you ever noticed that all of your limbs taper? What if they were not tapered? Does tapering have anything to do with how birds fly? For that matter, how does a boomerang fly? Why does a golf ball rise? Why does a golf ball have all of those dimples? What do those dimples have to do with swimmers? Is bat speed really the most important factor in hitting home runs? Is a "rising" fastball more difficult to hit because it actually rises? Does that fact have any relationship to corner kicks in soccer?

Through a comprehensive understanding of biomechanics, we can actually answer all of these questions. Sometimes we ask ourselves such questions simply because we would like to know, but usually we just dismiss them. The amazing thing is that many of the questions we ask out of curiosity often have some relationship to each other, but we may never make that connection. The ability to make those connections can further our understanding of the movement-related sciences, which in turn makes us more skilled practitioners. Putting practice aside, knowledge of biomechanics aids us in understanding what it is to be a human. Comprehending the connection between a human and the environment with which the human interacts can be deeply fulfilling. But comprehending is the difficult part; the connections are not always obvious. To fully appreciate biomechanics, one must first understand its relationship to other movement-oriented disciplines. One more question: What is biomechanics?

1.2 UNDERSTANDING THE DISCIPLINE OF BIOMECHANICS

Biomechanics is simply the physics (mechanics) of motion exhibited or produced by biological systems. Traditionally, biomechanics is sometimes considered synonymous with the term **kinesiology**, which is the study of human motion. In turn, some disciplines consider kinesiology to be applied or functional anatomy. In our field, kinesiology has been expanded to include all of the movement-related sciences: the anatomical, biomechanical, cultural, motor, pedagogical, physiological, psychological, and sociological aspects of motion. Of these subdisciplines of kinesiology, biomechanics is, of course, the major focus of this textbook. However, biomechanics should never be considered completely independent of any of the movement-related sciences.

More specifically, biomechanics is a highly integrated field of study that examines the forces acting on and within a body as well as those produced by a body (Figure 1.1).

The study of biomechanics also requires that we consider the consequences of the resultant motions produced by forces. Biomechanics is special in that it integrates biological characteristics with traditional **mechanics** (the branch of physics specifically concerned with the effect of forces and energy on the motion of bodies). Within mechanics, one may be concerned with **statics**, the study of systems in a state of equilibrium (at rest or in a constant state of motion); or **dynamics**, the study of systems that are in a state of accelerated or changing motion (Serway & Jewett, 2019). Whether a system is in a state of equilibrium or a state of acceleration, it may be analyzed from two perspectives: **kinetics** and **kinematics**. Kinetics is the study of forces that inhibit, cause, facilitate, or modify motion of a body. Some words in popular language that are actually examples of aspects of kinetics are *friction*, *gravity*, and *pressure*. In contrast, kinematics is the study or description of the **spatial** (direction with respect to the three-dimensional world) and **temporal** (motion with respect to time) characteristics of motion without regard to the causative forces (Serway & Jewett, 2019). Actually, most people are quite familiar with kinematic characteristics

Biomechanics Physics (mechanics) of motion exhibited or produced by biological systems.

Kinesiology Multidisciplinary study of human motion, including the anatomical, biomechanical, cultural, motor, pedagogical, physiological, psychological, and sociological aspects of motion.

Mechanics Branch of physics concerned with the effect of forces and energy on the motion of bodies.

Statics Branch of mechanics concerned with objects in a state of equilibrium (at rest or in a constant state of motion).

Dynamics Branch of mechanics concerned with objects in a state of accelerated or changing motion.

Kinetics Study of forces that inhibit, cause, facilitate, or modify motion of a body.

Kinematics Study or description of the spatial and temporal characteristics of motion without regard to the causative forces.

Spatial Relating to, or with respect to, the three-dimensional world.

Temporal Relating to, or with respect to, time.

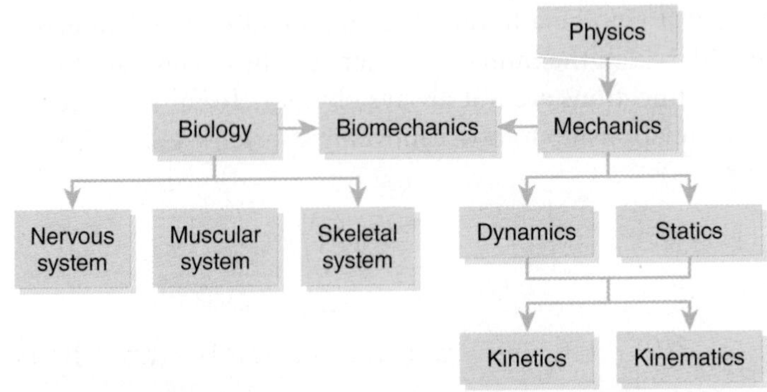

Figure 1.1 **The relationship of biomechanics to biology and physics (mechanics)**

such as displacement (distance traveled in meters or degrees) and velocity (displacement in a given time—for example, meters per second or degrees per second). Biomechanical analysis is applied to a variety of situations, some of which may be surprising. Therefore, we provide a small sampling of these situations.

The term *biomechanics* is sometimes immediately associated with the realm of sports. Indeed, biomechanical concepts are quite frequently applied to sports situations, so we begin with a couple of traditional sports examples. The sport of soccer has many areas to which biomechanical analysis can be applied. For example, a soccer player may injure a knee while trying to outmaneuver an opponent (Figure 1.2).

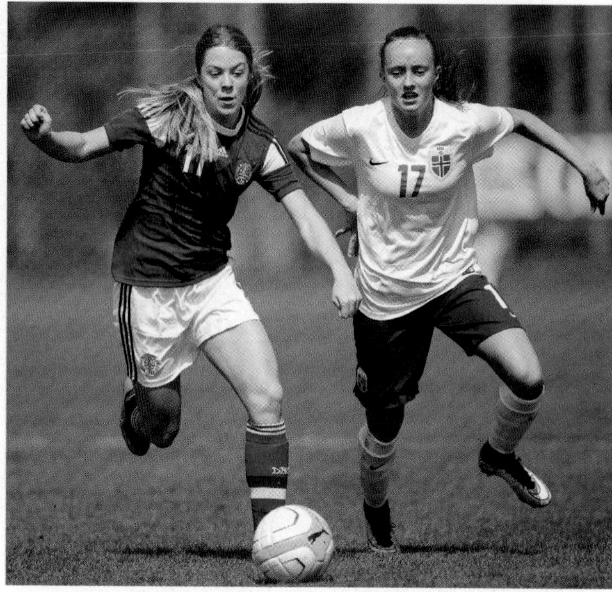

Figure 1.2 **A soccer player attempting to outmaneuver an opponent.** What elements of the situation are potential sources of injury?
© Herbert Kratky/Shutterstock

From a *kinematic* perspective, one may be interested in how fast the soccer player was moving at the moment of the injury (was the maneuver performed too quickly?). In *kinetic* terms, we may want to look to the forces involved in the situation. How much force is absorbed by the body when making a quick change in direction? What force makes the change of direction possible? How much force is required to tear an anterior cruciate ligament? Biomechanics can also help us understand whether or not this type of injury is more likely to occur on natural grass or on an artificial playing surface. Still other biomechanists may be interested in whether this injury is more common in females than in males. If so, what are the characteristics of the body that predispose females or males to this particular type of injury? Further, if one attempts to prevent or treat the injury with the use of taping or bracing, how much will this precaution hinder the soccer player's performance? These are just a few of the questions that biomechanical researchers attempt to answer.

In the sport of swimming many questions can also be answered through the use of biomechanical knowledge. We are definitely interested in how fast swimmers can swim and how quickly they can execute a flip turn (*kinematics*) (Figure 1.3), but we are also interested in the forces that act on a swimmer because the environment is a liquid (*kinetics*). Biomechanists study various strategies

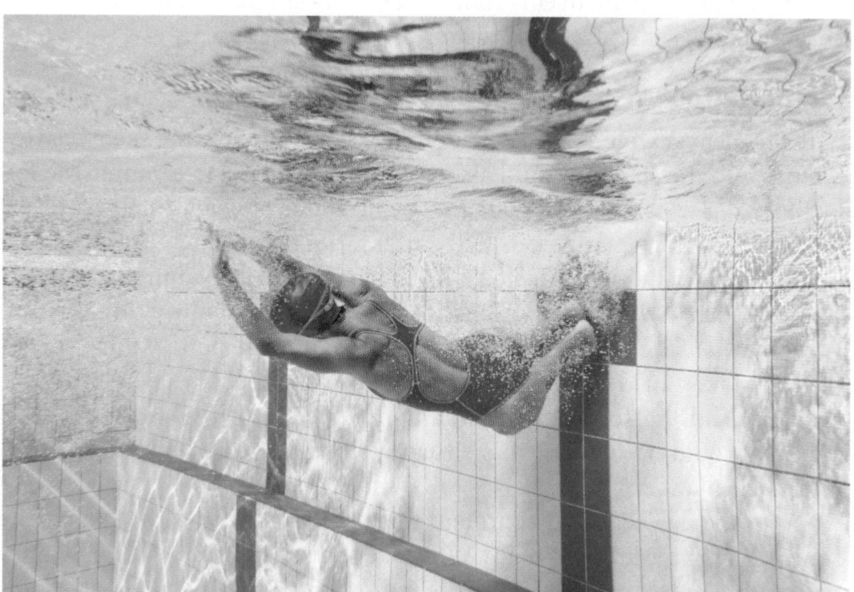

Figure 1.3 **A swimmer performing a flip-turn.** In what ways has this swimmer likely attempted to enhance performance?
© Microgen/Shutterstock

Figure 1.4 **The ollie skateboarding technique.** What causes the skateboard to rise?
© pio3/Shutterstock

used by swimmers to move easily though the water. For example, does shaving the body actually help? We now know that full body swimsuits provided such an advantage that they were banned after the Beijing Olympics in 2008. The use of use of advanced materials and design were considered "technical doping." One may also be interested in whether or not swimming close to another swimmer (*drafting*) is helpful. A biomechanist may also study a particular swimming stroke to determine if it is likely to result in a repetitive stress disorder. Still other researchers may be interested in the body type that is most suitable for swimming success.

In recent years, some nontraditional sports have become extremely popular (rock climbing, skateboarding, and snowboarding, for example). In the sport of skateboarding, many different maneuvers are performed that are fascinating from a biomechanical perspective. One extremely common skateboard maneuver is the "ollie," a technique used by skaters to hop over objects and onto or down from elevated surfaces (Figure 1.4).

Figure 1.5 **An infant walking.** Notice that the arms are held high and the feet are in a wide stance. In what situations would you adopt this same position?
© Ivanko80/Shutterstock

Although the maneuver is common, it is not simple. It must be coordinated precisely because the skater and the board must travel similar aerial paths without being tethered together. Kinematic studies of the ollie would examine such parameters as the height of the hop or how fast the skater was moving at the time of the ollie. Kinetic examination of the maneuver would reveal such information as to how the skater causes the board to bounce into the air. One could also use kinetic analysis to examine the forces absorbed by the skater when landing from an ollie. And one obvious kinetic issue associated with skateboarding is the numerous falls and injuries that occur.

Outside of the world of sport, biomechanists also examine more common human motions such as walking. Notice the characteristics of a walking infant (Figure 1.5). A biomechanist could examine the pattern for kinematic values such as the length of the infant's strides, the distance between the feet, the height at which the arms are held, and the speed of the infant's progress as a result of these factors. Kinetic analysis would be used in this situation to figure out why the infant walks in that particular manner.

Why are the feet so far apart and the toes pointed outward? What benefit is derived from carrying the arms so high? Why take such short strides? And by the way, haven't I noticed the same characteristics when observing a chimpanzee walk on two legs? Why would they share the same walking pattern?

In addition to developmental walking patterns, some biomechanists study movement in people who may have amputations or congenital abnormalities of the body. Designing prostheses, for example, requires close attention to biomechanical factors. If the kinetic parameters of one side of the body are not the same as those of the opposite side, then abnormalities in kinematic patterns emerge. These abnormalities not only cause self-consciousness in the user of the prosthetic but may also lead to further injury. But one should think beyond prosthetics being used to simply restore normal motion. Biomechanical principles have also been used to design prosthetic devices that aid in sports performance (Figure 1.6).

Common to all of these examples (maybe with the exception of the infant) are the numerous pieces of equipment used to train athletes and, in rehabilitation settings, to gain or regain muscular strength and endurance (Figure 1.7). Through the use of biomechanical principles, each piece of training equipment is designed to produce optimal results while minimizing the risk of injury to the user.

Figure 1.6 **An amputee running with prosthetics.** What do you think are the important aspects of prosthetic design for athletes?

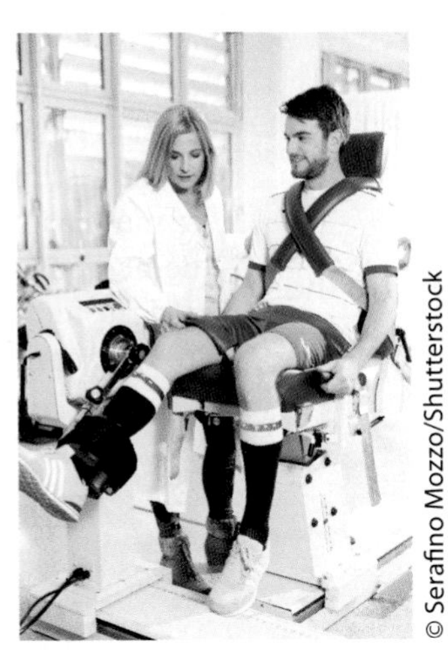

Figure 1.7 **Exercise and therapy equipment.** Notice the design elements of the various pieces of equipment. Why are there so many structural differences?

These are just a few of the numerous situations in which biomechanical analysis can be used to gain insight. As these examples show, many values labeled as *kinematic* or *kinetic* are actually familiar to most people in some way. Biomechanists simply study the parameters in greater detail. Even though many kinetic and kinematic values in biomechanics are somewhat familiar, the field of biomechanics is often misperceived as being isolated from the other movement-related subdisciplines. Therefore, a major goal of this textbook is to make the necessary connections between biomechanics and other disciplines. We begin with some explanation of related fields of study that are referred to throughout the textbook. We will also refer back to the previous examples to demonstrate the perspectives from which other disciplines might approach the same situation.

1.3 RELATIONSHIP OF BIOMECHANICS TO OTHER MOVEMENT DISCIPLINES

One inherent difficulty with the study of biomechanics is that connections to content in other courses may not be as readily apparent as in some other disciplines. In fact, a more integrated discipline is difficult to imagine (Figure 1.8).

At the most superficial level, biomechanics is about movement. Movement is caused by the contraction of skeletal muscle. To understand skeletal muscle contraction, one has to consider issues such as muscle fiber type and metabolism, topics that are traditionally studied in the discipline of **exercise physiology** (the study of physiology under conditions in which physical work has caused disrupted homeostasis). The student of biomechanics must also be concerned with the mechanisms used by the nervous system to control and coordinate the many intricate movements of the musculoskeletal system, which are specifically the interest of the field of **motor control**.

Exercise physiology The study of physiology under conditions in which physical work has caused disrupted homeostasis.

Motor control Mechanisms used by the nervous system to control and coordinate the movements of the musculoskeletal system.

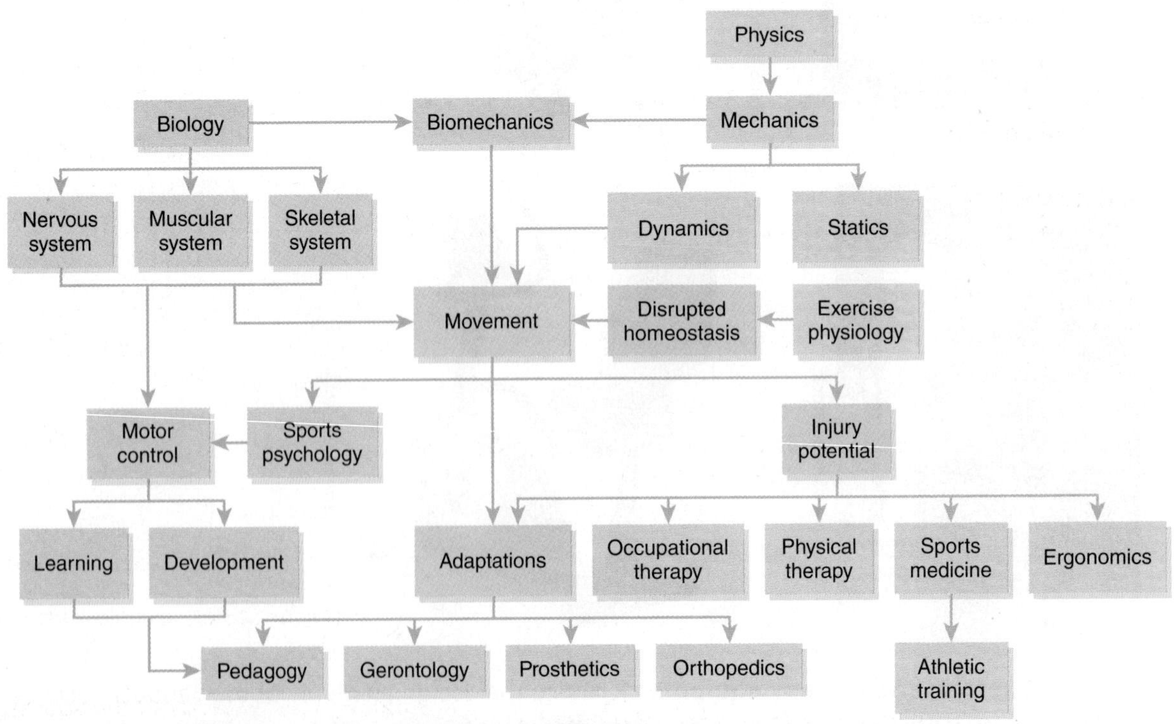

Figure 1.8 **Connections among movement-related disciplines**

Motor control progresses throughout the life span (**motor development**) and can undergo relatively permanent change to become more proficient through experience, practice, or both (**motor learning**). The changes in motor control are accompanied by changes in biomechanical movement patterns. Some movement fields are concerned with human motion as it applies to the work environment. **Ergonomics**, for example, is a discipline that examines human–machine interaction. Ergonomists use many biomechanical techniques in analyzing the work environment. Other related disciplines are primarily concerned with prevention, immediate treatment, and rehabilitation from both acute and chronic injuries that result from human motion. **Physical therapy** and **occupational therapy** are the fields dedicated to evaluating and treating movement abnormalities. Disordered movement may be caused by injury, lack of coordination, muscular imbalance, or congenital conditions. Physical and occupational therapists must be familiar with biomechanical principles to properly diagnose movement disorders and design the most appropriate intervention. In the field of **sports medicine**, practitioners such as athletic trainers focus on preventing and immediately treating injuries that occur during sports. Preventing injury may require such methods as bracing and taping, both of which can affect normal human motion. Movement patterns may be unusual because of temporary or permanent changes to the body that occur congenitally or because of injury or disease. If such changes are present, the biomechanist must be concerned with variation from and compensation in expected movement patterns; these alterations can collectively be called **adapted movement**. Of course, intertwined with all of these fields are the disciplines of **pedagogy** (the study of teaching) and **coaching**. Teachers and coaches work with people at different ages throughout the life span and must try to modify or improve movement behaviors while considering the various abilities of the population with whom they work. The variations in ability with which teachers and coaches contend arise from the interactions of all of these anatomical, biomechanical, motor-related, and physiological factors that are unique to each person at any given time. In common to all of the movement-related disciplines is an understanding of **functional anatomy**. No matter the movement-related discipline, the practitioner must always have an in-depth knowledge of the human body. One must know how the body moves when it is healthy to know when it is injured. Understanding the interrelationships among various body systems is also important to know how damage to one area may cause abnormalities in a seemingly unrelated area.

This section has covered some basic terminology of the disciplines associated with biomechanics. It has also presented some superficial connections to other fields of study. However, to fully comprehend the relationships between biomechanics and other movement-related sciences, a slightly more in-depth overview of some of these related disciplines is needed. The following fields are not the only ones to which biomechanics is related; they are simply the most directly related and are therefore the ones that will be discussed most often throughout the textbook.

Motor development Progression of motor control throughout the life span because of maturation.

Motor learning Relatively permanent changes in proficiency of motor control through experience and/or practice.

Ergonomics Discipline concerned with human–machine interaction.

Physical therapy Field dedicated to evaluating and treating movement abnormalities.

Occupational therapy Field focused on helping people to improve their ability to carry out activities of daily living and self-care tasks (i.e., "occupations") after an injury, disability, or other health condition.

Sports medicine Field dedicated to the prevention, immediate treatment, and rehabilitation of injuries that occur during sports participation.

Adapted movement Movement patterns that emerge because of compensation for changes to the physical body.

Pedagogy Study of principles and methods of instruction.

Coaching Study of principles and methods of instructing athletes.

Functional anatomy Study of the specific functions of individual structures that make up an organism.

Exercise Physiology

The beauty of the human body can best be appreciated when its systems are challenged, for example, during exercise. Exercise requires the cooperation of all of the body systems. The neurological system initiates muscle contraction to move the skeletal system and then orchestrates the coordination of different muscles for smooth motion throughout the exercise session. Increased activity of the muscular system requires that large amounts of energy be produced to fuel muscle contraction. The metabolic systems provide energy for muscle contraction in the form of adenosine triphosphate (ATP). For the metabolic machinery to make ATP, two things are needed: oxygen and food "fuel" such as fat and carbohydrate. The respiratory system brings oxygen in from the air (and removes the carbon dioxide that is produced during metabolism). Once inside the body, oxygen must have some method of transport (as must the fat and carbohydrate that are used as fuel). Oxygen and fuels are transported within the body by our cardiovascular system in the blood that travels through vessels, with the heart as a circulatory pump. The fuel—fat and carbohydrate—is, of course, brought into the body by the digestive system. However, the endocrine system controls whether those fuels are stored in the body tissues or released into the bloodstream for transport. Among its myriad other functions, the endocrine system also affects the rates of the metabolic pathways and various acute (immediate) and chronic (long-term) physiological adaptations to exercise. In the process of muscle contraction and metabolism, large amounts of heat are produced. To prevent heat illness, the heat produced during metabolism must be transported (once again by the cardiovascular system) to the skin (integumentary system) for dissipation. Much of our metabolic heat is eventually lost through evaporation of sweat. Sweat gland operation and the prevention of dehydration caused by loss of water in sweat are both moderated in part by the endocrine system. So the field of exercise physiology was born from the study of various acute and chronic changes to the human body systems that occur when homeostasis is disrupted. As this highly simplified series of events demonstrates, nothing creates homeostatic disruption better than simply exercising.

In our previous examples, we mentioned a soccer player, a swimmer, a skateboarder, and an infant walking. One might be tempted to say that the exercise physiologist approaches each situation by examining metabolic demands, training needs, etc. However, each situation is more complex than may be readily apparent. Muscles produce forces, and forces are in the realm of *kinetic* analysis. So biomechanics and exercise physiology are automatically connected through the muscular system. In the case of our injured soccer player, muscular fatigue may be a causative factor in the injury. Force produced by muscles is used not only to move the body but also to maintain joint integrity. So as muscles fatigue, body kinematics change; the capability to hold the joint together strongly enough may be reduced, resulting in an injury caused by a change in *kinetics*. In the case of our swimmer, kinetics arising from muscle force is also an issue. However, less obvious factors may be present. Performance in water can be affected by temperature. The muscular system helps to maintain proper body temperature by shivering when it is too cold. If the muscles are shivering, they cannot contribute to the swimming stroke as effectively as needed (kinetic change), and swimming kinematics can change. Muscular forces are also necessary to cause the skateboard to bounce off the ground to perform an ollie. When the skateboarder completes the ollie, muscular forces are necessary for a controlled landing. Finally, one factor needed for an infant to walk is muscular strength. Strength is needed to stand, to move one leg forward, and to maintain the weight of the body on one leg. Therefore, as muscular strength changes, so do the kinematics of walking.

As these examples show, the link between exercise physiology and biomechanics is the neuromuscular system. The muscles are the metabolic machines that cause motion of the skeletal system (and both muscles and skeleton are subject to various biomechanical factors). So the muscles are a *kinetic* factor that affects *kinematic* values. Further, those muscles are under control of the nervous system and rely on it for controlled, purposeful motion. The musculoskeletal and nervous (motor) systems are so intertwined that a student must have a deep understanding of the motor system to fully understand biomechanics.

Motor Behavior

Four similar terms are discussed in this section, and at times they can be confusing. The reason they belong together is the word that they all share: *motor*. In broad terms, *motor* means "movement"—muscles provide the force that drives human movement. Motor behavior is defined by three subareas: motor control, motor development, and motor learning. Each defines human movement in a different way, but they are all interrelated. Motor control is a theoretical area that describes how the nervous system controls muscle activation for coordinated skill performance. Motor development is concerned with how motor control changes over time. Finally, motor learning describes how humans learn and improve motor skills.

Figure 1.9 **An open-loop movement performed by a softball batter.** What is the key feature that defines it as open loop?
© Jon Osumi/Shutterstock

Motor Control

Theories of motor control attempt to explain how the nervous system controls the muscles during complex movements. Depending on the movement task, humans are believed to rely on one of two control systems—open loop or closed loop. **Open-loop** movements happen so quickly that the brain doesn't have enough time to receive feedback that can influence the current performance, whereas **closed-loop** movements can be changed during a performance as the brain receives sensory feedback from the eyes, ears, and **proprioceptors** (internal receptors that indicate force, velocity, and position) throughout the body.

An example of an open-loop activity is hitting a fast-pitch softball (Figure 1.9). As a ball leaves a pitcher's hand, a batter must decide quickly whether to swing or not swing. If the batter decides to swing, a motor program is selected and the information is sent to the muscles in one chunk. Once a forceful swing is initiated, it is almost impossible to stop, and a player can be injured in an attempt to hold the bat back during a "check swing." Although the batter receives sensory feedback during the swinging motion (i.e., she makes contact or misses the ball), the movement happens so quickly that she has no chance to change the performance. Sensory information gleaned from the swing can be used for the next performance but not the current one.

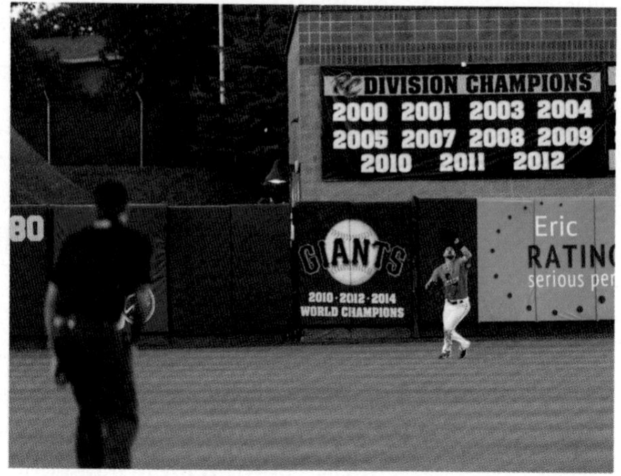

Figure 1.10 **A closed-loop movement: fly ball.** Can you think of an open-loop movement by an outfielder?
© Chris Allan/Shutterstock

Conversely, during closed-loop movements the performer receives feedback that can influence the current movement. In other words, a movement is initiated by the brain, and then the performer adjusts to a dynamic situation through the use of feedback. Continuing the softball example, assume the batter makes contact and the ball travels into the field of play. To understand closed-loop movement, assume that the ball has been hit high into the outfield. An outfielder begins moving in the direction of the ball but may need to adjust her position or speed, using visual feedback, to intercept the ball where it actually comes down. Environmental factors such as wind can make multiple corrections necessary, and these corrections are based on feedback received by the brain during the movement (Figure 1.10).

Open-loop Movements occurring too rapidly to be modified by sensory feedback.

Closed-loop Movements that can change during performance because of sensory feedback.

Proprioceptors Muscle, tendon, joint, and other receptors that provide information about position, force, velocity, and more.

All of the motions mentioned previously (playing soccer, swimming, skateboarding, and walking) are coordinated skills that must be controlled by the nervous system. Many different skills are required in the sports of soccer, swimming, and skateboarding. In some cases, the skills are open loop, whereas other skills are closed loop. Although the external *kinetics* of the situation may not change as a result of being open or closed loop, the internal *kinetics* (muscle forces) and resulting *kinematics* may vary. In other words, if external feedback is received by the performer during the movement, the way in which the movement is performed may vary from moment to moment. For example, the soccer player may have received some distracting feedback while performing a skill. The resulting lack of concentration during the motion may have changed the muscle *kinetics* and motion *kinematics* in such a way that the chances for injury were increased. Similarly, the *kinetics* and *kinematics* of the skills in swimming and skateboarding are affected by whether or not the skill is open or closed loop. If the skill is open loop, the *kinetics* and *kinematics* may change during the motion. These changes may be for the better, but they may also be for the worse.

FOCUS ON RESEARCH

Throughout this textbook, we include sections that serve to show the diverse and interdisciplinary nature of research in the field of biomechanics. This is the first chapter, so we have not yet introduced enough necessary concepts to allow full comprehension of the research studies themselves. Therefore, at this point we simply mention a couple of studies related to our introductory information.

In an issue of the *Journal of Strength and Conditioning Research*, we can find an article that looks at the actions and forces of one of the most common weightlifting movements: "Squatting Kinematics and Kinetics and Their Application to Exercise Performance" (Schoenfeld, 2010). Notice the words *kinematics* and *kinetics*; they indicate that the focus of this study is on both the spatial and temporal characteristics of lower-body movements and the forces that cause them. Using information gleaned during the study, the author provides recommendations to ensure optimal squatting performance and safety.

The cited study is just one example of the many different applications of biomechanics research. It is also used to increase safety of participants in sport. One current issue in sport is the incidence of concussion in American football. In an effort to design equipment that may reduce the likelihood of concussive injury, Broglio and colleagues collected real-time head impact telemetry data on high school and college athletes in a study titled "High School and College Football Athlete Concussions: A Biomechanical Review" (Broglio et al., 2012). Understanding the linear and angular acceleration forces applied to the head during a hit could lead to better helmet design and tackling rules that will reduce the incidence and severity of traumatic brain injury. As we cover biomechanical concepts in greater depth, we include more and more detail in our research focus sections. At this point, it is enough to begin to think about all of the topics that can be investigated through the use of biomechanical principles.

Motor Development

The human body begins as one cell that multiplies into trillions of differentiated cells that eventually form nervous, skeletal, muscle, and other tissues. From birth to advanced age, the body is in a dynamic state of change. The primary motor activities evident at birth are not voluntary; they are reflexive behaviors designed to gather information and to nourish and protect the body. Voluntary motion begins only when the nervous and muscular systems are ready, and reflexes are inhibited as voluntary control takes over. Growth is fairly steady from birth to puberty, when the body goes through immense changes. As a result of steady or rapid growth periods, the dynamics of motion change. Longer limbs and larger muscles allow the potential for increased performance in running and throwing; however, the nervous system must also learn to adapt to the new limb length.

Anyone interested in coaching mixed-gender youth sports will be happy to learn that girls and boys differ little in structure and physiology until they reach puberty. Few reasons exist to separate boys and girls in recreational and sport activities. All children have similar potential and mechanisms for gaining strength, aerobic endurance, and motor skills; differences in performance usually are a result of opportunity and practice. At the onset of puberty, dramatic changes begin happening relatively quickly and with dramatic consequences.

Changes that occur during and after puberty can have positive and negative effects on performance. Athletes who have learned to perform at the highest levels with pre-pubescent bodies can suddenly find their new body shape unaccommodating. Nadia Comăneci was an Olympic-level gymnast who never returned to the highest levels of competition once her body changed in size and weight. Michael Jordan was cut from his high school basketball team as a freshman, but he excelled once he adapted to his new size and strength. Males begin to have significant advantages over females in many physiological factors that affect performance, including height, shape (e.g., shoulder-to-hip width ratio), limb length, muscle-to-fat ratio, and the potential to build larger muscles through hypertrophy (Figure 1.11). All of these factors have the potential to affect biomechanical principles in the body.

The *kinetics* and *kinematics* of all skills change with motor development. In our example of the walking infant, we can identify several distinctive kinematic factors: arms held high, wide stance, short steps, flat footsteps, and little hip rotation. As the child develops physically and neurologically, these kinematic values change: arms are lowered and swinging in opposition to the legs, narrowed stance, longer steps, heel-then-toe footfalls, and greater hip rotation. The fields of motor

Figure 1.11 **Female and male body size before and after puberty.** How might the physical changes resulting from maturation affect body movement and sports performance?

development and biomechanics meet in an attempt to understand the kinetics of this developmental change in gait mechanics.

Motor Learning

The process of learning begins in the earliest stages of infancy when humans are basically "reflex machines" reacting to environmental stimuli. Early learning in movement is a trial-and-error process in which infants and toddlers attempt new activities. As any parent can attest after watching a child's first walking steps, failure is part of the learning process. As we get older, we learn new activities in varied ways. Sometimes we find old sports equipment in a basement and figure out how to use it through trial and error. People often emulate others, trying activities that look appealing and are taught new activities by relatives, friends, and in more formal settings by teachers and coaches.

Although failure is a common occurrence when learning new motor skills, teachers and coaches use a variety of tips and teaching techniques to increase the likelihood of success. The essence of physical education (PE) in schools is to help students become competent movers who learn a variety of skills. Ideally, it also instills in them a desire to pursue lifelong recreation and sport opportunities. Coaches endeavor to turn competent movers into higher-achieving athletes with specialized skills. For both teachers and coaches, understanding principles of motor learning can result in improved assessment, quicker and deeper learning, and better retention of skills.

Motor learning takes into account the structural and physiological changes through the life span but focuses primarily on neurological aspects of attaining and retaining motor skills. When learning and refining any motor skill, practice is imperative to success, but what kind of practice is best? Almost everyone has heard the adage "practice makes perfect"; in sports this adage is more accurately expressed as "perfect practice makes perfect." Thousands of golfers frequent local driving ranges each day in an attempt to become better players. They practice for hours and wonder why their scores never change. For some it could be the fact that they are using equipment that doesn't fit their body and allow biomechanical efficiency, but this is a topic for another area in this text. Most have equipment that is sufficient, but they practice incorrect techniques. Although a few golfers with no formal instruction eventually become skillful at the game through trial-and-error practice, most have little chance of attaining a higher level of proficiency without a knowledgeable teacher providing feedback. Practicing skills poorly results in learning poor skills. Depending on the task or even the level of a performer within a task (i.e., beginner, competent, expert), the type and amount of practice can differ in attaining maximal performance.

Practice is only one component of motor learning. Most sport and recreational activities require a person to pay attention to relevant stimuli, and strategies exist for learning to separate correct stimuli from irrelevant information. Certain methods of demonstrating tasks and providing oral instruction are clearer and more productive than others. These methods, along with appropriate teacher behaviors during practice, enhance understanding. Once students or athletes begin practicing a task, good teachers and coaches become "feedback machines," providing relevant skill feedback to encourage deeper learning and retention. Feedback comes in many forms and is delivered at different frequencies, depending on the performer or task. For example, beginners learn better with immediate feedback after each attempt. The same feedback frequency has the potential to hinder experienced athletes, who can become too dependent on it and not be able to perform well without it.

Successful teachers and coaches may not understand the exact mechanisms that result in developing high-skilled performers, but they rely on the neurological processes involved in motor learning. They know that tasks should begin with clear instructions and a demonstration. They understand that performers need feedback relevant to the skill that is delivered at the right time. They know that to perform successfully, players need to pay attention to relevant stimuli while ignoring irrelevant information. Finally, they understand that not all practice is good practice. Motor learning principles help coaches and teachers minimize the failure inherent in learning and refining motor skills and maximize the potential of each person.

Motor learning is a major factor in any of the skills that we have used as examples (playing soccer, swimming, skateboarding, and walking). Neurological control of all of these skills improves with practice and proper learning strategies—for example, more successful shooting in soccer, better flip turns in swimming, effortless completion of the ollie, and smooth strides while walking. Notice that all levels of motor behavior are intertwined with both the kinetics and kinematics of the resulting motion. So once again, our understanding of another movement-related field benefits from an understanding of biomechanics.

Ergonomics

As briefly mentioned earlier in this chapter, *ergonomics* is a discipline concerned with interaction of humans and machines and with the factors that influence that interaction (Bridger, 2018). Without proper analysis of the work environment, work tasks can be inefficient; but more importantly, potential exists for severe injury to the worker. So ergonomists attempt to improve the human–machine system. Ergonomists achieve this goal by "designing-in" a better human–machine interface or "designing-out" factors in the work task or environment that interfere with system performance. In general, the human–machine system is improved in the following ways (Bridger, 2018):

1. Designing the human-machine *interface* to make it more resistant to common human errors,
2. manipulating the work *environment* to enhance safety and appropriateness to the task,
3. changing the *task* itself to make it more compatible with the characteristics of the user, and
4. enhancing the *organization* of work tasks to better accommodate the psychological and social needs of the user.

Although biomechanics is a field often associated with sport, it is easy to understand how it applies to ergonomics. This association is natural because exercise is physical work, and physical work is exercise. Biomechanics and ergonomics are so highly related that ergonomics is sometimes referred to as **occupational biomechanics**. Many kinetic factors affect work tasks: muscle forces, weight of equipment, vibrations, surface textures, and so on. For example, should a worker keep an object close to or far away from the body when lifting (Figure 1.12)?

The kinematics of the work task can be analyzed by the ergonomists to estimate the effects of the above kinetic factors on the human–machine interface. These effects may be in the realm of potential injury or work efficiency. Based on the findings of the biomechanical analysis, the ergonomists can then make informed decisions to improve the human–machine system. Biomechanical analysis techniques can then be used to observe the resulting kinetic and kinematic changes to verify the effectiveness of the intervention. So the fields of ergonomics and biomechanics are inextricably related because no matter the movement situation, forces are involved and particular movement patterns result from those forces.

Injury Science

You are probably already familiar with some movement-related disciplines that are primarily concerned with prevention, immediate treatment, and rehabilitation from both acute and chronic injuries that result from human motion. However, you may not be as familiar with their

Occupational biomechanics Specialized area of biomechanics focused on human mechanics in work environments.

Figure 1.12 **An object being lifted using two different techniques.** Which do you believe to be safer, and why?

relationship to biomechanics. Many people have been treated by physical therapists. But how do they make their decisions about treatment?

Physical Therapy

Physical therapy is the field dedicated to preventing, evaluating, and treating movement abnormalities. Disordered movement may be caused by injury, disease, muscular imbalance, or congenital conditions. In addition, abnormal motion at one joint is often associated with abnormal motion at another joint. For example, a joint may exhibit movement abnormality as a result of structural defects that may be congenital or stem from injury or disease. This abnormal motion (as measured kinematically) is likely associated with abnormal forces acting on that structure (kinetics), leading to further motion abnormality (Oatis, 2017). In addition, abnormal motion at one joint may cause abnormal force application at another joint structure. For example, abnormal hip motion may lead to pain and dysfunction of the knee and ankle (Oatis, 2017). Physical therapists must be familiar with biomechanical principles to properly recognize and diagnose the underlying cause or causes of disordered movement (evaluation). Recognizing motion abnormality essentially requires kinematic analysis. The underlying cause of disordered movement is abnormal or excessive force application or distribution. The forces (kinetics) involved in abnormal movement may have originated outside the body (external) as a result of trauma or may be caused by abnormal kinetics at another joint (internal). Based on the biomechanical analysis, the physical therapist can then design the most appropriate intervention to maximize movement potential (treatment).

In the example of our soccer player with the injured knee, the physical therapist would view the incident from multiple levels. First, what was the underlying cause? Was it the immediate trauma of impact with the ground (external kinetics)? Or was it caused by abnormality at one joint making another more susceptible to injury (combination of external and internal kinetics)? After assessing the initial cause of the injury, the physical therapist must then design the appropriate intervention for effective rehabilitation from the injury (Figure 1.13).

In addition, if the cause was found to be partially associated with abnormal function at another (uninjured) joint, the physical therapist must also design an intervention to resolve the internal kinetic issue. So physical therapists use biomechanical concepts at every level from identification to treatment.

Occupational Therapy

Occupational therapists focus on helping people to improve their ability to carry out activities of daily living and self-care tasks (i.e., "occupations") after an injury, disability, or other health condition. Similar to physical therapists, occupational therapists educate people about injury prevention and the healing process, and they help people participate in meaningful activities. However, physical therapists tend to be more focused on the source of the problem such as the injured physical structures and tissues, whereas occupational therapists use a more "holistic approach to look not only at why a client's participation in activities has been impacted but also at the client's roles and environment" (National Board for Certification in Occupational Therapy).

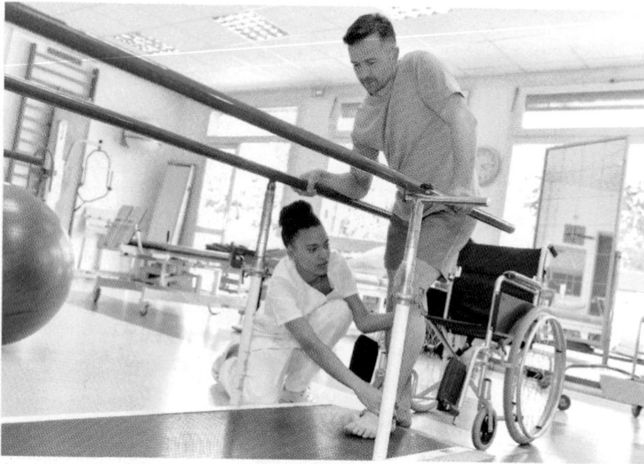

Figure 1.13 **Rehabilitation with use of handrails.** What are the kinetic and kinematic factors related to this specific situation?

For example, say a person has a stroke causing her to fall and sustain an injury such as a broken arm. She will most likely see a physical therapist to evaluate and diagnose the movement dysfunction. However, she will also see an occupational therapist to practice skills of daily life that may have been damaged by the stroke such as eating, walking, bathing, and dressing. Occupational therapists are also likely to make recommendations regarding the patient's home and work environments to enhance quality of life and increase independence. Therefore, occupational therapists must be familiar with the kinetic and kinematics of activities of daily living, as well as those related to the patient's interaction with the environment.

Sports Medicine

In the field of **sports medicine**, practitioners such as athletic trainers are focused on preventing and immediately treating injuries that occur during sports and on rehabilitating athletes after such injuries. Preventing injury may require such methods as bracing and taping, both of which can affect normal human motion. Bracing and taping are methods of manipulating the kinetic factors associated with a joint. In other words, a brace or tape can be used to prevent forces from causing excessive motion at a joint (Figure 1.14).

However, these treatment methods also affect the kinematics of motion. So a fine line exists between preventing injury and interfering with optimal performance. For these reasons, athletic trainers must be highly skilled in using these methods to maximize safety without excessively hindering movement.

When injuries do occur, the athletic trainer performs an immediate assessment and then physically examines the athlete (Pfeiffer, Mangus, & Trowbridge 2015). Biomechanics helps the athletic trainer understand the mechanism of injury (e.g., an impact force). The athlete must be treated as efficiently and effectively as possible, not only in an attempt to ensure immediate safety but also to avoid long-term movement abnormalities. Efficient assessment and examination requires knowledge of kinetics. What was the situation in which the injury occurred? In that type of situation, what is the most likely injury? These are actually questions of kinetics. What were the forces involved, and what is the likely result of those forces in this particular situation? The athletic trainer must also decide on the best immediate treatment and whether or not the athlete should return to the game. Without proper kinetic evaluation of the situation, the athlete may be prematurely allowed to reenter the game, and further injury can result. Because of the injury, the player may be referred to a physician, surgeon, or physical therapist who specializes in sports medicine. It is then the job of the athletic trainer to ensure that the athlete complies on a day-to-day basis with the prescribed treatment.

Sports medicine Field dedicated to the prevention, immediate treatment, and rehabilitation of injuries that occur during sports participation.

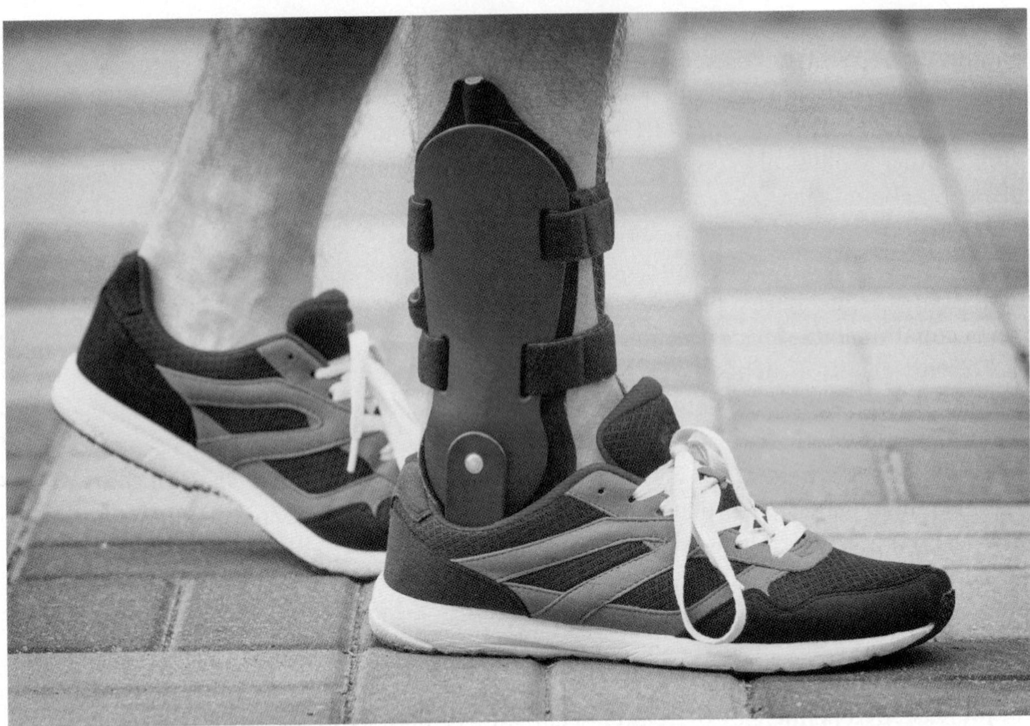

Figure 1.14 **An ankle brace.** What are the kinetic and kinematic changes that would be expected from use of the brace?

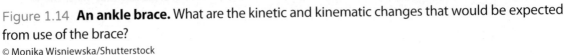

With our injured soccer player still in mind, the role of biomechanics in athletic training becomes clear. The athletic trainer sees the situation and then performs an assessment. Observing certain abnormalities of knee motion (kinematics), the athletic trainer diagnoses an anterior cruciate ligament injury. After the injury is repaired, the athletic trainer is responsible for the day-to-day care of the athlete to support rehabilitation and prevent future injury.

Pedagogy

Whether you aspire to teaching, coaching, or both, the principles of pedagogy form a foundation for success. The common objectives of quality teaching and coaching are to encourage learning and enhance performance. For the purpose of this text, *teaching* and *coaching* are used synonymously. Some believe that teachers and coaches are born for these roles. Although some evidence supports the notion that certain behavioral characteristics are common to a large percentage of successful teachers and coaches, intangible factors seem to exist that make some people better than others. A dictionary definition of *pedagogy* is "the principles and methods of instruction," but it has also become known informally as the "art and science of teaching."

Nobody would dispute that people with certain character and behavior traits have a higher aptitude for becoming good teachers, but evidence shows that all good teachers understand the scientific principles that encourage learning—you don't have to be born a teacher to become a competent one. Pedagogy in physical education and sport draws on several fields to form the basis of scientific principles for teaching and coaching: psychology, sociology, motor behavior, anatomy and physiology, biomechanics, and more.

Teachers first need to understand who their students are, both individually and as a group. This evaluation can include personality traits, physical limitations, and cognitive abilities of each student.

In addition, teachers need to understand social factors such as cultural and familial backgrounds. Designing and teaching a movement curriculum could differ dramatically between a rural and an inner-city school. Students at each of these schools would likely have different cultural backgrounds, expectations, and social concerns when they step into the gym.

Teachers and coaches must also have a sound understanding of the human body, including the ways it changes during normal maturation and in response to environmental influences such as training, equipment, injury, and weather. Children are not miniature adults and should not be taught or trained like adults. During childhood, windows of opportunity open, during which skills are more easily learned because the nervous system is ready. Critical periods in development also occur when overtraining can result in irreversible physical damage.

Beyond drawing from other fields of study, pedagogy has its own principles when it comes to encouraging learning. Teachers need a strong foundation of content knowledge, including skills, concepts, rules, tactics, and strategies. They also develop management skills that help them establish positive learning environments. Management begins with thorough planning, including appropriate curricular and lesson sequences designed to consider the ability level of students, available time, equipment, and space. Behavior management is another part of maintaining a learning environment—fostering a supportive atmosphere by individualizing tasks and planning for success.

Adapted Motion

A world-class runner sprints toward the finish line of a 100-meter dash, breaking the tape in 13.97 seconds. If you think that this time is good but not great, then consider the fact that she has no legs below the knees. With an extensive understanding of biomechanical principles and space-age materials, engineers helped a young girl who had both legs amputated below the knee become a world record holder. Children with varied disabilities attend almost every school, and the law mandates physical education for every child with an identified physical, cognitive, or behavioral challenge. The process of teaching movement activities to children with disabilities is called **adapted physical education**. In addition to receiving legislated PE, children with more severe physical disabilities also have a good chance of receiving physical and occupational therapy. These two disciplines rely heavily on the knowledge presented in this textbook. An understanding of biomechanical principles is a foundation for recognizing the changes and adaptations that can be made so that all people with disabilities have the chance to become successful movers.

In addition to school-based adapted physical education, programs such as Special Olympics and Paralympics offer additional opportunities for people with disabilities to be active in sport and leisure activities that range from recreational to elite-level competition. Given the general trend of this population toward sedentary lifestyles, it is imperative to promote lifetime physical activity beginning at an early age. The same principles that guide quality teaching in physical education and coaching are used to educate and enhance the performance of people with disabilities, whether they are students in schools learning basic motor skills or elite athletes who want to compete at the highest level. Instruction is individualized, the environment and equipment are modified (adapted), and, above all, the theme of *success* pervades all activities.

Success leads to a desire to continue physical activity. For some people, success could mean participating in a sport such as bowling (Figure 1.15).

For others, a definition of success may be developing enough strength to accomplish activities of daily living without help, like transferring from bed to a wheelchair, bathing, and getting dressed. This achievement can lead to a more independent lifestyle and higher self-esteem. Whatever the measure of success, it can develop a lasting appreciation for physical activity that benefits each person's physical and emotional well-being. Understanding how the human body moves

Adapted physical education The process of modifying equipment, the environment, or both in order to successfully teach movement activities to all populations.

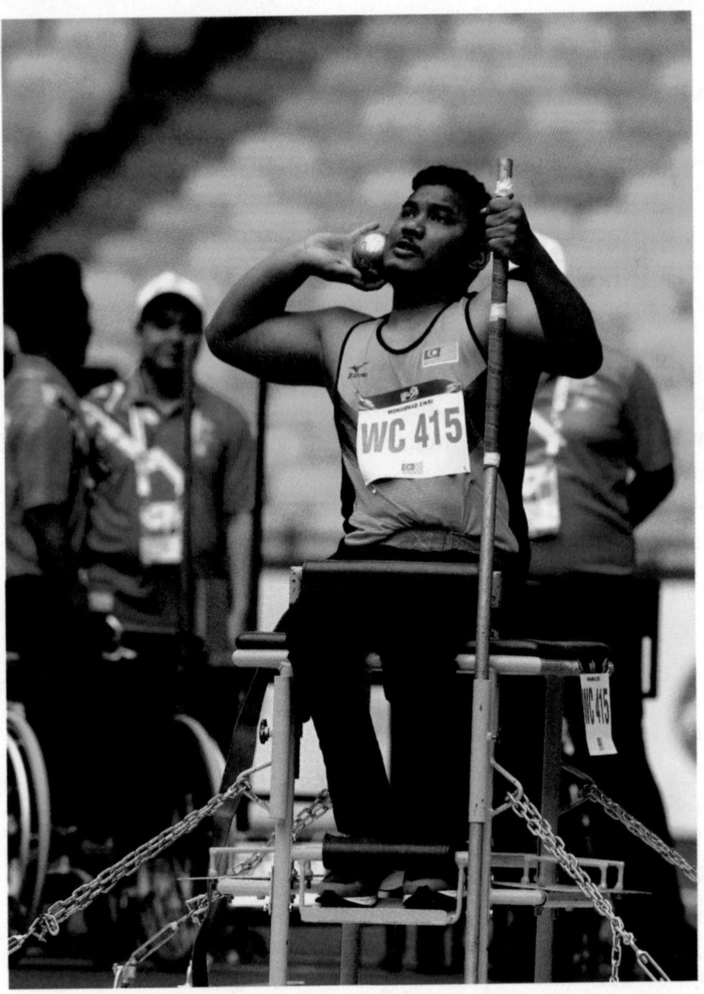

Figure 1.15 **Adapted seat for paralympic throwing.** Can you think of other equipment that allows people with disabilities to be active?
© Shahjehan/Shutterstock

and learns enables teachers to adapt movement activities that can help people with a variety of disabilities become competent, lifelong movers.

In addition to serving people with disabilities, adaptations are also becoming more common in youth sports. In the not-so-distant past, children used the same equipment that was designed for adults. Now youth soccer balls, baseball equipment, basketballs, and other equipment are smaller and lighter to match children's size and skills. Soccer fields for children are smaller, and the goals are more appropriately sized. Basketball rims in elementary schools are adjustable in height to promote more success and proper shooting form.

SUMMARY

Biomechanics is simply the physics (mechanics) of a living system's motion. Biomechanics is special in that it integrates biological characteristics with traditional mechanics (the branch of physics specifically concerned with the effect of forces and energy on the motion of bodies). Because biomechanics is a movement-related discipline, the student should approach the field from two perspectives: kinetics and kinematics. Kinetics is the study of forces that inhibit, cause, facilitate, or modify motion of a body. In contrast, kinematics is the study or description of the spatial (direction with respect to the three-dimensional world) and temporal (motion with respect to time) characteristics of motion without regard to the causative forces.

Biomechanics is a highly integrated discipline. At the most superficial level, biomechanics is about movement. Movement is caused by the contraction of skeletal muscle. Skeletal muscle contraction and metabolism are traditionally studied in the discipline of exercise physiology (the study of physiology under the conditions of disrupted homeostasis). The student of biomechanics must also be concerned with the mechanisms used by the nervous system to control and coordinate the many intricate movements of the musculoskeletal system, which are specifically the interest of the field of motor control. Motor control progresses with maturation throughout the life span (motor development) and can undergo relatively permanent change to become more proficient through experience or practice (motor learning) or both. The changes in motor control are accompanied by changes in biomechanical movement patterns.

Some movement fields are concerned with human motion as it applies to the work environment. Ergonomics for example, is a discipline concerned with human–machine interaction. Ergonomists use many biomechanical techniques in their analysis of the work environment. Other related disciplines are primarily concerned with prevention, immediate treatment, and rehabilitation from both acute and chronic injuries that result from human motion. Physical and occupational therapy are fields dedicated to evaluating and treating movement abnormalities. Therapists must be familiar with biomechanical principles to properly diagnose movement disorders and design the most appropriate intervention. In the field of sports medicine, practitioners such as athletic trainers focus on preventing and immediately treating injuries that occur during sports. Preventing such injury may require such methods as bracing and taping, both of which can affect normal human motion. Movement patterns may be different because of temporary or permanent changes to the physical body itself that occur congenitally or stem from injury or disease. If those changes to the body are present, then the biomechanist must be concerned with variation and compensation in expected movement patterns that can collectively be called *adapted movement.*

Intertwined with all of the above-mentioned fields are the disciplines of pedagogy (the study of teaching) and coaching. Teachers and coaches work with people at different ages throughout the life span and must try to modify or improve movement behaviors while considering the various abilities of the population with which they are involved.

In common to all of the movement-related disciplines is an understanding of functional anatomy. No matter the movement-related discipline, the practitioner must always have an in-depth knowledge of the human body. One must know how the body moves when it is healthy to know when it is injured. Understanding the interrelationships of the various body systems is also important to know how damage to one area may cause abnormalities in a seemingly unrelated area.

REVIEW QUESTIONS

1. A cyclist travels 100 kilometers at 50 km/hr. The wind is blowing in the same direction at 5 km/hr. Name as many kinetic and kinematic factors as you can for this situation.
2. Name the system that serves as the primary connection between biomechanics and exercise physiology.
3. The throwing pattern of a child changes with practice and age. Explain the relationships of the various movement-related disciplines involved with this change.
4. An office associate complains of back and wrist pain. Name some factors that could be of concern in this human–machine system.
5. A rugby player sustains a shoulder dislocation. Discuss the roles of the practitioners in various movement-related disciplines that may be involved.
6. Discuss the relationship of the fields of adapted motion and pedagogy.
7. Provide examples of both open-loop and closed-loop activities that could occur in a basketball game.

REFERENCES AND SUGGESTED READINGS

Bridger, R. S. 2018. *Introduction to ergonomics*, 4th ed. Boca Raton, FL: Taylor & Francis.

Broglio, S. P., J. T. Eckner, & J. S. Kutcher. (2012). Field-based measures of head impacts in high school football athletes. *Current Opinions in Pediatrics*, 24(6): 702–708.

Frederick, E. C., J. J. Determan, S. N. Whittlesey, & J. Hamill. 2006. Biomechanics of skateboarding: Kinetics of the ollie. *Journal of Applied Biomechanics*, 22(1): 33–40.

Gabbard, C. P. 2018. *Lifelong motor development*, 7th ed. Philadelphia, PA: Wolters Kluwer.

Kluka, D. A. 1999. *Motor behavior: From learning to performance.* Belmont, CA: Brooks/Cole-Thomson Learning.

Oatis, C. A. 2017. *Kinesiology: The mechanics and pathomechanics of human movement.* Philadelphia: Wolters Kluwer.

Pfeiffer, R. P., B. C. Mangus, & C.A. Trowbridge. 2015. *Concepts of athletic training*, 7th ed. Burlington, MA: Jones & Bartlett.

Powers, S. K., & E. T. Howley. 2018. *Exercise physiology: Theory and application to fitness and performance*, 10th ed. New York: McGraw-Hill.

Rogowski, I., K. Monteil, P. Legreneur, & P. Lanteri. 2006. Influence of swimsuit design and surface properties on the butterfly kinematics. *Journal of Applied Biomechanics*, 22(1): 61–66.

Schmidt, R. A., & T. D. Lee. 2011. *Motor control and learning: A behavioral emphasis*, 5th ed. Champaign, IL: Human Kinetics.

Schoenfeld, B. J. (2010). Squatting kinematics and kinetics and their application to exercise performance. *Journal of Strength and Conditioning*, 24(12): 3497–3506.

Serway, R. A., & J. W. Jewett Jr. 2019. *Physics for scientists and engineers*, 10th ed. Boston: Cengage Learning.

Serway, R. A., & C. Vuille. 2018. *College physics*, 11th ed. Boston: Cengage Learning.

CHAPTER 2

DESCRIBING THE SYSTEM AND ITS MOTION

LEARNING OBJECTIVES

1. Define the term *system*.
2. Understand basic anatomical terminology.
3. Explain how the system and the movement environment interact.

CONCEPTS

2.1 Introduction to the System

2.2 Anatomical Terminology

2.3 System Orientation

2.4 Motion at Segmental Links

2.5 The Movement Environment

2.6 Types of Movement

CONNECTIONS

2.7 Exercise Physiology

2.8 Motor Behavior

2.9 Pedagogy

2.10 Adapted Motion

Disagreements about observed events occur often in the world of sports, even in the presence of trained officials. Anyone who has watched a sporting event has probably seen two referees or umpires make opposite calls on the same play: A tennis umpire overrules a line judge, three football officials disagree on a call during the same play, a first-base umpire 20 meters away is asked to call a batter's motion by the home-plate umpire who is standing less than 2 meters from the batter. In sports such as diving, gymnastics, and ice-skating, panels of judges are asked to score sport skills that may have been performed at extremely rapid rates. With the advent of instant replay, viewers may find out that all of the officials were incorrect. However, veteran officials generally make the correct call, even when action occurs quickly. They are trained to look for critical elements while filtering out irrelevant information. What do you think are the critical elements to notice if a soccer player is injured during play? How would you describe the situation to a paramedic who arrives several minutes later?

Accurately observing and describing motion is especially difficult because the human body can perform so many different motions in extremely short periods. Careful observation and description are necessary in all movement-related disciplines. Exercise physiologists tend to observe skills in the context of muscle contraction and a task's metabolic needs. The student of motor control watches the same skill and thinks of the nervous system resources used to accomplish the highly complex motion. Motor development theorists are concerned with skill capability related to age, whereas motor learning experts think in terms of learning the task. Teachers and coaches must create methods for teaching the skill to other people and may have to adapt the motion for students with special needs. The common theme in all of these fields is observation and description.

As fledgling biomechanists, you will be introduced to a new language that will help you better understand and describe human movement and its complexity. This chapter will also help you recognize the salient features of motor skills, sport activities, and the motion of implements and objects that humans use and propel. With a common and consistent language and a trained eye for describing motion, you'll gain a deeper understanding of the role that biomechanics plays in sports and everyday life. The goal of this chapter is to give you the tools necessary to specifically describe the system of interest, the motion of the system, the location of the system within the environment, and the type of motion exhibited. With some practice, you will begin to look at motion in an entirely different way, noticing detailed aspects of movement that you may have previously ignored.

CONCEPTS

2.1 INTRODUCTION TO THE SYSTEM

To understand biomechanics and to begin biomechanical analysis, we must begin with an understanding of a clearly defined system of interest. A **system** is any structure or organization of related structures whose state of motion is of analytical interest. We tend to immediately assume that the system is an entire human, which may be true in some circumstances. However, the system may be only one part of the person or even some object that the human has kicked or thrown. A system is a simple concept, but the importance of defining it clearly should not be underestimated.

Let's use the following example for the remainder of this discussion: A biomechanist is interested in studying injury to find the most effective methods of prevention and treatment. In this situation, simply saying "people get injured playing soccer" is not especially useful. To be studied effectively, both the system and the circumstances must be clearly specified. In this particular case, we must begin with the system. What injury is the researcher studying? Is it damage to the knee, ankle, foot, or some other body part? If the researcher is concerned with the knee, then the system may be the tibiofemoral joint or simply the femur or the

System Any structure or organization of related structures whose state of motion is of analytical interest.

tibia. What is the age of the system? An injury may occur more frequently in a particular age group. For example, Osgood-Schlatter disease is an inflammation of the attachment site of the patellar tendon to the tibial tuberosity. This condition is typically observed in children and adolescents who regularly engage in activities that involve jumping. If age is a concern, then the system may be redefined as the tibiofemoral joint of 13-year-olds. What is the gender of the system? If the game is female soccer, then the quadriceps angle (Q-angle) may be important because females possess a wider pelvis than males. It could be that the femur and the tibia do not make appropriate contact (tibiofemoral incongruence). We may also be concerned with the relative lengths of various limbs forming the injured joint. The injury could also be the result of muscular imbalances around the joint of interest (e.g., the ratio of hamstring to quadriceps strength). Any of these concerns could change the specified system.

Remember that you may be the only person who has observed the system. Therefore, observing the system carefully and being as specific as possible when describing the system to others is critical. Every detail of the system will make a difference in identification. In the movement-related sciences, we must clearly describe the physical characteristics of the system and its state of motion because others may not have observed the event. Also, we must practice our observation skills because many of the motions that we witness happen at extremely rapid rates. We will begin with the discipline that studies the physical characteristics of systems; next, we'll discuss the motion of the systems.

Anthropometry

Anthropometry is the discipline that studies measurements of the body and body segments in terms of height, weight, volume, length, breadth, proportion, inertia, and other properties related to shape, mass, and mass distribution. Anthropometrics basically describe the shape of the system. A comprehensive presentation of the field of anthropometrics is beyond the scope of this text. However, the student of biomechanics should always be aware of how varying body shape and limb proportions affect motion. So the purpose of this section is to draw attention to the natural variability in human shape and illustrate how it can affect movement (Figure 2.1).

Anthropometry The discipline that studies measurements of the body and body segments in terms of height, weight, volume, length, breadth, proportion, inertia, and other properties related to shape, mass, and mass distribution.

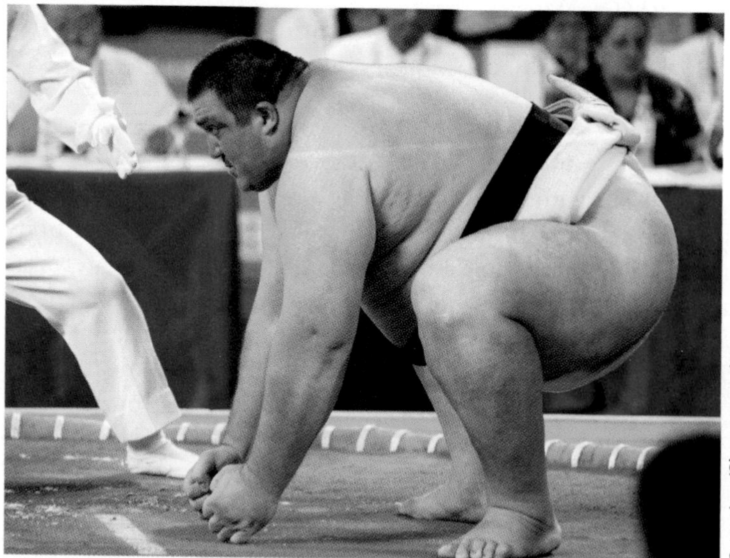

Figure 2.1 **Variability in human form.** How might these differences affect their sport performance?

For example, many competitive swimmers tend to have long torsos and short legs. Also, evidence suggests that great sprinters have short femurs in proportion to their tibia length. What body type makes for the most successful weight lifter? What about the length of the weightlifter's segments? What about great gymnasts? What is it about their bodies that aids in success? Inherent physical ability, even when combined with hours of intense training, can only go so far. Some of that physical capability comes simply from body shape. Because of the biomechanical effects of genetically determined body shape, particular activities are inherently easier for certain people.

For example, a majority of elite volleyball and basketball players are taller than the average, and the majority of elite female gymnasts are less than 5 feet tall. When people are comfortable with a particular sport activity, they tend to select that activity over others and potentially become successful within that sport. People also tend to opt out of sports in which they are not so comfortable. As a result of this sport self-selection by body type, we tend to see similarities in body features among athletes within a given sport. So any time that we are analyzing a particular movement, we must take into account the "build" of the person involved (i.e., the shape of the system).

A few anthropometric measurements students are usually familiar with even before a biomechanics course are height and weight, body mass index (BMI), somatotype, and waist-to-hip ratio. Height and weight are relatively simple measures but are still important, especially when considered along with other variables. But many pieces of information should be taken into account simultaneously. For example, if we are given only the information that the opponents in a wrestling match both weigh 47.5 kg, we can make no real judgments. However, if we also know that their heights are 1.5 m and 1.8 m, we can also assume some major differences in stature exist that could affect the match. If we also know that the wrestlers have similar body compositions, we can safely assume that the shorter of the two is probably stronger because more of his or her total body mass is composed of muscle tissue.

Statural Expressions

Equations and ratios have been developed that use multiple pieces of anthropometric data simultaneously to make assessments about health, stature, and athleticism. One such frequently used ratio is the **body mass index (BMI)**:

$$BMI = \frac{kg}{m^2} \tag{2.1}$$

where

kg = kilograms of body mass
m = height in meters

The most familiar use of BMI is its association with disease risk (i.e., as BMI increases, the risk of a variety of diseases also rises). One limitation to BMI is that it does not consider body composition—that is, relative amounts of fat and muscle. So care should be taken when interpreting the BMI value, because the numerator of the fraction can be affected by factors other than fat (i.e., other lean tissues such as muscle), which frequently leads to athletes being misclassified as having a high risk for disease. For the purposes of this text, BMI can still be a useful measure for giving some idea of an athlete's stature. For example, think about the BMI of a sumo wrestler versus that of a cross-country runner. In other words, particular BMI values are highly suitable for a given sport.

Body mass index (BMI) Ratio of body mass to height used to describe stature.

Ponderal index (PI) Ratio used to describe stature.

The **ponderal index (PI)** is a measure of stature similar to BMI but not as well known outside the fields of anthropometrics and biomechanics. Many methods for calculating PI have been suggested. The ratio used in this text is as follows:

$$PI = \frac{kg}{m^3} \tag{2.2}$$

where

kg = kilograms of body mass
m = height in meters

The PI yields a different numerical value, but it is used similarly to BMI (i.e., a person who weighs less relative to height has a lower PI). Again, in an era of athletics that encourages muscularity, one should use caution in making assumptions about the health of athletes using PI. However, we can use ratios such as BMI and PI to begin to take a closer look at the shape of the system. Some other expressions of PI that have been used are PI = $10^3 \times (\sqrt[3]{W}/H)$, where height is in centimeters; and PI = $H/\sqrt[3]{W}$, where height is in meters. A logical rationale exists for each recommended PI equation, and the semantics of the equations and their usefulness are controversial. However, normalization values such BMI and PI should be used only as guides when describing or interpreting the characteristics of a system. In terms of biomechanical analysis, parameters such as BMI and PI are meaningful only within a tightly narrow context. Highly technical kinematic and kinetic analyses of movement require more specific values such as segmental mass centers, mass distributions, and moments of inertia, all of which are covered in greater detail in subsequent chapters. For now, consider that expressions such as BMI and PI are associated not only with health status but also with success in some sporting events.

Somatotyping is another system for body classification that has been used for many years both quantitatively and qualitatively (Figure 2.2).

Somatotyping is a system of body-type description that uses the classification of people into three basic categories:

1. **ectomorphic**, being linear and relatively thin for height, which is also called *linear*;

Somatotyping System of body-type description based on weight and muscularity relative to height.

Ectomorphic Somatotype described as being linear and relatively thin for height.

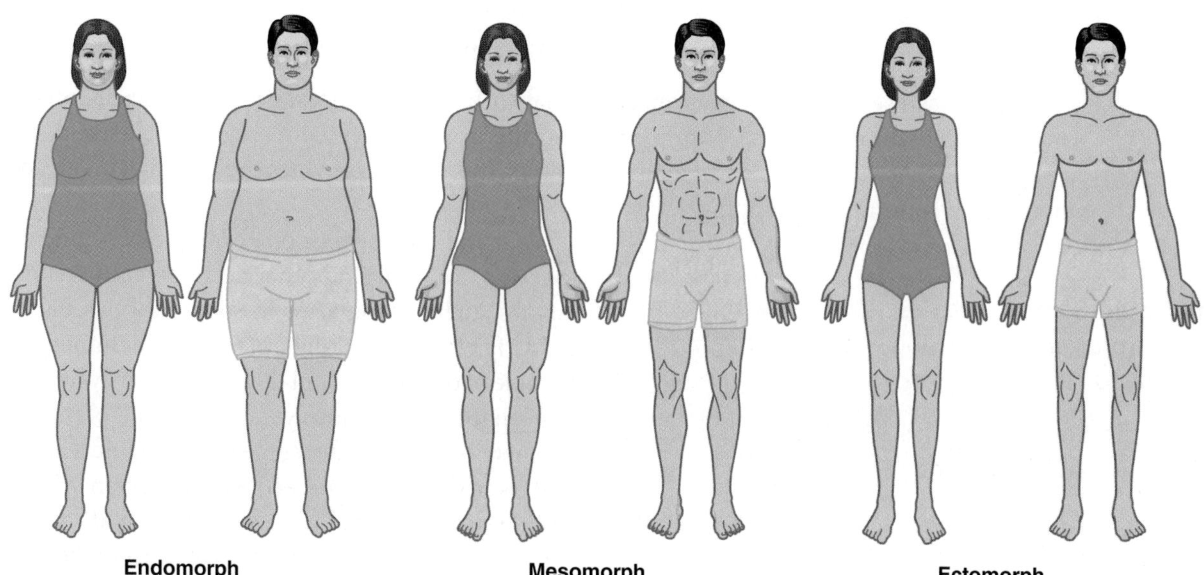

Figure 2.2 **The basic somatotypes: endomorph, mesomorph, and ectomorph.** In what sport can you imagine each individual participating?

2. **mesomorphic**, being muscular, strong, and possessing weight relatively proportional to height, which is also called *athletic*; and
3. **endomorphic**, which is rounder and relatively heavy for height.

A person does not have to fit perfectly into any given category but can possess varying degrees and combinations of traits from all three categories. As previously mentioned, a quantitative (points-based) system exists for classifying the degree to which a person possesses qualities of each category. However, most people tend to use this system in a more qualitative manner by visually categorizing a person. Even qualitatively, it can be assumed that most athletes are going to have a significant mesomorphic component blended with one of the other categories, depending on the sport. For example, an offensive lineman in football could be considered to have qualities of both the mesomorphic and endomorphic categories. In contrast, a combination of the ectomorphic and mesomorphic categories better describes a wrestler in a light weight class. In which category do you think a sumo wrestler would fall?

Rather than classifying or describing the entire body, biomechanical analysis is often more productive if we look at the anthropometrics of various individual parts of the body in relation to each other. Learning to pay close attention to individual body segments is important because of the variations in angular kinematics produced by varying limb proportions.

Body Segment Proportions

The ratio of waist to hip circumferences is another anthropometric measure that is often used to assess health risk. A person with a larger ratio (more "apple shaped") is at higher risk for disease than a person with a lower ratio ("pear shaped"). Even though the **waist-to-hip ratio** is associated with various disease states, the student of biomechanics should be aware of the difference in location of the whole body center of gravity that accompanies the change in distribution of weight from either above or below the waist. Consider the injured soccer player. What is the likely difference in waist-to-hip ratio for a female versus a male soccer player? Is this piece of information relevant in terms of the injury?

Many comparisons of system body-segment parameters are associated with various aspects of performance. One such comparison is that of limb proportion in the **crural index (CI)**:

$$CI = \frac{\text{length of the tibia}}{\text{length of the femur}} \times 100 \qquad (2.3)$$

As you can see from the equation, an animal with a large CI possesses a long distal segment (segment farther from the body) in proportion to the proximal segment (segment closer to the body) (Figure 2.3). As you will discover in later chapters, this quality provides the animal with an advantage in terms of running and jumping.

Crural index is just one example of segmental comparison. You can find other advantageous trends within sports such as long arms in successful baseball pitchers and high torso-to-leg-length ratios in swimmers (i.e., two people of the same height do not necessarily have the same leg length). As previously mentioned, all of these anthropometric variables have some biomechanical influence on a person's ability to succeed within a given sport.

Having looked at some ways of evaluating the system itself and its characteristics, we can turn to the movement situation and the environment in which the system is moving. In following sections of this chapter, we will learn how to clearly define the movement space. Here we also begin

Mesomorphic Somatotype described as being muscular, strong, and possessing weight relatively proportional to height.

Endomorphic Somatotype described as being rounder and relatively heavy for height.

Waist-to-hip ratio Ratio of waist circumference to hip circumference often associated with disease risk.

Crural index Ratio of the length of the tibia to the length of the femur.

Figure 2.3 **Two sprinters: human and cheetah.** Notice the proportions of the leg segments. Can you think of some other great sprinters in the animal kingdom?

to develop some understanding that the system of interest is not necessarily a separate entity from the environment with which it interacts. Using the example of the injured soccer player, our system may be defined as the tibiofemoral joint of teenage female soccer players. However, if we are researching types of injuries, we need to know, for example, whether the injury occurred on natural grass or an artificial surface. We also need to ask questions such as, What exactly was the situation? In what position was the system? Was it a cutting maneuver, a start, a stop? Was another player involved? Was the injury to the anterior or posterior cruciate ligament?

As can be seen from the preceding examples, the system is a specifically defined entity engaged in a particular movement situation within a specified space. Biomechanists must use common multidisciplinary terminology to perform research after the fact or to simply be able to specifically define the system, the movement situation, and the environment to someone who was not present. Now let's focus on specific definitions of movements and movement space.

2.2 ANATOMICAL TERMINOLOGY

To describe motion of the human body with consistency, researchers must have a common reference system and use consistent terminology. Much of the vocabulary used in biomechanics (and in most other movement-related disciplines) is derived from the field of anatomy. Comprehensive coverage of anatomical terminology is beyond the scope of this text, and anatomical terms vary slightly depending on the book. So the terms presented in this section represent the most common and widely used across disciplines (Floyd, 2018; Oatis, 2017; VanPutte, Regan, & Russo, 2017; Watkins, 2010).

Anatomical Position

The **anatomical position** is a commonly used reference point for the body itself; it refers to a person standing erect with all joints extended, feet parallel, palms facing forward, and fingers together (Figure 2.4).

Anatomical position Reference position defined by standing erect with all joints extended, feet parallel, palms facing forward, and fingers together.

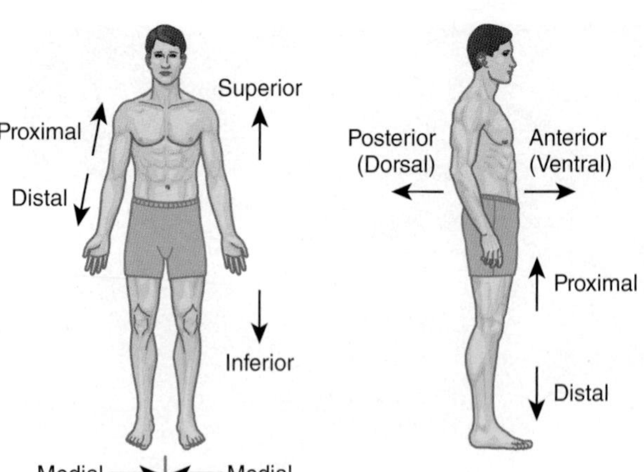

Figure 2.4 **The anatomical reference position with directions indicated.** When you give directions do you give a reference point?

Note that locations, movements, and relative positions of anatomical structures are always described according to the anatomical position whether or not the person is actually in the anatomical position at the time of description.

Directional Terms

Along with the anatomical position, directional terminology can be used for a more complete description of a movement or structural location. Remember that directional terms are usually used to describe *relative* location or position. The following is a list of the most commonly used directional terms:

- *Superior* and *inferior* are used to describe being toward or closer to the head and feet, respectively. For example, the knee is superior relative to the ankle but inferior to the hip. *Cephalo* or *cranial* may be used instead of *superior*, and *caudal* is sometimes used in place of *inferior*.
- *Anterior* means toward the front of the body, and *posterior* means toward the rear of the body. The pectoralis muscles are anterior to the heart. Alternative terms for anterior and posterior are *ventral* and *caudal*, respectively.
- *Medial* and *lateral* indicate position or movement toward and away from the midline of the body, respectively. One of the quadriceps muscles is closer to the midline, and one is farther away from the midline: thus, the names *vastus medialis* and *vastus lateralis*.
- *Proximal* means closer to the attachment of a limb to the body, and *distal* indicates having a position farther from the attachment of the limb to the body. The carpals are proximal to the phalanges.
- *Superficial* and *deep* describe relative proximity to the surface of the body. The gastrocnemius is superficial to the soleus.

2.3 SYSTEM ORIENTATION

Now that we have reviewed the anatomical reference position and directional terminology, we can introduce the fundamental orientation concepts of biomechanics. These particular concepts are useful for specifically describing motion of the body system and its segments in relation to the movement environment.

Planes of System Motion

Planar motion of a system or system segment is described as occurring in a plane. Geometrically, a plane is a flat two-dimensional (2-D) surface. So motion "in a plane" technically refers to the movement of that segment "describing" an imaginary plane. Movements can generally be classified as *uniplanar* (occurring in one plane, or two-dimensional) or *multiplanar* (occurring in more than one plane, or being three dimensional, or 3-D). Although most natural human motions occur in more than one plane, most segmental movements occur along one individual plane. Of course not all of the possible planes of motion are named, but an understanding of the concept of motion occurring in a plane is necessary for describing movements of the body as well as for comprehending many concepts you will encounter later. A **cardinal plane** is a plane that passes directly through the midline of the body (i.e., divides the mass of the body in half).

Cardinal plane Plane that passes directly through the midline of the body.

The three cardinal planes of motion are *sagittal*, *frontal*, and *transverse*. These planes are orthogonal, or perpendicular, to each other (Figure 2.5). The **sagittal plane** (also called *median* or *anteroposterior*) is vertical and divides the body in half along the midline into right and left masses. It runs superior to inferior and anterior to posterior. The **frontal plane** (also called *coronal* or *lateral*) is another vertical plane, but it divides the body in half along the midline into anterior and posterior masses. It runs superior to inferior and side to side. The **transverse plane** (also called *horizontal*) passes through the body horizontally and divides it into superior and inferior masses. It passes anterior to posterior and side to side.

Because the cardinal planes pass directly through the midline of the body and many body segments are lateral to the midline, most motions do not actually occur in one of the cardinal planes. Most motions should actually be pictured as taking place in one of the infinite number of planes parallel to the cardinal planes. In other words, a motion can be in a *sagittal* plane (such as knee extension) or a *coronal sagittal* plane (such as performing an abdominal crunch). In addition, a motion of the system may occur in a plane that is not parallel to one of the cardinal planes (e.g., a golf swing). These movements are said to describe *diagonal* or *oblique* planes (Figure 2.6).

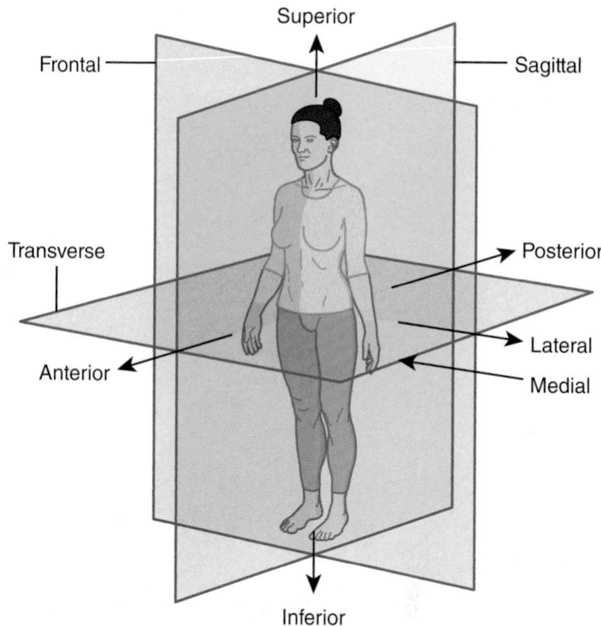

Figure 2.5 **The cardinal planes of motion: sagittal, frontal, and transverse.** Can you think of motions that occur in each of these planes?

Axes of System Motion

If we imagine the plane of motion as a wheel, the axis of rotation should be envisioned as the axle around which that wheel turns. In other words, a segmental movement describes a plane (*planar* motion) that rotates around a theoretical axis (*axial* motion). Just as there are three cardinal planes, there are three axes of rotation: *mediolateral*, *anteroposterior*, and *superoinferior*.

Each axis of rotation should be imagined as a line (axle) that is perpendicular to one of the described planes (Figure 2.7). Many variations of names for the axes are seen in textbooks. In this text, the axes are named for the anatomical directions in which they pass. Note, however, that some texts name an axis because it is parallel to a plane, or name it according to the intersection of the two planes by which it is formed. The **mediolateral axis** (also called *bilateral*, *frontal*, *frontal-horizontal*, *frontal-transverse*, and *transverse*) passes horizontally side to side and is perpendicular to the sagittal plane. The **anteroposterior axis** (also called *sagittal*, *sagittal-horizontal*, and *sagittal–transverse*) runs horizontally from front to back and is perpendicular to the frontal plane of motion. The **superoinferior axis** (also called *frontal-sagittal*, *longitudinal*, and *vertical*) passes up and down and is perpendicular to the transverse plane. Also, because there are diagonal (or oblique) planes of motion, diagonal axes of rotation exist perpendicular to each of those planes. Using our preceding examples, knee extension and flexion occur around a mediolateral axis, and a golf swing occurs around a diagonal axis.

Planes, Axes, and the Center of Gravity

Understanding the cardinal planes of motion and axes of rotation can aid elementary comprehension of certain

Sagittal plane Vertical plane dividing the body into right and left halves.

Frontal plane Vertical plane dividing the body into anterior and posterior halves.

Transverse plane Horizontal plane dividing the body into superior and inferior halves.

Mediolateral axis Axis that passes horizontally side to side and is perpendicular to the sagittal plane.

Anteroposterior axis Axis that runs horizontally from front to back and is perpendicular to the frontal plane of motion.

Superoinferior axis Axis that passes up and down and is perpendicular to the transverse plane.

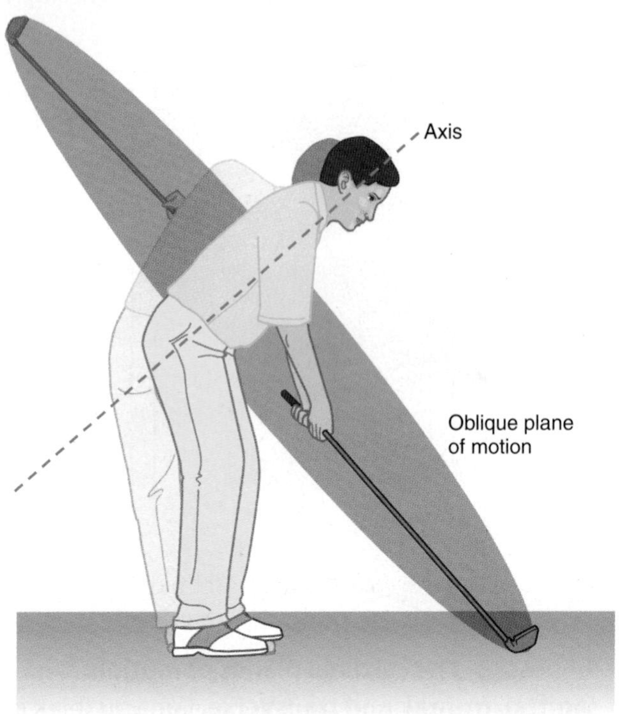

Figure 2.6 **Golf swing with an oblique plane.** What are some other sports motions that occur in oblique planes?

frequently used concepts in the field of biomechanics. As mentioned in the previous section, the cardinal planes divide the body into equal mass halves. Also, axes are formed by the intersections of two planes. Because the cardinal planes bisect the body, they must also pass through the **center of mass** (the point that represents the average location of a system's mass). Gravitational pull is concentrated at the center of mass. So at least in the vertical axis, the center of mass can be considered synonymous with the **center of gravity** (the point at which the force of gravity seems to be concentrated). In other words, the center of mass (or center of gravity) of the system is at the intersection of the three cardinal planes. As you will see in a Chapter 3, all forces can be represented with a line that possesses specific characteristics. Gravity can be represented with a line called the **line of gravity** that passes through the center of mass.

The line of gravity can be envisioned as the line at which the two vertical cardinal planes intersect (Figure 2.8). The center of mass and center of gravity are further elucidated as new concepts are introduced throughout the text.

Spatial Frames of Reference

To fully describe motion of the body system and its segments, we must be able to specifically define its position or location in space. This goal is achieved by establishing one or more frames of reference within a **Cartesian** or **rectangular coordinate system** (named for mathematician and philosopher René Descartes). An **origin (O)** and two or three orthogonal axes (each passing through the origin and defining one spatial dimension) are used to define a Cartesian coordinate frame of reference. For example, one of the most common rectangular coordinate systems uses two orthogonal axes that divide a plane into four quadrants (Figure 2.9).

The origin is a stationary point in the environment from which all measurements are made. For example, if you wanted to give someone directions to your home, you wouldn't begin from an arbitrary undefined point. The two of you would first agree on a starting point that is mutually understood and predefined. In biomechanics, we define the origin with the coordinates (0, 0) in two dimensions and (0, 0, 0) in three dimensions. Because most natural motions of the body occur in three dimensions, a 3-D coordinate frame of reference is used frequently throughout this text.

As previously mentioned, two or more perpendicular axes pass through the origin. Each axis represents a direction of motion. Many coordinate axis frames are used in different textbooks and research journals. The frame of reference we choose to use is merely a preference; the important factors are that it be clearly defined and that it is convenient for the specified purpose. For consistency, the convention adopted by the International Society of Biomechanics (ISB) is used throughout this text. We use a 3-D coordinate frame, so we need three axes: *x*-axis direction, *y*-axis direction, and *z*-axis direction (Figure 2.10).

Center of mass The point that represents the average location of a system's mass.

Center of gravity The point at which the force of gravity seems to be concentrated.

Line of gravity A vertical line representing gravity that passes though a system's center of mass.

Cartesian or rectangular coordinate system A frame of reference defined by an origin and two or three orthogonal axes, each passing through the origin and defining one spatial dimension.

Origin (O) A stationary point in the environment, from which all measurements are made.

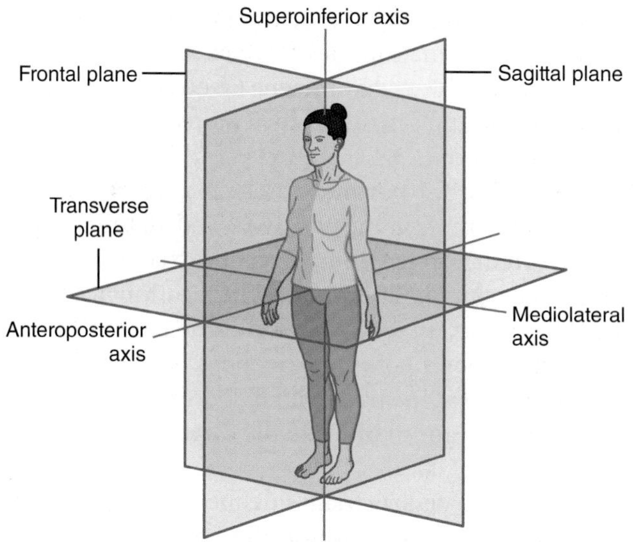

Figure 2.7 **The axes of rotation: mediolateral, anteroposterior, and superoinferior.** Nodding the head occurs around which axis?

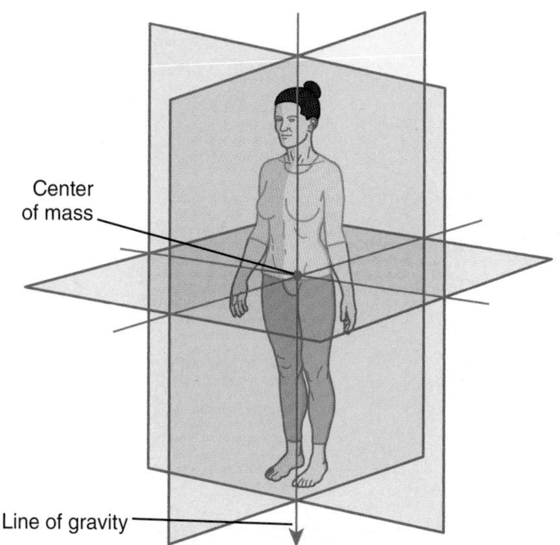

Figure 2.8 **The center of mass and line of gravity.** How could you locate your center of gravity?

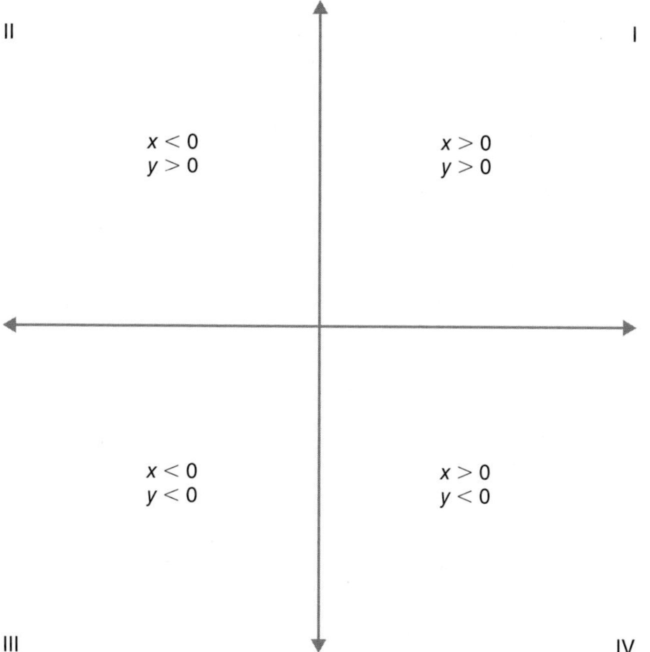

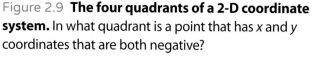

Figure 2.9 **The four quadrants of a 2-D coordinate system.** In what quadrant is a point that has *x* and *y* coordinates that are both negative?

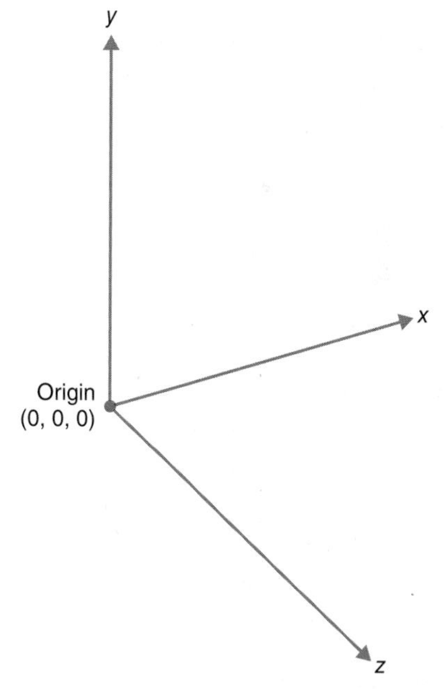

Figure 2.10 **A right-handed 3-D global coordinate frame.** Imagine the frame as a corner of the room in which you are located. Where are you located relative to the origin?

The x-axis direction (direction of progression) runs horizontally forward and backward relative to the system, with forward designated as positive and backward designated as negative. The vertical direction is the y-axis, which is perpendicular to the x-axis and runs superiorly (positive) and inferiorly (negative). The z-axis direction is horizontal and orthogonal to the x and y directions, running medially and laterally relative to the system. The positive z-axis direction is lateral or to the right, and

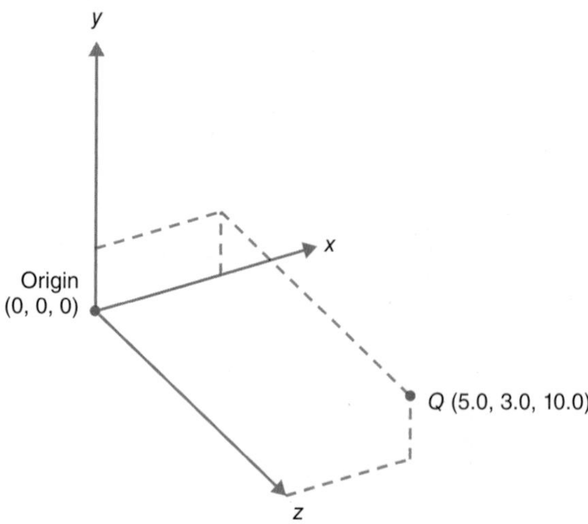

Figure 2.11 **The coordinates of a point Q located within the global coordinate frame.** Try to imagine the location of this point if you are located within the frame facing the *x* direction.

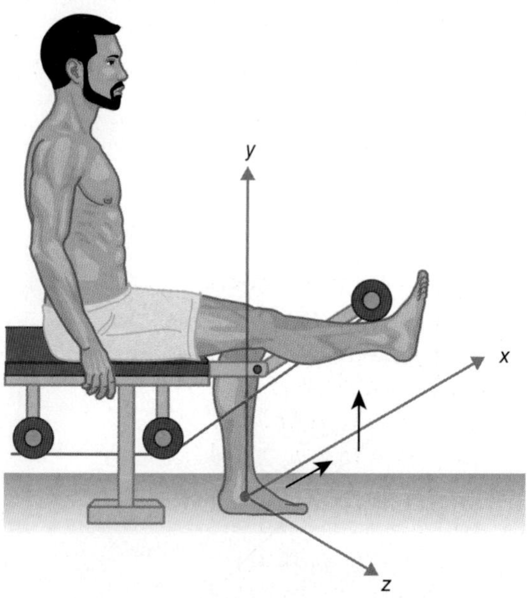

Figure 2.12 **Directions of motion of a point within a coordinate frame of reference.** In what directions would a runner move on a circular track?

Global reference frame A fixed frame of reference that allows the location of any point to be specified with respect to a defined origin.

Local reference frame A frame of reference attached to and moving with the system of interest.

the negative *z*-axis direction is medial or to the left of the system. So within a 2-D Cartesian coordinate frame of reference, a point Q with the coordinates (6, 8) can be located by moving 6 units in the positive *x* direction (forward) and 8 units in the positive *y* direction (upward).

If the frame of reference is 3-D (Figure 2.11), a point Q may possess the coordinates (5, 3, 10), which can be located by moving 5 units in the positive *x*, 3 units in the positive *y*, and 10 units in the positive *z* direction (right). So if a point is on one axis, its location can be fully designated with one coordinate. The location of a point in a *plane* formed by two axes requires two coordinates, and three coordinates are required for a point in 3-D *space*. So we can fully define the location of a point in space with a coordinate reference frame, or we may simply want to indicate the direction of motion of a point (Figure 2.12).

Notice that the coordinate axis frames defined above are *right-handed* (which is common). In other words, the *z*-axis is pointing to the right relative the plane formed by the *x*- and *y*-axes. You can further visualize this convention using your own hand. Begin by placing your right-hand palm up and making a fist. Now extend your index finger to point toward the positive *x* direction and extend (abduct) your thumb to point toward the positive *z* direction. Finally, extend your middle finger just to the extent that it is perpendicular to the other two. In other words, the middle finger points upward or positive in the *y* direction.

The same position can be used by your left hand to represent a *left-handed* coordinate frame, with the third axis pointing to the left of the plane formed by the other two axes. The reference frame is called *oblique* when the axes are not orthogonal. If we include oblique reference frames, an infinite number can be established. As previously stated, the ISB convention is used throughout this text for consistency. However, different reference frames are commonly used, not only within the field of biomechanics but also in engineering, mathematics, and physics. One common convention used in 3-D biomechanical analysis is one in which the *x*- and *y*-axes are orthogonal to each other in the transverse plane, and the *z*-axis is vertical instead of horizontal (Robertson et al., 2014). With all of the possibilities in mind, you can now understand the importance of clearly predefining the coordinate reference frame.

Fixed axes relative to the system, one of which is parallel to the ground, define the reference frame just described. This type of reference frame is called *global* (also *absolute*, *fixed*, *stationary*, and *inertial* or *Newtonian*). A **global reference frame** allows the position of any single point to be specified with respect to the defined origin. In other words, this frame is used to describe movement of the *entire* system as a whole relative to the start. However, to clearly describe the orientation of the entire system or the individual segments of the system (as opposed to defining the system as a single point moving in space), we must construct another reference frame within our global frame that moves with the system. This second frame is called the **local reference frame** (also *anatomical*, *cardinal, moving, relative, segmental,* and *somatic*) and has an origin and axes attached to the body. The origin of the local reference frame is established at the center of mass of the system or a system segment. The axes are orthogonal, have the same handedness as the global frame, and are aligned with those of the global frame when the system is in the anatomical position (Figure 2.13).

Depending on the needs of the observer, movement can be further defined by attaching multiple local coordinate frames to the system (Robertson et al., 2014; Zatsiorsky, 1998). For example, one local frame may be oriented similarly to the global frame and maintain that orientation while moving with the system. This type of frame is used to describe change in system orientation relative to its original orientation and is sometimes called *moving*. A second local frame (sometimes called *somatic* or body) may be fixed within the system that changes its orientation in space when system orientation is changed (describing change in position of one segment of the system relative to another).

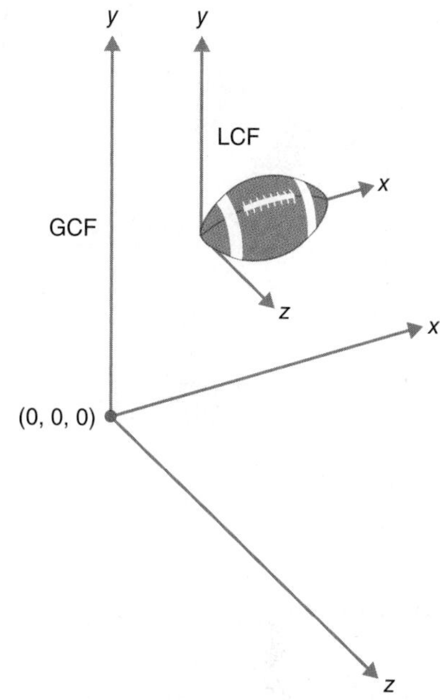

Figure 2.13 **Local coordinate frame located within the global coordinate frame.** Can you think of some sports in which a local coordinate frame would be necessary?

To gain insight into the difference between global and local reference frames and the information that each provides, let's use the example of observing a person driving a car. If you view the person driving the car from a global reference frame defined as the surface of Earth, you observe them to be moving in the positive x direction. However, the person is not necessarily moving in the positive x direction *relative to the car*. More specifically, viewed from the global frame, the driver's hands are moving in the positive x direction because they are attached to the driver. But relative to a local frame of reference, the driver's hands are on the steering wheel. The steering wheel allows the hands to move only up and down, not forward and backward. So, motion varies widely depending on your point of view. Therefore, establishing more than one reference frame provides multiple points of view that help to fully describe the event.

Tumbling in gymnastics is a great example of using multiple coordinate frames in sport to completely describe the motion of a system (Figure 2.14).

The only information relative to a global reference frame is that the system (gymnast) is moving forward (positive in the x direction), which is a huge understatement in terms of this skill. We need to further describe motion of the system in terms of rotation around the x-, y-, and z-axes of a local reference frame. The rotations around the x-, y-, and z-axes are also called **roll**, **yaw**, and **pitch**, respectively. In addition, the act of tumbling is being produced by changes in relative body segment positions. So a third axis frame is needed that helps to describe segmental motions. For example, the gymnast is moving forward and momentarily upward (global: $x+$ and $y+$ directions), may be rolling in midair (local moving: y and z directions around the x-axis), and have legs in a tuck because of knee flexion (local knee somatic: x and y directions around the z-axis). You can clearly see that motion of a system can be described with much greater detail and accuracy if multiple levels of reference frames are used.

Degrees of Freedom

Degrees of freedom (DOF) are literally the number of parameters that are "free" to vary. In biomechanical terms, DOF are the number of independent ways in which a system can move or the number of values required to completely describe system motion relative to the established

Roll A combination of rotation and translation in which each point on a surface contacts a unique location on the other surface.

Yaw Rotation of a system around the *y*-axis.

Pitch Rotation of a system around the *z*-axis.

Degrees of freedom (DOF) The number of independent ways in which a system can move.

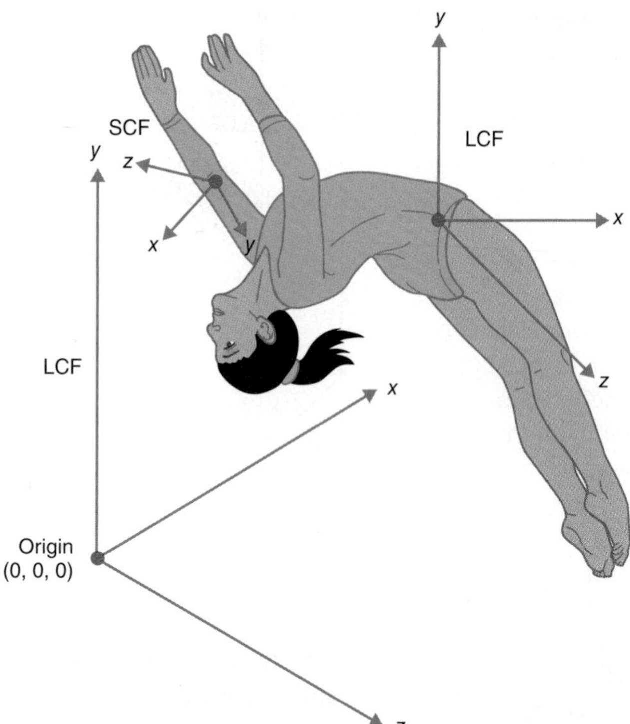

Figure 2.14 **Gymnast with global, local, and somatic reference frames indicated.** Which coordinate frame(s) are the most important in this situation?

coordinate reference frame. Two-dimensional (planar) motion has three degrees of freedom (some combination of movement in two possible directions and rotation around one axis). However, six DOF are required to describe 3-D system movements: a rigid segment freely suspended in space can move in three directions (three parameters required to specify the location of the segmental center of mass) and rotate around three axes (three values needed to describe segment orientation). If any constraints are imposed on the system, they must be subtracted from the six possible degrees of freedom. For example, a ball-and-socket joint such as the hip can rotate around any axis but cannot under normal circumstances move out along any of the three directions without dislocating (i.e., it is constrained in three of six possible movements types). Therefore, the DOF for a ball-and-socket joint is three (DOF = six minus three constraints). The human body is a system of segments. Any rigid segment in space has six DOF, therefore a system of n rigid segments has $6 \times n$ DOF. In systems such as the human body, joints connect the segments. These joint connections (and their skeletal features) impose constraints by limiting motion in certain planes or axes, so the number of DOF decreases (DOF in a jointed segmental system = $6n$ – number of constraints) (Zatsiorsky, 1998). So, even though the human system is made up of many segments (each of which would theoretically have six DOF), any particular segment or part of the system is constrained in one or more ways. Whereas planes, axes, and directions are ways of describing motion of the system (or a segment of the system), DOF give information as to *available* motion of the system or system segment.

2.4 MOTION AT SEGMENTAL LINKS

The human skeleton is linked together at various types of joints. So if the human body is the system of interest, then we must be able to describe the relative motion that occurs between two segments of the system. Once again, descriptions of joint motions vary slightly depending on the discipline or textbook. Therefore, the motions defined in the following sections represent generalizations of terminology used across various movement disciplines (Floyd, 2018; Oatis, 2017; VanPutte et al., 2017; Watkins, 2010).

Traditionally, intersegmental motions are termed *joint actions* and are introduced as new vocabulary to memorize. But one purpose of this book is to introduce as early as possible the use of biomechanical concepts to analyze movement. To make using these concepts easier and to aid understanding of skeletal movements, we will use a free-body diagram. A **free-body diagram** is a simplified representation of the system *free* of the movement environment. This type of minimalist rendering is often a stick figure or geometrical model of the system with the center of mass and points of contact with the environment depicted (Figure 2.15).

The center of mass must be represented because it is the point at which the force of gravity is concentrated. Points of contact with the environment are important because these are areas where the system is acted on by external forces. In situations in which a joint (segmental link) in the body is the system of interest (Figure 2.16), internal forces such as muscle forces and forces at the link itself (joint reaction forces) are usually represented along with the center of mass and environmental contact points (external forces).

Free-body diagram A simplified representation of a system free of the movement environment.

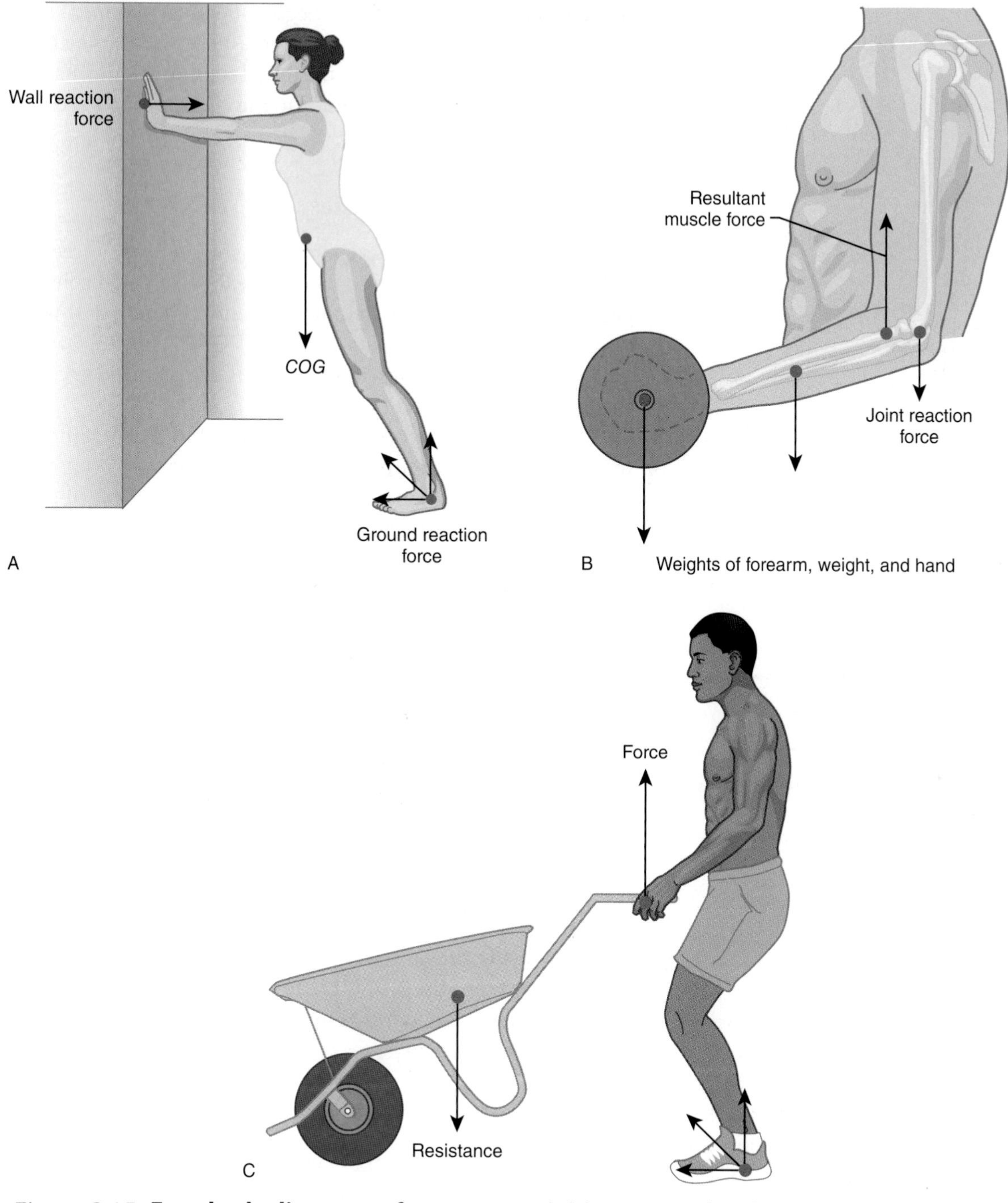

Figure 2.15 **Free-body diagrams of common activities.** Draw a free-body diagram of your current position.

Arrows showing the directions in which they act represent the forces in free-body diagrams. The specific characteristics of force representation with arrows are discussed in Chapter 3. We have not begun to elucidate all of the possible forces acting internally or externally to the system, so free-body diagrams in this chapter consist of the system alone. As you will see in subsequent chapters, the human body is a highly complicated system that can be engaged in highly complex movements. Free-body diagrams help visualize the many complex facets of a movement situation in a simplistic form for analysis. In essence, sometimes we must "suspend reality" and create a representation of only the most important aspects

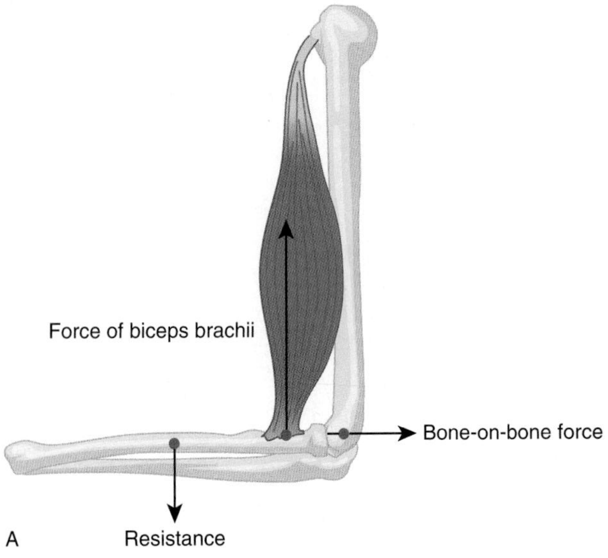

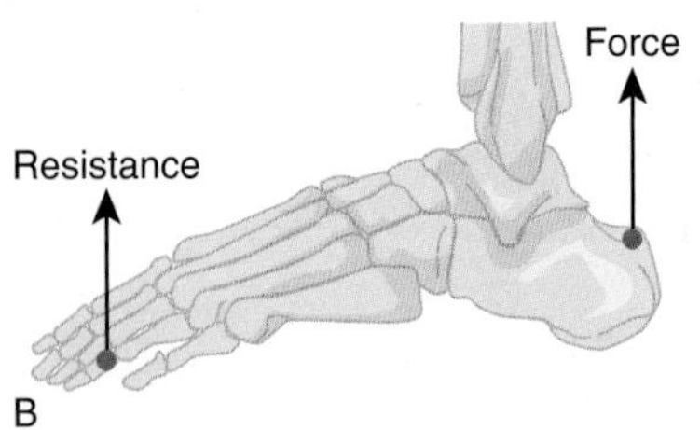

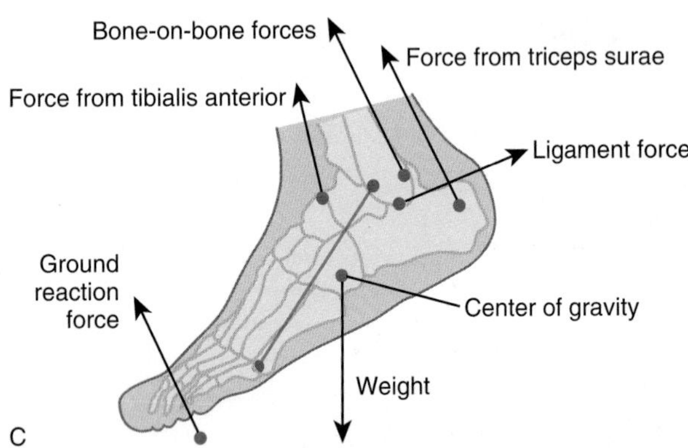

Figure 2.16 **Anatomical free-body diagrams.** Draw a free-body diagram of the hand during the performance of a push-up exercise.

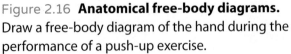

Flexion Segmental motion in a sagittal plane, around a mediolateral axis, and away from the anatomical position.

Extension Returns a segment to the anatomical position in a sagittal plane around a mediolateral axis and is described as increasing the angle at the joint.

of the motion. Even though the segmental link movements that follow are relatively simple, using free-body diagrams at this point enables you to engage in the initial stages of biomechanical analysis by becoming accustomed to the visualization necessary in the field of biomechanics.

Motions in Sagittal Planes

Most body segmental links (joints) possess degrees of freedom in the sagittal plane: ankle (talocrural), elbow, hip, interphalangeal, intervertebral, knee, shoulder (glenohumeral), and wrist. Of course, the relative motion of two segments in the sagittal plane occurs around a mediolateral axis and is in reference to the anatomical position (Figure 2.17).

For example, **flexion** of a joint results in segmental motion in a sagittal plane, around a mediolateral axis, and away from the anatomical position. Flexion is sometimes described as causing a decrease in joint angle. The angle that should be visualized in this convention is the angle formed by the segment of interest and the frontal plane while in the anatomical position. Specialized terms are used for flexion at the ankle: *dorsiflexion* deviates the foot from the anatomical position such that it moves in a sagittal plane toward the anterior tibia, whereas *plantar flexion* is the opposite motion in a sagittal plane away from the anatomical position resulting in the foot moving toward the posterior surface of the tibia (e.g., pointing the toe). **Extension** returns the segment to the anatomical position in a sagittal plane around a mediolateral axis and is described as increasing the angle at the joint. If extension is continued to the point that it exceeds the anatomical position, the movement is termed *hyperextension. Hyperflexion* is technically also possible but is prevented by joint structure (contact of one segment with another) in most non-injured joints. However, hyperflexion does occur in a healthy shoulder joint when it is flexed beyond vertical.

Motions in Frontal Planes

Body segmental links that are capable of frontal plane motion (around an anteroposterior axis) are the foot (subtalar and transverse tarsal joints), hip, intervertebral, metacarpophalangeal, shoulder, and wrist (Figure 2.18). Once again, all anatomical motions are described relative to the anatomical position.

Motion of the axial skeleton (at the intervertebral joints) in the frontal plane around an anteroposterior axis is termed *lateral flexion.* For example, leaning the trunk to one side and tilting the head are both lateral flexions.

The hip, metacarpophalangeal, and shoulder (glenohumeral) joints are capable of abduction and adduction.

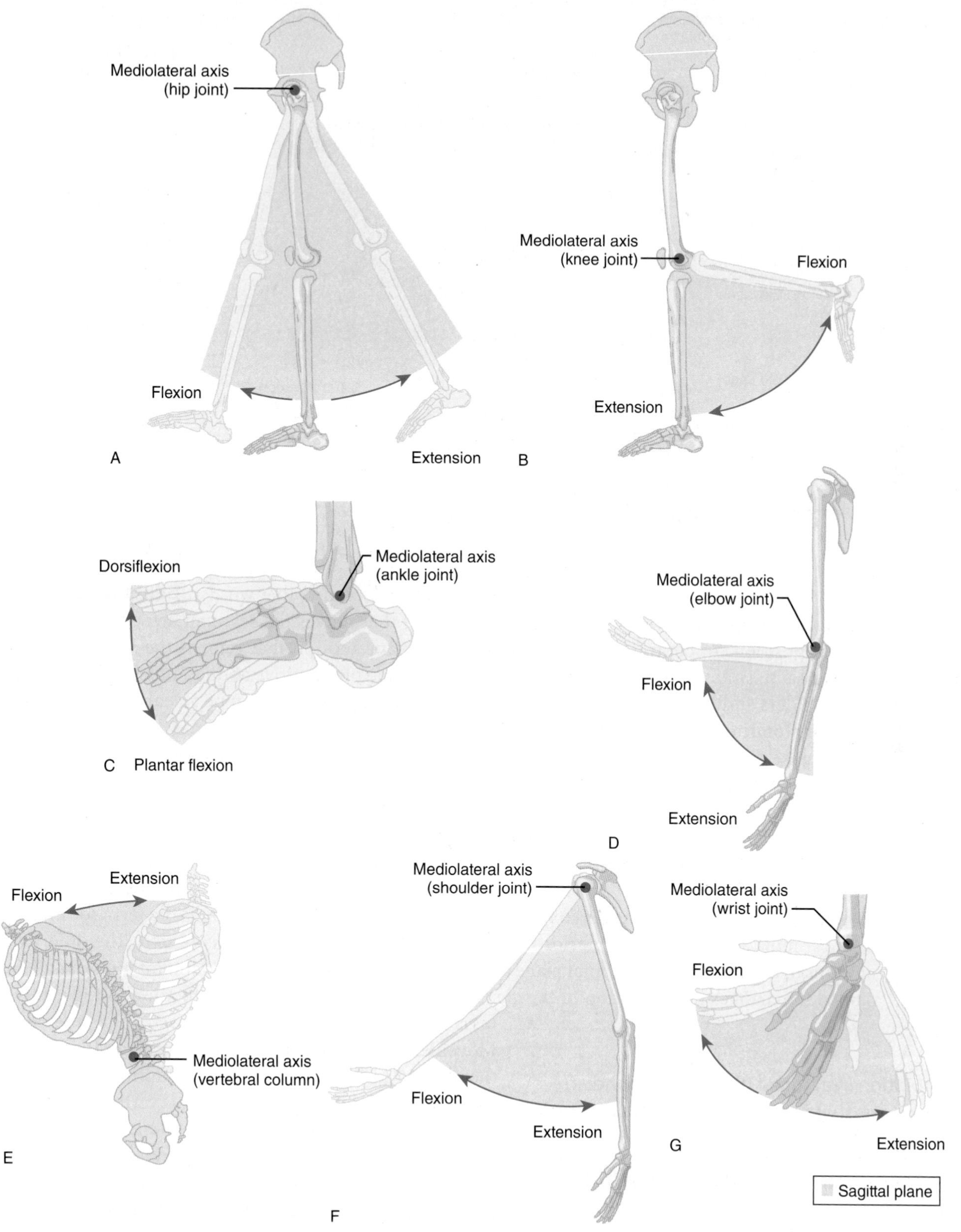

Figure 2.17 **Motions in sagittal planes.** Think of resistance training exercises that occur in sagittal planes.

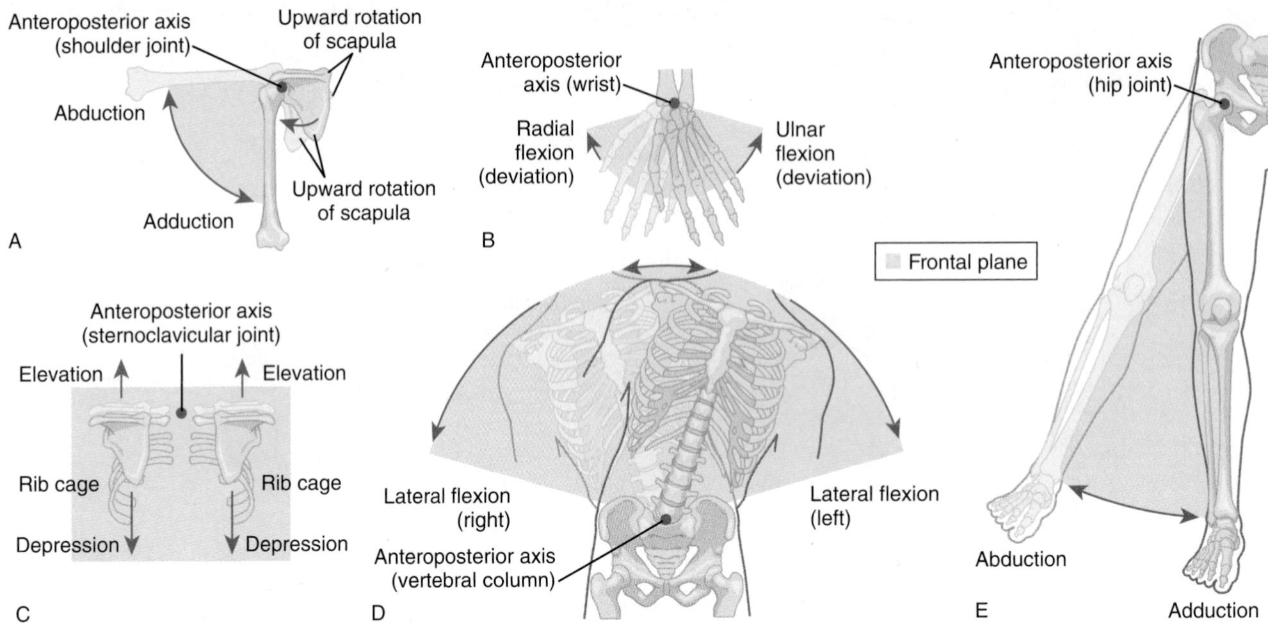

Figure 2.18 **Motions in frontal planes.** What exercises occur in frontal planes?

Abduction is the frontal plane movement analogous to flexion. In other words, **abduction** is motion in a frontal plane and around an anteroposterior axis that moves the segment away from the anatomical position. In contrast, **adduction** is frontal plane motion that returns the segment to the anatomical position. Jumping jacks is an exercise example of repetitive abduction and adduction of the shoulders and hips. Abduction of the metacarpophalangeal joints can be visualized by standing in the anatomical position and then spreading the fingers.

Because the number of degrees of freedom at the shoulder is large, many specialized movement terms are necessary to fully describe its motion. As previously mentioned, the glenohumeral joint of the shoulder is capable of flexion and extension in the sagittal plane and abduction and adduction in the frontal plane. However, when motions of the shoulder are being described, we must consider more than just the glenohumeral joint. For example, the scapula (shoulder blade) is capable of four motions in the frontal plane. Two of these motions are actually associated with sternoclavicular joint motion and are therefore considered motions of the shoulder girdle: *elevation* is the term for motion of the scapula superiorly, and inferior movement of the scapula is called *depression.* The other two motions of the scapula in the frontal plane are associated with acromioclavicular joint motion and are named for changes in orientation of the glenoid fossa of the scapula. The orientation of the glenoid fossa changes when the arms are raised above the head. Movement of the scapula in a frontal plane that causes the glenoid fossa to rotate upward (inferior angle of the scapula moving laterally away from the vertebral column) is called *upward rotation. Downward rotation* is movement of the shoulder that results in the glenoid fossa rotating downward (scapula rotating in a frontal plane, around an anteroposterior axis, with the inferior angle moving medially toward the vertebral column).

Specialized terms also exist for frontal plane motion of the foot at the subtalar and transverse tarsal joints. **Eversion** is frontal plane motion around an anteroposterior axis

Abduction Motion in a frontal plane and around an anteroposterior axis that moves the segment away from the anatomical position.

Adduction Frontal plane motion that returns the segment to the anatomical position.

Eversion Frontal plane motion around an anteroposterior axis such that the sole of the foot rotates outward or laterally.

such that the sole of the foot rotates outward or laterally. If the sole of the foot is rotated medially the motion is termed **inversion**. Take care with these terms, because they are often confused with ankle pronation and supination, which in this chapter are discussed with motions that occur in oblique or multiple planes.

Deviation is a term used for frontal plane motion around an anteroposterior axis at the wrist. If the thumb is moved closer to the radius (relative to the anatomical position), the movement is called *radial deviation* (also abduction and radial flexion). *Ulnar deviation* is frontal plane motion at the wrist that results in hand motion toward the ulna (adduction and ulnar flexion).

Motions in Transverse Planes

Skeletal segmental links with DOF in the transverse plane (around a superoinferior axis) directly from the anatomical position are the hip, intervertebral, shoulder, and radioulnar (Figure 2.19).

The axial skeleton (intervertebral joints) is capable of transverse plane motion at both the neck and trunk. *Rotation to the left* and *rotation to the right* are transverse plane motions accomplished by simply turning the head or twisting the torso to the left or right.

Internal rotation and external rotation are motions that are available at the hip and shoulder. For example, *internal rotation* at the shoulder (also called *inward* or *medial rotation*) is motion of the arm segment around a superoinferior axis that rotates it from the palms-forward anatomical position to a posture in which the palms are facing more medially and finally posteriorly. *External rotation* (outward or lateral) is the opposite motion in which the segment is returned to or beyond the anatomical position in a transverse plane (e.g., rotating the leg segment at the hip such that the toes point outward or laterally.

In addition to rotations, the shoulder is also capable of two other transverse plane motions. The shoulder girdle can be protracted and retracted. *Protraction* is movement of the shoulder girdle anteriorly away from the spine. Protraction is accompanied by abduction of

Inversion Frontal plane motion around an anteroposterior axis such that the sole of the foot rotates inward or medially.

Deviation Frontal plane motion around an anteroposterior axis at the wrist.

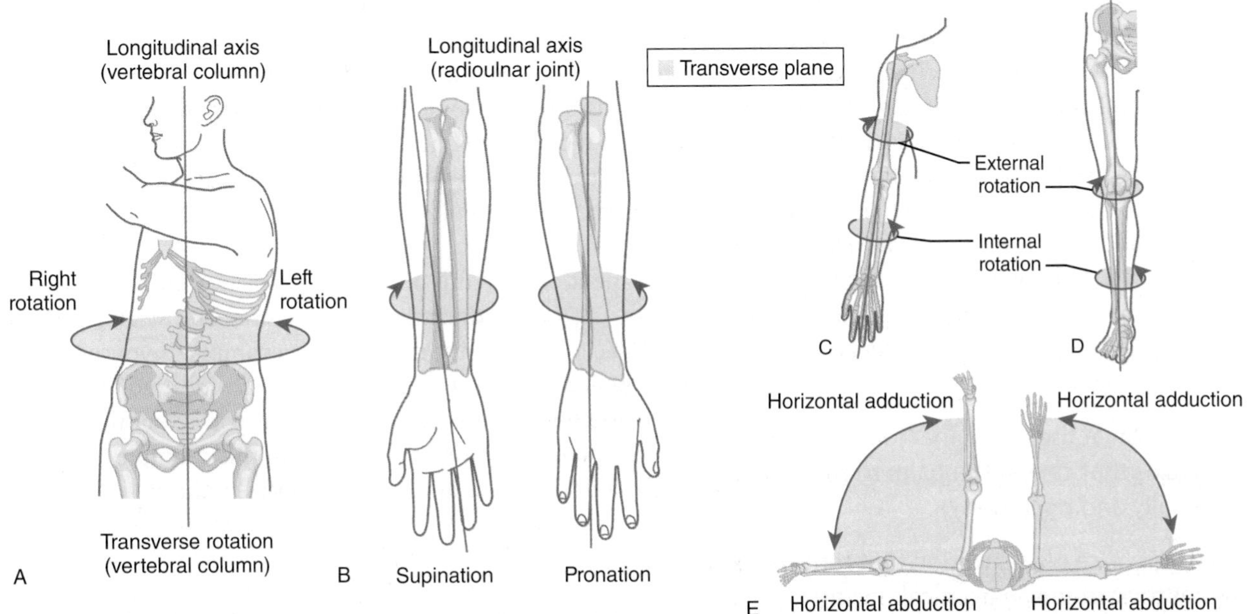

Figure 2.19 **Motions in transverse planes.** Think of daily activities that occur in transverse planes.

the scapula. *Retraction* is movement of the shoulder girdle posteriorly toward the spine and is accompanied by adduction of the scapula.

The radioulnar joint of the forearm is capable of pronation and supination. In the anatomical position, the radioulnar joint is in supination. **Pronation** is rotation at the radioulnar joint around a superoinferior axis that causes the palm to turn toward the body (medially and posteriorly). Do not confuse this action with internal rotation. Internal rotation occurs at the hip or shoulder; pronation occurs specifically at the radioulnar joint and is an independent action from rotation (i.e., you can rotate the palm medially without moving at the shoulder joint at all). Of course **supination** is the transverse plane motion that returns the radioulnar joint toward the anatomical position (palm moves anteriorly). Again, supination is independent of external rotation at the shoulder.

Rotation to the left and right, internal and external rotation, and pronation and supination can all occur directly from the anatomical position. Some transverse plane motions are possible if a segment is first moved into the transverse plane. For example, the frontal plane motions of abduction and adduction can occur in the transverse plane if the shoulder or hip is flexed first. If the hip or shoulder is flexed until a segment is moved into the transverse plane and then rotated laterally around a superoinferior axis, the movement is called *horizontal* or *transverse abduction* (also called *horizontal extension*). *Horizontal* or *transverse adduction* (horizontal flexion) is the opposite motion in which the segment is rotated around a superoinferior axis and through a transverse plane toward the midline of the body (medially).

Motions in Oblique and Multiple Planes

Some segmental link motions do not fit into the cardinal planes and axes described. For example, you have learned that planes and axes can be oblique. Therefore, if a segment is moved into an oblique plane before the motion of interest occurs, the segmental link motion is also oblique or diagonal (e.g., oblique abduction). In addition, movements can occur in more than one plane of motion, and some motions occur in planes but not around axes.

One segmental link motion that occurs in more than one plane (around more than one axis) is called *circumduction* (Figure 2.20).

Circumduction is possible at any joint in which the proximal end of the segment can remain fixed while the distal end of the segment is capable of describing a circle (so that the segment as a whole describes a cone). Because circumduction is a complex motion in multiple planes and axes, the joints that are capable of circumduction are the ones that have DOF in terms of flexion, extension, abduction, and adduction (hip, intervertebral, metacarpophalangeal, shoulder, talocalcaneal, and wrist).

Two other motions that do not quite fit into a clearly defined planar category are ankle and foot pronation and supination. Pronation of the ankle and foot requires simultaneous ankle dorsiflexion, forefoot abduction, and subtalar eversion. Supination is simultaneous ankle plantar flexion, forefoot adduction, and subtalar inversion (Floyd, 2018).

Although all of the segmental link motions mentioned to this point occur in planes and around axes, motion can occur in a plane but not around an axis. These nonaxial movements occur when a system segment is being moved as the result of an adjacent axial joint motion. For example, if you perform a bench press in weight lifting, the hands are moved in a plane but not around an axis. Can you name the plane? The planar motion of the hands in this example is caused by axial motions at the elbows and shoulders.

Pronation Rotation at the radioulnar joint around a superoinferior axis that causes the palm to turn toward the body (medially and posteriorly).

Supination Transverse plane motion that returns the radioulnar joint toward the anatomical position (palm moves anteriorly).

Use of Terminology

To avoid confusion, we need to refer back to the previous discussions of planes, axes, directions, and segmental motions. As can be seen, the cardinal planes (sagittal, frontal, and transverse) and axes (mediolateral, anteroposterior, superoinferior) are especially useful for describing musculoskeletal movement *from the anatomical position*. Because the anatomical position is a constant starting reference point, it is accompanied by cardinal axes (and planes) that are always the same. However, natural motion does not often begin in the anatomical position. For example, one can describe the movements of *individual body parts* of a tennis player in reference to the anatomical position, but it does not make sense to describe the position on the court of the *entire tennis player* (system) in reference to the anatomical position. So we need the ability to establish frames of reference outside of the body and independent of the anatomical position. In addition, we need to define multiple frames of reference. Cartesian coordinate frames of reference are versatile. The *x*, *y*, and *z* are not only *axes* that can be used in place of the cardinal axes but also can form a frame of reference in terms of *planes* and *directions* of motion of both individual segments and the entire system. For example, hip and knee extension and flexion occur around a mediolateral axis. This motion would be accompanied by motion of the femur and tibia in a sagittal plane (in reference to the anatomical position). What about motion of the foot or entire body (system) as a result of repetitive hip and knee flexion and extension? Hip and knee extension and flexion could be described in relation to the previously defined Cartesian coordinate frames of reference: hip and knee motion are occurring around *z*-axes, resulting in motion of the femur and tibia in planes formed by the *x*- and *y*-axes. Movement at the hip and knee results in foot motion in the $x\pm$ (forward and backward) and $y\pm$ (upward and downward) directions. The total resultant motion of the entire system center of gravity is $y\pm$ and $x\pm$ (Figure 2.21).

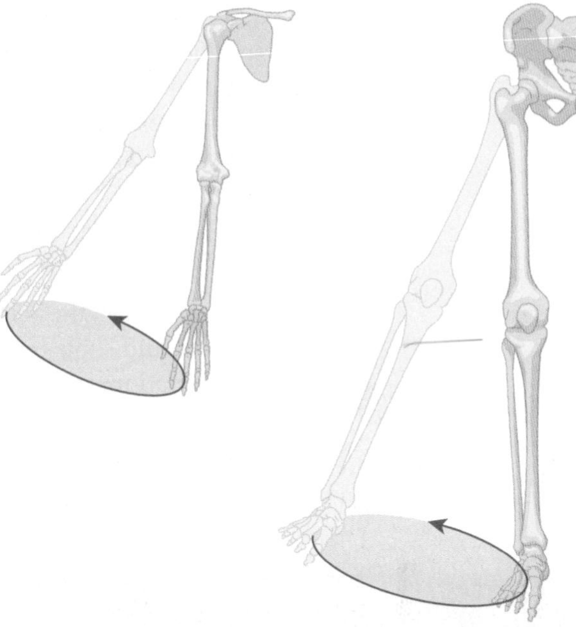

Figure 2.20 **Motion in oblique and multiple planes.** Name some oblique plane sports motions.

Figure 2.21 **Athlete and associated reference systems.** In what direction does the tennis ball move?

In addition to enhanced descriptive ability, Cartesian coordinate frames have an origin but do not require a specific starting body position to describe motion at any given point in time (i.e., the above description did not require any comparison to the anatomical position). So even though some terminology in the disciplines of anatomy and biomechanics is redundant, having several different conventions allows highly detailed interdisciplinary description of system movement.

2.5 THE MOVEMENT ENVIRONMENT

During a game of soccer, the ball is kicked for several different reasons and in different ways, including passes, clears, and shots on goal. At times the ball is in motion when struck by the foot, and at other times it is stationary. The movement—or lack thereof—of the ball is an environmental factor that must be taken into account when performing each kick. The nature of the skill is *open to change* when the ball is in motion, which is quite different than when the skill is performed with the ball at rest. The following section describes these two major environmental categories in which movements are performed. Note that some seemingly closed skills may take on properties of an open skill when external environmental factors have the potential to affect the outcome. In essence, a continuum of options exists between closed and open skills. Skills performed under the most constant conditions are at one end, and those with the most variable conditions are at the other.

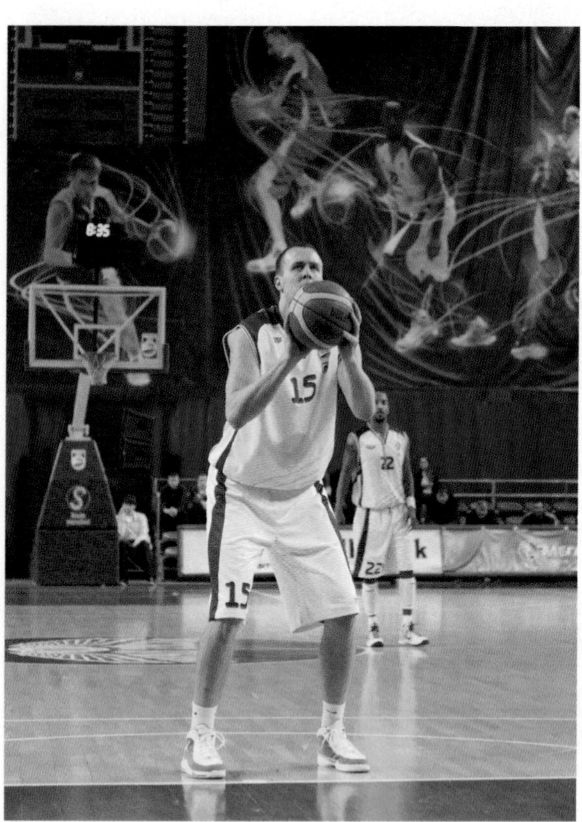

Figure 2.22 **Basketball free throw.** Think of some other closed sports skills.

Closed Skills

A skill is defined as **closed** when it is performed under standard environmental conditions (Figure 2.22). A basketball free throw is a good example; the goal is always at the exact same height, the free throw line is the same distance from the goal, and the ball is a standard size and weight.

Other examples include soccer penalty shots, putting in golf, bowling, archery, throwing horseshoes, and batting in T-ball. Notice that in each of these skills, the possibility of an external environmental influence is present. For instance, although the soccer penalty kick is taken from a standard distance and position, with a standard size ball at a standard size target, certain factors influence how the skill is performed. The playing surface (grass or artificial), position and tendencies of the goalkeeper, or wind may all affect how a player performs the skill. Putting on a golf green also presents factors that may affect the ball after the stroke is taken, such as the grain of the grass and changes in elevation, but the skill of striking the ball is constant. Few sport skills could be considered entirely closed or free from any environmental influence.

Open Skills

When movement must be altered because of the changing dynamics of the activity, environment, or object of interest, a skill is considered **open** (Figure 2.23).

Referring back to the game of soccer, you may have realized by now that most skills are not performed under

Closed skill Skill performed under standardized or predictable environmental conditions.

Open skill Skill performed in a changing or unpredictable environment.

static or standard conditions; the ball is in motion most of the time. To be successful, players must react to the ball in this constantly changing environment. Dribbling and passing could be considered closed skills when performed in isolation during practice, but these skills are not designed to be used in isolation. During a game, no two passes are likely to be identical, and dribbling is a dynamic skill that changes constantly depending on defensive pressure, open spaces, position of teammates, and the velocity of the player with the ball. Some sport and recreation skills are performed in less changeable environments, and some are performed in conditions that are highly unpredictable and constantly changing.

Figure 2.23 **Soccer header.** What are some closed skills in soccer?

2.6 TYPES OF MOVEMENT

Linear and Angular Motion

The term *motion* has been used many times thus far in our description of the system. We must now be more specific and clarify exactly what is meant by the word *motion*. **Motion** can be defined as a change in position with respect to both spatial (relative to points in space) and temporal (relative to time) frames of reference. In other words, a system moves in a given space and that movement requires a given amount of time. No change in motion occurs without force. For now, we will simply define a **force** as something that possesses the capability to cause a change in motion or shape of the system (i.e., a "pull" or "push"). Motion is relative, so the frame of reference is extremely important. **Relative motion** is the motion of one object with respect to a reference object. For example, if you are standing still it could be said that you are motionless relative to Earth. However, Earth is in orbit around the sun and moving through the galaxy. So although you may not be moving relative to Earth, you are in motion relative to other planets. A more specific example is two hockey players skating next to each other at 5 kilometers per hour (km/hr). They are both moving at 5 km/hr relative to the rink, but neither is moving relative to the other (0 km/hr). If another hockey player then skates past them at 15 km/hr, this third player is moving at 15 km/hr relative to the rink and 10 km/hr relative to the other two players (the numbers are subtracted if both objects are moving in the same direction). On the other hand, one player could be skating at 15 km/hr toward another player moving at 10 km/hr. In this case, their relative motion is 25 km/hr (the numbers are added if two objects are moving in opposite directions). In addition to motion being relative it can be defined as *translation*, *rotation*, or *general* (translation and rotation simultaneously).

Translation is motion *along* one of the *x*-, *y*-, or *z*-axes in which all points of the system move (change position) at the same time, in the same direction, and the same distance with respect to the defined reference frame despite their location on the system. During translation, all points on the system are moving at the same time, so their path can be represented as one point (e.g., the center of gravity) traveling along a line from one place to another. Therefore, translation is

Motion A change in position with respect to both spatial and temporal frames of reference.

Force Something that possesses the capability to cause a change in motion or shape of the system.

Relative motion The motion of one object with respect to a reference object.

Translation Motion along one of the *x*-, *y*-, or *z*-axes; linear motion.

also called *linear motion.* The linear path of a system in translation can be straight or curved. If the path of the system is a straight line (i.e., no change occurs in direction of motion), the motion is termed *rectilinear translation.* An example of rectilinear motion is the movement of an object when you drop it.

The object would travel in a straight line toward Earth. The change in linear position of the system in a straight line or its *linear displacement* (*d*) is described in terms of linear measurement units (e.g., millimeters, centimeters, meters, and kilometers). *Curvilinear translation* is the same as rectilinear in terms of the points on the system moving together, but in this case the path of the representative point is curved (in an arc) instead of straight (i.e., direction of system motion is changing). An example of curvilinear motion is a ski jumper. As soon as the jumper leaves the ramp, wind resistance slows him down and gravity pulls him toward Earth. Therefore, the resultant path of the ski jumper is an arc (curvilinear). Because the shortest distance between two points is a straight line, the actual path of the system in curvilinear motion is a greater distance than the linear displacement measured relative to the frame of reference.

Rotation occurs when the system is restricted to move *around* a fixed axis and therefore in a circular path (such as the segmental link motions described previously). Because the path of motion of one point on a rotating system or system segment describes a circle (and therefore the location of one segment relative to another forms an angle), rotation is also called *angular motion.* In the case of rotation, all points on the rotating segment have the same *angular displacement* (θ; the Greek letter theta), measured in degrees or radians. However, each point on a rotating segment has a different curvilinear displacement depending on its distance from the axis of rotation, which is called the *radius of rotation* (Figure 2.24).

For example, all points on the forearm and hand segments travel 45 degrees *angularly* (θ = 45°) if the elbow is flexed 45°; but individual points representing the center of mass of the forearm, the wrist, and the distal phalanges all describe arcs of varying lengths (i.e., travel different *curvilinear* distances).

Rotation Motion around a fixed axis and therefore in a circular path; angular motion.

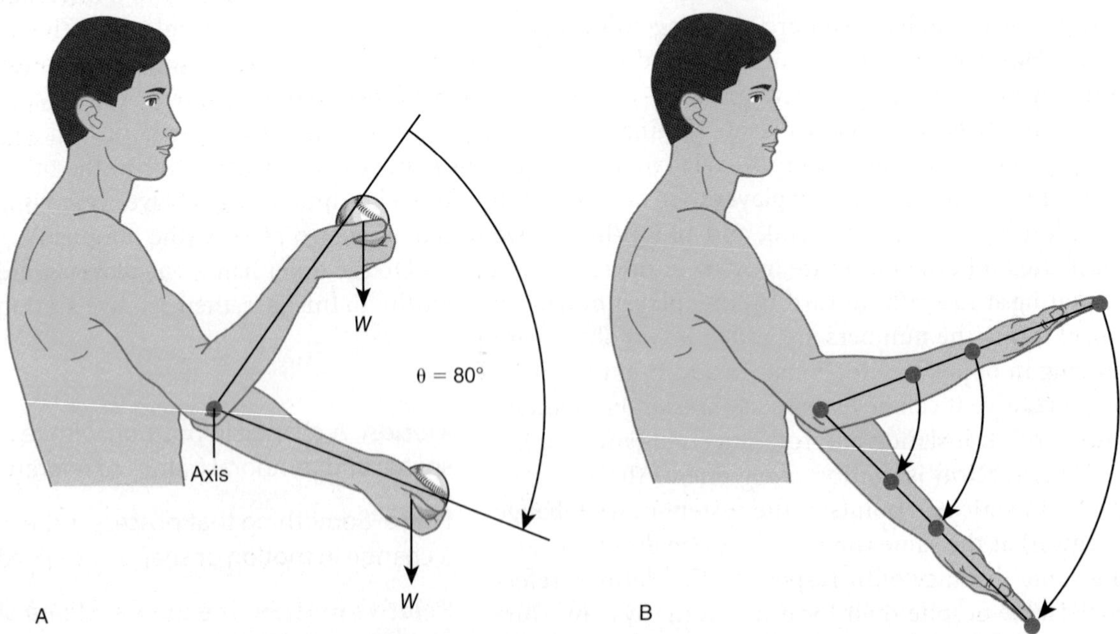

Figure 2.24 **Radius of rotation for various points on the arm.** What happens to the radius of rotation if an athlete uses an implement instead of the hand for striking?

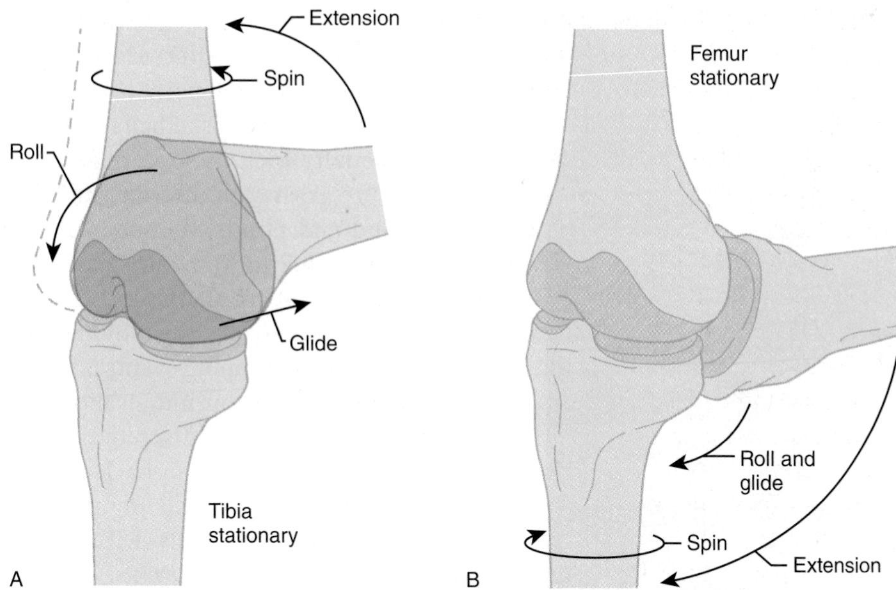

Figure 2.25 **Knee spin, roll, and glide.** Are there any joints in which pure spin occurs?

As previously described, nonaxial segmental motions are produced as a result of adjacent axial motions. In the human system, this motion is actually translation as a result of rotation. This combination of translation and rotation is called *general motion.* Specifically, the entire system or a segment of the system can be linearly displaced (translate) as a result of angular displacement at a segmental link (rotation at joints). For example, a person (the whole system) can run forward (translate) because of actions at the hips, knees, and ankles (joint rotation). Motion at segmental links (joints) is actually a great example of combining translation and rotation. We tend to think of joint movement as pure rotation. However, motion at most synovial joints is actually general in nature. If one segment in the skeletal system moves relative to another with pure rotation (no translation) it is called *spin.* **Spin** occurs if all points on one articulating surface (such as the ulna) come in contact with one point on another articulating surface (the humerus). Contrast this motion with **glide** (sliding) or pure translation, in which a point on one surface glides (or skids) over many points of an opposing surface. **Roll** (a combination of rotation and translation) occurs when each point on a surface contacts a unique location on the other surface (Figure 2.25).

Many natural joint motions are a combination of gliding and rolling. In the preceding tibiofemoral joint example, knee flexion causes the tibia to roll and glide posteriorly on the femur (the femur is in posterior roll and anterior glide). Knee extension causes the opposite situations of femoral anterior roll and posterior glide.

Discrete, Continuous, and Serial Skills

Skills can be classified into general categories by how they appear. Movement has traditionally been defined into three categories; *discrete, continuous*, and *serial.* This text identifies four categories of movement to alleviate some of the confusion that arises when skills don't seem to precisely fit the definition of one of the three traditional categories. A term that is often used synonymously with serial is *repeated discrete*, but it can be argued that these terms describe distinct categories.

Spin Occurs if all points on one articulating surface come in contact with one point on another articulating surface.

Glide Sliding or pure translation in which a point on one surface glides or skids over many points of an opposing surface.

Roll Rotation of a system around the *x*-axis.

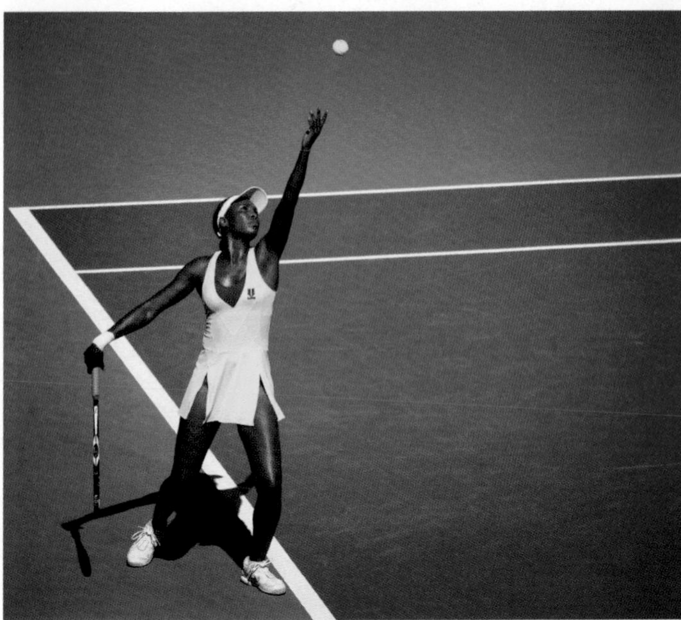

Figure 2.26 **A tennis player performs a discrete skill.** Why is pitching a discrete skill?
© Neale Cousland/Shutterstock

Figure 2.27 **Cycling.** Think of some other continuous skills.
© TORWAISTUDIO/Shutterstock

Discrete skill Motion that has a definite beginning and end-point.

Continuous skill Cycles of motion performed repeatedly with no well-defined beginning or end points.

Repeated discrete Cycles of motion that are seemingly continuous but require a recovery phase between propulsive actions.

Serial skill Movement that comprises a series of discrete motions.

Discrete

A movement is labeled a **discrete skill** when it has a definite beginning and end point (Figure 2.26). Examples of discrete skills are basketball free throws, baseball pitches, a penalty kick in soccer, and the volleyball or tennis serve. Although discrete skills have identifiable start and end points, they may also be initiated or followed by other movements. The long jump, high jump, and pole vault are all discrete skills that directly follow a moving approach. The moving approach to all three track and field jumping events is usually running, which is identified as a continuous movement. The flip turn in swimming is also a discrete skill that follows a continuous skill of freestyle or backstroke swimming.

Continuous skills are distinguished as cycles of motion performed repeatedly with no well-defined beginning or end points (Figure 2.27).

Examples include walking, swimming, bicycle riding, wheelchair "hand-cycle" racing, and in-line skating. Continuous movements can be stopped and started anywhere in the pattern. A critical identifier of continuous movements is the fact that at the finish of one cycle, the body is already in position to perform the next repetition.

Repeated Discrete

Some movements appear continuous but deserve a separate category of classification. Rowing is often labeled a continuous event, but at the end of a stroke the body is not in a position to perform the next stroke.

Between each pull there must be a recovery phase to put the rower in position to pull again. Therefore, a skill of this type is better labeled as **repeated discrete** (Figure 2.28). Another example in this category is traditional wheelchair racing. As in rowing, the propulsion phase leaves the athlete in a down-and-forward position from which she is not ready to push again without recovering to an upright position. Many assembly line jobs also fall into this category because workers rarely finish one repetition of their designated task in position to begin the next.

Serial

Some movements appear to be somewhat continuous in nature but are really a combination of discrete motions; these are called **serial skills**. In track and field, the triple jump is an excellent example of a serial movement Figure 2.29.

A continuous movement (run-up) initiates the activity; then the athlete performs three discrete actions without any stop between them (a hop, a step, and a jump), creating a new complex movement. An uneven parallel bar routine in gymnastics provides another good example. Points are deducted if a gymnast loses momentum during a routine, so it may seem to be a continuous activity. However, the routine is composed of series of discrete skills that are connected together into a constantly flowing pattern.

Figure 2.28 **A rower sculling.** Why is this skill repeated discrete instead of continuous?
© Stefan Ugljevarevic/Shutterstock

Gross and Fine Motor

Depending upon which textbook you are reading, movements are classified as gross or fine motor based on either the size of the muscle groups responsible for the action or for the precision of control necessary for the action. **Gross movements** generally result from major muscle group activity involving large muscles.

Movements classified as gross motor require little precision and include most fundamental skills such as walking, running, throwing for distance, skipping, and weight lifting. **Fine movements** are precise and generally controlled by small muscle groups or individual muscles (Figure 2.30). Examples include writing, playing the piano, threading a needle, and typing. Sometimes large muscle groups are asked to perform precise movements, and conversely, small muscles may be contracted with little regard for precision. For this reason, movement is classified as gross or fine *motor*, based on the precision of motion that the task requires.

Two examples that use common muscles or groups of muscles can help explain why precision of movement should be the best indicator for distinguishing between gross and fine motor skills. The quadriceps is a large four-muscle complex that performs multiple gross motor tasks during daily activity. Most actions that the quadriceps muscles perform require little precision: walking, climbing stairs, running, and jumping. However, the quadriceps muscles are asked to function with varying degrees of precision during some tasks. They function eccentrically

Gross movement Motion that is the result of large muscle group activity and requires little precision.

Fine movement Motion that is precise and generally controlled by small muscle groups.

Figure 2.29 **A triple jump.** Why is the triple jump not considered to be repeated discrete?
© Inspiring/Shutterstock

Figure 2.30 **Power lifting and texting.** Are there some precise skills performed by large muscle groups?

in lowering the body from standing to sitting position, a movement that requires more precise control to avoid falling heavily into the seat. Drivers of autos with manual transmissions take quadriceps precision to an even higher level. Any time the car is stationary, a driver must depress the clutch with his left leg (a gross motor skill requiring little precision), then *slowly* retract the leg to engage the clutch smoothly and avoid stalling the engine. This engagement point occurs in a relatively small range of motion and is controlled by a large muscle group.

At the opposite end of the spectrum are the small muscles that control side-to-side eye movement. These muscles are diminutive relative to the quadriceps and generally perform highly precise movements—such as allowing your eyes to move from word to word as you read this sentence. If you've never done it, watch a friend's eyes as he or she reads a paragraph; you'll be able to see each tiny movement as the eyes advance to a new word or line. These eye movements are called *saccades*, and they occur even when we are asked to follow an object in constant motion with our eyes such as watching a baseball travel from a bat into the outfield. These saccades require highly precise control to follow moving objects and are considered a fine motor skill. How could these muscles ever be considered gross motor effectors? Consider the case of a golfer who follows an instructor's advice: keep your head down and eyes on the ball until after you make contact. The ball leaves the club face traveling at high velocity, then the golfer must "find the ball" by moving her eyes rapidly in the approximate direction of ball flight. This saccade should be considered a gross movement, because the initial objective is to catch up to the ball. Once the ball is identified in flight, the muscle contractions become fine motor again as the eyes focus on and track it using small saccades.

Most sport and recreational activities require a combination of fine and gross motor skills. At times, the same skill may require different degrees of precision, depending on the environmental context. The car driver who had to let the clutch out smoothly from a stop does not have to be as precise when shifting between two gears while the car is in motion. An accomplished piano player can change the nature of a piece by playing the same notes, at the same tempo, but varying the pressure with which she depresses the keys. Complex tasks may include movements that require large and small muscle groups working together. For example, dart throwing combines fine finger control with gross control of the triceps to propel the dart toward the board. Therefore, movements can fall along a continuum from fine to gross, even within the same activity. This textbook identifies movements on this continuum based on the precision of movement rather than the size of the muscle or muscle group.

Kinetic Chain

In trying to describe motion of the system, using simplified models (such as the free-body diagrams referred to previously) is sometimes helpful. Thus far in the process of defining the system and describing its movements, we have modeled the human body system as having multiple segments connected at links (joints). If one imagines this system of segments and links as several interconnected chains, some specialized motion situations become easier to analyze and understand. For example, why does timing matter in throwing a baseball? Why does injury in one joint often lead to injury in another? Detailed understanding of the answers to these questions will come later. But even at this point, one can gain some basic insight into system motion if one understands the concept of a kinetic chain. At the basic structural and movement levels, a **kinetic chain** is simply a system of linked rigid bodies subject to force application. Remember that because all of the segments in the chain are linked together, motion at one link in the chain affects force transfer and therefore motion at one or more other links in the chain. For example, maximal medial rotation at the shoulder interferes with elbow flexion (i.e., degrees of freedom changes because of obstruction by the body). At the system performance level, a kinetic chain is an optimally timed sequence of link motions involving inter segmental force transmission. Kinetic chains vary in their particular characteristics because segmental links are all structurally different and movement situations are highly individualistic.

One type of classification for kinetic chains is their complexity. For example, in the arm kinetic chain, each segment participates in no more than two linkages (i.e., the hand is linked to the forearm; the forearm is linked to the hand and upper arm; and the upper arm is linked to the forearm and the shoulder) and is called a **serial** or **simple kinetic chain**.

A kinetic chain can also be complex (or branched). A **complex kinetic chain** is one in which a segment is linked to more than two other segments (Figure 2.31). For example, the human torso (if modeled as one segment) has two links at the hips, one at the neck, and two at the shoulders.

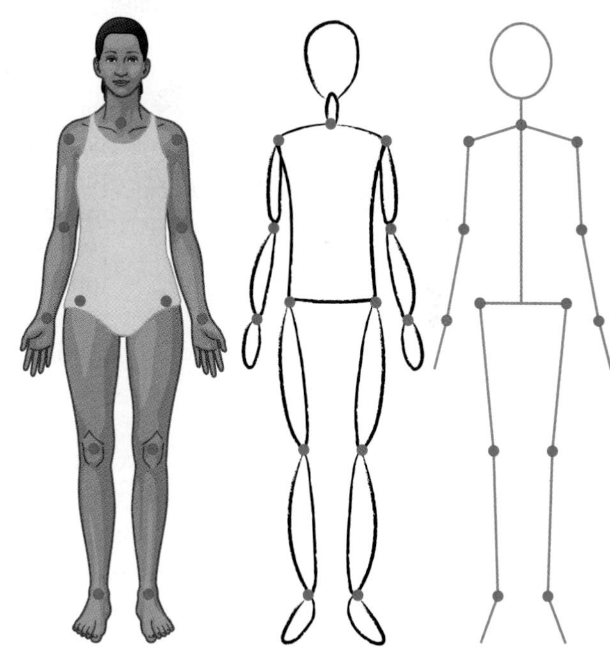

Figure 2.31 **A complex kinetic chain.**

Whether a kinetic chain is serial or complex is important in terms of total mobility of the chain and control of movement by the motor system. **Mobility** is the total DOF of a kinetic chain. Recall from our discussion of degrees of freedom that the number of segments and the number of links to other segments changes the total number of degrees of freedom (Zatsiorsky, 1998). Also, motion at one link can affect total chain DOF. In addition, more sophisticated methods of motor control are necessary as the chain becomes more mobile (gains degrees of freedom).

One of the most important considerations for kinetic chain classification, in terms of application to rehabilitation and sports training, is whether the chain is open or closed. Remember that, although the words are the same, this concept is not the same as an open or closed skill. The following are definitions of open and closed chain movements, which may or may not be performed in an open or closed environment.

Kinetic chain System of linked rigid bodies subject to force application.

Serial or simple kinetic chain Kinetic chain in which each segment participates in no more than two linkages.

Complex kinetic chain Kinetic chain in which a segment is linked to more than two other segments.

Mobility The total degrees of freedom of a kinetic chain.

Figure 2.32 **A baseball pitcher as a complex open kinetic chain.** Are there some sports situations that are not complex open chain?

Open Kinetic Chains

An **open kinetic (kinematic) chain** is one in which the most distal (or terminal) segment is free ("open") to move. Another classification is that motion can occur at one link (joint) in the chain without cooperative motion at other links. Therefore, open kinetic chains have greater mobility (more DOF). The force is usually applied at a more proximal segment (or proximal end of the same segment) to initiate motion. The force then can be transferred throughout other links in the chain. Of course, open kinetic chains vary in number of linked segments and complexity. An example of a serial kinetic chain motion with minimal segments is a barbell curl exercise.

The terminal hand segment holding the barbell is free to move. The force of the elbow flexors is applied to the proximal end of the forearm segment. In this example, the elbow is free to flex without motion at the shoulder or wrist. The advantage of such open kinetic chain exercise movements is that emphasis can be placed on one muscle or specific group of muscles (in this case, the elbow flexors—biceps brachii, brachialis, and brachioradialis). Throwing skills are open kinetic chain motions that are complex and have many participating segments (Figure 2.32).

Throwing is also an example of the need for optimal timing of force generation and intersegmental force transmission within a kinetic chain. For example, proficient throwing involves force generation and transfer from the base segments (legs) to the hips, then torso, humerus, forearm, and finally the hand. Notice in throwing that proper timing of segmental participation produces lag. In other words, the hips are rotating forward, while the upper torso is still rotating in the opposite direction (lagging behind). Then as the upper torso rotates forward, the humerus lags, and so on. The final result is a motion of the entire kinetic chain that is similar to that of a whip.

The proximal end of the whip is the area where force is initially applied. The initial force results in a wave of increasing velocity that travels toward the distal end of the whip. This sequence of actions, properly timed, results in rapid speed at the tip of the whip. Similarly, the goal of proficient throwing is extremely high speed in the distal hand segment of the chain, which can be transferred to the thrown object.

Figure 2.33 **A push-up exercise as a closed kinetic chain.** Name some other exercises that are closed chain.

Closed Kinetic Chains

In a **closed kinetic (or kinematic) chain**, the distal segment is stationary ("closed"), and therefore the total chain has less mobility (fewer DOF). In other words, the distal segment is in contact with the reference frame or portion of the reference frame (e.g., Earth or another object) that provides enough resistance to prohibit free motion. Therefore, motion at one joint is only possible with cooperative movement at other joints.

A push-up exercise is an example of a closed kinetic chain motion (both the toes and the palms are in contact with the surface of the environment). In this example, motion at the elbow can only occur if movement also occurs at the shoulder and wrist joints (Figure 2.33). But

Open kinetic chain Kinetic chain in which motion can occur at one link in the chain without cooperative motion at other links.

Closed kinetic chain Kinetic chain in which motion at one link is only possible with cooperative movement at other links.

classifying a movement as closed can sometimes be more difficult than it would seem. For example, squatting is a closed kinetic chain movement even though the head is free to move (and is in fact being moved up and down). The movement is classified as closed because the distal segment is the foot, and it is in contact with an immovable reference frame. This example is an instance in which thinking of cooperative joint motion is highly effective. During squatting, the knee can move only if motion occurs at the hip and ankle. Sometimes it is less confusing if one also considers the more proximal segments at which force is being applied to initiate the movement. Most of the force for movement in the squat example is applied to the femur and tibia by the hip and knee extensors (i.e., force application occurs proximal to the foot segment). It also helps to think of what *would* have happened if the foot had not been in contact with the immoveable reference frame. If the hip and knee extensors were contracted in the absence of foot contact with any surface, then the distal foot segment *would* have moved away from the body. But in the case of a squat, the body is forced to move away from the distal segment and is therefore classified as a closed kinetic chain.

The disadvantage in performance of closed chain motions is that one cannot focus so much on a specific muscle or muscle group. For example, in a squat exercise, both the hip and knee extensors are active. Closed kinetic chain motions have some advantages in terms of athletic training and rehabilitation. One advantage is that closed kinetic chain motions for training are more functional and sport specific because many sports skills are performed with the feet fixed.

Functional Kinetic Chains

The human system is a multisegmented complex kinetic chain during functional activities. Functional activities can basically be considered activities of daily living (or applied activities) as opposed to isolated motions performed in a closed environment (such as those observed in common exercise routines). If the system is a complex chain, then some links likely are engaged in open kinetic chain motions while other links are involved in closed kinetic chain motions. So a **functional kinetic chain** is a complex kinetic chain in which some links are involved in open chain motion and others are engaged in closed chain motion, which together produce a functional activity. One example of a functional activity is human locomotion (Figure 2.34).

Human gait can be divided into a stance (or weight-bearing) phase and a swing (or non-weight-bearing) phase. During locomotion, one foot is in contact with the ground (closed kinetic chain motion), while the other foot is free to swing forward (open kinetic chain motion). Locomotion is just one of the myriad activities classified as functional. Functional kinetic chains are also used in the performance of device and home management, self-care, and most sports skills.

Figure 2.34 **A runner as a functional kinetic chain.**

Compensatory Motion

Comprehending kinetic chain concepts is necessary to completely understand both normal and abnormal human movement. Remember that motion at one link in the kinetic chain affects motion at one or more other links. Therefore, musculoskeletal abnormality such as

Functional kinetic chain A complex kinetic chain in which some links are involved in open chain motion and others are engaged in closed chain motion.

injury or muscular imbalance at one segment or link in the chain can force other links to adapt their motion. These **compensatory movements** are adaptations at normal kinetic chain links as a result of abnormal motion at another link so that a task can still be accomplished. Awareness of compensatory movement is important for two reasons: (1) the task will now probably require more energy than before because it is no longer being carried out with use of the most efficient movement pattern (i.e., the pattern that minimizes muscle force requirements); and (2) changes in force-loading patterns on the compensating links can eventually lead to musculoskeletal abnormalities in other areas of the chain. In other words, compensatory motion arising from injury or muscular imbalance not only makes activities more difficult but also will possibly result in other injuries.

Compensatory movements Adaptations at normal kinetic chain links as a result of abnormal motion at another link.

FOCUS ON RESEARCH

As we proceed through the textbook, we hope it will become clear that the concepts in biomechanics are surprisingly related to many aspects of our daily lives. For example, millions of women wear high heels on a daily basis probably without the term *biomechanics* ever coming to mind. However, the relationship between wearing high heels, muscle activity, and dynamics of human walking was examined in a study by Simonsen et al. (2012).

First, we must explain several pieces of research equipment that are frequently used to study many aspects of biomechanics. One common tool for studying biomechanics is three-dimensional motion capture (3D MOCAP for short). The 3D MOCAP process records the motion of the system of interest using multiple cameras (often with high frame rates), each having a different view. Analyzing the videos via special software allows re-creation of the motion in a virtual 3-D Cartesian coordinate system and then the calculation of many static and dynamic parameters. Next, force platforms or force plates are instruments often recessed into the floor that are capable of measuring the forces exerted by a system moving on or across them. The force plates can measure the forces in all three directions of the Cartesian coordinate system discussed earlier. Simultaneous use of MOCAP with the force plates allows for calculation of many aspects of statics and dynamics that will be discussed throughout the book. Finally, electromyography (EMG) is the process of using a special instrument with electrodes to record the electrical signals generated by muscle fibers when activated. The signals can then be analyzed to evaluate muscle activation level, abnormalities, recruitment parameters, and so on.

Simonsen et al. (2012) used all of the just cited equipment to examine muscle activity and walking dynamics while wearing high heels (9 cm). Dynamics of the hip, knee, and ankle were measured using MOCAP along with force plates. EMG was used to measure activity of the following muscles of the leg: soleus, gastrocnemius medialis, tibialis anterior, vastus medialis, vastus lateralis, rectus femoris, biceps femoris, and semimembranosus. The researchers found that in the first half of the stance phase of walking on high heels, knee extensor moment peak was doubled. It was also found that the knee joint was flexed significantly more in the first half of the stance phase. A significant increase in knee and hip joint abductor moments in the frontal plane was also observed. Those differences, along with increases in several EMG parameters, led the authors to speculate about walking on high heels. The researchers conclude that the large increased knee joint extensor moment could be caused by the observed increased knee joint flexion during the stance phase along with increased EMG in the quadriceps muscle. Along with the significant increase in knee joint abduction moment in the frontal plane, the results may be indicative of a large increase in bone-on-bone forces in the knee joint while walking in high-heeled shoes. The authors also hypothesize that their findings may help explain the higher incidence of osteoarthritis in the knee joint observed in women when compared to men.

Notice the study involves planes of motion as well as directions of a Cartesian coordinate system. In addition, the study is directly related to the concept of compensatory movements in the kinetic chain. Because the ankle, knee, and hip are all links in the kinetic chain, a change in one will cause compensation in the others. In the case of high heels, changing the angle of the foot during walking leads to changes in knee joint dynamics and muscle activity patterns, resulting in potentially damaging forces at the knee joint. Unfortunately, compensatory motions often lead to or result from injuries.

CONNECTIONS

2.7 EXERCISE PHYSIOLOGY

At the most basic level, our field is about movement. Both biomechanists and exercise physiologists are interested in studying that movement, but the perspectives differ. As described in the previous sections, joint (segmental link) motions are described in terms of relative movement of two segments. Motions of the skeletal system are made possible through contraction of skeletal muscle. A biomechanist will tend to focus on the motion that results from muscle contraction, whereas an exercise physiologist will likely focus on the physiological processes (such as metabolism) occurring within the muscle that allow the contraction to take place. Now let's discuss some specific variations in perspective as they relate to the musculoskeletal system.

In general, a muscle has two points of attachment: a proximal attachment (sometimes called an *origin*) that tends to be at a relatively immoveable location, and a distal attachment (sometimes called an *insertion*) that tends to be on a relatively moveable segment. The two attachments of the muscle (and therefore two segments) are brought closer together during muscle contraction, with the less stable end being moved closer to the more stable (Figure 2.35).

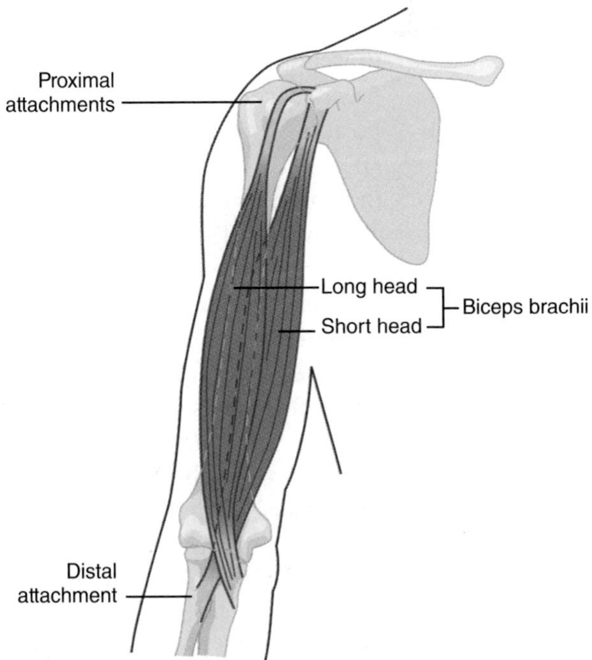

Figure 2.35 **The origin and insertion of the biceps brachii muscle.** Name some muscles that are only capable of one action.

The attachment that is most stable is not always obvious. For example, the distal attachment of the biceps brachii is brought closer to the more proximal attachments during the performance of a bicep curl exercise; however, the opposite is true when performing a pull-up exercise.

Biomechanical concepts can be used to study the resultant motion caused by contraction of a muscle with particular points of attachment. Biomechanics can also help explain the mechanical behavior of and potential injury to biological tissues such as bone, ligaments, and tendons that are affected by the muscle contraction. Principles of exercise physiology are also used to study biological tissues of the musculoskeletal system. However, the physiologist may be concerned with the specific cellular processes by which bones, ligaments, and tendons become stronger in response to muscle contraction. For instance, bone becomes stronger and denser in response to weight-bearing exercise. A biomechanist may be concerned with impact sustained by the musculoskeletal system during exercise sessions on varying surfaces, whereas a physiologist may be interested in how and what type of exercise is most effective for increasing the density of bone to prevent or treat osteoporosis.

It was already noted that joints of the body vary in their number of degrees of freedom. Therefore, having a wide range of muscles and muscle attachments available is necessary to move the segments of our skeletal system through all of the available degrees of freedom. In some cases, providing DOF at a segmental link is accomplished by more muscles. For example, the glenohumeral joint is classified as ball and socket and therefore has three degrees of freedom (rotation around any axis). However, do not mistake this fact to mean that the joint can perform only three actions. For example, flexion and extension are two rotations around a single axis. In addition, two different actions can occur around the same axis if the arm is not in the anatomical position at initiation of the observed motion. For example, both internal rotation (from the anatomical position) and horizontal adduction occur around a superoinferior axis. As a result of this increased mobility, 11 different muscles are necessary to provide all available degrees of freedom at the glenohumeral joint. The biomechanist may approach this situation by asking which muscle or muscles out of the many are most active at a given joint position, or which biomechanical factors are associated with a loss in mobility. An exercise physiologist might be more interested in whether fatigue in one of the muscles could predispose other muscles to injury.

In contrast to using multiple muscles, having a single muscle perform more than one function can also provide added DOF. For example, the biceps brachii can cause elbow flexion and forearm supination, as well as aid in shoulder flexion. Because many muscles act at each joint, and each one of those muscles has multiple functions, some redundancy of muscle action occurs. For example, four different muscles are involved in shoulder flexion; six muscles are used for wrist and phalangeal flexion, and seven muscles extend the cervical spine. Why so much redundancy? Having more than one muscle involved in an action strengthens the motion and enhances control. So not only does the human musculoskeletal system have many movement options, but also many of those motions can be highly precisely controlled. Once again, a biomechanist may be interested in exact movement patterns, joint positions, or velocities associated with contraction of a particular multifunctional muscle. In contrast, an exercise physiologist may be more interested in the exact cellular mechanisms that cause a change in that single muscle's function at different joint angles.

Finally, let's not forget that energy is needed for muscles to move the joints through all of the DOF necessary for exercise. The energy necessary for muscle contraction is produced by the metabolic systems, and the muscle contractions themselves are controlled by the neurological system. So the muscular system is the connection between the physiological, neurological, and skeletal systems. In fact, the systems are so interrelated that distinguishing whether a research project has been conducted by a biomechanist or exercise physiologist (or both) is often difficult. For example, in a study published in the journal *Medicine & Science in Sports & Exercise* (Roy & Stefanyshyn, 2006), the researchers observed that subjects running in shoes with a stiff midsole rather than a control (unmodified) midsole experienced a metabolic energy savings of approximately 1%. A subject such as midsole stiffness would often be found in the realm of biomechanical interest because it is related to forces acting in the foot. The metabolic cost of physical activity is frequently associated with exercise physiology because it is a cellular-level process. However, this study is a prime example of the need for interdisciplinary research in the movement-related disciplines. In biomechanics, the major concerns are the motion of the system itself and the forces involved in the movement situation. But one must always be aware that because the muscles cause the skeletal link movements, a physiological change within a muscle (e.g., fatigue) leads to a biomechanical outcome (e.g., a different running pattern).

2.8 MOTOR BEHAVIOR

As you have learned, the neurological system controls the muscles that are responsible for motion at the links of the skeletal system segments. So the observed biomechanical outcomes of the system are a coordinated effort under the control of the brain and spinal cord. Therefore, all of the different motions that have been described in this chapter are under the control of the motor system.

Motor Control

In terms of motor control, expressing how a motion is perceived is sometimes more effective than trying to describe the motion. For example, we have discussed Cartesian coordinate frames, which are constructed of orthogonal, or perpendicular, axes passing through an origin. These reference frames are a great way to describe a motion. However, we also discussed oblique reference frames in which the axes are nonorthogonal. In fact, an oblique reference frame is actually better representation of motion perception by the brain and sensory organs. For example, movement of the head can be *described* in terms of motion within a Cartesian coordinate frame. Motions (e.g., roll, yaw, and pitch) of the head are perceived by the semicircular canals of the vestibular system.

Although the semicircular canals are oriented in each of the major planes (sagittal, frontal, and transverse), they are not perfectly orthogonal to each other. This conformation means that the axes of the semicircular canals form an oblique coordinate reference frame within which motions are perceived. Therefore, the model of motor control of the head involves several transformations of Cartesian coordinates to oblique coordinates and oblique coordinates to Cartesian coordinates (Pellionisz, 1984; Zatsiorsky, 1998).

As the previous discussion illustrates, control of the system can be extremely complicated. Each human joint (link) possesses multiple degrees of freedom, and multiple muscles move each joint through those DOF. In fact, the human system possesses so many degrees of freedom that explaining the methods by which the motor system can learn, initiate, and control such smoothly coordinated motions, with so many variables to coordinate, can be difficult.

Russian biomechanist and physiologist Nikolai Bernstein (1897–1966) theorized that simply too many DOF must be controlled at the conscious level, and he put forth the idea that learning a motor skill is actually the process of solving the above "degrees of freedom problem." Bernstein divided learning into three stages (Schmidt & Lee, 2011):

1. *Freezing* degrees of freedom in which constraints are added to the movement so that fewer independent segments must be controlled, therefore simplifying the skill enough to allow crude performance;
2. *releasing* or removing constraints on degrees of freedom then occurs as the learner becomes more comfortable, which results in increased effectiveness of skill performance; and
3. *exploiting* the natural properties of degrees of freedom, a stage in which the learner becomes proficient in the movement by taking advantage of the various mechanical advantages inherent to the human system.

Motor Development

Development of walking patterns, from beginning to mature, provides evidence supporting Bernstein's theory that humans learn basic motor skills by decreasing the degrees of freedom during initial attempts, then freeing up additional degrees of freedom as they progress toward more mature and efficient patterns. When toddlers first attempt upright bipedal locomotion, they display several

Figure 2.36 **A beginning walker.** Restricting degrees of freedom simplifies the activity.
© Tanya Yatsenko/Shutterstock

characteristics that identify them as beginning walkers. They instinctively increase their base of support by spreading their legs and pointing their toes outward, flexing the knees minimally, rotating the pelvis as little as possible, and carrying the hands and arms stiffly in a "high-guard" position (Figure 2.36).

The lack of pelvic rotation, lack of flexion in the knees, and lack of arm swing evident in immature attempts at walking suggest that the human body restricts degrees of freedom to simplify the movement. As toddlers become more comfortable and efficient walkers through practice, additional degrees of freedom are added to allow longer and more accurate strides. Mature walkers demonstrate pelvic rotation and knee flexion and not only lower their arms to the side but also swing them contralaterally. This restricted to less-restricted movement pattern seems to be used when learning many fundamental skills such as hopping, skipping, striking, kicking, and more.

Motor Learning

When we learn complex movements later in life, excessive degrees of freedom are often the problem. In contrast to a toddler learning a fundamental skill, beginners attempting new and complex skills tend to use more motion than the skill requires. In essence, too many degrees of freedom are allowed at first but are reduced as performers become more efficient through practice. Juggling provides a good example of this process. The only body parts moving while a professional juggler performs a simple three-ball cascade are the lower arms, with slight movement at the shoulder joint. Contrast this technique to that of a novice juggler, who exhibits exaggerated shoulder motion in more than one plane, flexes and extends at the waist in more than one plane, and often moves her feet to correct for inaccurate tosses. As a novice improves, the movement, or degrees of freedom, seen in the lower body and waist diminishes, and the motion of the upper extremities becomes less exaggerated.

2.9 PEDAGOGY

Physical education teachers can improve performance of fundamental and complex motor skills by increasing or decreasing the degrees of freedom necessary for efficient movement. Children in preschool through second grade should be given plenty of opportunities to practice fundamental skills, with the teacher providing feedback designed to encourage increasing degrees of freedom of movement.

Older students learning more complex skills should also be allowed the maximum number of repetitions possible while being encouraged and given feedback to decrease the degrees of freedom. This approach promotes more efficient movement patterns that maximize power and minimize error. Secondary school students (middle and high school) and adults learning the game of golf can all benefit from formal instruction. Beginning golfers almost always add movements that are neither relevant—nor conducive—to sending a golf ball along an intended path with maximum velocity. Rather than rotating around the spine, beginners often sway laterally away from the target and then have to sway back toward the ball precisely the same amount if they want the club face to meet the ball exactly as it did when they addressed the ball. Amateur players who aspire to lower handicaps would do well to emulate professionals who have removed all but the essential degrees of freedom of motion in their swings.

2.10 ADAPTED MOTION

Most successful physical education teachers would agree that the word *adapted* is an adjective that describes everything they do, not just working with children who have formally identified disabilities. Each class that arrives in the gym will likely include students who range from beginners to advanced performers, and teachers must adapt their lessons, equipment, and teaching style to the needs of each one. Although the adapted physical education sections in this text focus primarily on those students with physical or cognitive challenges, the ideas and information are easily transferable to adapting physical education for any student that you teach.

Students with a variety of with physical challenges participate in physical education, physical therapy, and occupational therapy to increase their chances for success in sport, recreation, or activities of daily living. Muscle function lost because of amputation or diseases such as muscular dystrophy often require people to relearn movement patterns using different muscle groups, including modifying movement through different joints. In essence, degrees of freedom have been lost, and other links in the kinetic chain must learn to adapt to restore motion. Adapted physical education teachers often work closely with physical and occupational therapists to help people create new kinetic chains when normal movement becomes impossible. One example is learning to walk or run with a prosthetic after an amputation. Depending on the site of the surgery—above or below critical joints such as the knee and elbow—the muscles used to propel the body can differ significantly from those on the unaffected side. In essence, the prosthetic itself become part of the chain, and alterations to movement patterns are learned to compensate for it.

SUMMARY

Precise observation and accurate description of motion are necessities in all movement-related disciplines. A system is any object or group of objects whose motion is of analytical interest. The system can have numerous physical qualities, all of which can influence the characteristics of its motion. The physical qualities of the system are described by the field of anthropometrics. Directional terminology is shared among different fields to provide a common language for description of system motion.

Many movement terms are used to fully describe the movement of a system. System motion occurs in a plane and around an axis. The cardinal planes (frontal, sagittal, and transverse) bisect the body, and each has an associated orthogonal axis. Coordinate reference frames are used to describe motion of the system relative to the environment. The total number of movement options available to a system is its degrees of freedom.

Motion of the system within a predictable environment is called *closed*, whereas motion within a potentially changing environment is called *open*. Motion within that particular environment can be linear or rotational. Linear is motion *along* one of the x-, y-, or z-axes in which all points of the system move (change position) at the same time, in the same direction, and the same distance with respect to the defined reference frame despite their location on the system. Angular motion occurs if the system is restricted to move around an axis.

Skills performed by the system can be classified as *discrete*, *continuous*, or *serial*, depending on whether a definite beginning and end point can be identified and how many different movements make up the skill. A skill can be classified as gross or fine motor dependent on the level of precision required.

A kinetic chain is a system of linked rigid bodies subject to force application. Kinetics chains are classified according to complexity and whether cooperative motion is necessary among the links in the chain. The human system is a kinetic chain and must therefore adapt motions if one link in the chain becomes nonfunctional. The human kinetic chain has many degrees of freedom and therefore requires highly complex control mechanisms.

REVIEW QUESTIONS

1. What structures make up the system in the situation of a soccer injury? What about in a sumo wrestling match?
2. Sumo wrestler Akebono has a body mass of 236 kg and a height of 203 cm. What is his BMI? Find a classification table for BMI and classify Akebono in terms of disease risk. In what somatotype category does he fit? Why are these classifications not necessarily a bad thing for Akebono?
3. Explain the crural index and provide an example of an athlete or animal in which this value is high.
4. In what direction or directions does a runner move with reference to a global reference system?
5. In what direction or directions does the lower leg of a runner move in relation to the thigh?
6. What is the plane in which a runner's arms move?
7. As the runner's arms move in that plane, around which axis does the shoulder joint move?
8. How many degrees of freedom are available in planar motion? How many DOF are available in 3-D movements of the system? How many DOF does a ball-and-socket joint such as the shoulder possess?
9. What type of skill is a basketball free throw? Explain your answer.
10. If I drop an object, what type of motion will it exhibit? What if I throw the same object?
11. Classify running as one of the following types of motion: translation, rotation, or general. Explain your answer.
12. Provide examples of discrete, repeated discrete, continuous, and serial skills.
13. Provide simple and complex examples of open and closed kinetic chain motions. In which category is throwing?

EQUATIONS

Body mass index
$$BMI = \frac{kg}{m^2} \tag{2.1}$$

Ponderal index
$$PI = \frac{kg}{m^3} \tag{2.2}$$

Crural index
$$CI = \frac{\text{length of the tibia}}{\text{length of the femur}} \times 100 \tag{2.3}$$

REFERENCES AND SUGGESTED READINGS

Floyd, R. T. 2018. *Manual of structural kinesiology*, 20th ed. New York, NY: McGraw-Hill.

Oatis, C. A. 2017. *Kinesiology: The mechanics and pathomechanics of human movement*. Philadelphia, PA: Wolters Kluwer.

Pellionisz, A. J. 1984. Coordination: A vector-matrix description of transformations of overcomplete CNS coordinates and a tenorial solution using the Moore-Penrose inverse. *Journal of Theoretical Biology*, 110: 353–375.

Robertson, D. G. E., G. E. Caldwell, J. Hamill, G. Kamen, & S. N. 2014. Whittlesey. *Research methods in biomechanics*, 2nd ed. Champaign, IL: Human Kinetics.

Roy, J-P. R., & D. J. Stefanyshyn. 2006. Shoe midsole longitudinal bending stiffness and running economy, joint energy, and EMG. *Medicine & Science in Sports & Exercise*, 38(3): 562–569.

Schmidt, R. A., & T. D. Lee. 2011. *Motor control and learning: A behavioral emphasis*, 5th ed. Champaign, IL: Human Kinetics.

Simonsen, E. B., M. B. S. Svendsen, A. Nørreslet, H. K. Baldvinsson, T. Heilskov-Hansen, P. K. Larsen, T. Alkjær, & M. Henriksen. 2012. Walking on high heels changes muscle activity and the dynamics of human walking significantly. *Journal of Applied Biomechanics*, 28(1): 20–28.

VanPutte, C. L., J. L Regan, & A. F. Russo. 2017. *Seeley's anatomy and physiology*, 11th ed. New York, NY: McGraw-Hill.

Watkins, J. 2010. *Structure and function of the musculoskeletal system*. Champaign, IL: Human Kinetics.

Zatsiorsky, V. M. 1998. *Kinematics of human motion*. Champaign, IL: Human Kinetics.

CHAPTER 3

PARADIGMS FOR STUDYING MOTION OF THE SYSTEM

LEARNING OBJECTIVES

1. Describe the difference between qualitative and quantitative motion analysis.
2. Understand why different disciplines use different methods.

CONCEPTS

3.1 Qualitative Motion Analysis

3.2 Quantitative Motion Analysis

CONNECTIONS

3.3 Functional Anatomy

3.4 Injury Science

3.5 Motor Behavior

3.6 Pedagogy

3.7 Adapted Motion

When the only tool you have is a hammer, every problem begins to resemble a nail.

—Abraham Maslow

You've played some Frisbee in the backyard with friends and have become consistent in your throwing and catching skills. You read in the newspaper that tryouts will be held for a local Ultimate Frisbee team and decide that it sounds like fun. Players are warming up when you arrive, and you notice that they have throwing skills that you've never seen before. The first event of the day is a game of Ultimate

Frisbee, and the opposing team quickly finds your weakness—you only know how to throw the Frisbee using a backhand throw. From that moment on, you're helpless as they overdefend your backhand side. You make some good throws when the defense doesn't have time to cover you, but for you the game becomes an exercise in frustration against more skilled players. After the games, a skilled player offers advice on developing five different throws that will help you become more effective on offense. These new tools will come in handy at the next tryout. Different tools are necessary, depending on the job, the specifics of the job, and the desired outcome. In movement-based careers, assessing motion is critical in helping athletes, patients, and students improve performance, and the type of assessment tool varies dramatically among disciplines.

Biomechanics is a precise science that often uses quantitative measurements and mathematical formulas to describe and understand forces and motion within and created by the human body. Motion can be described in qualitative or quantitative terms. Unfortunately, some practitioners in the motion sciences keep the qualitative and quantitative tools in separate boxes. Quantitative and qualitative methods should be viewed not as competing but as complementary tools used to analyze movement.

Fundamentals is a term often used in teaching and coaching. Fundamentals are the general form and processes for movement activities that allow a performer the best chance at success. Fundamentals in Frisbee include grip, posture, and alignment. These are the first things that an experienced instructor looks at when observing a student. A throw performed using poor fundamentals has far less chance of being successful than one performed using sound fundamentals. Similarly, a physical therapist or athletic trainer may make an initial diagnosis based solely on visual (qualitative) inspection. Also, personal trainers analyze exercise form by watching the performance with particular key elements in mind.

Although most practitioners rely on observation and qualitative methods to identify errors and correct flaws, the information they use is based on data and physical principles proven through quantitative measurement. Recent developments in digital video and software provide great examples of the complementary nature of qualitative and quantitative analysis. Because a performer cannot "see" all the components while he performs, digital video analysis allows him to visually review the action. This method, primarily a qualitative one, can be paired with advanced software applications to gain more detailed and accurate information such as exact angles, velocities, and acceleration of body parts and implements. An example would be an app called Coach's Eye (www.coachseye.com). Biomechanical analysis can help reveal, for example, the optimum launch angle to propel a javelin. A physical therapist, on the other hand, could use video analysis to quantify an improvement in the range of motion of a patient undergoing a particular rehabilitative regimen.

This marriage of subjective observation and numerical data analysis provides movement specialists and elite performers with the specific feedback they need to diagnose and improve motion. As we progress through this chapter, keep in mind that qualitative methods of analysis are primarily deduced or supported through quantitative methods. So don't read this chapter thinking "qualitative versus quantitative"—each has its own place in teaching, coaching, physical and occupational therapy, athletic training, and other movement-based careers.

CONCEPTS

3.1 QUALITATIVE MOTION ANALYSIS

Qualitative analysis of motion describes how the human body "looks" on visual inspection as it performs skills, including its position in space, the position of body parts relative to each other, and in some cases the position of *segments* of body parts in relation to each other. When implements are required for the task, they may also be included in the analysis (racquets, bats, clubs, etc.). Qualitative analysis is subjective and typically based on visual observation, which can range from simple "eyeballing"—estimating—to video recording and evaluation, to sophisticated three-dimensional digital imagery. Presenting the range of available observation techniques that address qualitative analysis of motion is beyond the scope of this text. However, it's important to understand that qualitative analysis is built on a foundation of knowledge developed through quantitative measurement. The analytical tools presented in the following paragraphs are simply a few examples of the many qualitative methods that can be used by practitioners in movement-related disciplines.

This chapter presents two general approaches to qualitative analysis, which differ in approach and detail. The **composite approach** views the whole body as a system that progresses through stages or phases as it refines movement patterns. The **component approach** uses the same phase or stage method (actually defined as steps), but rather than looking at the whole body as a global system, it breaks it down into component sections, with each section progressing through more refined steps toward mature movement patterns. Both the composite and component approaches rely on observing movement changes across time, and information gathered through each technique can be used by a variety of practitioners.

Many assessment tools have been designed to identify ability levels in a variety of movement patterns. Some tools focus on fundamental skills such as walking, running, throwing, catching, jumping, and striking. Other tools are specifically designed to identify accurate performance of specific sport skills. Please note that using any tool properly requires training: Just as the beginning athlete learns and progresses in skill development through practice and instruction, valid and reliable qualitative assessment of human movement can be developed only through knowledge, practiced observation, and experience in using the tools. To provide you with a taste of the different qualitative approaches, we will use a single common motor skill (overhand throwing) to illustrate the commonalities and differences between these methods.

Composite Approach

The composite approach is also known as the total body approach and is often called *developmental biomechanics*. Although it breaks down movement patterns to primary body parts, the stages of skill progression are based on the product of all the body parts in combination. Each stage identifies important body parts used to perform the skill, and the number of stages can vary, depending on the requirements necessary to perform a task.

The Test of Gross Motor Development number 3 (TGMD-3) is a popular example of a composite approach for identifying fundamental motor skill ability. The TGMD-3 is a relatively easy-to-administer tool that includes locomotor activities (run, gallop, hop, leap, horizontal jump, and slide) and object-control skills (striking a stationary ball, stationary dribble, kick, catch, overhand throw, and underhand roll). It includes space and equipment requirements, directions for conducting each activity, and a simple scoring system based on fundamental criteria for successful performance.

Referring to Figure 3.1 (Ulrich, 2019), it's easy to see why this assessment is one of the most popular in physical education. The equipment needs and directions are clearly stated, the performance criteria are short and succinct, and the scoring system is based on only two possible scores for each criterion—a 1 or a 0. If the trained observer sees the criteria during the performance, she marks a "1" in the box; if not, she writes a "0." Standard subtest scores are calculated for the locomotor and object control sections, and an overall gross motor development quotient is derived from the sum of the subtest scores.

Another popular composite approach to evaluating the quality of fundamental movements is the total body approach presented by Seefeldt, Haubenstricker, and colleagues in a series of papers and presentations (1972, 1975, 1976, 1982, 1983). This series of evaluative tools

Composite approach Qualitative analysis approach that views the whole body as a system that progresses through stages or phases as it refines movement patterns.

Component approach Qualitative analysis approach that views the body in component sections, with each section progressing through more refined steps toward mature movement patterns.

Skill	Materials	Directions	Performance Criteria	Trial 1	Trial 2	Score
6. Overhand throw	A tennis ball, a wall, and 20 feet (6.1 meters) of clear space	Attach a piece of tape on the floor 20 feet from the wall. Have the child stand behind the tape line facing the wall. Tell the child to throw the ball hard at the wall. Repeat a second trial.	1. Windup is initiated with a downward movement of hand and arm			
			2. Rotates hip and shoulder to a point where the non-throwing side faces the wall			
			3. Steps with the foot opposite the throwing hand toward the wall			
			4. Throwing hand follows through after the ball release, across the body toward the hip of the non-throwing side			

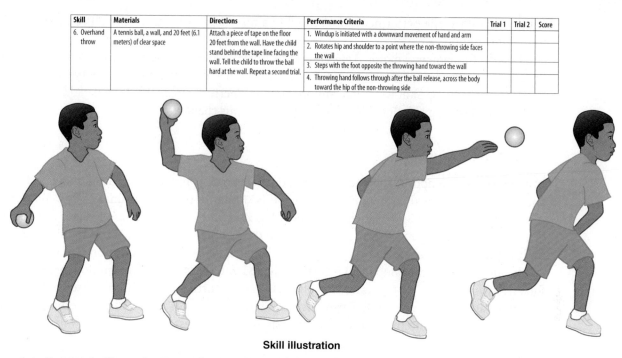

Skill illustration

Figure 3.1 **A sample from the Test of Gross Motor Development 3 for overhand throwing.** The TGMD number 3 is a popular example of a composite approach for identifying fundamental motor skill ability. Can you think of a similar device for physical therapists?

Courtesy of Dale Ulrich. Developer of the TGMD Assessment tool.

includes written descriptions of stages through which performers progress, and it provides serial drawings (similar to stop-frame photography) to show what each movement looks like. An important difference between this total body approach and the TGMD-3 is the lack of a formal scoring system.

Even without a formal scoring system, the total body approach can help movement specialists in identifying not only levels of ability (stages) but also areas where weaknesses in performance may originate. The illustrations add a visual interpretation of each written stage and provide additional information for evaluators. Notice in Figure 3.2 that although separate body parts are mentioned

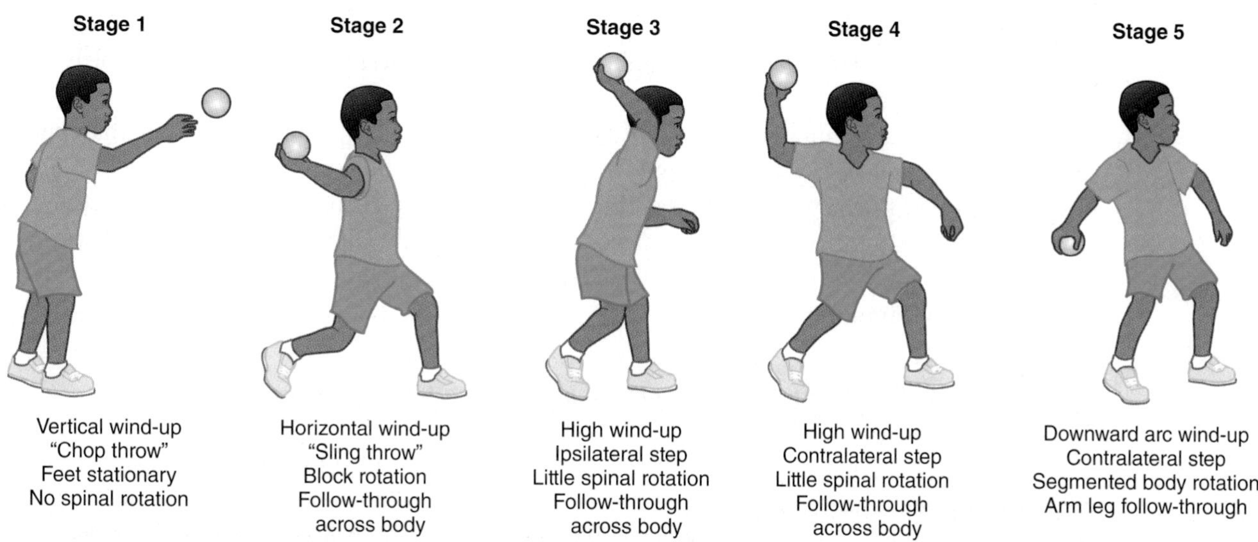

Figure 3.2 **Total body composite approach applied to throwing.** What are the advantages or disadvantages of this device compared to the TGMD-3?

within each stage, the performer is not scored in a more advanced stage until all component criteria in the previous one are met.

Beyond fundamental motor skills, composite analysis tools for specific sport skills have also been developed. From basketball shooting to soccer throw-ins, critical features have been identified for learning and refining skill technique and maximizing performance. This topic is addressed in more detail in the Pedagogy section of Connections later in this chapter.

Component Approach

In contrast to the composite approach, the component approach provides more detail about performance and is sometimes called *error analysis strategy*. It is similar in that each primary body component is observed, but it differs in that each component has its own evaluative series of stages or phases that can be advanced independently of one another. In the example, movement of overhand throwing, the body actions are broken down to component body parts: trunk, arms, and action of the feet.

In addition, the arm component is further divided into two phases of the motion: preparation and forward swing. Within the forward swing phase, the arm is further divided into two separate segments: the upper arm (humerus) and forearm. Five separate components are listed for overhand throwing, and each is divided into three or four distinct steps that progress from beginner to mature levels of skill (Figure 3.3).

In the composite approach, advancement to a later stage is based on the global sum of all the body components. Using the component approach, Roberton (1977) acknowledges that performers could show evidence of mature skill performance in one part of the body even as other body segments are at less mature levels; "one child might move ahead a stage in his trunk action while another child moved ahead a stage in his arm action" (p. 55).

Both the composite and component approaches, as well as other qualitative measurement devices, provide movement specialists with tools to subjectively evaluate human motion. The particular device you use will depend on several factors, including the nature of the task, the age and ability level of the performer, the setting (elementary school, high school, competitive athletics, rehabilitation setting), evaluator experience, and more. Many readers of this text will rely exclusively on qualitative analysis to evaluate and refine motion and performance. Qualitative analysis is highly effective and less time-consuming for teachers, coaches, and personal trainers, as well as for initial evaluation of movement disorders and injuries by therapists and athletic trainers. In fact, some of the great teachers and coaches in sport relied solely on this method. One example is Harvey Penick, a man who is regarded as a legend among golf instructors, someone who taught some of the best golfers in the world solely through subjective observation.

Remember that qualitative and quantitative analyses are complementary rather than competitive. We must qualitatively *observe* motion to give feedback, but some situations require a deeper understanding of the factors that *caused* the observed motion. Also, some performers require more accurate information that can only be measured quantitatively. The next section addresses some of these situations and begins to provide an understanding of the process of quantitatively analyzing a visually observed motion.

3.2 QUANTITATIVE MOTION ANALYSIS

The previous section helps us understand that motion analysis has a highly practical side. In fact, the methods already discussed are those most often used within the movement-related professions. Qualitative motion analysis also helps one develop the ability to *visually* analyze motion to rapidly isolate the various factors affecting the performance. However, a more specific analysis is often needed. In other words, we sometimes need to quantify the movement. Quantitative analysis may stem from the simple need for deeper understanding of why the system moves the way that it does. Also, the need for

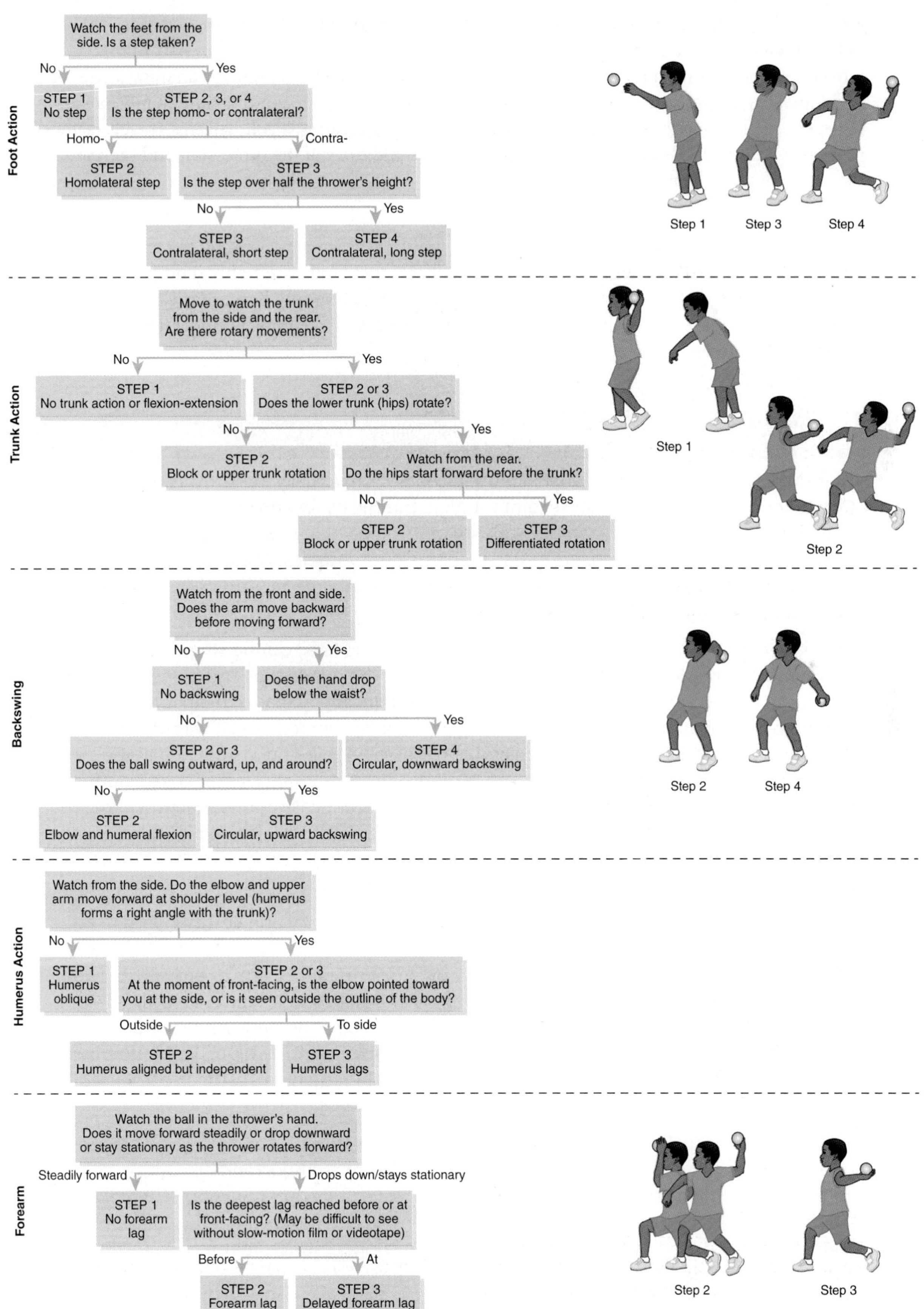
Foot Action
Watch the feet from the side. Is a step taken?
No
Yes
STEP 1
No step
STEP 2, 3, or 4
Is the step homo- or contralateral?
Homo-
Contra-
STEP 2
Homolateral step
STEP 3
Is the step over half the thrower's height?
No
Yes
STEP 3
Contralateral, short step
STEP 4
Contralateral, long step
Step 1
Step 3
Step 4
Trunk Action
Move to watch the trunk from the side and the rear. Are there rotary movements?
No
Yes
STEP 1
No trunk action or flexion-extension
STEP 2 or 3
Does the lower trunk (hips) rotate?
No
Yes
STEP 2
Block or upper trunk rotation
Watch from the rear.
Do the hips start forward before the trunk?
No
Yes
STEP 2
Block or upper trunk rotation
STEP 3
Differentiated rotation
Step 1
Step 2
Backswing
Watch from the front and side. Does the arm move backward before moving forward?
No
Yes
STEP 1
No backswing
Does the hand drop below the waist?
No
Yes
STEP 2 or 3
Does the ball swing outward, up, and around?
STEP 4
Circular, downward backswing
No
Yes
STEP 2
Elbow and humeral flexion
STEP 3
Circular, upward backswing
Step 2
Step 4
Humerus Action
Watch from the side. Do the elbow and upper arm move forward at shoulder level (humerus forms a right angle with the trunk)?
No
Yes
STEP 1
Humerus oblique
STEP 2 or 3
At the moment of front-facing, is the elbow pointed toward you at the side, or is it seen outside the outline of the body?
Outside
To side
STEP 2
Humerus aligned but independent
STEP 3
Humerus lags
Forearm
Watch the ball in the thrower's hand. Does it move forward steadily or drop downward or stay stationary as the thrower rotates forward?
Steadily forward
Drops down/stays stationary
STEP 1
No forearm lag
Is the deepest lag reached before or at front-facing? (May be difficult to see without slow-motion film or videotape)
Before
At
STEP 2
Forearm lag
STEP 3
Delayed forearm lag
Step 2
Step 3

Figure 3.3 **Developmental sequence of components of overhand throwing.**

a more specific analysis may be related to the desire to enhance the performance of elite athletes such as Olympians. Physical therapists and athletic trainers may need to quantify severity of injury, level of movement abnormality, or success of treatment methods. Quantitative motion analysis is also needed for biomechanical research. The research may have the goal of understanding human motion, optimizing sport performance, improving equipment design (e.g., creating a golf ball that travels farther), or preventing injury. We begin with some basic concepts and tools of the profession and then move into the actual process of analysis.

Scalars, Vectors, and Force

Throughout this textbook, we introduce many concepts that require not only representation but also manipulation of physical quantities. Physical quantities can simply be numerical or possess both numerical and directional properties. So at this point, gaining a complete understanding of how to classify and represent physical quantities is critical.

FOCUS ON RESEARCH

As is discussed in this chapter, when a muscle contracts it actually performs two functions. One function, of course, is to cause joint rotation, but the lesser known stabilization function is also vitally important. For example, stability of the knee joint is largely the responsibility of the combined actions of the quadriceps and hamstrings muscles. Therefore, it makes sense that if these muscles are fatigued there may be a risk to other structures of the knee joint such as the anterior cruciate ligament (ACL).

Thomas et al. (2010) studied the kinematics of the hip and knee during single-leg forward hops onto a force platform before and after fatigue of the quadriceps and hamstrings. Quadriceps and hamstring fatigue was associated with significant increases in hip internal rotation and knee extension and external rotation angle on initial contact. Also, it was observed that there were increases in knee extension and external rotation being maintained at the time of peak vertical ground reaction force. In addition, fatigue produced larger knee extension and smaller knee flexion and external rotation moments at peak ground reaction force. The researchers speculated that the changes in knee mechanics from fatigue may increase the risk of noncontact ACL injuries.

In addition, O'Connor et al. (2015) investigated the effects of hamstring fatigue on knee kinematics during single-leg stride landings followed by lateral and vertical jumps. During the stride landings followed by vertical jumps, fatigue of the hamstrings was found to be associated with reduced knee flexion angles, extensor moments, and energy absorption on landing. The researchers state that these findings indicate that fatigue is associated with a stiffer landing strategy, which is a pattern that would be consistent with increased ACL loading. In addition, the changes in kinetics and kinematics from fatigue were not observed in the lateral task, which may indicate that hamstring fatigue has greater effects in sagittal plane motions.

Although studies such as these cannot establish an absolute cause and effect for fatigue and ACL injury, they do provide evidence that is worthy of attention and consistent with our understanding of muscle force components. Many ACL injuries are noncontact. It makes sense that if the stabilizing components of the knee extensors and flexors are reduced from fatigue, then there will be greater stress transferred to other support structures of the knee that could lead to a noncontact injury of the ACL.

Scalars

Some physical quantities can be fully specified simply with a single numerical magnitude of appropriate units. A quantity that possesses only a magnitude but has no particular direction associated with it is called a **scalar quantity**. One example of a scalar quantity is mass. **Mass (m)** is the quantity of matter of which a body is composed. Mass is also a measure of a body's **inertia**, or resistance to having its state of motion changed by application of a force. The more massive the object, the more resistant it is to motion change. The appropriate Système International (SI) unit for mass is the kilogram (kg). For example, a body with a mass of 100 kg has more inertia than a body having a mass of 50 kg. In this example, notice that the mass of each object is specified simply with a single numerical value. Some other examples of scalar quantities and their SI units are a length of 2 meters (2 m), a speed of 5 meters per second (5 m/sec), and a temperature of 40 degrees Celsius (40 °C). Note that mass is not the same as volume (V), which is the three-dimensional space in cubic meters (m^3) occupied by an object. For example, 5 kg of muscle has the same mass as 5 kg of fat. However, the 5 kg of fat occupies a larger volume than the 5 kg of muscle (i.e., the fat occupies more space with the same amount of mass).

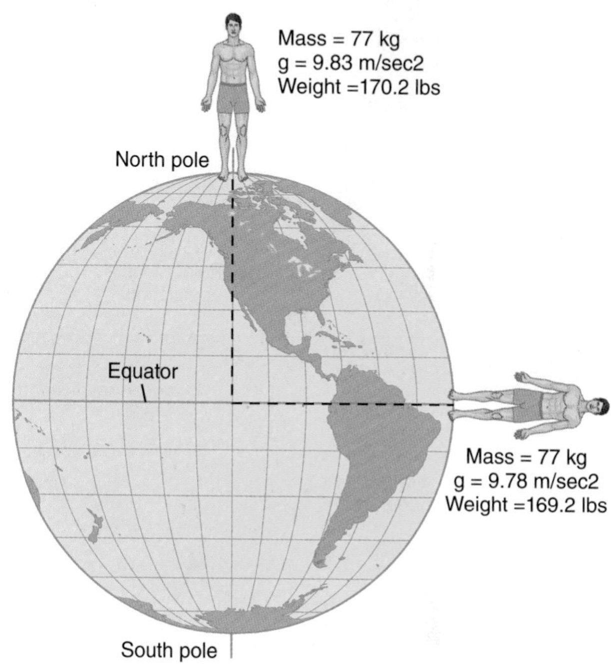

Figure 3.4 **Weight varies depending on proximity to the center of Earth, but mass does not.** What are some situations in which weight would change, but mass would not?

Vectors

Some physical quantities cannot be fully specified with a magnitude alone. A **vector quantity** can only be fully specified with a magnitude of appropriate units *and* a precise direction. Mass was used an example of a scalar quantity. Gravity pulls on an object's mass with a certain amount of force. A measure of the force with which gravity pulls on an object's mass is called **weight**. Because we also know the direction of gravity, weight is a vector quantity even though mass is a scalar. Many people speak of mass and weight synonymously, which is an easy mistake to make because mass and weight are directly proportional (i.e., a more massive object also weighs more). However, just because the two quantities are directly proportional does not mean they are the same quantity. As an extreme example, think of the difference in gravitational force on Earth versus the moon. Your mass is not affected by the difference in gravitational force (i.e., you are still composed of the same amount of matter in either location). But because weight is a measure of gravitational pull on your mass (and the moon has a lower gravitational pull than Earth), your weight will be less on the moon. That is an obvious example.

More subtle situations occur on Earth. For example, gravitational attraction between two bodies changes when the distance between their centers is changed. As the centers of two bodies are moved farther apart, the gravitational attraction between them goes down (and vice versa). So because planet Earth is not a perfect sphere (i.e., the poles are flattened, and the equator bulges), one can change one's weight without changing one's mass by traveling to areas of elevation that vary significantly from the norm (Figure 3.4).

Scalar quantity A quantity that can be fully specified simply with a single numerical magnitude of appropriate units.

Mass (m) The quantity of matter of which a body is composed.

Inertia A body's resistance to having its state of motion changed by application of a force.

Vector quantity A quantity that can only be fully specified with a magnitude of appropriate units and a precise direction.

Weight Measure of the force with which gravity pulls on an object's mass.

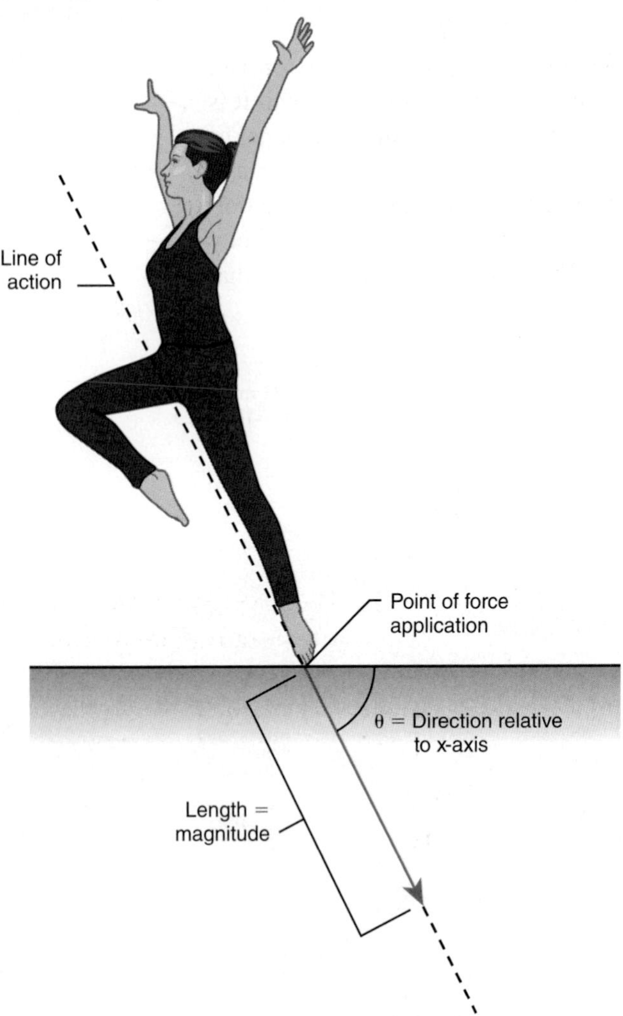

Figure 3.5 **A dancer leaping.** Draw a vector representing force being applied to a ball.

Direction Sense or way in which a force is applied; represented by the tip of a vector.

Orientation The alignment or inclination of the vector in relation to the cardinal directions.

Point of application The point or location at which a system receives an applied force; usually defined by the tail of the vector.

Magnitude Size or amount of an applied force; represented by the length of a vector.

Line of action An imaginary line extending infinitely along a vector through both the tip and tail, representing the path along which the vector would travel if moved forward or backward.

For example, weight is higher at Earth's poles because a person's center is closer to Earth's center. In contrast, a person weighs less in Denver, Mexico City, or at the equator because he is now farther from the Earth's center (as discussed later in this chapter, the spin of the Earth also has an effect on weight). These changes are not large, but the example helps to understand that mass (a scalar) and weight (a vector) are not the same. The SI unit of *force* is the newton (N). Because weight is a measure of gravitational *force,* weight is also measured in newtons. For the most part, 1 kg of mass weighs approximately 9.80 N. Of course, that is the number (9.80 N) that changes as one travels to higher elevations. Other examples of vector quantities are a velocity (v) of 3 m/sec north (speed with a specified direction), or a 1000 N force applied at 90° to a bone.

Vectors Representing Forces

Recall from the preceding chapter that arrows are used to represent forces in free-body diagrams. Arrows are appropriate because all forces (like gravity) are vector quantities. An arrow possesses certain specific characteristics that allow it to represent a vector quantity and therefore a force. An arrow has (1) a tip that points in a certain direction because of a given orientation, (2) a tail that defines where it began, (3) a certain distance between the tip and tail that defines a length, and (4) an imaginary path along which it *would* travel (Figure 3.5).

These qualities allow complete graphical representation of the following properties of a *force:*

- **Direction** is the way in which the force is applied (e.g., upward, downward, forward, backward, north, south, positive, negative) and is represented by the tip of the arrow or vector.
- **Orientation** is alignment or inclination of the vector in relation to the cardinal directions (e.g., vertical, 45° from horizontal), where θ is usually measured counterclockwise from the positive *x*-axis.
- **Point of application** is the point or location at which the system receives the applied force (e.g., at the toes, 2 cm from the axis of rotation of the elbow); it is usually defined by the tail of the vector.
- **Magnitude** is the amount or size of the applied force as depicted by drawing the length of the vector to scale (e.g., if the scale is 1 cm = 10 N, then a 10 cm vector represents a 100 N force).
- **Line of action** is an imaginary line extending infinitely along the vector through both the tip and tail (representing the path along which the arrow or vector would travel if moved forward or backward).

Though vectors are drawn as arrows, vector quantities can be indicated in handwriting by putting an arrow or line over or under the name of the vector ($\bar{A}$ or $\underline{A}$), or by boldfacing the vector name (**A**).

Vectors in Frames of Reference

Because of the special properties of vectors (being represented with lines or arrows having magnitude and direction), we must now revisit the idea of Cartesian coordinates and frames of reference. Recall from Chapter 2 that we can specifically define the position or location of an object (the system) in space by establishing one or more frames of reference within a Cartesian coordinate system. An origin (O) and two or three orthogonal axes (each passing through the origin and defining one spatial dimension) are used to define the coordinate frame of reference (Figure 3.6).

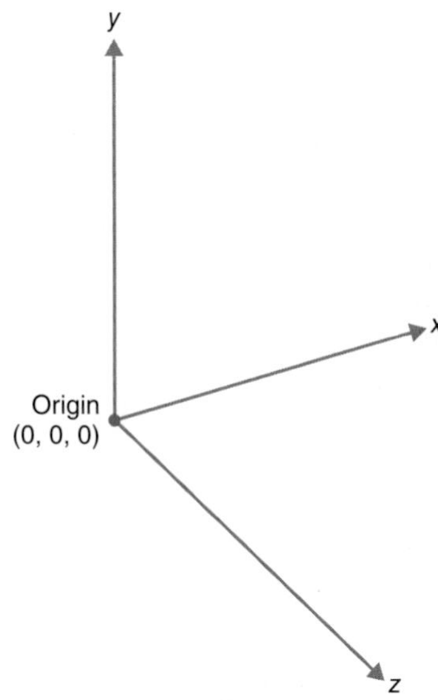

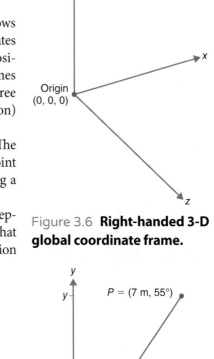

Figure 3.6 **Right-handed 3-D global coordinate frame.**

If a point is on one axis, its location can be designated with one coordinate. The location of a point in a plane formed by two axes requires two coordinates, and a point in space requires three coordinates. Any of these locations can be identified using a Cartesian (rectangular) coordinate system.

In biomechanics we need to describe not only single points but also segments (represented with lines connecting points) and forces (represented by vectors or arrows) that cause change in location of a particular point. For example, if we are interested in motion of the elbow joint such as the degree of flexion, we need to measure the joint angle. To form an angle, we have to have three points joined by lines instead of just the one point representing the elbow. In both cases, the key is that we are essentially using *lines* as opposed to single *points* as a means of representation. Because systems of interest are often composed of segments, we must be able to specify not only the location of one point on a segment but also the relative motion of multiple points on a single segment and motions of the segment as a whole. Forces cause change in motion of the system and its segments, so we must also have the capability of representing forces in the coordinate system with vectors. We will also see that various mathematical operations with vectors are necessary, for example to calculate the location of a point (system) or resultant motion of a segment after being acted on by more than one force (i.e., we can calculate the final location of a point if we know its initial location in the coordinate system and the magnitudes and directions of all of the forces acting on the point).

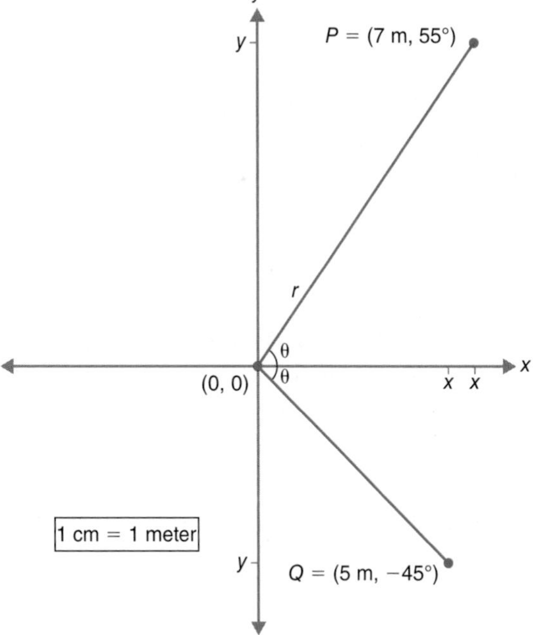

Figure 3.7 **Polar coordinates for two points.**

Because of these needs, it is sometimes more convenient and necessary to define a **polar coordinate system** and locate a point in space using its **plane polar coordinates** (Robertson et al., 2014). We must also be able to transform Cartesian coordinates into polar coordinates, and vice versa. In a polar coordinate system, there is still an origin and multiple reference axes (one for each dimension). As previously mentioned, the positive x-axis is often used as the reference axis. The location of the given point is then defined by its distance (radius) r from the origin and by the angle θ between the chosen reference axis and the line formed by connecting the given point to the origin (Figure 3.7).

As you have read, θ is often measured counterclockwise from the positive x-axis. So a point with the polar coordinates $(r, \theta) = (7.00 \text{ m}, 55°)$ is located 7 m away

Polar coordinate system A coordinate system in which the location of the given point is defined by its distance (radius) r from the origin and by the angle θ between the chosen reference axis and the line formed by connecting the given point to the origin.

Plane polar coordinates Coordinates (r, θ) representing the location of a point within a polar coordinate system.

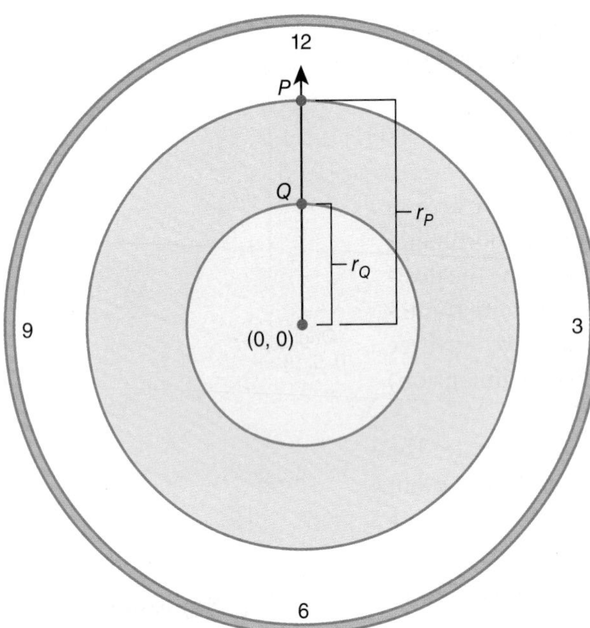

Figure 3.8 **Polar coordinates representing the radius of a circle.**

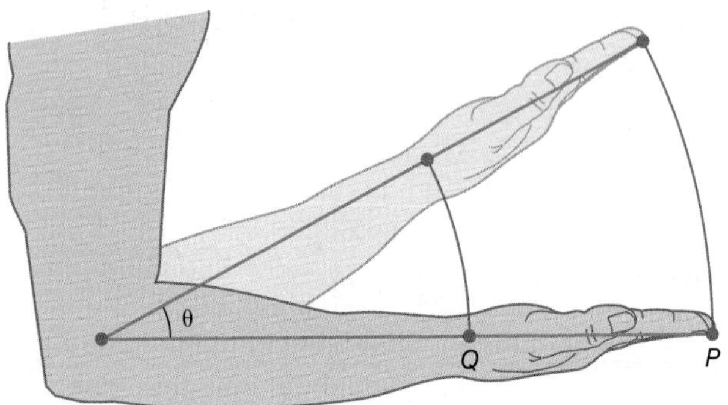

Figure 3.9 **Polar coordinate of the fingertip and wrist relative to the origin at the elbow.**

from the origin at an angle 55° above the reference axis. The angle is considered negative when measured clockwise from the reference axis. For example, a point that is located 5 m from the origin and 45° below the reference axis (i.e., clockwise from the reference axis) is specified with the polar coordinates (5.00 m, −45°). Notice that a radius (r) is used as one of the polar coordinates, because the distance r is the radius of the circle that is described if the line formed between the fixed origin and the point of interest rotated 360°. Because we used *counterclockwise* and *clockwise* above, think of a clock to further illustrate the point of polar coordinates.

The hand of a clock is fixed to the origin at the center. Imagine two points on the hand of the clock, one at the tip and one midway between the origin and the tip. Now imagine that the hand of the clock rotates all the way around from 12 o'clock to 12 o'clock with both points drawing a circle as they rotate (Figure 3.8). Both points have traveled −360°, but the circles are of two different sizes (i.e., have different radii). In other words, the points have the same *angular* displacement but two different *curvilinear* displacements. Now replace the image of the clock with a segment of the skeletal system (Figure 3.9).

If you imagine the origin as the elbow joint and the two points at the fingertip and wrist, you can understand why polar coordinates are necessary to describe motion of the human system. In other words, we are a collection of segments joined at points (or links), and we must be able to represent the relative motion of any point on any segment or motion of the entire segment.

Now let's apply these ideas to represent a vector in the coordinate reference frame. Basically, we use the same principles that are used to define a segment. The vector is drawn within the coordinate system such that the tail (representing the point of force application) is at the origin (0, 0), and the tip (indicating direction) has x and y coordinates. Of course, if we are discussing the point of force application, the O in this case is the origin of a local or somatic reference frame. Remember that the global reference frame has another O. In other words, we may be watching a soccer ball travel in reference to the origin of a global reference frame, with the center of the field being the O (Figure 3.10).

But a force that moves the ball is applied to the *ball*, not the field. So we place the tail of the force vector at the origin of the local reference frame attached to the ball and track the ball's motion relative to the *global* reference frame, with vectors that indicate the movements of the ball as a result of the applied forces (i.e., every time a force is applied to the ball, the O of the local reference frame has a new location relative to the O of the global reference frame). The magnitude of a vector within a coordinate reference frame is represented by r, and the orientation of the vector is given by θ relative to the positive x-axis. Therefore, a vector can also be represented with polar coordinates (r, θ).

So we have established that being able to transform Cartesian coordinates into polar coordinates and vice versa is sometimes necessary. Notice that if a second line is drawn from the point of interest (or from the tip of the vector) to the positive x-axis, a triangle is formed (i.e., r is one side, y is a side formed by drawing a line from the point to the x-axis, and x is the side formed by drawing a line from the O to the intersection of side y with the x-axis).

Because the resultant shape is a triangle (Figure 3.11), the coordinate transforms just discussed are possible with use of the most basic trigonometric functions. Before you panic at the mere mention of trigonometry, take a deep breath and realize that we simply need to remember a few basic elements that you are probably already familiar with: (1) "SOHCAHTOA"; (2) the **Pythagorean theorem**; and (3) the idea that trigonometry is based on simple ratios or proportions.

SOHCAHTOA is a mnemonic device used to remember the following equations:

$$\sin\theta = \frac{\text{side opposite } \theta}{\text{hypotenuse}} = \frac{y}{r} \quad \text{or} \quad y = r\sin\theta \quad \text{or} \quad r = \frac{y}{\sin\theta} \tag{3.1}$$

$$\cos\theta = \frac{\text{side adjacent to } \theta}{\text{hypotenuse}} = \frac{x}{r} \quad \text{or} \quad x = r\cos\theta \quad \text{or} \quad r = \frac{x}{\cos\theta} \tag{3.2}$$

$$\tan\theta = \frac{\text{side opposite } \theta}{\text{side adjacent to } \theta} = \frac{y}{x} \tag{3.3}$$

where r is the hypotenuse of a triangle that has x and y as the other two sides.

Do not overcomplicate any of this information; different forms of the same equation exist. Later in this chapter, we'll see that the form that is used depends on the variable for which we are solving.

The Pythagorean theorem expresses the relationships (ratios) between the lengths of the sides of a right triangle:

$$r^2 = x^2 + y^2 \text{ or } r = \sqrt{x^2 + y^2} \tag{3.4}$$

The final concept to remember is that trigonometry is based on ratios or proportions of the lengths of two sides in the given triangle. For example, if θ in the polar coordinates is 55°, then the ratio of y to r is 0.819 (i.e., $\sin 55° = 0.819$). In other words, the length of side y is 81.9% of the length of side r. The inverse function can be used to transform the ratio to the angle (i.e., $\sin^{-1} 0.819 = 55°$). The preceding concepts are all simple, but they are critical to understanding trigonometry.

Now that we have had a basic trigonometry refresher, we can perform other necessary functions. For example, if we have the Cartesian coordinates for the location of the wrist joint, we can describe the location of the wrist in space. However, locating a point in space is of little use because Cartesian coordinates alone do not provide any information as to the location of that point (the wrist) in relation to other parts of the body (e.g., the elbow or shoulder). In other words, do not forget that we are dealing with whole humans and not single points. To get further information from our Cartesian coordinates, we need to be able to perform coordinate transforms. Let's begin by transforming simple planar Cartesian coordinates (x, y) to polar coordinates (r, θ). For example, let's say that the location of a point Q in our Cartesian coordinate system is (−3.00 cm, −4.50 cm). We have the location of the point in space (Figure 3.12), but we would like to know the location of Q if it happened to be located on a rotating segment (line).

So we have to transform the Cartesian coordinates to find the radius (r) of the line formed by Q and the O of our coordinate system, as well as the angular distance traveled (θ) by the line on which our point of interest is located. We know the x and y coordinates, and we also

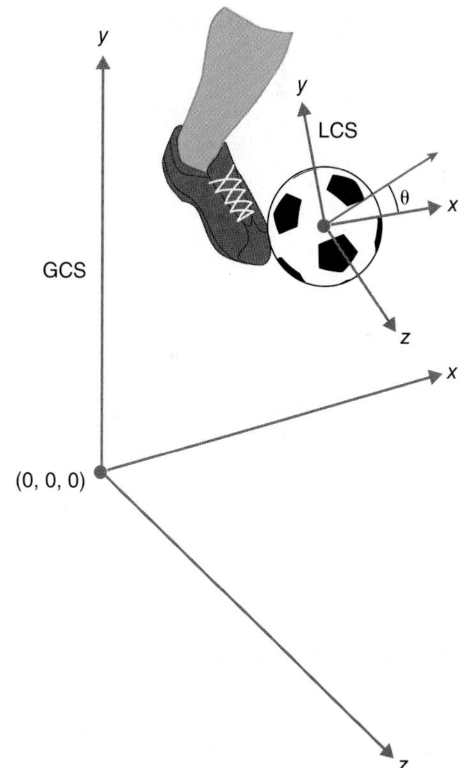

Figure 3.10 **A vector representing the force that is applied to a soccer ball.** Where would we place a vector representing the resultant path of the soccer ball relative to the global reference system?

Pythagorean theorem An expression of the relationships (ratios) between the lengths of the sides of a right triangle: $r^2 = x^2 + y^2$ or $r = \sqrt{x^2 + y^2}$

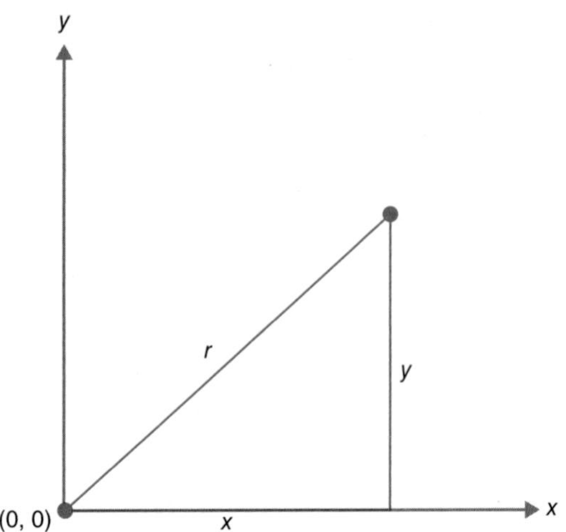

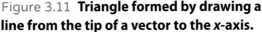

Figure 3.11 **Triangle formed by drawing a line from the tip of a vector to the *x*-axis.**

know from Equation 3.4 (Pythagorean theorem) that $r = \sqrt{x^2 + y^2}$. So in this case:

$$r = \sqrt{(-3.00\text{ cm})^2 + (-4.50\text{ cm})^2} = \mathbf{5.41\text{ cm}}$$

In other words, a line drawn (segment) from Q to the O is 5.41 cm long. Now we need to know how far this line rotated around the O. Again, we know x and y and we have an equation (3.3) that expresses the relationship (ratio) of the two coordinates in terms of θ:

$$\tan\theta = \frac{y}{x} = \frac{-4.50}{-3.00} = \mathbf{1.50}$$

We now have the ratio of y to x. Remember from the preceding that a ratio can be converted to an angle by using an inverse function:

$$\theta = \tan^{-1}(1.50) = \mathbf{56.3°}$$

However, we are measuring angles from the positive x-axis, so we must add 180° to any angle greater than positive 90° or negative 90° (i.e., in quadrants II or III, respectively) because the inverse tangent function on most calculators can only provide θ between +90° and −90°. In the midst of all of the mathematics, never lose contact with your common sense. We are measuring θ relative to the positive x-axis. Therefore, any point with a negative x coordinate must be located on a line that traveled greater than 90°. We got an answer of 56.3°, which we know is not large enough. So in our example, θ = 56.3° + 180° = 236.3°. In other words, the line (segment) formed by Q and the O rotated 236.3°. So our final polar coordinates for the point Q are (5.41 cm, 236.3°).

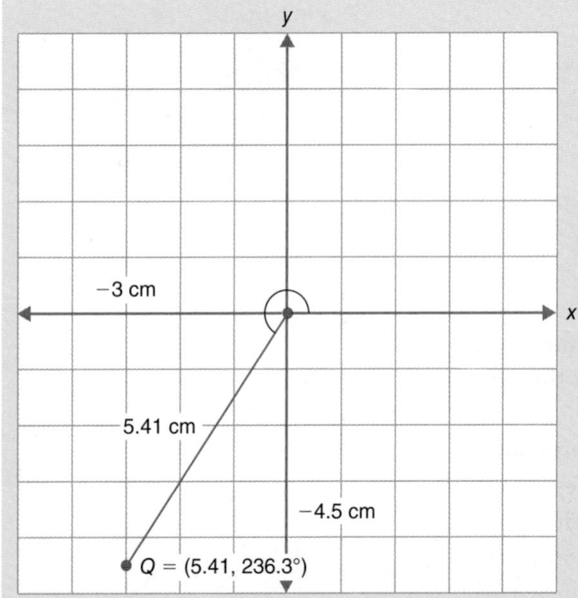

Figure 3.12 **Cartesian and polar coordinates for a point Q.**

Now let's transform the polar coordinates (5.41 cm, 236.3°) back to Cartesian using Equations 3.1 and 3.2:

$$y = r\sin\theta = (5.41\text{ cm})\sin 236.3° = \mathbf{-4.50}$$

$$x = r\cos\theta = (5.41\text{ cm})\cos 236.3° = \mathbf{-3.00}$$

As you can see, these transformations are simple with the use of basic trigonometric tools. In the following sections, you will see that basic trigonometry is highly useful for understanding biomechanics because all segments of the human body are linked together at axes of rotation. So, as you have read, the entire system undergoes *translation* because of joint *rotation*. These joint rotations are expressed in terms of θ. If θ changes, it means that a segment (represented by a line) was rotated from one location to another. For example, in a *somatic* reference frame, the joint (link) is the O; r is the distance from any point on the segment to the O; the positive x-axis is the original location of the segment (which is the axis from which θ is measured); and y is formed by drawing a line from any point on the segment to the x-axis. Therefore, a triangle is formed, and the use of trigonometry is necessary. These basic tools are used frequently throughout this text, especially when referring to joint motions and when working with vectors.

Special Properties of Vectors

We have already fully discussed the defining qualities of vector quantities: They can only be fully specified with a magnitude of appropriate units *and* a precise direction. We must understand some other properties of vectors before performing vector analysis (Robertson, 2014; Serway & Jewett, 2014; Serway & Vuille, 2018).

Vector Equality

One property of vectors is that two vectors are considered equal (**A** = **B**) as long as they possess the same magnitude and orientation. This concept of **vector equality** seems like a simple idea when written, but graphically it may not seem so obvious that all of the vectors in Figure 3.13 are, in fact, equal.

So the vectors are equal as long as the magnitudes are the same, and their lines of action are parallel, even though their tails may have different starting points within the frame of reference. As you'll soon learn, this concept is important in working with vectors graphically. The equality of vectors enables us to reposition a vector (maintaining its magnitude and orientation, of course) to perform a graphical analysis without actually affecting the vector itself.

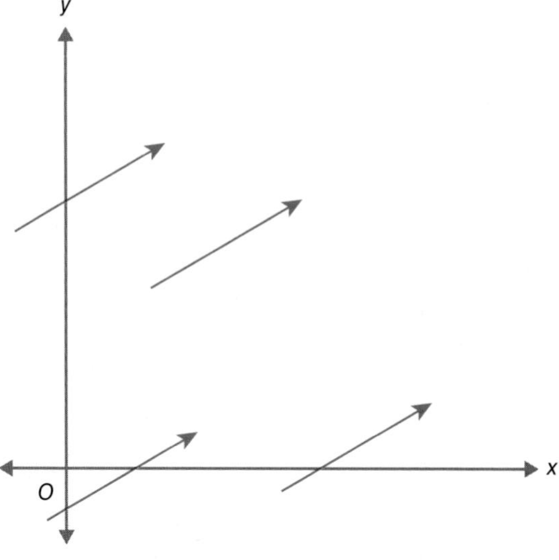

Figure 3.13 **Vector equality.**

Vector Addition

We will soon add vectors graphically, so you need to know that vector addition is both *commutative* and *associative* (you may recall these laws from one of your math classes). For our purposes, the **commutative law of addition** simply means that when we add vectors together the sum is independent of the order of addition:

$$\mathbf{A} + \mathbf{B} = \mathbf{B} + \mathbf{A} \tag{3.5}$$

The **associative law of addition** indicates that if we sum three or more vectors, the sum is independent of the grouping of the vectors for addition:

$$(\mathbf{A} + \mathbf{B}) + \mathbf{C} = \mathbf{A} + (\mathbf{B} + \mathbf{C}) \tag{3.6}$$

So far, so good: Two vectors are defined as equal if they have the same magnitude and orientation. Also, vector addition obeys commutative and associative laws. The last detail of addition that should be considered is basically a commonsense issue: as with scalars, the vectors being added must possess the same *type* of quantity (i.e., be of the same units). For example, you would never try to find the sum of 4 cm and 48 kg. The same applies to vectors of different units.

Vector Subtraction

To understand vector subtraction, we must first introduce the concept of the **negative of a vector**. We define the *negative of a vector* as simply another vector that, when added to the first, gives a sum equal to zero (i.e., the two vectors have the same magnitude but point in opposite directions):

$$\mathbf{A} + (-\mathbf{A}) = 0 \tag{3.7}$$

So to subtract one vector from another (**A** − **B**), we actually add vector −**B** to vector **A**:

$$\mathbf{A} - \mathbf{B} = \mathbf{A} + (-\mathbf{B}) \tag{3.8}$$

So vector subtraction is actually vector addition with use of the negative of a vector (i.e., the difference between **A** and **B** is the amount that would have to be *added* to **B** to equal **A**).

Vector equality Property that two vectors are considered equal (**A** = **B**) as long as they possess the same magnitude and orientation.

Commutative law of addition The sum of vectors added together is independent of the order of addition: **A + B = B + A**

Associative law of addition The sum of three or more vectors is independent of the grouping of the vectors for addition: (**A** + **B**) + **C** = **A** + (**B** + **C**)

Negative of a vector Another vector that, when added to the first, gives a sum equal to zero.

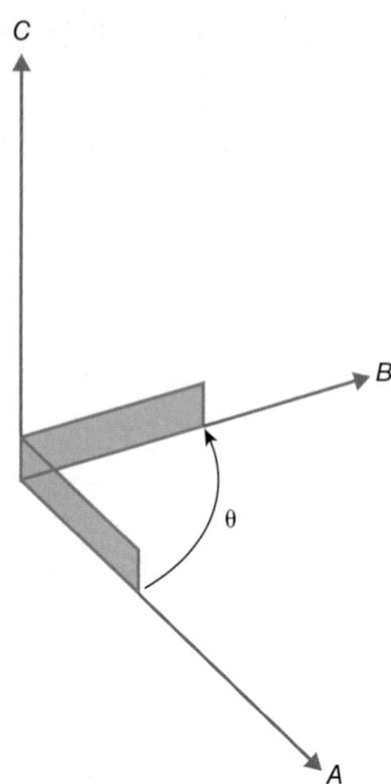

Figure 3.14 **Cross product of two vectors.** Notice that the cross product is perpendicular to the original two vectors.

Vector Multiplication

Two situations are possible with vector multiplication: (1) the product of a scalar and a vector, and (2) the product of a vector and another vector. Both are relatively straightforward.

When multiplying (or dividing) a scalar by a vector, the product is a vector quantity. For example, the product of the scalar number +9 and vector **A** is 9**A** (a vector of nine times the magnitude of **A** and possessing the same orientation because the scalar is positive). If the scalar number −9 were multiplied by **A**, then the product would be −9**A**. The resulting vector is still nine times the magnitude of **A** but is in the opposite direction because the scalar is a negative number.

Multiplying one vector by another (finding the *cross product*) results in another vector:

$$\mathbf{A} \times \mathbf{B} = \mathbf{C} \tag{3.9}$$

The orientation of **C** is perpendicular to the plane formed by **A** and **B**. Because **C** is perpendicular to the plane formed by **A** and **B**, the magnitude of vector **C** is calculated as follows:

$$\mathbf{A} \times \mathbf{B} \times (\sin \theta) = \mathbf{C} \tag{3.10}$$

where θ is the angle formed by **A** and **B**.

The concept of cross products is further elucidated in later chapters when calculating the turning effect that a muscle produces at a joint (Figure 3.14).

Graphical Methods of Vector Analysis

We now know that vectors are used in free-body diagrams to represent forces acting internally or externally to a system. Although we have not defined the forces that are represented in a free-body diagram, we can imagine there is always more than one (in fact, there are usually many). For example, let's look at an example free-body diagram of three acrobats and a teeter-totter (Figure 3.15).

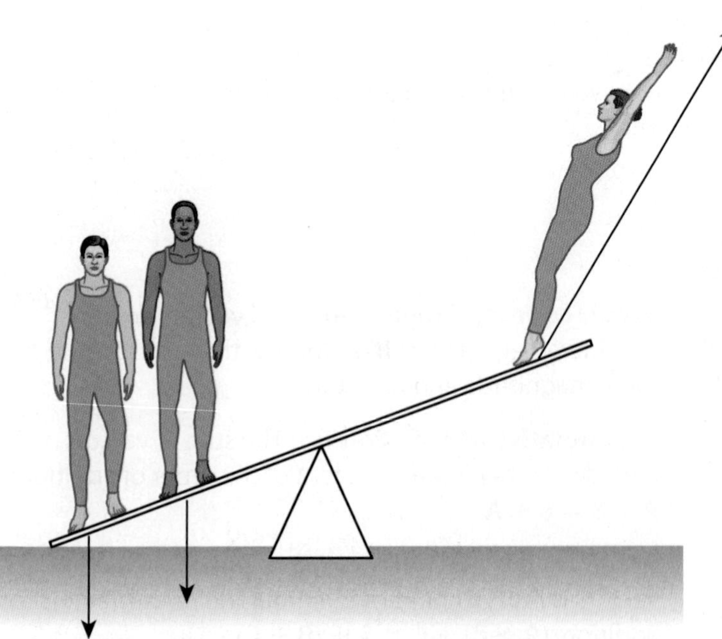

Figure 3.15 **Three acrobats on a teeter-totter.** Picture the path of the acrobat on the left.

Two of the acrobats are landing on one end of the teeter-totter, throwing the other acrobat into the air. We would need a vector to represent the force of gravity acting on each person and a vector representing the resulting force or "push" exerted by the teeter-totter on the third acrobat. Those are just a few of the forces in our diagram, but they are certainly the most important.

We must also be able to analyze multiple effects represented by a single vector. For example, in the free-body diagram, a vector represents the net force exerted by the teeter-totter on the acrobat. But that one force results in both upward (local $y+$) and backward (local $x-$) motion of the acrobat. Through vector analysis, we can calculate the portion of the teeter-totter force that results in upward motion of the acrobat and the portion that causes backward motion.

Also, vector analysis can be used to understand the resultant motion of the system that is acted on by many different forces simultaneously. In other words, the **resultant** is a vector that represents the sum of all forces (vector sum) acting on a system, and in this case representing the force with which, and the direction in which, the acrobat will fly. Remember that a vector quantity has both a magnitude and a direction. So instead of simply summing the force values (as we would with scalar math), we must use vector analysis to find the resultant, because all of the forces are not necessarily in the same direction.

At this point, we have actually already begun the biomechanical analysis process. We understand the system and can specifically describe the motion exhibited by the system within a coordinate reference frame. We can transform Cartesian coordinates of the system to polar coordinates. We know that forces cause changes in motion of the system and that those forces are represented by vectors. We can now use vector analysis to gain a greater understanding of the motion caused by forces.

We have mentioned that one force can cause movement in more than one direction simultaneously. We have also discussed the fact that a system is probably acted on by more than one force at the same time—and that the resultant is a vector representing the sum of all of those forces (vector sum) acting on a system. One method of calculating the directional effects of a single force (or the resultant effect of multiple forces) is graphical vector analysis. It is called graphical analysis because we literally draw diagrams (the easiest way is with use of graph paper). As you will see in the next section, trigonometry is often used when working with vectors. However, trigonometric methods are sometimes easier to understand if one has first visualized force vector effects by using graphical methods.

Vector Resolution

We begin with a look at the multiple effects that one vector actually represents. Remember from our example involving acrobats that, in the free-body diagram, a vector represents the net force exerted by the teeter-totter on the acrobat. Also recall that the one force exerted by the teeter-totter on the acrobat results in both upward (local $y+$) and backward (local $x-$) motion of the flying acrobat. So one force can have multiple directional effects. We use the process of **vector resolution** to *resolve* a single vector into its individual directional component vectors. **Component vectors** are the individual vectors that represent each of the multiple effects that one vector represents. A good example is music composition. One individual piece of music is composed of many individual notes (components). So resolution can be thought of as dividing that piece of music into those individual notes. A great way to visualize this process is to put your hands together palm to palm and point at something. You are pointing with one object, but that object is actually a combination of two hands.

Visualize vector resolution as spreading those hands apart (Figure 3.16).

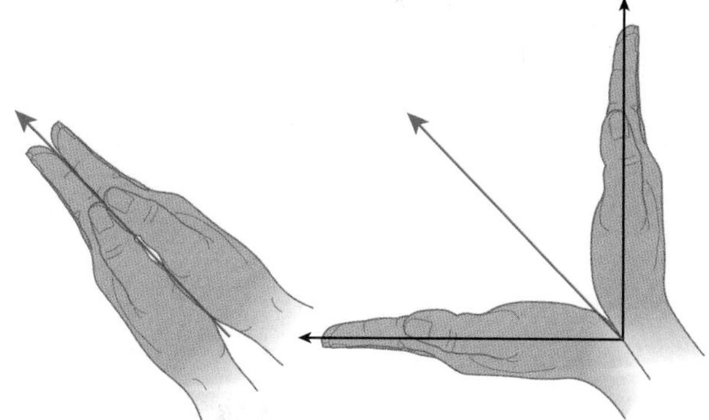

Figure 3.16 **Hands representing vector resolution.**

An infinite number of combinations of vectors could have combined to create the observed resultant vector. However, only one pair of the possible component vectors that are perfectly perpendicular to each could have combined to create one particular resultant. Therefore, our task is to find those two perpendicular components. In general, a force is capable of causing an object to have a particular "rise" and "run." In other words, the applied force can cause the system to travel a given distance both vertically and horizontally (as in the case of our flying acrobat). So for now let's call the two components the *vertical component* (also called *perpendicular*) and the

Resultant A single vector representative of the sum of multiple vectors.

Vector resolution The process by which individual directional component vectors of a single vector are determined.

Component vectors The individual vectors that are representative of each of the multiple effects that one vector represents.

horizontal component (also called *parallel*). Later, we will be more specific about the terminology of vector components.

We should mention before we begin that a vector may possess only one of these components. For example, if force is applied perfectly perpendicular to the ground (orthogonal to the x-axis), only a vertical component exists (i.e., all of the force is in the positive y direction). So when resolving a vector into its components, remember that the magnitude of one of the components may be equal to zero.

SAMPLE PROBLEM

Depending on your point of view, there are two ways to achieve our goal of resolving a vector into its vertical and horizontal components. Again, we will use examples of both to fully demonstrate the process. We begin with resolution using *vector parallelograms*. In other words, we are starting with one vector and would like to draw a parallelogram in which it fits corner to corner. Remember that the answer will always be correct if we follow some simple steps.

The first step is to draw the given vector to the established scale, being sure that it has the correct direction and orientation. Before proceeding to step two and while looking at the vector, remember the goal: We want to draw a parallelogram around our vector.

The corners of our parallelogram are defined by the tip and tail of the given vector (Figure 3.17). Also recall that the components for which we are solving (vertical and horizontal) are perpendicular to each other, so the parallelogram we are forming is always a square or rectangle. So our second step is to draw the sides of our parallelogram that represent the horizontal component. Do not be afraid to draw the lines longer than you know the final components will be (it actually helps).

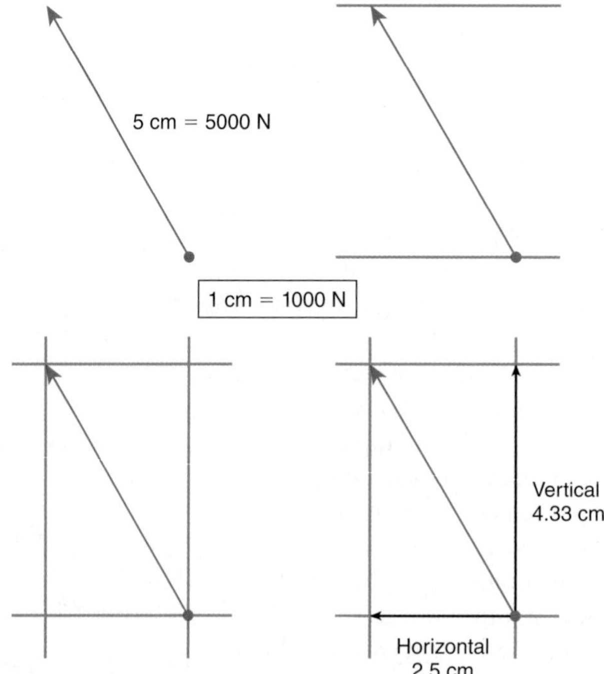

Figure 3.17 **Vector resolution using the parallelogram method.** Use your hands to define the tips of the methods.

Begin by drawing your first horizontal line touching the tail of the vector and parallel to the *x*-axis. Notice that in this case we are referring to the *global* coordinate frame of reference when drawing a line parallel to the *x*-axis, and this is not always the case. For example, when we resolve muscle force vectors, we use a somatic frame of reference. Next, we need to draw another horizontal line parallel to the first and touching the tip of the vector. We now have two sides of our parallelogram. The next step is to draw the vertical sides of the parallelogram. Again, it sometimes helps to draw the lines longer than we know is necessary. One side should touch the tail of the vector and be perpendicular to the horizontal lines drawn in the previous step (i.e., parallel to the *y*-axis). The next vertical line should be parallel to the first vertical line and touching the tip of the vector. The next step is simply to draw arrow heads on our lines, which actually creates two new vectors (our component vectors). In deciding where to draw the arrowheads, think of the direction and orientation of your initial vector. If the vector travels in the negative *x* direction and the positive *y* direction, then the horizontal component should have an arrow pointing in the negative *x* direction, and the vertical component should point in the positive *y* direction. These things may seem quite simple when they are explained, but they can be easily confused under pressure while you are staring at multiple lines and vectors on an exam. If you need an easy way to remember, think of our earlier example with the hands together and then apart. With the parallelogram drawn, you can put your hands together, placing them on the original vector and pointing in the same direction as the vector. Now split them apart while keeping them in contact at the wrist. One hand will point in the correct horizontal direction, and the other will point in the correct vertical direction. Once the new component vectors are drawn, you simply need to measure them and compare them to the established scale to calculate their individual magnitudes. So, in our example, the initial vector of 5000 N represents a vertically directed force of 4330 N and a horizontally directed force of 2500 N.

This is a good place in which to demonstrate the effect of a perfectly perpendicular force. In our preceding example, what would happen if the acrobats had applied a force to the third acrobat who was perfectly orthogonal to the *x*-axis? Keep in mind that the vector defines the corners of our parallelogram. So, in this case, there would be no defined *corners* and therefore the vector could not technically be *resolved.* So the vertical component would be equal to 5000 N, and the horizontal component would be equal to 0 N, causing the acrobat to fly straight up into the air.

Our final graphical technique is resolution with *vector chains* or *vector polygons*. This method can actually be thought of in several ways, so we will explain it as such. First, let's imagine creating a **vector chain**: a chain of vectors beginning at the origin of our reference frame and ending at the tip of our vector. Keep in mind that we are only looking for two components, so our chain only has two vectors. In addition, the vectors in this chain are perpendicular to each other. So essentially we are forming a vector polygon; it is just that the polygon is always a triangle.

To begin, we simply draw our vector to scale and imagine it as one side of a triangle (Figure 3.18). Next, we draw a horizontal line touching the tail and parallel to the *x*-axis (global). Then we draw a vertical line touching the tip of the vector and perpendicular to the horizontal line (parallel to the *y*-axis). Arrows are required as noted previously to indicate direction. Again, draw the arrows to indicate the direction and orientation of the original vector.

Again, notice the result if the given vector had been perfectly orthogonal to the *x*-axis. There is no way to define the horizontal component of a vector that points in the *y* direction at 90° to the *x*-axis, so all of the force would be vertical.

Notice that this last technique was simply one-half of the parallelogram method. That is because a parallelogram can be constructed from two triangles. So this method could be called the *vector triangle* method. If this method

Vector chain A form of vector analysis in which vectors are arranged tip to tail.

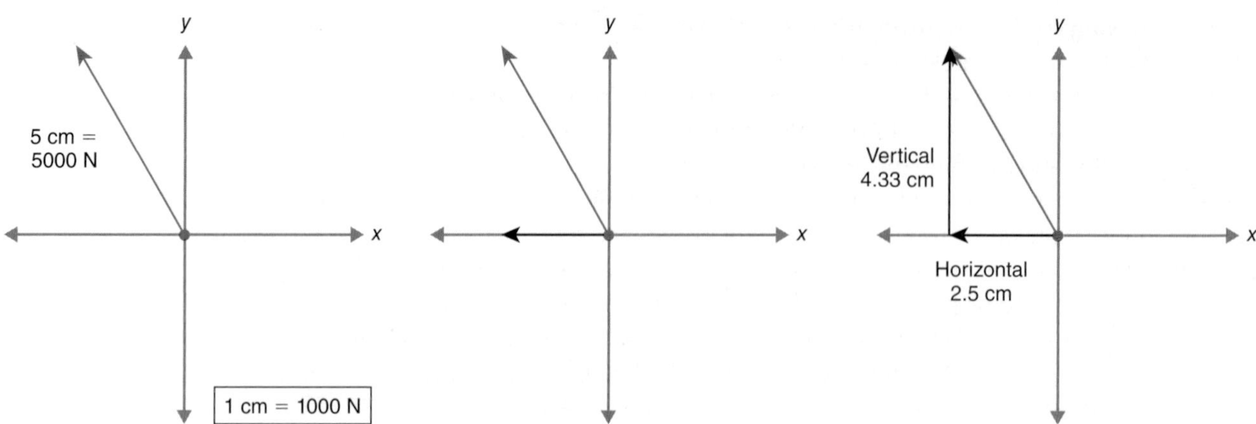

Figure 3.18 **Vector resolution using the chain method.** Notice that the chain forms a polygon.

has fewer steps, why not use it exclusively? We could ask the same question of any graphical method. If trigonometry is more precise and simple, why not always use it? The same answer applies to both questions. We cover graphical methods (as opposed to trigonometric) because they help us visualize what is actually happening during force application. In the same manner, we cover the parallelogram method because it helps us visualize the "splitting" of one force vector into its two directional components.

Vector Composition

Vector composition—the addition of two or more force vectors—is a situation in which multiple forces act on a system, and we would like to find the resultant force vector or vector sum (as opposed to our earlier work in which one force led to multiple effects). So, in this case, we are putting the hands back together. As with resolution, it can also help to think of the word *composition* in the context of other creative professions, such as composing one resultant piece of music from many notes or composing one resultant dance piece from many individual movement elements. In our profession, the resultant motion is composed of many individual forces. The forces involved may be simultaneous (concurrent) or sequential. In other words, we may want to know the resultant motion of a ball that is kicked by two people at the same time (simultaneous force application), or we may be interested in knowing the final position of the ball after it is kicked several different times (sequential force application). Either way, the process is the same. But as always, we must specify our system, the coordinate reference frame, and the situation completely.

Let's begin by understanding that two factors affect the complexity of vector composition: (1) the number of vectors, and (2) the relative directions and orientations of those vectors. The most simplistic case of vector composition is when only two forces are acting on the system. As for relative direction and orientation, the most basic case is that in which the involved forces (and their representative vectors) have the same line of action. When vectors have the same line of action, they are defined as **colinear vectors**. To be colinear, the involved forces do not necessarily have to be in the same direction; they just have the same line of action. Let's also recall our discussion of equality of vectors (Figure 3.13). We could, and often do, have a situation in which two concurrent forces are parallel—not colinear—to each other. But because of the equality of vectors, we can move a vector parallel to itself without changing its vital characteristics. So, for vector composition purposes, parallel vectors can be treated as colinear for analysis. But always keep in mind that they

Vector composition Process by which two or more vectors are summed to determine a single resultant vector.

Colinear vectors Vectors that have the same line of action.

are not colinear because this quality can lead to specialized types of motion that are covered in later chapters.

Let's look at two situations: (1) colinear forces with the same direction, and (2) colinear forces applied in opposite directions.

For our first situation, think of a car being pushed by two people as an example of colinear forces traveling in the same direction (Figure 3.19).

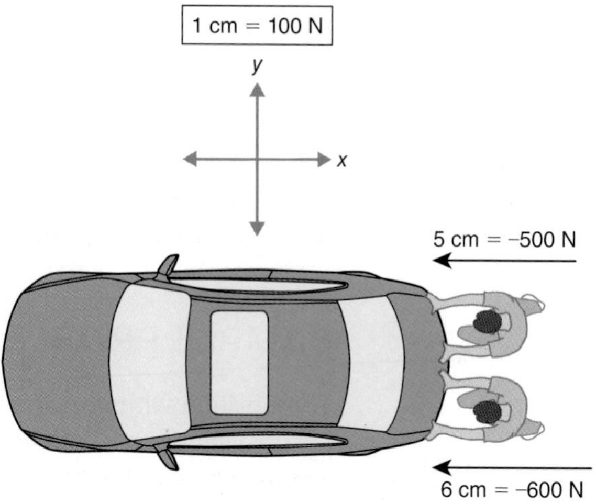

Figure 3.19 **Colinear vector composition of two vectors having the same direction.** Are these vectors equal?

One person pushes the car with a certain amount of force (600 N). The other person attempts to help by applying a force of 500 N to the car in the same exact direction. Recall that the magnitude of a vector is indicated by its length, which is drawn to scale (e.g., 1 cm = 100 N). So our first step is to draw both vectors to scale. Remember that we are working with vectors mathematically, which means that directions must not only be indicated with the tip of an arrow but also fully defined in our frame of reference. In this case, the vectors are colinear because they are parallel, so they are drawn in a row with the tip of one touching the tail of the other (forming a vector "chain"). Within a Cartesian coordinate frame of reference, notice that both forces are applied backward relative to the car, and therefore the directions of the vectors would be labeled negative (−) in the *x* direction. In the case of colinear forces, we can use simple algebraic addition. In our example, the magnitude of the resultant (net force) is (−600 N) + (−500 N) = −1100 N. Remember that the *negative* sign does not mean a *negative* force; it simply indicates direction. So the car experiences a force of 1100 N in the $-x$ direction. Notice that even though we know the value of the resultant, it is still good practice to always draw the resultant (to scale, of course) to better visualize the net effect of several forces.

As previously mentioned, colinear forces do not have to be in the same direction. Let's use a tug-of-war as an example. In this case, we would have four vectors in our chain: two vectors for one team and two vectors for the other team. Our first team has one person who pulls the rope with a force of 700 N and a second person pulling with a force of 450 N. Both of the force vectors for the first team are directed in the negative *x* direction (Figure 3.20).

Our other team has one player applying a force of 650 N and a second player applying a force of 600 N. The force vectors for the second team are directed in the positive *x* direction. So in this case, create a vector chain with two vectors pointed in the negative *x* direction and two pointed in the positive *x* direction. All of the vectors are still colinear, so we use simple algebraic addition: resultant (net force) = (−700 N) + (−450 N) + (+650N) + (+600) = +100 N. In our example, there is a net force of 100 N in the positive *x* direction. So our second team will win . . . slowly.

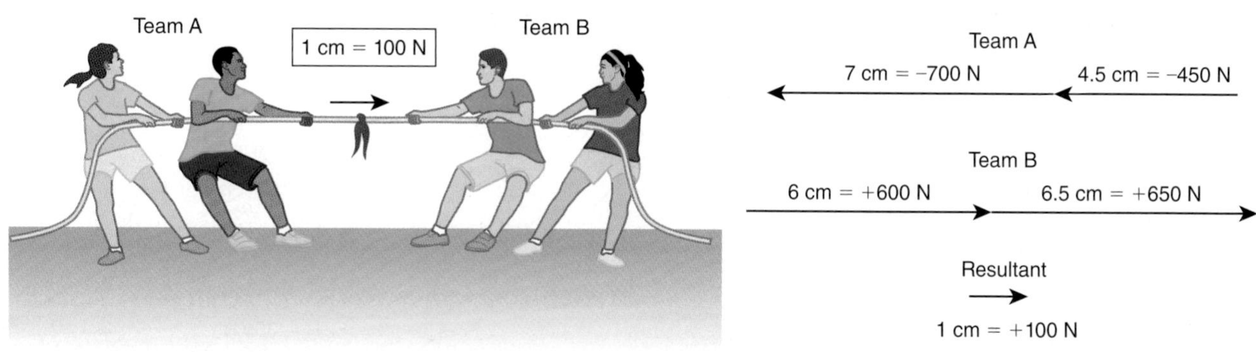

Figure 3.20 **Colinear vector composition of four vectors representing a tug-of-war.**

In the previous two examples, all of the vectors possessed the same line of action. However, this is not often the case. In most situations, many forces are acting simultaneously, all with different lines of action and in many different directions. Two basic graphical methods are used to compose noncolinear forces: vector parallelograms and vector chains. The use of one method over the other is for the most part a matter of personal preference, but it is also related to how many forces (and therefore vectors) are involved. Both are presented in this text to enhance understanding. As with all concepts, some people comprehend the process more effectively one way than another.

SAMPLE PROBLEM

Composition with vector parallelograms is just the opposite process of resolution with vector parallelograms. With vector resolution, we began with one vector and then constructed a parallelogram that fit corner to corner; with vector composition, we begin with two sides of the parallelogram and end by drawing a resultant from one corner to the other of the finished parallelogram. The process is relatively straightforward and can be accomplished with a series of simple steps. Just keep in mind that no matter the situation, constructing a vector parallelogram produces the correct resultant as long as the basic steps are followed. Let's start with two vectors and the idea that we are trying to construct a parallelogram.

Our first step is to position (draw) the vectors so that their tails are touching, being sure to retain their original length according to the established scale (Figure 3.21). Remember that a vector can be moved along its line of action without having its value changed, as long as it retains its initial magnitude and direction; so we achieve our first goal by "sliding" the vectors along their lines of action until their tails meet. The initial positions of the vectors are not going to change

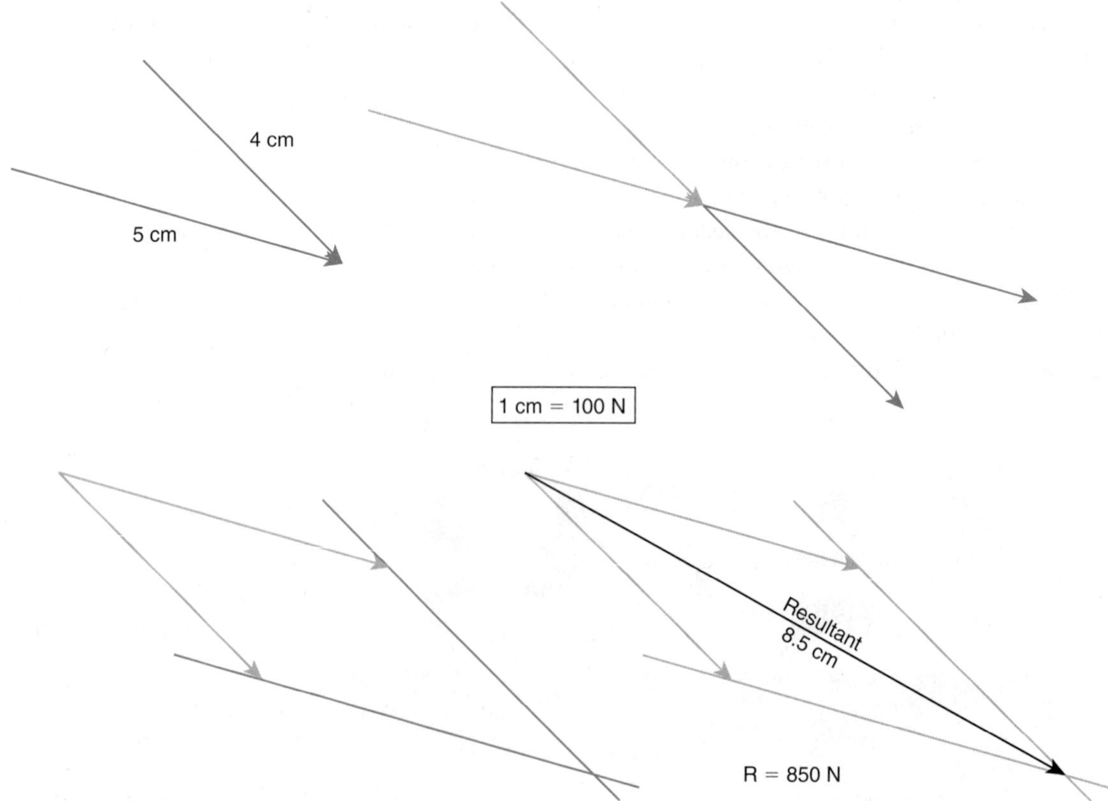

Figure 3.21 **Vector composition of two vectors using the parallelogram method.**

the desired outcome of our first step: They may be head to head, in which case we slide both; they could be head to tail, and we would only need to slide one; or they could already be tail to tail, and we would not need to slide either vector. After sliding along their lines of action, each vector still has its initial magnitude and direction, so no change has been made to the vector characteristics. This initial step should be imagined as drawing the first two sides of our vector parallelogram. Of course, our second step is to draw the other two sides of the parallelogram. Remember that the final shape is a parallelogram (opposite sides are parallel), so we need to draw one line perfectly parallel to the first vector and another line parallel to the other. Be careful: Optical illusion can sometimes eradicate good intention (and the answer will be correct only to the extent that the parallelogram is a true parallelogram). Our final step is to simply draw the resultant vector, which is an arrow reaching from one corner of our parallelogram to the other. The magnitude of the resultant is found by measuring the length of the line and then converting to newtons from the given scale. The tail of our resultant vector begins in the corner formed by the tails of the initial two vectors, and the head is in the opposite corner. Do not make the mistake of thinking that drawing the head of the vector is not important; vectors must have a specified magnitude of appropriate units *and* a specific direction and orientation. The orientation (θ) of the resultant vector can then be measured relative to the positive *x*-axis.

You can see that using vector parallelograms is effective and relatively simple when only two vectors are involved. The same process can be used for three or more vectors, but it is labor intensive and can be confusing. We need to try one example to demonstrate the process and gain some proficiency with the parallelogram method. We need to follow the exact same steps as just noted but work only with any two of the vectors in the situation at a time (Figure 3.22).

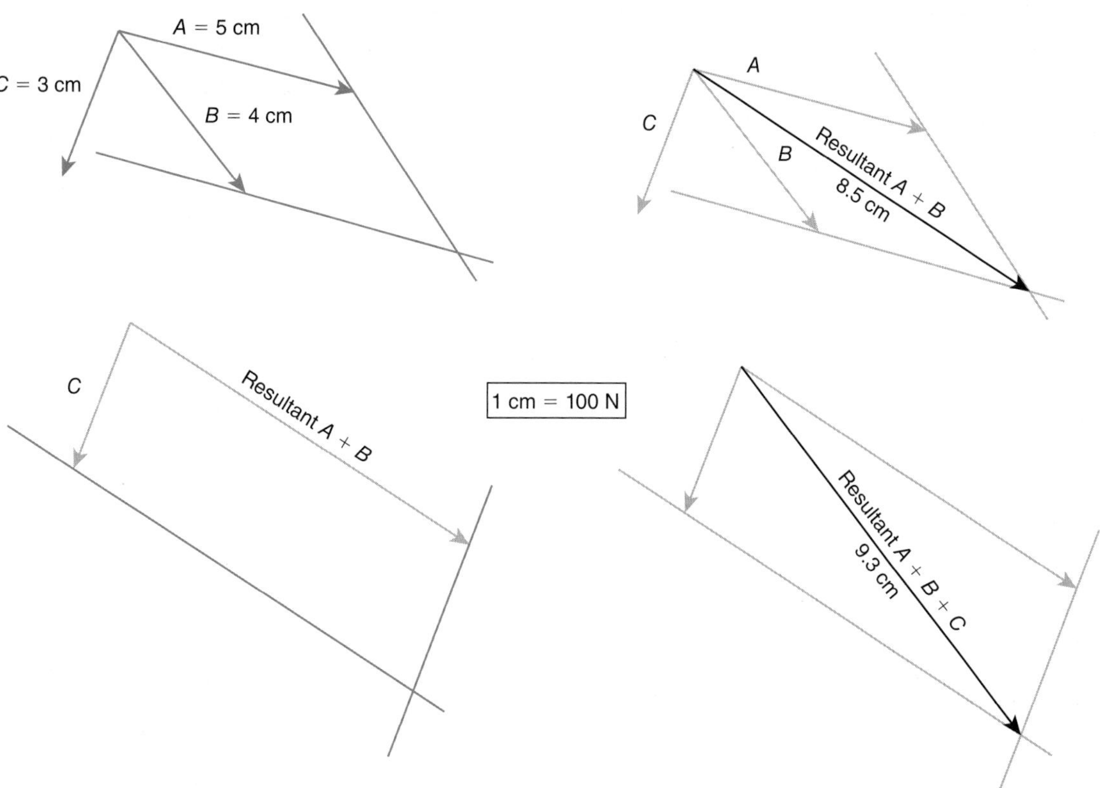

Figure 3.22 **Vector composition of three vectors using the parallelogram method.** What is a potential source of error when using this method?

For example, if we have vectors **A**, **B**, and **C**, we would simply choose any two of the vectors and compose them using the same steps from previously to find a resultant. Then the resultant from the first pair of vectors can be composed with the remaining vector. This process can be used with any number of vectors, two at a time. But as you can imagine, the process of composing vector parallelograms can be quite labor intensive when more than two vectors are involved.

SAMPLE PROBLEM

Composition with vector chains or vector polygons is a graphic method that is often used when three or more vectors must be composed. However, this method can be used with only two vectors. Actually, the concept is simple. Imagine that you are watching a hockey game (Figure 3.23).

One player hits the puck 10 m directly in the positive *x* direction; a teammate receives the puck and immediately strikes it so that it travels 15 m in a direction 30° above the positive *x*-axis (+*x*, +*y*) before it is intercepted by an opposing player. The opposing player, trying to clear the puck, strikes it so that it travels 20 m in a direction 170° relative to the positive *x*-axis (−*x*, +*y*). Each movement of the puck is the result of an applied force and has both a magnitude and direction, so each is a vector quantity. We could line up each vector end to end with the tip of one vector meeting the tail of the next and forming a vector chain. If we then wanted to know the distance the puck traveled relative to the initial location, we could draw a line from the starting point (the tail of the first vector) to its final location (the tip of the last vector) and measure the distance. Notice that drawing a line from start to finish would close the path of the puck and form the final side of a vector polygon. This final distance (magnitude) between start and finish is represented by a line that has a given direction and orientation relative to the initial position of the puck and is therefore a vector quantity (the *resultant* of all of the individual forces that acted on the puck).

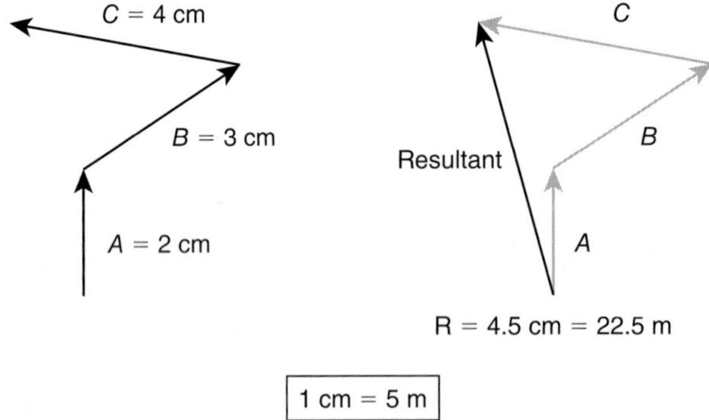

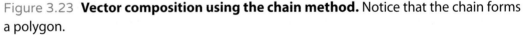

Figure 3.23 **Vector composition using the chain method.** Notice that the chain forms a polygon.

Notice that the order of the vectors in our vector chain does not matter because of the commutative nature of vector addition (Figure 3.24). Also notice that because vector addition is associative, the resultant is independent of the grouping of the vectors before summation (Figure 3.25).

Now let's cover the trigonometric methods of vector composition and resolution. If these methods are more appealing to you, then by all means use them. But always keep in mind the graphical representations of the following methods so that we do not lose touch with the visual aspects of trigonometry.

Trigonometric Methods of Vector Analysis

We do not want our trigonometric methods to seem completely disconnected from the graphical methods, so recall from our discussion of coordinate transforms that joint rotations are expressed in terms of θ. A change in θ means that a segment (represented by a line) was rotated from one location to another. So in a *somatic* reference frame, the joint (link) is the O; r is the distance from any point on the segment to the O; the positive x-axis is the original location of the segment (which is the axis from which θ is measured); and y is formed by drawing a line from any point on the segment to the x-axis. Drawing these lines constructs a triangle. Also recall that vectors are drawn within the coordinate system such that the tail is at the origin (0, 0), and the tip has x and y coordinates. The magnitude is represented by r, and the orientation of the vector will be given by θ relative to the positive x-axis. Again, drawing a line that connects the tip of the vector to the x-axis forms a triangle. In addition, constructing a parallelogram (during vector composition or resolution) with a vector drawn from one corner to the next actually forms two triangles. Finally, if one uses the vector polygon method with only two vectors, the polygon is actually a triangle. So now we understand: Graphic methods of vector math are actually just graphic methods of trigonometry. This text repeats many of these key points several times because trigonometry causes much undue anxiety for some people. The following section is simple if you understood the review of basic trigonometry and you comprehend that coordinate transforms and graphical vector analysis are both related to triangles (and therefore trigonometry).

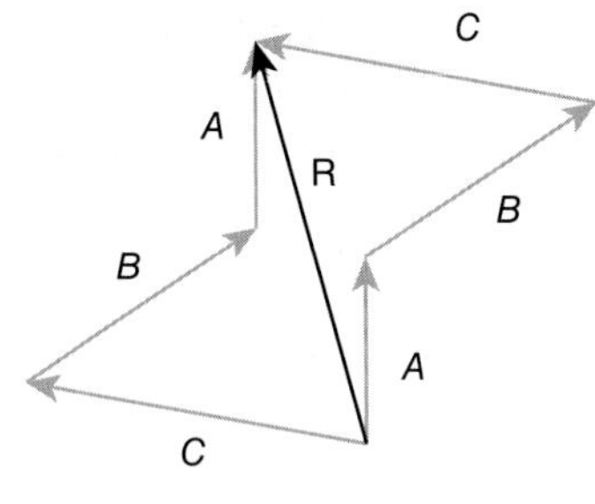

Figure 3.24 **Commutative property of vector addition.** Notice that the order of the vectors does not affect the resultant.

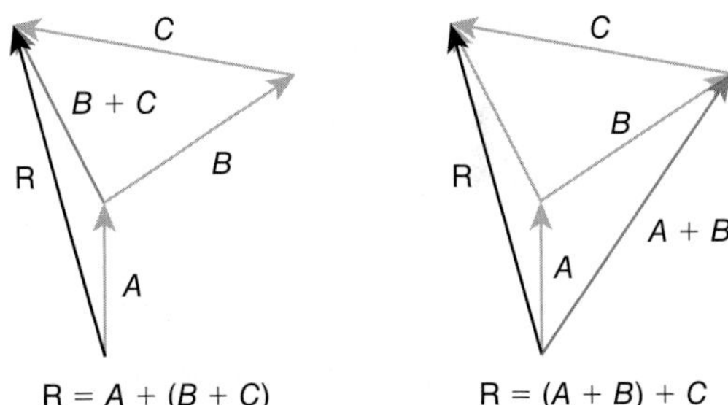

Figure 3.25 **Associative property of vector composition.** Notice that the grouping of the vectors does not affect the resultant.

Vector Resolution

Vector resolution is the process by which individual directional component vectors of a single vector are determined. Vector resolution with trigonometry is simple, and we have actually already used the concept when transforming polar coordinates (r, θ) to rectangular (Cartesian) coordinates (x, y). As we have stated several times, vectors are drawn within the coordinate system such that the tail is at the origin (0, 0), and the tip has x and y coordinates. The magnitude is represented by r, and the orientation of the vector is given by θ relative to the positive x-axis. If a drawn line connects the tip of the vector to the x-axis, it forms a triangle. Trigonometric vector resolution is simply the process of finding the x and y coordinates (components). In this case, the x and y coordinates represent the location of the tips of two component vectors.

SAMPLE PROBLEM

In our acrobat example (Figure 3.26), let's say that the teeter-totter applies a 5000 N force to the flying acrobat at an angle of 120° relative to the positive x-axis. Recall that Equations 3.1 and 3.2 can be used to transform the polar coordinates representative of our given vector (5000 N, 120°) to rectangular coordinates:

$$y = r\sin\theta = (5000\text{ N})\sin 120° = \mathbf{+4330.13\text{ N}}$$

$$x = r\cos\theta = (5000\text{ N})\cos 120° = \mathbf{-2500.00\text{ N}}$$

So the initial vector of 5000 N causes a vertically directed force on the acrobat of 4330 N and a horizontally directed force of 2500 N. Notice that the signs indicate the positive and negative x and y directions, respectively. It really is just that simple.

(continues)

(continued)

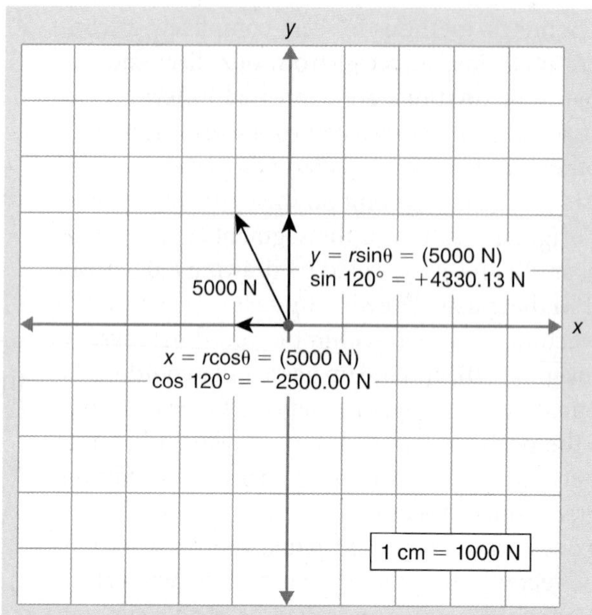

Figure 3.26 **Trigonometric vector resolution.** What is the benefit of this method versus a graphical method?

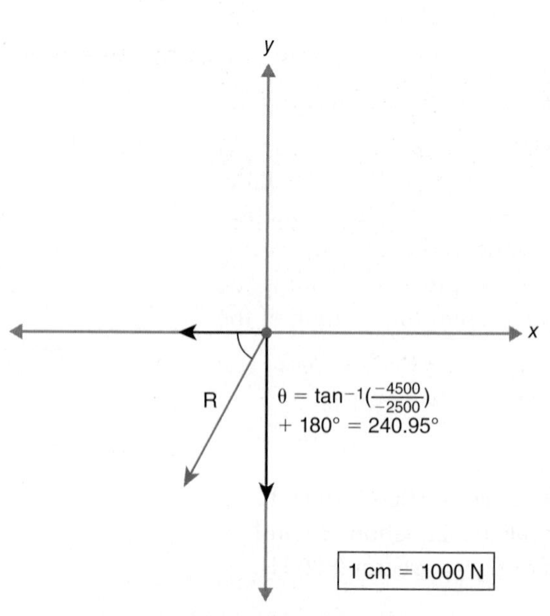

Figure 3.27 **Trigonometric vector composition.** What is the benefit of this method versus a graphical method?

Vector Composition

Vector composition is the process by which two or more vectors are summed to determine a single resultant vector. The simplest possible case of vector composition is one in which there are only two vectors to compose and they are *perpendicular* to each other (orthogonal). We have actually previously used the following method to perform a coordinate transform. The only difference is that instead of having x and y coordinates that represent the location of a point within our coordinate reference frame, the x and y coordinates represent the locations of the tips of our two vectors. Jumping is a great example (Figure 3.27).

Keep in mind that in this example we refer to force vectors, so the x and y coordinates are within a local reference frame. The jumper applies force to the ground so that his body will travel both vertically ($+y$ direction) and horizontally ($+x$ direction). We have discussed that these two components are perpendicular to each other. The jumper exerts a downward vertical force of 4500 N and a backward horizontal force of 2500 N. First we need to calculate the magnitude of the resultant. From our previous information, we know that the magnitude of a vector within a coordinate reference frame is defined by r in polar coordinates, and that $r = \sqrt{x^2 + y^2}$. So, in this example:

$$r = \sqrt{(4500\text{ N})^2 + (2500\text{ N})^2} = \mathbf{5147.82\ N}$$

Now we need to calculate the orientation of the resultant vector relative to the positive x-axis:

$$\tan \theta = \frac{y}{x} = \frac{-4500}{-2500} = \mathbf{1.80}$$

We now have the ratio of y to x (−1.80). The ratio can be converted to an angle by using an inverse function. Remember to add 180° to the answer, because the resultant vector is in the third quadrant:

$$\theta = \tan^{-1}(1.80) = 60.95° + 180° = \mathbf{240.95°}$$

So in this particular situation, the resultant force applied by the jumper to the ground has a magnitude of 5147.82 N and is oriented 240.95° relative to the positive x-axis. To avoid confusion, remember that this vector represents the force applied by the jumper to the ground. It is not the vector that would represent the final path of the jumper within the global reference frame. At this point, one can begin to understand the complexities of a complete biomechanical analysis.

Offensive football linemen applying forces to push a defensive lineman out of the way is an example of two vectors that are not perpendicular (Figure 3.28).

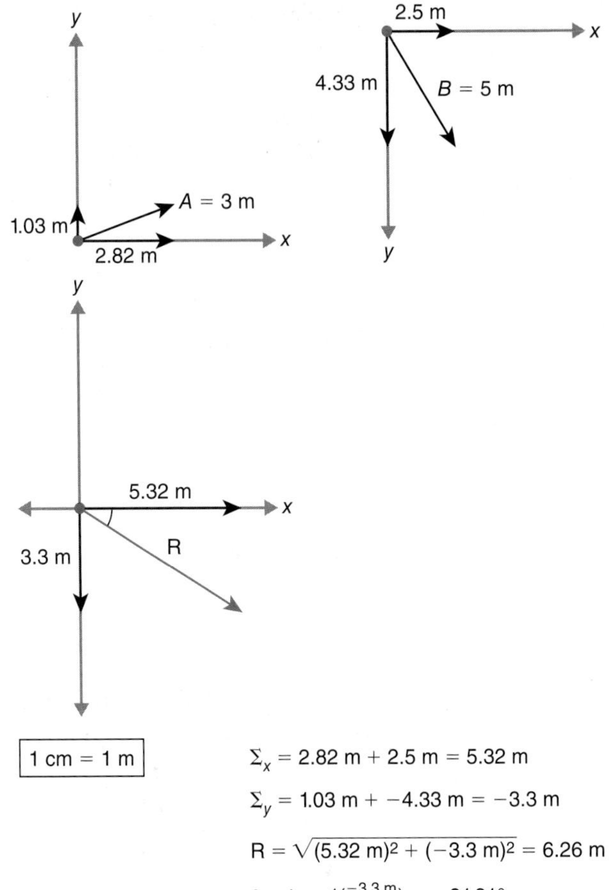

Figure 3.28 **Trigonometric vector composition of two nonorthogonal vectors.**

In this example, we are interested in the resultant motion of the defensive lineman, so our resultant vector is relative to the global reference frame. In the case of two vectors that are nonorthogonal, we simply resolve each vector, using the process from the previous section, and then add the x components and the y components to find the resultant. So, in this case, we find the x and y effects of each player's force and then add them together. In this example, let's say that offensive lineman A applies a force to the defensive lineman at an angle 20° relative to the x-axis that is capable of moving the defensive player 3 meters. Simultaneously, offensive lineman B applies a force to the defensive lineman at an angle 300° relative to the x-axis that is capable of moving the defensive lineman 5 meters. Our first step is to calculate the independent directional effects of each offensive lineman. To perform this calculation, we should recall Equation 3.1 ($y = r\sin\theta$) and Equation 3.2 ($x = r\cos\theta$). Basically, this process is the same as the one we used earlier to transform polar coordinates to rectangular (Cartesian) coordinates. Just recall that, in this case, the coordinates we are calculating define the location of the tip of our given vector. So we transform the polar coordinates given previously (3.00 m, 20°) to the rectangular coordinates of the tip of vector **A** as follows:

$$y = r\sin\theta = (3.00\text{ m})\sin 20° = \mathbf{1.03\ m}$$

$$x = r\cos\theta = (3.00\text{ m})\cos 20° = \mathbf{2.82\ m}$$

If player A acted alone, he would move the defensive player 1.03 m in the positive y direction and 2.82 m in the positive x direction. Next we transform the polar coordinates given for player B (5.00 m, 300°) to the rectangular coordinates of the tip of the representative vector:

$$y = r\sin\theta = (5.00\text{ m})\sin 300° = \mathbf{-4.33\ m}$$

$$x = r\cos\theta = (5.00\text{ m})\cos 300° = \mathbf{+2.50\ m}$$

If player B acted alone, he would move the defensive player 4.33 m in the negative y direction and 2.50 m in the positive x direction. However, both of the players acted on the defensive lineman, so we must find the combined effects. First we sum the effects in the y direction ($\Sigma y = 1.03\text{ m} + -4.33\text{ m} = -3.3\text{ m}$) and sum the effects in the x direction ($\Sigma x = 2.82\text{ m} + 2.50\text{ m} = 5.32\text{ m}$). Now we basically have rectangular coordinates and need to transform them to polar to define the vector representing the resulting effect of both players.

Several options exist for finishing our problem, and we could just choose one of those. However, because we may need some practice, we are going to exhaust all options. One option is simply to use the steps from our previous problem. We know that the *magnitude* of a vector within a coordinate reference frame is defined by r in polar coordinates and that $r = \sqrt{x^2 + y^2}$. So, in this example:

$$r = \sqrt{(5.32\text{ m})^2 + (-3.3\text{ m})^2} = \mathbf{6.26\text{ m}}$$

Now we need to calculate the orientation of the resultant vector relative to the positive x-axis:

$$\tan\theta = \frac{y}{x} = \frac{-3.3\text{ m}}{-5.32\text{ m}} = \mathbf{-0.62}$$

We now have the ratio of y to x (-0.62). The ratio can be converted to an angle by using an inverse function:

$$\theta = \tan^{-1}(-0.62) = \mathbf{-31.81°}$$

In our example, the combined effect of the two offensive linemen is to move the defensive lineman 6.26 m, with the orientation of the resultant motion being $-31.81°$ relative to the positive x-axis.

Another way to approach the problem is to solve for the orientation of the angle first. We know x and y and Equation 3.3 express the relationship (ratio) of the two coordinates in terms of θ:

$$\tan\theta = \frac{y}{x} = \frac{-3.30}{-5.32} = \mathbf{-0.62}$$

We now have the ratio of y to x. Remember from previously that a ratio can be converted to an angle by using an inverse function:

$$\theta = \tan^{-1}(-0.62) = \mathbf{-31.81°}$$

Now that we have θ, our final step is to calculate the *magnitude* of the resultant. This calculation can be accomplished by using one of the forms of either Equation 3.1 or Equation 3.2. Now we use both equations to further elucidate the trigonometric concepts:

$$r = \frac{y}{\sin\theta} = \frac{3.30}{\sin(31.81°)} = \mathbf{6.26\text{ m}} \qquad (3.1)$$

$$r = \frac{x}{\cos\theta} = \frac{5.32}{\cos(31.81°)} = \mathbf{6.26\text{ m}} \qquad (3.2)$$

No matter the method of choice, the combined effect of the two offensive linemen is to move the defensive lineman 6.26 m, with the orientation of the resultant motion being $-31.81°$ relative to the positive x-axis.

SAMPLE PROBLEM

The process just used is exactly the same no matter how many vectors are involved. We are going to perform one more example, this time composing three vectors using trigonometry. In this case, we track the motion of a puck relative to the rink, so all coordinates in this example are relative to a global reference frame (Figure 3.29).

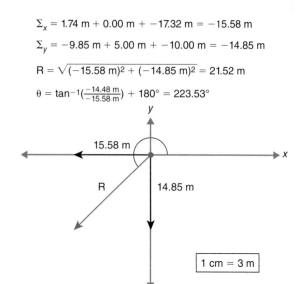

$\Sigma_x = 1.74 \text{ m} + 0.00 \text{ m} + -17.32 \text{ m} = -15.58 \text{ m}$

$\Sigma_y = -9.85 \text{ m} + 5.00 \text{ m} + -10.00 \text{ m} = -14.85 \text{ m}$

$R = \sqrt{(-15.58 \text{ m})^2 + (-14.85 \text{ m})^2} = 21.52 \text{ m}$

$\theta = \tan^{-1}(\frac{-14.48 \text{ m}}{-15.58 \text{ m}}) + 180° = 223.53°$

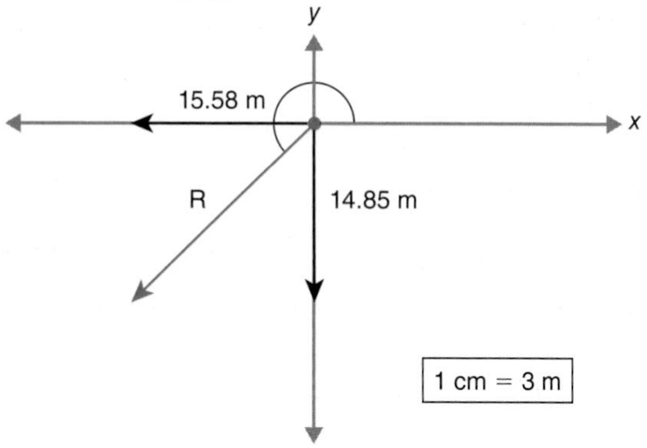

Figure 3.29 **Trigonometric vector composition of three nonorthogonal vectors.**

The puck's original location has the coordinates (−5 m, 3 m) within a global coordinate reference frame that has its O at the center of the rink. One player (A) passes the puck to another player 10 m away and oriented 280° relative to the *x*-axis. Player B then strikes the puck, causing it to travel 5 m to a player oriented at 90°. The last player (C) applies a force to the puck that moves it 20 m with a path oriented at 210°. We want to calculate the final total movement of the puck and the orientation of its resultant path. No matter how many vectors, we follow the same rules. Our first step is to resolve the individual vectors from each player into their *x* and *y* components.

First we transform the polar coordinates given previously (10.00 m, 280°) to the rectangular coordinates of the tip of vector **A** as follows:

$$y = r \sin \theta = (10.00 \text{ m}) \sin 280° = \mathbf{-9.85 \text{ m}}$$

$$x = r \cos \theta = (10.00 \text{ m}) \cos 280° = \mathbf{+1.74 \text{ m}}$$

If player A acted alone, then the puck would be moved 9.85 m in the negative *y* direction and 1.74 m in the positive *x* direction. Next, we transform the polar coordinates given for player B (5.00 m, 90°) to the rectangular coordinates of the tip of the representative vector:

$$y = r \sin \theta = (5.00 \text{ m}) \sin 90° = \mathbf{+5.00 \text{ m}}$$

$$x = r \cos \theta = (5.00 \text{ m}) \cos 90° = \mathbf{+0.00 \text{ m}}$$

If player B acted alone, the puck would move 5.00 m in the positive *y*-direction but not move at all in the *x*-direction. Notice that, in this case, a force has been applied orthogonally to one of our axes (the *x*-axis), so one of our components has a magnitude of zero because cos 90° = 0.00. If we had been using a graphical method, we would not have been able to resolve the given vector into two components and would have arrived at the same solution. Finally, we transform the polar coordinates given for player C (20.00 m, 210°) to the rectangular coordinates of the tip of the representative vector:

$$y = r \sin \theta = (20.00 \text{ m}) \sin 210° = \mathbf{-10.00 \text{ m}}$$

$$x = r \cos \theta = (20.00 \text{ m}) \cos 210° = \mathbf{-17.32 \text{ m}}$$

(continues)

(continued)

If player C acted alone, the puck would move 10.00 m in the negative *y*-direction and 17.32 m in the negative *x*-direction. As before, we are looking for the combined effects of the players. First, we sum the effects in the *y* direction ($\Sigma y = -9.85 \text{ m} + 5.00 \text{ m} + -10.00 \text{ m} = -14.85 \text{ m}$) and sum the effects in the *x* direction ($\Sigma x = 1.74 \text{ m} + 0.00 \text{ m} + -17.32 \text{ m} = -15.58 \text{ m}$). Now we basically have rectangular coordinates and need to transform them to polar to define the vector representing the resulting effect of both players.

Again, we know that the *magnitude* of a vector within a coordinate reference frame is defined by *r* in polar coordinates, and that $r = \sqrt{x^2 + y^2}$. So, in this example:

$$r = \sqrt{(-15.58 \text{ m})^2 + (-14.85 \text{ m})^2} = \mathbf{21.52\ m}$$

Now we need to calculate the orientation of the resultant vector relative to the positive *x*-axis:

$$\tan\theta = \frac{y}{x} = \frac{-14.85 \text{ m}}{-15.58 \text{ m}} = \mathbf{0.95}$$

We now have the ratio of *y* to *x* (0.95). The ratio can be converted to an angle by using an inverse function. Remember to add 180° to the answer because the resultant vector is in the third quadrant:

$$\theta = \tan^{-1}(0.95) = 43.53° + 180° = \mathbf{223.53°}$$

So, in our example, the combined effect of the three hockey players is to move the puck 21.52 m from the original location, with the orientation of the resultant motion being 223.53° relative to the positive *x*-axis.

Ideally, we have accomplished several goals in this section. First, we wanted to show the relationship between trigonometry and the graphical methods. Trigonometry is simpler and more precise (once one becomes proficient at using it). But we must also cover graphical methods because they help us visualize the effects of applying force. Finally, notice that all of the examples are actually lower-level motion analysis using quantitative methods. The quantitative methods are simply the mathematical representations of qualitative visual observations used by practical practitioners of biomechanics. So, even in light of all the trigonometry, never lose sight of the fact that all motion can be observed (analyzed) both qualitatively and quantitatively, depending on the needs of the situation. The section that follows connects these motion analysis techniques to other movement-related disciplines.

CONNECTIONS

3.3 FUNCTIONAL ANATOMY

We previously discussed that skeletal muscles produce the force required to move the joints through their many degrees of freedom. We have also mentioned that we usually represent muscle force with one vector, when in reality multiple muscles are responsible for the action at a given joint. For example, the *quadriceps femoris* muscle (*rectus femoris, vastus intermedius, vastus lateralis,* and *vastus medialis*) produces four forces with different lines of action simultaneously acting on a common point of application (the tibial tuberosity through the patellar tendon). So, in terms of our earlier discussion of vector composition, we would have four muscle force vectors, with one resultant vector producing the final motion of the tibia relative to the femur (Figure 3.30).

We should first revisit the idea that a single force can have multiple directional effects—that is, that it can be resolved into a vertical (perpendicular) component and a horizontal (parallel) component. A single muscle force vector can also be resolved to observe two different effects. The easiest way to understand the two effects of a muscle contraction is to first resolve a muscle force vector. We will use the parallelogram method, following the same preceding steps, with one exception: To resolve muscle force vectors, we use a somatic reference system (Figure 3.31).

In other words, your first horizontal line is drawn such that it is touching the tail of the vector and parallel to the *bone on which it has its insertion* (as opposed to the global x-axis as previously). All of the other steps are the same: Draw another horizontal line parallel to the first and touching the tip of the vector, draw the vertical sides of the parallelogram with one line touching the tail of the vector, and the other parallel to the first vertical line and touching the tip of the vector (both being perpendicular to the horizontal lines drawn in the previous step). Then carefully draw arrowheads on the lines to define two new vectors (our component vectors).

After resolving the vector, understanding the names of the individual components of a muscle force vector is much easier. Notice that we still have our vertical or perpendicular component, but in this case it is vertical or perpendicular to the bone. We name each muscle force vector component according to what action it would cause if allowed to act *alone.* For example, imagine a marionette on strings. Notice what would happen if you pull only the vertical (perpendicular) component "string." The forearm would rotate at the elbow relative to the humerus. So when resolving muscle force vectors, we define the vertical or perpendicular component as the **rotary component** because if it were allowed to act alone, it would cause joint rotation. Now notice what would happen if the horizontal (parallel) component acted *alone*—that is, pulled the horizontal (parallel) "string." The horizontal (parallel) component would cause the bones of the forearm to be pushed tightly into the humerus. So in this situation, we define the horizontal (parallel) component of a muscle force vector as the **stabilizing component** because it stabilizes the elbow joint. However, the horizontal (parallel) component can be slightly more complicated (Figure 3.32).

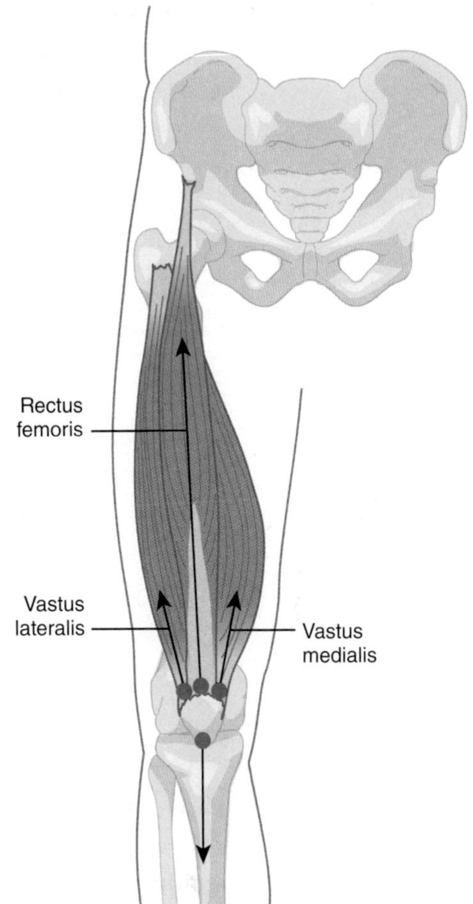

Figure 3.30 **Quadriceps force vectors.** Imagine the force vectors for the muscles of the lower leg.

Notice that if the elbow angle is changed to less than 90°, the horizontal (parallel) component has the opposite direction (i.e., the arrowhead is pointing away from the elbow joint). So, in this case, the horizontal (parallel) component acting alone would actually pull the forearm away from the humerus. When the horizontal (parallel) component is directed away from the joint. it is called the **destabilizing component** because it causes joint instability by pulling the links of the system apart. Visually, the resolution in this case can be extremely confusing, so be careful in resolving the vector and drawing the arrowheads in the correct directions. Find a set of rules that works for you and follow them in every case. Remember to look at the orientation of the original vector. If the original muscle force vector is directed upward and to the right relative to

Rotary component The vertical or perpendicular component of a muscle force vector representing the amount of force that would tend to cause joint rotation.

Stabilizing component The horizontal (parallel) component of a muscle force vector directed toward the joint, representing the amount of force that would tend to stabilize the joint.

Destabilizing component The horizontal (parallel) component of a muscle force vector directed away from the joint, representing the amount of force that would tend to destabilize the joint.

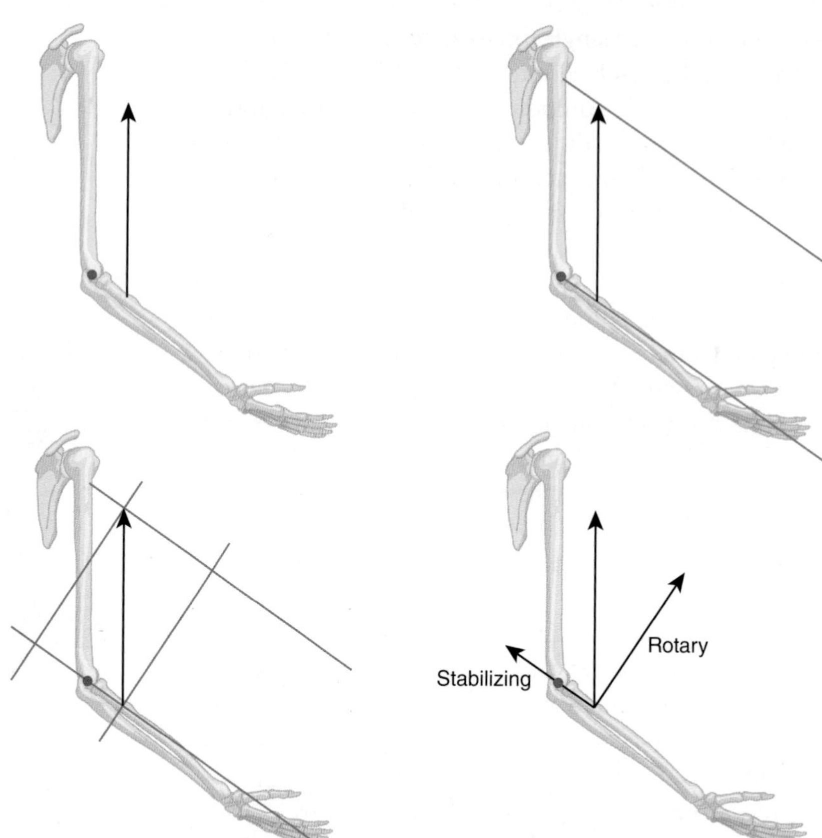

Figure 3.31 **Vector resolution of biceps brachii force vector into rotary and stabilizing components.**

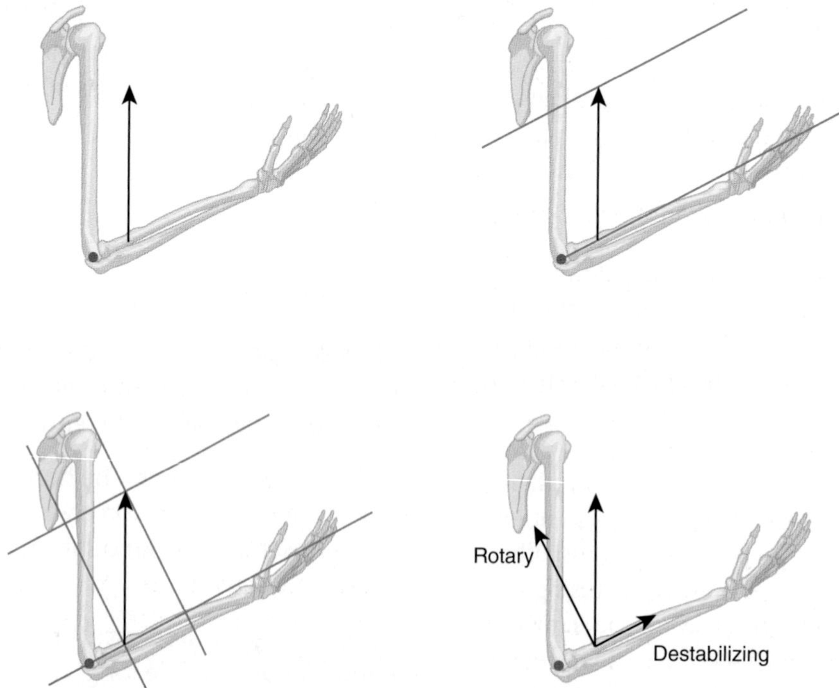

Figure 3.32 **Vector resolution of biceps brachii force vector into rotary and destabilizing components.** How is the elbow preventing from destabilizing?

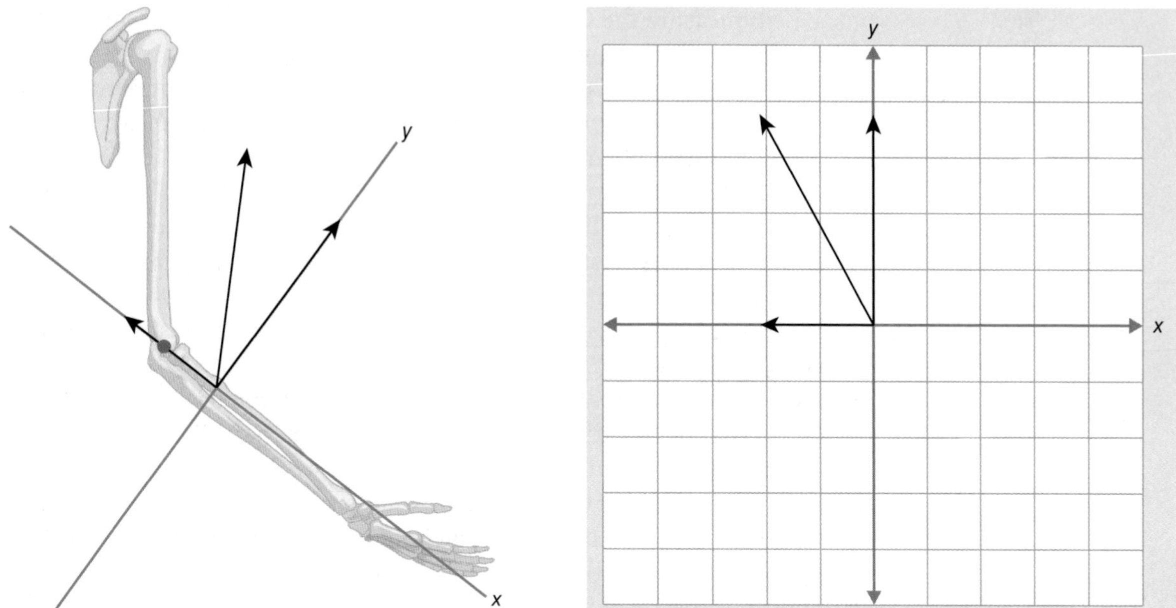

Figure 3.33 **Two methods for resolving an anatomical force vector.**

the parallel component (which in this case is parallel to the bone), the vectors representing the individual components must indicate upward and rightward directions. You can always use the two-hand method from earlier: Once the parallelogram is drawn, place both hands together pointing in the same direction as the vector and then split them apart while keeping them together at the wrist. The fingertips will point in the correct directions. Another option is to keep in mind that the source of the visual confusion is the use of a somatic reference system. So it is sometimes helpful to orient the bone in a free-body diagram with the *x*-axis or to remove the original vector from the free-body diagram altogether and resolve it according to one of our previous methods (Figure 3.33).

This concept of changing rotary, stabilizing, and destabilizing components at varying angles and the implications for motion are revisited in later chapters after some necessary concepts have been covered. For now, our purpose is to demonstrate the connection of muscle physiology and biomechanics and to engage in some basic quantitative motion analysis.

3.4 INJURY SCIENCE

Notice that the orientation of the vector representing muscle force is a function of the positions of the origin and insertion, as well as the alignment of the fibers in the muscle. Therefore, structural deviations have functional implications. Let's use the **Q angle** (quadriceps angle) as another example of the interconnectedness of the muscular aspects of exercise physiology and biomechanical concepts. Remember from the standpoint of vector composition that the quadriceps muscles all share a common insertion at tibial tuberosity by way of the patellar tendon. Therefore, their individual forces are composed into one representative vector. The Q angle is the angle formed by the longitudinal axes of the femur and tibia; it approximates the resultant line of action (sometimes called **line of pull** when referring to muscle) of the quadriceps muscles (Floyd, 2018; Oatis, 2017; Watkins, 2010) (Figure 3.34).

Q angle The quadriceps angle formed by the longitudinal axes of the femur and tibia, which approximates the resultant line of action of the quadriceps muscles.

Line of pull Resultant line of action of a muscle force vector.

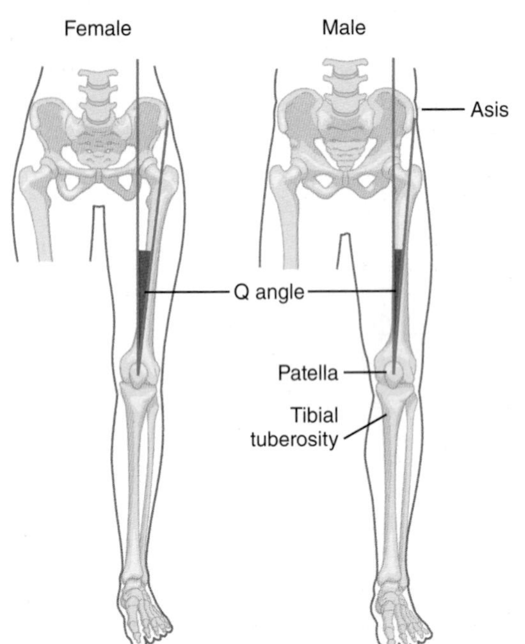

Figure 3.34 **Q angle for male and female.** Why is the angle different?

To measure the Q angle, two lines are drawn from the center of the patella: one to the anterior superior iliac spine of the pelvis (representing the femur), and another to the tibial tuberosity (representing the tibia). The Q angle is the smaller angle formed by these two lines (a rough estimate of femoral and tibial alignment). A "normal" Q angle is approximately 10° to 20°, with men ranging from 10° to 15° and women from 15° to 20° (Aglietti, Insall, & Cerulli, 1983). The gender difference is caused by the relatively wider pelvis of females. The reason for concern about Q angle is that the patella should track smoothly between the femoral condyles when the quadriceps contract to cause knee extension (Oatis, 2017, Watkins, 2010). If a person possesses a high Q angle, then the patella may not track normally during quadriceps contraction (i.e., it may be forced laterally). Though agreement is not universal (and allowances for individual variation must always be considered), belief is fairly consistent that Q angles that deviate significantly from normal (e.g., above 20°) are correlated with patellofemoral pain syndrome (caused by excessive forces on the border of the patella) and increased risk of knee injury (Aglietti et al., 1983; Boucher et al., 1992; Caylor, Fites, & Worrell, 1993; Lysholm & Wiklander, 1992; Messier et al., 1991; Moss, DeVita, &. Dawson, 1992; Neely, 1998; Zimbler et al., 1980).

In relation to Q angle, we can analyze patellofemoral pain from both anatomical and exercise training perspectives. Anatomically, we have already mentioned that the wider pelvis of a female leads to larger Q angles (by changing the location of the anterior superior iliac spine). However, the other line in the Q angle passes through the tibial tuberosity. So the location of the tibial tuberosity (as opposed to simply the shaft of the tibia) also affects the line of pull of the quadriceps muscle. From a muscle physiology training perspective, we must consider that other factors can potentially cause the abnormal patellar tracking and patellofemoral pain, although the Q angle may be within normal anatomical limits. For example, notice that the fibers of the vastus medialis possess an orientation that is extremely medial and lateral when compared to the other three heads of the quadriceps muscle. This orientation has implications for its major function: attempting to maintain normal patellar tracking by balancing the slightly lateral pull of the other three quadriceps muscle heads (Oatis, 2017). So it begins to become clear just how many other factors could cause patellofemoral pain because of abnormal patellar tracking. For example, the vastus lateralis or medialis may be excessively strong, weak, or tight relative to the other. This imbalance can change the resultant line of pull and therefore cause abnormal patellar tracking and result in pain.

There may not be a better example of the connection between qualitative and quantitative movement analysis than the fields related to rehabilitation. Doctors, therapists, and trainers rely on accurate quantitative measures to monitor the progress of their patients. These measures are used to guide therapeutic strategies and to decide when a patient is ready to advance or be considered "rehabilitated" and ready to resume activity. A patient entering rehabilitation is first put through a series of measures to gather baseline data. Quantitative measures such as strength, range of motion, balance, and joint laxity can all be accurately measured so that trainers have information to compare with measures taken during rehabilitation programs.

During the rehabilitation intervention (e.g., strength training, stretching, and balance activities), trainers often use qualitative techniques to assess and provide feedback on a patient's progress. At certain intervals, such as weekly or monthly, patients undergo the original battery of quantitative measures to validate the effectiveness of the intervention program. Completing the quantitative testing after each session would be too time consuming, so trainers rely on qualitative observation and feedback methods for daily routines—qualitative methods that are firmly grounded in the information provided by accurate quantitative analysis.

3.5 MOTOR BEHAVIOR

Motor Control

We have discussed the complex task of trying to control the many available degrees of freedom of the human musculoskeletal system. We have also mentioned that having a wide range of muscles available is necessary to move the segments of our skeletal system through all of those degrees of freedom. In motor control terms, keep in mind that the resultant motion is a vector sum of all of the forces involved. All of the muscular forces in the vector sum are being manipulated by the nervous system. Do not underestimate the complexity of this particular vector analysis problem. Many muscles are involved, but do not forget that each muscle is divided into motor units that can be recruited in various combinations. Each motor unit that is recruited (or fatigues) changes our vector sum by either adding force or becoming fatigued, which then lessens the force. At the same time, not all force from a given muscle contraction causes rotation; some stabilize (or destabilize) the joint. In short, human motion is a never-ending series of complex vector compositions and resolutions completely under the control of the nervous system. The volume and speed of mathematical calculations performed automatically by our nervous system are quite impressive, especially in light of the number of steps required to compose or resolve even the most basic problems presented in this chapter.

Motor Development

By what age should a child be able to walk? Do all motor skills develop naturally through normal maturation? Will every child eventually develop a mature throwing pattern? The answers to these questions are: it depends, no, and maybe. Most children develop the ability to walk between 9 and 13 months. Creeping, walking, and running develop through the course of normal maturation, but skills such as throwing, catching, kicking, and striking do not. Mature throwing patterns occur only through quality instruction and practice. How do we know these things? The primary basis for this understanding has been careful observation and qualitative evaluation. Infant and childhood development across cognitive, motor, and social domains has been studied for more than a century. In 1877, Darwin wrote "A Biographical Sketch of an Infant," a precursor to more recent developments in understanding human development. From the seminal work of Jean Piaget to current research in developmental biodynamics, psychologists, scientists, and others have been searching for patterns and sequences in cognitive and motor development in an attempt to understand and describe the process of change that children go through in the course of normal development.

Piaget developed his theoretical model of cognitive development through informal observation of his own children, proposing that infants and children progress through four distinct stages. He provided age ranges for each stage and suggested that the rate and degree of completion of any stage can vary by child; however, the sequence in which children pass through the stages is invariant. Gesell (1928) and Bayley (1936) also observed children but were more interested in the development of the physical organism. They focused on growth and development of human motion, describing changes over time and approximate age ranges that children should reach certain movement milestones. In fact, modified versions of Bayley's scales are still used to assess children's movement today. Although Piaget primarily focused on cognitive development, and Gesell and Bayley primarily identified movement changes, all share two significant commonalities—first, they all used qualitative methods (observation of changes over time); second, they all acknowledged that motor and cognitive development are intimately connected. More recent research in the area of motor skill development has also focused on the integrated nature of cognitive and structural development—in essence, coordinated movement can only result from the interaction and development of the brain and the body.

Motor Learning

New tasks tend to be inefficient, because the forces of the individual muscles are not acting together to produce one large resultant vector. For example, when learning to bench press, a novice is often observed to be "shaking" during the motion. That shaking is lack of coordinated effort among individual muscles. Therefore, each muscle force vector is not necessarily occurring in a pattern that is conducive to smooth motion or force production. Once the nervous system learns to contract the muscles in a more coordinated fashion, the force vectors compose a larger total resultant vector.

Weight training programs for women and prepubescent children rely on this coordinated muscle contraction. Among other neural mechanisms, this improved orchestration of individual muscle contractions leads to very rapid strength gains without a proportionate increase in muscle size. Post-pubescent males also gain strength through synchronization, but a more important factor is their increase in muscle size that results from higher levels of testosterone. Females and children do not gain significant muscle mass and must rely on the synchronization—through efficient recruitment and firing of muscles—to show noteworthy strength gains from training.

3.6 PEDAGOGY

Quantitative measurement (outcomes) of skill performance occurs often in physical education. Teachers assess fitness (number of push-ups or sit-ups, mile run times), keep score during games, and sometimes base grades on how well students perform isolated skills (percentage of tennis serves landing in the correct service area, percentage of free throws made). Valid reasons exist for collecting and using accurate quantitative information: (1) to provide data to monitor the progress of student improvement in skills and fitness; and (2) to provide accountability for a quality physical education or athletic program.

Although quantitative measures of studying human motion exist in the form of biomechanical analysis, performing this type of assessment in a school setting is neither relevant nor practical. This is not to say that biomechanics doesn't play a role in teaching and improving motor performance. To the contrary, the role of quantitative biomechanical analysis is quite important in qualitative evaluation and stems from identifying observable criteria that teachers use as guidelines.

Successful teachers are experienced observers; they constantly scan their classes as students practice skills. When they notice something wrong, they provide feedback to correct it. But how do they know what is right and wrong? Knudson (2013) mentions the importance of defining critical features for any given skill; in essence, these are the cues necessary to attain optimal performance. Knudson (2000) explains that "critical features are the most invariant technique points of a movement: They determine whether a movement is effective, efficient, and safe." He also validated the role of biomechanics in teaching by suggesting that "the science of biomechanics is most helpful in determining which technique points really contribute to successful execution and injury prevention" (p. 20).

These technique points or critical features are the cornerstone of observation instruments such as the TGMD-3, the total body approach, and the component approach of motion analysis. They are also the foundation of an important feature in quality pedagogy—**teaching cues**. Previously in this chapter, you learned that qualitative evaluation tools exist not only for fundamental motor skills but also for specific sport skills. The field of physical education in particular has been on the leading edge of developing assessment methods to help students learn and refine sport skills efficiently.

Pinheiro (2000) promotes the use of "criteria sheets" to perform qualitative assessment in elementary schools. The sheets include photos or pictures of the different phases of each skill (e.g., preparation, execution, and follow through), with written critical elements next to the photos. Similar assessment tools can be found in books and online for most common sports (Figure 3.35).

Teaching cues Single words or short phrases that identify critical elements of a skill.

Soccer

Observer's name: ______________________

Date: ______________________

Skills to be diagnosed: *Throw in* [specific skill]

Performer: ______________________

Illustration	Critical elements	Yes	No	Comment/Dx
 Prepatory phase	1. Face the target. 2. Feet shoulder width apart or stride. 3. Feet <u>behind</u> the side-line. 4. Secure the ball <u>overhead</u> with both hands. 5. Fingers spread to form "W" on the ball.	____ ____ ____ ____ ____	____ ____ ____ ____ ____	
 Execution phase	1. Bring ball behind the head. 2. Body <u>arched</u> backwards. 3. Body uncoils to release ball. 4. Ball released forward by both hands. 5. Release ball <u>"up in the sky."</u> 6. Feet in <u>contact</u> with the ground.	____ ____ ____ ____ ____ ____	____ ____ ____ ____ ____ ____	
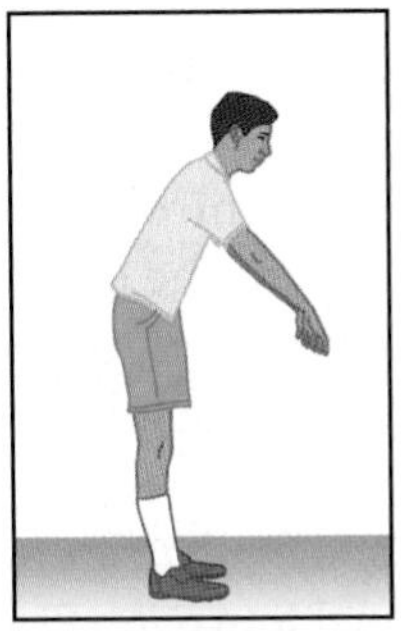 Follow through	1. Feet <u>still in contact</u> with the ground. 2. Arms extended forward. 3. Body <u>follows direction</u> of throw.	____ ____ ____	____ ____ ____	

(*continues*)

Figure 3.35 **Examples of task sheets for assessing soccer throw-in and volleyball passing.** Could you create a task sheet for a different sport skill?

Courtesy of Joey Feith, founder of ThePhysicalEducator.com.

FOREARM PASS
Ready Position
Player is relaxed with shoulders square to the ball.
Feet slightly wider than shoulder width with knees flexed.
Trunk is leaning forward slightly with arms in front of the body.
Pre Contact
Arms straighten and hand grip together to form a flat contact surface from wrists to elbows.
Knees are bent to 90°.
One foot is slightly in front of the other.
Contact
Arms are straight and form a 90° angle with trunk.
Trunk is straight with a forward lean.
The ball contacts both forearms at a point slightly above the wrists.
Arms stay together and straight as they swing slightly upward and forward as contact is made.
Follow Through
The body moves forward towards the target.
Player then returns to ready position in preparation for the next action.
Learn More!
Scan to learn more about this awesome skill!
See It In Action!
Scan to see this skill being performed live!
Time To Practice!
Scan to access a fun activity that will help you master this skill!
ThePhysicalEducator.com
Volleyball

Figure 3.35 *(continued)*

The criteria sheets are designed for students to use while watching each other perform, resulting in six benefits in the teacher–learner setting: (1) promoting learning of the critical elements, (2) forcing students to concentrate on specific problem areas, (3) acting as a control guide, (4) helping instructors perform a systematic observation, (5) helping instructors record the errors of the skill, and (6) motivating students. As students observe each other, they learn through qualitative evaluation. The critical elements are read by the evaluator and used to give feedback to the performer, which helps both the performer and the evaluator to remember them.

The importance of identifying and teaching using critical elements is evident in the expanding volume of resources available to practitioners. From articles in professional journals (Pinheiro, 2000), to Internet resources such as PE Central (http://www.pecentral.org/lessonideas/cues/cuesmenu.asp), to books (Fronske & Wilson, 2002), teaching cues based on critical features help teachers identify the most important components of motor skills and enhance student learning. Whether you are teaching fundamental or sport-specific skills, qualitative analysis tools identify the critical elements to help you understand *what* to observe, *how* to evaluate the performance, and what kind of relevant feedback to provide to improve it.

One crucial aspect of learning and refining any skill is the quality of feedback received by the performer. Successful teachers and coaches are not only experienced observers of human movement but also understand what must occur as a result of the observation to help the performer improve. In most cases, this intervention involves the use of relevant, skill-related feedback. Teachers and coaches rely primarily on informally observing performers to evaluate and refine skills, and qualitative assessment techniques present criteria they can use in providing relevant feedback.

3.7 ADAPTED MOTION

When teaching children known to have cognitive and physical challenges, assessing movement is no different than with nondisabled populations except for expectations. Most students with disabilities are able to learn and refine motor skills, but the rate and pace of change is unique to each person, depending on the type and severity of the disability. Children with challenges will progress through the same stages of fundamental movements and sport skills, but they may need more time to acquire the skills because of difficulties with strength, coordination, or comprehension of the task.

Qualitative analysis of movement plays an important role in providing children with disabilities the best chance of success in physical education. Formal qualitative assessment tools such as the TGMD-3, the Bruininks-Oseretsky Test of Motor Proficiency, and the Peabody Developmental Motor Scales derive information used to enhance educational opportunities. Results from formal testing are used to (1) identify children with delays who may benefit from additional support or services, (2) place children in the appropriate setting, (3) develop individualized education plans (required by law), (4) plan content and curricular needs, and (5) provide valid and reliable data on current performance and future progress. Beyond basic measures of motor proficiency in fundamental skills, many other assessment tools are available for measuring other areas related to movement.

Each year a larger number of people with disabilities are participating in organized sports through Special Olympics (www.specialolympics.org). Special Olympics, Inc. produces program guides for 25 sports, including those that are part of their summer and winter games, and also several popular national sports. The guides don't contain formal assessment tools, but they do have checklists that describe critical elements of sport skills that athletes should demonstrate. Most athletic events require some level of fitness, and several assessment tools have been designed for specific populations with disabilities. The Brockport Physical Fitness Test and FitnessGram are two examples that can be used for students with and without disabilities.

Not all people with disabilities choose to participate in sports, and sedentary lifestyles are common for many. The U.S. Surgeon General suggests that everyone should participate in 60 minutes of physical activity most days of the week to maintain a healthy level of fitness. One popular assessment tool that works well for providing valid information on activity levels is ACTIVITYGRAM. Individuals use a computer self-report system to record their activity levels throughout each day, and the software generates a report showing the amount and level of activity. The information gleaned from these reports is used to identify deficits and aid in programming activities for each person.

SUMMARY

Qualitative and quantitative methods of measurement are often viewed as competing, but each has its own place in teaching, coaching, physical therapy, athletic training, and other movement-based careers. This marriage of subjective observation and numerical data analysis provides movement specialists and elite performers with the specific feedback they need to diagnose and improve motion.

Qualitative analysis of motion describes how the human body "looks" on visual inspection as it performs skills, including its position in space, the position of body parts relative to each other, and in some cases the position of *segments* of body parts in relation to each other. Two general approaches to qualitative analysis are the composite approach and the component approach. The former views the whole body as a system that progresses through stages or phases as it refines movement patterns; the latter approach uses the same phase and stage method but, rather than looking at the whole body as a global system, breaks it down into component sections, with each section progressing through more refined steps toward mature movement patterns.

Quantitative methods of motion analysis arise from the need to further elucidate mechanisms underlying observed events or a requirement for highly precise numerical data. Although advanced software programs are often used to analyze motion data, the theoretical base for quantitative motion analysis can be understood through graphical and trigonometric methods of vector analysis.

REVIEW QUESTIONS

1. When can quantitative measures provide valuable information for physical education teachers?
2. What are "critical features" of a skill?
3. What is the most common qualitative assessment tool used by teachers?
4. What types of qualitative assessment tools are used by both physical therapists and athletic trainers?
5. In what units are mass and weight measured? Why are they not the same units?
6. Where are some places that you could go where your weight would change, but your mass would stay the same?
7. Name the four properties of a force and give brief definitions of each.
8. a. What is the difference between a scalar and a vector?
   b. Which one is mass?
   c. Which one is weight?
9. a. What does the length of a vector represent?
   b. The arrow point?
   c. The tail?
10. a. A muscle force can be resolved into which two components?
    b. What function does each of these perform?
11. Explain the Q angle, its normal values, and the consequences for an abnormally large value.

PRACTICE PROBLEMS

1. What are the values of the two muscle force vector components in the following problems? Interpret the answers to this question. 1 cm = 100 N.

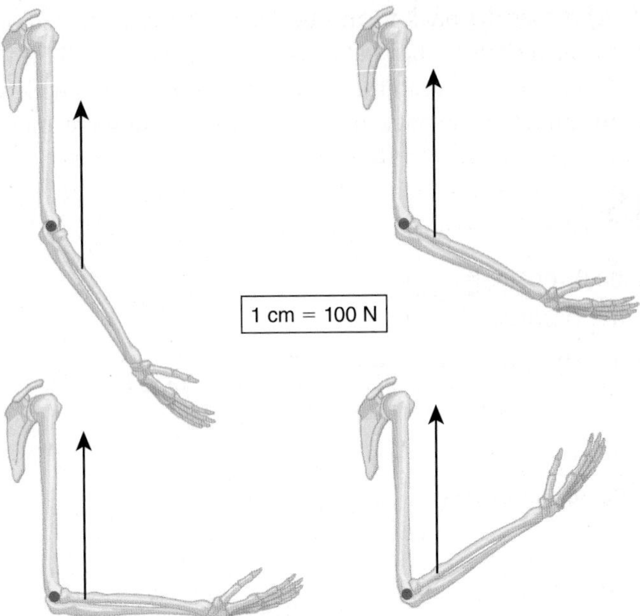

Figure 3.36 **Practice your vector resolution skills.**

2. What is the magnitude of the resultant in the following problems? 1 cm = 200 N.

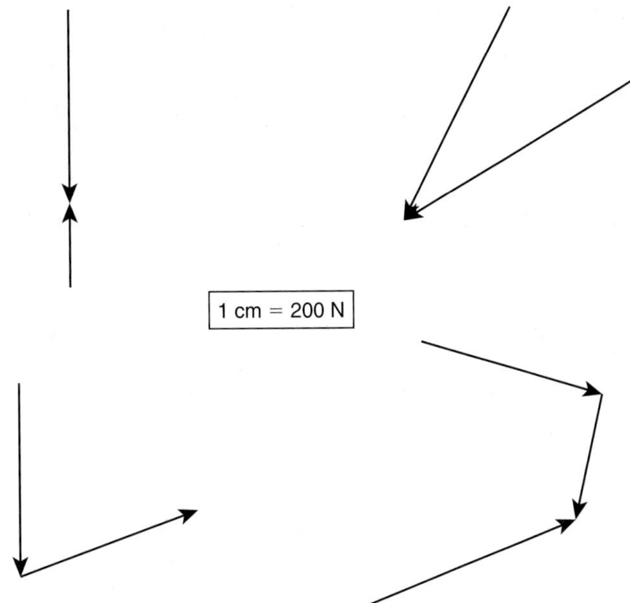

Figure 3.37 **Practice your vector composition skills.**

3. Compare (both numerically and with a written discussion) the vertical and horizontal components in the following situations. Show all work.
   a. A force of 600 N is applied to a football at an angle of 45°.
   b. A force of 600 N is applied to a football at an angle of 85°.
   c. A force of 600 N is applied to a football at an angle of 25°.

4. One hockey player (A) passes the puck to another player 10 m away and oriented 280° relative to the positive x-axis. Player B then strikes the puck, causing it to travel 5 m to a player oriented at 90°. The last player (C) applies a force to the puck that moves it 20 m with a path oriented at 210°. Calculate the total movement of the puck and the orientation of its resultant path. Show all work.

EQUATIONS

Sine $$\sin\theta = \frac{\text{side opposite } \theta}{\text{hypotenuse}} = \frac{y}{r} \quad \text{or} \quad y = r\sin\theta \quad \text{or} \quad r = \frac{y}{\sin\theta} \tag{3.1}$$

Cosine $$\cos\theta = \frac{\text{side adjacent to } \theta}{\text{hypotenuse}} = \frac{x}{r} \quad \text{or} \quad x = r\cos\theta \quad \text{or} \quad r = \frac{x}{\cos\theta} \tag{3.2}$$

Tangent $$\tan\theta = \frac{\text{side opposite } \theta}{\text{side adjacent to } \theta} = \frac{y}{x} \tag{3.3}$$

Pythagorean Theorem $$r^2 = x^2 + y^2 \tag{3.4}$$

Commutative law of addition $$\text{A} + \text{B} = \text{B} + \text{A} \tag{3.5}$$

Associative law of addition $$(\text{A} + \text{B}) + \text{C} = \text{A} + (\text{B} + \text{C}) \tag{3.6}$$

Negative of a vector $$\text{A} + (-\text{A}) = 0 \tag{3.7}$$

Vector subtraction $$\text{A} - \text{B} = \text{A} + (-\text{B}) \tag{3.8}$$

Vector cross product $$\text{A} \times \text{B} = \text{C} \tag{3.9}$$

$$\text{A} \times \text{B} \times (\sin\theta) = \text{C} \tag{3.10}$$

REFERENCES AND SUGGESTED READINGS

Aglietti, P., J. N. Insall, & G. Cerulli. 1983. Patellar pain and incongruence. I: Measurements of incongruence. *Clinical Orthopedics and Related Research*, 176: 217–224.

Bayley, N. 1936. *The California Infant Scale of Motor Development.* Berkeley: University of California Press.

Boucher, J. P., M. A. King, R. Lefebvre, & A. Pepin. 1992. Quadriceps femoris muscle activity in patellofemoral pain syndrome. *American Journal of Sports Medicine*, 20: 527–532.

Caylor, D., R. Fites, & T. W. Worrell. 1993. The relationship between quadriceps angle and anterior knee pain syndrome. *Journal of Orthopedic & Sports Physical Therapy*, 17:11–16.

Darwin, C. 1877. A biographical sketch of an infant. *Mind*, 2: 285–294.

Floyd, R. T. 2018. *Manual of structural kinesiology*, 20th ed. New York: McGraw-Hill.

Fronske, H., & R. Wilson. 2002. *Teaching cues for basic sport skills for elementary and middle school students.* San Francisco: Benjamin Cummings.

Gesell, A. 1928. *Infancy and human growth.* New York: McMillan.

Haubenstricker, J., C. Branta, & V. Seefeldt. 1983. Preliminary validation of developmental sequences for throwing and catching. Paper presented at the annual conference of the North American Society for the Psychology of Sport and Physical Activity, East Lansing, MI.

Haubenstricker, J., V. Seefeldt, & C. Branta 1983. Preliminary validation of developmental sequences for the standing long jump. Paper presented at the annual convention of the American Alliance for Health, Physical Education, Recreation, and Dance, Minneapolis, MN.

Haubenstricker, J., V. Seefeldt, C. Fountain, & M. Sapp. 1982. Preliminary validation of developmental sequences for kicking. Paper presented at the Midwest district convention of the American Alliance for Health, Physical Education, Recreation, and Dance, Chicago, IL.

Knudson, D. V. 2000. What can professionals qualitatively analyze? *Journal of Physical Education, Recreation, and Dance* 71(2): 19–23.

Knudson, D. V. 2013. *Qualitative diagnosis of human movement.* Champaign, IL: Human Kinetics.

Lysholm, J., & J. Wiklander 1992. Injuries in runners. *American Journal of Sports Medicine*, 15:168–171.

Messier, S. P., S. E. Davis, W. W. Curl, R. B. Lowery, & R. J. Pack. 1991. Etiologic factors associated with patellofemoral pain in runners. *Medicine & Science in Sports & Exercise*, 23: 1008–1015.

Moss, R. I., P. DeVita, & M. L. Dawson. 1992. A biomechanical analysis of patellofemoral stress syndrome. *Journal of Athletic Training*, 27(1): 64–66, 68–69.

Neely, F. G. 1998. Biomechanical risk factors for exercise-related lower limb injuries. *Sports Medicine*, 26:3 95–413.

Oatis, C. A. 2017. *Kinesiology: The mechanics and pathomechanics of human movement.* Philadelphia: Wolters Kluwer.

O'Connor, K. M., C. Johnson, & L. C. Benson. 2015. The effect of isolated hamstrings fatigue on landing and cutting mechanics. *Journal of Applied Biomechanics*, 31(4): 211–220.

Pinheiro, V. E. D. 2000. Qualitative analysis for the elementary grades. *Journal of Physical Education, Recreation, and Dance*, 71(1): 18–21, 25.

Roberton, M. A. 1977. Stability of stage categorizations across trial: Implications for the "stage theory" of overarm throw development. *Journal of Human Movement Studies*, 3:49–59.

Robertson, D. G. E., G. E. Caldwell, J. Hamill, G. Kamen, & S. N. Whittlesey. 2014. *Research methods in biomechanics*, 2nd ed. Champaign, IL: Human Kinetics.

Seefeldt, V. 1972. Developmental sequence of catching skill. Paper presented at the annual convention of the American Association for Health, Physical Education, and Recreation, Houston, TX.

Seefeldt, V., & J. Haubenstricker. 1975. Developmental sequence of punting. Rev. ed. Unpublished manuscript, Michigan State University, East Lansing, MI.

Seefeldt, V., & J. Haubenstricker. 1976. Developmental sequence of throwing. Rev. ed. Unpublished manuscript, Michigan State University, East Lansing, MI.

Seefeldt, V., & J. Haubenstricker. 1982. Patterns, phases, or stages: An analytical model for the study of developmental movement. In J. A. S. Kelso & J. E. Clark (Eds.), *The development of movement control and organization.* New York: Wiley.

Serway, R. A., & C. Vuille. 2018. *College physics,* 11th ed. Boston: Cengage Learning.

Serway, R. A., & J. W. Jewett Jr. 2019. *Physics for scientists and engineers,* 10th ed. Boston: Cengage Learning.

Thomas, A. C., S. G. McLean, & R. M. Palmieri-Smith. 2010. Quadriceps and hamstring fatigue alters hip and knee mechanics. *Journal of Applied Biomechanics*, 26(2): 159–170.

Ulrich, D. A. 2019. Test of Gross Motor Development: Examiner's Manual, 3rd ed. Austin: Pro-Ed publisher.

Watkins, J. 2010. *Structure and function of the musculoskeletal system*, 2nd ed. Champaign, IL: Human Kinetics.

Zimbler, S., J. Smith, A. Scheller, & H. H. Banks. 1980. Recurrent subluxation and dislocation of the patella in association with athletic injuries. *Orthopedic Clinics of North America*, 11: 755–770.

CHAPTER 4

INTERACTION OF FORCES AND THE SYSTEM

LEARNING OBJECTIVES

1. Understand Newtonian laws.
2. Describe the types of forces that act on a system.
3. Explain force application and material properties.

CONCEPTS

4.1 Properties of Force

4.2 Introduction to Newtonian Laws

4.3 Types of Forces Affecting System Motion

4.4 Force, Force Application, and Material Properties

4.5 Resultant Force

CONNECTIONS

4.6 Human Performance and Injury Science

4.7 Motor Behavior

4.8 Pedagogy

4.9 Adapted Motion

Two divers prepare for their next dive in the 3-meter springboard event. On a signal from the lead partner, both begin their carefully choreographed walking approach, and each of their steps causes increasingly more height as they near the end of their boards. As if they were two videotapes of the same diver side by side, they take off simultaneously and perform two-and-a-half forward flips with two twists and enter the water at the same instant. This event is called *synchronized diving*. Imagine the difficulty for an individual diver of performing this dive correctly—then add a second diver who must exactly match each step and attain the same relative height off the board so that both divers enter the water at exactly the same moment and execute each twist and flip with precise timing!

During these dives, several forces are at work outside and within each diver. The diving board provides an external motive force to gain additional vertical height, which gives the divers more time to execute elements to score points. What material properties must various diving boards possess? Once the divers clear the board, the only force acting on the system (the divers) is gravity, which pulls them downward toward the water. To perform the twists and flips, the divers contract and relax muscles to move body segments closer or further from axes of rotation. These changes in the shape of the system result in different orientations and rates of spin.

This chapter focuses mainly on forces, their application, and various material properties. Knowledge of forces and material properties is of great importance in all movement-related disciplines, sometimes with surprising variety. In the field of physical therapy or athletic training, the area of interest may be the effects of forces or behavior of material properties within the human body itself. For example, human biological materials such as ligaments and cartilage can be damaged when force is applied. In sport science, we may be more interested in an implement being used by the performer. A motor behaviorist will more likely be concerned with the relationships between forces, material properties, and the developmental stages of life. Still another point of view may be that of overcoming challenging situations by using material properties and forces to your advantage. No matter the discipline, the information in this chapter is of great importance because forces cause changes in motion and affect biological materials. In addition, those changes in motion are affected by environmental forces and material properties.

CONCEPTS

4.1 PROPERTIES OF FORCE

So far, we have defined **force** as something that possesses the capability to cause a change in the motion of a system. Basically, we can think of a force as a "pull" or "push." Now it is time to become more specific and gain a deeper understanding of force. One important aspect of the definition of force is that force possesses the *capability* to change motion. So we need to understand clearly that a force does not have to change motion, but it could. Otherwise, we would not be able to sit still while being acted on by the force of gravity. Another important part of the definition is that force *changes* motion. Forces do not cause motion, they cause change in motion (Serway & Jewett, 2019; Serway & Vuille, 2018). In other words, motion can exist without force, but there must be a force to change the state of a system's motion. In addition, some definitions of force also state that it can cause a change in the *shape* of the system. But in actuality, a shape change is the result of force acting to change the motion of an individual part of the system. Although these details may seem trivial, they are important for a deeper understanding of the detailed characteristics of force, a core concept of biomechanics. The nuances of force will become clear as Newtonian laws are introduced and applied in this and subsequent chapters.

Because the concept of force may seem abstract, previous chapters covered motion analysis and vector composition to develop the habit of closely observing motion and to form a visual representation of force. You have learned that vectors can be used to graphically represent the following *properties of a force:*

- Direction is the way in which the force is applied (e.g., upward, downward, forward, backward, north, south, positive, negative) and is represented by the tip of the arrow or vector.

Force Something that possesses the capability to cause a change in motion of a system.

- Orientation is the alignment or inclination of the vector in relation to the cardinal directions (e.g., vertical, 45° from horizontal), where θ is usually measured counterclockwise from the positive x-axis.
- Point of application is the point or location at which the system receives the applied force (e.g., at the toes, 2 cm from the axis of rotation of the elbow); it is usually defined by the tail of the vector.
- Magnitude is the amount or size of the applied force; it is depicted by drawing the length of the vector to scale (e.g., if the scale is 1 cm = 10 N, then a 10-cm vector represents a 100-N force).
- Line of action is an imaginary line extending infinitely along the vector through both the tip and tail (representing the path along which the arrow or vector would travel if moved forward or backward).

Now that the basic tools are in place, it is time to introduce the basic laws of motion. These laws of motion help provide a more concrete idea of the properties of force and force–motion relationships that to this point in the text have merely been represented by vectors in free-body diagrams. Though vector analysis and free-body diagrams are highly useful tools in understanding biomechanical analysis, we must gain greater insight as to the causes of the observed changes in motion.

4.2 INTRODUCTION TO NEWTONIAN LAWS

Recall that *motion* can be defined as a change in position with respect to both spatial (relative to points in space) and temporal (relative to time) frames of reference. In other words, a system moves in a given space, and that movement requires a given amount of time. You have also learned that no change in motion occurs without force. Now we need to briefly introduce what have come to be known as Newton's laws of motion (Cajori, 1934). Isaac Newton (1642–1727) formulated the basis of classical mechanics in his writing *Philosophiæ Naturalis Principia Mathematica* (*The Mathematical Principles of Natural Philosophy*), or *The Principia.* At this point, we will only introduce the laws along with brief explanations of each. However, in later chapters each law will be reintroduced and explored in the context of biomechanical situations.

Newton's First Law: Law of Inertia

Every body perseveres in its state of rest—or uniform motion in a right line—unless it is compelled to change that state by forces impressed upon it. This law explains the motion of a system in the absence of externally applied force. Simply stated, a system at rest will remain at rest, and a system in motion will remain in motion in a straight line until acted on by an external force. So it is the natural tendency of a system to resist change in motion. Recall that *inertia* is the resistance of a system to having its state of motion changed by application of a force. Therefore, this law is also called the *law of inertia*. We also mentioned that *mass* is a measure of a system's inertia. So, the more massive an object is, the greater its tendency to remain in its current state of motion.

One misconception about this law is that it means that there is no motion without force. This interpretation is incorrect; Newton's first law simply tells us that the current state of rest or motion of the system will be maintained unless, and until, there is interaction with an external force. So motion can occur without force, but no *change* in motion occurs without force. Remember that a force is something that possesses the capability to cause a change in motion of a system. So, once again, Newton's first law is a statement of what happens in the absence of an applied force. Newton's second law of motion explains situations in which the system is acted on by an external force.

Newton's Second Law: Fundamental Law of Dynamics or Law of Acceleration

The alteration of motion is ever proportional to the motive force impressed and is made in the direction of the right line in which that force is impressed. In other words, change in motion is directly proportional to the magnitude of the applied force. According to Newton's first law, if the state of motion of the system changes, a force must have been applied. The observed change in motion is called **acceleration**. Newton's second law states that the acceleration (change in the state of motion) of the system will be directly proportional to the applied force. For example, if 1 N of force is applied to an object, that object will accelerate by a certain amount. If 2 N of force is then applied, the object will experience double the acceleration. This law also means that the acceleration of the object will be proportional to the sum of the forces (or net force) acting on it. In other words, two 1-N colinear (having the same line of action) forces may be acting in the same direction instead of one force of 2 N. Also, colinear forces of equal value could act on the system in opposite directions. In that case, the sum of the forces (or net force) is zero, and the resulting acceleration is zero. But no matter whether one force or many is applied, the acceleration (change in motion) of the system will be directly proportional to the net applied force. However, this law holds true only as long as the mass of the object is constant. Because mass is a measure of resistance to change in motion (or resistance to acceleration), changing the mass of the system will change the motion that results from applying a given force. In our previous example, 1 N of force is applied to a given object, and a certain amount of acceleration is observed. But if that same amount of force (1 N) is applied to an object that possesses twice the mass, then half the acceleration will be observed. This result occurs because an object with double the mass has twice the inertia (or twice the resistance to having its state of motion changed). With all of this in mind, Newton's second law is often stated as, "The acceleration of a system is directly proportional to the sum of the forces (or net force) acting on it and inversely proportional to its mass."

Newton's Third Law: Law of Reciprocal Actions or Law of Action–Reaction

To every action there is always an opposed equal reaction, or the mutual actions of two bodies on each other are always equal and directed to contrary parts. Simply stated, forces exist in pairs. Newton's third law is probably the most misunderstood because of the abstract way in which it is presented. For example, one common way of stating this law is, "For every action there is an equal and opposite reaction." However, that statement is overly simplistic and can be misleading. For example, if a person exerts a force on Earth that causes that person to jump 1 meter (action), will Earth also move 1 meter (reaction)? We know that Earth will not display a perceptible "reaction" equal to that of the jumper. Also, overly simplified statements of Newton's third law leave some of us pondering the confusing question of how Earth knew that the jumper was going to push on it so that it could push back at precisely the same moment. In fact, in our example the jumper and Earth do experience the same *force*—but in opposite directions. Although this statement also is an oversimplification, it can be the key to grasping Newton's third law. The muscles of the jumper generate forces that pull on the bones to extend the hips and knees. The resultant force comes in the form of the feet pushing on Earth. Remember from Newton's second law that the acceleration of a system is directly proportional to the net force acting on it and *inversely proportional to its mass.* In the example of our jumper, both systems (the person and Earth) experience the same force produced by the muscles. But Earth's acceleration is much less than that of the jumper because of its extremely large mass. As a result, the muscular forces that attempt to move the segments of the leg away from the torso are actually redirected and cause the torso to move away from the segments of the leg. In essence, the feet attempt to move away from the body, but Earth is too massive to be pushed out of the way. So the body, possessing much less mass than

Acceleration A change in the state of motion of the system caused by an applied force.

Figure 4.1 **A Newtonian frame of reference.**

Earth, is accelerated away from the feet. When this series of actions is studied in the context of Newton's second law, one can begin to understand that although every action *force* is met with an equal and opposite reaction *force*, every action does not produce an equal and opposite reaction.

Newton's Law of Universal Gravitation: Law of Gravitation, or Law of Attraction

Every body in the universe attracts every other body with a force directed along the line of centers for the two objects that is directly proportional to the product of their masses and inversely proportional to the square of the separation between the two objects. This is, of course, the law that was purportedly inspired by Newton's observing a falling apple while sitting under a tree on his family's farm in Lincolnshire (Figure 4.1).

Whatever the inspiration, Newton was attempting to explain gravity mathematically. At this point, we will discuss the law of universal gravitation only theoretically. The mathematical expression will be explained in greater detail in later sections. At its most basic level, Newton's law of universal gravitation attempted to explain the interaction of objects even when they are not in contact (action at a distance). The ramifications are enormous when we consider that this law actually explains the forces that are acting between all of the objects in the universe, including the planetary bodies. In general, every particle (body) in the universe is attracted to every other particle (body). The force of that attraction increases proportionally with the masses of those bodies (i.e., the more massive the bodies, the greater the force of attraction) and inversely with the distance between the bodies (i.e., the closer the bodies, the greater the force of attraction). For the most part, the attractions between objects that we deal with in everyday life are so small that they can be ignored—their masses are relatively small compared to that of a planet. This is why, even though Newton's law of universal gravitation applies to all objects in the universe, we tend to think of gravitational attraction as existing only between *Earth* and other objects. Another reason for this misconception is that we are in daily contact with Earth, which is extremely massive and exerts a significant gravitational force on us and other objects. Because it is such a major force in our lives, we tend to forget that all bodies are attracted to each other to some degree.

4.3 TYPES OF FORCES AFFECTING SYSTEM MOTION

Now that we have introduced Newton's laws of motion and have some understanding of the involvement of force in that motion, we can be more specific as to the types of forces that a system may encounter. Note that we tend to think there are many types of forces. We use phrases such as "muscular force," "the force of gravity," and "ground reaction force." We read terms such as "lift force," "drag force," and "buoyant force." All of these uses of the word *force* make it easy to think that there must be an infinite number. The reality is that only a few fundamental forces actually are known in nature. Of those fundamental forces, only two are of concern in the field of biomechanics. However, the use of many types of classification systems is sometimes necessary for reasons of convenience and thorough understanding of the process of human motion analysis. We will also be using multiple classification systems in this text. But as you read along, keep in mind that any force being discussed fits into one of a few basic categories.

Noncontact (Field) Forces

Newton's law of universal gravitation explains the interaction of objects even when they are not in contact. Obviously, this is a highly abstract concept. For most people, perceiving that one object can affect another without being in contact with that object is difficult. Because of this difficulty, Michael Faraday (1791–1867) introduced the idea of a *field* (Serway & Vuille, 2018). For example, Earth (or any other object possessing mass) creates an invisible field of influence stretching throughout space. The moon interacts with this field created by Earth, as opposed to interacting directly with Earth. Therefore, the force of gravitational attraction discussed by Newton is one example of a **field force** or *noncontact force* (Serway & Vuille, 2018). Sometimes it is easier to conceptualize field forces by thinking of magnets. When two magnets are held in close enough proximity, one can actually feel the force of attraction (or repulsion) between them, even though they are not in contact (i.e., action at a distance).

All known fundamental forces in nature are field forces. The following are the four types of field forces, and they are in order of strongest to weakest:

1. The **strong nuclear force** occurs between subatomic particles and prevents the nucleus of an atom from exploding because of the repulsive electric force produced by its protons;
2. the **electromagnetic force** between electric charges (such as in the preceding magnet example;
3. the **weak nuclear force**, which is a product of some radioactive decay processes and plays an important part in the nuclear reactions by which the sun produces energy; and
4. the **gravitational force**, which exists between bodies of mass (Serway & Vuille, 2018).

Of these four fundamental forces, only electromagnetic and gravitational forces are of serious concern in biomechanics because all forces experienced by the body are a combination of those two.

People are often surprised to learn that gravity is relatively weak among all fundamental forces. But if we compare the speed at which various forces can change the motion of a body, we can gain a deeper understanding. For example, let's use the situation of an egg dropped from a 10-story building to the concrete sidewalk below. Compare the length of time that it takes gravity to bring that egg to the concrete (many seconds) with the length of time it takes the concrete (held together with stronger forces) to stop the egg's fall (less than a second). Another example is a simple jump into the air. We apply a force to the concrete surface with our feet, which according to Newton's third law results in equal and opposite forces acting on our body and the concrete. The resultant force causes us to overcome the force of gravity for a short time, but it does not crack the concrete. So, although we tend to see gravity as a strong force, the reality is that it just seems that way because our mass is so small compared to that of Earth (as discussed in the explanation of Newton's second law). By the way, we would not want gravity to be an extremely strong force because we would have no way to overcome it to jump (or move in any way, for that matter).

Field force A force that acts at a distance without making contact with the object that it is affecting.

Strong nuclear force Force that occurs between subatomic particles and prevents the nucleus of an atom from exploding because protons produce a repulsive electric force.

Electromagnetic force Force that occurs between electric charges.

Weak nuclear force Force that is a product of some radioactive decay processes.

Gravitational force Force that exists between bodies of mass.

Contact Forces

Contact forces are exactly what their name implies: the result of physical contact between two bodies. In our previous example of jumping, the feet contact the ground to apply the force. If you hit a baseball with a bat, there is physical contact. Physical contact also occurs between a sled and snow. Contact force is seemingly straightforward. However, at the atomic level contact forces are really electric (field) forces (Serway & Jewett, 2019). But this is one of those cases in which using both classifications makes more sense in terms of motion analysis. For example, a complete free-body diagram of jumping would necessitate the inclusion of both field forces (such as gravity) and contact forces (such as friction and ground reaction force). Besides, biomechanically analyzing movement from an atomic perspective is impractical.

Figure 4.2 **An astronaut engaging in an extravehicular activity.** Think of some ways astronauts maintain control of their movements.

External and Internal Forces

A force, whether noncontact or contact, can be classified as external or internal relative to the system. **External forces**, or *extrinsic forces*, are those that interact with the system from the outside (i.e., the external movement environment). In contrast, **internal forces**, or *intrinsic forces*, act within the defined system (i.e., internal to the system). The confusion comes about in understanding that only external forces can cause a change in the *motion* of a system (Newton's first law of motion), whereas internal forces can change only the *shape* of the system. As we proceed through the following discussion, never assume that the system is the entire person. Remember that the system is any object whose motion is of interest.

The following sections describe various forces. People often attempt to classify each force as exclusively external or internal. However, recall that a force is defined as external or internal *relative to the system*. Within the movement environments that are common to us, gravity is clearly an *external* force (it acts from outside the system). If you jump from a diving platform, gravity changes your state of motion and pulls you toward Earth. In this example, the system is defined as the whole body. As a result, muscle forces in this case are *internal* (acting inside the system). Because the only forces that can change the state of motion of a system are external, muscle forces can change only the shape of the system in our example. In other words, once the system is no longer in contact with the platform, the only external forces are gravity and air resistance. Therefore, muscle contractions (internal to the system) can cause various individual body parts to move, but the path of the center of gravity of the diver is under the influence of external forces only and will fall in a straight line toward Earth. So a diver can flap his arms all he wants in an attempt to fly, but flapping arms merely change the shape of the system rather than its motion.

To further illustrate this point, think of an astronaut. Microgravity is the ultimate case of humans working outside their natural environment. Notice that astronauts working in microgravity (Figure 4.2) must always remain in contact with some stationary object. Humans must have external forces to translate our center of gravity. Without contact with an object, an astronaut can change *shape* but is for the most part helpless in terms of moving from one place to another.

Contact forces Forces that are the result of physical contact between two bodies.

External forces Forces that interact with the system from the outside.

Internal forces Forces that act within the defined system.

How can we possibly walk around—a clear change in motion of the system—if muscle forces are internal and therefore incapable of causing change in the motion of the system? One can analyze this particular situation from a couple of perspectives. Technically, muscle forces *are* internal to the human system in the case of walking. However, muscle forces are *external* to the skeletal system. Recall that external force (according to Newton's first law) can cause a change in system motion. So even if a force technically originates internal to the human body system, it can change the motion of a body segment (and therefore the entire system) as long as the ultimate force is external to that segment. To expand on that aspect of the explanation, let's analyze this movement situation from another perspective. Muscle forces (being internal to the human system) can only change the shape of the system. In this case, that shape change is in the form of an attempt to move the leg segment away from the center of gravity of the body. However, recall from Newton's second and third laws that forces are paired, and that the smaller of two masses will be affected the most if acted on by an equal amount of force. Because Earth is so much more massive than the system in our example, it does not move out of the way and allows the shape change in terms of the leg segment moving away from the center of gravity of the system. Therefore, the resulting shape change is in the form of the center of gravity of the system moving away from the leg segment. Of course in this explanation of walking, there was a force applied downward and backward to Earth by the system. According to Newton's third law, there must have also been an equal force simultaneously applied forward and upward by Earth to the system. So even though the muscles that applied the force to Earth are internal to the system, the equal and opposite force applied by Earth is external and can therefore change the motion of the entire human system. This explanation of how we move requires that we classify forces not only as external or internal but also as *action* and *reaction* (according to Newton's third law).

Action and Reaction Forces

Again, this classification of forces is straightforward but somewhat relative in definition because the action and reaction are simultaneous and initiated by the same force. The **action force** is what we often think of as *the* force (i.e., it is the initially applied force). In our walking example, the action force is the force applied by the system to Earth by its change in shape (which was initiated by an internal force). The **reaction force** is the simultaneous equal counterforce acting in the opposite direction to the action force. In the walking example, the reaction force is the equal and oppositely directed force applied by Earth to the system. This reaction force is external to the system and therefore capable of changing the motion of the center of gravity of the system. So, because muscles are internal to the human system, reaction forces are the major means of our locomotion (Figure 4.3).

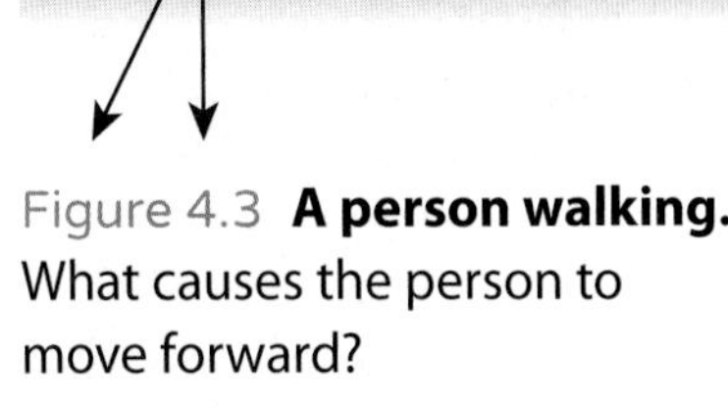

Figure 4.3 **A person walking.** What causes the person to move forward?

4.4 FORCE, FORCE APPLICATION, AND MATERIAL PROPERTIES

In the preceding sections, the concept of *force* (and its relationship to motion) is discussed thoroughly and in many contexts, and we're starting to understand the various categories into which forces are classified. We now must focus on specific forces, their application, and their relationship to material properties.

Action force The initially applied force.

Reaction force The simultaneous equal counterforce acting in the opposite direction to the action force.

Gravity

Gravity has been discussed in many different contexts throughout the text so far. Gravity is the focus because it is one of the most pervasive external forces in the everyday life of the human system. In this section, we will review concepts from previous chapters and then further elucidate the concept of gravitational force and its interaction with the system.

Recall that the cardinal planes bisect the body and therefore must also pass through the *center of mass* (the point at which all of a system's mass is concentrated). Gravitational pull is concentrated at the center of mass. In other words, the center of mass of the system is at the intersection of the three cardinal planes. Gravity can be represented with a line called the *line of gravity* that passes through the center of gravity (Figure 4.4). Therefore, the center of mass is often considered synonymous with the *center of gravity* (the point at which the force of gravity seems to be concentrated). However, *center of mass* is a more general term. The center of gravity is actually a reference to the center of mass, but in only one axis: the one representing the direction of gravity (the vertical axis).

So the line of gravity can be envisioned as the line at which the two vertical cardinal planes intersect (Figure 4.4). *Center of mass* is the correct term for referencing the two horizontal axes (Winter, 2009).

You have also read that gravity pulls on an object's mass with a certain amount of force. A measure of the force with which gravity pulls on an object's mass is called *weight.* Because we also know the direction of gravity, weight is a vector quantity, even though mass is a scalar. Many people speak of mass and weight synonymously. However, just because the two quantities are directly proportional does not mean they are the same quantity. An object's mass is not affected by differences in gravitational force (i.e., you are still composed of the same amount of matter in any location). Recall that the *weight* of an object is a measure of the gravitational force with which Earth pulls on the object's mass. Because weight is a measure of gravitational pull on your mass (and the moon has a lower gravitational pull than that of Earth), your weight can change on the moon without a concomitant change in mass. Changes in weight at various locations on Earth are not large, but the knowledge is useful for understanding mass and weight and their relationship to gravity. Newton's law of universal gravitation further elucidates these concepts in terms of mass, weight, and the distance between two bodies. Newton's law of universal gravitation attempted to explain the interaction of objects even when they are not in contact (action at a distance). Every particle (body) in the universe is attracted to every other particle (body). The force of that attraction increases proportionally with the masses of those bodies (i.e., the more massive the bodies, the greater the force of attraction). However, that force of attraction decreases if the same two bodies are farther apart (the force is inversely proportional to the distance between the two objects) and increases if they are closer together.

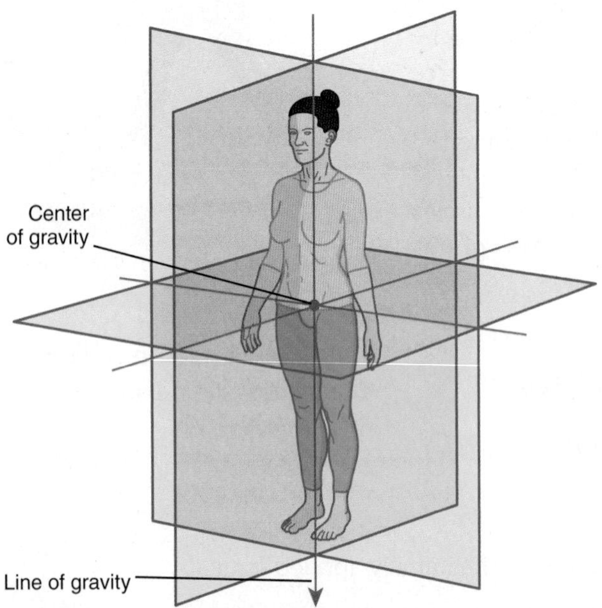

Figure 4.4 **Line of gravity.** What is the difference between the center of gravity and the center of mass?

Therefore, Newton's law of universal gravitation helps us understand why our weight changes as we get closer, to and farther from, the center of Earth's mass. In other words, these changes in force explain why a person will weigh slightly less at the equator (where that person is farther from the center of Earth) and slightly more at the poles (closer to the center of Earth's mass). This same law, taken in conjunction with Newton's second law, explains why we fall at a faster and faster rate if we jump from an extreme height (and no air resistance is present).

Newton's second law says that the acceleration of an object is directly proportional to the applied force. As we fall, we are getting closer and closer to Earth. According to the law of universal gravitation, this means that the force of attraction between us and Earth is gradually increasing, and therefore so is our rate of descent (according to Newton's second law). Gravity will be discussed in further detail in subsequent sections.

Because of gravitational force, objects are pushed (or pulled) together. Whenever objects are in contact, the potential for friction and pressure arises.

Friction

At the most basic level, **friction** is the force that resists the sliding of two objects in contact. Therefore, friction exists whenever two objects are in contact and have the potential to slide across each other. Because friction resists sliding, the vector that represents friction force has a direction opposite that of the potential sliding direction and is parallel to the two surfaces in contact. We tend to not think of friction on a regular basis, even though (and probably because) it is a constant force in our lives. Because we cannot see it in action, we tend to notice only its *absence* rather than its *presence*. For example, we may not think consciously of friction being present when we walk. However, we do notice the *absence* of friction force if we unexpectedly step on a patch of ice. This example illustrates the importance of friction in human locomotion. Without friction force, we could not move from one place to another. So now we must gain some insight into the exact role of friction in human locomotion.

Let's use some tools we previously learned along with our recently gained knowledge of Newton's laws. First, remember that in a free-body diagram of this situation, a vector will be oriented downward and backward, which represents the force being applied by the foot to the ground. Also recall that the vector can be resolved into a perpendicular (vertical) component and a parallel (horizontal) component (Figure 4.5).

Figure 4.5 **Vector resolution of a person walking.** Imagine all of the different surface textures that people walk on every day.

According to Newton's third law, vectors that represent reaction forces of equal magnitude will also be present, oriented in opposite directions to the components previously described (upward and forward as opposed to downward and backward). So when contact is made between the foot and the ground, part of the total applied force will be oriented perpendicularly to the two contacting surfaces (the foot and the ground). This component of the force that acts downward on the ground (and upward on the walking system) is called **normal force** (or *normal contact force*). Because this force acts downward on one surface and upward on the other, normal force is often defined as the force that presses two surfaces together. The normal force is responsible for the system's center of gravity being temporarily elevated during locomotion. So the system applies a normal force to the ground that is met with an equal and opposite normal force from Earth, which elevates the system's center of gravity. This oppositely directed normal force of Earth is often called **ground reaction force**.

The other component of the force applied by the system to Earth is backward and parallel to the contacting surfaces. Thus, an equal and oppositely directed force

Friction The force that resists the sliding of two objects in contact.

Normal force Force that acts downward on one surface and upward on another.

Ground reaction force An equal and oppositely directed normal force from Earth.

exists that is forward and parallel to the contacting surfaces. This oppositely directed parallel force is friction, and it is responsible for forward motion of the system's center of gravity. So friction not only prevents our foot from slipping backward on a contacting surface but also provides the reaction force that leads to locomotion. Always be careful in identifying the direction of friction force. Notice in the preceding example that friction acts in a direction opposite to that in which the foot would potentially slide on the ground, which is actually in the same direction as the resultant motion of the system.

Now that we have a good concept of how friction behaves, let's gain a better understanding of the factors involved. We will begin with the simplest case, one in which two objects have the potential to slide, but are not actually sliding, and the surfaces are dry.

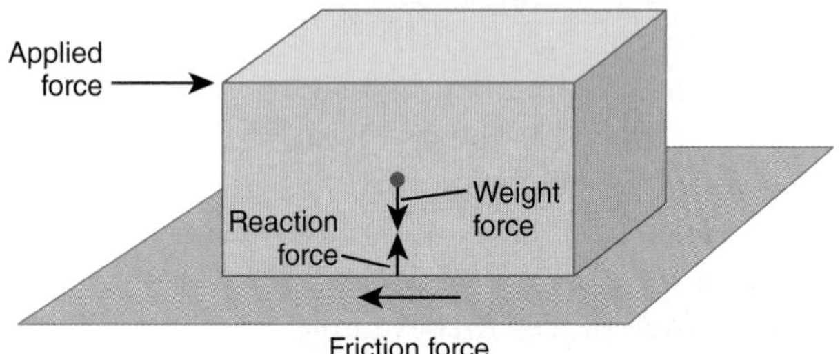

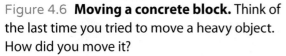

Figure 4.6 **Moving a concrete block.** Think of the last time you tried to move a heavy object. How did you move it?

Static Friction

As you have read, **static friction** exists when two contacting surfaces are not currently sliding relative to each other but do possess the *potential* for movement. As you already know from experience, objects differ in the ease with which they slide across one another. Let's explain the reason for these differences. Start by remembering the last time you tried to slide a heavy object. We will use a concrete block as an example (Figure 4.6).

Let's say that the block is heavier than you imagined. In an attempt to make the block slide, you apply a force in the direction you want it to move. But because the block is heavier than you thought, it doesn't initially move. That's because the static friction force holding it in place is greater than the force you applied. So you gradually apply greater amounts of force until sliding begins. At the point of sliding, you have applied a force large enough to overcome the opposing static friction force. Because static friction resists the start of sliding motion, it is often referred to as *starting friction.*

Intuitively, you probably already have some idea of the factors affecting static friction. First, you know that the heavier the block, the more difficult it will be to move. You probably also know that the difficulty will depend on the texture of the block surface, as well as the texture of the surface over which the block will potentially slide (e.g., a wooden floor, carpet, etc.). The relationship of these factors can be expressed in an equation:

$$\mu_s \leq \frac{f_s}{F_n} \tag{4.1}$$

where

μ_s = coefficient of static friction (measured experimentally)
f_s = static friction force (or force required to start the slide of the object)
F_n = normal force (in this case the weight of the block because of gravity)

The **coefficient of friction** (mu) is an experimentally measured, dimensionless value that represents the proportion of the friction force resisting the sliding motion of the object to the normal force holding the objects together. The coefficient is a constant for any two surfaces and depends on their surface textures and hardness. In other words, the coefficient of friction represents the difficulty of sliding any given surface over another because of their textures. This difficulty comes from interactions among the molecules at the surfaces of the objects. No matter how smooth a surface appears, it

Static friction Friction that exists when two contacting surfaces are not currently sliding relative to each other but do possess the *potential* for movement.

Coefficient of friction An experimentally measured dimensionless value representing the proportion of friction force resisting sliding motion of the object to the normal force holding the objects together.

has some texture. When viewed microscopically, even the surface of glass can be seen to have grooves and scratches. Those grooves, scratches, and pores on the surface of one object interacting with those on the other are what impede motion (i.e., increase friction). So when two surfaces are in contact, the grooves and imperfections are "interlocked" and provide resistance (i.e., friction) against moving the objects past one another. The more molecular interactions between two surfaces, the higher the coefficient of friction. For example, as Table 4.1 shows, the coefficient of friction would be much lower for a skate on ice than for rubber-soled shoes on a wooden floor. Typical values are from 0.01 to 1.00 (Grigoriev & Meilikhov, 1997).

A value of 0.00 would indicate a totally frictionless interaction. Some would argue that a value of 1.00 is the maximum for friction and that values greater than 1.00 are actually a coefficient of adhesion rather than friction. Either way, a lower value indicates greater ease of sliding between two

Table 4.1

Common Coefficients of Friction for Various Pieces of Athletic Equipment on Different Surfaces

Coefficients of Friction[a]	μ_s	μ_k
Steel on steel	0.74	0.57
Aluminum on steel	0.61	0.47
Copper on steel	0.53	0.36
Rubber on concrete	1.0	0.8
Wood on wood	0.25–0.50	0.2
Glass on glass	0.94	0.4
Waxed wood on wet snow	0.14	0.1
Waxed wood on dry snow	—	0.04
Metal on metal (lubricated)	0.15	0.06
Ice on ice	0.1	0.03
Teflon on Teflon	0.04	0.04
Synovial joints in humans	0.01	0.003

[a]All values are approximate. In some cases, the coefficient of friction can exceed 1.0.

SOURCE: From Serway & Jewett, 10th ed., 2014 p. 116.

Skates on ice	0.003–0.007
Bobsled runners on ice	0.01–0.05
Skis on snow	0.05–0.20
Tennis balls on wood	0.25
Tennis balls on artificial surfaces	0.50–0.60
Tennis shoes on artificial surfaces	1.3–1.8
Basketball shoes on wooden floor	1.0–1.2
Basketball shoes on wooden floor (dusty)	0.3–0.6
Cleated shoes on astroturf	1.2–1.7

SOURCE: Nigg et al., 2000, p. 56.

given surfaces. Notice that the coefficient of friction depends upon surface textures and is therefore independent of normal force. The role of normal force is sometimes easier to understand if we rearrange Equation 4.1:

$$f_s \leq \mu_s F_n \tag{4.2}$$

where

f_s = static friction force (or force required to start sliding of the object)
μ_s = coefficient of static friction (measured experimentally)
F_n = normal force (in this case the weight of the block attributable to gravity)

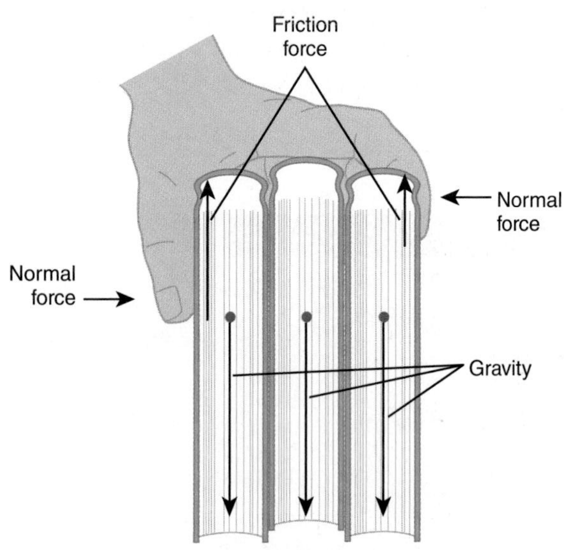

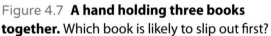

Figure 4.7 **A hand holding three books together.** Which book is likely to slip out first?

Equation 4.2 shows not only the relationship of the textures of the contacting surfaces (μ_s) but also the fact that friction force is proportional to the normal force holding the objects together. For example, if you grasp three books in one hand, you must pinch them together with enough normal force to prevent sliding (Figure 4.7). Without a strong enough normal force, static friction force will not be great enough to prevent slipping.

The same example helps explain the interaction of normal force and coefficient of friction. Think about which book will most likely slip out first. It will probably be the one in the middle. In this situation, friction is present not just between the surface of the middle book and the surfaces of the other two books, but also between the surface of the skin and the outer two books. Most likely, the coefficient of friction is lower for textbook cover on textbook cover rather than for skin on textbook cover. So even with the same applied normal force, the middle book will most likely fall first because it has a lower coefficient of friction with its contacting surfaces.

SAMPLE PROBLEM

Equation 4.2 also helps demonstrate that, as greater force is applied parallel to the surfaces, static friction force increases to a maximum value just before slippage. At that moment immediately before the object begins to slide (**impending motion**), $f_s = \mu_s F_n$ (as opposed to $f_s < \mu_s F_n$). Let's use Equation 4.2 in a mathematical example. If the concrete block in our previous example weighs 100 N and the coefficient of friction for concrete on wood is 0.80, then $f_s = 0.8 \times 100$ N. The force required to start the block sliding is 80 N. In other words, the force of static friction at impending motion is 80 N. At any point before impending motion, the force of static friction is less than 80 N. To fully understand this situation, we must first think of Newton's third law. If I apply a force to the block, it is met with an equal force in the opposing direction. So if I apply a force of 30 N in the positive *x*-direction, it will be met with a static friction force of 30 N in the negative *x*-direction. If I then apply 45 N, it will be met with 45 N in the opposite direction, and so on (in all of these cases, $f_s < \mu_s F_n$). So as more parallel force is applied, it is met with a greater static friction force all the way until impending motion. At this point, the parallel force is equal to the total capability of the object to resist sliding motion (which is calculated as $f_s = \mu_s F_n$). If any further parallel force is applied, the object will begin to slide. Once that object does begin to slide, a different form of friction is present.

Impending motion The moment immediately before an object begins to slide because of the application of a force.

Kinetic friction is sometimes referred to as *dynamic friction* and is the term for friction in cases when the two surfaces are already sliding relative to each other (Figure 4.8). In the case of kinetic friction, the grooves and pores on the surfaces of the objects can be envisioned to be "bumping" into each other rather than being locked together (as in a situation of static friction).

Because the surface molecules are just bumping across one another and do not have to be "unlocked," the coefficient of kinetic friction is lower than the coefficient of static friction. Therefore, as soon as the applied parallel force exceeds the static friction force (f_s) and the object begins to slide, friction force actually decreases. The lower coefficient of kinetic friction is one of the reasons why less effort is required to maintain the sliding of an object compared to the effort required to initiate the slide. Even though the size of the coefficients of friction in these two situations is slightly different, the calculations are similar:

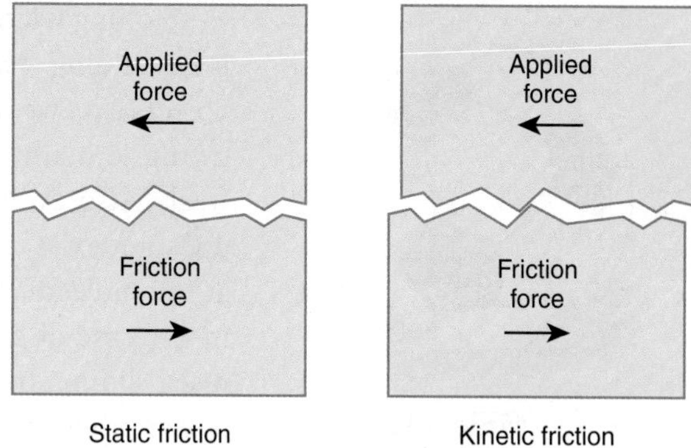

Figure 4.8 **Molecular interactions causing friction.** Why is the coefficient of kinetic friction lower than that for static friction?

$$\mu_k = \frac{f_k}{F_n} \tag{4.3}$$

where

μ_k = coefficient of kinetic friction (measured experimentally)
f_k = kinetic friction force (or force required to continue the sliding of two objects)
F_n = normal force (in this case the weight of the block because of gravity)

and

$$f_k = \mu_k F_n \tag{4.4}$$

where

f_k = kinetic friction force (or force required to continue the sliding of two objects)
μ_k = coefficient of kinetic friction (measured experimentally)
F_n = force (in this case, the weight of the block because of gravity)

Notice the only difference in the formulas is that no situation of inequality exists in the case of kinetic friction. In the case of static friction, the inequality holds until the point of impending motion (i.e., all situations in which the applied parallel force is less than the maximum value of static friction). At the point of impending motion (the greatest degree of static friction before sliding begins), the equality in the static friction equations holds. Because in situations of kinetic friction the object is already sliding, there is no point at which kinetic friction force can be less than the product of the coefficient of kinetic friction and the normal force.

Rolling friction exists whenever one surface is rolling over another but not sliding. Rolling friction, just like static and kinetic friction, depends on a coefficient of friction and normal force. We have seen that the coefficient of kinetic friction is less than that of static friction. Similarly, the coefficient of rolling friction is less than that of either static or kinetic friction. However, the difference is not slight. As you have doubtless experienced, it is *much* easier to move a heavy object by placing it on wheels and rolling it than it is to drag it across the ground. Rolling is easier not because friction does not exist when you roll an object. Also, the object is not any lighter (i.e., the normal force is the same). Rolling is easier because the coefficient of rolling friction is significantly less than even the coefficient of kinetic friction.

Kinetic friction Friction that exists when two surfaces are already sliding relative to each other.

Rolling friction Friction that exists whenever one surface is rolling over another but is not sliding across it.

Figure 4.9 **Molecular interactions causing rolling friction.** Why is the coefficient of rolling friction lower than that for static or kinetic friction?

Rolling friction is actually a specialized case of static friction because the object is not sliding. As a matter of fact, sliding is usually avoided in the case of rolling an object (e.g., we tend to actively prevent our tires from skidding). So we are actually trying to maintain the static friction of the rolling object while allowing it to roll freely. Also, even though one object is moving by rolling over another, kinetic friction does not exist unless one of the objects actually begins to slide across the other. So at the molecular level, the source of rolling friction is a little different. Remember in the case of static friction, imperfections in object surfaces are interlocked and resist sliding. In the case of kinetic friction, the imperfections are "bumping" across one another. In situations of rolling friction, one molecular surface is being "peeled away" from another (Figure 4.9). In other words, molecular interaction still occurs between the two surfaces, but there is no attempt to "unlock" them by sliding them across one another. Also, the imperfections are not "bumping" into each other. One surface is essentially being pulled away from another, and the molecular surfaces resist. An excellent visual image would be that of a wheel rolling across a surface when both are covered with Velcro®.

There will be resistance to rolling, but not because the wheel is sliding across the other surface. Instead, resistance will be caused by the molecular interaction of the material surfaces. Velcro is an extreme example. But once again, molecules (imperfections) of one surface adhere to molecules on the other surface, even when both are apparently flawless. So the coefficient of rolling friction depends upon the same factors as the coefficients of static and kinetic friction. Just as smooth hard surfaces slide across each other more easily than irregular soft surfaces, smooth hard objects roll across (or are rolled on) more easily than rough soft surfaces.

As can be inferred from what you have just read, creating surfaces that roll easily without sliding requires a delicate balance. The same characteristics of surfaces that reduce rolling friction also reduce both static and kinetic friction (i.e., surfaces that roll easily also slide easily). For example, we do not want a lot of rolling friction on our tires because it requires more energy to "peel" the surface of the tire away from the pavement. However, manufacturing a tire that is extremely hard and smooth is not feasible because it will slide too easily. In the case of tires, we do want to maintain static friction, or the tires slide and we lose control. This balance is the idea behind antilock brakes. If the tire stops rolling too abruptly, static friction is overcome and sliding begins. Once sliding begins, friction decreases because the coefficient of kinetic friction is lower. This decrease in friction in turn makes it even more difficult to stop or control the car. So the purpose of antilock brakes is to slowly bring the wheels to a stop by generating a precise amount of braking force to keep the tires just below the point of completely coming to a stop. This strategy maintains the static friction between the tire and the pavement to allow maximum braking force while also letting the driver maintain directional control of the vehicle.

As you can see, friction plays a role in almost every aspect of our lives. It is there any time two surfaces are in contact. We have used examples such as concrete blocks and tires, but think of the many sports situations in which friction is a factor. What about trying to throw a wet football accurately? Think about the choice of skis for particular snow conditions. How about the difference between playing soccer on natural grass and on artificial turf? What about tennis on a clay court versus on grass? Think of the grip on an implement such as a racquet, bat, or club. An infinite number of surface interactions occur in our daily lives, workplaces, and sports situations. Therefore, friction is always a factor.

Pressure Versus Force

Friction is not the only concern when forces are exerted and objects are in contact; pressure is also an important issue. The terms *pressure* and *force* are sometimes used interchangeably. However, **pressure** is the magnitude of applied force acting over a given area. One of the best examples of using pressure to one's advantage is the famous trick in which a magician lies on a bed of nails. We have all seen the trick and know that somehow the magician avoids being punctured by the nails. There is really no trick at all; the magician is merely manipulating pressure. In our definition, the weight of the magician is the applied force. The area over which it is applied is the total area of the tips of the nails. You can truly understand the difference between force and pressure if you imagine changing the trick so that the magician is required to lie on only *one* nail instead of hundreds. In this case, the force has not changed (i.e., the magician still weighs the same). However, the area over which the force is applied has been reduced significantly.

Now that we have some understanding of the difference between pressure and force, let's actually calculate the pressure in these two situations. First, we need the equation for calculating pressure:

$$P = \frac{F}{A} \tag{4.5}$$

where

P = pressure
F = applied force
A = area over which the force is applied

Pressure has units of force divided by units of length squared. Because force is measured in newtons and area is usually measured in square meters, pressure has Système International (SI) units of newtons per square meter (N/m^2). Another term for the SI unit of pressure is the *pascal* (Pa). One newton per square meter is equal to one pascal ($1\ N/m^2 = 1\ Pa$). However, in our example, we are working with the tips of nails. So, for our example, we will be using area in millimeters.

Pressure The magnitude of applied force acting over a given area.

SAMPLE PROBLEM

Let's begin our example with the situation of the magician lying on a standard bed of nails (Figure 4.10).

Say the magician is of average size and weighs 758 N. We will estimate that the tip of one nail has an area of 0.25 mm². We will also estimate that a typical bed of nails consists of 2,500 nails. This would mean that the total area to which the magician's weight will be applied is 625 mm² (2,500 nails × 0.25 mm² of area per nail). Using Equation 4.5, the pressure can be calculated as follows:

$$P = \frac{758\ N}{625\ mm^2} = \frac{1.213\ N}{mm^2}$$

Thus, even though the applied force is 758 N, the force is spread out over such a large area that the magician is not actually experiencing that much force per unit area. To give some comparison,

(continues)

(continued)

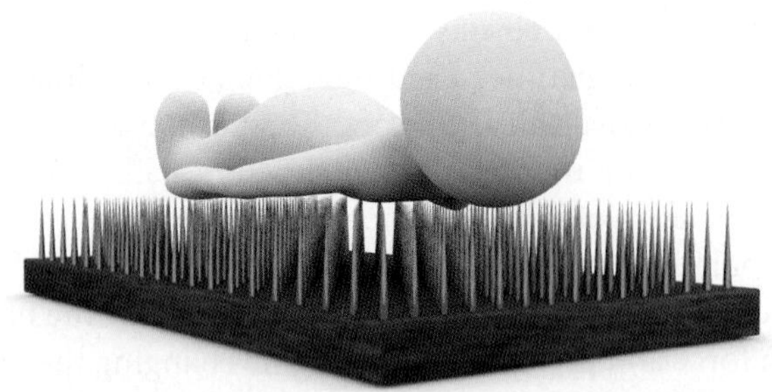

Figure 4.10 **A magician on a standard bed of nails.** Why is his skin not punctured?

the **tensile strength** (maximum stretch that a material can withstand without rupture) of skin is estimated to average 9.25 N per square millimeter (between 2.5 and 15 N/mm^2, depending on the location on the body). As our calculation shows, the magician is probably not comfortable, but he is also not in grave danger. Now we can calculate the pressure in our extreme example of lying on one nail.

The magician's weight is the same, so the applied force is still 758 N. But with the use of only one nail, the total area is significantly reduced to 0.25 mm^2 (1 nail × 0.25 mm^2 of area per nail). Using Equation 4.5, the pressure can be calculated as follows:

$$P = \frac{758 \text{ N}}{0.25 \text{ mm}^2} = \frac{3032 \text{ N}}{\text{mm}^2}$$

A pressure of 3,032 N/mm^2 is obviously greater than the 9.25 N/mm^2 that is required to puncture the skin of the magician. So as you can clearly see, pressure can change drastically even in the presence of the same applied force.

Although the pressure applied by the tip of a nail is visually obvious, don't ignore the less obvious pressure being applied to you at this moment. You are currently experiencing atmospheric pressure (or barometric pressure) from the weight of the atmospheric fluid (gases) above you. Because we are adapted to the atmospheric pressure in our climate, we tend not to think of it as *weighing* anything. However, if you recall the last time you swam to the bottom of a pool, you can probably vividly remember the pressure exerted by the weight of the water above you. The increased pressure may have even caused some discomfort in your ears. In both situations, pressure is being exerted on you by the weight of the fluid in the atmosphere (either gases or water). Both air and water pressures are measured in N/m^2 (or Pa). But you may also be familiar with measurement of pressure in millimeters of mercury (1 mm Hg = 133.3 Pa), another commonly used unit of pressure. Millimeters of mercury are often used when measuring barometric pressure (for instance, during a weather report) and blood pressure (for example, a systolic blood pressure of 120 mm Hg).

If you are currently at sea level, you are experiencing one atmosphere of pressure (approximately 1.013×10^5 Pa or 760 mm Hg). As you travel to areas of higher elevation, there is less atmospheric gas piled up on you. At an altitude of 5,486 m, the atmospheric pressure is approximately half that at sea level. So at an extremely high altitude, such as the summit of Mt. Everest (8848 m), the atmospheric pressure is only 250 mm Hg (approximately 0.333×10^5 Pa)!

As mentioned, we tend to not think of the atmosphere as having *weight*. However, the force of the atmosphere is approximately 133,500 N (if we assume the area of the body to be 5,080 cm^2)! Our bodies would collapse under the weight of the atmosphere if it were not for the equal and opposite reaction force provided by the fluid in our tissues and body cavities (Serway & Vuille, 2018).

Tensile strength The maximum stretch that a material can withstand without rupture.

Table 4.2
The Relationship of Water Depth and Pressure

Depth		Pressure			Inspired Air (mm Hg)	
ft	m	atm	mm Hg	Hypothetical Lung Volume (mL)	PO_2	PN_2
Sea level		1	760	6000	159	600
33	10	2	1520	3000	318	1201
66	20	3	2280	2000	477	1802
99	30	4	3040	1500	636	2402
133	40	5	3800	1200	795	3003
166	50	6	4560	1000	954	3604
200	60	7	5320	857	1113	4204
300	90	10	7600	600	1590	6006
400	120	13	9880	461	2068	7808
500	150	16	12,160	375	2545	9610
600	180	19	14,440	316	3022	11,412

atm = atmospheres
PO_2 = partial pressure of oxygen
PN_2 = partial pressure of nitrogen

McArdle et al., 2014

Just as pressure varies with elevation above sea level, pressure also varies with depth under the surface of the water. More specifically, a column of fresh water exerts a pressure equal to one sea level atmosphere (760 mm Hg) for each 10 m of descent below the surface of the water (Table 4.2).

Therefore, a diver at a depth of 10 m will experience two atmospheres of pressure (1520 mm Hg): one atmosphere from the weight of the air at the surface of the water, and one atmosphere from the weight of the column of water (McArdle, Katch, & Katch, 2014). The weight of the ambient air at the surface of the water must be accounted for in accordance with **Pascal's law**: pressure applied to a fluid is transmitted undiminished to every point of a fluid and to the walls of the container. Therefore, all points at a given depth will experience the same pressure. So you can see that a submerged body experiences considerable pressure, even at shallow depths. This is one reason why a person cannot simply use an extraordinarily long snorkel when diving. One problem is that the person would simply be rebreathing his or her own air because the snorkel would contain a large amount of dead space. A lesser-known obstacle to using a superlong snorkel is that at a depth of even 1 m, the pressure against the chest cavity is often great enough to prevent inspiration by the respiratory muscles (McArdle et al., 2014). Above water, you find it easy to breathe because the air that the chest cavity displaces during inhalation is the same pressure as the air that you are attempting to inhale. Underwater, the difficulty for a diver with a long snorkel would be attempting to breathe air that is only pressurized to one atmosphere by trying to inflate the chest cavity against the pressure of the surrounding water, which is at a pressure greater than one atmosphere. When you breathe air from a scuba tank, you are inhaling air at the same pressure as the surrounding water, which allows you to breathe with the same ease as you do above the surface of the water.

Pascal's law Pressure applied to a fluid is transmitted undiminished to every point of a fluid and to the walls of the container.

Stress, Strain, and Elasticity

So far, we have discussed forces in terms of systems or surfaces being pushed together. However, the human body contains many connective tissues that must resist being pulled apart. We know that applied forces can come in the form of a "push" or a "pull." In speaking of the musculoskeletal system, the externally applied forces are often "pulls." For example, muscles generate forces that pull on tendons. The tension in that tendon is transferred into a pulling force on the bone. Ligaments prevent bones from being pulled apart. In all of these cases, an external force is acting to deform a given material (biological tissue). Depending upon the structural characteristics of the material, internal forces resist that deformation to varying degrees. The external force acting to deform the material is called **stress**. Stress is strongly similar to pressure in that it is calculated as an external force acting over a given cross-sectional area:

$$\sigma = \frac{F}{A} \tag{4.6}$$

where

σ = stress
F = applied force
A = area over which the force is applied

Stress (like pressure) has units of force divided by units of length squared, and therefore it has SI units of newtons per square meter (N/m^2). Depending on the author, the difference between pressure and stress is usually described either in terms of the direction of externally applied force or whether or not the object is a solid or a fluid. Often stress is defined as "internal pressure"—that is, the same quantity as pressure—but the external force acts along the long axis of the object and potentially can cause deformation by pulling (tension). Some experts define pressure as a specialized form of stress in which an object is subjected to uniformly distributed force applied perpendicularly over its entire surface. For example, an object submerged in water would experience stress (more specifically pressure) from the evenly distributed force of water over its entire surface. We will simply define pressure as *compression stress.*

Whereas stress is the force applied over a given area in an attempt to cause deformation, **strain** is the resulting magnitude of deformation as a result of the applied stress. In other words, strain is the percentage change in length from an applied stress and can be calculated as follows:

$$\varepsilon = \frac{\Delta l}{l_i} \tag{4.7}$$

where

ε = strain
Δl = change in length ($l_{final} - l_{initial}$)
l_i = initial or original length

So if we apply a particular amount of tensile stress to a section of elastic tubing, it will stretch. The amount of stretch (total change in length) in proportion to its original length is the strain. Notice that the units for strain actually cancel and give a percentage. However, strain is often presented in units of mm/mm, cm/cm, or m/m.

Stress The external force acting to deform a material.

Strain The resulting magnitude of deformation as a result of an applied stress.

Notice also that, in addition to the strain in the direction of applied stress, there is also strain orthogonal to the direction of loading (Figure 4.11). In other words, the diameter of the tube will become smaller as it is lengthened (tensile strain is accompanied by transverse

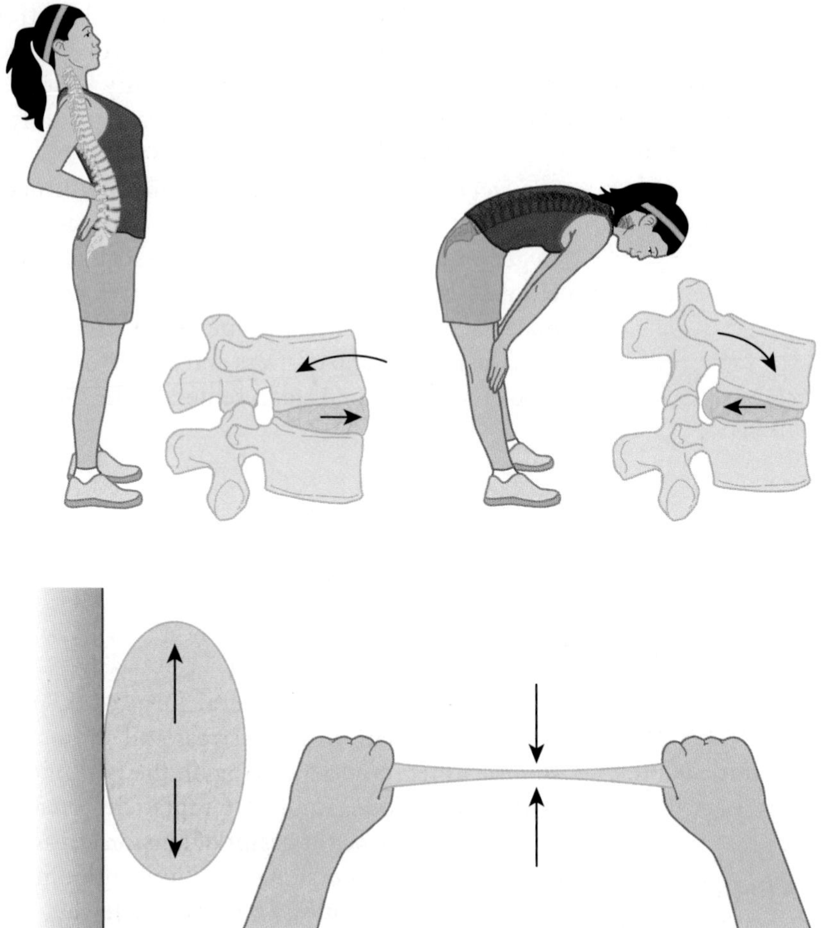

Figure 4.11 **Diameter changes as materials are stretched or compressed.**

strain). The same phenomenon can be observed in a racquetball as it hits the wall. The ball is compressed in one direction, and it expands in the opposite direction. In the human body, intervertebral discs bulge under compressive loads (sometimes too much).

The tendency of a material to exhibit transverse (lateral) strain simultaneously with axial (longitudinal) strain is expressed as **Poisson's ratio**:

$$\nu = \frac{\varepsilon_t}{\varepsilon_a} \quad (4.8)$$

where

ν = Poisson's ratio
ε_t = transverse strain (at an angle orthogonal to the stress)
ε_a = axial strain (in the direction of the stress)

Poisson's ratio has values between 0.00 and 0.50, with 0.50 being a perfectly incompressible material. Rubber is almost 0.50, steel is approximately 0.30, and cork is near 0.00. Keep in mind that because the geometry of a material changes with loading, actual stress and actual strain change from moment to moment. For example, actual stress changes during loading because *area* in the stress calculation is changing in accordance with Poisson's ratio.

Poisson's ratio An expression of the tendency of a material to exhibit transverse (lateral) strain simultaneously with axial (longitudinal) strain.

Real-time measurement of actual stress and strain of biological materials is extremely difficult, so reported magnitudes for mechanical properties tend to be nominal (or engineering) values taken before actual loading. Knowing the actual value of any given quality is not as important as understanding that the complex conditions within the human body are constantly changing and highly interrelated.

Depending on the material being deformed, the relationship of stress and resulting strain also varies (Figure 4.12). In general, strain is proportional to stress (in accordance with Hooke's law) and the relationship is a constant for a given material and a particular type of deformation.

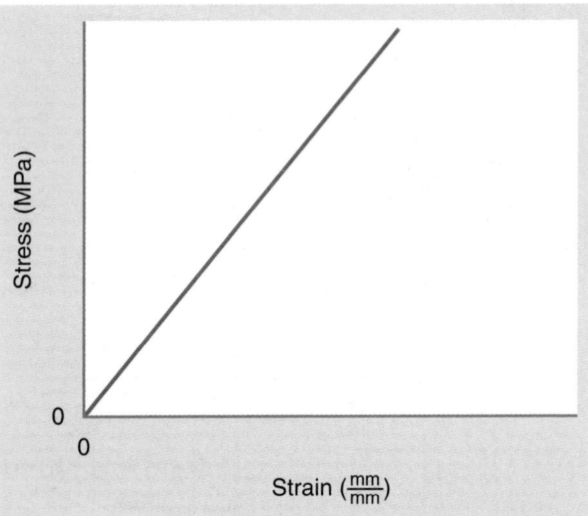

Figure 4.12 **Relationship of stress and strain.**

The relationship of stress and strain for a given material and type of deformation is represented by the **elastic modulus**:

$$E = \frac{\sigma}{\varepsilon} \quad (4.9)$$

where

E = elastic modulus
σ = stress
ε = strain

Note that this is a general form of the elastic modulus. The specific relationship of stress and strain can be calculated slightly differently, depending on the type of deformation that is of interest. For example, attempted deformation of the material may be the result of tension stress (Young's modulus), compression stress (bulk modulus), or shear stress (shear modulus) (Serway & Jewett, 2019; Serway & Vuille, 2018). **Tension stress** occurs when two forces are applied to a system in opposite directions away from each other (Figure 4.13a). In other words, the forces are colinear or have parallel lines of action but are in opposing directions so that the system is being "pulled apart."

In contrast, **compression stress** is the result of two forces being applied to the system in opposite directions toward each other (Figure 4.13b). **Shear stress** occurs from the application of two parallel forces that tend to simultaneously displace one part of a system in a direction opposite another part of the system (Figure 4.13c). Two special cases of loading that also cause deformation are bending and torsion. **Bending** occurs when two off-axis forces are applied such that tension stress is caused on one side of the system and compression stress occurs on the other side (Figure 4.13d). **Torsion** is caused by two forces being applied in such a way that part of the system is rotated around its longitudinal axis in a direction opposite rotation of another part of the system (Figure 4.13e).

Elastic modulus An expression of the relationship of stress and strain for a given material and type of deformation.

Tension stress Occurs when two forces are applied to a system in opposite directions away from each other.

Compression stress The result of two forces being applied to the system in opposite directions toward each other.

Shear stress Occurs from application of two parallel forces that tend to simultaneously displace one part of a system in a direction opposite another part of the system.

Bending Occurs when two off-axis forces are applied such that tension stress is caused on one side of the system and compression stress occurs on the other side.

Torsion Caused by two forces being applied in such a way that part of the system is rotated around its longitudinal axis in a direction opposite of the rotation of another part of the system.

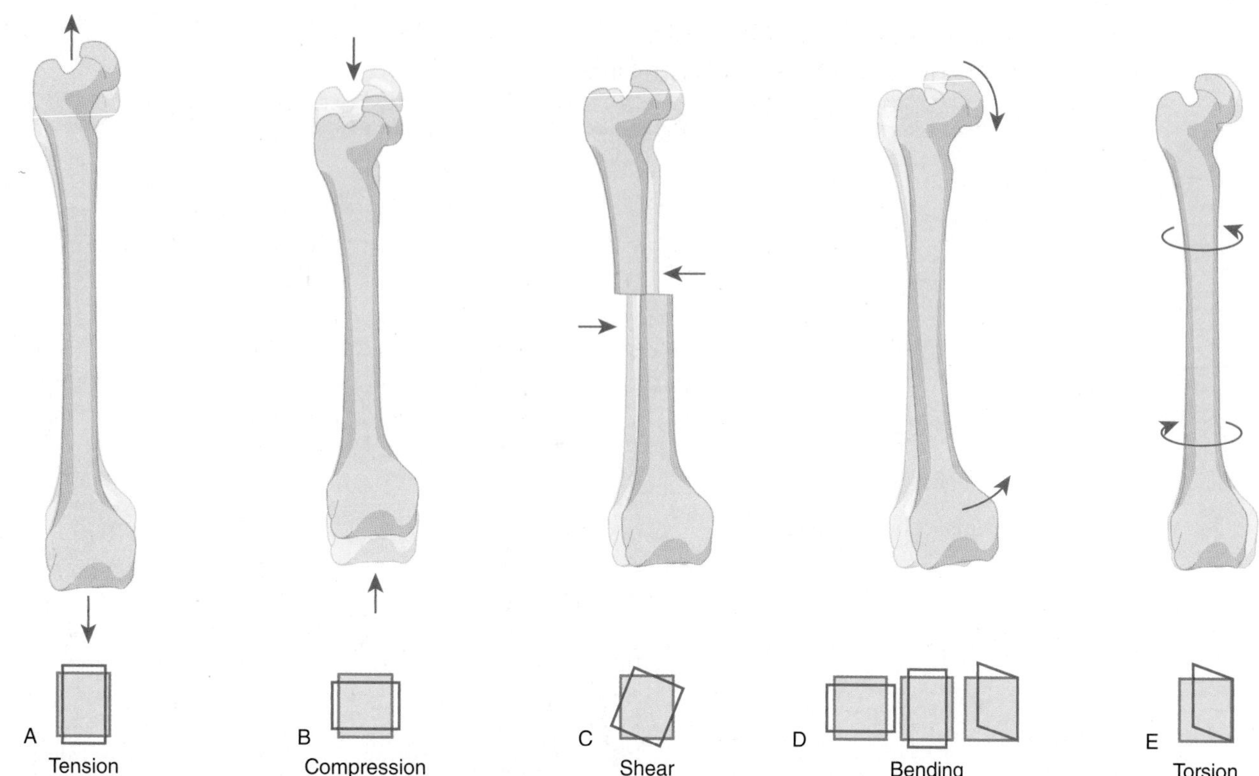

Figure 4.13 **Tension, compression, shear, bending, and torsion.** Think of some injuries that are caused by each type of loading.

Young's Modulus

When specifically referring to a condition of tension stress, the elastic modulus is called **Young's modulus** and is calculated as a more specific form of Equation 4.9:

$$Y = \frac{F/A}{\Delta l/l_i} \tag{4.10}$$

where

Y = Young's modulus
F/A = tensile stress
$\Delta l/l_i$ = tensile strain

Let's calculate the Young's modulus of a cord with a cross-sectional area of 5 cm^2 that changes in length from 100 cm to 150 cm when 50 N of force is applied. First we calculate the stress as 50 N/5 cm^2 = 10 N/cm^2. Next we calculate strain as (150 cm − 100 cm)/100 cm = 0.5. Finally, we can calculate Young's modulus as 10 N/cm^2/0.5 = 20 N/cm^2. Table 4.3 compares the Young's modulus for various materials.

Graphic representation of Young's modulus (and other stress–strain curves) is essential for understanding the response of biological materials to applied forces. For example, Figure 4.16 compares the stress–strain curves of various types of materials.

Young's modulus Term for the elastic modulus specifically referring to a condition of tension stress.

Table 4.3
Young's Modulus for Various Materials

Material	Young's Modulus (Y) in GPa	Material
Cartilage	0.0005–0.0008	
	0.01–0.1	Rubber
Ligament	0.3–0.4	
Tendon	0.8–1.2	
	2–4	Nylon
	11	Oak
Osteoporotic compact bone	12	
Normal compact bone	16	
Normal compact bone (6th decade)	17–20	
	30	Concrete
	72	Glass
	105–120	Titanium
	170	Stainless steel
	190–210	Wrought iron and steel
	400–410	Tungsten
	1,050–1,200	Diamond

Elastic region The linear portion of any given stress–strain curve; a material will return to its original shape if the tensile stress is removed within this range.

Yield point The point at which an applied stress can lead to permanent deformation.

Plastic region The nonlinear response of the material after the yield point; some degree of deformation will persist after removal of the stress.

Strength The maximum stress or strain that a material can withstand without permanent deformation.

Ductility The force per unit area required to deform a material and is represented by the steepness (slope) of the stress–strain curve.

Toughness The total energy required to cause material failure.

Yield strength Stress at the yield point of a material.

Ultimate strength The maximal stress that a material can withstand before failing.

Failure strength The stress at which a material actually breaks or ruptures.

First, notice that the stress–strain curves of all materials are not linear. The linear portion of any given stress–strain curve is called the **elastic region** because the material will return to its original shape if the tensile stress is removed within this range (Watkins, 2010. However, there is a **yield point** at which the applied stress can lead to permanent deformation. The nonlinear response of the material after the yield point is called the **plastic region** because some degree of deformation will persist after removal of the stress. Notice that each curve represents three other qualities of the material: **strength**, **ductility**, and **toughness** (Nigg, MacIntosh, & Mester, 2000). In general, the *strength* of a material is the maximum stress or strain that it can withstand without permanent deformation. More specifically, **yield strength** is stress at the yield point of a material. Beyond that point, permanent deformation will occur. However, yield strength is not the highest point on the stress–strain curve. The maximal stress that a material can withstand prior to the initiation of failure is called **ultimate strength**. **Failure strength** is the stress at which the material actually breaks or ruptures. Ultimate strength and failure strength often, but do not always, have the same value. Think of a rope that is being stressed. The rope may rupture suddenly (ultimate strength = failure strength), or it may fray and then rupture. While it is fraying it is not as strong, but it hasn't completely ruptured. Rupture may be the result of a single acute bout of loading or perhaps the result of accumulated microtrauma from cyclic loading and unloading. In the case of accumulated microtrauma, the stress required to produce failure gradually decreases, and failure may occur at a stress level lower than ultimate strength. For example, bending a piece of metal over and over again gradually weakens it until it eventually ruptures at a stress that it could previously withstand.

The *ductility* of a material is the force per unit area required to deform a material and is represented by the steepness (slope) of the stress–strain curve (Figure 4.14). Materials that fail at low stress but can withstand a large strain are called **ductile** (or pliant). A material that can withstand high stress but fails with relatively low strain is defined as **brittle** (or stiff). For example, bone is less ductile (more brittle) than tendon, and ligament is more ductile (or pliant) than tendon (Figure 4.15).

One of the most useful qualities displayed by a stress–strain curve is toughness, the total energy required to cause material failure (Nigg et al., 2000). In other words, it takes a lot of energy to rupture a tough material. Toughness can be estimated by observing the total area under the stress–strain curve. It can be seen in the figure that a material can possess high strength; if at the same time it has low ductility, however, it will not be particularly tough.

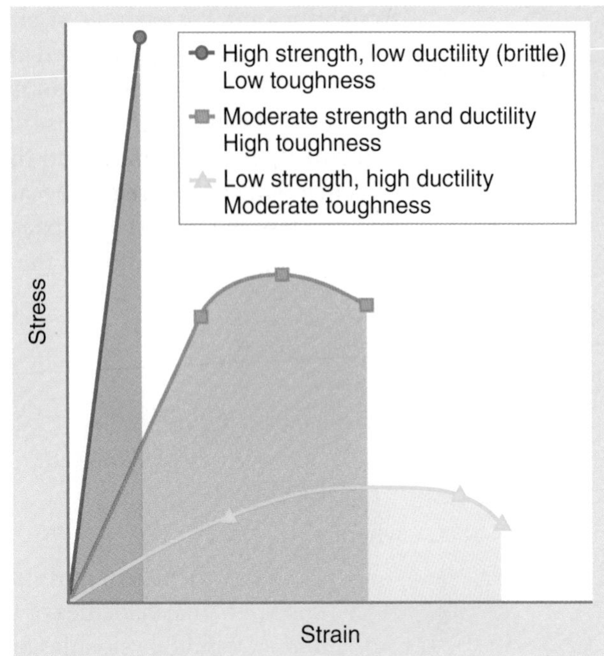

Figure 4.14 **Comparison of various stress–strain curves.** Identify each part of the curve.

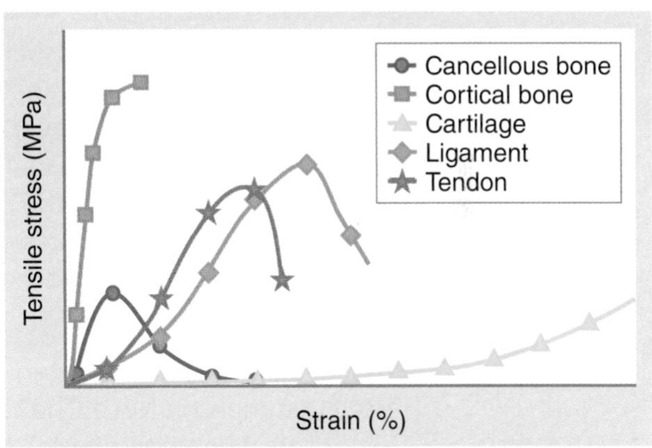

Figure 4.15 **Comparison of stress–strain curves of biological materials.** Why are the curves so different?

Coefficient of Restitution

We have defined the elastic modulus of a material as stress divided by strain, providing some indication of a material's resistance to being stretched. We have discussed that there is an elastic region of a material's stress–strain curve in which the material will reform to its original shape once the stress is removed. Materials vary in their ability to reform after deformation. The **coefficient of restitution** (or coefficient of elasicity) is a parameter observed after reformation that indicates the ability of an object to return to its original shape after deformation. A common application in sports is to calculate a coefficient of restitution for a particular type of ball by dropping the ball from a known height and observing the rebound. The coefficient of restitution is calculated as follows:

$$e = \sqrt{h_{rebound}/h_{drop}} \tag{4.11}$$

where

e = coefficient of restitution
$h_{rebound}$ = height of the rebound
h_{drop} = height of the drop

The coefficient of restitution ranges in value from 0.00 to 1.00 (with 0.00 being perfectly inelastic and 1.00 being perfectly elastic). The calculation is relatively

Ductile The quality of a material that fails at low stress but can withstand a large strain; also known as *pliant*.

Brittle The quality of a material that can withstand high stress but fails with relatively low strain; also known as *stiff*.

Coefficient of restitution A parameter observed after reformation that indicates the ability of an object to return to its original shape after deformation; also known as *coefficient of elasticity*.

straightforward. For example, if a ball is dropped from a height of 3.0 m and rebounds 1.8 m, the coefficient of restitution is calculated as $\sqrt{1.8\text{ m}/3.0\text{ m}} = 0.7746$.

The coefficient of restitution may also be calculated using the change in velocities of the objects involved in a collision. In the case of the ball from above, the collision would be with a particular sports implement. The implement and the ball have individual velocities before impact. After impact, the velocities will have decreased because of energy lost on collision. The higher the coefficient of restitution, the less energy lost at impact. Therefore, energy lost at impact can be estimated by calculating the coefficient of restitution in the form of change in relative velocities:

$$e = \frac{v_{af} - v_{bf}}{v_{bi} - v_{ai}} \tag{4.12}$$

or

$$-e = \frac{v_{af} - v_{bf}}{v_{ai} - v_{bi}} \tag{4.13}$$

where

e = coefficient of restitution
v_{af} *and* v_{bf} = final velocities of the objects (after impact)
v_{bi} *and* v_{ai} = initial velocities of the objects (before impact)

FOCUS ON RESEARCH

During a collision in sports such as American football, spectators (and often researchers) tend to focus on the impact itself in terms of potential for injury. This is especially true with regard to recent concerns over concussion risk. However, other factors involved might not be so obvious to the casual observer. For instance, many people do not give a second thought to the textures of the materials from which uniforms are made. Remember that the interaction of the textures of two materials is reflected in the coefficient of friction (μ_s). Changing the materials that make up uniforms changes the μ_s when there are physical interactions.

Rossi et al. (2016) investigated the influence of friction between football helmets and jersey materials on force. The researchers discuss the often overlooked "pocketing effect" that occurs when one player's helmet hits another player. The player's head compresses the padding and tissues of the other player, creating a "pocket" that holds the head in place. This pocketing effect is positive in that it helps to dissipate forces. However, if the head stays in the pocket there may be an increased risk of cervical spine injury because of the increased time of exposure to compressive forces. One factor affecting whether or not the head can come out of the pocket is friction between the helmet and the other materials that it's contacting. With this in mind, the investigators measured the coefficients of friction between glossy and matte helmet surfaces with youth, high school, and collegiate jerseys, as well as silk screened, stitched on, and sublimated jersey numbers. To measure the μ_s, the investigators used an inclined plane, a 1.4-kg aluminum block, and a means of increasing and measuring the angle of inclination. The researchers found that the highest μ_s for helmet combinations was glossy versus glossy. Glossy helmet versus collegiate jersey was the highest μ_s among the helmet-to-jersey combinations. And finally, the highest μ_s for helmet-to-jersey number combinations was that of glossy versus silk screened. It was also calculated that the highest μ_s combination was glossy helmet versus silk-screened numbers and that it could increase the force on the player's helmet by 3,553.88 N compared to the other combinations. The investigators point out that these findings are clinically relevant because forces in football range from 3,922 to 7,845 N and neck failure can occur at forces ranging from as low as 3,340 to 4,450 N (Ivancic, 2012; Nightingale et al., 1996; Swartz, Floyd, & Cendoma, 2005).

Currently, many organizations and researchers are focused on preventing traumatic brain injury in sport. Based on the findings of this study, the scope of research may need to be broadened to include some less obvious contributors to injury. Many observers see uniforms and padding as decorative and protective, but this study provides evidence that more research is needed to find combinations of materials that could actually reduce risk of cervical and perhaps other types of injuries.

A ball traveling 20 m/sec strikes an area on an implement that is traveling 25 m/sec. The velocity of the ball off the implement is 25 m/sec, and the velocity of the implement after the impact is 5 m/sec. Equation 4.12 can be used to calculate the coefficient of restitution for this particular ball and implement:

$$e = \frac{v_{af} - v_{bf}}{v_{bi} - v_{ai}} = \frac{5 - 25}{-20 - 25} = 0.4444$$

Care should be taken with this particular situation, however, because directions are indicated with signs. For example, the ball is initially given a negative sign because it is traveling the opposite direction ($-x$) of the implement ($+x$) relative to the person swinging the implement. However, the signs are the same after impact, because the ball and implement are traveling in the same direction. Notice that even though Equations 4.11 and 4.12 (or 4.13) differ, the concept remains the same: The higher the coefficient of restitution, the greater the energy remaining after impact in the form of either bounce height or velocity. No matter the calculation, coefficient of restitution depends on the material properties and temperature of the objects and on the relative velocities at impact. For example, graphite bats have higher coefficients of restitution, and warm materials are "livelier" than cold materials.

Viscoelasticity

We have discussed material properties in some depth. However, one other factor can affect the behavior of biological materials. If a material is **viscoelastic**, its deformation is also affected by both the rate of loading and the length of time it is subjected to a constant load. Most biological materials are viscoelastic to some degree because of their fluid components. Bone, for example, responds to high loading rates with increased strength and stiffness (Figure 4.16).

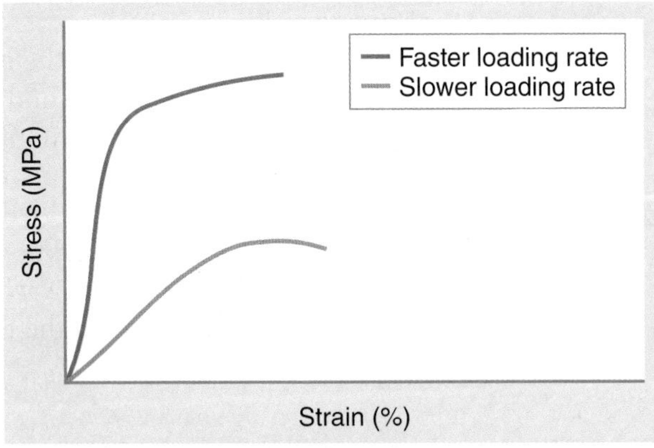

Figure 4.16 **Relationship of stress–strain to loading rate.**

This is an advantage in human locomotion because loading rates can be extremely high during running. However, some ligaments do not respond in the same way and may fail in situations of high loading rate. In terms of injury, this means that material failure will depend on the rate of loading. So injury may occur because of the

Viscoelastic The quality of a material whose deformation is affected by both the rate of loading and the length of time that it is subjected to a constant load.

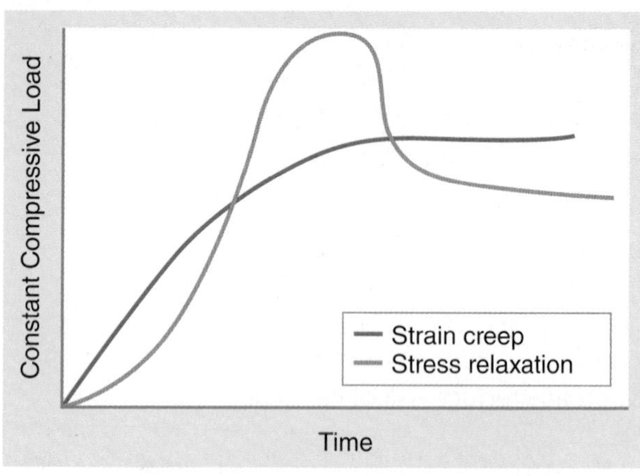

Figure 4.17 **Creep of an intervertebral disc.**

rupture of the ligament at high loading rates, whereas avulsion (rupture of the bony attachment of the ligament) may occur at lower rates of loading (Oatis, 2017). Also, intervertebral disks are subjected to a compressive load throughout the day. In response to this constant load, the intervertebral discs lose vertical height and bulge (in accordance with Poisson's ratio) as fluid is slowly squeezed out of the disc. As fluid is lost, the disc will experience increasing strain until no further fluid is exuded (and strain will become constant). **Creep** is the term for this property of experiencing increasing strain (continued deformation) under a constant stress (Figure 4.17). **Stress relaxation** is the corresponding eventual decrease in stress that will occur as fluid is no longer exuded.

The disc will then gain vertical height as the person sleeps at night because the compressive load of standing has been removed.

Fluid Forces

Two types of forces are exerted by a fluid on the system: (1) force from being submerged in a fluid and (2) force from moving through the fluid. At this point, we will only give a basic introduction to these forces because they will be discussed in much greater detail in Chapter 9.

Buoyant Fluid Force

Buoyant force is the vertical, upward-directed force acting on an object that is submerged in a fluid. Two key concepts from our discussion of pressure are needed to understand buoyant force. First, Pascal's law states that pressure applied to a fluid is transmitted undiminished to every point of a fluid and to the walls of the container; therefore, all points of a submerged body at a given depth will experience the same pressure. Second, pressure increases in large increments with relatively small changes in depth (Figure 4.18). Now let's talk about water itself. If we imagine a pool of water and then outline 1 cubic meter of that water, we can see that the water is neither rising nor falling.

Therefore, we know that the net force on the cubic meter of water must be zero. Otherwise, in accordance with Newton's second law, the section of water would be accelerated. Let's clearly define the forces involved. First think of Pascal's law and realize that every point on a given level of the section of water is experiencing the same pressure. Start with the sides of the section: All of those force vectors are colinear and will therefore sum to zero (i.e., no acceleration from side to side). Besides, we are talking about buoyant force, so we are primarily concerned with vertical forces. We know that there will be downward force exerted on the area by the water above. We also know that there will be force acting on the bottom of the area of water. Because the bottom of the section of water is at a greater depth, the pressure is higher than at the top. But if the force is higher on the bottom, how is it that the defined area of water is not rising? Don't forget that the water itself weighs something because of gravity, so it exerts a second downward force. Because of the weight of the water itself, the forces are balanced and sum to zero (preventing acceleration). *Buoyant force* is the difference between the downward and upward forces caused by fluid pressure. As the last example shows, the difference between the downward and upward forces is the weight of the water in that defined area.

Creep The term for the property of experiencing increasing strain (continued deformation) under a constant stress.

Stress relaxation The eventual decrease in stress that will occur as fluid is no longer exuded.

Buoyant force The vertical, upward-directed force acting on an object that is submerged in a fluid.

Archimedes (287–212 B.C.) understood this concept and stated it in what has become known as **Archimedes' principle**: a body submerged in a fluid will be buoyed up by a force that is equal in magnitude to the weight of the displaced water. For example, let's imagine that we replace the defined area of water in a pool with an empty container that fills the same amount of space (i.e., has equal volume to the displaced water).

The water surrounding the empty container will behave the same as if the displaced water is still there. In other words, the water pressure at a given depth is a constant. So the empty container will experience the same downward force from the water pressure produced by the weight of the water above; it will also experience the same upward force from the water pressure at the bottom of the container. In this case, the difference is that the container is empty and will not weigh enough to balance the upward and downward forces. So because there is now a net force acting vertically upward, the container will be accelerated upward. The opposite situation will occur if the weight of the object is greater than the weight of the water it displaces (i.e., the net force will be downward and the object will sink). Buoyant force will be discussed in greater detail in Chapter 9.

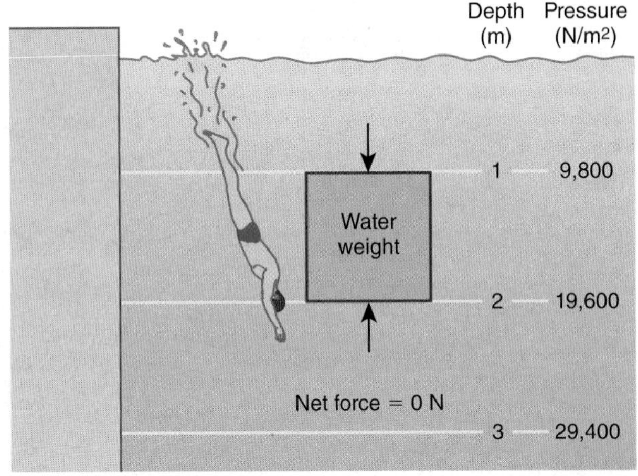

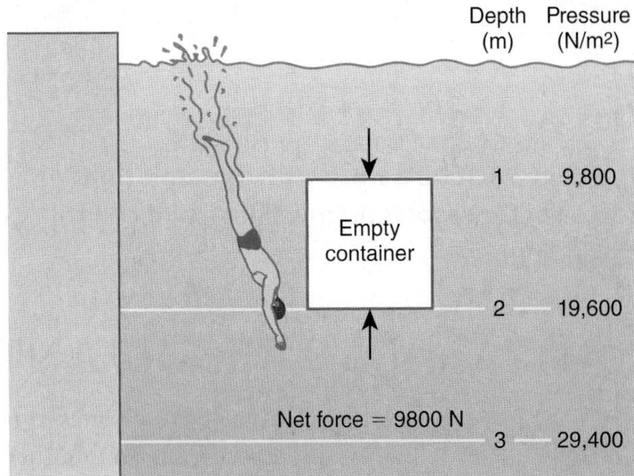

Figure 4.18 **One cubic meter of water in a swimming pool.** Why is the pressure higher at the bottom?

Dynamic Fluid Force

As you have read, there is force acting on any object that is submerged—buoyant force. In contrast, **dynamic fluid force** acts on a system that is moving through a fluid. When moving through a fluid, the system (whether an airplane, a bird, a car, a dolphin, or a human) causes particles of the fluid to be deflected so the system can move forward. In other words, the system applies a force to the fluid particles that moves them along the outside of the system. If the system is tilted (such as the wing of an airplane or bird), the deflection is even greater.

According to Newton's third law of motion, there must be an equal and opposite directed force applied to the system by the fluid particles. Dynamic fluid force is the equal and opposite directed force of the fluid particles in reaction to the applied force of the system moving through the fluid. Similar to other forces, the dynamic fluid force vector can be resolved into two components: a parallel component and a perpendicular component (Figure 4.19).

The parallel component of dynamic fluid force acts in the opposite direction of system motion with respect to the fluid and is called **drag force**. Drag force tends to resist motion of the system through the fluid (or decelerate the object). The force of drag can easily be experienced by holding your hand out the window while riding in a car. If your palm is forward, you will experience the force of

Archimedes' principle A body submerged in a fluid will be buoyed up by a force that is equal in magnitude to the weight of the displaced water.

Dynamic fluid force The equal and oppositely directed force of the fluid particles in reaction to the applied force of the system moving through the fluid.

Drag force The parallel component of dynamic fluid force that acts in the opposite direction of system motion with respect to the fluid; tends to resist motion of the system through the fluid.

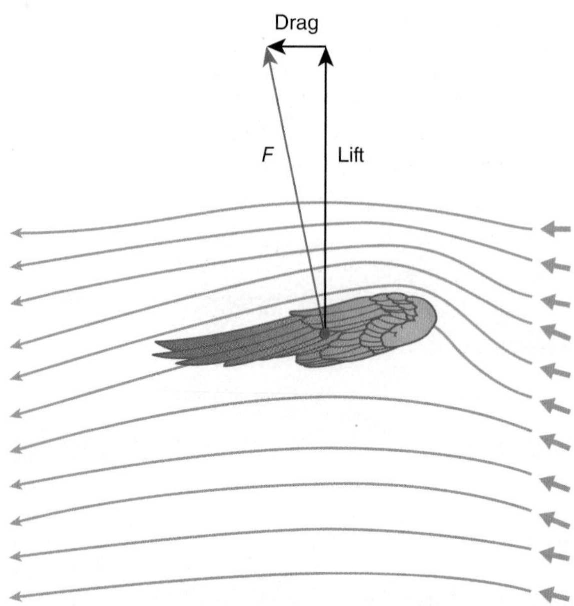

Figure 4.19 **A bird wing with dynamic fluid force vector resolved.** How does a bird control its landing?

drag attempting to push your hand in the direction opposite to that of the motion of the car.

The perpendicular component of dynamic fluid force can act in any direction that is perpendicular to system motion with respect to the fluid, even though it is termed **lift force** (which implies upward motion). Therefore, lift tends to change the direction of system motion. For example, air particles tend to be deflected downward if a wing is tilted with the leading edge up. This means that the reaction force applied by the particles to the wing will cause upward motion. You can also experience this effect with your hand as it moves through oncoming air. Simply change the position of your hand from completely palm forward to a slanted position with the leading edge up. You will experience lift force (along with the drag force) that tends to cause the hand to rise. This is the same force that allows both airplanes and birds to fly.

Dynamic fluid force, like buoyant fluid force, will be discussed in much greater detail in Chapter 9. The objective at this point is to gain some understanding of all of the forces to which the system is subjected.

4.5 RESULTANT FORCE

To this point we have discussed noncontact and contact forces, as well as the fact that forces can be classified according to whether or not they originate from within the system or external to the system. Only external forces can cause changes in system motion (Newton's first law of motion). In accordance with Newton's third law of motion, forces are paired and can therefore also be classified as action or reaction. Because action and reaction forces are simultaneous, there is sometimes confusion in their classification (i.e., the *reaction* is sometimes misinterpreted as *the action*). Some would say that there are further classes into which individual forces can be categorized. However, it is probably clear at this point that forces are not lone entities. In lieu of further categorization of individual forces, our approach will be that there is an ultimate effect from the simultaneous application of multiple forces. So in this section, we will classify forces according to *resultant* force. In Chapter 3, vector analysis was used to calculate the *resultant* vector in situations of multiple force application. We will now revisit vector composition and use the resultant to classify a force according to its ultimate effect on the system.

Lift force The perpendicular component of dynamic fluid force can act in any direction that is perpendicular to system motion with respect to the fluid; tends to change the direction of system motion through the fluid.

Motive force A force that tends to cause a change in motion in the form of increased velocity or change in direction of the system; also known as *propulsive force.*

Motive and Resistive Forces

One way to classify forces according to their attempted resultant is simply to observe whether or not the force was motive or resistive. If a force tends to lead to a change in motion in the form of increased velocity or change in direction of the system, then we can define the force as **motive** (or *propulsive*) **force**. When a skydiver jumps from an airplane, gravity is the motive force accelerating the diver back to Earth. In the body, muscle forces are considered motive as long as the situation is one in which the applied muscular force is greater than the force provided by the external load (e.g., in the concentric phase of a resistance training exercise).

In contrast to motive force, **resistive force** tends to prevent changes in motion by other external forces or decrease the velocity of a system that is already in motion. In our resistance-training example, muscle forces are motive as long as they are greater than the force of an external load. However, muscle forces are *resistive* and gravity is motive when one is lowering an object. In the skydiving example, gravity is motive while air resistance provides a resistive force. If one rolls a ball down a hill, gravity is motive while both air resistance and friction are resistive. In the human body, resistive forces are created by the friction within joints (which articular cartilage and synovial fluid attempt to reduce), resistance to stretch provided by connective tissues, and the resistance that liquid provides as it flows.

Remember that at any given instant a force can be motive and then resistive—or even motive and resistive simultaneously. For example, one characteristic of a motive force is that it can cause a change in direction of a system. Resistive forces can decrease the velocity of a system. Think of friction force between the surface of the floor and the sole of a shoe when a person attempts to rapidly change his or her direction of motion. In this case, friction is motive in that it provides the force for an acceleration and resistive in that it keeps the foot from slipping while changing direction.

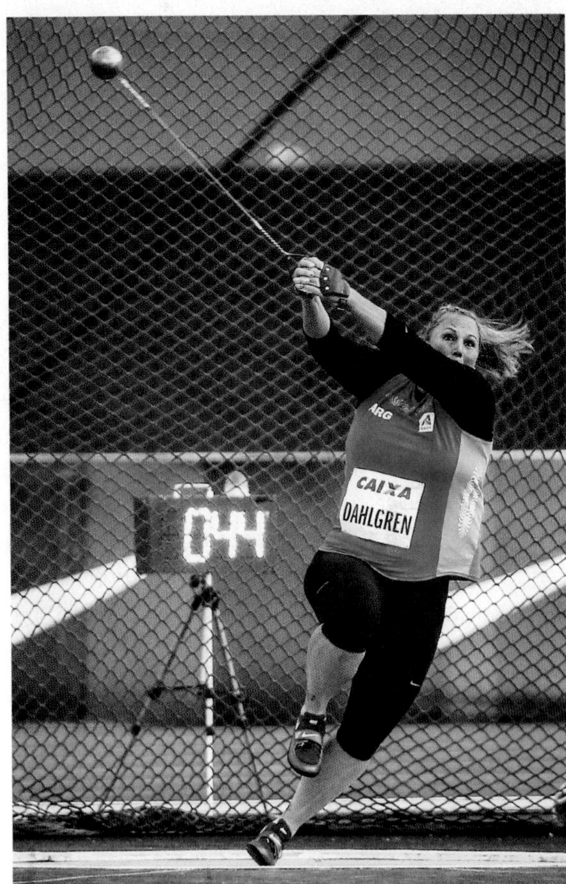

Figure 4.20 **The hammer throw.** What keeps the hammer moving in a circle?

Centripetal and Centrifugal Forces

Although the words *centripetal* and *centrifugal* are often paired with *force*, in fact the two words refer to the resultant *effects of forces*. To clarify, **centripetal force** is any force that causes a system to exhibit circular motion, which is often called "center seeking." This center-seeking motion could potentially be caused by any force, depending upon the situation. To clarify, think of attempting to push a rope in a circle. It simply won't work. Instead, we have to pull the rope in a circle. That pulling force is centripetal, acting to angularly accelerate the rope. Let's use the hammer throw as an example (Figure 4.20).

The hammer thrower spins in circles while swinging (pulling) the hammer from a cable. According to Newton's first law of motion, objects tend to move in right lines unless acted on by an external force. So because the hammer is not moving in a straight line, there must be some external force causing it to move in a circle. The force causing angular acceleration (i.e., circular change in motion) is called *centripetal*. In the case of the hammer throw, the tension in the cable caused by the pull of the athlete's hands is the force that has a centripetal result. Gravity can be a centripetal force in that it forces the moon to remain in orbit around Earth. Friction acts as a centripetal force any time a bicycle or car travels in a circle. So think of centripetal force as any force that causes a system to move in a circular path.

The term *centrifugal force* is even more frequently misused. Centrifugal force, which is often called "center

Resistive force A force that tends to prevent changes in motion by other external forces or decrease the velocity of a system that is already in motion.

Centripetal force Any force that causes a system to exhibit circular motion.

Figure 4.21 **A hammer throw at the time of release.** What makes the hammer move in a straight line?

fleeing," is sometimes defined as the force that is equal to, and opposite, the centripetal force. The reality is that what is often called "centrifugal force" is simply the absence of sufficient centripetal force to maintain a circular path of the system. For example, the hammer in our example of the hammer thrower will no longer travel in a circle when the thrower releases the handle. Instead, the inertia of the hammer will cause it to travel in a path tangent to the circle and in the direction it was moving at the time of release (Figure 4.21). This result is in keeping with Newton's first law; the natural tendency of the object is to move in a straight line.

However, notice that there is no force pulling the hammer out of the circular path. With the disappearance of the centripetal force, the hammer is no longer constrained in a circular path and simply travels in a straight line. Think of a car making a sharp turn. Friction is the centripetal force keeping the car in a circular path. Imagine what would happen if friction suddenly disappeared while making a turn over a patch of ice. The car will slide in a tangential direction. But again notice that the inertia of the car causes it to travel in a straight line; there is no force pulling it out of the circular path. Therefore, centrifugal force is a fictitious or false force. A false force appears to act on the object, but one cannot actually identify that force. Think back to the example of pushing a rope in a circle. You actually have to pull the rope. You apply a force, accelerating the rope in direct proportion to the size of the applied force. The natural tendency of the rope is to move in a right line in the direction of the acceleration. However, you haven't let go of the rope so it is forced to travel in a circle. Once you do let go of the rope, it is free to move in the direction in which you accelerated it. So no force is pulling the rope out of the circle. In other words, there is no "centrifugal force." There is simply the lack of a centripetal force.

CONNECTIONS

4.6 HUMAN PERFORMANCE AND INJURY SCIENCE

Exercise Physiology

Even though we discussed *pressure* in a biomechanical context, atmospheric pressure can have large ramifications for physical performance. Traveling to high altitudes, where atmospheric pressure is low, has many health-related consequences. However, these consequences are beyond the scope of this text, so we will focus on the performance aspects of altitude. The problem for physical performance at high altitude is oxygen delivery. More specifically, the bonding of oxygen to hemoglobin depends on the pressure of oxygen in the alveoli of the lung (i.e., less oxygen pressure in the atmosphere, less bonding of oxygen to hemoglobin). We use oxygen (delivered via hemoglobin in our red blood cells) in metabolic processes to convert the fuel substrates we eat (carbohydrate, fat, and protein) into the high-energy compound used within the body (adenosine triphosphate, ATP). Our maximal capability to use oxygen in metabolic processes to make ATP is called **maximal oxygen uptake** (or VO_2 max—maximal volume of oxygen consumed). Therefore, maximal oxygen uptake is an indicator of the capacity to perform aerobic work (endurance exercise).

We mentioned that as you travel to areas of higher elevation, there is less atmospheric gas piled on you. So if you are currently at sea level, you are experiencing one atmosphere of pressure (approximately 1.013×10^5 Pa or 760 mm Hg), whereas the atmospheric pressure at an altitude of 5,486 m is approximately half that at sea level. So at an extremely high altitude such as the summit of Mt. Everest (8,848 m), the atmospheric pressure is only 250 mm Hg (approximately 0.333×10^5 Pa). Total atmospheric pressure results from a mixture of gases. Our atmosphere is approximately 20.93% oxygen, 0.03% carbon dioxide, and 79.04% nitrogen. That means that at sea level, 20.93% of the total 760 mm Hg pressure is the result of oxygen. So the pressure of oxygen at sea level is approximately 159.068 mm Hg ($0.2093 \times$ atmospheric pressure). The *percentage* of oxygen in the atmosphere is the same at high altitude as it is at sea level. But remember, that's a percentage of a lower total amount of pressure. So the pressure provided by oxygen at the summit of Mt. Everest is only about 52.325 mm Hg.

Maximal oxygen uptake does not seem to be affected to a significant degree below altitudes of 1,500 m (McArdle et al., 2014). However, at altitudes above 1,500 m, a linear decrease in maximal oxygen uptake of approximately 10% occurs for every 1,000-m increase in altitude (McArdle et al., 2014). This means that at an altitude of 6,248 m, maximal oxygen uptake is only around 50% of that at sea level. Imagine climbing a mountain with half your VO_2 max tied behind your back!

Functional Anatomy

Many of the topics in this chapter are related to material properties. The ability of biological materials to sustain repeated applications of stress and strain has direct implications for the health of the musculoskeletal system. For example, the coefficient of friction in synovial joints is approximately 0.01–0.003 (Mow, Ratcliffe, & Poole, 1992; Serway & Jewett, 2019). This extremely low coefficient of friction allows for the gliding of one bone across another without erosion of the joint.

Another example of force application and its relationship to biological material design is that of compression stress applied to intervertebral discs. First, remember that when an object is compressed in one direction, it expands in the opposite direction. This tendency of a material to exhibit transverse (lateral) strain simultaneously with axial (longitudinal) strain is expressed as *Poisson's ratio*. Also recall *Pascal's law*: pressure applied to a fluid is transmitted undiminished to every point of a fluid and to the walls of the container. During compressive loading, pressure increases within the nucleus pulposus of the intervertebral disc (Figure 4.22a). Because the intervertebral disc contains fluid, the nucleus pulposus exerts pressure against the surrounding annulus fibrosus in accordance with Pascal's law. This causes the disc to bulge in accordance with Poisson's ratio (Figure 4.22b).

Stress from compressive forces applied to intervertebral discs is sometimes referred to as **hoop stress**, and the resultant bulging is called **radial expansion** (Oatis, 2017). Through this process, the intervertebral disc is capable of withstanding extremely high loads. Radial expansion from hoop stress also helps explain why people tend to be 1–2 cm taller in the morning. However, if the compressive load is extreme, the disc can bulge to the point of compressing nerves (causing pain) or even rupture. If the intervertebral disc is dehydrated

Maximal oxygen uptake Maximal capability to use oxygen in metabolic processes to make ATP; expressed as VO_2 max.

Hoop stress Stress caused by compressive forces applied to intervertebral discs.

Radial expansion The bulging of an intervertebral disc in accordance with Poisson's ratio.

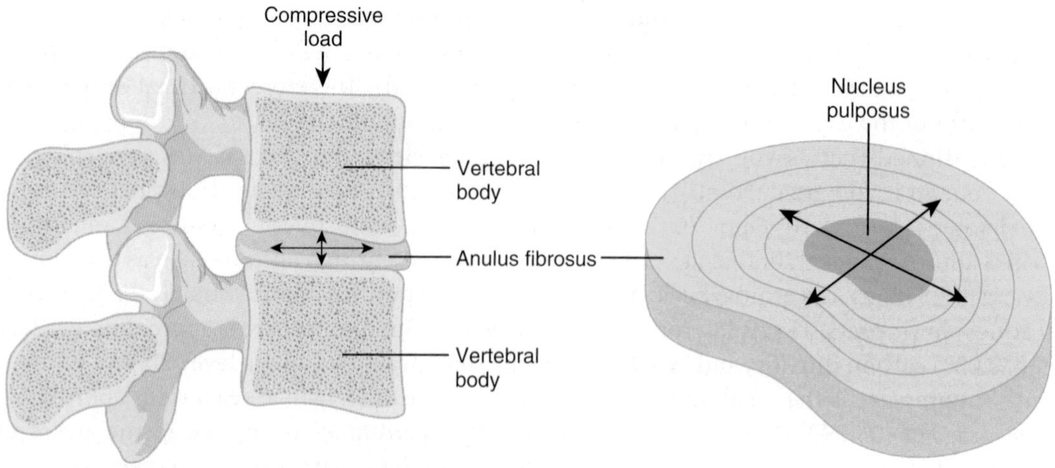

Figure 4.22 **Radial expansion of an intervertebral disc.**

or surgically incised, then there are changes in the ability of the disc to withstand compressive loads. If the disc is unable to transform the compressive load radially to the annulus fibrosus, then higher than normal loads are sustained by the vertebral bodies, which can lead to degeneration (Oatis, 2017).

Sport Science

Technology has undoubtedly changed sport dramatically within the past two decades. New materials and the advancing design of equipment have brought changes in performance and safety, sometimes to the detriment of competitive events. On the positive side there are advances in helmet technology that reduce the acceleration forces on the head and brain by spreading impact loads during hits. At the other end are swimsuits that gave athletes such an advantage that they were considered "technical doping" and banned from competition. All of the changes in sport equipment and rules can be traced back to manipulating forces.

Baseball, softball, and golf have all undergone a revolution in equipment design, with newer bats and clubs constructed of metal and composites rather than traditional wood. If you watch any baseball game with players below the professional level, you will likely miss the traditional sound of a wooden bat making solid contact with a baseball. Except in Major League Baseball and affiliated minor leagues, standard bats are now constructed from aluminum and composite materials such as carbon fiber. The coefficient of restitution of aluminum and newer composite materials is much higher than that of wood, so a baseball or softball leaves these new bats at a much higher velocity. This is an advantage for younger athletes who desire to hit the ball farther. However, older venues and Major League Baseball parks that were designed to be played with wooden bats would become "Home Run Derby" facilities if professional players were allowed to use metal bats. Smaller parks already experience this as a new generation of players becomes bigger and stronger.

An additional reason that professional baseball does not allow metal bats is that they're concerned about pitcher safety. Pitchers are in a vulnerable position at the end of their throwing motion, and a line drive coming back off a metal bat would arrive in a shorter time. Although aluminum bats are not allowed in professional baseball, wooden bats are now constructed from maple, instead of the traditional ash. Maple is denser, and collisions between these bats and baseballs means less energy loss, which also increases ball speed—but not as much as with metal and composite bats. This same safety issue has now trickled down to Little League baseball, which adopted rules in 2018 related to coefficient of restitution limitations.

A similar situation is happening currently in golf. In the late 20th century, equipment manufacturers began experimenting with different materials in an effort to improve driving distance and accuracy. For the same reason that wood bats were replaced by newer materials, traditional wood clubs were replaced by a variety of materials with a higher coefficient of restitution and lower weight, including metal, graphite, and carbon. The result was higher swing speeds and less energy lost at impact, which led to higher ball velocity and more distance. At the same time, ball manufacturers were also experimenting with different materials and design to increase their coefficient of restitution. When these changes are put together, the average golfer was able to gain a significant advantage in distance.

The added distance is even more significant for professionals, and the rule makers of golf have set maximal standards for clubs and balls so that courses don't become obsolete. Many classic golf venues such as Augusta National, home of the Masters Tournament, have had to lengthen their courses to accommodate players hitting the ball so far. Certainly one factor in golf balls traveling farther is the size, strength, and fitness of the recent generation of athletes, but critics still suggest that the newer clubs and balls were taking away from the skill of the game, and the PGA instituted strict testing procedures at tournaments.

As with golf, new materials and design have added to the enjoyment of many tennis players, with lighter and larger rackets creating more power and less fatigue. For tennis players at higher elevations where the air is less dense, there is now an approved ball that makes the game more consistent with traditional equipment. The 6% to 8% larger ball has more aerodynamic drag and travels through the thinner air in a similar way to a standard ball at sea level, allowing players more time to react.

When the first edition of this textbook was written, there was controversy in the sport of swimming. World records were being broken at an amazing rate—and not just by small time increments. The reason for the change was a revolutionary full-body swimsuit made by Speedo called the LZR Racer. This new suit reduced pressure drag by using compression to give swimmers a more streamlined shape, reduced viscous drag through specially designed surface textures, and added buoyancy by trapping air inside within the suit. The suit was banned in 2010 because it was deemed to negatively affect the spirit of the sport.

From striking implements and balls to clothing to safety gear, equipment is changing sport through a thorough understanding of, and application of, physics. Advanced materials and design allow lighter and larger implements that add power and accuracy, decrease (or increase) drag forces, and spread the force of impact loads to reduce injury. Manufacturers know that recreational players will buy and use equipment that will give them an advantage, so they encourage their engineers to find ways to accomplish this.

4.7 MOTOR BEHAVIOR

Motor Development

Gravity is the force we understand well because it plays a significant role in our daily lives. The human postural system is governed subconsciously and is responsible for orienting the body relative to gravity. During the first years of life, infants and toddlers progress through a series of milestones that prepare the body for upright locomotion. The earliest milestone is the ability to lift the head off a surface, and we progress to sitting, standing, and eventually walking. Every parent understands that gravity sometimes exceeds a toddler's ability to remain upright against it, but as the child's nervous and muscular systems mature the incidence of falling becomes less frequent. Humans have developed the instinctive ability to interact positively with a force that surrounds us at all times. This adaptability is especially noteworthy when considering the extensive changes that the human form evolved through when upright bipedal locomotion was established.

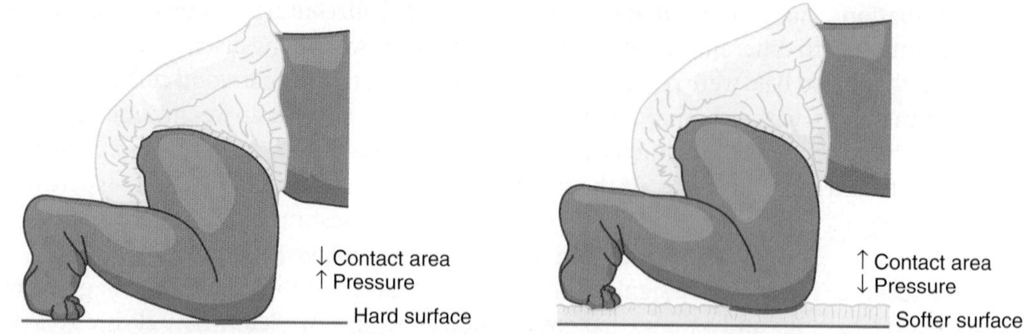

Figure 4.23 **Infant crawling on two different surfaces.** How does the surface relate to motor development?

Adaptability is also common in relation to development within different environments. Newell (1986) suggested that individual movements arise from the interaction of organismal constraints, movement environment constraints, and task constraints (Haywood & Getchell, 2014). For example, some children prefer to crawl on all fours (hands and feet) instead of on their hands and knees. There are several constraints that could lead to this behavior, including muscular strength and reinforcement from caregivers. But in the context of our chapter, a special type of movement environment may also play a role. For example, if the house in which the child lives has only hardwood floors, crawling could be extremely uncomfortable because of the extra pressure on the knees (Figure 4.23a).

Carpet and other soft floors help distribute the weight of the infant's body over a larger area, thereby decreasing pressure (Figure 4.23b). So, according to Newell's model, the environmental constraint of the crawling surface (which changes pressure) could play a part in what is considered to be a nontraditional motor pattern. A newer product available for infants, but used for years by volleyball players, mimics the effects of a carpeted floor. For the same reason that volleyball players wear kneepads, infants now have a way to spread the force load over a larger area when crawling on harder floors.

4.8 PEDAGOGY

Teaching

This text's early mention of "forces" reflects the timing with which some pedagogists believe the concept of force should be taught in physical education (Graham, Holt-Hale, & Parker, 2013). The authors of one of the most widely used textbooks for teaching elementary physical education recommend a two-part approach for students in early grades: skill themes and movement concepts. Skill themes are the "verbs" of movement, or the things that we can do such as walking, running, throwing, kicking, and more. Movement concepts are "adverbs" that can modify the skill themes.

Graham and colleagues believe that the foundation of successful human movement is derived from a solid functional understanding of movement concepts, or *how* we move. Included in this category are the concepts of time, relationships, and effort. Effort is associated with the notions of force, with subcategories of dualisms such as fast–slow, bound–free, and strong–light. In the beginning, it is important for children to understand these concepts by using extreme contrasts—for example, by moving on tiptoes to sneak up on someone versus stomping to scare someone away (light versus strong force). Once students understand the extremes of these concepts, they begin to explore the continuum that exists between the extremes.

In later movement experiences, students and athletes with knowledge of these continuums can be more effective. One example of this occurs during fitness testing of young children. Teachers who conduct the mile run as a measure of cardiovascular fitness often don't realize that this test may not be measuring the cardiovascular component at all. Everyone gathers on the starting line, and at the "Go" signal, they take off at different speeds. Unless students are taught the concept of pacing themselves, some will inevitably run as fast as they can for a short distance, then have to stop and walk to catch their breath. These "rabbits" only understand the extremes of fast and slow. Eventually, the students who understand pacing will catch and pass the rabbits, who tire from their "burst and stop" method of running.

The concept of an infinite continuum of force can be applied to successful participation in a variety of sports. Soccer players have to vary the force they apply to the ball, depending on the speed of their dribble, the length of a pass, or a shot on goal. Softball and baseball players vary the force of their swings to hit the ball to specific locations in the field. Golfers apply different amounts of force when striking every putt, and basketball players have to adjust the force they apply to each shot, depending on their distance from the goal. Force is an important concept to understand; it is presented early in this text for the same reason that physical educators teach it early to their students—a foundation of understanding force helps students be more successful in the future.

Coaching

Because of heavy rain before a soccer match, a coach preparing a team to play on a natural grass field needs to take more information into account. Cleat height may need to be altered because the top layer of soil will be damp and loose. Longer cleats will reach deeper into the ground (compensating for less traction) and give players the traction they need to accelerate and change direction quickly. The players will also need to be reminded that the ball will react differently when it comes into contact with the field because of the water. With less friction from wet grass, the ball will skid more and bounce less. The rolling friction will change also; as the surface of the ball pulls away from the ground while rolling, it lifts water with it. This will result in the ball rolling more slowly than on a dry field.

In the 1970s, several new baseball stadiums were built with artificial turf, and a few professional baseball stadiums actually replaced grass fields with artificial turf. Visiting teams often looked inept while fielding balls because this human-made surface resulted in different ball reactions because of significant differences in friction and the coefficient of restitution compared to natural grass. This gave the home team a real advantage but also resulted in the possibility of different injuries to players. The surface was not as soft as natural turf, so diving to make a play became a more painful endeavor. The coefficient of friction was also higher than grass, and players sliding along the surface often experienced "carpet burn." Many professional teams have returned to natural grass fields, or newer and softer human-made fields.

In the sports of rugby and American football, violent contact is inherent in the game. Avoiding injury during tackling for each sport is based on the concept of pressure, although the two sports have different means of dissipating pressure forces. In American football, players wear specially designed pads that spread the forces of impact over a greater surface area of the body, thereby decreasing the relative load in any one area (remember the magician and the bed of nails). Rugby players do not wear special equipment and therefore must rely on a different method of dissipating the forces of contact between players. This is where good coaching comes in. Tackling strategies to avoid injury include attacking from the side to decrease the overall collision force, aiming for softer parts of the opponent's body, and wrapping both arms around a player's legs (this technique spreads the force over a large area). Even when using equipment in American football, good tackling technique can decrease the pressure created by violent contact by spreading it across more surface area, with the benefit of greatly reduced risk of injury.

4.9 ADAPTED MOTION

Adapted Physical Education

There are individuals with disabilities who rely on external forces to participate in physical activities. One example is the use of gravity to effectively propel bowling and bocce balls. Quadriplegics and other individuals with limited upper body mobility use a specially designed ramp at a bowling alley. A ball is set onto the ramp by a helper, and its elevated status above the floor results in it having potential energy. The athlete then guides the position of the ramp physically or with outside help to aim at a specific target. When the ramp is aimed properly, the athlete gives the ball a slight nudge from its initial position on a flat portion of the ramp to the declined rails that run to the floor. Gravity pulls the ball downward, and the ball generates enough velocity to travel the 20 meters to the pins. The same type of ramps are often used with young children who don't have the strength to physically send a ball down the alley.

In the sport of bocce, a similar gravity assist is the basis for propulsion. However, because of the dynamic nature of the game, the ramp height varies, and players can position the ball within the ramp to adjust for higher or lower velocity. For athletes with no motor control below the neck, a specially designed headpiece is used to hold the ball in place until a player is ready to send it toward the target (Figure 4.24).

The force of gravity allows individuals with no ability to propel objects through internal muscular forces the opportunity to play highly competitive sports.

A second example of adapting equipment is using friction force. Being able to catch and control a thrown object is a key element for success in a variety of activities. The ability to pick up and move objects is also an integral part of independent living. Velcro and similar "hook and loop" materials help individuals in both sport and daily life by providing friction force that cannot be developed by the body, and it can also help those with perceptual difficulties such as visual impairments. A simple example of one product contains two Velcro-covered discs with a strap on the back of each and a light ball covered with the complementary Velcro partner material.

Two players each fasten one disc to their catching hand using the strap, and almost any throw that touches the Velcro on the disc will result in a successful "catch." This product has even become popular for individuals with no disabilities as a recreational game as the added friction allows spectacular catches that would be nearly impossible without the benefit of Velcro (see Figure 4.25). The friction that Velcro offers also benefits individuals with limited grip strength. Aids designed around its use include devices for writing, shaving, cooking, cleaning, and maintaining control of objects.

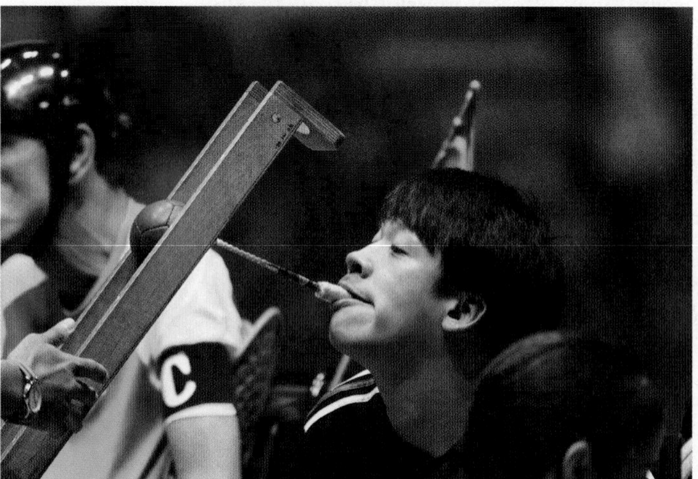

Figure 4.24 **Equipment adapted for bocce.** Can you think of some ways to adapt equipment for your favorite sport?
© A.RICARDO/Shutterstock

Prosthetics

In terms of material properties, mechanical concepts are used in the development of effective prostheses and orthotic devices. Although nonusers may focus on the visual aspect, a user of the prosthetic is more likely to be concerned with functional realism: Does the prosthetic produce the feeling of natural motion? In reference to this chapter, the concern for a prosthetic would be issues such as coefficient of restitution and pressure. To mimic a natural footstep, the prosthetic must have a coefficient of restitution that allows proper recoil after deformation (Figure 4.26).

Also, the prosthetic must be fit to the person in such a way that it does not place excessive pressure on the area of natural tissue. For patients who undergo amputation, the type of prosthetic to be employed in the future has implications in the nature of the surgery. Two factors that need to be carefully considered are the site of the amputation (above or below a joint), and the desired shape of the remaining limb for maximum effectiveness and comfort of the prosthetic. For knee- and hip-replacement devices, the coefficient of friction must be variable to replicate natural gait mechanics (i.e., greater at the beginning and end of the motion to replicate the natural resistance of connective tissues that surround a healthy joint). Finally, orthotic devices are often used in shoes to relieve excessive amounts of pressure in one particular area of the sole of the foot. So many mechanical principles are used in the design and development of prosthetic and orthotic devices that we can't cover them all here. We will see many other principles applied throughout the book. At this point, the key is to focus on the material properties of the devices.

Figure 4.25 **Velcro-covered products as adapted equipment.**

In contrast to devices used on land, we should also discuss adapted activity in water. One of the more beneficial, motivational, and enjoyable settings for adapted physical activity is a pool. Therapists prescribe aquatic programs for several reasons, but the primary one is buoyancy. The force applied by water (upward) counteracts the force of gravity, a force that on land requires adequate muscular strength to maintain posture or move successfully. For individuals with limited mobility or strength, an aquatic environment becomes a place where they can function independently, and the warm temperature of the water has additional benefits beyond the low-gravity atmosphere.

Buoyancy is based on Archimedes' principle: a body submerged in a fluid will be buoyed up by a force equal in magnitude to the weight of the displaced water. Buoyancy has the net effect of making the body seem like it weighs less than on land. A majority of individuals who require mobility assistance on land (walkers, wheelchairs, crutches) are able to move without assistance in a pool. This can be a liberating experience for a population that has to rely on others for mobility outside the pool setting, and it can lead to higher self-esteem, better morale, and an increase in motivation toward other aspects of rehabilitation (Skinner & Thompson, 1983). In addition, individuals who have little chance of competing in land-based sports and recreational activities often find events that they can excel in with the help of a buoyant environment.

Figure 4.26 **An athlete with prosthetics running.** What topics from this chapter are related to running with prosthetics?

Although the human body will generally float or maintain buoyancy when the lungs are full of air, it will sink without air in the lungs. There are other factors such as body-fat percentage that will affect buoyancy, but let's look at the case described in the first sentence. If we want to make sure that a child will maintain positive buoyancy, how can we guarantee it? Information in this chapter provides the answer. All you need to do is make sure that the weight of the displaced water is always greater than the weight of the individual. We do this through the use of assistive devices such as inflatable armbands—remember "floaties" from your earliest pool experiences? But how do relatively small inflatable objects provide so much buoyancy? A beach ball example will help explain the answer. Let's assume that a small beach ball weighs 1.112 N and is filled with 0.0283 cubic meters of air. Using Archimedes' principle, if we submerge the ball into water, an upward buoyant force will be exerted on the ball equal in magnitude to the weight of the displaced water. A cubic meter of water weighs approximately 9800 N. If the beach ball is completely submerged in water, the water will exert an upward force of 277.34 N against a 1.112 N object. A pair of inflatable armbands that can be inflated with a total of 0.0566 cubic meters of air can provide more than 533 N of buoyant force.

SUMMARY

This chapter has more clearly defined the concept of force, its application, and the various properties of materials that pertain to force application. Force is necessary to cause changes in motion, and each force is paired with an equal and opposite reaction force. Only external forces can serve to change the motion of the entire system, whereas internal forces change the shape of a system. Gravity is the most pervasive force in our lives because it constantly acts on us. Because of gravity and other forces, objects are often in contact. Whenever objects are in contact, there is the potential for them to slide across one another. Friction is the force of concern in any situation during which sliding could occur. Also, when forces are applied, there is the potential for pressure, stress, and strain. Biological materials vary in their ability to withstand stress and strain and also vary in their response to different rates of stress and strain. If the system is in a fluid medium, it will be subjected to fluid forces. If immersed in a fluid, the system will be buoyed up by buoyant force. If the system is moving through a fluid, it will be subjected to the dynamic fluid forces of drag and lift. The resultant motion caused by a force can be motive or resistive. Force application and material properties have implications of human performance, injury, and prosthetic design.

REVIEW QUESTIONS

1. When we exert a force on Earth, its mass is too large to allow our segment (leg) to move away from the body. Therefore, what must the body be doing in relation to the segment in order for us to move?
2. How is it possible to lie on a bed of nails without injury?
3. Why are spike-heeled shoes not as effective as snowshoes for walking on snow without sinking?
4. Explain the difference between stress and strain. Also explain how they are used in the calculation of Young's modulus and discuss the interpretation of that modulus value.
5. Explain viscoelasticity and its relationship to loading rate.
6. How does loading rate affect the safety of exercise performance?
7. Give the equation for determining the coefficient of elasticity and then put it into your own words.
8. What would the coefficient of restitution be for a rubber ball compared to a ball of clay?
9. Friction depends on what two factors? (Do not just give the pieces of the equation above; define the pieces of the equation for this answer.)

10. What does it mean if the coefficient of friction for two surfaces is equal to 0.0? What if it is equal to 1.0?
11. The coefficient of friction for a synovial joint is 0.001. Why is this fact important?
12. How can friction be both resistive and motive during running?
13. In terms of friction, why is it often easier to pull rather than push an object?
14. What provides the centripetal force in a centrifuge?
15. What is the cause of the centripetal force in the case of riding a bike in a circle?
16. What happens if a body's weight is greater than the weight of the fluid it displaces, and what is this relationship called? What happens if a body's weight is less than the weight of the fluid it displaces? If they are equal?
17. If the whole body is the system, of what type are muscle forces?
18. What type of force is the only one that can cause motion of a system?
19. If the above answers are true, then how does the body move (such as in walking)?
20. When a diver leaves the board, muscle forces can only serve to change her or his ________, because they are internal to the system.

PRACTICE PROBLEMS

1. If the concrete block weighs 200 N and the coefficient of friction for concrete on wood is 0.80, what is the force of static friction at impending motion?
2. If a person weighs 600 N, and the surface area of the bottom of both feet combined is 200 cm^2, what is the pressure under the person's feet? What is the pressure if the person stands on only one foot?
3. What is the stress on the Achilles tendon if 4,700 N of force are applied and the cross-sectional area of the Achilles tendon is 90 mm^2?
4. If the tendon in the previous problem has a length of 8 cm in the resting state and a Young's modulus of 1,044.4 N/cm^2, what will be the strain and the absolute change in length when the force is applied?
5. Three balls are dropped from a height of 3 m. The balls have coefficients of restitution of (a) 0.9, (b) 0.6, and (c) 0.3. How high is the rebound for each ball?
6. A baseball traveling 25 m/sec strikes an area on a bat that is traveling 30 m/sec. The velocity of the ball off of the bat is 40 m/sec, and the velocity of the bat after the impact is 10 m/sec. Calculate the coefficient of restitution for this particular bat and ball.

EQUATIONS

Coefficient of static friction	$\mu_s \leq \frac{f_s}{F_n}$	(4.1)
Static friction force	$f_s \leq \mu_s F_n$	(4.2)

Coefficient of kinetic friction $$\mu_k = \frac{f_k}{F_n} \quad (4.3)$$

Kinetic friction force $$f_k = \mu_k F_n \quad (4.4)$$

Pressure $$P = \frac{F}{A} \quad (4.5)$$

Stress $$\sigma = \frac{F}{A} \quad (4.6)$$

Strain $$\varepsilon = \frac{\Delta l}{l_i} \quad (4.7)$$

Poisson's ratio $$\nu = \frac{\varepsilon_t}{\varepsilon_a} \quad (4.8)$$

Elastic modulus $$E = \frac{\sigma}{\varepsilon} \quad (4.9)$$

Young's modulus $$Y = \frac{F/A}{\Delta l/l_i} \quad (4.10)$$

Coefficient of restitution $$e = \sqrt{h_{rebound}/h_{drop}} \quad (4.11)$$

or

$$e = \frac{\nu_{af} - \nu_{bf}}{\nu_{bi} - \nu_{ai}} \quad (4.12)$$

or

$$-e = \frac{\nu_{af} - \nu_{bf}}{\nu_{ai} - \nu_{bi}} \quad (4.13)$$

REFERENCES AND SUGGESTED READINGS

Cajori, F. 1934. *Sir Isaac Newton's mathematical principles.* Trans. Andrew Motte, 1729. Berkeley: University of California Press.

Graham, G., S. A. Holt-Hale, & M. A. Parker. 2013. *Children moving* (9th ed.). Reston, VA: McGraw-Hill.

Grigoriev, I. S., & E. Z. Meilikhov (Eds.). 1997. *Handbook of physical quantities.* Boca Raton, FL: CRC Press.

Haywood, K. M., & N. Getchell. 2014. *Life span motor development*, 6th ed. Champaign, IL: Human Kinetics.

Ivancic, P. C. 2012. Biomechanics of sports-induced axial-compression injuries of the neck. *Journal of Athletic Training*, 47(5): 489–497.

McArdle, W. D., F. I. Katch, & V. L. Katch. 2014. *Exercise physiology: Energy, nutrition, and human performance*, 8th ed. Alphen aan den Rijn, Netherlands: Wolters Kluwer.

Mow, V. C., A. Ratcliffe, & A. R. Poole. 1992. Cartilage and diarthrodial joints as paradigms for hierarchical materials and structures. *Biomaterials*, 13: 67–97.

Newell, K. M. 1986. Constraints on the development of coordination. Pp. 341–361 in M. G. Wade & H. T. A. Whiting (Eds.), *Motor development in children: Aspects of coordination and control.* Amsterdam: Martin Nijhoff.

Nigg, B. M., B. R. MacIntosh, & J. Mester (eds). 2000. *Biomechanics and biology of movement.* Champaign, IL: Human Kinetics.

Nightingale, R. W., J. H. McElhaney, W. J. Richardson, T. M. Best, & B. S. Myers. 1996. Experimental impact injury to the cervical spine: Relating motion of the head and the mechanism of injury. *Journal of Bone and Joint Surgery (American)*, 78(3): 412–421.

Oatis, C. A. 2017. *Kinesiology: The mechanics and pathomechanics of human movement*. Baltimore: Wolters Kluwer.

Rossi, A. M., T. L. Claiborne, G. B. Thompson, & S. T. Todaro. 2016. The influence of friction between football helmet and jersey materials on force: A consideration for sport safety. *Journal of Athletic Training*, 51(9): 701–708.

Serway, R. A., & J. W. Jewett Jr. 2019. *Physics for scientists and engineers with modern physics*, 10th ed. Boston: Cengage Learning.

Serway, R. A., & C. Vuille. 2018. *College physics*, 11th ed. Boston: Cengage Learning.

Skinner, A. T., & A. M. Thompson (Eds.). 1983. *Duffield's exercises in water*. London: Bailliere Tindall.

Swartz, E. E., R. T. Floyd, & M. Cendoma. (2005). Cervical spine functional anatomy and the biomechanics of injury due to compressive loading. *Journal of Athletic Training*, 40(3): 155–161.

Winter, D. A. 2009. *Biomechanics and motor control of human movement*, 4th ed. Hoboken, NJ: John Wiley & Sons.

CHAPTER 5

LINEAR MOTION OF THE SYSTEM

LEARNING OBJECTIVES

1. Define linear kinematics.
2. Explain the relationship of linear kinetics to Newtonian law.
3. Understand the relationship between linear kinetics and energy transfer.

CONCEPTS

5.1 Linear Kinematics

5.2 Linear Kinetics and Newtonian Laws

5.3 Linear Kinetics and Energy Transfer

CONNECTIONS

5.4 Human Performance and Injury Science

5.5 Motor Behavior

5.6 Pedagogy

5.7 Adapted Motion

The skip points out the exact spot where she'd like the stone to arrive. The lead, second, and third nod and plan their strategy. Using the proper weight and curl, the skip sends the hammer toward the house. The plan is a double takeout of opponents' stones while leaving the hammer on the button. As the stone careens toward the house, the sweepers keep it on course by listening to the skip's commands. The hammer crashes into the first opposing stone at an acute angle and sends it out of the house while maintaining velocity toward the second stone. It collides with the second stone directly in the center, sending it out of the house and stopping the hammer in the house to earn a point.

If the previous paragraph made sense to you, it's likely that you're a fan of curling, a sport played on ice using granite stones. The ice surface provides a nearly frictionless surface, and the heavy stones display consistent direction changes through elastic collisions. Elite curlers understand the nature of these collisions and propel the stones at different velocities (weight) and spin rates (curl), depending on the nature of the shot. To see an animation of this scenario, visit the Curling Basics website (https://www.curlingbasics.com/en/double-takeout.html). This chapter will help you understand why the stones react as they do during a curling match.

This chapter focuses on linear kinematics and kinetics such as velocity, acceleration, momentum, work, power, and energy. It also covers situations in which objects collide. You are probably familiar with most of those terms and have certainly witnessed collisions. However, linear kinematics and kinetics are important to various disciplines in different ways. For example, knowledge of linear kinetics can help an athletic trainer understand concussion. We can also gain knowledge of why strength and conditioning specialists should pay attention to the playing surface. Principles in this chapter can help teachers and coaches understand the physical capabilities of students of different ages. We can also understand the design of various pieces of safety equipment such as airbags, helmets, boxing gloves, and shoes. Boxing gloves and shoes are safety equipment? We will see.

CONCEPTS

5.1 LINEAR KINEMATICS

In Chapter 1, you learned that a system of interest may be analyzed from two perspectives: kinetics and kinematics. *Kinematics* is the study or description of the *spatial* (direction with respect to the three-dimensional world) and *temporal* (motion with respect to time) characteristics of motion without regard to the causative forces. Issues such as displacement, velocity, direction, and acceleration are all kinematic concepts of interest when describing motion of the system. In this particular chapter, we are concerned with *linear* kinematics. Recall that linear motion (or translation) is motion along an axis (such as the x, y, or z axis) in which all points of the system move (change position) at the same time, in the same direction, and the same distance with respect to the defined reference frame despite their location on the system. Also remember that the linear path of a system in translation can be straight (rectilinear motion) or curved (curvilinear motion).

Linear Distance and Displacement

In Chapter 2, you read that we can specifically define the position or location of an object (the system) in space by establishing one or more frames of reference within a Cartesian coordinate system. The location of a point in a plane formed by two axes requires two coordinates, and three coordinates are required for a point in space. The specific position of the system gives us a starting point from which to measure various kinematic parameters. One of those parameters is linear distance traveled. **Linear distance traveled** (l) is simply the total *length* of the path traveled by the system of interest. Linear distance is described in terms of linear measurement units (e.g., millimeters, centimeters, meters, and kilometers). For example, we may have the piece of information that a soccer player traveled 30 m. However, notice that direction is not specified. Because we only have a magnitude, linear distance is a scalar quantity.

In contrast, **linear displacement** (d or Δp) is the change in linear position of the system in a *straight line.* Linear displacement is also described in terms of linear measurement units, and the terms *linear distance traveled* and *linear displacement* are often used synonymously.

Linear distance traveled The total length of the path traveled by the system of interest.

Linear displacement The change in linear position of the system in a straight line.

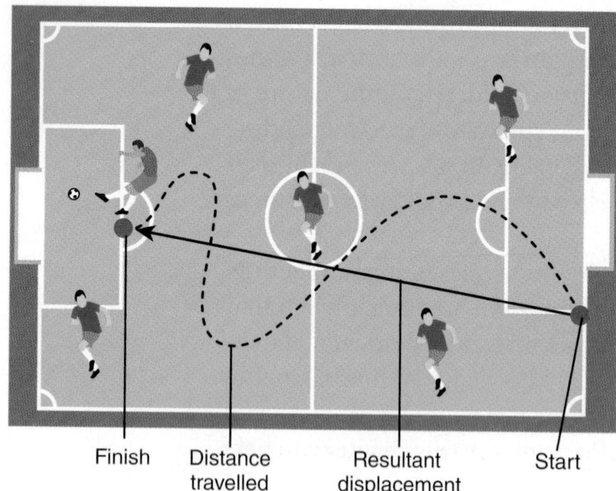

Figure 5.1 **Displacement of a soccer ball.** Notice the difference between linear distance traveled and resultant displacement.

However, linear displacement is in a specified direction from the initial position to the final position of the system (i.e., the shortest distance between the two points) and is therefore a vector quantity. The length of the vector represents total linear displacement in a straight line (magnitude), and the tip (arrowhead) of the vector represents the resultant direction of motion. Remember that, as a vector quantity, the displacement vector can be resolved into total distance traveled in the *x* direction and total distance traveled in the *y* direction.

As you can see in Figure 5.1 that linear distance traveled (30 m in our soccer player example) may be greater than displacement (maybe only a total displacement of 10 m). Also note that the two quantities may be equal, but displacement (being a straight-line distance) will never be greater than distance traveled.

Linear Speed and Velocity

In most situations, simple distance traveled or even displacement is not an especially useful quantity. We often want to know how fast a system changes position—that is, the *rate of motion*. To calculate the rate of motion, we need to know the interval during which the position change took place. However, we must also consider whether or not we are concerned with the interval of a linear distance traveled (l) or the interval of a linear displacement (d). If our only pieces of information are linear distance traveled and the duration, then we are technically calculating speed. **Speed** is the scalar rate of motion and is calculated as follows:

$$s = \frac{l}{\Delta t} \tag{5.1}$$

where

s = the speed, or rate of motion
l = length (or linear distance traveled)
Δt = time$_{final}$ − time$_{initial}$ (the change in time, or interval)

The units for describing speed are a unit of length divided by a unit of time. So if we know that the 30 m covered by a soccer player took 10 sec, then the player's speed was 3 m/sec.

However, if we know the linear displacement of the soccer player, velocity can be calculated. **Velocity** is the vector rate of motion, or rate of motion in a specific direction:

$$v = \frac{d}{\Delta t} \tag{5.2}$$

where

v = the velocity, or rate of motion in a specific direction
d = position$_{final}$ − position$_{initial}$ (displacement, or change in position)
Δt = time$_{final}$ − time$_{initial}$ (the change in time, or interval)

Speed The scalar rate of motion.

Velocity The vector rate of motion, or rate of motion in a specific direction.

The units for describing velocity are also a unit of length divided by a unit of time. In our soccer example, the linear displacement was 10 m, so linear velocity would be 1 m/sec if the interval was 10 seconds.

We should also consider that the soccer player may not have had the same rate of motion throughout the position change. There was probably a period of speeding up and a period of slowing down (or several changes, depending on the situation). During these times, the rate of motion would not be as great as the **peak rate of motion**, or maximum rate of motion achieved. In other words, the rate of motion is not necessarily a constant throughout the movement. So if $time_{initial}$ in our equation is 0 sec (the start time) and $time_{final}$ is 10 sec (the total elapsed time during the motion change), we are actually calculating an **average speed** or **average velocity**. We may be interested in the rate of motion at one given instant in time, **instantaneous speed** or **instantaneous velocity**. For example, our soccer player may have traveled 8 m of the total 30 m in the time between second 4 and second 8 ($\Delta t = 4$ sec). During that particular interval, the rate of motion was 2 m/sec (different from both our average speed and average velocity calculated in Equations 5.1 and 5.2). So a kinematic value such as the peak rate of motion is an instantaneous speed or instantaneous velocity measured during the period in which the greatest distance was covered in the shortest time.

Now we need to turn our attention to the kinematic value that describes the changes in rate of motion observed during movement.

Linear Acceleration

We have defined the scalar quantities of linear distance traveled and speed as well as their vector counterparts, displacement and velocity. We have also discussed that instantaneous velocity is not necessarily the same as average velocity. This difference implies that velocity may be changing throughout the motion. In other words, the system goes through periods of speeding up and slowing down, and these periods may be associated with changes in direction of the system. Previously, we stated that acceleration is the term for a change in state of motion. More specifically, **acceleration** is a change in magnitude or direction (or both) of the velocity vector with respect to time. Mathematically, acceleration is calculated as:

$$a = \frac{\Delta v}{\Delta t} \tag{5.3}$$

where

a = the acceleration, or rate of change in linear velocity
$\Delta v = velocity_{final} - velocity_{initial}$ (change in velocity)
$\Delta t = time_{final} - time_{initial}$ (the change in time, or interval)

The units for describing linear acceleration are a unit of length divided by a unit of time divided by a unit of time. For example, in 2 seconds ($time_{initial} = 0$ sec and $time_{final} = 2$ sec) a cheetah can change its state of motion from a complete standstill ($v_{initial} = 0$ m/sec) to a velocity of approximately 18 m/sec (v_{final}). This ability means that the cheetah is capable of accelerating 9 m/sec/sec or 9 m/sec^2. We specifically used the example of a cheetah to make a point. In many instances (e.g., sports and predator–prey relationships), peak velocity is not the most important issue; rather, how fast you can *change* your rate of motion is often the difference between success and failure (i.e., how fast a sprinter gets out of the blocks is sometimes more important than the sprinter's top velocity).

One should notice from the equation that acceleration can be equal to zero ($velocity_{final} = velocity_{initial}$), a positive number ($velocity_{final} > velocity_{initial}$), or a negative number ($velocity_{final} < velocity_{initial}$). If acceleration is equal to zero, it doesn't mean that the system is no longer in motion,

Peak rate of motion Maximum rate of motion achieved.

Average speed or average velocity The average rate of motion.

Instantaneous speed Speed at one given instant in time.

Instantaneous velocity The rate of motion at one given instant in time.

Acceleration A change in magnitude or direction (or both) of the velocity vector with respect to time.

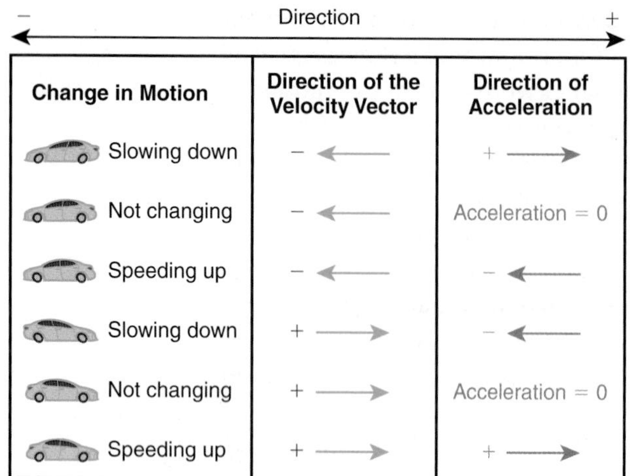

− Direction +

Change in Motion	Direction of the Velocity Vector	Direction of Acceleration
Slowing down	− ←	+ →
Not changing	− ←	Acceleration = 0
Speeding up	− ←	− ←
Slowing down	+ →	− ←
Not changing	+ →	Acceleration = 0
Speeding up	+ →	+ →

Figure 5.2 **Direction of motion versus direction of acceleration.**

only that the system is not changing its current state of motion. For example, your acceleration is equal to zero when you are driving and the speedometer in your car is reading a constant velocity. Your car is still in motion, but there is no acceleration because velocity is constant. If acceleration is positive, then the system is changing its rate of motion to a greater value (speeding up). In the cheetah example, the animal changed from a velocity of 0 m/sec to a velocity of 18 m/sec. In the case of negative acceleration (such as pushing the brake pedal of your car), the system is changing its rate of motion to a lesser value (slowing down). Some refer to negative acceleration as *deceleration*. The term is acceptable, but always remember that any change in state of motion is acceleration.

Also be aware that because acceleration has a magnitude and occurs in a specific direction, it is a vector quantity. However, do not confuse the direction of system motion with the direction of acceleration. Let's use driving your car as an example (Figure 5.2). If you increase the velocity of your car while driving in the positive *x* direction, then the direction of both system motion and acceleration is positive. If you then slow down but are still traveling in the positive *x* direction, then the direction of system motion is still positive but the acceleration is negative. Acceleration is negative because $\text{velocity}_{\text{final}}$ is less than $\text{velocity}_{\text{initial}}$. Simple enough, but this issue can become confusing. Recall that velocity has a direction and is on the top of the acceleration equation. So if you stop the car and begin to accelerate in the negative *x* direction, the direction of system motion is negative and acceleration is still negative even though you are speeding up. In this case, acceleration is negative because $\text{position}_{\text{final}}$ is less than $\text{position}_{\text{initial}}$ (along the *x*-axis) in the velocity equation, so $\text{velocity}_{\text{final}}$ in the acceleration equation is negative. One more example: Let's say you now begin to slow down while traveling in the negative *x* direction. Now $\text{velocity}_{\text{final}}$ is positive because $\text{position}_{\text{final}}$ is greater than $\text{position}_{\text{initial}}$ in the velocity equation. $\text{Velocity}_{\text{initial}}$ is negative because that was the previous direction of travel. So the direction of system motion is still negative, but the resultant acceleration is positive.

Finally, acceleration (like speed and velocity) is not always a constant value. It can change over the interval in which it is measured. Therefore, **average acceleration** is the rate of change in velocity divided by the entire interval over which it changed; the **instantaneous acceleration** is the rate of change in velocity at one specific instant in time.

5.2 LINEAR KINETICS AND NEWTONIAN LAWS

Now that we have discussed the linear *kinematics* of system motion, we must introduce the linear *kinetics* of system motion. *Kinetics* is the study of forces that inhibit, cause, facilitate, or modify motion of a body. The forces discussed in Chapter 4 (such as gravity and friction) are all of interest in studying the *kinetics* of system motion. Again, in this chapter our attention will be focused on kinetics during linear motion.

Average acceleration The rate of change in velocity divided by the entire interval over which it changed.

Instantaneous acceleration The rate of change in velocity at one specific instant in time.

Application of the Newtonian Laws

When we speak of kinematics, we must always remember that each value is simply a measurement of the motion exhibited by the system as a result of force application (or the lack of force application). The forces involved are studied in the realm of kinetics. Newton attempted to explain the behavior of systems in the absence and presence of forces, so we return to Newton's laws to better understand the resultant motion of systems (Cajori, 1934; Serway & Jewett, 2019; Serway & Vuille, 2018).

Newton's First Law: Law of Inertia

Every body perseveres (remains) in its state of rest, or of uniform motion in a right line, unless it is compelled to change that state by forces impressed thereon. According to our kinematic values above, this law means that if an object is at rest, it will not undergo linear displacement without the application of an external force. Without linear displacement, velocity is equal to zero. If the object is already in motion and traveling at a given velocity, there is no change in that linear velocity without an externally applied force. If linear velocity does not change, then the system does not accelerate. So Newton's first law helps us understand the relationship between kinematics and kinetics. Without the forces (kinetics) discussed in Chapter 4, the spatial and temporal characteristics of motion (kinematics) remain constant.

Remember that this law explains what happens to a system in the absence of applied forces. So in most *biomechanical* situations, technically no direct application of Newton's first law is possible because we are always being acted upon by the force of gravity. In addition, gravity and air resistance both act on the system when it is airborne. However, if we look at these two forces in isolation, or ignore them altogether, we can develop a deeper understanding of Newton's first law as it applies to the linear motion of the system. First, let's use the example of jumping from a diving board. We know that gravitational force acts downward on the system and that air resistance acts upward on the system. But in the case of jumping from a diving board, no other *external* forces act on the system. Let's also assume that the system is a human. Any movements that the system makes while airborne are caused by *internal* forces. Remember that internal forces can only serve to change the shape of the system. Without the application of any other external forces, the current state of motion of the diver will be maintained; that is, no matter what movements of the body are made, the diver will continue falling in a straight line toward Earth until she makes contact with the water. The water then applies a force to the diver, which changes the state of motion.

Another example is that of projectile motion. If a person applies a force to project an object through the air, both gravity and air resistance act immediately on that object, causing it to slow and eventually fall to Earth. These forces account for the curvilinear path of the projectile downward to Earth. But if we ignore gravity and air resistance (i.e., no forces are present), we would see an entirely different path of projectile motion. According to Newton's first law, the motion of the projectile would be maintained in its current state and the projectile would travel in a straight line with a constant velocity and never fall to Earth. Of course we know that forces are acting on the projectile, but the example does help to illustrate the relationship of Newton's first law to system motion. Even though these forces are present, many athletes will experience Newton's first law on a regular basis (normally in a negative context). For example, many pitchers, quarterbacks, and kickers have uttered the phrase "I wish I had that one back!" This statement is an admission that once the projectile leaves the athlete's hand or implement, nothing can be done to change its path. Once contact is lost with the projectile, no further force can be applied to accelerate or change the direction of projectile motion.

However, keep in mind that Newton's first law explains what happens in the *absence* of externally applied forces. So these examples are not direct applications of Newton's first law. They merely serve to illustrate what happens in the absence of forces *other than* gravity and air resistance. Let's examine what happens when forces are applied.

Newton's Second Law: Fundamental Law of Dynamics

The alteration of motion is ever proportional to the motive force impressed and is made in the direction of the right line in which that force is impressed. In this case, the system's state of motion changes. Therefore, according to the first law there must have been an applied force. The observed change in motion

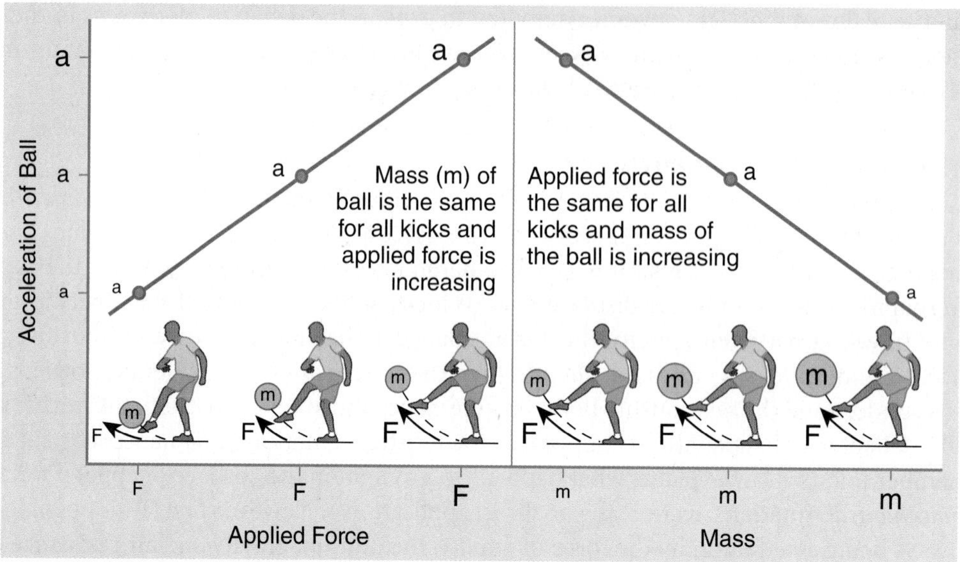

Figure 5.3 **Relationship of force, mass, and acceleration.** What does this relationship have to do with the size of offensive linemen in football?

is called *acceleration*, which we defined mathematically in Equation 5.3. Newton's second law is stating that the acceleration (change in state of motion) of the system is directly proportional to the sum of the forces (or net force) acting on it and inversely proportional to the mass of the system to which the force is applied (Figure 5.3).

Mathematically, the relationship of applied force and the resulting acceleration can be stated as follows:

$$a = \frac{\Sigma F}{m} \tag{5.4}$$

where

a = the acceleration, or change in motion of the system
ΣF = the vector sum of the forces applied to the system
m = the mass of the system to which the force is applied

If we multiply through by m, Equation 5.4 becomes:

$$\Sigma F = ma \tag{5.5}$$

where

ΣF = the vector sum of the forces applied to the system
m = the mass of the system to which the force is applied
a = the acceleration, or change in motion of the system

Remember that forces are vector quantities. As such, Equation 5.5 can be resolved into three component equations:

$$\Sigma F_x = ma_x \qquad \Sigma F_y = ma_y \qquad \Sigma F_z = ma_z$$

Simply stated, multiple forces may be acting in the x direction, the y direction, or the z direction singly or in combination. The *resultant* acceleration of an object is the vector sum of all of the forces acting on the system (i.e., many forces act simultaneously to produce one resultant acceleration). Let's return to projectiles as an example. Once the projectile has left the hand of the athlete, it exhibits a curvilinear path. If we resolve that path into its component vertical and horizontal vectors, we can explain the path of the projectile with Newton's second law of motion. The parallel component of dynamic fluid force acts in the opposite direction of system motion with respect to the fluid and is called *drag force* (see Chapter 4). So drag force is acting in the negative *x* direction with respect to projectile motion. Therefore, the sum of the forces in the *x* direction is acting to decrease the ball's velocity (deceleration). Simultaneously, gravity is acting in the negative *y* direction to accelerate the ball toward Earth. The resultant motion of the projectile is the vector sum of all of the forces in the *x*, *y*, and *z* directions. So the ball slows down horizontally and is accelerated toward Earth vertically, creating the observed curvilinear path (Figure 5.4).

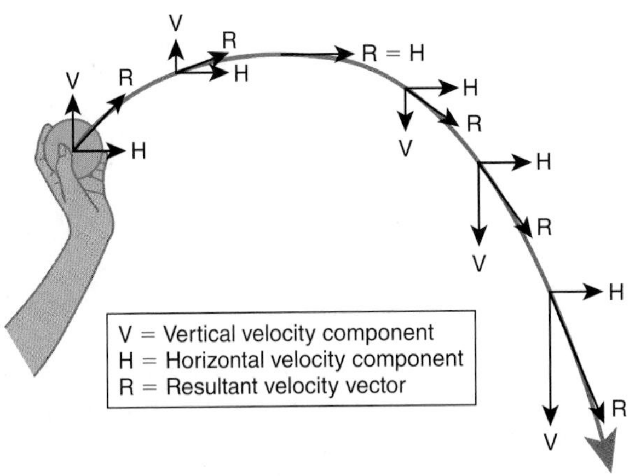

Figure 5.4 **Motion of a projectile.** What forces cause the path to be curvilinear?

Note that the forces summed as vectors in Equation 5.5 are forces that we have already discussed; the product of mass and acceleration do not create a force. The acceleration of the system is directly proportional to the vector sum of the applied forces and inversely proportional to the mass. So, for example, *ma* in Equation 5.5 simply means that a large acceleration of a large mass must have been caused by a large force application. Also observe that Equation 5.5 represents an *instantaneous* acceleration. In other words, the net forces at that instant in time are summed as vectors to create a single instantaneous acceleration. Remember from our previous sections that velocity and acceleration are not necessarily constant over a given interval. The reason for the variations in acceleration is that the net applied force is not necessarily a constant value.

Remember that Newton's first law explains motion in the absence of force application. Notice that Newton's second law explains situations in which a system is under the influence of forces but in equilibrium. A system in **equilibrium** is either at rest or moving with a constant velocity. According to Equation 5.5, the sum of the forces in this situation is equal to zero. More specifically, an object at rest is exhibiting no linear displacement and therefore no velocity or acceleration. An object moving with a constant velocity also has no acceleration. In both cases acceleration is equal to zero. Therefore, in accordance with Equation 5.5, there must be no *net* force ($\Sigma F = 0$). So even though a system is not in motion, it does not mean that no force has been applied.

Now that we have discussed the application of force, we must further elaborate on the idea that forces are paired.

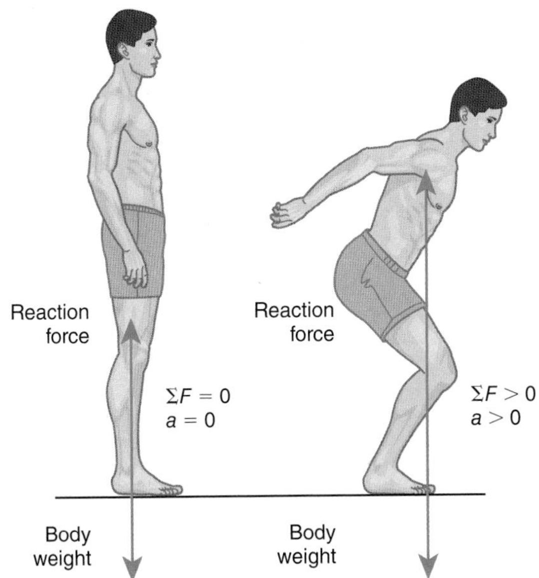

Figure 5.5 **The vertical jump.** What causes the net force to propel the jumper forward?

Newton's Third Law: Law of Reciprocal Actions

To every action there is always opposed and equal reaction; or the mutual actions of two bodies on each other are always equal and directed to contrary parts. Newton was simply stating that forces exist in pairs. To further illustrate this point, we can return to the idea of systems that are in equilibrium. A system in equilibrium is experiencing applied forces that sum to zero. A person standing still in preparation to perform a vertical jump is in equilibrium (Figure 5.5).

The force of gravity acting to accelerate the person in the negative *y* direction is balanced with an equal and opposite reaction force in the positive *y* direction provided by Earth. So the sum of the forces is equal to zero,

Equilibrium State in which the system is either at rest or moving with a constant velocity.

and no acceleration occurs. If our person then generates muscle forces to jump, both systems (the person and Earth) experience that force to the same degree. Because the jumper is accelerated in the positive y direction, we know there must be a net force. What causes the net force? The equal and opposite force of Earth in response to the force applied by the jumper is greater than the force of gravity, so our jumper is temporarily accelerated in the positive y direction. Once the jumper loses contact with the ground, only the force of gravity is acting. So the net force is then in the negative y direction, and the jumper is accelerated back toward Earth by the force of gravity. Once the jumper is back in contact with the ground, Earth applies an equal and opposite force to that of gravity and equilibrium is returned.

Don't forget that Newton's second law of motion is also at work in this example. Remember that acceleration is inversely proportional to the mass of the system to which force is applied. So even though the jumper and Earth experience the same force, the resultant acceleration of Earth is much less than that of the jumper because of its extremely large mass. As a result, the muscular forces that are attempting to move the segments of the leg away from the torso are actually redirected and cause motion of the torso away from the segments of the leg. In essence, the feet are attempting to move away from the body, but Earth is too massive to be pushed out of the way. So the body, possessing much less mass than Earth, is accelerated away from the feet. This example helps illustrate a couple of concepts. Forces are paired in that one force is met with an equal and opposite reaction force. However, because of variations in mass, equal and opposite forces can produce unequal reactions (accelerations) in accordance with Newton's second law.

Newton's Law of Universal Gravitation

Every body in the universe attracts every other body with a force directed along the line of centers for the two objects that is directly proportional to the product of their masses and inversely proportional to the square of the separation between the two objects. With all of the preceding concepts in mind, we now can become more specific about gravitational force. Remember that Newton was attempting to explain gravity mathematically. The algebraic representation of the law is as follows:

$$F = \frac{Gm_1m_2}{r^2} \tag{5.6}$$

where

F = the magnitude of the gravitational force
G = the universal gravitational constant (measured experimentally)
m_1 and m_2 = the masses of the two bodies (in our case, some object and Earth)
r^2 = the distance between the centers of mass of the two bodies (i.e., the distance from Earth's center to its surface)

We have mentioned that the *weight* of an object is a measure of the gravitational force acting on that object's mass. We have also discussed Newton's second law, which states that the acceleration (change in motion) of a body is proportional to the force impressed. In Equation 5.6, all of the factors are constants except the mass of the object interacting with Earth. In addition, Earth's gravitational force causes the object to accelerate toward it. Therefore, we can define another factor using the existing constants:

$$g = \frac{Gm_2}{r^2} \tag{5.7}$$

where

g = the acceleration of the object caused by application of Earth's gravitational force
G = the universal gravitational constant (measured experimentally)
m_2 = the mass of Earth
r^2 = the distance from Earth's center to its surface

Using Equation 5.7, Equation 5.6 can be restated as:

$$F = mg \tag{5.8}$$

where

F = the magnitude of the gravitational force
m = the mass of the body interacting with Earth
g = the acceleration of the object caused by gravitational force

In addition, because weight is a measure of gravitational force, Equation 5.8 can be used to calculate the weight of an object (i.e., *weight* = *mg*). For example, an object that has a mass of 1 kg weighs approximately 9.81 N (1 kg × 9.81 m/sec^2 = 9.81 kg · m/sec^2 = 9.81 N). The acceleration of an object caused by Earth's gravitational force (g) is measured experimentally, and for most situations it can be assumed to be a constant value of 9.81 m/sec^2 (or 9.81 m/sec/sec) downward. This means that if no other forces act on a body, it will fall at a rate of 9.81 m per second for the first second and then fall an additional 9.81 m per second faster for each additional second (i.e., it will be falling at 19.62 m/sec after two seconds, etc.). Recall that the Système International (SI) unit of *force* is the newton (N). Because weight is a measure of gravitational *force*, weight is also measured in newtons. Referring to Equation 5.7, you can see that 1 kg of mass weighs approximately 9.81 N. Keep in mind that the assumed constant of 9.81 m/sec^2 changes as one travels to differing elevations because r^2 in Equation 5.7 changes.

Linear Momentum and Linear Impulse

In common language, we often use phrases such as "gaining momentum" or "a shift in momentum." These phrases are often used synonymously with the difficulty level of stopping or prevailing over an object, a person, or a group of people. Newton did not use the term **momentum**; instead he referred to a system's quantity of motion. The greater the quantity of motion possessed by a system, the more difficult it is to stop that motion. In technical terms, more force would have to be applied to stop a system that possesses a large amount of momentum. Momentum can be gained in a couple of ways. Most people would agree that a charging elephant has a lot of momentum. One obvious reason for this is the size of the elephant. So we know that mass is one factor in momentum. However, we can also list some objects of small mass that possess large amounts of momentum: a bullet shot from a gun or a baseball thrown by a major league pitcher. In both cases, the mass isn't large but the objects are traveling with great velocity. So velocity is the other factor in momentum. Therefore, linear momentum (M; designated p in some disciplines) can be calculated as follows:

$$M = mv \tag{5.9}$$

where

M = the momentum of the system (or quantity of motion)
m = the mass of the system in motion
v = the linear velocity of the system

Momentum is expressed in units of mass multiplied by units of velocity (kg · m/sec). So a system having a mass of 70 kg and traveling at 4 m/sec has a linear momentum of 280 kg · m/sec. To illustrate this point, a system with a mass of 4 kg traveling at 70 m/sec also has a momentum of 280 kg · m/sec. You can see from Equation 5.9 that an object at rest (v = 0 m/sec) has no momentum. In addition, momentum is a vector quantity and has the same direction as the system's velocity. It can also be reasoned that in most movement situations mass is constant, so changes in momentum are normally the result of changes in velocity. In most situations, those changes in momentum are of particular interest. Changes in momentum are directly related to Newton's first two laws of motion.

Momentum A system's quantity of motion.

Linear Momentum and Newton's First Law

Considering momentum in the context of Newton's first law of motion, especially as it applies to systems colliding, is important. According to Newton's first law, there will be no change in state of motion without an externally applied force. An object in motion has both a mass and a velocity and therefore a given amount of momentum. If no change in state of motion is possible without an externally applied force, then no change in momentum can occur without an externally applied force. Therefore, Newton's first law may be restated as the **principle of conservation of linear momentum**:

> *In the absence of a net externally applied force, the total momentum of a system that comprises multiple bodies remains constant in time.*

One application of this principle is in predicting the outcome of collisions. If two objects collide, the total momentum of the system just before collision equals the total momentum just after collision. One of the keys to understanding the concept is a clear definition of the system. The *system* in this case must be made up of two or more bodies (e.g., a person jumping and Earth, two cars, a catcher in baseball and a ball, two rugby players). There is no guarantee that the momentum of one individual body in the system will be conserved because it is interacting with the other body or bodies in the defined system. For example, the momentum of a baseball changes drastically when it is caught. However, the total momentum of the catcher and the baseball remains the same because momentum of one body (pitcher) is gained as momentum of another body is lost (ball). Another important point to remember is that the law of conservation of linear momentum (similar to Newton's first law) applies in the *absence* of forces (i.e., isolated systems). Therefore, external forces such as air resistance, friction, and gravity are often ignored when discussing collisions in the context of the principle of conservation of linear momentum. Of course, these forces are present and do act to change the momentum to a certain degree during collisions.

Figure 5.6 **Bowling ball and pin.** Is this collision elastic or inelastic?

Because collision is such a great example of conservation of momentum, let's now be more specific about the phenomenon. Collisions can be classified into two general categories: elastic and inelastic. If two objects collide and bounce off of one another, the event is said to be an **elastic collision**. For example, when a bowling ball collides with a pin, the pin bounces off of the ball (Figure 5.6). The ball and the pin make up the system in this elastic collision. We know that in the absence of any external forces, the total momentum of the system is the same before and after the collision. However, this equivalence doesn't mean that the ball and the pin will be traveling at the same velocity as before the collision. The ball will travel more slowly, and the pin will increase its velocity.

Therefore, the momentum of one object is reduced and the momentum of the other object is increased proportionately to keep the total momentum of the system constant. Mathematically, the principle of conservation of linear momentum in an *elastic* collision can be defined as:

$$(m_1 v_{1i} + m_2 v_{2i}) = (m_1 v_{1f} + m_2 v_{2f}) \quad (5.10)$$

where

m_1 and m_2 = the masses of the objects involved in the collision

v_{1i} and v_{2i} = the initial velocities of the objects (velocities before impact)

v_{1f} and v_{2f} = the final velocities of the objects (velocities after impact)

Principle of conservation of linear momentum In the absence of a net externally applied force, the total momentum of a system that comprises multiple bodies remains constant in time.

Elastic collision A collision in which two objects collide and bounce off of one another.

Notice that the equation allows for differing velocities after the collision but implies that the total momentum before and after the collision will be constant. Again, remember that this relationship assumes the absence of external forces such as friction that can change momentum. However, the relationship holds well if we look at the collision immediately after impact (before other forces have time to significantly affect momentum).

Also recall Equation 4.12, in which the change in relative velocities of two objects that collide was used to calculate the coefficient of restitution:

$$e = \frac{v_{af} - v_{bf}}{v_{bi} - v_{ai}} \tag{4.12}$$

where

e = coefficient of restitution
v_{af} and v_{bf} = final velocities of the objects (after impact)
v_{bi} and v_{ai} = initial velocities of the objects (before impact)

Notice that Equation 4.12 simply uses the concept of conservation of linear momentum to calculate a material property. The masses of the objects are assumed to be constant, and the differences in velocities between the objects before and after collision are observed. If the objects in the collision are highly elastic, then their individual velocities will change less. Therefore, the final velocities will be similar to the initial velocities, and the coefficient of restitution will be closer to 1.00.

If the objects are not especially elastic, the coefficient will approach 0.00, and the collision will fall into our other broad category termed *inelastic*. In an **inelastic collision**, the objects collide and stick together. The key concept is that if the objects are stuck together after the collision, they will be traveling at the same velocity (e.g., a wide receiver in football successfully catching a pass). So in this case, total momentum will again be conserved and velocities will change. But the velocities of the objects after the collision will be equal. The mathematical relationship will change only slightly:

$$(m_1 v_{1i} + m_2 v_{2i}) = (m_1 + m_2) v_f \tag{5.11}$$

where

m_1 and m_2 = the masses of the objects involved in the collision
v_{1i} and v_{2i} = the initial velocities of the objects (velocities before impact)
v_f = the final velocity of the objects (velocity after impact)

We can see that momentum will still be conserved, but notice the change in this equation. The velocities of the objects before impact can be different values, but after impact they will share a common velocity because they are stuck together. Because of these known values (i.e., momentum is conserved, and velocity of the objects after an inelastic collision is a common value), the final velocity of two objects after an inelastic collision is a predictable value calculated by dividing through by $(m_1 + m_2)$:

$$v_f = \frac{m_1 v_{1i} + m_2 v_{2i}}{m_1 + m_2} \tag{5.12}$$

where

v_f = the final velocity of the objects (shared velocity after impact)
m_1 and m_2 = the masses of the objects involved in the collision
v_{1i} and v_{2i} = the initial velocities of the objects (velocities before impact)

Inelastic collision A collision in which two objects collide and stick together.

SAMPLE PROBLEM

A prime example that demonstrates Equations 5.11 and 5.12 is a person who jumps to catch a ball (assuming that he actually catches it; Figure 5.7).

For example, the person may have a mass of 70 kg (m_1) and the ball a mass of 1 kg (m_2). Before the catch, the person has jumped into the air but is not traveling horizontally. Therefore, the horizontal linear velocity of the person before the catch is 0 m/sec (v_{1i}). Let's say that the ball is traveling at 10 m/sec (v_{2i}). This would mean that the total momentum of the system before impact is:

$$[(70 \text{ kg})(0 \text{ m/sec}) + (1 \text{ kg})(-10 \text{ m/sec})] = -10 \text{ kg} \cdot \text{m/sec}$$

Notice that the sign indicates the direction of the total momentum of the system. We can then use Equation 5.12 to calculate the velocity that both parts of the system will be traveling after the catch:

$$v_f = \frac{[(70 \text{ kg})(0 \text{ m/sec}) + (1 \text{ kg})(-10 \text{ m/sec})]}{70 \text{ kg} + 1 \text{ kg}}$$

$$v_f = -0.1408 \text{ m/sec}$$

This result means that after the inelastic collision, the person and the ball will be traveling a common velocity of 0.1408 m/sec. Once again, the sign indicates the final direction of travel. Now let's insert this final velocity into Equation 5.11 to see if momentum was conserved:

$$[(70 \text{ kg})(0 \text{ m/sec}) + (1 \text{ kg})(-10 \text{ m/sec})] = (70 \text{ kg} + 1 \text{ kg})\ (-0.1408 \text{ m/sec})$$

$$-10 \text{ kg} \cdot \text{m/sec before collision} = -10 \text{ kg} \cdot \text{m/sec after collision}$$

So, one can see that the total linear momentum was conserved in the collision and the final velocity of two objects in an inelastic collision is predictable.

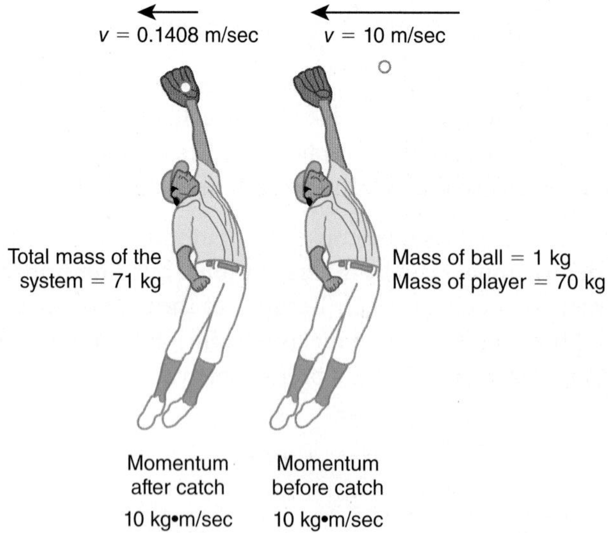

Figure 5.7 **Person catching a ball.** How could this situation be either elastic or inelastic?

Note that all collisions cannot be perfectly categorized as *elastic* or *inelastic*. In other words, two objects bounce off each other or stick together because of a collision to varying degrees. Also, don't forget the forces that we have ignored to create an isolated system. So the actual result of a collision in terms of velocity change and momentum conservation depends on the degree to which the objects stick together during the collision, the elasticity of the objects, friction (whether or not one or both of the objects are in contact with the ground), air resistance, and gravity.

Linear Momentum and Newton's Second Law

Once again, Newton's first law of motion assumes the absence of external forces. In the event that external forces are applied to the system, a relationship exists, of course, between linear momentum and Newton's second law of motion. First recall that Equation 5.5 ($\Sigma F = ma$) is an instantaneous acceleration. In other words, the net forces at that instant in time are summed as vectors to create a single instantaneous acceleration. Remember from our previous sections that velocity and acceleration are not necessarily constant over a given interval. The reason for the variations in acceleration is that the net applied force is not necessarily a constant value (i.e., you push a wheelchair 10 m but not with the same force at every given instant over the course of the entire 10 m). This also means that the amount of time that a given force is applied is not a constant value. For example, you might apply 1,000 N of force for 2 sec and then 1,500 N for 3 sec. So it is not usually the instantaneous acceleration that is of interest because an instantaneous acceleration doesn't really provide a large amount of information as to the result of the force application. We tend to place more importance on the outcome of an event—for example, how fast was the car going when it hit the wall? So, for most purposes we need to know the average net force and therefore the average acceleration caused by that force. Now what does that have to do with momentum?

Remember from Equation 5.3 that average acceleration is:

$$a = \frac{\Delta v}{\Delta t}$$

and (from Equation 5.5) that:

$$\Sigma F = ma$$

If we replace instantaneous acceleration in Equation 5.5 with average acceleration from Equation 5.3, we form Equation 5.13:

$$\Sigma F = \frac{m\Delta v}{\Delta t} \tag{5.13}$$

We can multiply both sides by Δt and form:

$$\Sigma F\,(\Delta t) = m\Delta v \tag{5.14}$$

or

$$\Sigma F\,(\Delta t) = m(\text{velocity}_{\text{final}} - \text{velocity}_{\text{initial}})$$

We have already defined momentum as mv. So with distribution we eventually come to what is known as the *linear impulse equation*:

$$\Sigma F\,(\Delta t) = \Delta M \tag{5.15}$$

where

ΣF = average net force applied to the system
$\Delta t = \text{time}_{\text{final}} - \text{time}_{\text{initial}}$ (interval of force application)
$\Delta M = \text{momentum}_{\text{final}} - \text{momentum}_{\text{initial}}$ (change in momentum caused by the applied force)

So **linear impulse** is the product of applied force and the interval of force application. The linear impulse equation is often written as $\Sigma F\,(\Delta t) = J$. We have left it in the form shown in Equation 5.15 as a reminder of what

Linear impulse The interval during which force is applied.

is being accomplished, a change in momentum. When you do see *J*, pay close attention to whether or not it means "impulse" or "joules." So the linear impulse equation is another expression of Newton's second law. There must be an applied force to accelerate a body (ΣF). That force must be applied for a certain time (Δt). The more massive the body (m), the more force must be applied to accelerate it (a). Acceleration is a change in velocity (v). Mass and velocity together represent the momentum (M) or quantity of system motion. If mass is constant and the velocity of the mass changes, then the mass is accelerated. That acceleration is caused by external force application. This takes us back to Newton's second law of motion. So we can infer that Newton was speaking of changing *momentum* when he referred to changing *motion*.

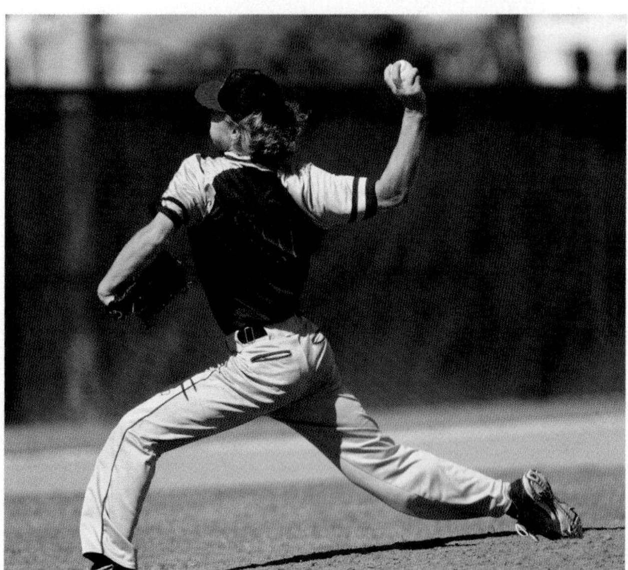

Figure 5.8 **Windup of a baseball pitcher.** Why is the windup needed?

It may seem to be a matter of semantics, but the implications of the linear impulse equation are quite far-reaching. First, the basics: the greater the force or time the force is applied, the greater the change in linear momentum of the object. Next we should draw attention to the fact that the impulse equation is of great importance in situations of both increasing and decreasing momentum. We tend to only think of accelerating objects (increasing their momentum). But a critical application of the impulse equation is decelerating objects (decreasing their momentum).

Let's start with an example of increasing momentum. Most sports involve at least some element of throwing or striking. In either case, applying a force to increase the linear momentum of an object is often the goal. Of course athletes train their muscles to enhance the ability to apply force. But because a person can apply only so much force, many athletes also manipulate the Δt factor of the impulse equation (i.e., a longer time over which the force is applied). For example, why do we use a windup before throwing a ball or a backswing before using an implement (Figure 5.8)? The windup of a pitch leaves the arm (and the ball) behind the body.

So force will be applied to the ball from the time it is behind the body until it is released. That period is longer than it would be if force application had begun from next to the body. Of course other factors are involved that will be discussed in Chapter 11 on muscle physiology. But one can see that maximal change in linear momentum of the ball is the result of maximal force application and maximal time of force application. We can use a mathematical example to further demonstrate this point.

SAMPLE PROBLEM

Let's say we have two people who are both capable of applying an average force of 40 N to a 1 kg ball during a pitch. Our first pitcher applies the average force of 40 N for 1.00 seconds. The resulting change in momentum (from 0 kg · m/sec) would be 40 kg · m/sec. If we divide the resulting momentum by the mass of the ball (1 kg), we find that the velocity of the ball at release will be 40 m/sec. If our second pitcher applies the same amount of force for only 0.50 seconds, the resulting change in momentum will be 20 kg · m/sec, and the final velocity of the ball will be 20 m/sec. What if our second pitcher wants to cause the same change in momentum as the first pitcher but still only apply force for 0.50 seconds? If we divide the desired change in momentum (40 kg · m/sec) by the allotted time (0.50 sec) to solve for force, we find that the second pitcher must apply 80 N (80 kg · m/sec/sec) of force to accomplish the same change in momentum. Once again, you can apply more force or apply force for a longer time to cause a given *increase* in momentum.

Increasing momentum is relatively straightforward. Now what about using linear impulse to decrease momentum? This application of the equation has numerous ramifications of which people are often not fully aware. What if I asked you to jump from a height of 3 meters? What would you do to make your landing more comfortable? Would you land stiffly, not allowing your hips or knees to bend on impact? Intuitively, most people would say the opposite: to "give" on impact. That is the most comfortable (and safest) way to land. The reason is found in the impulse equation.

SAMPLE PROBLEM

For example, our person who will be landing has a mass of 70 kg. Let's say that his velocity on landing is 9 m/sec. Therefore, he has a linear momentum of 630 kg · m/sec. The momentum will be changed to 0 kg · m/sec by the ground reaction force provided by Earth as the person lands. The key to comfort is how long it takes Earth to cause a change in momentum of 630 kg · m/sec. First, let's have the person land stiffly (Figure 5.9). This kind of landing means that the momentum change will be almost immediate (fractions of a second). Let's use an arbitrary value of 0.1 sec for the time of momentum change. If we solve for force by dividing the change in momentum by the allotted time, we find that Earth will apply 6,300 N of force to change the momentum of the person to 0 kg · m/sec in 0.1 sec. Now let's have our person "give" on landing so that it takes just a fraction of a second longer for Earth to change momentum to 0 kg · m/sec (Figure 5.9). The change in momentum will still be 630 kg · m/sec, but this time we will allow 0.2 sec for the change to take place. With even this slight change in time, we find that the person will only absorb an average force of 3,150 N. Compared to 6,300 N, that's quite a difference.

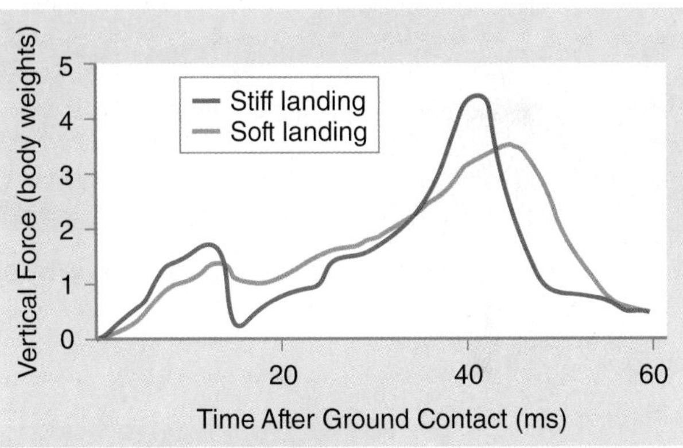

Figure 5.9 **Ground reaction force during stiff and soft landing.** Why does it help to "give" when landing?

Landing is not the only situation in which the linear impulse equation applies to decreasing linear momentum. The impulse equation changed the entire shoe industry. Would you jog on a regular basis in tap shoes? Why not?

Notice the soles of shoes used for various activities (Figure 5.10). They vary in both material quality and thickness. They may even contain "shock absorbers." *Shock*, an important word, in this instance is synonymous with *impulse*. The key to shock absorption is to try to reduce the amount of force applied during impact. Because the mass of the person is a constant, once again Δt is manipulated. In the case of shoes, the change comes in the form of increasing the time between the instant the sole of the shoe touches the ground and the instant the momentum of the body is brought to zero. This is accomplished by changing the thickness or material qualities of the sole of the shoe. A shoe with a thin, hard sole is not good for shock absorption because the interval for momentum change will be particularly short. As we discussed previously, short intervals for change in momentum result in high average force application. In contrast, a thicker sole made of more elastic material increases the time between contact and zero momentum (therefore decreasing average force absorbed by the person). Shoes are just one example; think of how many devices in our lives (e.g., padding in football, helmets, and airbags) are designed for shock absorption.

Also recall that viscoelastic materials exhibit stress and strain relationships that are loading-rate dependent. So in terms of safety, loading the body rapidly means that the resulting large forces absorbed by the body will be transferred to biological materials that are viscoelastic. This can be a dangerous combination in situations of impact (which occur in almost every sport).

Figure 5.10 **Comparison of various shoes.** Why are the soles so different?

FOCUS ON RESEARCH

The sport of running has been gaining popularity, with millions participating as a form of exercise and in competitive events such as 5Ks, marathons, and triathlons. With the increase in participation also comes an increase in the number of injuries in runners even with advances in running shoes and information about proper form and training. A recent trend for some runners is a move to barefoot running as a way to decrease collision forces (decreased linear impulse), reduce running cost, and increase muscle strength. However, runners who have moved from shod to barefoot are still reporting injuries after the transition (Giuliani et al., 2011). It is interesting to note that humans run barefoot using a forefoot strike and run shod using a heel strike. It seems that the padding in running shoes allows humans to fundamentally alter the natural running gait.

A recent study examining kinematic and kinetic data of runners during shod and barefoot running reveals that it may be the foot-strike pattern that is more important for injury reduction than the presence of footwear. In a study titled "Is the Foot Striking Pattern More Important Than Barefoot or Shod Conditions in Running," Shih, Lin, and Shiang (2013) recruited healthy habitually shod runners to participate in four different running conditions: (1) barefoot with heel strike, (2) barefoot with forefoot strike, (3) shod with heel strike, and (4) shod with forefoot strike. Investigators used 3-D cameras to collect kinematic data and load cells to collect kinetic data.

Researchers found that participants had significantly reduced loading rates when using a forefoot strike pattern whether barefoot or shod. The loading rate was significantly higher when barefoot using a heel strike pattern versus shod condition, which may explain why people who move to barefoot from shod running without changing their heel strike pattern report similar rates of injury. Kinematic analysis revealed that there was more knee flexion during the forefoot strike pattern that provided a larger cushioning effect. The researchers concluded that striking pattern plays a more important role in reducing force loads than whether the runners were wearing shoes or not.

5.3 LINEAR KINETICS AND ENERGY TRANSFER

In popular language, the words *work* and *energy* are used in many different contexts. This section explores these concepts specifically within the realm of kinetics. We know that a force must be applied to accelerate an object of constant mass. In mechanical terms, work was performed if the object is linearly displaced by that force application. Energy is the capacity to perform that work. Therefore, work is an energy transfer (Serway & Jewett, 2019). Many forms of energy exist in the universe, including chemical, electromagnetic, mechanical, and nuclear. In mechanics, our primary focus is on mechanical energy. Mechanical energy is composed of energy associated with position (potential energy) and energy associated with motion (kinetic energy) (Serway & Vuille, 2018).

In our previous section, we discussed linear kinetics in terms of Newton's laws. Now let's use an energy-transfer approach to kinetics.

Work

As previously mentioned, **work** is performed when an object is displaced by the application of a force. Work is a transfer of energy. More specifically, work is the product of applied force and the magnitude of displacement in the direction of the applied force:

$$W = F(d) \tag{5.16}$$

where

W = work performed on the object
F = force applied to the object to perform the work
d = displacement caused by the applied force ($\text{position}_{\text{final}} - \text{position}_{\text{initial}}$)

The units for describing work are units of force (N) multiplied by units of distance (m). The SI unit for work is the *joule* (J); 1 J = 1 N · m = 1 kg · m^2/s^2. For example, if a force of 100 N is used to displace an object 3 m, the work performed was 300 J or 300 Nm. As with our other linear quantities, assume that the force applied to perform the work may not be constant. So it must be specified whether or not one is referring to a constant force or the average force.

The concept of work is straightforward. However, some confusion does occur in two areas: (1) work can be positive or negative and (2) no matter how much physical effort was involved on the part of the person, technically no *mechanical* work was performed if the object wasn't displaced. We can actually demonstrate both concepts with one example. Let's consider the work performed while walking up and down stairs (Figure 5.11).

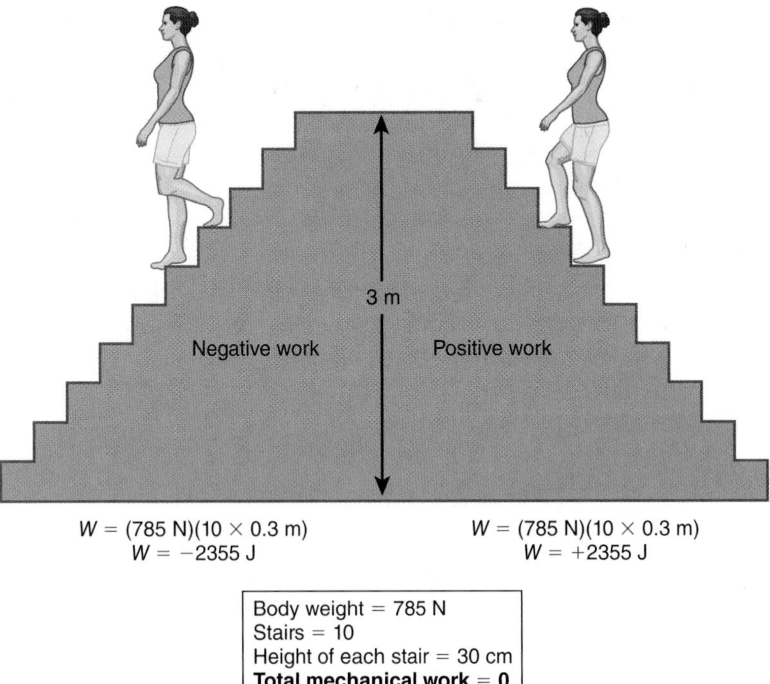

Figure 5.11 **Work while walking stairs.** What is the difference between mechanical work and physiological work?

Work Displacement of an object caused by the application of a force; a transfer of energy.

In our example, a person weighing 785 N walks up 10 steps, each having a height of 30 cm. When our subject reaches the top of the stairs, 2,355 J of work will have been performed, and our subject will have been displaced upward by 3 m:

$$W = (785 \text{ N})(10 \times 0.3 \text{ m}) = 2355 \text{ J}$$

In this case, the direction of displacement is the same as that of the applied force. Because displacement is in the same direction as the applied force (ground reaction force), the work is positive. If the person then turns and walks down the steps, the work performed will be the same. However, the person is now moving downward and the ground reaction force is still upward. Because the direction of displacement is now in the opposite direction of the applied force, the work is negative. If we then notice that the person is in the same place at which he started (i.e., displacement is 0), we can see that the total mechanical work performed by the person was 0 J ($W_{total} = 2355 \text{ J} + [-2355 \text{ J}] = 0$). It seems counterintuitive because we tend to think of work as being synonymous with energy expenditure. Although there was some *physiological* work required to walk up and down the stairs, there was no mechanical work performed.

We mentioned that work was a transfer of energy. If you think in terms of positive and negative work, one can understand the energy transfer. For example, the act of successfully lifting an object is positive work (applied force is in the same direction as displacement). If work is positive, then energy is transferred to the system. In other words, energy is being transferred to the object to displace it upward. When the object is then returned to the original position, the work is negative (a force is applied upward by the person but the object is displaced downward). If work is negative, energy is transferred *from* the system. In this case, energy is transferred from the object to the person applying the force.

A final point to consider is that in both of our examples multiple entities are performing work simultaneously. In both instances, gravity is also performing work. In the case of walking up the stairs, gravitational force is performing negative work on the person (i.e., the force of gravity is downward but the person is displaced upward). The opposite is true when walking down the stairs (gravity performs positive work). Therefore, one should always be aware of the *specific* force that is causing displacement of the object.

Power

Although the ability to perform work is an important quality, in most sports the most important factor is *how quickly* one can perform work. For example, football players must possess great strength because they exert forces against each other throughout the game. But it is also important for them to be able to exert those forces in a short time. In mechanics, **power** is the term used to express the amount of mechanical work performed in a given interval. Because work is a transfer of energy, power is the time rate of energy transfer (Serway & Jewett, 2019). Power, or the rate of performing work, is expressed mathematically as:

$$P = \frac{W}{\Delta t} \quad (5.17)$$

where

P = power
W = mechanical work performed
Δt = $\text{time}_{final} - \text{time}_{initial}$ (interval of work performed)

Power The amount of mechanical work performed in a given time interval.

Because $W = F(d)$, power can also be expressed as:

$$P = \frac{F(d)}{\Delta t} \quad (5.18)$$

where

P = power
$F(d)$ = mechanical work performed
$\Delta t = \text{time}_{\text{final}} - \text{time}_{\text{initial}}$ (interval of work performed)

and because $v = d/\Delta t$, power is also:

$$P = Fv \quad (5.19)$$

where

P = power
F = force applied to the object to perform the work
v = velocity, or rate of motion in a specific direction

Figure 5.12 **Olympic weightlifter.** How would power be measured in this situation?

The units for describing power are units of work (J) divided by units of time (sec). The SI unit for power is the *watt* (W); 1 W = 1 J/s = 1 N · m/s. If the person in our previous example walks up the stairs (2,355 J of work) in 5 seconds, then the power is 471 W or 471 J/s. We can understand the importance of the interval if the person runs up the stairs in 3 seconds, in which case the power is 785 W. So power is greater if the same amount of work performed in a shorter time. In fact, stair sprinting is often used to measure power. In the Margaria-Kalamen test (Margaria et al., 1966), a person climbs a staircase three steps at a time as fast as possible. Power is then calculated using the mass of the person, the vertical distance between the third and ninth steps, and the interval between contacting the third and ninth steps (calculated by using switch mats). Note also that, similar to velocity and acceleration, power can be calculated as an average value or an instantaneous value. For example, the Wingate power test (Bar-Or, 1987) is a popular power test that entails a person cycling as fast as possible for 30 sec on an ergometer against a resistance that is based on her body mass (e.g., 7.5% of body mass). The ergometer is interfaced with a computer, and the software can provide an average power output for the entire 30 sec or an instantaneous value such as peak power.

The ability to produce power is of great importance in sports that are considered explosive (e.g., Olympic weight lifting, Figure 5.12, and various track and field events such as sprinting and putting the shot).

But one should notice that the performance of any explosive or powerful sports event takes place in a brief period. This is because human power production is generated by muscle, which has metabolic limitations. Because limitations to human power are at the muscular level, let's return to the topic of power in Chapter 11 dealing with the biomechanics of skeletal muscle.

Potential Energy

Recall that mechanical energy is composed of energy associated with position (potential energy) and energy associated with motion (kinetic energy). The capacity of an object to perform work because of its position, deformation, or configuration is called **potential energy** (or **stored energy**). The term *potential energy* refers to the

Potential energy (or stored energy) The capacity of an object to perform work based on its position, deformation, or configuration.

potential for conversion to kinetic energy (energy of motion). So when thinking of potential energy, it is helpful to realize that an object possessing potential energy has the capacity to perform work but is not yet in motion.

Gravitational Potential Energy

One form of potential energy is gravitational potential energy. **Gravitational potential energy** is potential energy that an object has because of its position relative to a reference surface (often Earth). Gravitational potential energy is expressed mathematically as:

$$GPE = wth \tag{5.20}$$

where

GPE = gravitational potential energy (or potential energy because of position)
w = weight of the object
h = height or position of the object relative to a reference surface

Recall from Equation 5.8 that we calculate the weight of an object as follows:

$$F = mg$$

where

F = the magnitude of the gravitational force
m = the mass of the body interacting with Earth
g = the acceleration of the object caused by gravitational force

Therefore, GPE can also be calculated:

$$GPE = mgh \tag{5.21}$$

where

GPE = gravitational potential energy (or potential energy because of position)
m = the mass of the body interacting with Earth
g = the acceleration of the object caused by gravitational force (9.81 m/sec^2)
h = height or position of the object relative to a reference surface

The units for describing gravitational potential energy are units of force (N) multiplied by units of length (m). Similar to the units of work, the SI unit for GPE is the joule (J); $1 J = 1 N \cdot m = 1 kg \cdot m^2/s^2$. As a matter of fact, returning to our previous example of work will clarify many points. In our example, a person weighing 785 N walked up 10 steps, each having a height of 30 cm. The work was calculated to be: $W = (785 N)(10 \times 0.3 m) = 2{,}355 J$. Because the person is now at the top of the stairs, he has GPE as a result of being displaced above Earth. Let's calculate the GPE, this time starting with the person's mass rather than his weight. The person has a mass of 80.02 kg and has been displaced 3 m above the ground (h = 3.00 m). Therefore, GPE is calculated as follows:

$$GPE = (80.02 kg)(9.81 m/sec^2)(3.00 m) = 2355 J$$

Notice that the work and the gravitational potential energy are the same. The person now has potential energy because of the work performed during displacement. We have mentioned that work is a transfer of energy. In this case, energy was being *stored* during performance of the initial work. So while the person is at the top of the stairs, the energy *stored* during the performance of the work has the *potential* to become kinetic energy if the person then jumps off. Therefore, energy was transferred by work.

Gravitational potential energy The potential energy that an object has based on its position relative to a reference surface (often Earth).

Elastic Potential Energy

The second type of potential energy is called **elastic potential energy** (or **strain energy**); it is potential energy stored in a deformed object (e.g., a spring). For example, when an archer deforms a bow by pulling on the string, the bow then possesses the potential to perform work on an arrow. Also, stored elastic energy in the musculoskeletal system is used in the performance of plyometric exercises. In Chapter 4 we discussed stress, strain, and elasticity. Because *strain* energy is produced by deforming an object, we must realize that the mechanical properties of that object play an important role in its ability to store energy. We can see this in the mathematical definition of strain energy:

$$SE = \tfrac{1}{2}k\Delta x^2 \qquad (5.22)$$

where

SE = strain energy (or potential energy arising from deformation)

k = stiffness constant (spring constant) of the material representing its ability to store energy on deformation

Δx^2 = deformation (or change in length) of the object from its nondeformed state squared

Figure 5.13 **An archer.** How is the bow similar to a pole used by a vaulter?

The stiffness (spring) constant has units of N/m, so strain energy has units of N/m multiplied by units of length squared (m^2). Therefore, the SI unit for SE is the joule (N · m). Notice that the same units are used for both work and GPE. In terms of sport, we tend to think of implements such as poles used by vaulters and bows used by archers (Figure 5.13).

But inside the human body, muscles and tendons that are stretched also store strain energy, which can be used to enhance the performance of work as the muscle is subsequently contracted. This book covers this idea in greater detail in Chapter 11 on skeletal muscle mechanics. For now, we will simply use an example with which most people are familiar . . . shoes. We have mentioned that the type of shoe sole can affect the change in momentum as one plants his foot from a stride. Now let's look at the same situation from a different perspective—that is, *return* of some of the energy absorbed by the sole of shoe.

The typical spring constant of the sole of a shoe used for marathon running is between 10^6 N/m and 10^7 N/m (Nigg et al., 2000). On planting the foot, let's assume the sole deforms 3.00 mm. In this case, the strain energy is calculated as follows:

$$SE = \frac{(10^6\ \text{N/m})(0.003\ \text{m}^2)}{2} = 4.5\ \text{J}$$

This result means that 4.5 J of energy will be stored in the sole of the shoe during ground contact. Remember that this is potential energy that can then be used to perform work as the sole of the shoe reforms. As the sole of the shoe returns to its original shape, the potential energy associated with deformation is converted into energy associated with motion (kinetic energy).

Kinetic Energy

Objects in motion have the potential to perform work. This energy associated with motion is **kinetic energy**. One can observe this capacity to perform work by observing collisions. We discussed collisions in the context of

Elastic potential energy (or strain energy) The potential energy stored in a deformed object.

Kinetic energy The energy (potential to perform work) associated with motion.

momentum; now we must observe them from a different perspective. Let's assume a car has stopped at a traffic light. Another driver, not seeing the car stop, hits the car from behind. You will notice that the car that was not moving is set into motion by the moving car. According to Newton's first law, a force must have been applied by the moving car to the motionless car. In other words, one car performed work on the other car; work requires energy, and the energy was generated by motion. Kinetic energy is expressed mathematically as:

$$KE = \tfrac{1}{2}mv^2 \qquad (5.23)$$

where

KE = kinetic energy (or energy associated with motion)
m = mass of the object in motion
v^2 = velocity of the object in motion squared

The units for describing kinetic energy are units of mass (kg) multiplied by units of velocity squared (m^2/sec^2). Therefore similar to the units of work, GPE, and SE, the SI unit for KE is the joule (J); $1\ J = 1\ N \cdot m = 1\ kg \cdot m^2/s^2$. You get the point with the units. To perform work, something must possess energy. Therefore, the units are always the same. Notice two aspects of the equation: (1) an object that is not moving ($v = 0$ m/sec) has no kinetic energy and (2) velocity is squared in the equation, so changes in velocity have large effects on kinetic energy.

SAMPLE PROBLEM

We can demonstrate the importance of velocity by comparing two projectiles, a baseball and a bullet. Let's say a baseball having a mass of 0.145 kg is thrown at 45 m/sec. The bullet has a mass of 0.010 kg and is fired from a gun at 1000 m/sec. Clearly, the baseball is more massive. But if we calculate the kinetic energy of both objects, we see the exponential effect of velocity:

$$\text{Baseball KE} = \frac{(0.145)(45\ \text{m/sec})^2}{2} = 146.81\ \text{J}$$

$$\text{Baseball KE} = \frac{(0.010)(1000\ \text{m/sec})^2}{2} = 5000\ \text{J}$$

So, although the mass of an object affects its kinetic energy, a small object such as a bullet has the ability to perform large amounts of work if it is moving at ultrahigh velocity.

We have mentioned both momentum and kinetic energy in reference to collisions. In the absence of a net externally applied force, the total momentum of a system that comprises multiple bodies remains constant. This is not necessarily the case for kinetic energy. During impact, some kinetic energy is usually converted into internal energy, sound energy, thermal energy, and work (Serway & Vuille, 2018). For example, if a player hits a ball with a racquet (Figure 5.14), the energy used to deform the ball and the racquet will be returned during reformation.

However, energy is also released in the form of sound and heat. It's easy to understand that some degree of kinetic energy usually is lost during a collision if we imagine the contrary. If there wasn't some degree of kinetic energy loss during collision, then a dropped ball would bounce up and down infinitely. Theoretically, conservation of kinetic energy depends on whether the collision is *elastic* or *inelastic*. In an elastic collision, both momentum and kinetic energy are conserved. In other words, the objects collide and bounce off of each other, retaining both their precollision masses and their velocities. During an inelastic collision, the objects stick together to some degree, and at least some of the kinetic energy is converted to a different form. Therefore, during inelastic collisions momentum is conserved, but kinetic energy is not. Most collisions are not purely elastic or inelastic, so varying degrees of kinetic energy conservation are observed.

One final topic of interest is a comparison of momentum and kinetic energy. Both are made of the same quantities (mass and velocity). However, velocity is of greater importance in terms of kinetic energy. Momentum is the quantity of motion possessed by an object; the more momentum, the more difficult it is to change the state of motion of the object. Kinetic energy is the ability of an object to perform work based on its motion; the greater the kinetic energy, the more capability of the object to perform work. Therefore, one should see kinetic energy as the ability of a system to perform work to change momentum.

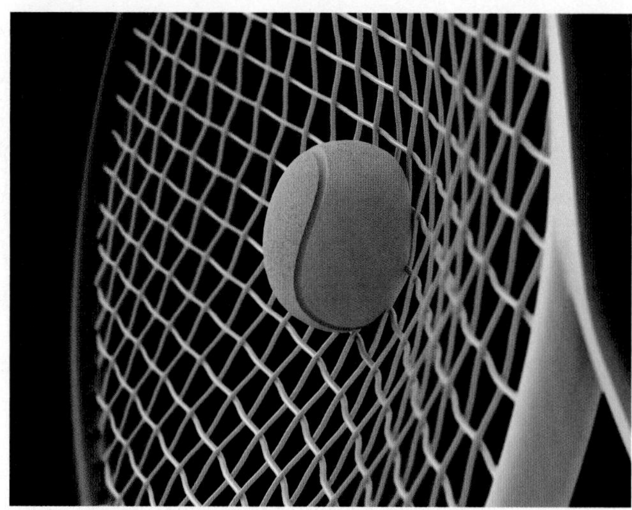

Figure 5.14 **Racquet hitting a ball.** What are the sources of energy return in this situation? What about energy loss?

SAMPLE PROBLEM

Figure 5.15 **Diving board.**

For example, let's say that two performers run toward an apparatus such as a diving board as preparation to bounce into the air (Figure 5.15). The first performer has a mass of 65 kg and is running at 4.5 m/sec. The second performer has a mass of 75 kg and is running at 3.9 m/sec.

The performers in this case have the same momentum (292.5 kg · m/sec), but different amounts of kinetic energy (KE of first performer = 658.125 J; KE of second performer = 570.375 J). So our first performer has a greater ability to perform work on the diving board. This work will be stored in the form of strain energy, which will be recovered to subsequently perform work on the performer as the diving board reforms. The performers have the same momentum, so the one that has the greatest work performed on him (has the greatest change in momentum) will be thrown the farthest by the diving board. Because the first performer has deformed the diving board the most, he will experience the greatest change in momentum on reformation.

Conservation of Mechanical Energy

We have clearly demonstrated that a relationship exists between work and energy: Energy is the capacity to perform work, and work is a transfer of energy. Now let's use two mathematical expressions to summarize the relationships previously discussed.

You have probably heard the statement "Energy cannot be created or destroyed; it can only be transferred from one form to another." This is a general statement about energy, but it also applies to the mechanical energy with which we are concerned. Therefore, one important concept to highlight is the **law of conservation of mechanical energy**: *In the absence of externally applied forces other than gravity, the total mechanical energy of a system remains constant* (Serway & Jewett, 2019; Serway & Vuille, 2018). This law can be expressed mathematically as follows:

$$(KE + PE)_i = (KE + PE)_f \quad (5.24)$$

where

KE_i and PE_i = initial kinetic and potential energies
KE_f and PE_f = final kinetic and potential energies

Let's use the example of a diver on a platform (Figure 5.16). While standing at the top of the platform, the diver has potential energy because she is separated from Earth. If the person is not moving, she has zero kinetic energy.

If the diver then jumps from the platform, potential energy gradually decreases as the diver gets closer to the water. Simultaneously, kinetic energy is gained as the velocity of the diver increases. Therefore, the sum of kinetic and potential energy remains a constant as one form of energy is transferred into another.

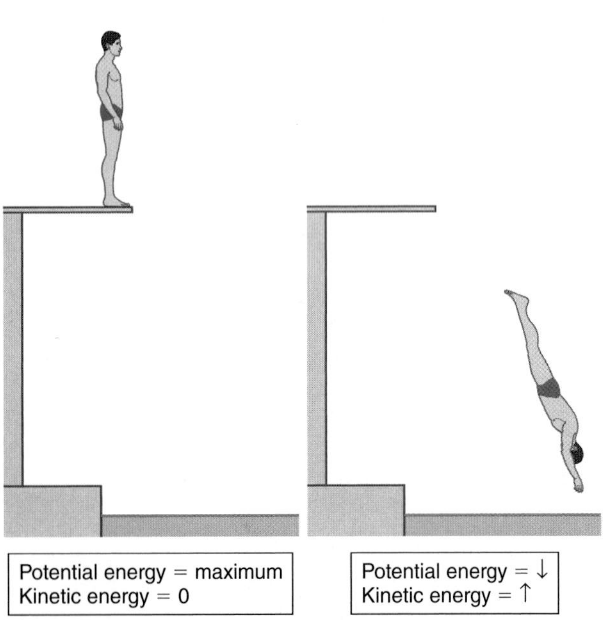

So we know that the mechanical energy is conserved. This conservation is attributable to energy transfer. We have stated that energy transfer from one system to another takes place in the form of work. So another important concept is the **principle of work and energy**: *The work performed by externally applied forces other than gravity causes a change in energy of the object acted on* (Serway & Jewett, 2019; Serway & Vuille, 2018). This principle is represented as:

$$W = \Delta E = \Delta KE + \Delta PE + \Delta TE \quad (5.25)$$

where

W = mechanical work performed
$\Delta E = E_{final} - E_{initial}$ (change in energy)
$\Delta KE = KE_{final} - KE_{initial}$ (change in kinetic energy)
$\Delta PE = PE_{final} - PE_{initial}$ (change in potential energy)
$\Delta TE = TE_{final} - TE_{initial}$ (change in thermal or heat energy)

So the change in the sum of the kinetic, potential, and thermal forms of energy produced by the application of externally applied force equals the mechanical work performed. As an example of the work–energy relationship, let's analyze the energy change during a golf drive. Before

Potential energy = maximum
Kinetic energy = 0

Potential energy = ↓
Kinetic energy = ↑

Figure 5.16 **Platform diving.** Why does this diver have potential energy?

Law of conservation of mechanical energy In the absence of externally applied forces other than gravity, the total mechanical energy of a system remains constant.

Principle of work and energy The work performed by externally applied forces other than gravity causes a change in energy of the object acted on.

being struck by the club, the golf ball has a tiny amount of potential energy because of its position on the tee. The ball has zero kinetic energy because it is stationary (has zero velocity). However, the club does have kinetic energy as it approaches the tee. The kinetic energy of the club gives the club head the ability to perform work on the ball. At contact, the energy of the club is used to deform the ball and propel it forward and upward. The work performed on the ball (energy change) can be observed in the changes in both its potential and kinetic energy. The potential energy of the ball increases as it gains height, and its kinetic energy increases because of its velocity change. So again, the total work performed on the ball is equal to the change in the sum of its kinetic and potential energy.

Finally, note that all of these concepts are related to our previous discussion of the linear impulse equation (5.15):

$$\Sigma F\,(\Delta t) = \Delta M$$

Most sports situations involve changing the momentum of a system. This requires the application of a force that simultaneously causes a change in kinetic energy of the system through a change in velocity. The application of a force can also change the potential energy of the system by causing an elevation change. The greater the force applied or the longer the interval over which the force is applied, the greater the momentum change but also the greater the energy transfer. So the important realization is an understanding of the interrelationship of all of the concepts related to force, work, and energy.

CONNECTIONS

5.4 HUMAN PERFORMANCE AND INJURY SCIENCE

Sport Science

Strength and conditioning specialists (as well as athletic trainers) must consider biomechanical factors when training (and treating) athletes. One of these factors could be as simple as the surface on which the game is played. For example, the degree of hardness of a field can have major implications for the linear impulse equation ($\Sigma F\,(\Delta t) = \Delta M$). Using artificial turf on football playing fields has been a subject of controversy for years. The original artificial turf was usually constructed by placing simple carpeting over concrete.

In terms of linear impulse, this construction means that the athletes experience great forces over short intervals because the surface provides little or no give on impact. Because the injury rate among players on the original artificial turf was high, newer forms have been designed to mimic natural grass. These newer materials consist of an artificial grass with two layers of nap. One layer is longer and re-creates the look and feel of natural grass (Figure 5.17). The shorter nap is tightly curled and used to trap rubber pellets at a level below the longer nap. Also, the entire surface is placed over dirt as opposed to concrete. The result is that newer artificial turf (similar to natural grass over dirt) increases the time over which the force is absorbed by giving on impact. The forces absorbed during sport occur in two forms in relation to the playing surface. The first is running, during which force is absorbed by impact of the foot with the ground. The second occurs when the athlete makes sudden starts and stops or changes in direction (accelerations).

Figure 5.17 **Newer artificial turf design.** Why is this design better than original Astroturf?

Dirt and grass (or rubber pellets in the newer artificial turf) can slide out of the way, again increasing the interval during which force is applied. Artificial turf doesn't exhibit this behavior to the same extent. Therefore, if the foot stops abruptly when it would normally slide, the potential to sprain the ankle or damage the knee increases. The ramifications for the strength and conditioning specialist are in the area of recovery. After a game on a hard playing surface, athletes will need more recovery time before their next training session to prevent injury because they have absorbed many large impulses during the game that lead to microtrauma in the joints. The direct implication for the athletic trainer is the need to be prepared for more treatment than if the athlete played on a natural surface.

Several sports involve elastic collisions as a strategic concept. Bowling, curling, bocce, and shuffleboard are target games that involve changing the dynamics of velocity and direction, depending on the situation of the game. The discussion will begin with bowling because this game is slightly different than the other three. Bowling consists of 10 targets (pins) set up in a triangle about 20 m away, with one point of the triangle facing the bowler. The intent is to knock all of the pins over with one or two shots using a standard circumference ball. The first ball is propelled down the alley and collides with the pins, in turn creating several collisions between the ball and pins as well as between pins and other pins. With luck, all the pins will fall with only one throw. If not, the game gets more complicated because almost any combination of the 10 pins may remain standing.

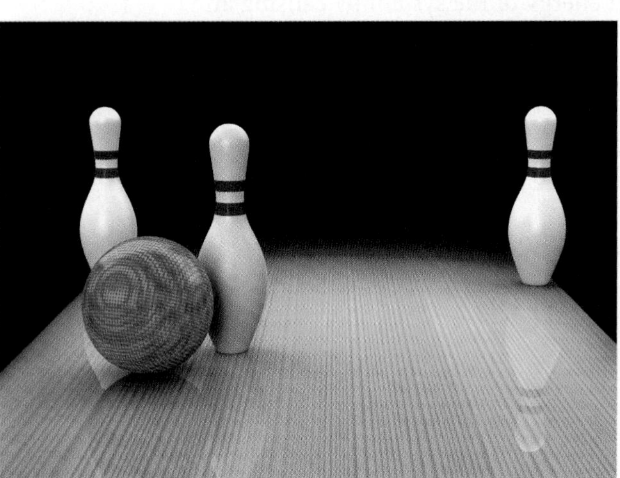

Figure 5.18 **Bowling pins.** Think of as many principles as possible from this chapter that are related to this situation.
© billdayone/Shutterstock

When the pins are separated by more space than the width of the ball (Figure 5.18), the only way to knock them all over is to deflect one pin into another or to have the ball deflect off one pin toward another. To do this, the bowler must understand the concept of elastic collisions. For example, a bowler may send a ball to the right of a pin that lies forward of a single pin to the left. When the ball strikes the pin at the proper angle, the pin is accelerated toward the one on the left, and the ball changes direction slightly to the right. One other variable comes into play during each roll—the mass of the ball. The pins are a standard size and weight, but bowling balls can vary in weight from 2 kg to 7 kg. Conventional teaching suggests that a bowler use the heaviest ball that she can control. Why do you think this is? The answer lies in the section about momentum. A bowler would have to be able to propel the 2 kg ball at more than three times the velocity to arrive at the pins with the same momentum of the 7 kg ball. More momentum equals less deflection of the ball during the collisions with pins.

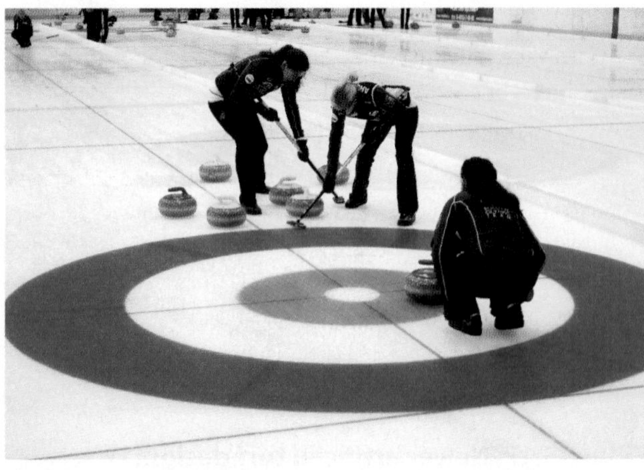

Figure 5.19 **Curling.**
© Anton Gvozdikov/Shutterstock

In the games of bocce, curling, and shuffleboard, collisions take on a more tactical role than bowling. These games involve multiple objects being sent into a target area by two different teams with the objective being to get your team's objects closer to a target (Figure 5.19). Once an object enters the target area, an opponent has several choices. First, he may choose to send one of his objects closer to the target without striking anything along the way. Second, he may decide to knock an opponent's object away from the target area. Finally, he may attempt to hit one of his own objects to move both (the thrown and resting ball, stone, or puck) closer to the target.

In all of these games, the objects are made of hard materials, which results in highly elastic collisions. Skilled players learn that the angle and velocity of each collision will depend on the nature of the task. A collision that is centric will tend to transfer most of the energy from the arriving object to the stationary one, thereby leaving the thrown one close to the location of the collision while the previously stationary ball is sent away. The farther that the thrown objects hit from centric contact, the less energy will be transferred from one object to the other, and both will move away from the location of the collision. Using a deft touch in both angle and velocity, skilled players are able to deflect opponents' objects away while leaving their own closer to the target. They might also choose to strike their own objects so that both will end closer to the target. A great example of this process is the hammer stone in curling. This is the final stone sent toward the target area during each end of play during a match. The hammer gets its name from the high velocity and violent collision force it imparts on opposing stones in the target area (see the opening scenario in this chapter).

Injury Science

The linear impulse equation ($\Sigma F\,(\Delta t) = \Delta M$) has huge implications for injury, not only in sport but also in activities of daily living. Remember that to change momentum in a short time, high forces must be used (and vice versa). So think of all of the instances in everyday life in which the dissipation of force in a safe way in necessary. Padding in athletic uniforms, air bags and crumple zones or bumpers in cars, and boxing gloves are all designed to prevent the body from experiencing high forces caused by quick changes in momentum (Figure 5.20).

© dean bertoncelj/Shutterstock

© Attapon Thana/Shutterstock

Figure 5.20 **Athletic equipment, air bags, and a car bumper.** What equation do all of these pieces of safety equipment have in common?

In other words, all of these devices increase the interval over which the momentum is brought to zero. According to the linear impulse equation, this increase in duration decreases the force sustained by the person. So increasing the time during which the impulse is received aids in absorbing and dissipating the impact forces. The less the force sustained by the body, the lower the risk of injury.

The brain itself also has a protective mechanism against sudden changes in motion (momentum) of the head. The brain floats in a cushion of cerebrospinal fluid, so it is not in direct contact with the skull. Once the head is in motion, it has a given amount of linear momentum. The brain also has momentum. When the head is suddenly decelerated by impact forces, its momentum is brought to zero rapidly by the applied force. The brain is also decelerated because of the force of the skull applied to it. However, for a few milliseconds, the brain is still traveling in the original direction of the head even though the head is no longer moving. The brain doesn't stop moving at the same time as the head because of the layer of cerebrospinal fluid. So the cerebrospinal fluid increases the interval over which its momentum is brought to zero. However, in some instances the cerebrospinal fluid cannot dissipate the impact force of the brain hitting the skull to a large enough degree to prevent injury. This type of injury, in which the brain impacts the skull, is called a *concussion*. Concussion can also occur if the head is still but suddenly accelerated by an impact force (such as being hit with an object). In this case, the brain is accelerated by an impact of the skull. In either case, the cause of the damage is a collision of the skull and the brain. Repeated concussion can lead to permanent brain damage. In boxing, this permanent damage is sometimes called "punch drunk" syndrome. Over time, boxing gloves have been made thicker and thicker in an attempt to minimize the damage caused by repeated impact. Again, the padding in the glove increases the interval over which the force is applied. Just think about the origins of boxing: It was barefisted!

5.5 MOTOR BEHAVIOR

Motor Development

The human body is a dynamic adaptable organism. As we change in size, systems that are responsible for producing force change, depending upon the requirements of the body. As we get larger, it requires more energy to produce motion. The process of change can be either positive or negative in terms of ability to produce motion.

First we'll discuss positive changes in motive force production or, in simple terms, getting stronger. As an infant grows, her weight increases. With the increase in weight comes a requisite need for increased force production to initiate and maintain movement. This increased force is produced in two ways: (1) the muscles adapt and become larger and stronger and (2) the bones also grow longer, providing longer levers for the muscles to act upon. Muscles increase their strength as a reaction to the increasing needs of the infant, toddler, or growing child. As the child grows larger, the muscle and bones adapt by getting longer and stronger.

The same phenomenon can be seen in the mother and is a basic principle of weight training known as *overload*. When muscles are strained, the body adapts by making them stronger. Just as the infant's or toddler's muscle adapt as the body grows larger and heavier, the muscles of a mother who lifts and carries her child must also adapt and get stronger as the child grows. Some of the strength gains are attributable to neural adaptations such as increased firing rates and more efficient recruitment of motor units. In post-pubescent males, strength gains are the result of the same neural adaptations, plus hypertrophy—an increase in muscle size that is regulated by the presence of testosterone. Whatever the mechanism of strength increase, the body has the ability to adapt to increasing stress and strain by becoming stronger.

The opposite is also true: The human body can adapt to decreased strain by decreasing muscle size when it's not necessary. The best example of this phenomenon is disuse of a limb because of a broken bone. When a cast is applied to a broken limb, and medical prescription is to avoid placing any weight on that limb, the muscles inside the cast lose the natural strain placed on them through normal use. This disuse results in muscle atrophy—the muscle loses both size and strength. Anyone who has experienced the sight of a person immediately after a cast is removed can attest to the significant difference in size between the affected and unaffected limbs. Something that cannot be seen is the remodeling that the bone in the affected limb has also undergone. Just as muscles adapt to the strain—or lack thereof—placed on them, so do bones (Wolff, 1892). As muscles become weaker and strain is removed from a limb, the bones are remodeled to a smaller size to compensate. There's no reason to have a large, strong bone in the absence of strain, so the body adapts.

The good news for anyone who suffers an injury that results in disuse atrophy is that the effects can be reversed. Therapy includes increasing the strain on the limb in a progressive manner. The increased strain on the limb will result in adaptation of both muscle and bone as each become stronger in response to the requirements of the exercise.

Motor Learning

In the realm of motor learning, we see definite implications of linear impulse and kinetic energy as they apply to the proficiency of throwing (ballistic) skills (Figure 5.21). Early throwing patterns are typically characterized by minimal leg or trunk motion. In contrast, proficient throwers have a backswing

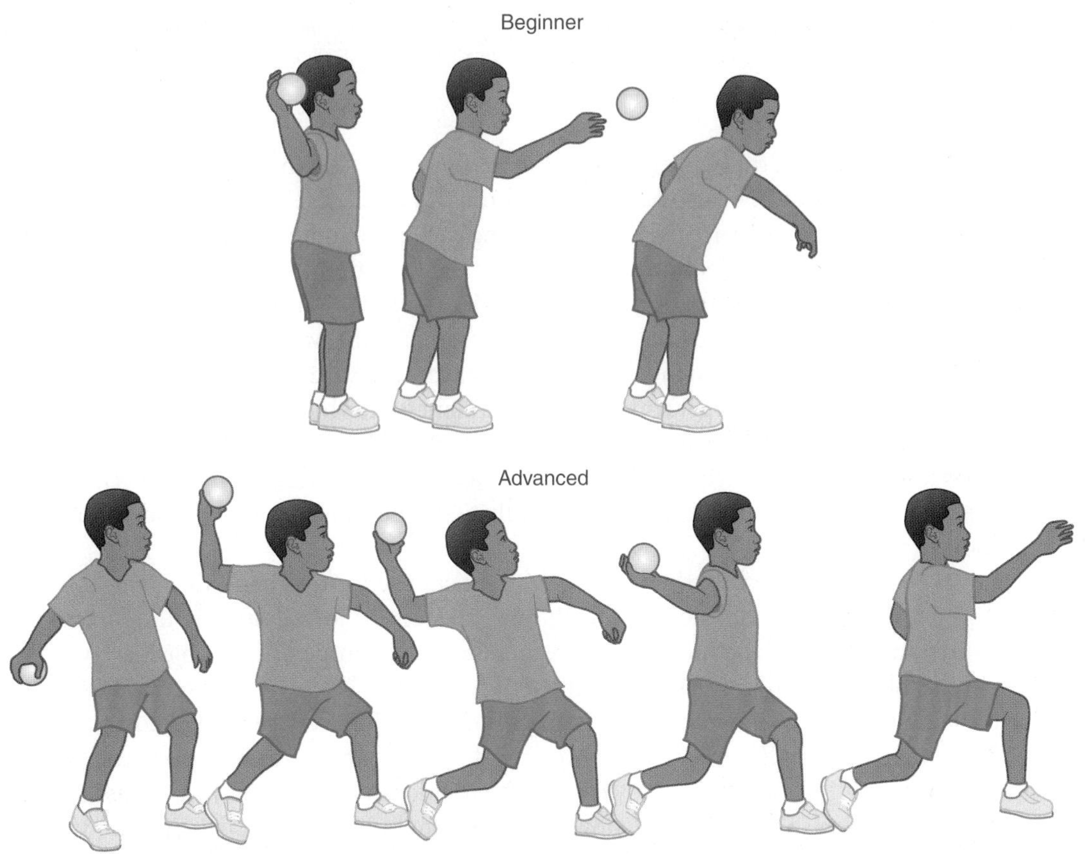

Figure 5.21 **Throwing sequence of early versus proficient thrower.** What concepts from this chapter explain the differences between the throwers?

or windup with the throwing arm and simultaneously step forward with the leg opposite the throwing arm (Haywood & Getchell, 2014).

We mentioned proficient throwing in Chapter 2 as an example of the open kinetic chain concept. Proficient throwing involves force generation and transfer from the base segments (legs) to the hips, then torso, humerus, forearm, and finally the hand. Proper timing of segmental participation produces lag. In other words, the leg opposite the throwing arm steps forward. Then the hips are rotated forward while the upper torso is still rotating in the opposite direction (lagging behind). Then as the upper torso rotates forward, the humerus lags, and so on. The final result is a motion of the entire kinetic chain that is similar to that of a whip. The proximal end of the whip is the area of initial force application. The force then travels in a wave toward the distal end of the whip. As most people are fully aware, this properly timed sequence of actions results in ultrahigh speed of the tip of the whip—in fact, the crack of a whip results from the end surpassing the speed of sound and creating a sonic boom. Similarly, the goal of proficient throwing is to achieve high speed in the distal hand segment of the chain that can be transferred to the thrown object. In terms of linear motion concepts, all of the motions of a proficient thrower allow force to be applied to the ball for a longer time. This longer time of force application can be analyzed in two ways. First, in terms of linear impulse, the momentum of the ball is changed by applying a force over the greatest duration possible. Alternately, from a work–energy transfer point of view, the total work performed on the ball is demonstrated by the large change in its kinetic energy. Developing proficient throwing motion is a learned task and is used as an example in the following section.

5.6 PEDAGOGY

Teaching

The development of proficient throwing skills—as well as any other skill in which a performer projects an object away from the body—can also be presented through a teacher's eyes. Four skill cues, or critical elements, that teachers use to enhance overhand throwing each contribute to applying additional linear momentum to an object. First, students are taught to turn their nondominant side toward the intended target. This cue encourages trunk rotation that is not evident in early throwing attempts. The second cue is to bring the ball to a point next to the ear of the dominant (throwing hand) side. In putting the ball next to the ear, the thrower has to move the throwing arm rearward and bend his elbow. The third cue involves stepping toward the target with the nondominant foot, which is the first motion in a series that creates the whip effect previously mentioned. When sequenced correctly, the force generated by the forward step is transferred to the trunk, which rotates forward adding its own impulse. This summed force generated by the legs and trunk is then carried through the shoulder, which adds its own force, then the elbow to the hand.

As you can see, each of the first three cues that teachers use for overhand throwing contributes to adding linear impulse to an object. The fourth cue is necessary for dissipating the forces created by the first three cues, and it is the most common cue used when propelling an object away from the body—this is the follow-through. The momentum of the performer has to be controlled after release of the object. The follow-through is designed to dissipate the forces that were generated during the throw over a relatively long time. In the case of a thrower, the dominant arm continues along the same path after release and crosses in front of the body toward the nondominant side, all while muscles are acting like automobile brakes to slow the limb down. Forward momentum is also dissipated by the large quadriceps muscles working eccentrically as the knee is allowed to flex.

Now imagine adding the momentum of a heavy object to the performer—for example, in T-ball, a baseball game played by young children before they are introduced to hitting a pitched ball. A player stands with his side to the target in preparation for hitting the ball; he uses his arms and trunk rotation to move the bat away from the ball, and starts the forward swing by stepping toward the target with his nondominant foot. Notice that these are similar cues to throwing. Whether he contacts the ball or not, the heavy bat is moving at high speed with a lot of momentum as it passes through the hitting area. With sufficient strength, the child will be able to slow the bat down in a controlled manner, but some children swing so hard that they develop more momentum in the bat than they have the strength to slow down effectively. These batters continue their follow-through, dissipating the force of the swing simply by spinning in a circle as far as it takes to slow down. Although this can be entertaining to watch, it's not the most effective way to bat.

Now let's consider what happens at the other end of a throw or hit because children can also be taught to receive an object more efficiently. Early attempts at catching a thrown object are characterized by fear reactions (the head turning away as the object nears) and the object bouncing off the receiver. Fear reactions can be lessened by using light, nonthreatening objects such as yarn balls, foam balls, and inflatable balls. To decrease the chance of a ball bouncing off the receiver, teachers use cues that are designed to allow a student to increase the time that the object can be slowed by the body, thereby decreasing the relative force.

The first cue for catching is to face the object with the feet staggered slightly back and front, which provides a solid foundation while allowing the rear leg to absorb force. The second cue is to extend the arms and hands forward and then give with the ball as it contacts the hands. This method effectively allows the student more time to dissipate the momentum of the object and more time to successfully grip the object with his hands.

In simple terms, a volleyball set is a combination of a catch and throw in quick succession. Beginners learning the skill often slap the ball or lose control trying to strike the ball upward as it is coming down. Experienced players have learned that a quiet set is a good set and they use skills based on linear impulse and momentum to successfully accept a moving ball and send it away. The same cues that help students become efficient catchers also make for quiet set shots. First, the setter extends her arms upward, cupping each hand so the ball will contact only her fingertips. As the ball contacts her fingertips, both finger extension and elbow flexion cushion the ball as it continues to move toward the player's body. This is where high- and low-level players differ most significantly—beginners often try to send the ball upward without first slowing it down and controlling it. Now that the experienced setter has slowed the ball down, she can apply controlled force over a longer time, using both her arms and fingertips. When performed correctly, a set shot combines the skills of catching and throwing in one continuous motion and cannot be heard by spectators.

Coaching

A trend in American-style football since the early 1990s has been the tendency of linemen who are larger and heavier people, especially on the offensive line. The reason for this, and similar changes for certain positions in rugby, can be traced to Newton's laws. It's harder to move an object that is more massive. American football coaches often sacrifice mobility for added mass in certain positions.

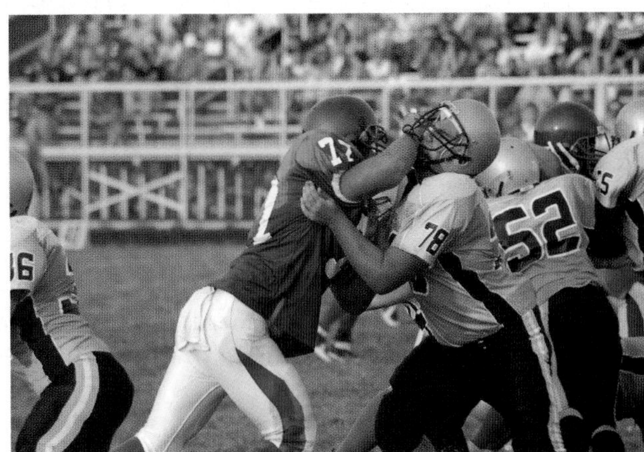

Figure 5.22 **Offensive linemen in football.**

Linemen (Figure 5.22) are generally the largest men on a football field when it comes to sheer mass. It's not uncommon to see players weighing 150 kg in the National Football League and average weights of all interior offensive linemen in the range of 135 kg. More mass takes more force to move, and the role of offensive linemen generally involves not letting defensive players through the line. Similarly, the position of prop in rugby is relegated to the largest players on a team. During a scrum, it is more difficult to move the larger mass of the heaviest players. In football and rugby, the trade-off for this size is the necessity to substitute players to keep them fresh—for just as it takes a lot of force for an opposing player to move a heavier player, so too does it take tremendous amount of energy for the players to move themselves.

5.7 ADAPTED MOTION

Adapted Physical Education

The benefits of lighter equipment cannot be overlooked in the area of physical education, especially in adapted PE. A child struggling to perform with adult-sized equipment is both inappropriate and counterproductive. Equipment that is too heavy and large to move with proper form encourages compensatory movement and teaches the child bad habits. The child also risks frustration from a lack of success and might give up on the sport or activity altogether. Nothing is more frustrating for a teacher than watching a child "give up" because he isn't successful. Success equals *fun*, and fun activities tend to be repeated.

Good teachers in physical education design success into their lessons, and one way to do this is by using modified equipment. For children lacking adequate strength to use "standard size" equipment—even standard size children's equipment—adapting different materials can afford success in skill performance. For example, a student with a muscular strength deficiency may have a difficult time learning to control a volleyball. A good progression for success might begin with volleyball skill development using a balloon. Although it may not travel in the same way that a real volleyball will, the student learns the proper form for striking without having to apply excessive force. As the student becomes more successful at the skill, the next step is to introduce a "trainer" volleyball. These balls come in a variety of sizes and weights. Trainer volleyballs bridge the gap between the lightest balloons and standard volleyballs, allowing the student to experience ball flight appropriate for game play while still being successful in spite of limited strength.

Any sport or activity that involves an object being propelled away or received by a player is a candidate for adapting equipment. If a traditional bat or hockey stick is too heavy, then hollow plastic implements are available. If those are too heavy, Styrofoam striking implements are also available. Is a heavy ball causing a lack of success? Even youth sports teams understand the nature of successful play through modified equipment. Soccer balls, footballs, basketballs, and more come in child-appropriate sizes. In learning the skill of receiving an object, lighter equipment serves two important functions. First, it takes less energy to slow down and control an object with less mass. Second, it removes one of the primary fears of beginning catchers, the fear of the object hitting them and causing pain. A walk through a toy store will reveal a variety of equipment designed for children to throw and catch, and you might not be surprised now that most of these objects are made of soft, light foam (e.g., Nerf® products).

Newer equipment designed for young children is smaller, lighter, and shorter than traditional adult-sized equipment. The change in size and weight is based on an understanding of motor and nervous system limitations. By designing equipment that fits the needs of children in regard to strength, visual system development, and evolving neuromuscular control, students can enjoy more success. Much of the success comes from students being able to accomplish the skills with proper form. For teachers working with children with disabilities, there sometimes is a need for creativity in adapting equipment for specific motor deficiencies, but with a thorough understanding of the principles involved in imparting force to send or receive an object, modifications in equipment can help *all* to become successful in a variety of activities.

Prosthetics

In Chapter 4, we mentioned that to mimic a natural footstep, a prosthetic must have a coefficient of restitution that allows proper recoil after deformation. In the context of this chapter, we need to think of how the coefficient of restitution relates to potential to perform work. As the prosthetic device is deformed during impact, it stores potential energy (more specifically, strain energy) that can then be used to perform work as it recoils. Therefore, the prosthetic device that is designed properly can actually help to propel the person forward using stored energy.

One of the limiting factors in prosthetic foot design when it comes to energy transfer is the dynamics of the natural gait. When walking forward, one foot propels the body by pushing rearward with the leg and foot muscles, which provide the propulsive force. As the opposite foot strikes the ground in front of the body, its job is to first slow the momentum generated by the push off, in essence dissipating energy through eccentric muscle action. Designers of prosthetic feet had to design products that would not only dampen the energy of the foot striking the ground but also find a way to store the energy to be released later in the gait cycle. If the only factor in successful prosthetic design was energy return, then the device could essentially be a spring—as some high-performance athletic prosthetics currently demonstrate (Figure 5.23). As the user specifications move from sport to everyday activity, designers add shock absorption into the prosthetics.

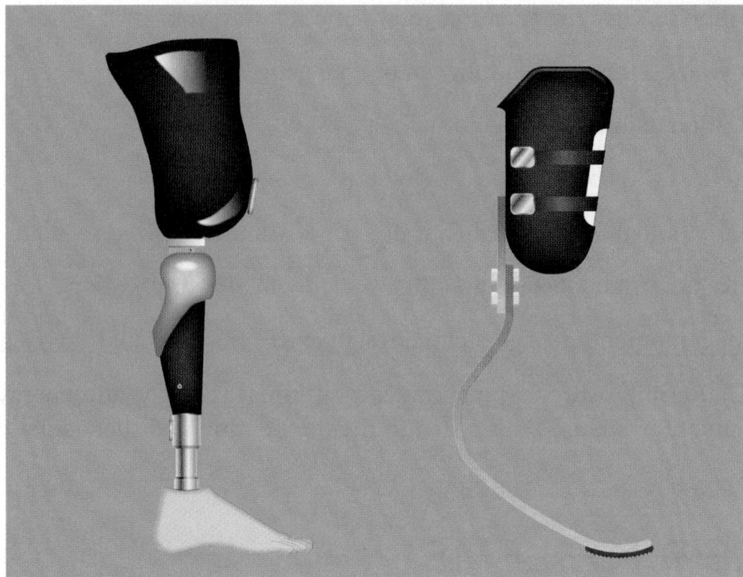

Figure 5.23 **Comparison of prosthetic devices.** Which device would be better for sprinting?

SUMMARY

In this chapter, we looked at the concepts of linear kinematics and kinetics. *Kinematic values* describe motion without concern for the forces that cause that motion. Basic kinematic values are displacement, velocity, and acceleration. *Kinetics* is the branch of mechanics concerned with the relationship between motion and the forces that cause the changes in motion. Newton's laws help us understand the kinetics of motion in linear terms. Once an object is in motion, it has a certain amount of *momentum* (quantity of motion). To change that momentum, a *force* must be applied over a given interval. The shorter the interval, the greater the force required to change momentum. This concept of *linear impulse* has widespread implications for sport, safety, and injury. We can also describe kinetic concepts from the point of view of *energy transfer*. *Work* is performed whenever an object is displaced by an applied force. The rate at which that work is performed is called *power*. Objects have the ability to perform work, depending on their position (*potential energy*) or their motion (*kinetic energy*). Mechanical energy is conserved; therefore, total work performed can be observed in terms of the sum of the changes in potential and kinetic energy of the objects involved.

REVIEW QUESTIONS

1. Explain the difference between speed and velocity.
2. What does it mean if acceleration is equal to zero?
3. Define "g" and then give its numeric value.
4. Explain variation in the acceleration caused by gravity (g) if a person travels from the north pole to the equator.
5. Define linear momentum. Give the equation and then put it into your own words.

6. Define linear impulse. Give the equation and then put it into your own words.
7. What are the two ways that momentum can be changed?
8. Define conservation of linear momentum.
9. What is the safest way to absorb shock? Why?
10. Why does a foam football hurt less when caught than a similarly sized regulation ball?
11. If a person walks up a flight of stairs and then back down, was any work performed?
12. In what situation is momentum of greater interest than KE? In what situation is KE more important?
13. Explain the relationship of work, potential energy, kinetic energy, and energy conservation in the following situation: a child pulls a sled to the top of a hill and then slides down the hill to a complete stop.

PRACTICE PROBLEMS

1. If a car goes from 0 km/hour to 40 km/hour in 4 seconds, what is the acceleration?
2. Calculate the weight at the north pole (g = 9.83 m/sec^2) and at the equator (g = 9.78 m/sec^2) of a person who has a mass of 65 kg.
3. What is the momentum of a 100-kg system moving at 5 m/sec? Show all work.
4. Using $F(t) = \Delta M$, what is the magnitude of force that would have to be applied to stop the motion of the system (i.e., momentum changes to 0) in question 3 within 2 seconds? Show all work.
5. How long would it take to stop the motion of the system (i.e., momentum changes to 0) in question 3 by applying a force of 100 N? Use $F(t) = \Delta M$. Show all work.
6. A body that has a mass of 1 kg and is moving 2 m/sec has an inelastic collision with a body that has a mass of 100 kg and is moving 0 m/sec. What is the total momentum of the system before the collision? What is it after the collision? What is their final velocity? Which body is affected most? Why?
7. If a barbell that weighs 300 N is moved a distance of 3 m, how much work is performed? If the work takes place in 1 sec, what is the power? What is the power if it takes 0.5 sec for the work to take place?
8. What is the kinetic energy of a 5-kg ball traveling at 60 m/sec?
9. Jack has a mass of 65 kg and hits a diving board at 4.5 m/sec. Jill has a mass of 75 kg and hits the diving board at 3.9 m/sec. Which diver will jump the highest off the board? Show all work. Two concepts are necessary for a complete answer.

EQUATIONS

Speed

$$s = \frac{l}{\Delta t} \tag{5.1}$$

Linear velocity

$$v = \frac{d}{\Delta t} \tag{5.2}$$

Linear acceleration

$$a = \frac{\Delta v}{\Delta t} \tag{5.3}$$

Relationship of acceleration and force	$a = \frac{\Sigma F}{m}$	(5.4)
	$\Sigma F = ma$	(5.5)
Gravitational force	$F = \frac{Gm_1m_2}{r^2}$	(5.6)
Acceleration caused by gravity	$g = \frac{Gm_2}{r^2}$	(5.7)
Gravitational force	$F = mg$	(5.8)
Linear momentum	$M = mv$	(5.9)
Linear momentum in an *elastic* collision	$(m_1v_{1i} + m_2v_{2i}) = (m_1v_{1f} + m_2v_{2f})$	(5.10)
Coefficient of restitution	$e = \frac{v_{af} - v_{bf}}{v_{bi} - v_{ai}}$	(4.12)
Linear momentum in an *inelastic* collision	$(m_1v_{1i} + m_2v_{2i}) = (m_1 + m_2)v_f$	(5.11)
Final velocity after an inelastic collision	$v_f = \frac{m_1v_{1i} + m_2v_{2i}}{m_1 + m_2}$	(5.12)
Derivation of linear impulse	$\Sigma F = \frac{m\Delta v}{\Delta t}$	(5.13)
	$\Sigma F\,(\Delta t) = m\Delta v$	(5.14)
	or	
	$\Sigma F\,(\Delta t) = m(\text{velocity}_{\text{final}} - \text{velocity}_{\text{initial}})$	
Linear impulse	$\Sigma F\,(\Delta t) = \Delta M$	(5.15)
	or	
	$\Sigma F\,(\Delta t) = \Delta m$	
Work	$W = F(d)$	(5.16)
Power	$P = \frac{W}{\Delta t}$	(5.17)
	$P = \frac{F(d)}{\Delta t}$	(5.18)
	$P = Fv$	(5.19)
Gravitational potential energy	$GPE = wh$	(5.20)
	$GPE = mgh$	(5.21)
Strain energy	$SE = ½k\Delta x^2$	(5.22)
Kinetic energy	$KE = ½mv^2$	(5.23)
Conservation of mechanical energy	$(KE + PE)_i = (KE + PE)_f$	(5.24)
Relationship of work and energy	$W = \Delta E = \Delta KE + \Delta PE + \Delta TE$	(5.25)

REFERENCES AND SUGGESTED READINGS

Bar-Or, O. 1987. The Wingate anaerobic test: An update on methodology, reliability, and validity. *Sports Medicine*, 4: 381–394.

Cajori, F. 1934. *Sir Isaac Newton's Mathematical Principles*. Trans. Andrew Motte, 1729. Berkeley: University of California Press.

Chen, W. L., J. J. O'Connor, & E.L. Radin. 2003. A comparison of the gaits of Chinese and Caucasian women with particular reference to their heel strike transients. *Clinical Biomechanics* (Bristol, Avon), 18(3): 207–213.

Cole, G. K., B. M. Nigg, A. J. van Den Bogert, & K. G. Gerritsen. 1996. Lower extremity joint loading during impact in running. *Clinical Biomechanics* (Bristol, Avon), 11(4): 181–193.

Giuliani, J., B. Masini, C. Alitz, B.D. & Owens. 2011. Barefoot-simulating footwear associated with metatarsal stress injury in 2 runners. *Orthopedics*, 34 (7): 320–323

Haywood, K. M., & N. Getchell. 2014. *Life Span Motor Development*, 6th ed. Champaign, IL: Human Kinetics.

Margaria, R., P. Aghemo, & E. Rovelli. 1966. Measurement of muscular power (anaerobic) in man. *Journal of Applied Physiology*, 21: 1662–1664.

Nigg, B. M., B. R. MacIntosh, & J. Mester (eds). 2000. *Biomechanics and Biology of Movement*. Champaign, IL: Human Kinetics.

Radin, E. L., R. B. Martin, D. B. Burr, B. Caterson, R. D. Boyd, & C. Goodwin. 1984. Effects of mechanical loading on the tissues of the rabbit knee. *Journal of Orthopedic Research*, 2: 221–234.

Radin, E. L., K. H. Yang, C. Riegger, V. L. Kish, & J. J. O'Connor. 1991. Relationship between lower limb dynamics and knee joint pain. *Journal of Orthopedic Research*, 9: 398–405.

Serway, R. A., & J. W. Jewett Jr. 2019. *Physics for scientists and engineers with modern physics*, 10th ed. Boston: Cengage Learning.

Serway, R. A., & C. Vuille. 2018. *College physics*, 11th ed. Boston: Cengage Learning.

Shih, Y., K-L. Lin, & T-Y. Shiang. 2013. Is the foot striking pattern more important than barefoot or shod conditions in running? *Gait and Posture*, 38(3): 490–494.

Simon, S. R., I. L. Paul, J. Mansour, M. Munro, P. J. Abernathy, & E. L. Radin. 1981. Peak dynamic force in human gait. *Journal of Biomechanics*, 14: 817–822.

Wolff, J. *Das Gesetz der Transformation der Knochen*. Berlin: A. Hirschwald, 1892. Trans. P. Maquet and R. Furlong as Wolff, J., 1988. *The law of bone modeling*. New York: Springer Verlag.

CHAPTER 6

ANGULAR MOTION OF THE SYSTEM

© technotr/Getty Images

LEARNING OBJECTIVES

1. Define angular kinematics.
2. Explain the relationship of angular kinetics to Newtonian law.
3. Understand the relationship between angular kinetics and energy transfer.

CONCEPTS

6.1 Torque and Angular Motion

6.2 Torque and the Center of Gravity

6.3 Angular Kinematics

6.4 Angular Kinetics and Newtonian Laws

6.5 Angular Kinetics and Energy Transfer

CONNECTIONS

6.6 Human Performance and Injury Science

6.7 Motor Behavior

6.8 Adapted Motion

It's the ninth inning of a baseball game during the World Series. A right-handed power hitter is coming to the plate with two outs and the bases loaded, and his team is behind by one run. The opposing coach has a choice of bringing in a left-handed or right-handed relief pitcher. Which does he choose? Consider that the strike zone is identical for either pitcher, and both throw nearly the same velocity and have the same variety of pitches to choose from. If the coach plays the percentages, he'll choose the right-handed pitcher.

One primary concern in this batting situation is the movement of the ball. Even with a straight pitch such as a fastball, a right-handed pitcher naturally throws a ball that travels slightly away from the batter. Add to this small deviation the fact that the right-handed pitcher can throw a curve ball that moves laterally away from the batter to a higher degree, and getting a hit becomes even more difficult. Good baseball coaches understand that batters hit the ball more effectively if it's traveling toward them, so they will generally call in a pitcher who will be able to throw the ball away from the batter in a critical situation.

This chapter will help you understand the nature of rotation and its effects on the human body and other objects. An understanding of angular motion is a key component in the study of biomechanics. Because the human body consists of segments connected at joints, almost all of our motions involve rotation. When we swing an implement, we are rotating it. When we throw or strike a ball, it is often rotating (which is sometimes desired and sometimes not). Muscles cause these rotations and can therefore be injured during the motion. The nervous system must adapt as we develop physically during childhood and into adulthood; it has to control the muscles that rotate the ever-changing body segments. Those are some of the more obvious topics related to rotation. Some topics, however, are not so obvious: Why is the knob of a door so far from the hinges? How does a cat rotate to land on its feet? Why does a diver spin faster in a tuck position? How is prosthetic design related to angular kinetics? Angular motion is present in all aspects of our lives. Therefore, it is of interest to biomechanists, physiologists, physical therapists, prosthetic designers, and motor behaviorists.

CONCEPTS

6.1 TORQUE AND ANGULAR MOTION

In Chapter 5, you read about the concepts related to linear motion. As you are probably aware, systems rarely move in purely linear motion. Usually, linear motion is accompanied by some degree of rotation. In this chapter, we look at the concepts related to angular (or rotary) motion. Recall from Chapter 2 that *rotation* occurs when the system is restricted to move around a fixed axis and therefore in a circular path (such as the segmental link motions described previously). For example, the forearm is rotated about an axis at the elbow joint during elbow flexion. During this elbow flexion, the path of the fingertips would describe an arc. Because the path of motion of one point on a rotating system or system segment describes a circle (and therefore the location of one segment relative to another forms an angle), rotation is also called *angular motion.* To this point, we have always referred to *force* when discussing changes in motion. We have done so because, in accordance with Newton's first law, there is no change in motion without externally applied force. This law still applies when the type of motion is angular. Indeed, all of the underlying concepts of linear and angular motion are the same (keep this fact in mind as we cover the material). However, we must be aware that some specific details related to force and angular motion do vary.

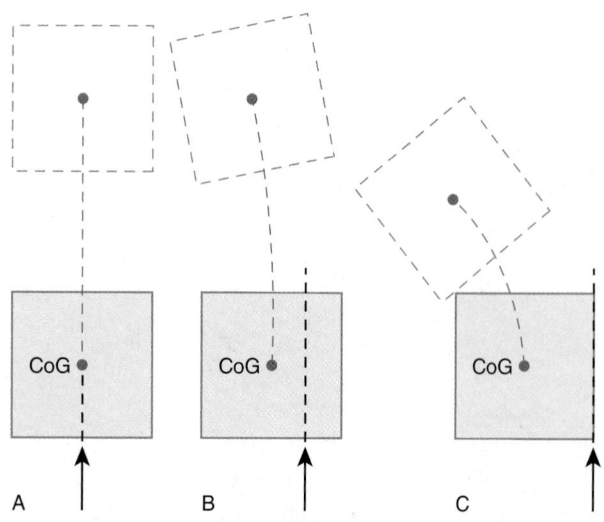

Figure 6.1 **Forces applied to a box.** What determines whether or not the box will rotate?

One concept that requires further specification is that of the point of application of the force. First, let's consider the simple case of a force applied to a box we wish to move. In our first scenario (Figure 6.1a), the force is applied directed through the center of the box. A force of this type (directed through the center) is called a **centric force**. In other words, a centric force has a line of action that passes through the center of mass of an object.

In this case, the box will slide linearly because of the externally applied *centric* force. But what happens if a force is applied in such a way that the line of action does not pass through the center of mass (Figure 6.1b)? In this situation, the box may still slide, but some rotation will also occur. In other words, an *off-center* force is necessary if a change in angular motion is desired. An off-center force is called an **eccentric** ("away from center") **force. Torque** is the tendency of an eccentric force to rotate an object around an axis. Torque is also called **moment of force** or simply **moment.** Although not everyone agrees on the specific differences and usages of the term *torque* versus *moment*, there are some consistencies. Torque is sometimes defined as the rate of change of angular momentum of an object (a movement force). In other words, usage of the term *torque* would imply that either angular velocity or the moment of inertia of an object (or both) is changing. In contrast, *moment* is the term often used for the tendency of one or more applied forces to rotate an object around an axis but without necessarily changing the angular momentum of the object (a static force). Either way, one should notice that the relative degree of linear and angular motion of the box depends on how far away from center the eccentric force is applied. In other words, if the eccentric force is applied only slightly away from the center, the box will exhibit some degree of both linear and rotary motion. But if the force is applied well away from the center, the box may display almost pure rotary motion (Figure 6.1c). In the case of pure rotary motion, the center of the box becomes the **axis of rotation** (the point about which a body rotates). So torque depends on two factors: (1) the magnitude of externally applied force and (2) the distance that the force is applied from the axis of rotation. The perpendicular distance between the axis of rotation and the line of action of the applied force is called the **force arm, moment arm,** or **lever arm.** Recall that the *line of action* is an imaginary line extending infinitely along the vector through the point of application and in the direction of the applied force.

Mathematically, torque is calculated as follows:

$$T = F \times r \times \sin\theta \quad (6.1)$$

where

T = torque (or moment of force)
F = magnitude of the externally applied force
r = distance between the axis and the point of force application (or radius of rotation of the point of force application)
θ = angle at which the force acts relative to the horizontal (positive x-axis)

The *perpendicular* distance between the axis of rotation and the line of action of the applied force is given by $r \times \sin\theta$. So if the moment arm is measured with a ruler as the perpendicular distance ($\theta = 90°$) between the line of action of the force and the axis of rotation, then $\sin\theta = 1.00$ and Equation 6.1 can be restated as:

$$T = F \times d \quad (6.1)$$

where

T = torque (or moment of force)
F = magnitude of the externally applied force
d = moment arm (*perpendicular* distance between the axis of rotation and the line of action of the applied force)

Please keep in mind that the quantity d (or moment arm) is $r \times \sin\theta$ and not simply the distance between the

Centric force A force having a line of action that passes through the center of mass of an object.

Eccentric force A force having a line of action that does not pass through the center of mass of an object.

Torque (moment of force or moment) The tendency of an eccentric force to rotate an object around an axis.

Axis of rotation The point about which a body rotates.

Force arm (moment arm, or lever arm) The perpendicular distance between the axis of rotation and the line of action of the applied force.

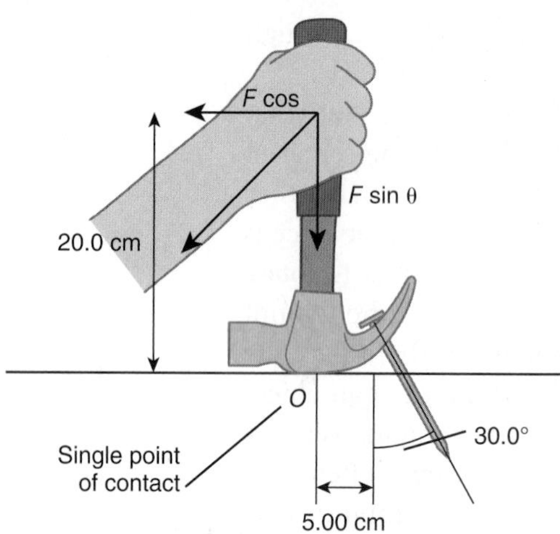

Figure 6.2 **Hammer pulling out a nail.** How would a longer handle on the hammer help to pull out the nail?

axis and the point of force application. Also, be careful not to confuse this equation with the one for work. The units for describing torque are units of force (N) multiplied by units of distance (m). So the Système International (SI) unit for torque is the N · m. For example, if a 200-N force is applied to the handle of a hammer to pull out a nail and the moment arm is 0.2 m, then the torque = 40 N · m.

Notice in Figure 6.2 that the force vector has been resolved into its vertical component (F sin θ) and its horizontal component (F cos θ). Two pieces of information require further attention. First, notice that the line of action of the vertical component passes through the axis of rotation at O. This means that there is no moment arm for the vertical component because there is no distance between the axis of rotation and the line of action. Therefore, the vertical component cannot produce torque ($T = F \times 0 = 0$ N · m). This is the same reason that the centric force in our box example caused only linear motion.

Notice that the first form of Equation 6.1 ($T = F \times r \times \sin \theta$) is similar to Equation 3.10 ($\mathbf{A} \times \mathbf{B} \times \sin \theta = \mathbf{C}$). Equation 3.10 was used to demonstrate multiplying one vector by another. Multiplying one vector by another (finding the *cross product*) results in another vector, with the orientation of **C** being perpendicular to the plane formed by **A** and **B**. Torque is the cross product of the force vector and the moment arm vector. Therefore, the vector representing torque can be visualized as an arrow perpendicular to the plane formed by the force vector and the force arm (moment arm). Torque is a vector quantity; as such, it has a magnitude and a direction. Don't confuse the direction of the torque with the direction of the applied force. Torque has a turning effect. So the direction of torque is that associated with the rotation that it causes clockwise or counterclockwise (Figure 6.3).

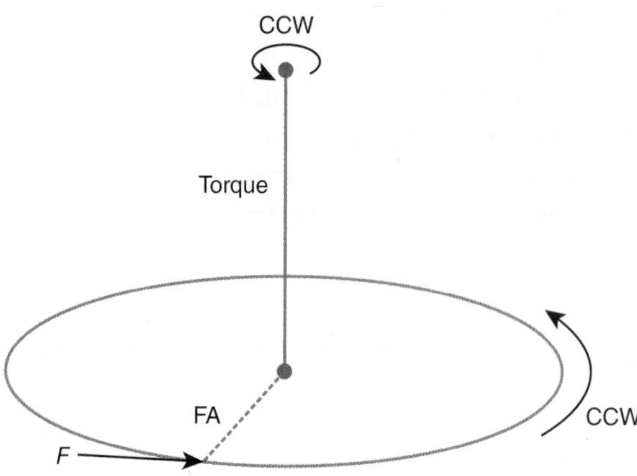

Figure 6.3 **The vector representing torque.**

So if a person were able to observe the torque vector from the preceding, it would be turning either clockwise or counterclockwise. As is the case with force, direction of torque is specified with a positive (+) or negative (−) sign. In the case of a force vector, we indicate direction relative to a defined reference frame and then assign a negative or positive value accordingly. In the case of torque, we must first identify the axis around which the rotation will occur. We then assign a positive or negative sign according to whether the rotation about that axis will be counterclockwise or clockwise. A torque that causes counterclockwise rotation is assigned a positive value, and a clockwise rotation is assigned a negative value. The approach to remembering this convention is called the *right-hand thumb rule.* First, identify the axis around which torque will cause rotation (x, y, or z). Once the axis is known, we can identify the plane of motion that is perpendicular to the axis. Imagine that you wrap your right hand around the axis of rotation, with your thumb pointing in the positive direction associated with that axis (Figure 6.4).

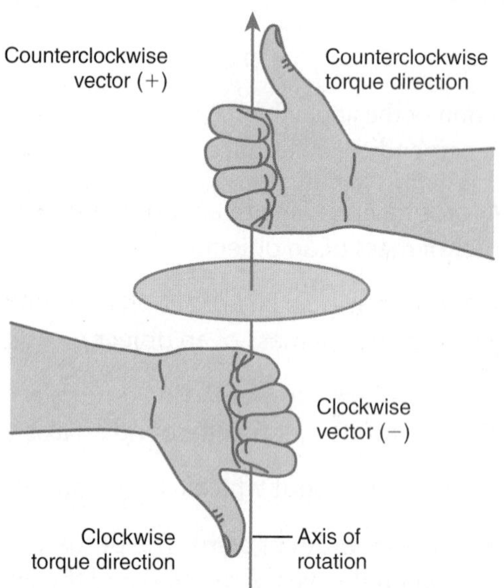

Figure 6.4 **Demonstration of the right-hand thumb rule.**

Your fingers will be in the *plane* of motion and will be wrapped in the direction of rotation. An example of the right-hand thumb rule in everyday living is the tightening or loosening of jar lids, screws, faucets, and so on. Most of these threaded objects follow the right-hand thumb rule. People often try to think of a way to remember which way to turn these objects. Many people use the phrase "righty-tighty, lefty-loosey." Of course right and left have no relationship to rotation, but the idea is the same as the right-hand thumb rule. For example, which way do we turn a screw to make it go into a board? According to the "righty-tighty" rule, we would turn it to the right in order to tighten it. However, "right" and "left" have no meaning in terms of rotation, because the head of the screw could be said to move to the "right" for part of the time and "left" for part of the time (i.e., it depends on how you look at it).

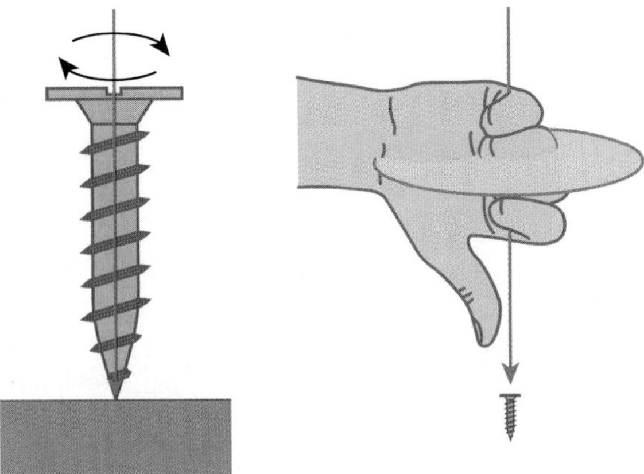

Figure 6.5 **The right-hand thumb rule indicates in which direction the screw will rotate.**

The right-hand thumb rule is much more reliable. First, we define the axis of rotation and plane of motion. With a screw and a board, the axis of rotation will likely be defined as the y-axis if the board is on the ground. The plane of motion will be the plane in which the head of the screw moves and is perpendicular to the y-axis (in this case, a transverse plane). Next, we simply define the direction in which we would like the screw to move. In this case, the desired direction of motion is *into* the board; this would be in the *negative y* direction according to our Cartesian reference frame (Figure 6.5a). If we now point our thumb in the direction of desired motion (Figure 6.5b), we will see that the head of the screw has to be rotated clockwise (which corresponds to the negative direction of motion along the y-axis).

So the right-hand thumb rule provides some consistency in defining the turning effect of a torque. Remember that the torque itself is not "positive" or "negative." The sign simply indicates the direction or sense of the rotation around the defined axis.

Similar to forces, once torque directions have been established, the torques can be summed to indicate the final magnitude and direction of motion (i.e., net torque). For example, if the sum of the torques in in one of the diagrams of Figure 6.6 equals +300 N · m, the system will rotate counterclockwise with a magnitude of 300 N · m of torque. If the sum is negative, we know that the system will rotate clockwise. To find the net torque, we simply follow these steps:

1. establish the magnitude of the applied force according to the given scale (e.g., 1 cm = 10 N of force),
2. find the length of the moment arm by measuring the perpendicular distance between the axis of rotation and the line of action of the applied force,
3. torque for each force is calculated according to the equation,
4. establish the direction of each torque using the right-hand thumb rule, and
5. sum the torques to find the net magnitude and direction of rotation caused by the torques.

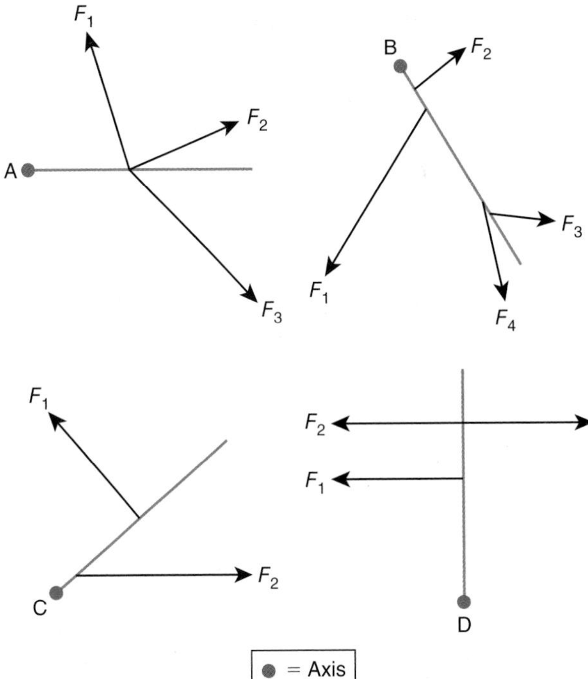

Figure 6.6 **Calculating net torque.** How are these situations similar to muscle contractions that cause rotation?

Torque is a critically important biomechanical concept. It not only applies to rotating objects in general but also directly applies to motion of the human body. Our muscles apply forces to bones in order for us to move. Those muscle forces are eccentric, therefore we move because of torque (Figure 6.7).

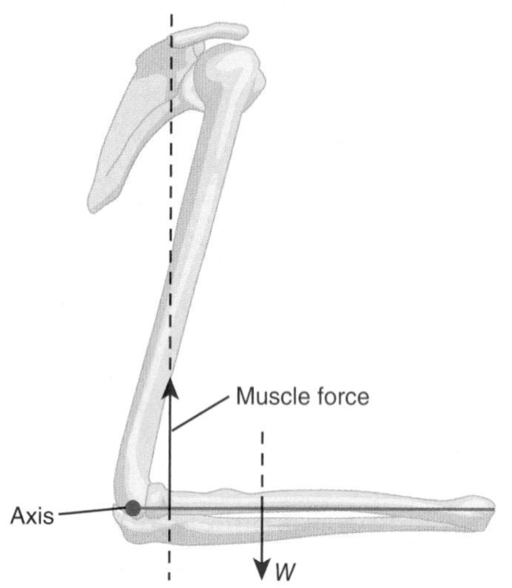

Figure 6.7 **Torque produced by the biceps brachii muscle.** What will be the direction of rotation?

So we will refer to torque in the following sections related to angular motion but also throughout the rest of the book. Keep in mind as we go along that all change in motion is caused by force. If the force is centric, the motion will be linear. If the force is eccentric, a moment arm is defined, and the motion will, at least to some extent, be angular. We will also be referring back to the concept of net torque because it applies to several of our following sections. Remember that if net torque is positive, the system will rotate counterclockwise in accordance with the right-hand thumb rule. If net torque is negative, rotation caused by the applied torque will be clockwise. We now turn our attention to the center of gravity, a direct application of net torque.

6.2 TORQUE AND THE CENTER OF GRAVITY

In Chapter 2 we defined the center of gravity in terms of its relationship to the cardinal planes of motion. The cardinal planes bisect the body, and as such they must also pass through the center of mass (the point at which all of a system's mass is concentrated or the point at which any plane passing through it divides the mass of the system in half). Gravitational pull is concentrated at the center of mass. So, at least in the vertical axis, the center of mass can be considered synonymous with the center of gravity (the point at which the force of gravity seems to be concentrated) (Winter, 2009). In other words, the center of mass (or center of gravity) of the system is at the intersection of the three cardinal planes. With an understanding of torque, we can gain deeper insight into the definitions of the center of mass and the center of gravity.

First, let's begin with this definition of the center of mass: the point at which any plane passing through it divides the mass of the system in half. According to this definition, the center of mass of an object possessing uniform mass and mass distribution (i.e., a symmetrical object made of the same material throughout its structure) is directly in the middle of the object. Let's use the board of a teeter-totter as an example. The teeter-totter is made of a consistent material throughout and is symmetrical in shape. So its center of mass is located directly in its geometrical center, and any plane passing through that point will divide the mass of the teeter-totter in half.

Now, to understand the center of gravity, we must remember that gravitational force will be acting on every particle of the mass of the object. Also recall that gravitational force acts downward in the vertical direction. Now, to understand the center of gravity, let's place the teeter-totter board on its platform (Figure 6.8).

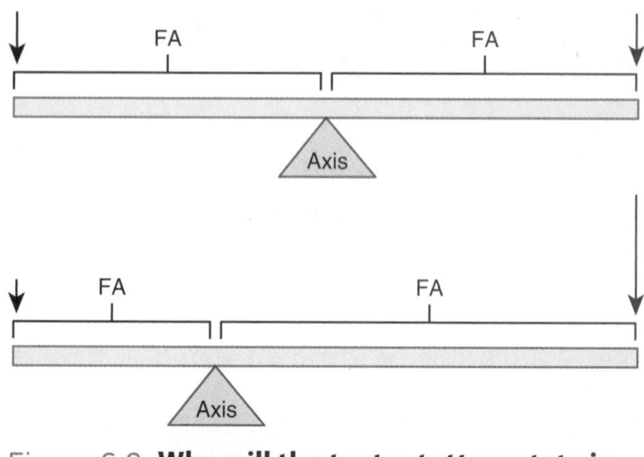

Figure 6.8 **Why will the teeter totter rotate in the second situation?**

Notice that a teeter-totter balances as long as the platform is directly in the middle. What if the platform is not placed in the middle? Notice that the board of the teeter-totter rotates around the platform, which acts as an axis of rotation. We observe rotation, so torque must have been present. Did it suddenly appear, or was it there all the time? The board is the same, and the platform (axis of rotation) was always there. So what changed? The difference is that the mass of the board is not evenly distributed on each side of the axis of rotation. Technically, this unevenness means that two values changed. First, there is more mass on one side of the axis of rotation. And because weight depends on mass ($F = mg$), there is now more force on one side of the axis of rotation. Second, there is now a greater perpendicular distance between the axis of rotation and the line of action of the force on one side of the teeter-totter. Greater force and a longer force arm (moment arm) lead to greater torque on one end of the board. So if we return to the idea of net torque, we see that the net torque in this case is not equal to zero, so rotation occurs.

So let's return to the situation in which the teeter-totter was balanced and did not rotate. In this case, the force on each end was the same, and the moment arms were the same. Therefore, the torque on each end of the teeter-totter is the same. If allowed, one end of the teeter-totter would rotate clockwise, and the other would rotate counterclockwise. This tendency means that one of the torques is a negative value and the other is positive. Therefore, the sum of the torques is zero and no rotation takes place. So another definition of the center of gravity is the point around which all of the torques (or moments) sum to zero. This is why the term *center of gravity* only applies in the vertical direction (the direction of gravity). In the two horizontal directions, the term *center of mass* is used because gravity isn't acting in those directions (Winter, 2009).

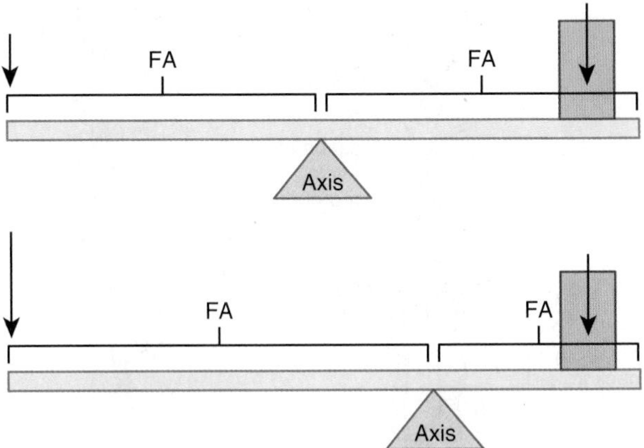

Figure 6.9 **Torque acting on a teeter-totter.** What causes the teeter-totter to balance in the second situation?

What about the center of gravity in an object that is not of uniform mass or mass distribution? For example, we could place some weight on one end of the teeter-totter (Figure 6.9). Once we have done that, what must be done to balance the teeter-totter again? We must relocate the axis of rotation to a point closer to the added mass.

This move will balance the torques and show the location of the center of gravity. The same concept applies to the human body (Figure 6.10). For example, the human torso contains many cavities and therefore does not consist of as much mass as the lower body. The head is relatively solid and locates above the torso. The lower body tapers such that more mass is located near the hips. This distribution means that we cannot simply locate the center of gravity by measuring equal distances from the top of the head to the bottom of the feet. But the principle is the same; we must find the point around which all of the torques sum to zero. The final result of this complicated mass distribution is that the center of gravity varies between approximately 53% and 59% of standing height measured from the soles of the feet, depending on gender

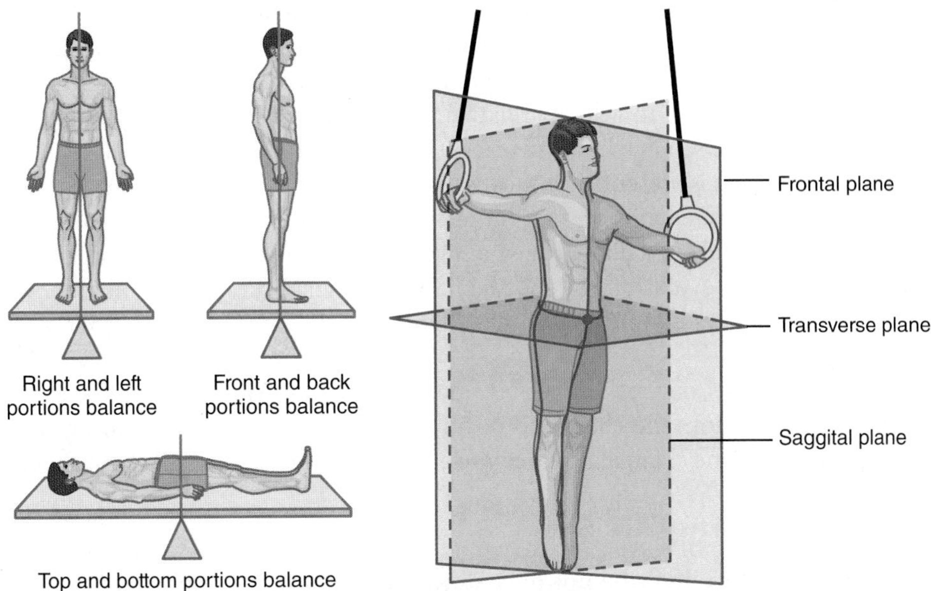

Figure 6.10 **Location of the center of gravity in a human.**

Figure 6.11 **Possible locations of the center of gravity.** How could you locate your center of gravity?

(Braune & Fischer, 1889; Dempster, 1955; Hay, 1973; Heinrichs, 1990; Plagenhoef, Evans, & Abdelnour, 1983; Winter, 2009).

Females tend to have a center of gravity that varies between 53% and 55% of standing height, whereas males have a range of 54% to 59%. This difference arises because females tend to have relatively more mass distributed in the lower body than males. For example, females tend to have a larger pelvis, narrower shoulders, and greater fat and muscle deposits below the waist. In addition, a child's center of gravity is higher because the child's head is relatively large in comparison to her height. Remember that any time mass is shifted, the location of the center of gravity changes because we must find a new location for the axis of rotation to balance the torques.

So understanding that any change in body position redistributes a portion of the mass is important. This redistribution in turn changes the torque about a given axis. Therefore, to find the center of gravity, the axis of rotation must be moved in the direction of the most massive area (Figure 6.11). In fact, the center of gravity may actually be located outside of the body. For example, where is the center of gravity of a donut?

Knowing the location of the center of mass or center of gravity during human motion is useful in events in which the path of the center of gravity is critical to successful performance of the skill. The location of the center of mass and center of gravity also has implications for both balance and floating position in fluids. These implications are discussed in subsequent sections and chapters.

6.3 ANGULAR KINEMATICS

In Chapter 5, we described the linear kinematics of motion. Kinematics is the study or description of the spatial (direction with respect to the three-dimensional world) and temporal (motion with respect to time) characteristics of motion without regard to the causative forces. We described motion in terms of displacement, velocity, and acceleration. *Angular kinematic* values are described in the same way. We will again be discussing displacement, velocity, and acceleration. Displacement is still defined as change in position, velocity is change in position relative to the interval in which it takes place, and acceleration is change in velocity relative to the interval in which the change takes place. So the basic concepts are the same, with some exceptions. One difference is that an eccentric force is applied, and therefore torque is present. This means that all of our quantities are now angular in nature, which means there is a specified axis of rotation. Also, a segment rotating around an axis may have the ability to travel completely in a circle (maybe several times). Therefore, some angular quantities require that the number of revolutions be specified. For the most part, in the human body, this isn't a problem. But the human system may be riding a bicycle, in which case the number of revolutions is certainly a concern.

As with linear kinematics, our first concern is a starting position. We will then proceed to displacement from that initial position, the change in position relative to the interval during which it took place (velocity), and finally the change in velocity relative to the interval of that change (acceleration).

Angular Distance and Displacement

In biomechanics we need to describe not only single points but also segments (represented with lines connecting points). The key is that we are using *lines* as opposed to single *points* as a means of representation. Because systems of interest are often composed of segments, we must be able to specify not only the location of one point on a segment but also the relative motion of multiple points on a single segment and motions of the segment as a whole. Again, we need a specific position of the system to provide a starting point from which to measure various kinematic parameters. So first we define the *position* or location of the object (or system) that is rotating. You learned in Chapter 3 that it is convenient and necessary to define a *polar coordinate system* and locate a point in space using its *plane polar coordinates.* We also described the steps needed to transform Cartesian coordinates into polar coordinates and vice versa. Recall that a polar coordinate system has an origin (O) and multiple orthogonal axes (one for each dimension), and the positive x-axis is often used as the reference axis. In the case of angular kinematics, the origin is the axis of rotation. The **angular position** of the given point is then defined by its distance (radius) r from the origin, and the angle (θ; the Greek letter *theta*) between the chosen reference axis and the line formed by connecting the given point to the origin. So a point with the polar coordinates $(r, \theta) = (7.00\text{ m}, 55°)$ is located 7 m away from the origin at an angle 55° above the reference axis (Figure 6.12).

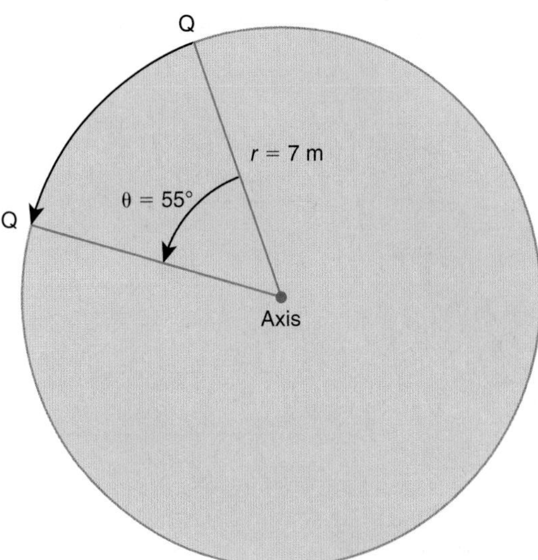

Figure 6.12 **Polar coordinates of a point located 7 meters from the axis of rotation.**

If the reference axis is also capable of moving, then the angular position is called *relative.* If the reference axis cannot move relative to the segment of interest, then the angular position is *absolute.*

Once the angular position is established, we have a reference position from which to measure angles and changes in motion. **Angular displacement** (Δθ) is the change in angular position of a segment or any point on the rotating segment. In the case of rotation, the segment and all points on the rotating segment have the same angular displacement. The angular displacement of the segment or any point on the segment can be measured in degrees or radians. Degrees are straightforward; most people know that a circle has 360°. But we can also measure angles in radians (radius units). A **radian** (rad) is the ratio of the length of a circular arc to the length of the circle's radius. Remember that any point on a rotating segment will describe an arc. When the length of that arc is equal to the radius of the circle, then one rad of angular displacement has been traveled. So θ in radians is calculated as:

$$\theta = \frac{l}{r} \tag{6.2}$$

where

θ = angular measurement in radians
l = length of the described arc
r = radius of the circle

Because a radian is a ratio of length to length, it is sometimes described as a "unitless" quantity. Technically, a radian is meters along the circumference/radial meters. The circumference of a circle = 2π rad (or 6.28 rad) = 360°. Therefore, 1 rad = 57.3°. So π rad (or 3.14 rad) = 180°, and π/2 rad (or 1.57 rad) = 90°. Both degrees and radians are acceptable units of measure for angular displacement. However, some mathematical situations are simplified by the use of radians. Our next equation is an example of the convenience of radians.

Angular position The location of the given point, defined by its distance r (radius) from the origin, and the angle (θ; the Greek letter *theta*) between the chosen reference axis and the line formed by connecting the given point to the origin.

Angular displacement The change in angular position of a segment or any point on the rotating segment.

Radian The ratio of the length of a circular arc to the length of the circle's radius; 1 rad = 57.3°.

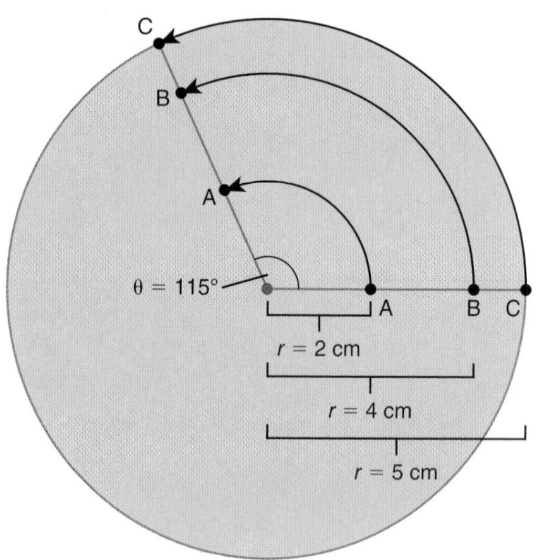

Figure 6.13 **Radius of rotation and curvilinear displacement.** Each radius of rotation produces a different curvilinear path.

As you have read, each point on a rotating segment describes an arc. That arc is the curvilinear displacement of the point on the segment. Although every point on a rigid segment will travel the same angular distance, each point has a different curvilinear displacement (Figure 6.13), depending on its distance from the axis of rotation (**radius of rotation**).

This curvilinear displacement is the length of the arc (l) as shown in Equation 6. 2. Therefore, we can calculate the linear displacement of a point on a rotating segment by rearranging Equation 6.2:

$$l = \Delta\theta r \tag{6.3}$$

where

l = linear displacement of a point on a rotating segment
$\Delta\theta$ = angular displacement of the line or segment measured in radians
r = radius of rotation of the point on the rotating segment

Because radians are unitless, linear displacement will be in linear distance units. For example, if a point is on a segment 1 meter from the axis of rotation (radius of rotation = 1 m) and the segment is angularly displaced 3 rad (172°), the linear distance traveled by the point is 3 meters. Be careful not to confuse the angular and linear displacements of the point. All points on the line will be *angularly* displaced 3 rad, but the *linear* displacement of each point will depend on its radius of rotation. Let's use a different type of situation to fully illustrate this point.

SAMPLE PROBLEM

Consider two wheels, one with a diameter of 1 m, and the other with a diameter of 2 m. What will be the linear distance traveled by each wheel on completion of one full revolution? First, one complete revolution would be 6.28 rad (i.e., 2π) or 360°, so all points on both wheels travel an *angular* distance of 6.28 rad or 360°. However, the wheel having a diameter of 1 m will travel a *linear* distance of 3.14 m, whereas the wheel with the diameter of 2 m will travel a *linear* distance of 6.28 m. So although all of the points on a rotating segment travel the same angular distance, linear distance of a point on the segment depends on the radius of rotation of that point.

Angular Speed and Velocity

As with linear kinematics, we need to calculate the rate of motion. **Angular speed** is the scalar rate of angular motion and is calculated as follows:

$$\delta = \frac{\varphi}{\Delta t} \tag{6.4}$$

where

δ = angular speed, or rate of angular motion
φ = angular distance traveled
$\Delta t = \text{time}_{\text{final}} - \text{time}_{\text{initial}}$ (the change in time, or interval)

Radius of rotation The distance from the axis of rotation to a given point on the rotating segment.

Angular speed The scalar rate of angular motion.

The units for describing angular speed are a unit of angular distance divided by a unit of time. So if we know that 85° of angular distance was traveled by a segment in 5 sec, then the angular speed of the segment was 17°/sec.

Angular velocity is the vector rate of angular motion. To calculate the rate of change of angular displacement, we need to know the angular displacement and the interval in which the angular displacement took place. Don't forget that because angular velocity is a vector quantity (i.e., based on displacement), direction must be specified according to the right-hand thumb rule. Angular velocity or the rate of change of angular displacement is calculated as:

$$\omega = \frac{\Delta\theta}{\Delta t} \tag{6.5}$$

where

ω = the angular velocity, or rate of angular motion in a specific direction
$\Delta\theta$ = angular position$_{final}$ − angular position$_{initial}$ (angular displacement)
Δt = time$_{final}$ − time$_{initial}$ (the change in time, or interval)

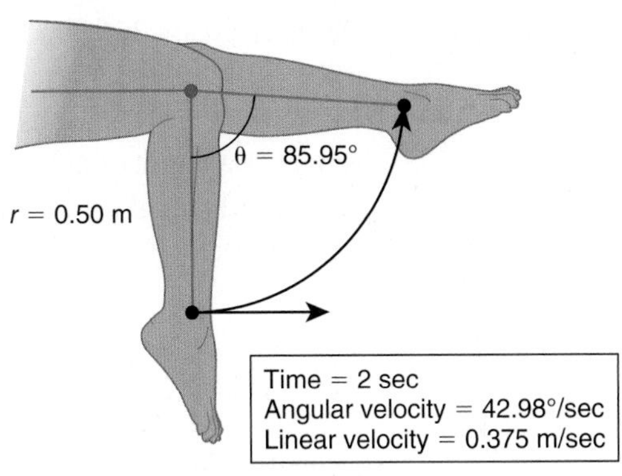

Figure 6.14 **Angular versus linear velocity.**

The units for describing angular velocity are a unit of angular distance divided by a unit of time. Therefore, if a segment of the body travels 45° and the interval is 1 sec, then the angular velocity would be 45°/sec (or 0.785 rad/sec).

Because Equation 6.3 allows us to calculate the linear distance traveled by a point on a rotating segment, we can also calculate the linear velocity of a point on a rotating segment (Figure 6.14). Linear velocity of a point on a rotating segment is called **tangential linear velocity** because the linear velocity vector of the point on the segment is tangent to the path of the object and perpendicular to the radius of the circular path (Serway & Jewett, 2019).

Tangential linear velocity is calculated mathematically as follows:

$$v_T = \frac{\Delta\theta r}{\Delta t} = \omega r \tag{6.6}$$

where

v_T = linear velocity of a point on a rotating segment (tangential linear velocity)
ω = the angular velocity, or rate of angular motion
$\Delta\theta$ = angular position$_{final}$ − angular position$_{initial}$ (angular displacement in radians)
r = radius of rotation of the point on the rotating segment
Δt = time$_{final}$ − time$_{initial}$ (the change in time, or interval)

Because angular displacement in Equation 6.6 is in radians and the radius of rotation is a linear distance, the units are the same as linear velocity (m/sec). For example, if the knee joint travels through 1.5 rad (85.95°) of angular displacement in 2 sec, and we are tracking a point that is 0.50 m from the knee (axis of rotation), then angular velocity of all points on the segment is 0.75 rad/sec (42.98°/sec), but the linear velocity of the point of interest is 0.375 m/sec.

As with linear quantities, the rate of motion may vary throughout the change in angular position. During these variations, the rate of motion is not as great as the **peak rate of motion**, or maximum rate of motion achieved. In other words, the rate of motion is not necessarily a constant throughout the movement. So again, we are actually calculating an *average angular speed* or *average angular velocity.* We may be interested in the rate of motion at one given instant in time, **instantaneous angular speed** or **instantaneous angular velocity.** So a kinematic value

Angular velocity The vector rate of angular motion, or rate of angular motion in a specific direction.

Tangential linear velocity The linear velocity of a point on a rotating segment.

Peak rate of motion Maximum rate of angular motion achieved.

Instantaneous angular speed The rate of angular motion at one given instant in time; also called *instantaneous angular velocity.*

such as the peak rate of motion is an instantaneous speed or instantaneous velocity measured during the period in which the greatest distance was covered in the shortest interval. For example, the angular velocity of a golf club is not a constant value throughout the swing. The angular velocity of the club starts at a value of zero, then increases to a given maximum value, and then is brought back to zero. We could calculate an average angular velocity for the entire interval, or we could calculate the peak angular velocity.

Angular Acceleration

As with linear velocity, angular velocity is not always a constant value: There are periods of speeding up and slowing down, and these periods may be associated with changes in direction of the rotating system. Previously, we stated that acceleration is the term for a change in state of motion. More specifically, **angular acceleration** is a change in magnitude or direction of the angular velocity vector with respect to time. Mathematically, angular acceleration is calculated as:

$$\alpha = \frac{\Delta\omega}{\Delta t} \tag{6.7}$$

where

α = angular acceleration, or rate of change in angular velocity
$\Delta\omega$ = angular velocity$_{final}$ − angular velocity$_{initial}$ (change in angular velocity)
Δt = time$_{final}$ − time$_{initial}$ (the change in time, or interval)

The units for describing angular acceleration are a unit of angular distance divided by a unit of time divided by a unit of time. For example, in 3 seconds (time$_{initial}$ = 0 sec and time$_{final}$ = 3 sec) an arm segment changes its state of motion from 45°/sec ($v_{initial}$ = 45°/sec) to an angular velocity of 90°/sec (v_{final}). This change means that the angular acceleration is 15°/sec/sec. Keep in mind that angular acceleration is a vector quantity, with both a magnitude and direction of change in angular velocity. The direction is specified according to the right-hand thumb rule. Also, as with linear acceleration, angular acceleration can be equal to zero (angular velocity$_{final}$ = angular velocity$_{initial}$), a positive number (angular velocity$_{final}$ > angular velocity$_{initial}$), or a negative number (angular velocity$_{final}$ < velocity$_{initial}$). If angular acceleration is equal to zero, it doesn't mean that the system is no longer rotating. It simply means that the system is not changing its current state of motion.

Also, angular acceleration (like angular speed and angular velocity) is not always a constant value. It can change over the interval in which it is measured. Therefore, **average angular acceleration** is the change in angular velocity divided by the entire interval over which it changed; the **instantaneous angular acceleration** is the change in angular velocity at one specific instant in time. As with our example of swinging a golf club, there is a period during which the golf club is accelerated, an interval during which peak acceleration is reached, and then a period of deceleration. We could therefore calculate an average acceleration or an instantaneous acceleration (such as peak acceleration).

Equation 6.6 allows us to calculate the tangential linear velocity of a point on a rotating segment. If the magnitude of the velocity vector changes, then a linear acceleration has occurred. We can calculate this **tangential linear acceleration** of a point on a rotating

Angular acceleration A change in magnitude or direction of the angular velocity vector with respect to time.

Average angular acceleration The change in angular velocity divided by the entire interval over which it changed.

Instantaneous angular acceleration The change in angular velocity at one specific instant in time.

Tangential linear acceleration The linear acceleration of a point on a rotating segment.

segment. Again, it is called *tangential* because the velocity vector is tangent to the path of the object and is calculated as follows:

$$a_T = \frac{\Delta \omega r}{\Delta t} = \alpha r \tag{6.8}$$

where

a_T = linear acceleration of a point on a rotating segment (tangential linear acceleration)
α = angular acceleration, or rate of change in angular velocity
$\Delta\omega$ = angular velocity$_{\text{final}}$ − angular velocity$_{\text{initial}}$ (change in angular velocity)
r = radius of rotation of the point on the rotating segment
Δt = time$_{\text{final}}$ − time$_{\text{initial}}$ (the change in time, or interval)

Again, because of the unitless nature of radians, the units are the same as linear acceleration (m/sec/sec).

Centripetal Acceleration

We discussed centripetal force in Chapter 4 as simply a term to describe any force that causes a system to exhibit circular motion, often called *center seeking*. This kind of motion could potentially be caused by any force, depending on the situation. For example, to continuously swing a rope in a circle, one must pull the rope toward the center of the circle. This action causes the rope to exhibit a somewhat uniform circular motion. If the motion of the rope is uniform, then there is no *angular* acceleration. However, because the path of a point on the rope is circular, the point is always changing direction. Remember that acceleration is a change in state of motion, and that change can be in the form of change in magnitude or direction of the velocity vector. In this case, the magnitude of the velocity vector is not changing, but the direction is constantly changing. So in the case of circular motion, the point of interest is constantly being accelerated toward the center of the circle (i.e., the acceleration vector is perpendicular to the path and points toward the center of the circle) (Serway & Jewett, 2019). Acceleration caused by change in *direction* of the velocity vector is called **centripetal acceleration** (center-seeking acceleration). Be careful not to confuse centripetal acceleration with tangential linear acceleration, which is caused by a change in *magnitude* of the velocity vector. Mathematically, centripetal acceleration is calculated as:

$$a_c = \frac{v_T^2}{r} \tag{6.9}$$

where

a_c = centripetal acceleration
v_T^2 = tangential linear velocity
r = radius of rotation of the point on the rotating segment

So if a car is driving around a track, the faster it travels the more centripetal acceleration there must be to keep it on the track. By the same token, if the track is smaller, there will also have to be a greater centripetal acceleration (Figure 6.15).

Because acceleration is caused by applied force, more friction force is needed to accelerate the car toward the center of the track (i.e., to keep it from sliding off). The friction force in this example is a centripetal force, causing centripetal acceleration. The relationship of centripetal force and centripetal acceleration will be discussed in further detail in a subsequent section.

Centripetal acceleration An acceleration caused by change in *direction* of the velocity vector; a center-seeking acceleration.

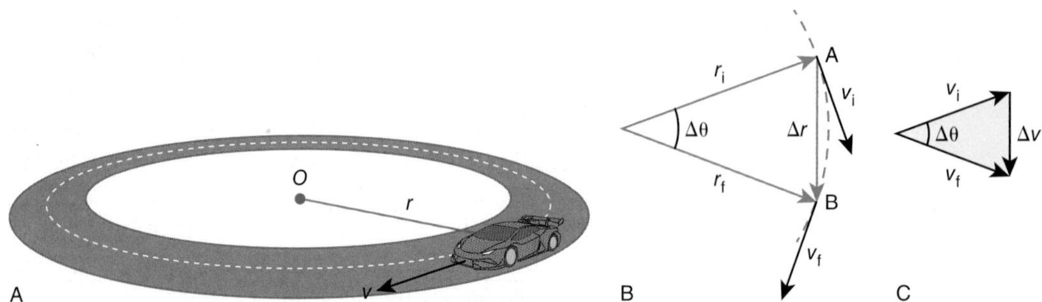

Figure 6.15 **A car on a race track.** What force is acting centripetally in this situation?

6.4 ANGULAR KINETICS AND NEWTONIAN LAWS

Kinetics is the study of forces that inhibit, cause, facilitate, or modify motion of a body. Again, in this chapter our attention will be focused on kinetics during *angular* motion. The forces discussed in Chapter 4 (such as gravity and friction) are still the forces of interest in studying the *angular kinetics* of system motion.

Application of the Newtonian Laws

Kinetics explores the forces involved with system motion. Newton attempted to explain the behavior of systems in the absence and presence of forces. So we will return to Newton's laws in an attempt to better understand the resultant motion of systems (Cajori, 1934; Serway & Jewett, 2019; Serway & Vuille, 2018). However, remember that in the case of angular kinetics, the forces are applied eccentrically. Eccentric forces cause angular motion, which means there is a defined axis of rotation.

Newton's First Law: Law of Inertia

Every body perseveres in its state of rest or of uniform motion in a right line unless it is compelled to change that state by forces impressed thereon. In terms of the kinematic values prevously described, this law means that if an object is at rest it will not undergo angular displacement without the application of an external eccentric force (torque). Without angular displacement, angular velocity is equal to zero. If the object is already rotating with a given velocity, there will be no change in that angular velocity without an externally applied eccentric force. If there is no change in angular velocity, then there is no angular acceleration of the system. So again, Newton's first law helps us understand the relationship between kinematics and kinetics. Without the torques (kinetics) discussed previously, the spatial and temporal characteristics of motion (kinematics) remain constant. Remember that this law explains what happens to a system in the absence of applied torques. So once again, in most *biomechanical* situations, Newton's first law cannot be directly applied because we are subjected to eccentric forces (torques) in almost any movement situation. However, we do need to revisit the concept of inertia.

So, we will now return to the basic concept of inertia (which isn't so basic when referring to angular motion). This law explains the motion of a system in the absence of externally applied force. In terms of angular kinetics, it explains what happens in the absence of an externally applied eccentric force. Simply stated, a system at rest will remain at rest, and a system in motion will remain in motion in a straight line until acted on by an external torque. So the natural tendency of a system is to resist change in motion. We have described *inertia* as the resistance of a system to having its state of motion changed by application of a force. Therefore, this law is also called the *law of inertia*. We also mentioned that *mass* is a measure of a system's inertia. So, the more massive an object is, the greater its tendency to remain in its current state of motion. This section is concerned with **rotational inertia** (or **angular inertia**): the resistance of an object to having its state of angular motion changed. As with linear kinetics, mass is still a good indicator of a system's angular inertia. However, when a system is rotating, mass is not the only indicator of resistance to motion change; mass *distribution* also plays a role. For example, try swinging a wooden bat from the "fatter" end. You should notice a difference in the force necessary to rotate the bat. The bat has less rotational inertia. But the mass of the bat didn't change; we simply held it in a different position. What did change?

The difference is that much of the mass of the bat is now closer to the axis of rotation (i.e., a change in mass distribution). Every particle of mass in the bat provides resistance to change in motion. But the amount of resistance to change in motion exhibited by a particle varies depending on how far the particle is from the axis of rotation. Don't forget that torque is also called *moment* and depends on force and the distance that the force is applied from the axis of rotation. The mathematical relationship of mass, mass distribution, and rotational inertia is called **moment of inertia**:

$$I_a = \Sigma m_i r_i^2 \tag{6.10}$$

where

I_a = moment of inertia about a given axis
m_i = mass of a given particle i
r_i^2 = radius of rotation of the particle i to the axis of rotation

The units for describing moment of inertia are units of mass (kg) multiplied by units of distance (m) squared. Therefore, the SI unit for moment of inertia is $kg \cdot m^2$. For example, let's calculate the moment of inertia for a mass of 20 kg being swung in a circle at the end of a 5-meter rope: $20 \text{ kg} \times (5 \text{ m})^2 = 500 \text{ kg} \cdot \text{m}^2$. In the case of our previous example of a wooden bat swung from the fat end, the number of particles is the same, but more of them are closer to the axis of rotation. This arrangement lowers the moment of inertia, and the bat is less resistant to having its state of angular motion changed.

Not all cases are as simple as this example. In the case of extended objects (such as cones, cylinders, disks, or spheres), we cannot simply calculate the moment of inertia using Equation 6.10. Instead, variations of Equation 6.10 are used according to principles of calculus (Robertson et al., 2014; Serway & Vuille, 2018).

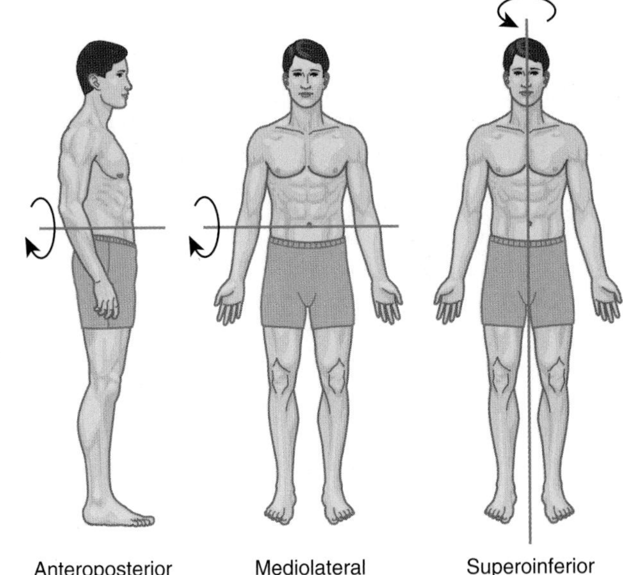

Figure 6.16 **Rotation around each of the principle axes.**

Notice that we calculate the moment of inertia about a given axis. We do so because the moment of inertia for an entire system (for example, the entire human body) varies with respect to the axis of rotation (i.e., we have one linear inertia, our mass, but we have many moments of inertia). This variability arises from the fact that the human body can change its shape (e.g., muscles cause flexion, extension, abduction, etc.) about a given axis and therefore change its moment of inertia about that given axis. For example, a human body rotating in the air is exhibiting angular motion around one of the three principle axes (Figure 6.16).

When referring to the moment of inertia with respect to one of the principle axes of rotation, the moment of inertia is called the **principal moment of inertia**. Therefore,

Rotational inertia (angular inertia) The resistance of an object to having its state of angular motion changed.

Moment of inertia The mathematical relationship of mass, mass distribution, and rotational inertia.

Principal moment of inertia Term used when referring to the moment of inertia with respect to one of the principle axes of rotation.

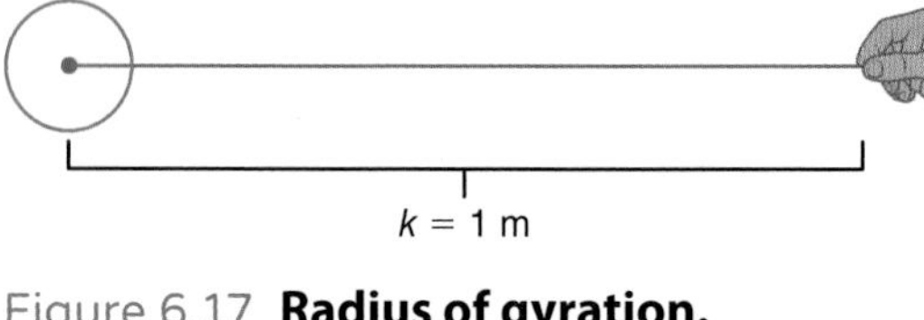

Figure 6.17 **Radius of gyration.**

the moment of inertia depends on the mass of the object, the shape of the object (mass distribution), and the axis of rotation (Robertson et al., 2014; Serway & Vuille, 2018; Winter, 2009).

Moment of inertia is sometimes expressed as the following:

$$I_a = mk_a^2 \quad (6.11)$$

where

I_a = moment of inertia about a given axis
m = mass of the entire object
k_a^2 = radius of gyration about the given axis

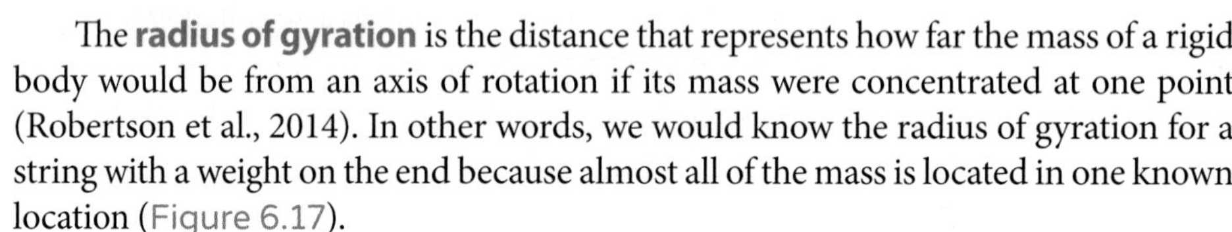

The **radius of gyration** is the distance that represents how far the mass of a rigid body would be from an axis of rotation if its mass were concentrated at one point (Robertson et al., 2014). In other words, we would know the radius of gyration for a string with a weight on the end because almost all of the mass is located in one known location (Figure 6.17).

Notice in either equation, the *distribution* of the mass is much more important than the quantity of mass itself in the determination of resistance to change in angular motion. People manipulate the moment of inertia in many situations without knowing the exact mechanism behind it. For example, if a child wants to swing a bat that is too massive, we often tell her to "choke up on it." Translated into biomechanical terminology, "choke up" is the same as saying "bring the mass closer to the axis of rotation." The human body itself is constructed with attention to moment of inertia. Human limbs tend to taper overall and from one joint to the next (Figure 6.18). For example, the upper arm tends to taper towards the elbow, and the upper arm is larger than the forearm.

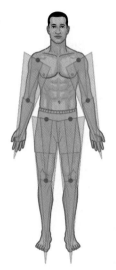

Figure 6.18 **Distal tapering of human limbs.** How is the shape of human limbs a benefit?

This tapering keeps most of the limb mass closer to the axis of rotation, which keeps the moment of inertia of the limb as small as possible. Can you imagine if it were the other way around (Figure 6.19)?

The lower rotational inertia of the limb makes it less resistant to changes in rotational inertia by applied forces. Newton's first law of motion explains what happens in the absence of applied forces (or in this case, torques). Newton's second law of motion deals with motion changes in the presence of force application.

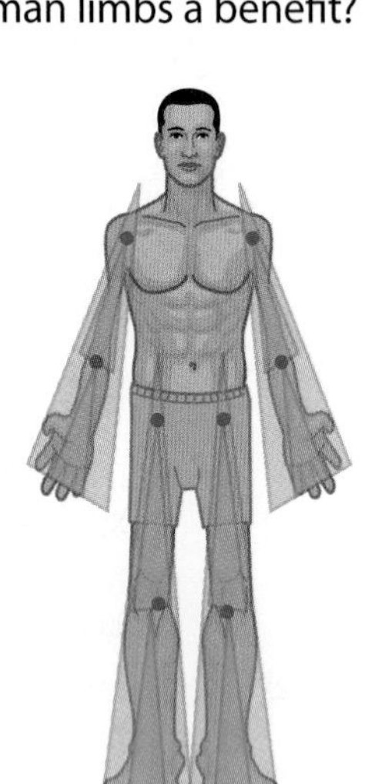

Figure 6.19 **Human limbs tapered proximally.** Why would this body be more difficult to move?

Newton's Second Law: Fundamental Law of Dynamics

The alteration of motion is ever proportional to the motive force impressed and is made in the direction of the right line in which that force is impressed. We have discussed that if there is a change in the state of motion of the system, according to Newton's first law there must have been an applied force (assuming mass is constant). The observed change in motion is called *acceleration*, which we have already defined mathematically in both linear and angular terms. In linear terms, Newton's second law states that the acceleration (change in state of motion) of the system will be directly proportional to the sum of the forces (or net force) acting on it and inversely proportional to the mass of the system to which the force is applied. Mathematically, the relationship of applied force and the resulting acceleration can be stated as follows:

$$a = \frac{\Sigma F}{m} \quad (5.4)$$

Radius of gyration The distance that represents how far the mass of a rigid body would be from an axis of rotation if its mass were concentrated at one point.

where

a = the acceleration, or change in motion of the system
ΣF = the vector sum of the forces applied to the system
m = the mass of the system to which the force is applied

If we multiply through by m, Equation 5.4 becomes:

$$\Sigma F = ma \tag{5.5}$$

where

ΣF = the vector sum of the forces applied to the system
m = the mass of the system to which the force is applied
a = the acceleration, or change in motion of the system

Notice the three components of these two equations: (1) Acceleration is the change in motion caused by an externally applied force, (2) acceleration is proportional to the magnitude of force, and (3) acceleration is inversely proportional to the mass of the object. These same equations can be expressed in angular terms if one understands the basics. In this chapter, we are discussing angular kinematics and kinetics. So in Equations 5.4 and 5.5, we need to refer to *angular* acceleration. For angular motion to occur, the applied force must be eccentric (i.e., possess a moment arm). So *force* must be changed to *torque*. And mass is simply an indicator of inertia. We found in our last section that rotational inertia takes into account mass and mass distribution. So *mass* must be changed to rotational inertia. If we make these simple changes, we have the angular analogues of Equations 5.4 and 5.5:

$$\alpha_a = \frac{\Sigma T_a}{I_a} \tag{6.12}$$

where

α_a = angular acceleration about a given axis, or change in angular motion of the system
ΣT_a = the sum of the torques (net torque) about a given axis
I_a = moment of inertia about a given axis (rotational inertia)

If we multiply through by I, Equation 6.12 becomes:

$$\Sigma T_a = I_a \alpha_a \tag{6.13}$$

where

ΣT_a = the sum of the torques (net torque) about a given axis
I_a = moment of inertia about a given axis (rotational inertia)
α_a = angular acceleration about a given axis or change in angular motion of the system

So in angular situations, Newton's second law is simply a statement of the fact that angular acceleration is directly proportional to the net torque applied and inversely proportional to the rotational inertia of the segment being accelerated.

As an example, think back to our previous discussion of how human limbs taper. Keeping the mass closer to the axis of rotation lowers the rotational inertia of the limb. According to Equation 6.12, a limb with lower rotational inertia will be accelerated to a greater degree with a given amount of torque. So our bodies gain an advantage in angular acceleration by keeping the rotational inertia to as small a value as possible. Another example of how we manipulate rotational inertia during angular acceleration is the bending of the knee in order to run. Try to run with your legs straight. You will probably notice that (besides being awkward) it requires a lot of effort (Figure 6.20).

Figure 6.20 **Radius of gyration during running.** Which of these is more difficult to perform?

Normally when you run, you bend the knee before you accelerate the limb forward. Bending the knee lowers the radius of gyration of the limb, which lowers its rotational inertia. With a lower rotational inertia, the limb can be angularly accelerated with lower torque application. Of course, many factors are involved in running, but the theme is obvious. In most situations, the most efficient way to move the human body is to maximize acceleration while minimizing torque requirements. One way to accomplish this task is to manipulate rotational inertia. Because the mass is often a constant value, changes in inertia often come in the form of changes in mass distribution (radius of gyration).

Previously in the chapter, we discussed that torques, similar to forces, can be summed to indicate the final magnitude and direction of motion (i.e., net torque). For example, if the sum of the torques is positive, then the system will rotate counterclockwise with a given magnitude. If the sum is negative, then we know that the system will rotate clockwise. To find the net torque, simply follow these steps:

1. Establish the magnitude of the applied force according to the given scale (e.g., 1 cm = 10 N of force).
2. Find the length of the moment arm by measuring the perpendicular distance between the axis of rotation and the line of action of the applied force.
3. Calculate the torque for each force according to the equation.
4. Establish the direction of each torque using the right-hand thumb rule.
5. Sum the torques to find the net magnitude and direction of rotation caused by the torques.

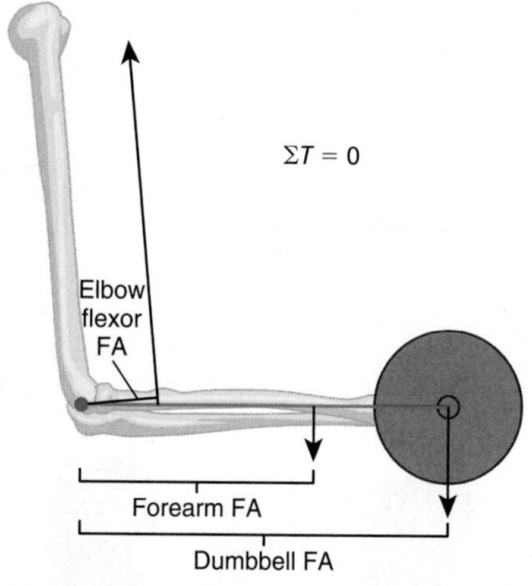

Figure 6.21 **Force vectors for a dumbbell curl exercise.**

Remember that Newton's first law explains motion in the absence of torque application. Notice that Newton's second law explains situations in which a system is under the influence of torques but is in equilibrium. A system in **equilibrium** is either at rest or rotating with a constant angular velocity. According to Equation 6.13, the sum of the torques in this situation is equal to zero. More specifically, an object at rest is exhibiting no angular displacement and therefore no angular velocity or angular acceleration. An object rotating with a constant angular velocity also has no angular acceleration. In both cases, acceleration is equal to zero. Therefore, in accordance with Equation 6.13, there must be no *net* torque ($\Sigma T = 0$). So even though a system is not in motion, it does not mean that there is no applied torque. For example, if a person is holding a dumbbell in her hand in a constant position (zero acceleration), then the torque produced by muscle force application must be equal to the torques produced by the weight of the forearm and dumbbell (Figure 6.21).

One application in which we use the idea of net torque is in solving for unknown forces. For example, if we know that the system isn't rotating, then we know that the net torque is equal to zero. If we know that the sum of the torques is zero and we have all of the forces and force arms except one value, then we can solve for the missing value. Let's revisit Figure 6.21 with more detail to elucidate this concept. The dumbbell is being held in a constant position, so the sum of the torques equals zero. The torques in this case are the torque produced by the elbow flexors (which we will represent with one force vector), the torque of the weight of the forearm, and the torque of the dumbbell itself. First, notice that we have a lateral view of the right arm of the person. From this point of view, the elbow flexor torque will be a positive value because it would cause counterclockwise rotation if it acted alone. In other words, clockwise or counterclockwise depends upon your perspective, so the point of view must be clearly specified. The torques produced by the weight of the forearm and the dumbbell are both negative values because they would cause clockwise rotation if they acted alone.

Equilibrium A state of being either at rest or rotating with a constant angular velocity.

SAMPLE PROBLEM

Let's say that the forearm segment weighs 17.84 N and its center of gravity is 0.127 m from the axis of rotation at the elbow (moment arm is 0.127 m). The dumbbell weighs 111.47 N and is located 0.381 m from the axis of rotation (moment arm is 0.381 m). The insertion of the elbow flexors is approximately 0.025 m from the axis of rotation, but we do not know the muscle force (Figure 6.22).

We can solve for muscle force if we remember that in this case $\Sigma T = 0$ and $T = F \times d$. Therefore, $\Sigma(F \times d) = 0$. So we need to calculate the torques:

Torque caused by weight of the forearm $= -17.84\ \text{N} \times 0.127\ \text{m} = -2.266\ \text{N} \cdot \text{m}$

Torque caused by weight of the dumbbell $= -111.47\ \text{N} \times 0.381\ \text{m} = -42.470\ \text{N} \cdot \text{m}$

Torque from elbow flexor force (F_m) $= + F_m\ \text{N} \times 0.025\ \text{m} = + F_m(0.025)\ \text{N} \cdot \text{m}$ and then solve for the unknown muscle force:

$$(-2.266\ \text{N} \cdot \text{m}) + (-42.470\ \text{N} \cdot \text{m}) + F_m(0.025\ \text{m}) = 0$$

$$F_m(0.025\ \text{m}) = +\ 44.736\ \text{N} \cdot \text{m}$$

$$F_m = \frac{+44.736\ \text{N} \cdot \text{m}}{0.025\ \text{m}}$$

$$F_m = +1789.44\ \text{N}$$

So it requires 1,789.44 N of elbow flexor force to hold up a total weight of 129.31 N. Seems a little one-sided, doesn't it? But if you look at the length of the force arm for the muscle in comparison to the force arm for the dumbbell, you can see the reason for the large force requirement. It seems to be a disadvantage for us to have such short moment arms, but we will return to this issue in Chapters 8 and 11 and discover that there are benefits.

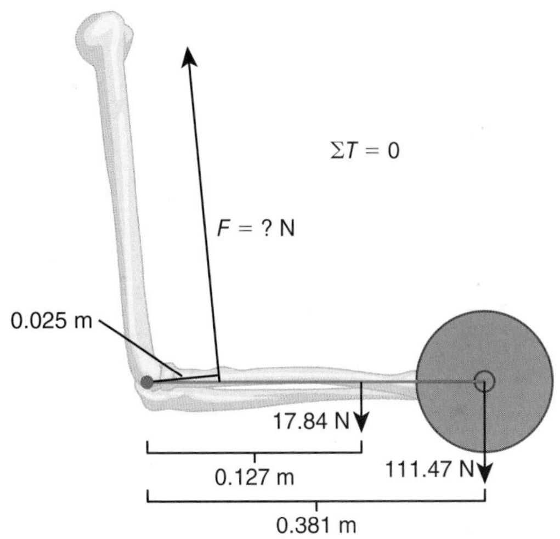

Figure 6.22 **Force vectors for a dumbbell curl exercise.**

We should mention that the torques that are summed in Equation 6.12 are created by forces that we have already discussed. The difference is that the forces are applied eccentrically. So the product of rotational inertia and angular acceleration do not create a torque. The angular acceleration of the system is directly proportional to the sum of the applied eccentric forces (torques) and inversely proportional to rotational inertia. So for example, *Ia* in Equation 6.13 simply means that a large angular acceleration of a segment possessing a large rotational inertia must have been caused by applying a lot of torque.

Finally, we should mention that the distribution of mass also has implications for behavior of an object when struck by another object. You have read that applying a centric force causes linear acceleration of an object and that eccentric forces cause angular acceleration of an object. Now that we have further information, we can be more specific about the application of these forces. For example, an implement such as a bat has a center of mass. If the bat is suspended by the grip and struck by a ball at the center of mass, the bat attempts to move in pure linear motion with no rotation. The attempted linear

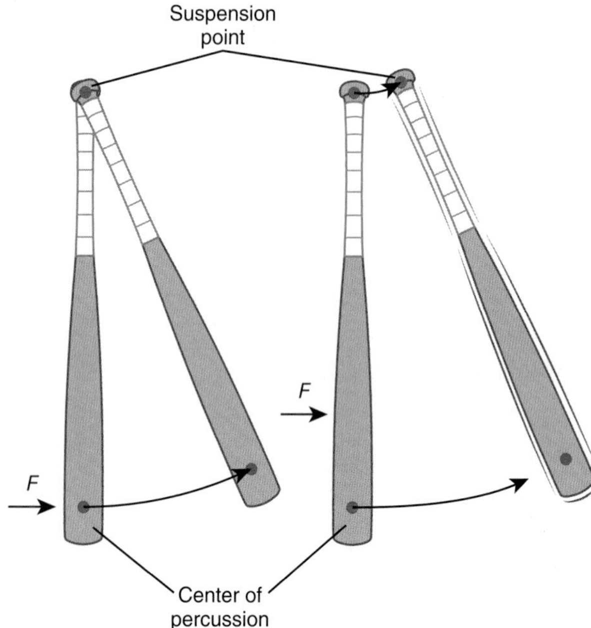

Figure 6.23 **Linear and angular motion of an object struck at different points.**

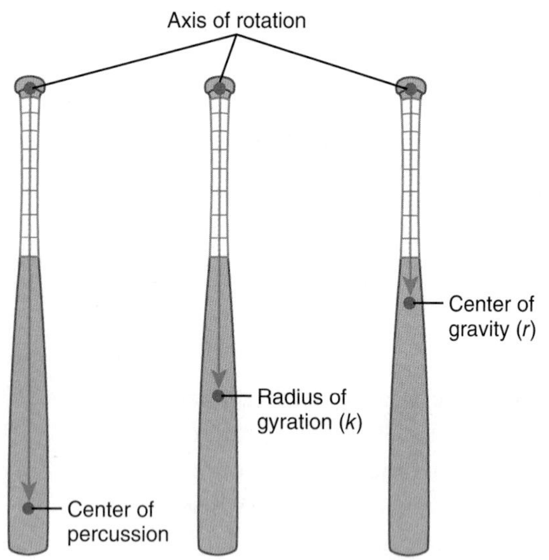

Figure 6.24 **Relationship of the center of mass, radius of gyration, and center of percussion.**

Center of percussion Spot on an implement that, when struck, will cause the implement to experience pure rotation about the suspension point without vibration.

motion will be felt as vibration. If the bat is stuck at the **center of percussion**, however, it experiences pure rotation about the suspension point and no vibration is felt (Figure 6.23). Therefore, the user of the implement experiences minimal force at the hands (suspension point) if the center of percussion is struck with another object. Any other striking point results in some combination of linear and angular motion.

The center of percussion is sometimes called the "sweet spot." However, other areas of an implement produce special results when struck. For example, we discussed the *coefficient of restitution*. All implements have a location in which this value is the highest. Notice that unless a body is of uniform mass and spherical in shape, there is a relationship between the locations of the center of mass, radius of gyration, and the center of percussion (Figure 6.24).

The center of gravity will be the most proximal to the axis of rotation, and the center of percussion will be the most distal, and the radius of gyration will fall somewhere between (Robertson et al., 2014). A complete discussion of sports implements is beyond the scope of this text, but the key concept here is that forces are applied eccentrically to create torque. In angular kinetics, the distribution of the mass of an object has implications for the rotational inertia of the object, as well as the behavior of the object when struck.

Similar to linear motion situations, mass is often constant. So we tend to think of accelerations as being caused only by applied forces or torques. But understand that Newton's laws refer to acceleration being proportional to the applied force or torque. Don't forget that acceleration is inversely proportional to mass or rotational inertia. Because a body tends to have a given mass (a single linear inertia), we tend to focus on the force application when referring to change in linear motion. However, a body (especially a human body) may have many different moments of inertia because many different axes of rotation exist, and we can change the distribution of mass about an axis. For example, we have discussed that flexing or extending a limb changes the mass distribution and therefore the rotational inertia of the limb. This change in rotational inertia affects acceleration even if torque is constant. We will return to this concept in the section on angular momentum.

One other application of Newton's second law to angular kinetics is the idea of centripetal acceleration. Recall that centripetal acceleration is caused by change in *direction* of the velocity vector. For example, the linear velocity vector of a car traveling in circular motion around a track is tangent to the track. And according to Newton's first law, because the car is not traveling in a straight line (which is an object's tendency), there must be a force applied. The force in this case is a centripetal force (directed toward the center of

FOCUS ON RESEARCH

Recall that the moment of inertia around a given axis depends on the mass of the object and the distribution of the mass relative to the axis of rotation. The moment of inertia around a limb axis is changed with amputation and subsequent use of a prosthetic device. For years, it has been known that the metabolic cost of walking is increased by loss of a limb (or portion of a limb). The increase in energy expenditure varies, depending on the level of the limb loss. There has been some disagreement in the research literature as to whether reducing or adding mass to a prosthetic will decrease the level of metabolic cost and walking asymmetry observed in amputees. Because of this delicate relationship of mass, mass distribution, and moment of inertia, prosthetic limb design can be a highly complex issue.

Smith and Martin (2013) used metabolic analysis and motion capture to investigate the effects of prosthetic mass distribution on metabolic cost and walking asymmetry in unilateral, transtibial amputees. One novel aspect of this study is that the mass added to the prosthesis was not an absolute amount as in some previous studies. Instead, added mass was calculated individually for each subject and was equal to the estimated difference in mass between the intact and prosthetic legs. Contrary to some speculation, symmetry in kinematic and temporal parameters did not improve with added mass at different locations on the limb. With loads positioned distally on the limb, asymmetries in stance and swing time increased by 3.4% and 7.2%, respectively. It was also found that when mass was added to the thigh, maximum knee angular velocity asymmetries increased by 6% versus approximately 10% with mass positioned near the ankle of the prosthesis. Energy costs of walking was found to increase by 12% by adding 100% of the estimated mass difference between intact and prosthetic legs to the ankle of the prosthesis. Adding that same 100% difference in mass to the prosthesis center of mass increased metabolic cost by approximately 7% and by 5% when added to the thigh center of mass.

The researchers advise caution when deciding to add mass to a prosthetic device as an alternative to current lightweight designs. Based on their finding that added mass did not improve asymmetry or lower metabolic cost, they warn that mass should only be added to a prosthesis distally if some other benefit is to be gained.

the circular path) that causes the centripetal acceleration (change in direction of the velocity vector). Remember that mathematically, centripetal acceleration is calculated as:

$$a_c = \frac{v_T^2}{r} \tag{6.9}$$

where

a_c = centripetal acceleration
v_T^2 = tangential linear velocity
r = radius of rotation of the point on the rotating segment

If we revisit Equation 5.5 ($\Sigma F = ma$) and make a substitution, we can see the relationship between Newton's second law and centripetal force:

$$\Sigma F = ma_c = m\frac{v_T^2}{r} \tag{6.14}$$

where

ΣF = net force causing centripetal acceleration
m = mass of the object being centripetally accelerated
a_c = centripetal acceleration
v_T^2 = tangential linear velocity
r = radius of rotation of the point on the rotating segment

Notice that there is always some relationship to the three Newtonian laws. Some call Equation 6.14 the equation for *centripetal force needed.* This name is appropriate because the equation demonstrates the relationship between mass, acceleration, and the amount of center-seeking force needed to maintain the circular path of the object. For example, friction provides the centripetal force in the case of a car traveling around a track. If it suddenly disappeared, the car would no longer travel in a circular path (it would travel in some linear path tangent to the circle). The *centripetal force needed* is how much friction force is needed to maintain that circular path. As you can see, the mass of the object has an effect, but the tangential velocity plays a much greater role in determining the quantity of force needed to maintain the circular path of an object. One can witness the intuitive understanding of this equation that children have by observing them riding bicycles. When on a bike and approaching what appears to be a slick surface, a child will slow down and make a wider turn. Slowing down has a dramatic effect on the amount of friction (centripetal force) necessary to maintain the circular path of the bike. Also, a wider radius means that the velocity vector doesn't experience as much change in direction, therefore lowering the force needed to maintain the circular path. In this text, we do not refer to Equation 6.14 as the "equation for calculating *centripetal force needed*" because this name sometimes gives the false impression that centripetal force is a separate force rather than a force that is center seeking.

Eccentric forces create torque, and angular acceleration is directly proportional to the net torque. Is the applied torque subject to Newton's third law of motion?

Newton's Third Law: Law of Reciprocal Actions

To every action there is always an opposite and equal reaction—or the mutual actions of two bodies on each other are always equal and directed to contrary parts. So we know that forces exist in pairs. Because torques are produced by eccentric forces, they are subject to the same conditions. In other words, for every torque there is a torque of equal magnitude directed in the opposite direction. A simple example is the swinging of an implement. Torque is applied to the implement by the person. The implement exerts a torque of equal magnitude to the person. The key to understanding equal and oppositely directed torques is to realize that the objects (the person and the implement) producing the torques share a common axis of rotation. Because the axis of rotation is shared, both objects are affected by a common torque.

Notice that Newton's second law of motion is also at work in this example. In other words, the torques are equal and opposite but the accelerations of the objects aren't. Remember that, in linear terms, acceleration is inversely proportional to the mass of the system to which force is applied. In the case of angular motion, angular acceleration is inversely proportional to rotational inertia. So, as with forces, torques are paired in that one torque will be met with an equal and opposite reaction torque. However, because of variations in rotational inertia, equal and opposite torques can produce unequal reactions (angular accelerations) in accordance with Newton's second law. In our example, the implement has lower rotational inertia than the person and is therefore accelerated to a greater degree.

Angular Momentum and Angular Impulse

You have learned that the term *momentum* refers to a system's quantity of motion. The greater the quantity of motion possessed by a system, the more difficult it is to stop that motion. In linear terms, more force has to be applied to stop a system that possesses a large amount of linear momentum.

In angular motion situations, possessing large amounts of momentum means that a large torque must be applied to stop rotation. Linear momentum is affected by a system's mass and velocity and is calculated as follows:

$$M = mv \tag{5.9}$$

where

M = the momentum of the system (or quantity of motion)
m = the mass of the system in motion
v = the linear velocity of the system

Notice that the two quantities are mass (a measure of inertia) and velocity. We can replace these linear quantities with their angular analogues and calculate **angular momentum**:

$$L_a = I_a \omega_a \tag{6.15}$$

where

L_a = angular momentum of the system (or quantity of angular motion) about a given axis
I_a = moment of inertia of the system in motion about a given axis
ω_a = the angular velocity of the system about a given axis

Angular momentum is expressed in units of rotational inertia (kg · m^2) multiplied by units of angular velocity (rad/sec). Therefore, the SI unit for angular momentum is kg · m^2/sec. So an object having a mass of 7 kg being rotated at the end of a 10-m rope at 3 rad/sec has an angular momentum equal to 2100 kg · m^2/sec. You can see from Equation 6.15 that an object at rest (ω = 0 rad/sec) has no angular momentum. In addition, momentum is a vector quantity and has the same direction as the system's angular velocity. Similar to linear motion situations, mass is often constant. However, it was prevously mentioned that a body may have many different moments of inertia because many different axes of rotation exist. For example, we have discussed that flexing or extending a limb changes the mass distribution and therefore the rotational inertia of the limb. Therefore, changes in angular momentum are normally the result of changes in radius of gyration (which changes the moment of inertia) or changes in angular velocity. Those changes in angular momentum are of particular interest in most situations. Changes in angular momentum (or the lack thereof) are directly related to Newton's first two laws of motion.

Newton's First Law: Law of Inertia

According to Newton's first law, there will be no change in state of motion without an externally applied force. An object in motion has both a mass and a velocity and therefore a given amount of momentum. If there is no change in state of motion without an externally applied force, then there can be no change in momentum without an externally applied force. In Chapter 5, Newton's first law was restated as the principle of conservation of linear momentum:

> *In the absence of a net externally applied force, the total momentum of a system comprised of multiple bodies remains constant in time.*

We have already covered the concept that Newton's first law also applies to angular motion, so it can also be stated as the **principle of conservation of angular momentum**:

> *In the absence of a net externally applied torque, the total angular momentum of a system comprised of multiple bodies remains constant in time.*

Because angular momentum is conserved, it has implications for the performance of many sports skills.

Angular momentum A system's quantity of angular motion.

Principle of conservation of angular momentum In the absence of a net externally applied torque, the total angular momentum of a system comprised of multiple bodies remains constant in time.

Figure 6.25 **A diver and a skater.** Why do the athletes spin faster when mass is moved closer to the axis of rotation?

These implications are especially applicable for skills in which external forces can be ignored (e.g., aerial events or events with low friction such as events performed on ice), because the less external eccentric forces are present, the more angular momentum is conserved. For example, in diving and ice skating, we observe increased angular velocity as limb segments are brought close to the body (Figure 6.25). How can angular momentum remain constant if angular velocity is increased? A diver brings the limb segments closer to the body, reducing the moment of inertia. If the moment of inertia is decreased, angular velocity must increase if angular momentum is to be conserved.

So even though the angular velocity in our example changed, the overall *quantity of motion* (momentum) was conserved. Another example of this phenomenon is the spin of an ice skater. On the ice, we can virtually ignore external torques. During a spin, the skater's angular velocity can be seen to increase as limb segments are brought closer to the body, and angular velocity decreases as the limbs are moved farther from the axis of rotation. In both cases, the moment of inertia is being manipulated to cause change in angular velocity. The angular velocity changes to conserve angular momentum. The reason for manipulating the moment of inertia is that there are no external torques to angularly accelerate the skater during the spin. The situation is the same as with the diver: in the air, there are no external torques because gravity doesn't have a force arm (i.e., it is a centric force).

Notice that in these examples we have the seemingly contradictory notion of acceleration without an externally applied torque. Again, we tend to focus on the applied force or torque in reference to acceleration. However, recall that angular acceleration is proportional to applied torque but is inversely proportional to rotational inertia. So we can have a change in velocity of the system and still obey Newton's first law as long as the total quantity of motion (angular momentum) is conserved.

Even though angular momentum is conserved in the absence of an externally applied torque, angular momentum can, to some degree, be transferred from one principal axis to another. **Cat rotation** or the **cat twist** is a prime example of this maneuver. We have all heard that cats tend to land on their feet. This statement holds if the cat is dropped from a height of at least 0.50 to 1.00 meters (to give it time to perform its aerial maneuver). From Figure 6.26, one can see that the cat first assumes a pike (bent position).

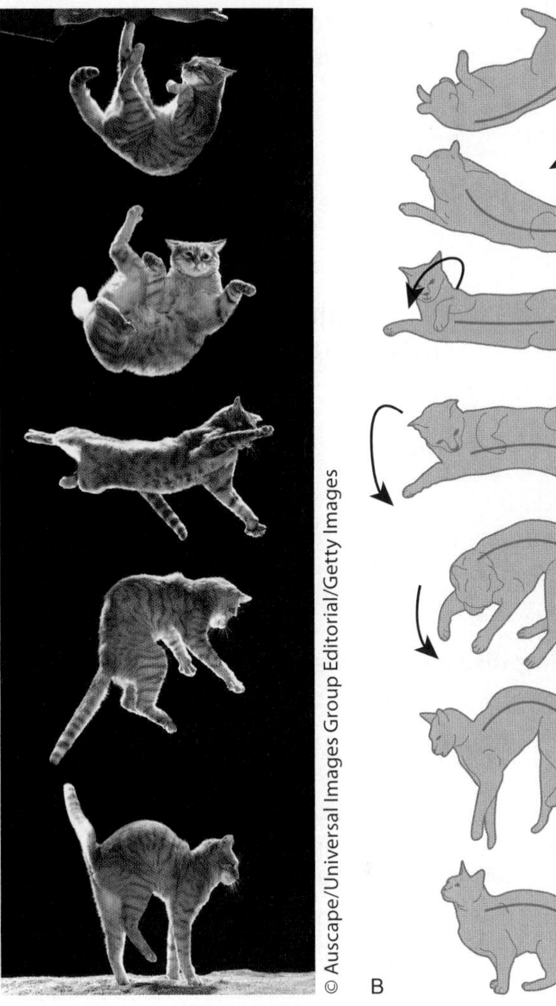

Figure 6.26 **Cat twist technique.** Can you think of some situations in which humans perform this technique?

This is an important position because it creates two axes of rotation, one for the upper body and one for the lower body. The first step in cat rotation is to lower the moment of inertia for the upper body by bringing the forelimbs closer to the axis of rotation (lower radius of gyration, k). Next, internal torques are used to turn the upper body in one direction toward the ground. This rotation of the upper body can be completed with little reaction of the lower body because of the relative difference between the moments of inertia about the two axes. In other words, the forelimbs are close to the body (decreased k), but the rear limbs are still extended (large k), so there is little reaction of the lower body. It makes sense from a survival standpoint to rotate the upper body first to make eye contact with the ground. Once the upper body has completed its rotation, the forelimbs are extended, which increases the moment of inertia about the upper body axis (because of increased k), and the rear limbs are pulled closer to the body (decreased k around the lower-body axis). Now that the upper-body moment of inertia is large compared to the lower-body moment of inertia, the cat can rotate the lower body toward the ground with little reaction from the upper body. That maneuver completes the cat twist. By the way, it's called a *cat twist*, but some aerial maneuvers by divers require the same type of rotation. Also, because this twist depends on the radius of gyration around one axis relative to the other, the more exaggerated the pike the better. A 90° pike is preferable because it maximizes the radius of gyration around each axis relative to the other.

So we have discussed what happens in the absence of externally applied torque. If the quantity of motion changes because of torque, we must discuss Newton's second law of motion.

Newton's Second Law: Fundamental Law of Dynamics

Once again, Newton's first law of motion assumes the absence of external forces (or torques in this case). In Chapter 5, you read about the relationship between linear momentum and Newton's second law of motion. We have also discussed that velocity and acceleration are not necessarily constant over a given interval. The

Cat rotation or cat twist A maneuver in which upper and lower body moments of inertia are manipulated in order to rotate while airborne.

reason for the variations in acceleration is that the net applied force is not necessarily a constant value. Using these concepts we derived the *linear impulse equation*:

$$\Sigma F\,(\Delta t) = \Delta M \qquad (5.15)$$

where

ΣF = average net force applied to the system
$\Delta t = \text{time}_{final} - \text{time}_{initial}$ (interval of force application)
$\Delta M = \text{momentum}_{final} - \text{momentum}_{initial}$ (change in momentum caused by the applied force)

As with our previous sections, we can substitute angular analogs for the values in Equation 5.15 and derive the **angular impulse** equation:

$$\Sigma T_a\,(\Delta t) = \Delta L_a \qquad (6.16)$$

where

ΣT_a = average net torque applied about an axis
$\Delta t = \text{time}_{final} - \text{time}_{initial}$ (interval of force application)
$\Delta L_a = \text{angular momentum}_{final} - \text{angular momentum}_{initial}$ (change in angular momentum)

Angular impulse The interval of torque application.

A large change in angular momentum will require the application of a large amount of torque or a large interval of torque application (or both).

SAMPLE PROBLEM

Let's use a merry-go-round to demonstrate this concept and relate it to some others (Figure 6.27). First we must calculate the moment of inertia of the merry-go-round. The merry-go-round has a mass of 455 kg and a diameter of 3 m. Note that a merry-go-round is actually an extended object, so as previously discussed the moment of inertia *cannot* simply be calculated as $I_a = \Sigma m_i r_i^2$. In this case, we model the merry-go-round as a disc and use the equation $I_a = ½mr^2 = (0.50 \times 455\text{ kg}) \times (1.5\text{ m})^2 = 511.88\text{ kg} \cdot \text{m}^2$. If we want to impart a lot of angular momentum to the merry-go-round, we not only apply a lot of torque (the force is eccentric) but also might run around with the merry-go round a few times to impart the torque for as long a time as possible. For example, we could apply a torque of 383.91 N · m for 4 sec and cause the angular momentum of the merry-go-round to change from 0 kg · m²/sec to 1,535.64 kg · m²/sec (this change would mean that the merry-go-round has a speed of 3 rad/sec). However, if I wanted the momentum to increase to 2,559.4 kg · m²/sec (angular speed of 5 rad/sec) with application of the same torque (383.91 N · m), I would have to apply that torque for an additional 2.67 sec. One simply has to solve for time after setting up the problem as $\Sigma T_a\,(\Delta t) = \Delta L_a$: 383.91 N · m (time) = 2,559.4 kg · m²/sec−1,535.64 kg · m²/sec.

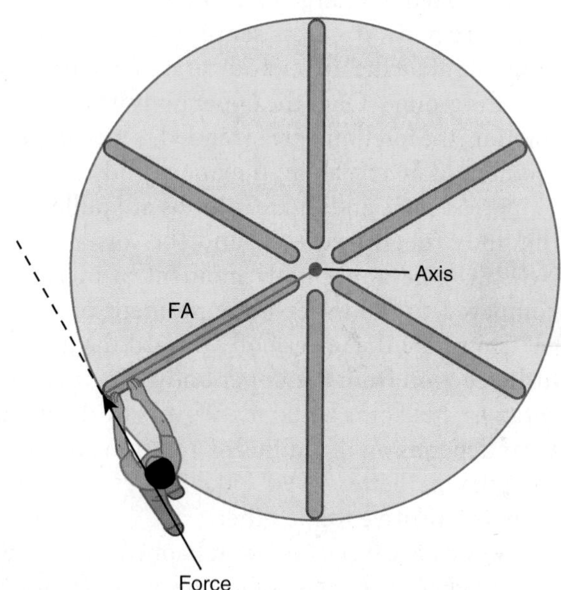

Figure 6.27 **A merry-go-round.**

By maximizing torque and time of torque application, we impart maximal angular momentum to the merry-go-round. What if you want to stop the merry-go-round? You can choose to stop it abruptly, in which case a large torque will have to be applied. Or you can stop it slowly by applying less torque for a longer time. For example, if the merry-go-round has an angular momentum of 2,559.4 kg · m²/sec and we choose to stop it in 2 seconds, it will require a torque application of torque (2 sec) = 0 kg · m²/sec−2,559.4 kg · m²/sec = 1,279.7 kg · m²/sec² or 1,279.7 N · m. However, if we have 4 seconds to stop the merry-go-round, it will only require a torque of torque (4 sec) = 0 kg · m²/sec−2,559.4 kg · m²/sec = 639.85 kg · m²/sec² or 639.85 N · m.

No matter the object, you will notice that the faster it spins, the more difficult it is to stop. The stability caused by rotation is called **gyroscopic stability**. Angular momentum is the product of rotational inertia and angular velocity. If the angular velocity of the object is greater, the angular momentum will be greater. Large angular momentum requires large angular impulse to stop. So increasing angular velocity increases gyroscopic stability. A toy top is a good example (Figure 6.28).

You will notice that the faster it is spun, the longer it stays upright. This is the case for a couple of reasons. First, the increased velocity has given the top a large amount of gyroscopic stability. But second, gravity is applying a torque in the negative y direction. The key is that the top has angular momentum in the x–z plane (Serway & Jewett, 2019). So it takes that much longer for the top to fall over. Similarly, footballs are thrown and punted with a spiraling motion to increase their gyroscopic stability. A spiraling football or an arrow with fletching is more likely to maintain its intended course because they possess more angular momentum.

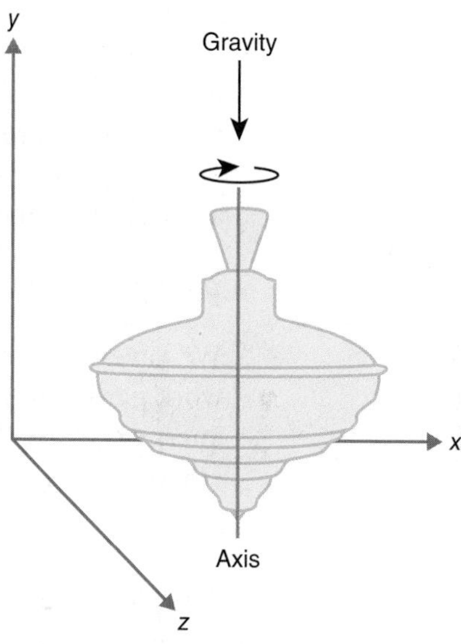

Figure 6.28 **A toy top.** Why does the top spin longer if spun faster?

6.5 ANGULAR KINETICS AND ENERGY TRANSFER

Angular acceleration of an object is directly proportional to the torque applied. Also, in Chapter 5, you learned that work is performed if an object is displaced by the application of force and that energy is the capacity to perform that work. Therefore, work is an energy transfer. We now need to examine the angular analogues of the linear expressions of energy transfer.

Work

As mentioned in Chapter 5, *work* is performed when an object is displaced by the application of a force and is a transfer of energy. More specifically, work is the product of applied force and the magnitude of displacement in the direction of the applied force:

$$W = F(d) \tag{5.16}$$

where

W = work performed on the object
F = force applied to the object to perform the work
d = displacement caused by the applied force ($\text{position}_{\text{final}} - \text{position}_{\text{initial}}$)

If we realize that the application of a torque also causes displacement, we can express the equation for work in angular terms:

$$W_a = F_T\,(r\Delta\theta) = T_a\,(\Delta\theta) \tag{6.17}$$

Gyroscopic stability Stability of an object caused by its rotation.

where

W_a = work performed on the object around a given axis
F_T = tangential force applied to perform the work
r = radius of rotation of the point at which the force is applied (moment arm)
$\Delta\theta$ = angular displacement caused by the applied force (measured in radians)
T_a = torque applied to perform the work about a given axis

In other words, **rotational work** is the angular displacement of an object around an axis caused by the application of a torque.

The units for describing work are units of force (N) multiplied by units of distance (m). The SI unit for work is the *joule* (J); 1 J = 1 N · m = 1 kg · m²/s². For example, if 500 N of force is applied 3 meters from an axis of rotation to rotate an object 3 radians, then the work performed around the axis is 500 N or 500 kg · m/s² × 3 m × 3 rad = 4500 J = 4500 N · m = 4500 kg · m²/s². So again, we have an angular analogue to a linear concept.

Power

We have also discussed the concept of *power*, the term used to express the amount of mechanical work performed in a given interval. Because work is a transfer of energy, power is the time rate of energy transfer (Serway & Jewett, 2019). Power, or the rate of performing work, is expressed mathematically as:

$$P = Fv \tag{5.19}$$

where

P = power
F = force applied to the object to perform the work
v = velocity, or rate of motion in a specific direction

By substituting our angular expressions, we can calculate the rate of performing work by applying an eccentric force:

$$P_a = F_T \frac{(r\Delta\theta)}{\Delta t} \tag{6.18}$$

or

$$P_a = T_a \frac{(\Delta\theta)}{\Delta t} \tag{6.19}$$

or

$$P_a = T_a \omega \tag{6.20}$$

where

P_a = power, rate of work performed on the object around a given axis
F_T = tangential force applied to perform the work
r = radius of rotation of the point at which the force is applied (moment arm)
$\Delta\theta$ = angular displacement caused by the applied force (measured in radians)
T_a = torque applied to perform the work around a given axis
Δt = $time_{final}$ − $time_{initial}$ (interval of work performed)
ω = angular velocity

Rotational work Angular displacement of an object about an axis caused by the application of a torque; a transfer of energy.

So **rotational power** is the amount of angular mechanical work performed during a given interval.

Again, the units for describing power are units of work (J) divided by units of time (sec). The SI unit for power is the *watt* (W); 1 W = 1 J/s = 1 N · m/s. So if 450 N of force is applied 5 meters from an axis of rotation to rotate an object 3 radians and the work is performed in 2 sec, then the power around the axis is (450 N or 450 kg · m/s² × 5 m × 3 rad)/2 sec = 3375 W = 3,375 J/s = 3,375 N · m/s.

Kinetic Energy

Once an object is in motion, it has the potential to perform work. This energy associated with motion is *kinetic energy.* Kinetic energy is expressed mathematically as:

$$KE = \tfrac{1}{2}mv^2 \tag{5.23}$$

where

KE = kinetic energy (or energy associated with motion)
m = mass of the object in motion
v^2 = velocity of the object in motion

A rotating object also has energy of motion. That **rotational kinetic energy** is associated with its rotational inertia and its angular velocity as follows:

$$KE_R = \tfrac{1}{2}\, I_a \omega_a^{\,2} \tag{6.21}$$

where

KE_R = rotational kinetic energy (potential to perform work associated with angular motion)
I_a = moment of inertia around a given axis
$\omega_a^{\,2}$ = angular velocity of an object around a given axis

The units for describing kinetic energy are units of mass (kg) multiplied by units of velocity squared (m²/sec²). Therefore, the SI unit for KE_R is the joule (J); 1 J = 1 N · m = 1 kg · m²/s². For example, let's calculate the moment of inertia for a mass of 20 kg being swung in a circle at the end of a 5-meter rope: 20 kg × (5 m)² = 500 kg · m². If we add the additional information that the rope is being swung with an angular velocity of 5 rad/sec, we can calculate rotational kinetic energy: (0.50 × 500 kg · m²) × (5 rad/sec)² = 6,250 J = 6,250 N · m = 6,250 kg · m²/s². Note that this is *not* a new kinetic energy. Kinetic energy is simply energy of motion. Each particle of mass on a rotating segment has kinetic energy. The difference in the equations arises simply from the fact that rotational inertia takes mass distribution into account. And once again, notice the large contribution of velocity to kinetic energy. In other words, a more massive bat would wield more kinetic energy when swung, but it would be more effective to swing the bat faster.

Conservation of Mechanical Energy

In Chapter 5, you learned the law of conservation of mechanical energy: *In the absence of externally applied forces other than gravity, the total mechanical energy of a system remains constant.* Now we need to discuss conservation of mechanical energy as it applies to angular motion. It's really quite simple. In Chapter 5, we only considered linear quantities:

$$(KE + PE)_i = (KE + PE)_f \tag{5.24}$$

Rotational power The amount of angular mechanical work performed during a given interval.

Rotational kinetic energy The energy (potential to perform work) associated with angular motion.

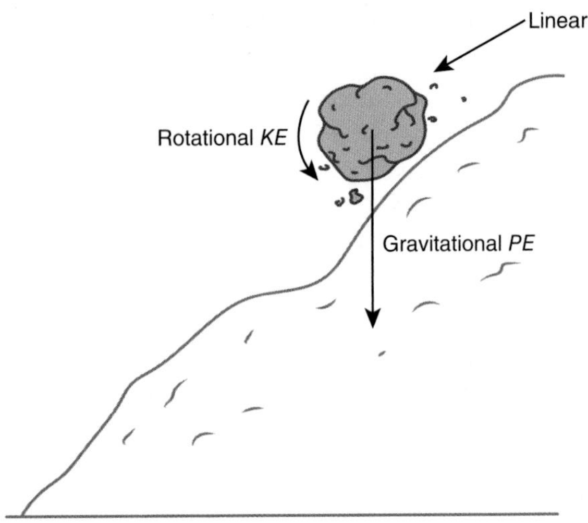

Figure 6.29 **A boulder rolling down a hill.**

where

KE_i and PE_i = initial kinetic and potential energies
KE_f and PE_f = final kinetic and potential energies

But what if an object is in angular and linear motion simultaneously (for example a rotating ball)? We simply need to include any energy contributed by rotation. So our equation becomes:

$$(KE_L + KE_R + PE)_i = (KE_L + KE_R + PE)_f \quad (6.22)$$

where

KE_{Li}, KE_{Ri}, and PE_i = initial linear kinetic, rotational kinetic, and potential energies
KE_{Lf}, KE_{Rf}, and PE_f = final linear kinetic, rotational kinetic, and potential energies

So, for example, a boulder rolling down a hill has potential energy because of its height, linear kinetic energy because of its linear motion, and rotational kinetic energy because it is rolling (Figure 6.29).

Also, remember that we are ignoring externally applied forces other than gravity (e.g., in this case, friction) (Serway & Vuille, 2018).

We have stated that energy transfer from one system to another takes place in the form of work. So our final task is to revisit the principle of work and energy: *The work performed by externally applied forces other than gravity causes a change in energy of the object acted on.* This principle is represented as:

$$W = \Delta E = \Delta KE + \Delta PE = \Delta TE \quad (5.25)$$

where

W = mechanical work performed
$\Delta E = \mathrm{E_{final}} - \mathrm{E_{initial}}$ (change in energy)
$\Delta KE = \mathrm{KE_{final}} - \mathrm{KE_{initial}}$ (change in kinetic energy)
$\Delta PE = \mathrm{PE_{final}} - \mathrm{PE_{initial}}$ (change in potential energy)
$\Delta TE = \mathrm{TE_{final}} - \mathrm{TE_{initial}}$ (change in thermal or heat energy)

Now that we know work is performed by eccentric forces, we must include rotational kinetic energy in the equation:

$$W = \Delta E = \Delta KE_L + \Delta KE_R = \Delta PE + \Delta TE \quad (6.23)$$

where

W = mechanical work performed
$\Delta E = \mathrm{E_{final}} - \mathrm{E_{initial}}$ (change in energy)
ΔKE_L = Linear $\mathrm{KE}_{L\mathrm{final}}$ − Linear $\mathrm{KE}_{L\mathrm{initial}}$ (change in linear kinetic energy)
ΔKE_R = Rotational $\mathrm{KE}_{R\mathrm{final}}$ − Rotational $\mathrm{KE}_{R\mathrm{initial}}$ (change in rotational kinetic energy)
$\Delta PE = \mathrm{PE_{final}} - \mathrm{PE_{initial}}$ (change in potential energy)
$\Delta TE = \mathrm{TE_{final}} - \mathrm{TE_{initial}}$ (change in thermal or heat energy)

So the change in the sum of the linear kinetic, rotational kinetic, potential, and thermal forms of energy produced by the application of externally applied force equals the mechanical work performed. For example, our rolling boulder has the potential to perform work on your car parked at the bottom of the hill. That work will be equal to the sum of the changes in forms of energy.

CONNECTIONS

6.6 HUMAN PERFORMANCE AND INJURY SCIENCE

Functional Anatomy

The structure of the human body is such that we sometimes must use concurrent torques to prevent actual linear displacement. For example, the scapula is not directly attached to the thorax, so no bony structures are present to inhibit unwanted motion. Therefore, the muscles of the posterior thorax must stabilize the scapula. Upward rotation of the scapula is achieved by contraction of both the upper and lower trapezius muscles. The torques produced by these muscles are equal in magnitude, opposite in direction, and noncolinear.

A **force couple** consists of two forces that are not colinear but have parallel lines of action and act in opposite directions (Robertson et al., 2014). Two muscles co-contracted in opposite directions in an attempt to accomplish one action are called an **anatomical force couple** (Oatis, 2017). Anatomical force couples attempt to accomplish a desired action but also help to prevent unwanted motions. So in the case of the trapezius muscles contracting concurrently, there is no linear displacement of the scapula; there is only rotation upward (Oatis, 2017). For example, if the upper trapezius acted alone, the scapula would exhibit rotation (because the force is eccentric) but there would also be some degree of superior linear displacement. Similarly, if the lower trapezius acted alone, the scapula would exhibit rotation, but there would be some degree of inferior linear displacement (Figure 6.30). So, by working together, the muscles of the anatomical force couple achieve the task of rotating the scapula while minimizing linear displacement.

Sport Science

In any sport that uses implements for striking, rotational inertia of the implement is an important issue. If the length of an implement is increased, it gains advantages and disadvantages. The longer radius of rotation means there will be greater linear velocity at the striking end of the implement. This greater velocity would

Force couple Two forces that are not colinear but have parallel lines of action and act in opposite directions.

Anatomical force couple Two muscles co-contracted in opposite directions in an attempt to accomplish one action, prevent an unwanted motion, or both.

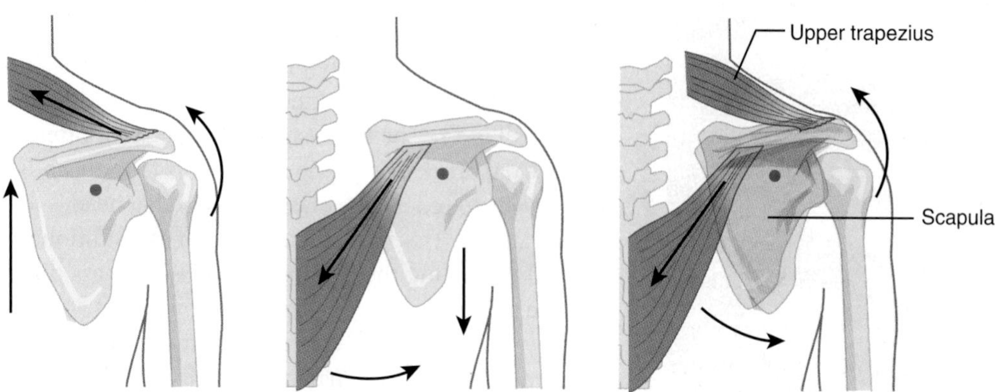

Figure 6.30 **Upper and lower trapezius muscles.** Can you think of other anatomical force couples?

be an advantage in imparting kinetic energy to the object being struck. However, a longer implement requires greater application of force to accelerate because of its greater rotational inertia. So it could be that the athlete cannot swing the longer implement with the same angular velocity as the shorter one.

However, new technology must also be taken into account. Wooden bats are made from hardwoods and by professional rules must be solid in construction. College and youth league baseball have different rules when it comes to bat design. Aluminum or composite bats are now the standard in every league except professional baseball. Aluminum bats can be manufactured lighter than an equivalent size wood bat, and an added benefit is that the material provides a higher coefficient of restitution than wood. This property means that the aluminum bats can be accelerated more quickly and impart more energy into the ball.

In the sports of tennis and racquetball there has been a similar revolution in equipment. Lighter materials allow players to accelerate racquets more quickly while maintaining the potential energy return of heavier traditional materials. In racquetball especially, the new racquets were made so light that an increase in length created no detrimental effects on swing speeds. Racquets became longer as the rules changed, and today's equipment is both lighter, longer, and more powerful and has changed the game because players now can reach previously unplayable shots.

Golf club designers have tried to create the best of both worlds by increasing the length of the club while simultaneously decreasing the mass by using lighter materials. In addition, some designers have manipulated the moment of inertia around the longitudinal axis of the club. Because the ball is contacted off-center of the club shaft, a torque is created around the longitudinal axis of the club. If the torque is large enough, the club could twist on impact. Twisting of the club on impact is not exactly conducive to accurate ball placement. Designers now distribute more of the mass of the club head toward the heel and toe (the areas closest and farthest from the player), thereby increasing the moment of inertia around the longitudinal axis of the club (Figure 6.31).

Figure 6.31 **Two club heads of different mass distributions.**

Baseball is another sport in which rotational inertia of implements is manipulated, both legally and illegally. One illegal practice is called *corking*. In an effort to lighten a wood bat, some players drill out part of the mass of the bat and replace it with a lighter cork material. This corking effectively reduces the mass of the bat and thereby lowers its rotational inertia. With lower rotational inertia, the bat can be swung with a greater angular velocity. Unfortunately (or fortunately depending on your perspective), a couple of other factors are at work in this case that the batter may not have considered. First, the structural integrity of the bat is compromised because of being drilled. Second, the ball has kinetic energy of its own and is therefore capable of performing work on the bat. The combination of a weakened bat with greater kinetic energy and the kinetic energy of the ball is a recipe for destroying a bat. So it often happens that the bat breaks, revealing its contents to thousands of fans. A legal way to reduce the mass of a wood bat head is to remove some material. This is called *cupping* and has additional benefits beyond lightening the bat, including better mass distribution and the ability to use a denser piece of wood.

Several throwing events at track meets rely on rotational acceleration to generate maximal velocity and therefore the best chance to maximize distance. Three throwing events require the participants to remain within a circular boundary during and after a throw: the hammer, the discus, and the shot put.

The hammer is a heavy ball attached to a short length of chain with a handle at the other end. Athletes hold the handle and slowly begin accelerating the ball while spinning their bodies around a vertical axis. Using the friction forces between their feet and the ground, the muscles of the lower body generate torque to continue this angular acceleration. The ball is held in a circular arc by centripetal force until the athlete has generated the maximal rotational velocity possible within the boundaries of the throwing circle, usually three to four turns. At that point, he releases the handle; with no centripetal force acting on the hammer, it leaves the circular arc on a tangential path. Complicating the matter, the handle must be released so that the hammer flies into a specific boundary within a 35–40 degree arc.

Rotating the body enables the athlete to impart force on an object for a longer time, resulting in more velocity at release (Figure 6.32). The same rotation strategy is employed during the discus throw and, more recently, the shot put. With the slightly larger throwing circle and the extended throwing arm acting as a counterweight, the spin delivery in the discus event is easier to master. The traditional glide approach in shot put is easier to learn and master, but the 540-degree spin delivery has been found to be highly effective. The advantage comes from the longer time that the thrower is able to impart force on the shot, thereby increasing the velocity of the ball at release.

While we're on the subject of flying objects, the concept of angular inertia may help you understand how a discus and Frisbee fly so far. First, aerodynamic factors come into play; these are described in detail in Chapter 9. However, one important concept that allows the aerodynamics to act on these objects over a large distance relates to this chapter. To have any chance of covering maximal distance, the Frisbee and discus have to be spinning. The faster the spin, the more rotational inertia the object will have. Remember that inertia is resistance to change—an object with more inertia will take more force to change its orientation, or in this case the stability of the discs in flight. When thrown properly with high spin rates, the effects of aerodynamic lift can stay consistent over a longer time, thereby increasing the distance the discs travel.

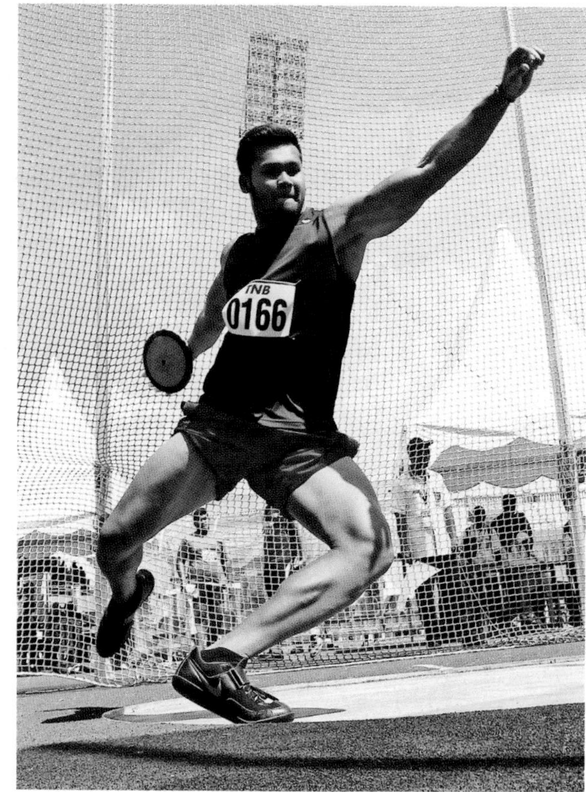

Figure 6.32 **Hammer throw and discus throw.**

From the flying objects to the rolling sports, angular physics is full of lessons. Lawn bowling is a popular game that is played competitively on manicured grass rinks and is similar to the game of bocce played in many backyards. However, there is one salient difference between the two games that is relevant to this chapter. In both sports the object is to roll your balls closer to a target ball than your opponents'. However, the strategy of each game is different because of the nature of the equipment. In bocce, the balls are perfectly round and therefore roll straight along the manicured grass surface. If an opponent's ball was between a thrower and the target ball, the only option was to knock the blocking ball away and ideally leave your own ball closer. Lawn bowling equipment offers more options. Lawn bowling balls have changed in design but still incorporate a feature that causes them to curve. Original bowling balls had a weight inserted into a round ball, and the weight was not placed near the center. This structure created an eccentric force as the ball rolled along the "rink," and this force produced a curved path that could be controlled by the player. Off-center weights were banned recently, but today's equipment counts on the same concept of eccentric weight to allow the balls to curve. New equipment is not perfectly round, so it is *biased* toward one side. Bowlers can change the orientation of the ball before rolling it to make it curve in either direction. Rather than knocking an opponents ball out of the way, experienced bowlers can curve their own balls around an opponent's ball that may be blocking a straight shot.

Until the advent of 20th-century technology, golf balls were also manufactured with eccentric weighting. The cores of golf balls could range from a little to a lot off-center. Unfortunately, this offset wasn't desirable or beneficial to golfers. For the average golfer, it wasn't too much of a problem, but it was an important issue for professionals. A ball with eccentric weighting will wobble in the air because of the spin imparted by a club, resulting in a loss of distance and accuracy. Unlike players using lawn bowling equipment, golfers had no idea where the center of mass lay, and this uncertainty could affect the ball's lift or lateral movement, depending how the ball was oriented when struck. On the greens, a ball that was weighted off-center would react in a similar way to a lawn bowling ball, curving in an unintended way. Professionals counted on manufacturers to provide them with the most centric golf balls they could produce because these balls improved the consistency of their play. Today, quality control in golf ball manufacturing has improved the consistency of flight and roll with mechanized processing and testing that ensures the center of mass is in the center of the ball. In 1932, the president of a rubber company became frustrated with the consistency of his shots and wondered if the golf balls had something to do with it. According to company lore, he took some golf balls to his dentist's office to x-ray them and found that the cores were indeed off-center. The rubber company was named Acushnet, and its president and an engineer friend spent the next three years developing a ball design and manufacturing process that ensured consistent placement of the core at the center of the ball. This ball was named Titleist®, a name that still maintains a reputation as one of the highest quality golf balls in the world.

For racing cars that compete on circular or oval tracks, drivers and engineers have found that increasing the circumference of the tires on the outside of the car increases performance. The correct term for this technique in racing is called *tire stagger* and is a result of understanding angular motion. Remember that for a given change in angle, a point closer to the axis of rotation travels less linear distance than a point farther away from the axis. The outside of the car is farther from the center of the track and therefore travels a longer distance. If tires of the same circumference were placed on both sides of the car, the outside tires would be turning slightly faster than the inside tires. The larger circumference of the outside tires when using stagger balances wheel speed side to side, and this balance seems to reap benefits in terms of how well a racing car handles.

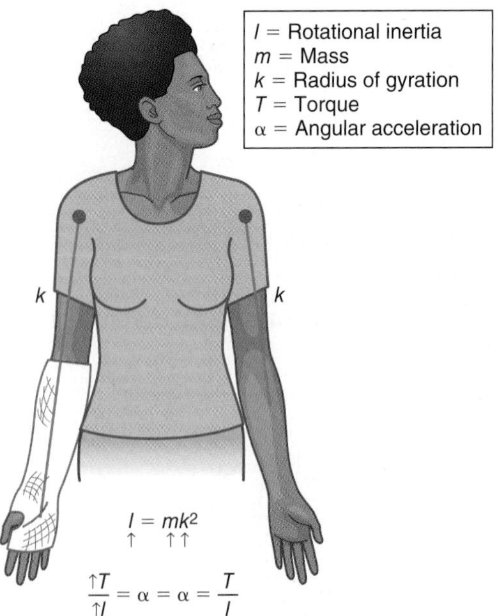

Figure 6.33 **Person in a cast.** Are there any benefits to muscle atrophy?

Injury Science

Consider how the moment of inertia changes around a limb when that limb is placed in a cast. Casts are the worst of both worlds in terms of rotational inertia. Not only is a large amount of mass added to the limb, but also that mass is often located distal to the axis of rotation (Figure 6.33). That positioning increases both quantities in the rotational inertia equation (m and k^2).

In terms of angular kinetics, it will require a greater amount of torque to rotate the segment wrapped in the cast. An interesting observation is that the body tends to respond to the immobilization with muscular atrophy. We, of course, judge the atrophy to be a negative side effect. But looking at the situation from the point of view of rotational inertia is interesting. Could it be that atrophy is the body's way of decreasing rotational inertia by lowering the mass of the segment? Cast design has changed over the years, and casts are now much lighter. But one should always consider the inertial properties of a limb when applying casts or bracing devices. Also consider the implications for torque. For example, if more torque is required on one segment than the other, then the situation could easily lead to muscular imbalance and potential injury of some other part of the body stemming from compensation.

Another interesting phenomenon in the realm of angular kinetics is an injury called *radial head subluxation*, which is also known as "pulled elbow" or "nursemaid's elbow." Pulled elbow is an injury to the elbow that usually occurs in preschool-aged children, often as a result of swinging them in circles by their hands (Figure 6.34). The cause of the injury is usually innocent enough, and it often happens in the course of play or when preventing a child from falling (Kunkler, 2000). The specific injury is caused by the head of the radius being pulled through the ring of the annular ligament (Oatis, 2017). The injury is more common in children younger than age six because the annular ligament is not yet fully developed, meaning that it is weaker and more susceptible to tearing.

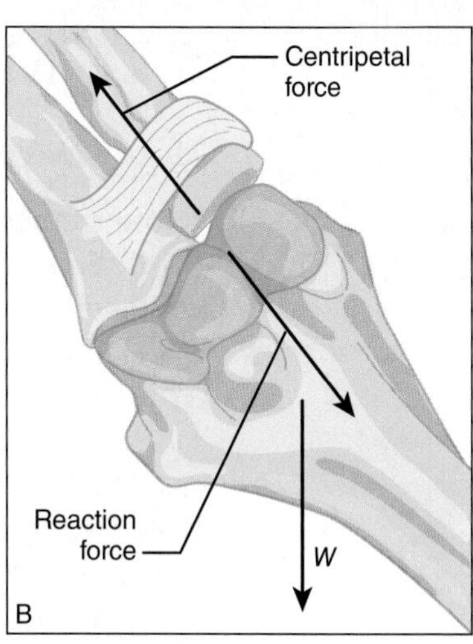

Figure 6.34 **Radius and annular ligament.**

One can understand the cause of the injury if it is put in the context of a centripetally applied force. Remember that the natural tendency of an object is to move in a linear path. So to maintain the circular path of the child, a center-seeking force must be applied to keep the child from traveling tangentially to the circle. The force needed to maintain the circular path is directly proportional to the tangential velocity squared. So as the velocity increases, there is a substantial increase in the amount of center-seeking force that must be applied to the child's hands. Therefore, one should always be careful when swinging or pulling a child by the hands. Also, in accordance with the linear impulse equation, one should be especially careful if the applied force is large and applied quickly. If one does want to swing a child during play, it may be a better idea for the child to grasp your arms as opposed to the other way around. Recall from Chapter 3 that a muscle force vector is resolved into a rotary component and a stabilizing component. If the child is holding onto your arms, she will likely be contracting her elbow flexor muscles. Some of the force of muscle contraction will be acting to stabilize the elbow, therefore helping to prevent radial subluxation.

Another common injury caused by forces generated during rotation is *tennis elbow* in people who participate in racquet sports. Actually, tennis players make up only 5% of the total cases of this injury, which is also known as *lateral epicondylitis*. Manual laborers, office workers, and other athletes account for most cases. However, it is still called *tennis elbow* because as many as 50% of tennis players experience this injury at some point. The exact nature of this disorder isn't well understood, but it's thought to be caused by small tears in the tendon attaching forearm muscles to bones near the elbow (lateral epicondyle). The cause of tennis elbow is thought to be the frequent, powerful extension of the elbow joint. Manual laborers who swing hammers daily are a good example of forceful extension, and athletes who play racquet sports are no less forceful. The good news is that lighter equipment and better design make newer racquets easier to accelerate; they also transmit less vibration and torque to the arm of the player who hits off-center. It will be interesting to see whether the frequency of tennis elbow declines among tennis players and other athletes who participate in racquet sports.

Shoulder injuries are a common injury for baseball pitchers. When you take into account the forces involved in moving and stabilizing the shoulder joint while delivering a high velocity pitch, it's easy to understand why. Consider this fact: A pitcher who throws a baseball nearly 42 m/sec exceeds internal shoulder instantaneous angular velocities greater than 7,500°/sec (Fleisig et al., 1996). Put in different terms, if the shoulder maintained this rate of angular velocity for one second, it would complete 20 revolutions during that one second! To accelerate, stabilize, and decelerate this motion takes immense forces generated by shoulder muscles. Over the course of a game, pitchers may generate these forces more than 100 times. It's no wonder they need four to five days of rest between games.

6.7 MOTOR BEHAVIOR

Motor Development

Young runners tend to have little flexion in the extremities while running (Haywood & Getchell, 2014). From the perspective of rotational inertia, the radius of gyration is huge. If the rotational inertia of the limbs is large, then more torque must be applied. Because children are not capable of producing much torque, the resultant running motion is awkward, slow, and requires a lot of effort (Figure 6.35a).

As the child becomes more proficient, the recovery leg is accelerated forward after assuming a flexed position. In addition, the elbows of a proficient runner tend to be flexed at a 90° angle (Figure 6.35b). These two simple actions lower the rotational inertia of the limbs, which reduces the effort required for running (Haywood & Getchell, 2014). As children learn to run with higher speeds, the heel of the recovery side can nearly come into contact with the gluteus muscles—a position that moves the mass of this leg as close to the axis of rotation as possible.

Motor Learning

As children grow in length, growth occurs primarily because of length changes in the long bones of the legs and arms. With bone growth, the length of the lever arm changes over time, and the distal ends of limbs will move at higher velocities. The angle of insertion does not change appreciably relative to the

Figure 6.35 **Early versus proficient running.**

changes in bone length from infancy to adulthood. If you have ever pushed against the wrong side of a door in an attempt to open it, you'll get a better idea of this concept (the "wrong" side being the one close to the hinges). If you push near the hinges, it takes greater force to move it, but you don't have to push far to create a larger change of position at the side away from the hinges.

As limb length changes, children and teens often go through a clumsy stage as their neural and muscular systems relearn movement skills that once required different amounts of force for identical movements. When growth spurts occur, parents, teachers, and coaches need to be aware that children may regress in sport skills as their brain and muscles reorganize to accommodate new forces created by and exerted on longer moment arms.

Golf clubs are interesting tools for the task for which they are designed. Longer clubs equate to higher head velocity and are designed to send the ball farther, whereas shorter clubs generally send the ball higher and less distance. In an interesting twist, one young professional golfer with a background in physics is changing the way we think about the length of golf clubs, and it might have as much to do with motor learning as with gaining an advantage from the varying the size of each club. Bryson DeChambeau had his entire set of irons modified to have identical shaft lengths; typically, each iron will be about 0.5″ inch different than the next. He chose one club in the middle of the range of irons, which means that some are longer and some shorter than traditional sets. His reasoning is that he can make the exact same swing with each club rather than having a slightly different posture and swing for each one. In theory, this could shorten the amount of time it takes to learn and practice the game of golf and removes the need to constantly change body segment positions for each iron shot.

6.8 ADAPTED MOTION

In the area of prosthetic design, there is usually a problem of balancing aesthetics with functionality. In other words, it is especially difficult to design a prosthesis that emulates a living leg while re-creating the necessary mass and mass distribution requirements for normal gait. If the rotational inertia characteristics of the prosthetic are not the same as the living leg, torque production requirements for acceleration of the limbs will vary from one side of the body to the other. This variation will eventually lead to muscular imbalance, as well as abnormal gait. Therefore the fitting (and refitting) of prosthetic devices is of utmost importance.

For children with disabilities that lead to limited torque production such as cerebral palsy, multiple sclerosis, and muscular dystrophy, the use of traditional sport and recreation equipment can be frustrating. The amount of torque necessary to lift and swing a bat, racquet, and stick can be difficult or impossible to generate, so teachers and therapists use one of two options to get students into the game. The first is changing the rules of the activity to let children with inadequate strength use a body part to strike or send an object. The second is to use lighter equipment that takes less strength to hold and swing. These strategies allow students with disabilities a chance to participate rather than sit on the sidelines.

SUMMARY

The concepts in this chapter were concerned with *angular kinematics* and *kinetics. Kinematic values* describe motion without concern for the forces that cause that motion. Basic kinematic values are *angular displacement*, *angular velocity*, and *angular acceleration. Kinetics* is the branch of mechanics concerned with the relationship between motion and the forces that cause the changes in motion. In the case of angular motion, the applied forces must be *eccentric* (torque producing). Newton's laws help us understand the kinetics of motion in angular terms. Once a rotating object is in motion, it has a certain amount of *angular momentum* (quantity of motion). To change that momentum, an eccentric force must be applied over a given interval. The shorter the interval, the greater the torque required to change momentum. This concept of angular impulse has widespread implications for sporting implements, casts, and bracing, as well as prosthetics. We can also describe angular kinetic concepts from the point of view of energy transfer. Work is performed whenever an object is displaced by an applied eccentric force. The rate of performing that work is called *power.* Objects in angular motion have the capability to perform work because of their motion. Mechanical energy is conserved, therefore total work performed can be observed in terms of the sum of the changes in potential and kinetic energy of the objects involved.

REVIEW QUESTIONS

1. Give the equation for torque and then state it in your own words.
2. Why is the knob of a door located across from the hinges?
3. Where are the center of mass and center of gravity of a donut?
4. Define angular velocity and angular acceleration. Give the equations and then put them into your own words.
5. Define rotational inertia. Give the equation and then put it into your own words.
6. Provide two equations for the expression of rotational inertia. In what situation is each expression preferable?
7. Why is it easier to swing your leg forward (angularly accelerate it) when it is bent at the knee? Why is it easier to perform jumping jacks with your arms bent?
8. Why is it important (from a rotational inertia point of view) to be sure that a prosthetic device is the correct mass and mass distribution?
9. Human limbs tend to taper distally. Using two equations, fully explain why this arrangement is a distinct advantage.
10. In what ways are the equations for linear and angular momentum similar?
11. In what ways are the equations for linear and angular impulse similar?
12. Why must a torque be applied to change angular momentum, whereas linear momentum can be changed by application of a force?
13. Define gyroscopic stability. Use an equation to help explain.
14. Using two equations, explain why a toy top spun faster is more gyroscopically stable.
15. What is the purpose of fletching on an arrow?
16. Once a diver leaves the board, does angular momentum change? Why?
17. Once a diver leaves the board, does linear momentum change? Why?
18. Can gravity change angular momentum? Why?
19. Using two equations, explain what happens to angular velocity when a diver moves from a pike position to a layout position.
20. When a gymnast lands from a vault, how does the ground cause the body to stop in a vertical position?

PRACTICE PROBLEMS

1. For systems shown in Figure 6.36, calculate the individual torque magnitudes (indicate direction with the appropriate sign) and then calculate the net torque and indicate the direction of system motion with the appropriate sign. 1 cm = 20 N.

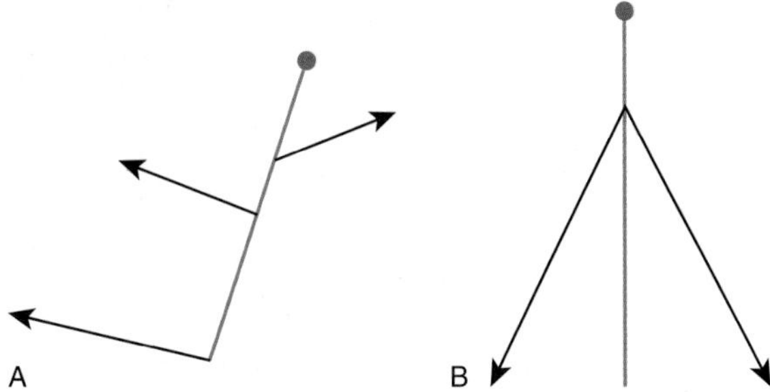

Figure 6.36 **Practice calculating net torque and resultant direction of rotation.**

2. A point is 1 meter from the axis of rotation of a segment. The segment rotates 3 radians. What is the linear distance traveled by the point?
3. Calculate the linear and angular velocity of the point in question 2 if the segment rotates the 3 radians in 0.5 seconds.
4. Calculate the linear and angular acceleration of the point in question 2 if its velocity is changed to 4 rad/sec in 1 sec.
5. What is the rotational inertia of a 10-kg ball being swung at the end of a 10-m rope? What about a 20-kg ball on a 10-m rope? What about a 10-kg ball on a 20-m rope?
6. What is the angular momentum of the ball in question 5 if we know the angular velocity is 5 rad/sec?
7. What is the magnitude of torque that would have to be applied to stop the motion of the system (i.e., angular momentum changes to 0) in question 6 in 3 seconds? Show all work.
8. How long would it take to stop the motion of the system (i.e., angular momentum changes to 0) in question 6 by applying a force of 100 N · m? Show all work.
9. Calculate the work performed around an axis if 650 N of force is applied 5 meters from an axis of rotation to rotate an object 4 radians.
10. Calculate the power around an axis if 550 N of force is applied 6 meters from an axis of rotation to rotate an object 3 radians and the work is performed in 5 seconds.
11. What is the rotational kinetic energy of a 10-kg ball at the end of a 10-m rope that is being swung at 10 rad/sec?

EQUATIONS

Torque	$T = F \times r \times \sin \theta$	(6.1)
Theta in radians	$\theta = \frac{l}{r}$	(6.2)

Linear displacement of a point on a rotating segment	$l = \Delta\theta r$	(6.3)
Angular speed	$\delta = \frac{\varphi}{\Delta t}$	(6.4)
Angular velocity	$\omega = \frac{\Delta\theta}{\Delta t}$	(6.5)
Tangential linear velocity	$v_T = \frac{\Delta\theta r}{\Delta t} = \omega r$	(6.6)
Angular acceleration	$\alpha = \frac{\Delta\omega}{\Delta t}$	(6.7)
Tangential linear acceleration	$a_T = \frac{\Delta\omega r}{\Delta t} = \alpha r$	(6.8)
Centripetal acceleration	$a_c = \frac{v_T^2}{r}$	(6.9)
Moment of inertia	$I_a = \Sigma m_i r_i^2$	(6.10)
	$I_a = mk_a^2$	(6.11)
Relationship of acceleration and force	$a = \frac{\Sigma F}{m}$	(5.4)
	$\Sigma F = ma$	(5.5)
Relationship of angular acceleration and torque	$\alpha_a = \frac{\Sigma T_a}{I_a}$	(6.12)
	$\Sigma T_a = I_a \alpha_a$	(6.13)
Relationship of Newton's second law and centripetal force	$\Sigma F = ma_c = m\frac{v_T^2}{r}$	(6.14)
Linear momentum	$M = mv$	(5.9)
Angular momentum	$L_a = I_a \omega_a$	(6.15)
Linear impulse	$\Sigma F(\Delta t) = \Delta M$	(5.15)
Angular impulse	$\Sigma T_a(\Delta t) = \Delta L_a$	(6.16)
Work	$W = F(d)$	(5.16)
Work performed about an axis	$W_a = F_T(r\Delta\theta) = T_a(\Delta\theta)$	(6.17)
Power	$P = Fv$	(5.19)

Power about an axis	$P_a = F_T \frac{(r\Delta\theta)}{\Delta t}$	(6.18)
	or	
	$P_a = T_a \frac{(\Delta\theta)}{\Delta t}$	(6.19)
	or	
	$P_a = T_a \omega$	(6.20)
Kinetic energy	$KE = ½mv^2$	(5.23)
Rotational kinetic energy	$KE_R = ½\, I_a \omega_a^{\,2}$	(6.21)
Conservation of mechanical energy	$(KE + PE)_i = (KE + PE)_f$	(5.24)
	$(KE_L + KE_R + PE)_i = (KE_L + KE_R + PE)_f$	(6.22)
Relationship of work and energy	$W = \Delta E = \Delta KE + \Delta PE = \Delta TE$	(5.25)
	$W = \Delta E = \Delta KE_L + \Delta KE_R = \Delta PE + \Delta TE$	(6.23)

REFERENCES AND SUGGESTED READINGS

Braune, W., & O. Fischer. 1889. *The Center of gravity of the human body as related to the German infantryman*. Leipzig. Available as report ATI 138 452, National Technical Information Service, Springfield, VA.

Cajori, F. 1934. *Sir Isaac Newton's mathematical principles*. Trans. Andrew Motte, 1729. Berkeley: University of California Press.

Dempster, W. T. 1955. *Space requirements of the seated operator: Geometrical, kinematic and mechanical aspects of the body with special reference to the limbs*. Technical Report WADC-TR-55-159, Wright-Patterson Air Force Base, OH: Wright Air Development Center.

Fleisig, G. S., N. Zheng, S. W. Barrentine, R. F. Escamilla, J. R. Andrews, & L. J. Lemark. 1996. Kinematic and kinetic comparison of full-effort and partial effort baseball pitching. *Proceedings of the 20th Annual Meeting, American Society of Biomechanics*, Atlanta, Georgia, Oct. 17–19, pp.151–152.

Hay, J. G. 1973. The center of gravity of the human body. *Kinesiology*, III: 20–44.

Haywood, K. M., & N. Getchell. 2014. *Life Span Motor Development*, 6th ed. Champaign, IL: Human Kinetics.

Heinrichs, R. N. 1990. Adjustments to the center of mass proportions of Clauser et al. (1969). *Journal of Biomechanics*, 23: 949–951.

Kunkler, C. E. 2000. Did you check your nursemaid's elbow? *Orthopaedic Nursing* 19(4):49–52; quiz 53–5.

Oatis, C. A. 2017. *Kinesiology: The mechanics and pathomechanics of human movement*. Baltimore: Wolters Kluwer.

Plagenhoef, S., F. G. Evans, & T. Abdelnour. 1983. Anatomical data for analyzing human motion. *Research Quarterly for Exercise and Sport*, 54(2): 169–178.

Robertson, D. G. E., G. E. Caldwell, J. Hamill, G. Kamen, & S. N. Whittlesey. 2014. *Research methods in biomechanics*, 2nd ed. Champaign, IL: Human Kinetics.

Serway, R. A., & J. W. Jewett Jr. 2019. *Physics for scientists and engineers with modern physics*, 10th ed. Belmont, CA: Brooks/Cole-Thomson Learning.

Serway, R. A., & C. Vuille. 2018. *College physics*, 11th ed. Boston: Cengage Learning.

Smith, J. D., & P. E. Martin. 2013. Effects of prosthetic mass distribution on metabolic costs and walking symmetry. *Journal of Applied Biomechanics*, 29(3): 317–328.

Winter, D. A. 2009. *Biomechanics and motor control of human movement*, 4th ed. New York: John Wiley & Sons.

CHAPTER 7

SYSTEM BALANCE AND STABILITY

LEARNING OBJECTIVES

1. Describe the differences between equilibrium, stability, and balance.
2. Describe factors that increase or decrease linear and rotational stability.
3. Understand the relationship between stability and energy transfer.

CONCEPTS

7.1 Equilibrium, Stability, and Balance

7.2 Linear Stability

7.3 Rotational Stability

7.4 Stability and Energy Transfer

CONNECTIONS

7.5 Injury Science

7.6 Motor Behavior

7.7 Pedagogy

7.8 Adapted Motion

A gymnast is preparing for a tumbling pass in the last event of the meet. After a deep breath, she performs a back handspring followed in quick succession by a round-off back flip and another back flip in the tuck position. The crowd applauds wildly, because it understands that this was the part of the routine that involves the most risk and might cause the gymnast to fall off the 4-inch wide beam. That's correct; she performed two back flips and landed precisely back onto a 4″-wide landing zone!

The balance beam event is one of the ultimate tests of the human vestibular and neuromuscular systems in sport.

The human body contains a variety of sensors that are designed to send information to the brain that will keep us upright against gravity and let us know how we're oriented and moving in space. The brain sorts all of the information and sends signals to the motor cortex to fire muscles in a coordinated way to maintain the orientation desired. In most cases, the muscle actions are voluntary, but protective reflex mechanisms can still cause involuntary movements, especially in the case of losing balance. If the gymnast lands one of the back flips with her center of gravity outside of the 4″ beam, her body will usually compensate by leaning to the other side or reflexively throwing her arm to the other side. In this chapter, you'll learn about the forces that can enhance and inhibit maintaining balance.

Many important concepts such as force, gravity, torque, and center of gravity were introduced in previous chapters. In this chapter, these concepts are applied to situations in which stability and balance are of the utmost importance. Understanding stability requires a base knowledge of both linear and angular concepts. Balance and stability are important concepts in sports and in activities of daily living. All sports require the manipulation of balance and stability. Most people tend to think of maximizing balance and stability, but minimizing them is the key in some sports. In daily living, we can observe the importance of stability in human movement by watching an infant learn to creep and walk. At the other end of the life span continuum, we observe the loss of stability as people age. Because balance and stability are ever-present concepts in any movement-related field, applications of this material and connections to other disciplines are numerous.

CONCEPTS

7.1 EQUILIBRIUM, STABILITY, AND BALANCE

The previous two chapters discussed the idea of *equilibrium* to some extent. In general, a state of equilibrium is achieved if the resultant force and net torque (moment) acting on an object are equal to zero. Therefore, in accordance with Newton's first law of motion, the object in equilibrium is either at rest or moving with a constant velocity because no external force has been applied. In reference to Newton's second law of motion, equilibrium requires that the object is either motionless or its acceleration is equal to zero (constant velocity) because there is no *net* force. More specifically, equilibrium can be either linear (translational) or rotational. **Linear** or **translational equilibrium** is achieved when the net of the external forces acting on a system is equal to zero ($\Sigma F = 0$). For example, in a tug-of-war, the rope will not move linearly one direction or the other if all of the opposing pulling forces are similar in magnitude and acting in opposing directions ($\Sigma F = 0$). Notice that forces are being applied, but they sum to zero, so the object (rope) remains in linear equilibrium.

In a tug-of war, the forces are applied linearly. We could have a case in which off-axis forces (or torques) are being applied. **Rotational equilibrium** is achieved when the net of the external torques (moments) acting on a system is equal to zero ($\Sigma T = 0$). Any combination of linear and rotational equilibrium can be achieved by an object. For example, an object may be completely motionless (i.e., rotating and traveling with zero linear or angular velocity), or an object may travel linearly with a constant velocity and rotate with constant velocity (zero linear or angular acceleration). And as we have seen, an object can exhibit purely linear or purely angular motion, depending on the degree to which the applied force is eccentric. We will explore all of these options in the following sections.

Linear or **translational equilibrium** A state achieved when the net of the external forces acting on a system is equal to zero ($\Sigma F = 0$).

Rotational equilibrium A state achieved when the net of the external torques (moments) acting on a system is equal to zero ($\Sigma T = 0$).

Static Equilibrium

As mentioned in the previous section, objects are often in linear and angular motion simultaneously (i.e., general motion). However, we will begin our discussion with the situation of **static equilibrium**, a situation in which the system is in linear and rotational equilibrium ($\Sigma F = 0$ and $\Sigma T = 0$) *and* possesses zero linear or rotational velocity. So an object in static equilibrium is completely at rest, which requires satisfying all of the following six equations (in three-dimensional space) (Serway & Jewett, 2019; Serway & Vuille, 2018):

$$\Sigma F_x = 0 \qquad \Sigma F_y = 0 \qquad \Sigma F_z = 0$$

$$\Sigma T_x = 0 \qquad \Sigma T_y = 0 \qquad \Sigma T_z = 0$$

where

ΣF is the sum of the forces in the x, y, and z directions and ΣT is the sum of the torques (moments) around the given axis.

Satisfaction of the above six equations is necessary for static equilibrium in three-dimensional (3-D) analysis. However, in biomechanics the condition of static equilibrium is often used for two-dimensional (2-D) analysis. We actually used this method in the previous chapter to solve for an unknown force. The condition of static equilibrium in 2-D analysis is satisfied with the following three equations (relating to the three degrees of freedom in planar motion) (Serway & Jewett, 2019; Serway & Vuille, 2018):

$$\Sigma F_x = 0 \qquad \Sigma F_y = 0 \qquad \Sigma T_a = 0$$

where

ΣF is the sum of the forces in the x and y directions, and ΣT is the sum of the torques (moments) around any single given axis.

In 2-D analysis, we are concerned with coplanar forces. For example, an upward-directed muscle force is used to counteract the gravitational force (weight) of a limb segment. These two forces lie in the same plane. If the sum of the coplanar forces is equal to zero, then the linear equations of 2-D static equilibrium are satisfied. These coplanar forces are acting around a given axis. If the sum of the torques about the given axis is zero, then the rotational equation of 2-D static equilibrium is satisfied. Notice that we need specify only one axis. Only one is needed because if an object is in linear static equilibrium and the sum of the torques around one axis perpendicular to the specified plane is equal to zero, then the net torque around any other axis is also equal to zero (Serway & Jewett, 2019).

Static equilibrium A situation in which the system is in linear and rotational equilibrium ($\Sigma F = 0$ and $\Sigma T = 0$) *and* possesses zero linear or rotational velocity.

SAMPLE PROBLEM

In Chapter 6, we used only the torque equation to solve for an unknown force. Now we will use all three of the preceding equations to solve for unknown quantities. In this situation, we will use the example of the quadriceps muscles contracting to hold the lower leg in static equilibrium (Figure 7.1). The lower leg is in static equilibrium in this situation because it is in both linear and rotational equilibrium ($\Sigma F = 0$ and $\Sigma T = 0$) *and* possesses zero linear or rotational velocity. We need some anatomical information for this problem. Most information of the type we will be using can be found in tables from various studies that use cadavers to gain anatomical characteristics. The quadriceps tendon is inserted on the tibia, approximately 5 cm from the axis of rotation of the knee joint, and the force vector representing quadriceps force is oriented at approximately 30° (Buford et al., 1997). The weight of the lower leg of an 80-kg male is approximately 48 N (Plagenhoef, Evans, & Abdelnour, 1983). If the person is 1.78 m in height, then the center of gravity of the lower leg is approximately 0.20 m from the knee joint (Braune & Fischer, 1889; Dempster, 1955; Plagenhoef, et al. 1983).

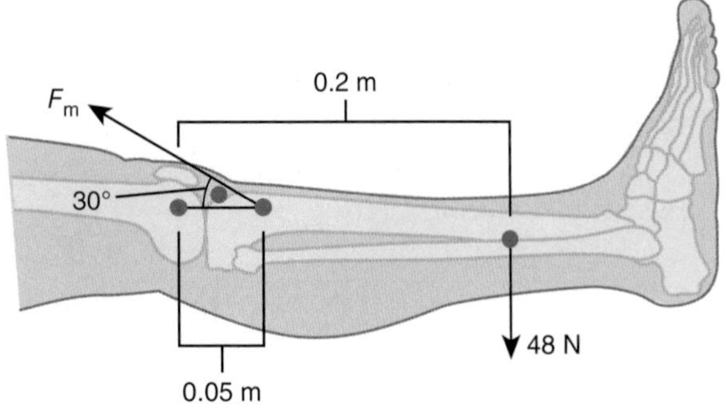

Figure 7.1 **Solving for quadriceps force.** Why is the necessary muscle force so large?

As before, our first task is to solve for quadriceps force necessary to hold the lower leg in static equilibrium. We can use the $\Sigma T_a = 0$ static equilibrium equation to accomplish this calculation. Keep in mind that the actual equation for torque is $T = F \times r \times \sin\theta$. In addition, in this analysis we will also consider that as the tibia is pulled into the femur by the muscle contraction, the femur will exert a reaction force on the tibia. So we also want to solve for the magnitude and orientation of the reaction force vector of the femur. Using the other static equilibrium equations ($\Sigma F_x = 0$ and $\Sigma F_y = 0$), we can also solve for the horizontal and vertical components of this joint reaction force. We need to calculate the torques (directions are according to the right-hand thumb rule):

$$\text{Torque caused by weight of the lower leg} = +48.00 \text{ N} \times 0.20 \text{ m} = +9.60 \text{ N} \cdot \text{m}$$

$$\text{Torque from quadriceps force} = -\text{F}_\text{m} \sin 30° \times 0.05 \text{ m} = -\text{F}_\text{m} \sin 30° \,(0.05) \text{ N} \cdot \text{m}$$

where

F_m = muscle force

and then solve for the unknown muscle force:

$$\Sigma T_a = 0$$

$$(+9.60 \text{ N} \cdot \text{m}) + (-\text{F}_\text{m} \sin 30° \,(0.05) \text{ N} \cdot \text{m}) = 0$$

$$\text{F}_\text{m} = \frac{-9.60 \text{ N} \cdot \text{m}}{-0.025 \text{ m}}$$

$$\text{F}_\text{m} = +384.00 \text{ N}$$

The quadriceps muscle force required to hold the lower leg in a static position is +384.00 N. Notice the large amount of force that must be generated by the muscle because of its short force arm. Now we need to solve for the vertical and horizontal components of the joint reaction force (directions are in reference to an *x–y* coordinate frame). First, we sum the forces in the vertical direction to solve for the vertical component of the joint reaction force (R_v):

$$\Sigma F_y = 0$$

$$\text{R}_\text{v} + (\text{F}_\text{m} \sin 30°) + (-\text{weight of the lower leg}) = 0$$

$$\text{R}_\text{v} + (+384.00 \sin 30° \text{ N}) + (-48.00 \text{ N}) = 0$$

$$\text{R}_\text{v} = -192 \text{ N} + 48.00 \text{ N}$$

$$\text{R}_\text{v} = -144.00 \text{ N}$$

So the vertical component of the joint reaction force is −144.00 N. Notice that in this case R_v must have a negative value because it is directed opposite to the upward force of the tibia.

(*continues*)

(continued)

Next we sum the forces in the horizontal direction to solve for the horizontal component of the joint reaction force (R_H):

$$\Sigma F_x = 0$$

$$R_H + (F_m \cos 30°) = 0$$

$$R_H + (384.00 \cos 30° \text{ N}) = 0$$

$$R_H = -332.55 \text{ N}$$

So the horizontal component of the joint reaction force is −332.55 N. Notice that in this case R_v must have a negative value because it is directed opposite to the force of the tibia. Also notice that the weight of the lower leg is not included because we are assuming that it is acting perpendicular to the tibia and therefore doesn't have a horizontal component.

Now that we have the vertical and horizontal components of the joint reaction force vector, we can solve for its magnitude and orientation. We need to calculate the magnitude of the resultant reaction force vector (F_r). From our previous information we know that the magnitude of a vector within a coordinate reference frame is defined by *r* in polar coordinates and according to the

Pythagorean theorem $r = \sqrt{x^2 + y^2}$. So in this example:

$$F_r = \sqrt{(-144 \text{ N})^2 + (-332.55 \text{ N})^2} = +362.39 \text{ N}$$

Now we need to calculate the orientation of the resultant vector:

$$\tan\theta = \frac{y}{x} = \frac{-144}{-332.55} = 0.433$$

We now have the ratio of *y* to *x* (0.481). The ratio can be converted to an angle of orientation by using the inverse tangent function. Remember to add 180° to the answer because the resultant joint reaction vector is in the third quadrant:

$$\theta = \tan^{-1}(0.433) = 23.41° + 180° = 203.41°$$

Now we know that the joint reaction force will have a magnitude of 362.39 N and that its representative vector will be oriented at 203.41° relative to the positive *x*-axis.

Note that some assumptions are made when using static equilibrium equations to solve for unknowns. All forces except for gravity and muscle force are ignored. Second, we can only solve for one muscle force vector. For example, the quadriceps muscles are modeled as one muscle. If this were an example of elbow flexion, we would have to choose one elbow flexor or model all of them as producing one collective force. These assumptions must be made or the equation will have more than one unknown value, a condition referred to as **statically indeterminate**, in which there are an infinite number of solutions (Oatis, 2017; Robertson et al., 2014). Even with these assumptions, 2-D analysis based on static equilibrium can give a remarkably good idea of musculoskeletal force requirements and reactions.

Statically indeterminate A situation in which there is more than one unknown value in an equation, causing an infinite number of solutions.

Dynamic Equilibrium

One should not necessarily assume that a zero net force or torque acting on an object means that the object is not moving; it simply means that the object is not accelerating at the time of observation. If an object has an initial velocity that is not equal to zero and is acted on by zero net force (or torque), it will continue to move with a constant velocity (Serway & Vuille, 2018). Therefore, a system in **dynamic equilibrium** is in motion, but it is experiencing no change in velocity or direction. This steady state means that all of the forces acting to cause the motion are balanced by equal and oppositely directed inertial forces. The condition of dynamic equilibrium in 3-D analysis is satisfied with the following six equations (Serway & Jewett, 2019; Serway & Vuille, 2018):

$$\Sigma F_x - ma_x = 0 \qquad \Sigma F_y - ma_y = 0 \qquad \Sigma F_z - ma_z = 0$$

$$\Sigma T_x - I_x\alpha_x = 0 \qquad \Sigma T_y - I_y\alpha_y = 0 \qquad \Sigma T_z - I_z\alpha_z = 0$$

where

ΣF is the sum of the forces in the given direction
ma is the product of the body's mass and its acceleration in the given direction
ΣT is the sum of the torques (moments) around the given axis
$I\alpha$ is the product of the body's rotational inertia and angular acceleration around the given axis

As with static equilibrium, the equations of dynamic equilibrium can be reduced to three for 2-D analysis (Serway & Jewett, 2019; Serway & Vuille, 2018):

$$\Sigma F_x - ma_x = 0 \qquad \Sigma F_y - ma_y = 0 \qquad \Sigma T_a - I_a\alpha_a = 0$$

where

ΣF is the sum of the forces in the given direction
ma is the product of the body's mass and its acceleration in the given direction
ΣT is the sum of the torques (moments) around any given axis
$I\alpha$ is the product of the body's rotational inertia and angular acceleration about any given axis

To fully understand this idea, it is sometimes helpful to use an example of imbalance in the sum of the forces and the inertial force (i.e., $\Sigma F - ma \neq 0$). The typical example of this phenomenon is the apparent change in weight of an object as measured by a scale on an elevator that is accelerating (Figure 7.2).

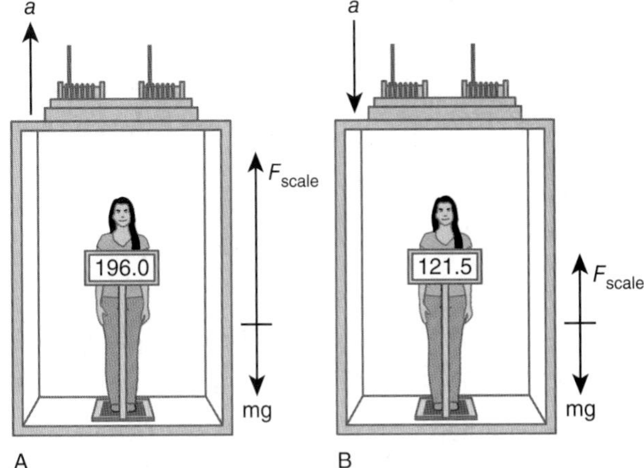

Figure 7.2 **Person on a scale in an elevator.** Why does weight change on an elevator?

For example, as the elevator accelerates upward, an oppositely directed (downward) inertial force is created. This force causes the weight of the body as measured by the scale to increase, even though the actual weight of the body has not changed. The forces to be summed in this situation are the weight of the object and the upward inertial force caused by the elevator. Because the acceleration of the elevator is upward, it has a positive value. This positive force exerted by acceleration is added to the weight of the object, and the scale shows a higher reading. If the elevator accelerates downward, it has a negative value. This negative force exerted by acceleration is subtracted from the constant weight of the object, and the scale shows a decrease in weight (Serway and Jewett, 2004). Note that the actual weight of the object did not change; only the weight as measured by the scale varies. The cause of this phenomenon is an imbalance between the downward force of the object and the upward inertial force.

Now that we have a full understanding of equilibrium, we need to define *stability* and *balance*. **Stability** is simply the resistance of an object to having its equilibrium disturbed. The more stable the object, the more resistance it has against forces or torques that attempt to disrupt its equilibrium. In a mechanical sense, balance is often used

Dynamic equilibrium A situation in which a system is in motion but is experiencing no change in velocity or direction.

Stability The resistance of an object to having its equilibrium disturbed.

synonymously with the term *equilibrium*. And if we are referring to an inanimate object, that definition is appropriate. However, if we are referring to an organism, *balance* takes on a slightly different connotation. **Balance** is the ability to control the current state of equilibrium, and it implies conscious effort and coordination. So stability is more of a *mechanical* term, whereas balance is more of a *neuromuscular* reference. Let's examine equilibrium in terms of an object's stability.

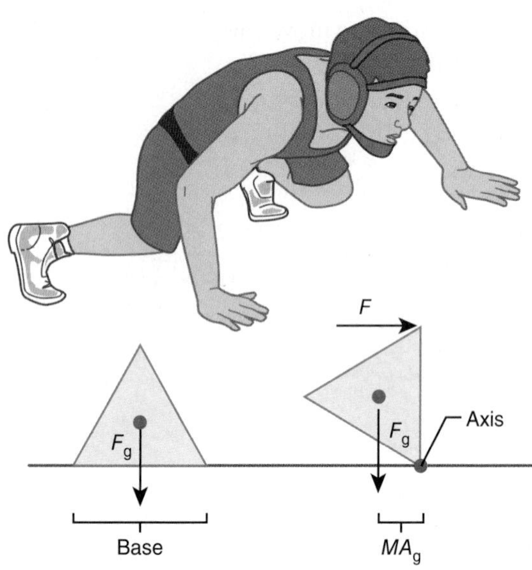

Figure 7.3 **A pyramid-shaped object and a wrestler in a defensive sprawl.** What do the pyramid and the wrestler have in common?

Stable Equilibrium

In the sport of wrestling, a defender will sometimes sprawl to prevent being turned over and pinned (Figure 7.3). A large amount of torque is necessary to rotate a person in a sprawled position. Any object placed in such a way that a large force or torque is required to disrupt its current position is said to be in **stable equilibrium**. A pyramid-shaped object placed on its largest side would be an example of stable equilibrium.

If we were to apply a small force to the tip of the pyramid, a torque would be created around an axis at one edge of the base of the pyramid. Notice that the torque would cause the pyramid's center of gravity to be raised. So the lower the center of gravity, the longer a torque must be applied to completely topple an object. If the applied eccentric force is removed before the pyramid falls over, a torque caused by gravity will return the object to its original position because the line of gravity is still on the original side of the axis of rotation (i.e., the force arm for gravity is on the original side). So a force applied to an object in stable equilibrium allows gravity to act as a **restorative force** that attempts to return the object to its original position. The return to the original position is caused by the force arm for gravity remaining on the side of the applied force. If, however, the applied force caused the line of gravity of the pyramid to fall opposite the original side of the axis of rotation, gravity would produce a torque and cause the object to fall over. In this case, the restorative force actually becomes a **disruptive force**. The torque caused by gravity is now disruptive because its force arm is now on the opposite side of the applied force. So the more torque required to overcome the restorative torque, the more stable the object is considered to be.

Unstable Equilibrium

Sports such as sprinting require an athlete to become mobile as fast as possible. In these sports, the highly stable position of the wrestler in our previous example would be a disadvantage because a large torque would be necessary to disrupt the current state of equilibrium. So in some instances, athletes must adopt a stance that enables a fast transfer from stance to motion (i.e., less force or torque is required to disrupt equilibrium). A runner in the starting blocks is a prime example. The runner is in a state of equilibrium that can be changed rather abruptly by simply lifting the hands (Figure 7.4). At this point, you can probably guess the definition of unstable equilibrium. An object in **unstable equilibrium** requires little force or torque to disrupt its current state. Let's say that we somehow managed to balance the pyramid on its tip.

Balance The ability to control the current state of equilibrium; implies conscious effort and coordination.

Stable equilibrium A state in which a large force or torque is required to disrupt a body's current position.

Restorative force A force that acts to return an object to its original position.

Disruptive force A force that acts to move an object away from its original position.

Unstable equilibrium A state in which little force or torque is required to disrupt a body's current position.

If we were to apply even a small force to the pyramid, the torque created around the axis at the tip would be enough to cause the pyramid to fall over. Notice in this case that the center of gravity didn't have to be lifted at all; it can only fall. So any force at all causes a torque that is disruptive to the original position. Again, the torque is disruptive because the force arm for gravity is on the opposite side of the applied force. So a force applied to an object in unstable equilibrium causes gravity to act as a disruptive force that attempts to move the object farther from its original position of equilibrium. One could also see this from the point of view that the disruptive force in this case is actually restorative because it moves the object to a position of greater stability.

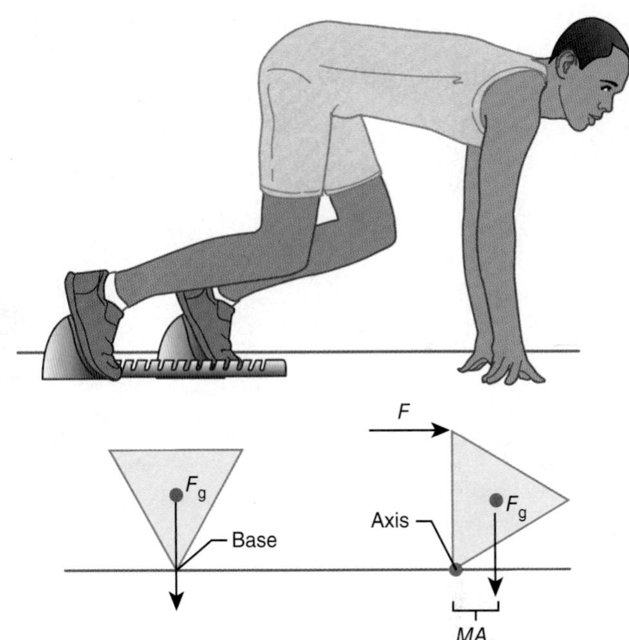

Figure 7.4 **A pyramid-shaped object and a sprinter in the starting blocks.** What do the pyramid and the sprinter have in common?

Neutral Equilibrium

An object in a state of **neutral equilibrium** has no tendency to fall in one direction or the other. In other words, moving the object from its original position has no tendency to produce either a restorative or disruptive force (Serway & Jewett, 2019). An example of this type of object would be a sphere (such as a ball) on a flat surface (Figure 7.5).

As in the preceding examples, notice that a sphere on a flat surface has a line of gravity that passes directly through the base. But unlike force applied to the pyramid, a force applied to a sphere does not cause the line of gravity to fall to one side or other of that base. Therefore, the ball has no tendency to fall in one direction or the other. So, once friction force stops the motion caused by the original force application, the ball will regain its position of neutral equilibrium. The key is that the ball is on a flat surface. To disrupt neutral equilibrium, one would need to tilt the surface on which the ball is placed. Tilting the surface would cause the line of gravity to fall outside the base of the ball, creating a force arm for gravity. The ball would then roll downhill because of the disruptive torque of gravity around an axis at the ball's point of contact on the surface.

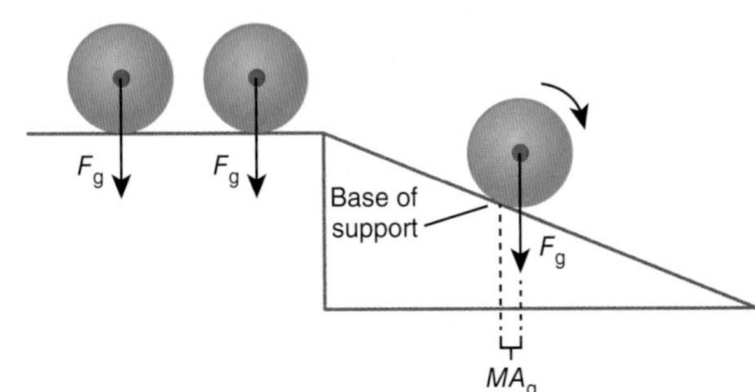

Figure 7.5 **A ball on flat surface and ball on sloped surface.** What is the specific difference that causes the ball to roll?

7.2 LINEAR STABILITY

We have discussed equilibrium and stability in a general sense. It is now necessary to be more specific about the situations of interest. We have said that linear or translational equilibrium is achieved when the net external force is equal to zero ($\Sigma F = 0$). Also, recall that stability is the resistance of an object to having its equilibrium disturbed. Therefore, **linear stability** is the resistance of a body to having its linear equilibrium (i.e., linear inertia) disrupted.

In accordance with Newton's second law of motion, the more massive the object (more inertia) the more force must be applied to produce a given acceleration. So in linear terms, one of the major factors affecting stability is the mass of the object. One of the key words is *linear.* Try not to think in terms of an object falling

Neutral equilibrium A state in which there is no tendency to fall in one direction or the other.

Linear stability The resistance of a body to having its linear equilibrium disrupted.

Figure 7.6 **Sumo wrestlers.**

over; focus on movement in a straight line. Think of the difficulty of pushing a large football blocking sled loaded with football players. The need for linear stability depends on the situation or sport. For example, a sumo wrestler is likely to have high body mass index (Figure 7.6). The same would be true for an offensive lineman in football.

In both situations, the advantage of being massive is that it makes it extremely difficult (more force is required) for other athletes to move that massive person linearly out of the way or to stop his forward progress. There is a tradeoff; increased stability is often accompanied by decreased mobility. For example, we would probably never see a 230-kg sumo wrestler succeed at a sport such as gymnastics. Gymnastics requires rapid directional changes. The less massive a gymnast is, the less force is required to produce a given acceleration. Therefore, successful gymnasts tend to be smaller than athletes in many other sports (McArdle, Katch, & Katch, 2014). However, this small mass also means that gymnasts are not as stable as much larger athletes. So the average gymnast would probably never be a champion sumo wrestler. Imagine a sumo wrestler or offensive lineman on ice skates trying to make a quick directional change.

Then again, try to imagine a world class figure skater on a professional football field attempting to defend the quarterback. So, maximizing linear stability is critical if the situation requires maintaining a given position or state of linear motion. In contrast, situations that require large accelerations also require minimization of linear stability. Note that the stability requirements are not necessarily the same for every position within a given sport. For example, many players on a football field are not nearly as massive as an offensive lineman.

Linear Stability and External Forces

Maximizing or minimizing linear stability is not all about manipulating mass. If we want to maintain linear equilibrium, we can make a conscious attempt to minimize external forces that could change the current state of motion. For example, in sports such as cycling, rowing, and swimming, maintaining a straight path is critical. But cycling, rowing, and swimming are examples of sports in which linear stability must be increased by some means other than adding excess mass. In these sports, athletes attempt to minimize the external forces that could disrupt linear equilibrium. For example, in all three sports, reducing drag force maintains linear equilibrium better than adding mass does.

An additional example is ice skating. A skater moves easily on ice not only because of the smaller mass but also because the friction force provided by the ice is minimal. Friction can be an important factor in linear stability. The ice is an obvious example of a surface that decreases the friction force that would normally act to prevent motion or slow the skater once motion has begun. But many examples of linear stability are related to friction in less obvious ways. For example, our offensive lineman is massive, which means that his weight will push him into the playing surface. Recall that friction force is proportional to the normal force holding the objects together. Thus, the greater the mass of the athlete, the greater the reaction force, and the greater the resulting friction force between the athlete and the surface of play. But also think of the contacting surfaces between the person and the ground. If the offensive lineman is wearing cleats, the footwear will further enhance linear stability by increasing friction force. Also consider the playing surface itself. Is the surface natural grass or artificial turf? We know that the coefficients of friction are probably not the same for those two surfaces. Therefore, changing the mass of an object actually affects linear stability in two ways: (1) Greater force is required to linearly accelerate a more massive object, and

(2) greater mass leads to increased friction force between the object and the surface it is contacting because of an increase in normal force. In addition, linear stability can be affected by the coefficient of friction of the surfaces that are in contact.

7.3 ROTATIONAL STABILITY

As you have read, rotational equilibrium is achieved when the net external torque (moment) is equal to zero ($\Sigma T = 0$). Thus, **rotational stability** is the resistance of a body to having its rotational or angular equilibrium disrupted. In other words, rotational stability is resistance to having equilibrium disrupted by being rotated about an axis by a net torque (or moment).

In accordance with Newton's second law of motion, the more angular inertia the object possesses, the more torque must be applied to produce a given angular acceleration. As with linear stability, one major factor affecting rotational stability is the inertia (mass) of the object. The key word here is *angular.* Think of objects falling or being tipped over around an axis. For example, imagine pushing a box over instead of just pushing it across the ground. If you tip it over, you are applying a torque about an axis that is formed by one edge of the box. Most situations of stability are actually concerns of rotational stability, especially in sports. Even in the examples of the sumo wrestler and the offensive lineman, rotational stability issues are present simultaneously with the linear stability issues. For example, the linear stability of the offensive lineman may be high because of a large friction force (Figure 7.7). But all that this fact really means is that he is resistant to being slid across the ground. In terms of rotational stability, another player may rotate the lineman around an axis at his feet (i.e., push him over even though his feet are planted).

Because rotational equilibrium and rotational stability are involved to some degree in most stability situations, let's examine determining factors in detail.

Figure 7.7 **Football linemen.**

Rotational Stability and the Center of Gravity

The location of the center of gravity relative to the base of support is an important factor in rotational stability. So first we must define the base of support. An object's **base of support** is the entire area bounded by a perimeter formed by the object's points of contact with a resistive surface that provides a counterforce (Figure 7.8).

Fully understanding this concept is critical. Many people mistakenly consider the base of support to be only the points of contact with the surface. If you are hanging by a bar with both hands, your base of support includes the part of the bar between your hands. If you are leaning against a wall, one point of the perimeter of your base is formed by the contact area with the wall. If you are standing on one foot, then your base of support is the area under the foot.

Remember that as the rotational equilibrium of an object is disrupted by an applied torque, one edge of the base becomes the axis of rotation. Also recall that if the center of gravity remains on its original side of the axis of rotation, then gravity exerts a restorative torque when the force attempting to move the object is removed. If, however, the center of gravity is pushed to the side opposite its original position relative to the axis of rotation, it will become a disruptive torque. As you have read, it becomes disruptive because the relative position of its force arm (moment arm) has changed (Figure 7.9).

So in reality, rotational stability is directly related to the location of the center of gravity. If the center of gravity

Rotational stability The resistance of a body to having its rotational or angular equilibrium disrupted.

Base of support The entire area bounded by a perimeter formed by the points of contact with a resistive surface that provides a counterforce.

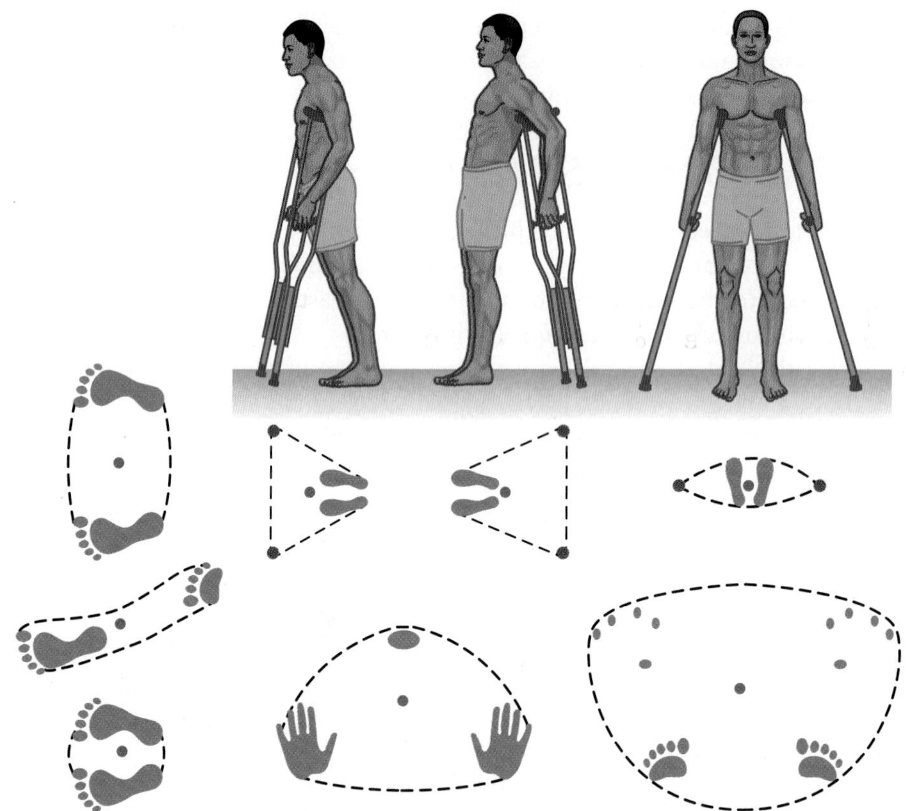

Figure 7.8 **Various possibilities for the base of support.** Match a sport skill to each base.

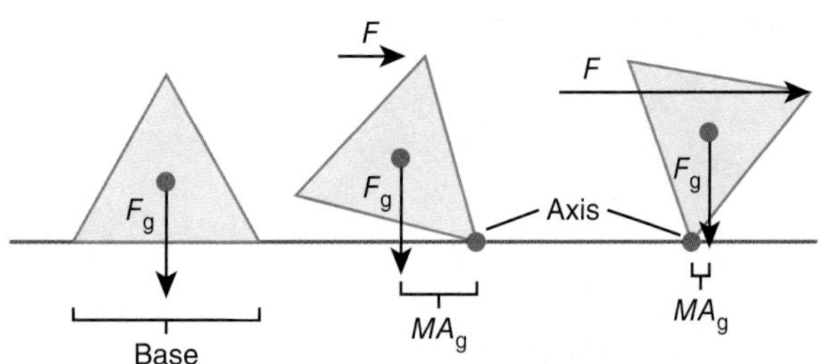

Figure 7.9 **The force arm for gravity acting on a pyramid-shaped object.** Why will the pyramid eventually fall over?

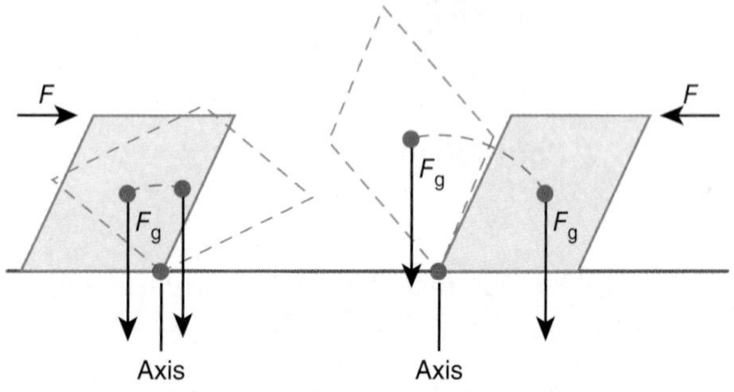

Figure 7.10 **Differing locations of the center of gravity.** Why is it easier to push one of these objects over?

is in the middle of the base of support, then gravity has no force arm to be either restorative or disruptive. In this case, the object is in rotational equilibrium. If the applied torque is large enough to rotate the object around an axis formed by one of its edges, the line of action of the center of gravity will get closer to the edge of the base. The center of gravity at this point is torque producing, but unless it crosses to the other side of the axis it will remain a restorative torque. The closer it gets to the edge, the more likely it is to cross over the axis of rotation. Therefore, *one way to increase rotational stability is to keep the center of gravity in the middle of the base of support.* In other words, if the line of gravity is already near the axis of rotation, little torque or time of torque application will be needed to move it outside the base (Figure 7.10).

The effect of the position of the center of gravity is actually attributable to the change in length of the force arm for the gravitational force. So basically, in terms of understanding stability, it's torque against torque. If you want to rotate the object around an axis, you must apply a torque. Remember that gravity also applies a torque (in the opposite direction). If gravity has a large force arm, more torque is required on your part to rotate the object. Keeping the center of gravity in the middle of the base simply maximizes the length of gravity's force arm (Figure 7.11).

Visualizing rotational stability during suspension is actually a great way to understand the importance of keeping the center of

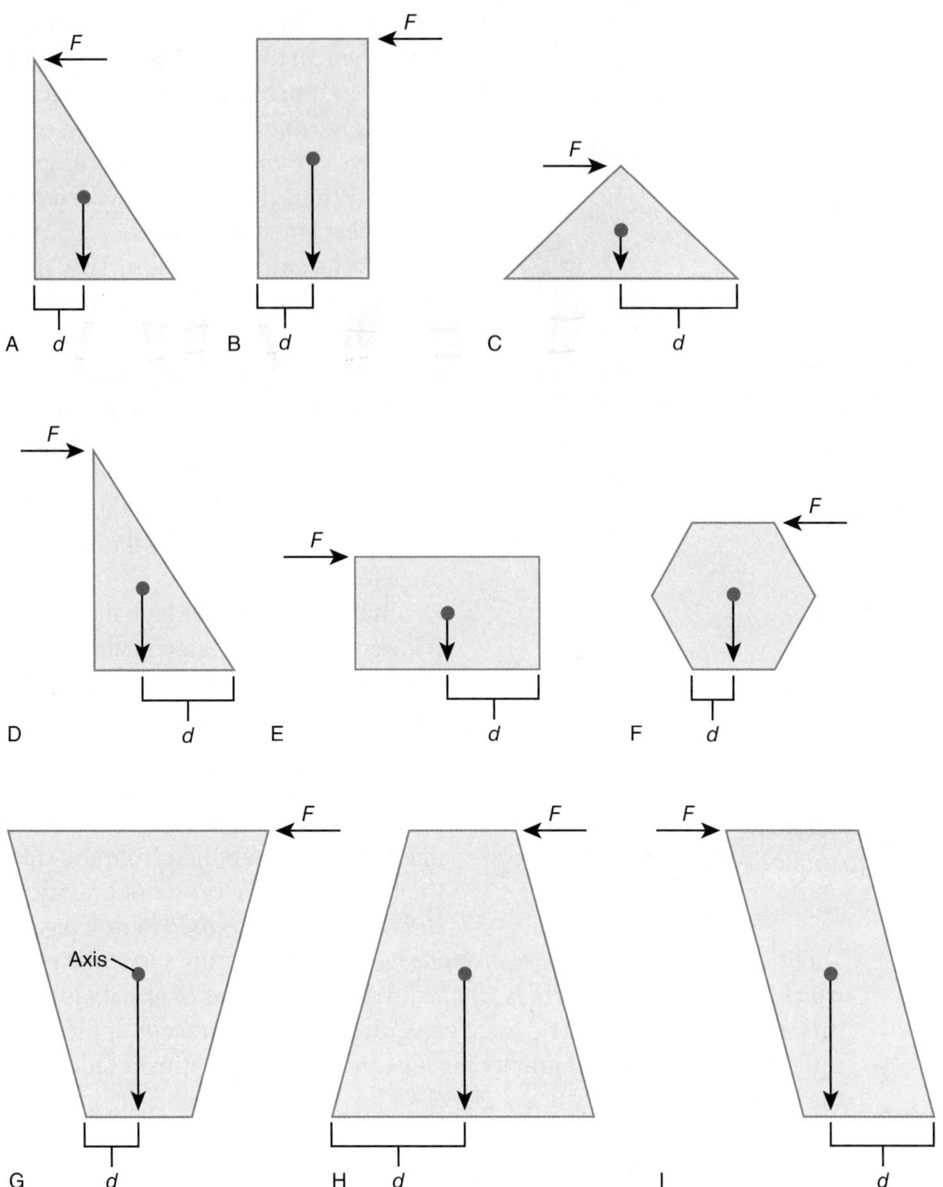

Figure 7.11 **The length of gravity's force arm for objects of various shapes.** Which of the objects will be easiest to topple? Why?

gravity in the middle of the base. Imagine hanging by two hands from a bar in a position of equilibrium. The base of support is outlined by the points of contact with the bar, and the center of gravity would be directly in the middle of that base. Now imagine releasing one hand (Figure 7.12).

A new base of support is now formed by the single hand still in contact with the bar, and your center of gravity is well outside that base. Because the line of gravity is outside the base, gravity has a force arm and becomes torque producing. This torque will cause the center of gravity to rotate around the new axis formed by the contact hand. Notice that the body will come to a new position of equilibrium when the line of gravity once again passes through the base of support. Equilibrium is reestablished because the force arm for gravity is no longer present, so gravity can no longer apply torque, and therefore rotation will stop. Therefore, *a suspended object will come to rotational equilibrium in a position in which the center of gravity is at its lowest point possible, given the particular constraints of the situation.*

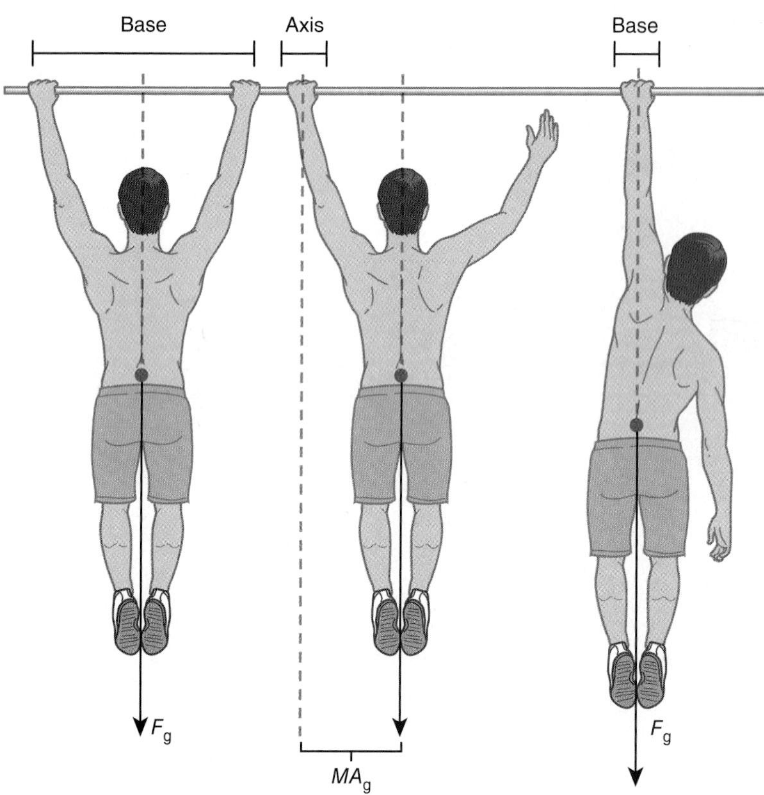

Figure 7.12 **Equilibrium during suspension.** What is the specific difference that causes the person to rotate? Why does rotation stop?

Recall from the discussion of equilibrium that when an object is rotated around one edge of its base, the center of gravity is actually lifted. Therefore, *another way to increase rotational stability is to lower the center of gravity of the object*. When the center of gravity is lowered, the object must be rotated a greater distance around the axis of rotation. This situation will require greater torque, or torque will have to be applied for longer, to move the line of gravity outside the base (Figure 7.13).

Now, of course, the exception to increasing stability by lowering the center of gravity would be a situation in which the center of gravity is lowered but moved toward the edge of the base of support opposite that of the externally applied force (Figure 7.14).

In this case, even though the center of gravity is lowered, the effect is attenuated because the line of gravity is now closer to the edge. Being closer to the edge, the line of gravity requires little force to move it outside the base of support.

The exceptions also have exceptions. The exception just described assumes that the external force will be applied from the side opposite the location of the center of gravity. *Rotational stability can be increased by moving the center of gravity to one side of the base of support as long as it is the same side of the base on which the external force will be applied.* When the center of gravity is positioned in this way, a greater eccentric force is required, or force must be applied for a longer time, because the line of gravity must be moved to the opposite side to fall outside the base of support (Figure 7.15).

You have read that rotating an object around an axis requires a torque to be applied. This torque is countered by a gravitational torque in the opposite direction. The same idea applies to this concept. Moving the center of gravity toward the applied force actually maximizes gravity's force arm. Without an externally applied force, this change actually destabilizes the object because gravity in this position can become a disruptive torque. If an externally disruptive torque *is* applied, we must maximize the torque caused by gravity to maintain equilibrium.

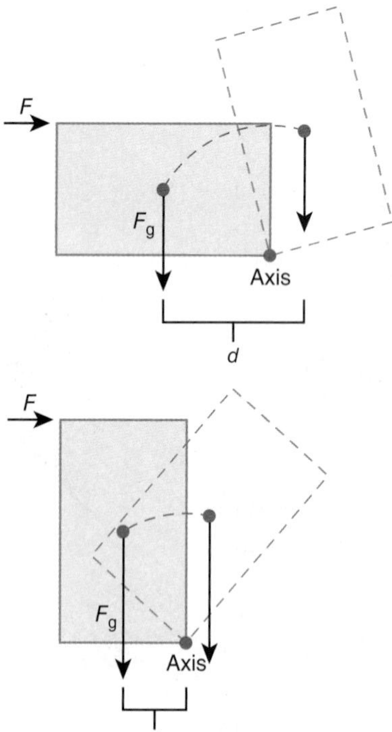

Figure 7.13 **Change in height of the center of gravity.** How does the location of the center of gravity affect stability?

Rotational Stability and the Base of Support

So far, we have focused on keeping the line of gravity within the boundaries of the base of support. To this point, you have read only about positioning the center of gravity to achieve that goal. However, a primary determinant of rotational stability is the characteristics of the base of support itself. The characteristics are important because the size and shape of the base of support affect the ease with which the line of gravity falls outside the boundaries of the base.

The size of the base of support is one of the most important factors in maintaining rotational equilibrium. As soon as the line of gravity falls outside the base, gravity becomes a disruptive torque producer. When the base of support is as wide as

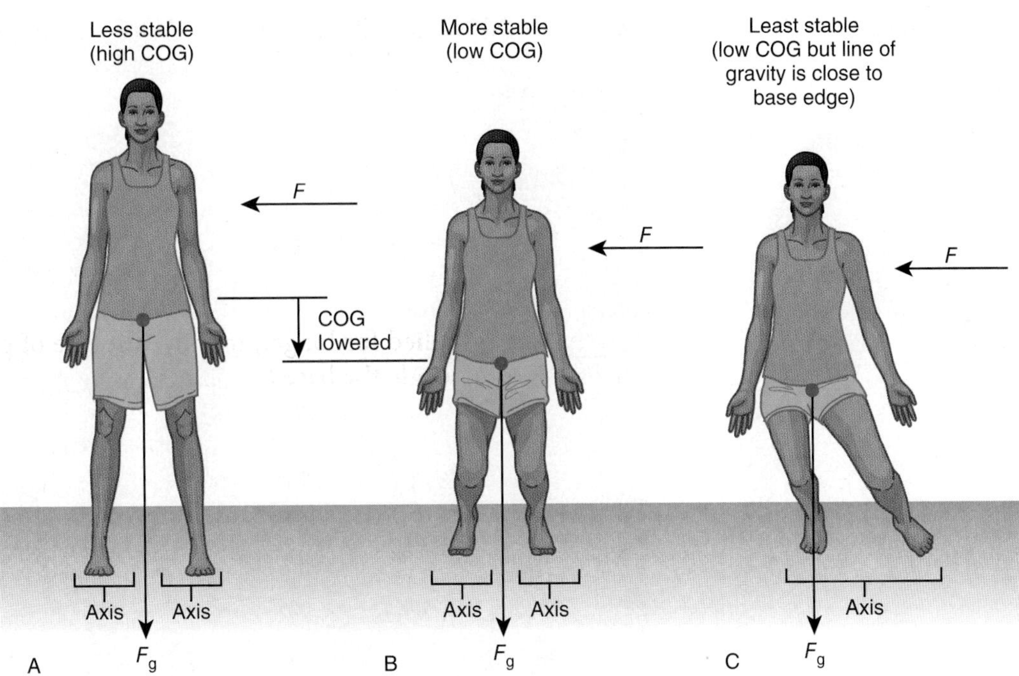

Figure 7.14 **Person in three positions.** Which position is the most stable?

possible, it becomes extremely difficult for the line of gravity to fall outside that base. It is simply a matter of the distance that the center of gravity will have to be moved. The greater the horizontal distance to the edge of the base, the greater the applied force or time of force application required to disrupt rotational equilibrium.

Therefore, *another way to increase rotational stability is to increase the size of the base of support.* Remember what you have read about stable and unstable equilibrium. A pyramid positioned on its widest side is much more stable than a pyramid balanced on one of its corners. We can see the importance of this concept by observing the body position of almost any athlete in the stance he or she takes before initiating movement (Figure 7.16).

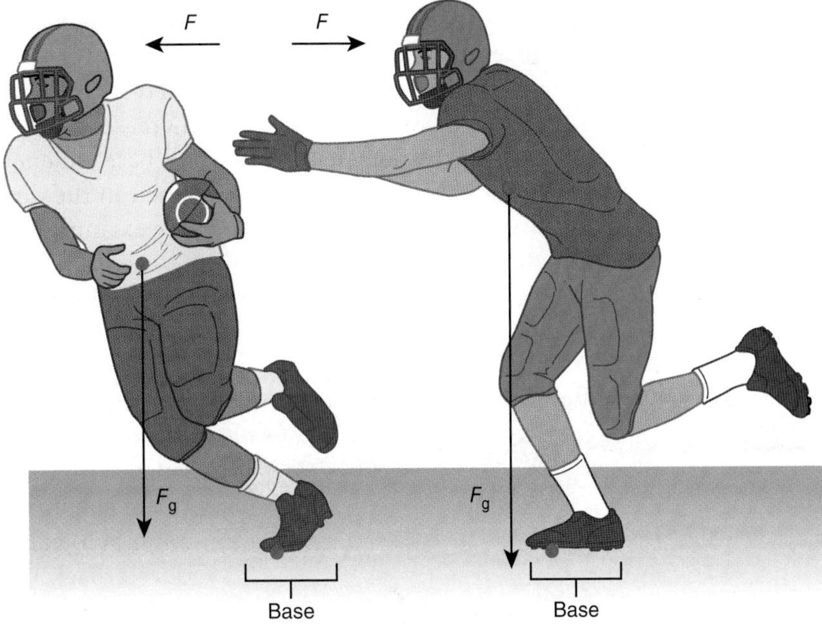

Figure 7.15 **Locations of the center of gravity relative to the location of the applied force.**

While observing various stances in sports, however, notice that the base of support is not always the same shape. Preparatory stances are normally wider than a natural stance, but the width tends to be exaggerated in one direction more than in the other (e.g., medially and laterally, or anteriorly and posteriorly). This imbalance is necessary because to *maximize rotational stability, the*

Directional (or **planar**) **stability** A state in which a body is more stable in the direction that the base is widened.

© Gino Santa Maria/Shutterstock

© Ahturner/Shutterstock

Figure 7.16 **Several sports stances that require stability.** Can you think of other situations that require similar levels of stability?

base of support should be widened in the direction of the line of action of the externally applied force. This concept is similar to that of moving the center of gravity to the same side as the externally applied force. The key is to put as much distance between the line of gravity and the potential axis of rotation as possible. Doing so will increase the amount, or duration, of torque necessary to move the line of gravity outside the base. So a base widened in one direction does not ensure stability in all directions. The most stability is gained in the direction in which the base is widened—that is, *stability is directional or planar*. A prime example of **directional** or **planar stability** is a gymnast on a balance beam (Figure 7.17).

A gymnast on a balance beam usually has one foot in front of the other; therefore, the base is widened anteriorly and posteriorly. This foot position comes at the expense of narrowing the base medially and laterally. Therefore, the gymnast tends to be relatively directionally stable in the forward and backward direction (sagittal plane), but at the same time may have trouble maintaining balance from side to side (frontal plane). This concept of directional stability poses a precarious problem for the center on a football team. To snap the football between his legs to the quarterback, the center must assume a wide base medially and laterally. Doing so causes a directional stability problem because the applied force will more than likely come from the anterior direction.

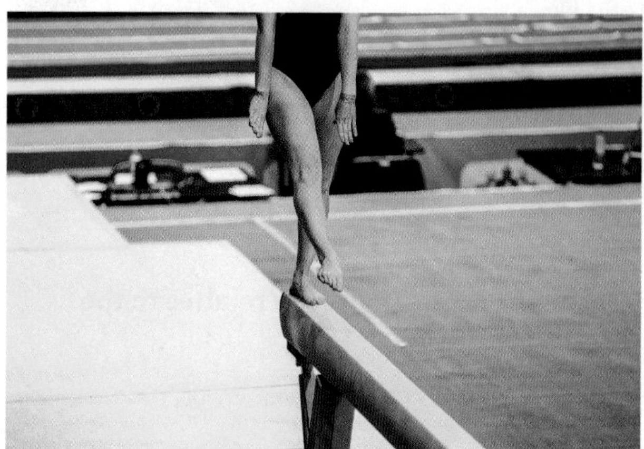
Figure 7.17 **Gymnast on balance beam.** What is the width of her base of support?
© sportpoint/Shutterstock

Notice that the center can compensate for this loss of stability by lowering the center of gravity as much as possible and simultaneously moving it toward the front of the base. As you have learned, sprinters in starting blocks take advantage of directional stability (or instability, depending on the point of view), as well as some other concepts of rotational stability (Figure 7.18).

While in the starting position, the sprinter is directionally stable anteriorly and posteriorly. But why adopt this position to begin with? Why not just start from a standing position? Remember that stability comes with a cost in terms of mobility. A simple standing position is not overly stable or unstable. The sprinter needs to be as mobile as possible. The most mobile position would be highly unstable. The sprinter does need to be stable enough to remain in position until the start signal is given but then she needs to be instantly unstable. How do you get the best of both worlds? A hint to the answer is that many directionally stable positions can easily become directionally *unstable* in a short time. Notice what happens to the base of support when the tips of the fingers are removed from the ground at the starting signal. The base is now formed only by contact of the toes with the ground, whereas before it included all of the area between the tips of the fingers and the toes. With this new base of support, the line of gravity of the sprinter is outside the base and is torque producing. With the split-second removal of the fingertips from the ground, the sprinter has assumed a highly unstable (i.e., mobile) position: (1) the center of gravity is outside the base of support, and (2) the base of support has been minimized. This change causes an unbalanced force, which results in accelerating the person into the race.

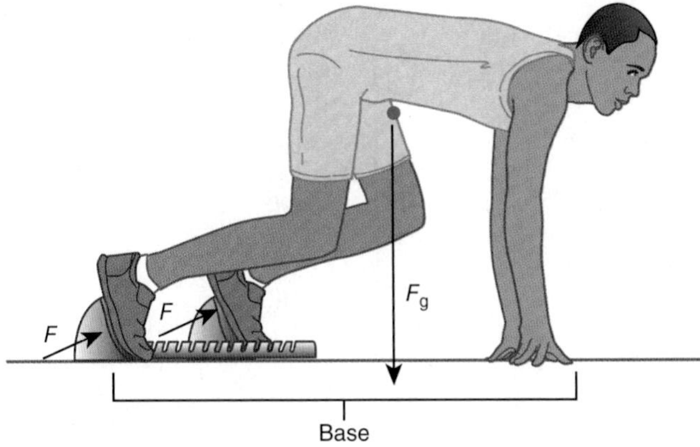

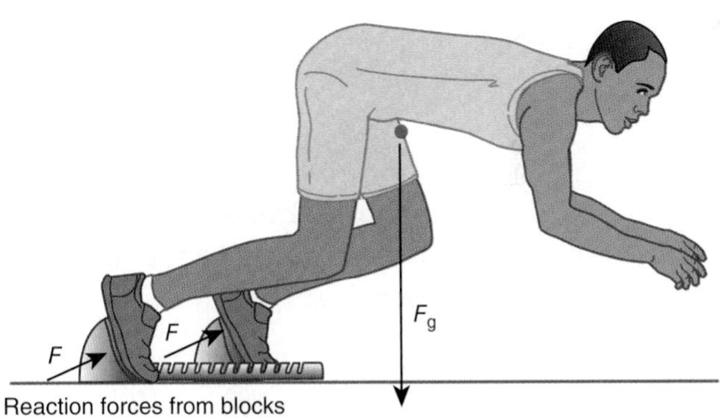

Figure 7.18 **A runner in starting blocks showing base of support.** Why is this takeoff position an advantage?

Unfortunately, some common misunderstandings are related to the concept of directional stability. For example, many novice (and not so novice) weightlifters fail to analyze one of the most commonly uttered phrases in resistance training: "Your feet should be shoulder width apart." The underlying advice is good because we do want stability while lifting weights. However, the base should not always be widened in the same direction. For example, let's analyze a biceps curl exercise (elbow flexion; Figure 7.19). The common mistake seen with this exercise is "rocking" the entire body while lifting the barbell.

This rocking motion is caused by directional instability. A person following the "feet shoulder width apart" advice has widened the base of support in the frontal plane and is therefore stable in that plane but directionally unstable in the other planes. The motion of the barbell is in the sagittal plane. The instability in the sagittal plane is observed as the rocking motion of the body begins. A full understanding of the concepts of stability should lead the weightlifter to two conclusions: (1) The weight is not being moved in the frontal plane, so stability in that plane is not of primary importance; and (2) it may be more useful to make sure that one foot is slightly in front of the other to provide directional stability in the sagittal plane (Figure 7.20).

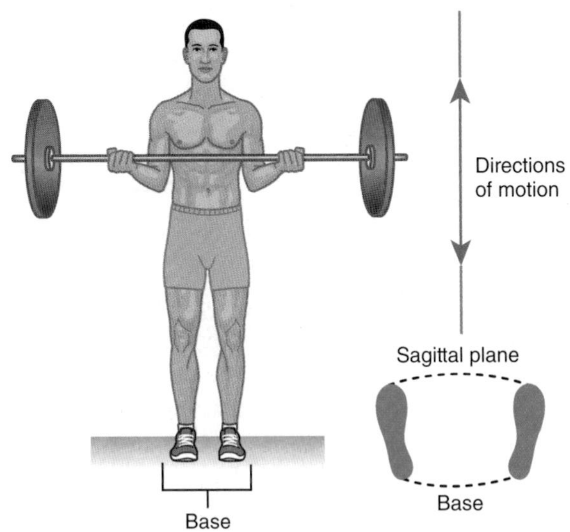

Figure 7.19 **Biceps curl exercise.**

So as we have seen, what matters when attempting to maximize rotational stability is not only the size of the base of support but also its shape. Besides the position of the center of gravity and the shape of the base of support, one other factor affects rotational stability.

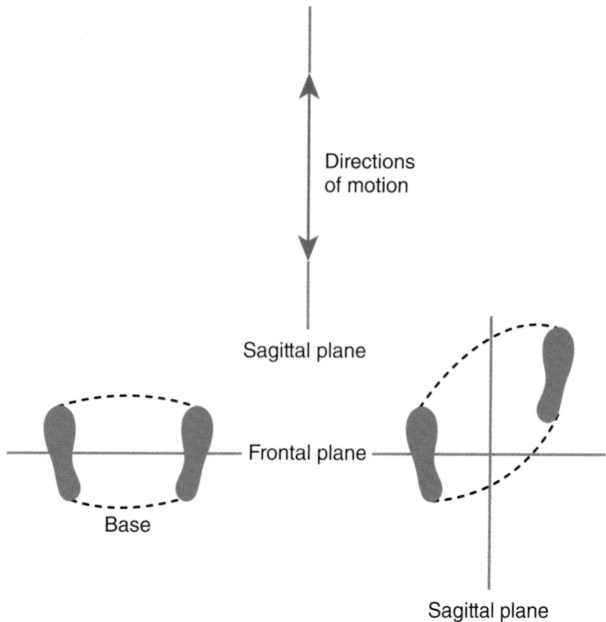

Figure 7.20 **Bicep curl foot placement.** Why does the change in foot placement enhance the effectiveness of the exercise?

Rotational Stability and Mass

Of course mass is always a factor in any form of stability, and rotational stability is no exception. Again, according to Newton's second law, a greater force must be applied to a more massive object to achieve a given acceleration. In rotational situations, larger mass translates into a larger rotational inertia, and a greater torque must be applied to an object of greater rotational inertia to cause a given angular acceleration about a given axis. We discussed the concepts of rotational inertia and angular acceleration in Chapter 6:

$$I_a = mk_a^2 \tag{6.11}$$

where

I_a = moment of inertia. about a given axis
m = mass of the entire object
k_a^2 = radius of gyration about a given axis squared

$$\alpha_a = \frac{\Sigma T_a}{I_a} \tag{6.12}$$

where

α_a = angular acceleration about a given axis, or change in angular motion of the system
ΣT_a = the sum of the torques (net torque) about a given axis
I_a = moment of inertia about a given axis (rotational inertia)

But one should also think of rotational stability in terms of some previous information about the center of gravity and its relationship to stability. Remember that the center of gravity must be raised to rotate it around an axis at one edge of the base of support. More torque, or longer torque application, is required to raise (rotate) the center of gravity of a more massive object to disrupt its equilibrium. Therefore, *rotational stability can be enhanced by increasing the mass of the object.* Think back to the sumo wrestler example. Because of the mass of the sumo wrestler, he is both linearly and rotationally stable (that is, as long as he assumes the typical wide stance with a low center of gravity).

7.4 STABILITY AND ENERGY TRANSFER

In previous chapters you have read about work (or energy transfer), potential energy, and kinetic energy. One can also analyze the stability of a system from an energy-transfer perspective. Because the center of gravity must be raised to rotate it around an axis, work is performed ($W = F \times d$). As work is performed, energy is transferred. In this case, the energy transfer changes the potential energy of the object (i.e., as the center of gravity is raised by the work performed, the potential energy of the object is increased).

Recall that if the center of gravity remains on its original side relative to the axis of rotation, it will produce a restorative torque on release. However, if the center of gravity is displaced to the opposite side of the axis of rotation, its torque is considered disruptive. In either case, this restorative or disruptive torque produced by gravity is actually the conversion of the potential energy to kinetic energy as the objects falls to a position of equilibrium (Figure 7.21). In terms of energy transfer, the more work that must be performed to disrupt an object's equilibrium, the more stable the object. You have learned several ways to increase stability. One simple way is to *lower the center of gravity.* The lower the center of gravity, the more work must be performed to displace that center of gravity upward (a large change in potential energy). If the object's center of gravity is high, then little work is needed to disrupt equilibrium (a small change in potential energy). We also mentioned *keeping the center of gravity in the middle of the base*. This concept also relates to increasing the vertical displacement necessary (increased work) to disrupt equilibrium. All of these factors affect the distance (d) portion of the work equation ($W = F \times d$). Also, because force = mass × acceleration, *increasing the mass of the object* affects stability by increasing the "force" factor in the work equation. Therefore, the more stable the object, the more work must be performed to disrupt its equilibrium. Potential energy is changed during performance of the work. The potential energy of the displaced object is then converted to kinetic energy as the object falls around an axis at the edge of the base of support.

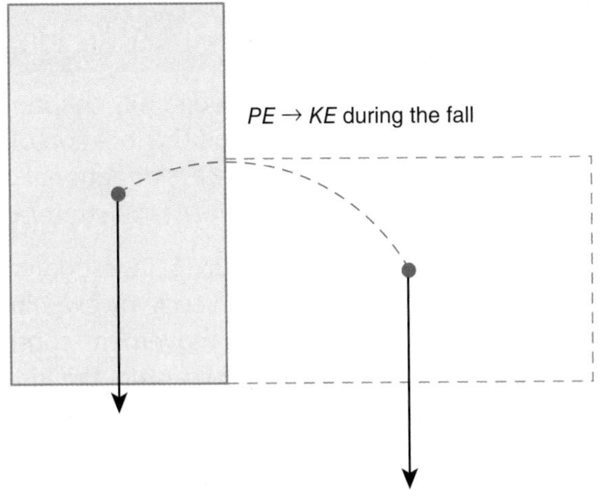

Figure 7.21 **Block in two positions.**

CONNECTIONS

7.5 INJURY SCIENCE

Physical and Occupational Therapy

Much like people with chronic disabilities, those who sustain acute trauma can also lose the ability to maintain balance and equilibrium for a variety of reasons. Physical and occupational therapists are trained to assess the causes of balance dysfunction and to help people regain the ability to maintain equilibrium. Factors that can cause balance problems include loss of sensory information, loss of integrative function within the brain, loss of motor control, and lack of sufficient strength to counteract forces acting on the body. Therapists have protocols and equipment designed to determine the causes of postural, static, and dynamic instability.

Spinal cord injury and neuropathy, such as with advanced diabetes, are two possible reasons for loss of sensory input. **Proprioceptors** are responsible for providing information concerning joint position, velocity, muscle tension, rate of change in length, and other important information related to maintaining equilibrium. When sensory structures or delivery pathways are compromised, the brain doesn't receive needed information to make sense of where the body is located in space or where and how specific body segments are moving (or not moving) in relation to each other.

In some cases such as stroke, concussion, or other head injury, sensory systems are functioning properly, but the integrative brain centers that control balance and movement may be affected. Depending on the site of injury, the brain may not be able to process the information provided by sensory systems or make proper adjustments through coordinated muscular contractions to counteract unstable situations. When motor areas of the brain are affected, sensory and integrative systems can function normally, but the ability to contract muscles in a coordinated manner is affected. The brain may understand that the body is in a state of instability but be unable to do anything about it.

As with muscle, bone, and other tissues, the brain is considered plastic: They all have the potential ability to sustain and recover from damage, but they do so in different ways. A broken bone rebuilds through a process of cell genesis. Tears and other trauma to muscle cells result in the proliferation of myofibrils and other connective tissues to repair damaged areas. Plasticity in the nervous system

Proprioceptors Responsible for providing information concerning joint position, velocity, muscle tension, rate of change in length, and other important information related to maintaining equilibrium.

FOCUS ON RESEARCH

As discussed in this chapter, stability is affected by factors such as the width of the base of support and the location of the center of gravity in relation to that base of support. In addition, balance is directional or planar. Simple enough, as long as you are standing still on two feet. But what's the strategy for riding a unicycle?

Lee et al. (2016) investigated the coordination of balance and propulsion processes in learning to ride a unicycle by examining the contribution of the arm, leg, and torso motions to the structure of the movement-coordination pattern. A motion-digitizing system along with cameras was used to capture the kinematics of the motion and the cycling performance. The process was divided into three learning stages: early, middle, and advanced. In the early stage, all participants used a handrail as an aid for balance. The middle stage was trials without a handrail. Participants were considered to be in the advanced stage if they could ride the unicycle with a success rate of more than 70% (three of the six participants reached this stage). Changes in the associations of the upper- and lower-body degrees-of-freedom (DOF) were observed over the three stages. Participants were more afraid of falling in the early stage and therefore relied more on the handrail to maintain balance. This slower speed could not provide enough angular momentum to maintain balance. Because the arms are being used to hold a rail, a smaller number of upper-body joints are participating in pedaling-related principal components such as counteracting or following the pedaling movement. The researchers also found that once the task was learned, the contribution of the upper-body segments increased as well as the segmental involvement of pedaling-related principal components. The researchers speculate that this finding is an indication that the processes of balance and propulsion became more coupled as a result of success and the riding movement became composed of a larger number of pedaling components. The researchers state that these findings provide evidence that while learning to ride a unicycle, balance and propulsion must be made more coordinated to make the system more controllable while simultaneously mastering the redundant DOF.

As can be seen from this study, balance can become a highly complex task dependent on the situation. Riding a unicycle involves both linear and angular momentum, a small base of support, as well as coordinating balance in the sagittal plane with propulsion forces.

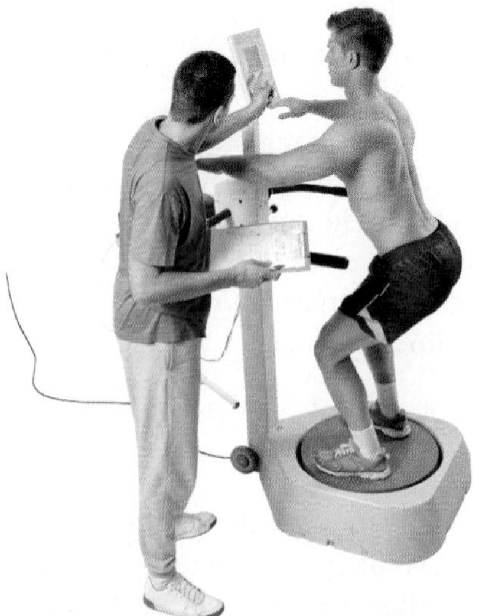

Figure 7.22 **Computer aided balance assessment equipment.**

relies not only on neurogenesis (regeneration of brain cells) but also on the brain relearning through the process of making new connections between remaining neurons.

Physical and occupational therapy for nervous system damage is based on the brain's ability to "rewire" around damaged areas and sometimes retrain other areas in the brain to assume the function of the damaged areas. Therapy for relearning or improving balance focuses on practicing balance in a safe, controlled environment. Much of the same equipment used for children with disabilities is also used by physical and occupational therapists—from simple floor exercises to balance boards, balance balls, and newer inflatable cushions in a variety of shapes and sizes. To assess balance and to help develop dynamic balance in a controlled environment, products such as the Biodex® Balance System SD™ offer computer-based evaluation and rehabilitation protocols (Figure 7.22).

An additional benefit of using equipment is that it can help develop proprioceptive awareness and also increase muscular strength. Strength is another important area for therapists who work with patients who have balance problems. Maintaining balance takes a coordinated effort of both the nervous and muscular systems. If the muscles can't produce enough strength to counteract forces acting on the body, or they fail to produce it quickly enough, the result is instability and falls. For physical and occupational therapists, enhancing balance often takes the form of strength training. Developing stronger and more powerful muscles is sometimes the only therapy necessary for patients with stability problems.

7.6 MOTOR BEHAVIOR

Motor Development

At both ends of the spectrum of life, humans are challenged in their quest to attain (after birth) and maintain (in old age) the ability to move through space. The key in both cases is creating a state of equilibrium and stability. Early in life we learn how to gain stability through trial and error, and with gains in muscular strength and development of neural systems, coordinated upright locomotion becomes subconscious. As we begin losing neural capacity and muscular strength later in life, we revert to the same strategies a toddler uses to maintain upright locomotion.

Have you ever wondered why a newborn horse is able to stand and run within hours after birth, but a human baby doesn't acquire this ability for months? Some of this discrepancy has to do with the horse being more physically mature at birth than a human. A foal's muscles and nervous system are in a more advanced state of development (partly because of a longer gestation period). Human babies do not have sufficient muscular strength or neural control at birth. Two other important factors hinder a human baby's ability to support herself in an upright position: (1) a high center of gravity and (2) a relatively small base of support.

At birth, humans are widely different in shape, proportion, and center of gravity than in childhood or later years. We know that humans develop in a cephalocaudal direction (head to tail), so the head is the largest part of the body at birth. A newborn's head constitutes 25% of his height, and through the age of two years this percentage does not change appreciably (Gabbard, 2018). The result of this general body shape is a high center of gravity (Figure 7.23).

Any time the relatively large mass of the head moves away from the base of support, maintaining an upright orientation requires a lot of muscular force. During the first two years of life, we progress through sitting, standing, and walking; and as every parent knows, the slow development of the muscular and nervous systems often cannot keep up with the infant's desire to be upright. The battle between the force of gravity, center of gravity, base of support, and top-heaviness results in greater disruptive torque than an infant can effectively counter. Topples and falls are part of the developmental process, but as the vestibular system and motor cortex mature and muscles gain strength through repeated attempts, infants become increasingly capable of counteracting this torque and thus become more successful at upright locomotion.

Developing upright locomotion occurs in a highly predictable sequence and has more to do with physics than you may imagine. Through repeated use in spoken language, *crawling* has become the word commonly used to describe the skill of traveling on the hands and knees with the torso clear of the floor. To remain consistent with motor development literature and dictionary definitions, this book uses the more appropriate term for this action: *creeping*. Creeping is a later stage in voluntary locomotion; the first step is crawling, in which the infant drags the torso across the floor by pushing off with the arms and legs. This technique allows a large base of support for the raised head—the mass of the entire body against the floor easily counteracts the mass of the head. As the infant gains strength, crawling becomes creeping,

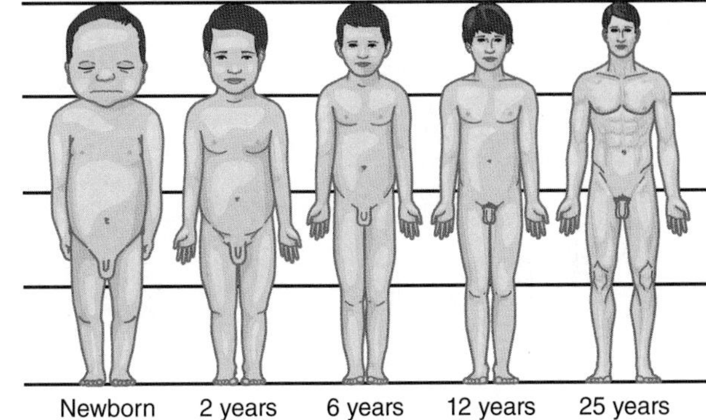

Figure 7.23 **Comparison of relative size of a baby's head to an adult's.** How does the relative size of the head affect stability?

Figure 7.24 **Moving mass over the base of support and then widening the base of support for balance.** How does the base of support change?
© Oksana Kuzmina/Shutterstock

and things get a little more interesting. As long as both hands and feet are in contact with the floor, an infant has essentially the same base of support as crawling. However, the intent of creeping is locomotion, and this requires at least one of the four points of support to leave the ground momentarily, creating a state of disrupted equilibrium as the base of support changes shape (Figure 7.24).

Initial attempts at forward motion see the infant barely raising a hand or knee off the floor and returning it quickly to reacquire balance. With more practice, infants instinctively learn that when a hand or knee is raised, they can more easily maintain balance by shifting their weight to the opposite side. Notice how the head has shifted to put this mass over the supporting arm in Figure 7.24. This change allows her to lift the motive body part for a longer amount of time, resulting in longer "strides" and faster speeds.

As infants master prone locomotion, their attention turns to upright locomotion. This step presents a great challenge because their base of support is dramatically reduced. To compensate for the lack of stability, toddlers demonstrate several behaviors that help them maximize their base of support: legs are spread shoulder width apart, toes are pointed outward, and arms are carried in a high guard position (Figure 7.25a). The arms in high guard position may protect infants in case of falls, but they also act much like the long pole that tightrope walkers use to maintain stability. With strength gained through practice and maturation of the nervous system, the stability strategies used in initial attempts are no longer required. The base of support narrows as the legs come closer together, the toes decrease their outward turn, and the arms come down from the high guard position (Figure 7.25b). By the time a child reaches kindergarten, she is easily able to walk heel to toe along a line with her arms at her sides.

The strategies that infants use to increase their stability as they acquire upright locomotion are instinctive and subconscious. At the opposite end of the life span, older adults consciously apply the same strategies as the aging process advances. Falls among older adults become more frequent as various biological systems reach advanced stages of regression. Falls among the elderly are a leading cause of injury. Injuries caused by falls often lead to the loss of independent lifestyles and even death in extreme cases (Greenhouse, 1994). The increased frequency of falls among older adults can be attributed to changes in sensory, vestibular, and muscular function and in the distribution of mass and the density of materials. These topics are covered in more detail in the next section.

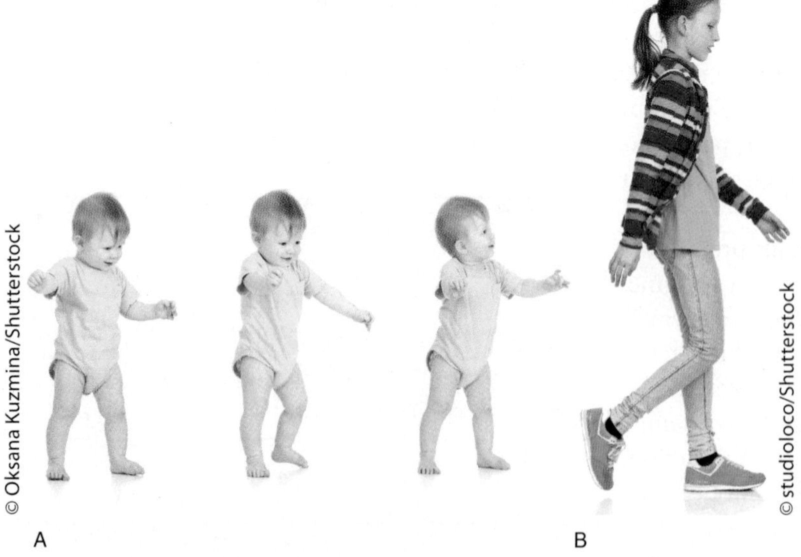

© Oksana Kuzmina/Shutterstock © studioloco/Shutterstock

Figure 7.25 **An early walker, and a proficient walker.** What are similarities in these situations?

Motor Control

Several structures within the brain coordinate to help us maintain postural (unconscious) and static (conscious) balance. Neurons in balance areas of the brain begin dying off after age 40 (Gabbard, 2018), and both muscle mass and contractile properties diminish as well (McArdle et al., 2014). The end result is a nervous system that is no longer able to maintain equilibrium, and muscles that don't react quickly enough or with enough force to counteract the fall. The good news is that with an active lifestyle, balance practice, and assistive devices (e.g., canes, walkers) if needed, independent mobility can continue, and the risk of falls can be reduced.

Postural sway is a term used to describe oscillations of the center of mass within the base of support during normal erect standing posture because of variations in muscle tension. Toddlers display postural sway, which decreases as their nervous and muscular systems mature. Through adolescence and adulthood, postural sway is almost nonexistent unless pathology is involved. However, with deterioration of nervous system and muscle function, postural sway becomes more evident again in older age. Signals that we are out of balance are generated and understood more slowly by the brain, and corrective muscle reactions to this information are also slowed. The result is that the body sways side to side and front to back like a tree in the wind. Postural sway is an indicator of deficiency in the body's ability to maintain equilibrium, and it is highly correlated with falls in older adults (and, for that matter, falls while children learn to walk).

Whether compromised equilibrium is caused by pathology or by regression of biological systems because of advanced age, humans find ways to fight its effects. We make a conscious effort to increase the base of support that reverses the pattern followed by children learning to walk. The angle of toeing out increases, stride width increases, step length decreases (to shorten the time spent in an unstable position), and arm swing decreases. When these changes are no longer enough to maintain upright posture, assistive devices make independent mobility possible by increasing the base of support. Canes add a second point (or more with newer quad canes) of support during the swing phase of walking. Crutches create a tripod, and walkers generally have four points on the ground during the swing phase. Each device can be used to counteract neural and strength deficiencies.

Before the need arises for assistive devices, people approaching old age should do all they can to maintain strength and should also practice balance activities. Research has reported that both measures reduce postural sway and increase stability. Rogers, Fernandez, and Bohlken (2001) conducted a study with older adults using elastic bands for resistance exercises while balancing on air-filled exercise balls. Their conclusions suggest that by challenging the physiological systems involved in balance, the participants improved both static and dynamic balance. Formal exercise forms such as Tai Chi, Pilates, and yoga have also been reported to be highly effective methods for older adults to maintain strength and improve balance (Ni et al., 2014).

7.7 PEDAGOGY

In its simplest form, walking is purposeful "falling" followed by the application of muscular strength to avoid falling completely. In essence, it's controlled loss of balance. Try this: Stand and fall forward, moving your center of gravity away from your base of support with your feet stationary. After a half second, move one leg forward to stop falling; that was one step in a walking pattern. This is the process that a toddler progresses through as he develops more mature walking patterns, and it's a good example of how balance is a learned skill rather than an inherent ability.

Do good gymnasts have exceptional balance because they are naturally good balancers, or do they become good balancers through doing gymnastics? Research shows that the latter is probably true. Successful performance of almost every motor skill relies

Postural sway Oscillations of the center of mass within the base of support during normal erect standing posture from variations in muscle tension.

on the ability to establish and maintain balance, or equilibrium, and requires the successful integration of several biological systems. Sensory, motor, muscular, skeletal, and vestibular systems must work in harmony to counteract forces that attempt to upset a state of equilibrium—the most common force being gravity.

Sports and physical education share a common stance called *ready position*. The feet are a little more than shoulder width apart, the toes are angled out slightly, the knees are bent, and the arms are lifted parallel to the floor. This stance is called *ready position* because athletes are ready for most situations. They are ready to move in any direction quickly and are also in the most effective position to absorb force. By this point in the chapter, you should understand the two most important reasons why the ready position is better in most situations than other body positions. The base of support is widened, and the center of gravity is lowered, resulting in a shape that is more resistant to being knocked over, can accelerate quickly in any direction, and can apply force in a more effective manner.

Teachers should understand that balance is an acquired ability and is activity specific. Having excellent postural or static balance does not necessarily indicate good dynamic balance, and good dynamic balance in one activity does not indicate good dynamic balance in all activities. Curriculum should be designed across all grade levels to include balance practice for most recreational and sport activities.

7.8 ADAPTED MOTION

Adapted Physical Education

People with cognitive and physical disabilities often have problems with balance because of their disability or a lack of opportunity to be active. Because balance is a learned behavior and is the key to success in most movement activities, including activities of daily living, it becomes an important topic in the adapted physical education curriculum. Activities to enhance balance include tumbling, educational gymnastics, and the use of specially designed equipment (Figure 7.26).

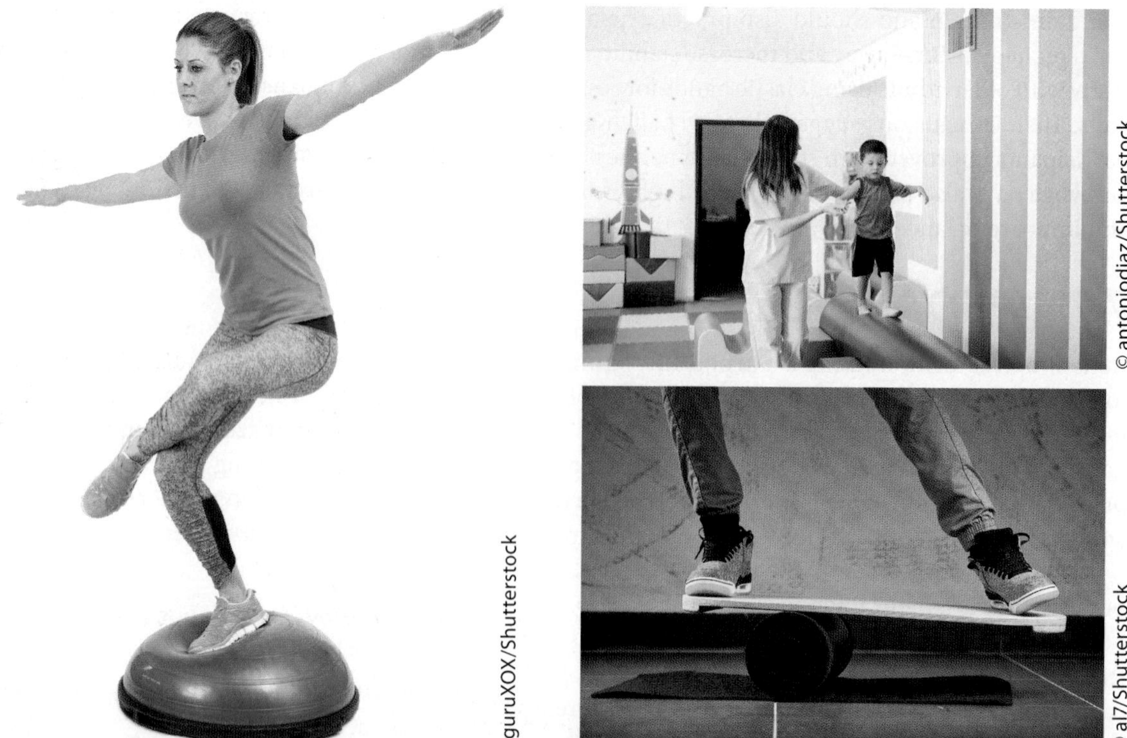

Figure 7.26 **Equipment used for balance training.**

Use of such specially designed equipment enables participation in physical education activities that are designed to enhance balance.

Pregnancy

The third trimester of pregnancy is accompanied by substantial weight gain. This gain is obviously from the rapidly growing fetus, the increased mass of the uterus, and the large amount of amniotic fluid. In biomechanical terms, this change in weight leads to an anteriorly displaced center of gravity. Recall that moving the center of gravity closer to the edge of the base of support can lead to a state of unstable equilibrium. To compensate for this change, the posture is adapted to incorporate a backward lean and exaggerated anterior curvature of the lumbar spine. This adaptation is accomplished by tilting the pelvis forward, and the resulting posture is referred to as **lordosis** (Figure 7.27).

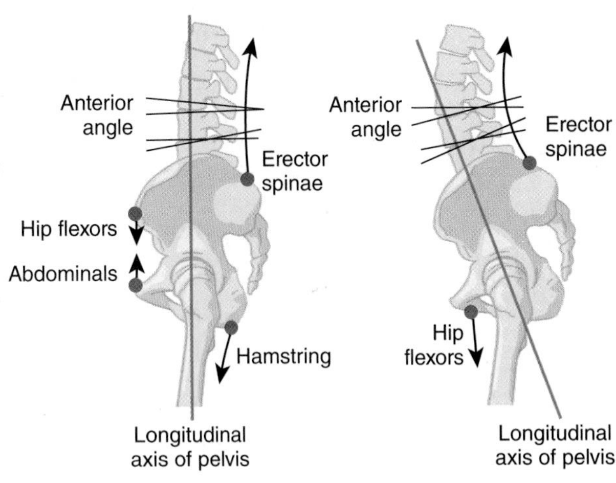

Figure 7.27 **Lordosis.**

This lordotic posture becomes progressively more exaggerated as the pregnancy progresses (Franklin & Conner-Kerr, 1998; Otman et al., 1989). The backward lean maintains the center of gravity in the middle of the base of support. However, this posture also increases lumbar stress and may lead to back pain. One may also see a widened walking stance during pregnancy, which is a clear attempt to enlarge the base of support. These changes in posture and gait may seem to be obvious in origin but do not minimize their importance. Stability must be maintained to prevent falling and injury of the soon-to-be mother and her unborn child.

Lordosis A posture in which there is exaggerated anterior lumbar curvature accompanied by excessive anterior tilt of the pelvis.

SUMMARY

Equilibrium, stability, and balance are important in all sports and activities of daily living. Equilibrium is achieved if the resultant force and resultant torque (moment) acting on an object are equal to zero. Linear or translational equilibrium is achieved when the net of the external forces acting on the system is equal to zero, and rotational equilibrium is achieved when the net of the external torques (moments) is equal to zero. Any combination of linear and rotational equilibrium can be achieved by an object. Equilibrium can be either static or dynamic. Static equilibrium is a situation in which the system is in linear and rotational equilibrium *and* possesses zero linear or rotational velocity. A system in dynamic equilibrium is in motion, but it is experiencing no change in velocity or direction. Stability is simply the resistance of an object to having its equilibrium disturbed and applies to both linear and rotational situations. The more stable the object, the more resistance it has against forces or torques that act to disrupt its equilibrium. Balance is the ability to control the current state of equilibrium, and it implies conscious effort and coordination. Some situations require maximal stability, whereas others require instability. Stability is affected by location of the center of gravity, size of the base of support, and the mass of the object.

REVIEW QUESTIONS

1. Why does rotation occur if the line of gravity falls to one side of the axis of rotation?
2. What is required to disrupt neutral equilibrium? Why?
3. Why is a four-point stance used for takeoff in a sport such as sprinting?
4. Why does the mass of a body affect the amount of torque necessary to cause loss of equilibrium?

5. Why does lowering the center of gravity increase stability? What is the exception?
6. When the body is suspended by two hands, what happens to the center of gravity if one hand is removed? Why? When will it stop? Why?
7. List four ways to increase stability and explain how each one works.
8. In what direction will a segment move if the sum of the torques is greater than zero? What about less than zero?
9. In what ways does the defensive sprawl of a wrestler affect stability? How is the wrestler's mobility affected by this position?
10. Name some skills that require an athlete to be suddenly mobile in any direction. How does this mobility requirement affect the choice of a base of support?
11. What do infants who are learning to walk and older adults have in common related to balance?
12. In what ways does a tightrope walker's pole increase stability?

PRACTICE PROBLEMS

The moment arm of the shoulder abductor muscles is approximately 5 cm from the axis of rotation of the shoulder joint, and the force vector representing shoulder abductor muscle force is oriented at approximately 150° relative to the positive *x*-axis. The weight of the upper extremity of an 80-kg male is approximately 45 N. If the person is 1.78 m in height, then the center of gravity of the upper extremity is approximately 0.25 m from the shoulder joint (Braune & Fischer, 1889; Buford et al., 1997; Dempster, 1955; Plagenhoef et al., 1983).

1. Draw a free-body diagram of the person described. The person should be facing you and have the left arm abducted parallel to the ground.
2. Calculate the muscle force needed to hold the arm in the static position.
3. Calculate the vertical component of the joint reaction force.
4. Calculate the horizontal component of the joint reaction force.
5. Calculate the orientation of the vector representing joint reaction force.

EQUATIONS

Equations satisfying the condition of static equilibrium in 3-D space:

$$\Sigma F_x = 0 \qquad \Sigma F_y = 0 \qquad \Sigma F_z = 0$$

$$\Sigma T_x = 0 \qquad \Sigma T_y = 0 \qquad \Sigma T_z = 0$$

Equations satisfying the condition of static equilibrium in 2-D space:

$$\Sigma F_x = 0 \qquad \Sigma F_y = 0 \qquad \Sigma T_a = 0$$

Equations satisfying the condition of dynamic equilibrium in 3-D space:

$$\Sigma F_x - ma_x = 0 \qquad \Sigma F_y - ma_y = 0 \qquad \Sigma F_z - ma_z = 0$$

$$\Sigma T_x - I_x\alpha_x = 0 \qquad \Sigma T_y - I_y\alpha_y = 0 \qquad \Sigma T_z - I_z\alpha_z = 0$$

Equations satisfying the condition of dynamic equilibrium in 2-D space:

$$\Sigma F_x - ma_x = 0 \qquad \Sigma F_y - ma_y = 0 \qquad \Sigma T_a - I_a\alpha_a = 0$$

Moment of inertia about a given axis

$$I_a = mk_a^2 \tag{6.11}$$

Angular acceleration about a given axis

$$\alpha_a = \frac{\Sigma T_a}{I_a} \tag{6.12}$$

REFERENCES AND SUGGESTED READINGS

Braune, W., & O. Fischer. 1889. *The center of gravity of the human body as related to the German infantryman*. Leipzig. Available as report ATI 138 452, National Technical Information Service, Springfield, VA.

Buford, W. L. Jr., F. M. Ivey, J. D. Malone, R. M. Patterson, G. Peare, D. Nguyen, & A. A. Stewart. 1997. Muscle balance at the knee—Moment arms for the normal knee and the ACL minus knee. *Transactions on Rehabilitation Engineering*, 5:367–379.

Dempster, W. T. 1955. *Space requirements of the seated operator: Geometrical, kinematic and mechanical aspects of the body with special reference to the limbs*. Technical Report WADC-TR-55–159, Wright-Patterson Air Force Base, OH: Wright Air Development Center.

Franklin, M. E., & T. Conner-Kerr. 1998. An analysis of posture and back pain in the first and third trimesters of pregnancy. *Journal of Orthopaedic and Sports Physical Therapy*, 28: 133–138.

Gabbard, C. P. 2018. *Lifelong motor development*, 7th ed. Philadelphia: Wolters Kluwer.

Greenhouse, A. H. 1994. Falls among the elderly. Pp. 611–626 in Albert, M. L. & J. E. Knoefel (Eds.), *Clinical Neurology of Aging*, 2nd ed. New York: Oxford University Press.

Lee, I. C., Y.T. Liu, & K. M. Newell. 2016. Learning to ride a unicycle: Coordinating balance and propulsion. *Journal of Motor Learning and Development*, 4(2): 287–306.

McArdle, W. D., F. I. Katch, & V. L. Katch. 2014. *Exercise physiology: Energy, nutrition, and human performance*, 8th ed. Philadelphia: Wolters Kluwer.

Ni, M., K. Mooney, L. Richards, A. Balachandran, M. Sun, K. Harriel, M. Potiaumpai, & J. F. Signorile. 2014. Comparative impacts of Tai Chi, balance training, and a specially designed yoga program on balance in older fallers. *Archives of Physical Medicine and Rehabilitation*, 95(9): 1620–1628.

Oatis, C. A. 2017. *Kinesiology: The mechanics and pathomechanics of human movement*. Philadelphia: Wolters Kluwer.

Otman, A. S., M. S. Beksac, & O. Bagoze. 1989. The importance of "lumbar lordosis measurement device" application during pregnancy, and post-partum isometric exercise. *European Journal of Obstetrics, Gynecology, and Reproductive Biology*, 31:155–162.

Plagenhoef, S., F. G. Evans, and T. Abdelnour. 1983. Anatomical data for analyzing human motion. *Research Quarterly for Exercise and Sport*, 54: 169–178.

Robertson, D. G. E., G. E. Caldwell, J. Hamill, G. Kamen, & S. N. Whittlesey. 2014. *Research methods in biomechanics*, 2nd ed. Champaign, IL: Human Kinetics.

Rogers, M. E., J. E. Fernandez, & R. M. Bohlken. 2001. Training to reduce postural sway and increase functional reach in the elderly. *Journal of Occupational Rehabilitation*, 11: 291–298.

Serway, R. A., & J. W. Jewett Jr. 2019. *Physics for scientists and engineers with modern physics*, 10th ed. Boston: Cengage Learning.

Serway, R. A., & C. Vuille. 2018. *College physics*, 11th ed. Boston: Cengage Learning.

CHAPTER 8

THE SYSTEM AS A MACHINE

LEARNING OBJECTIVES

1. Understand the similarities between the human body and machines.
2. Describe lever, pulley, and wheel-and-axle systems in the human body.
3. Describe the advantage of each system to human motion.

CONCEPTS

8.1 Musculoskeletal Analogy of Machines

8.2 Lever Systems

8.3 Pulley Systems

8.4 Wheel-and-Axle Systems

CONNECTIONS

8.5 Human Performance

8.6 Ergonomics

8.7 Adapted Motion

The art of warfare is an amazing study in humans' understanding of energy systems. For each new defensive advance, some clever inventor designed a weapon to overcome it. Before the advent of explosives, massive fortifications built from meters-thick stone were impervious to human siege. Impervious, that is, until those outside the walls designed machines that could propel massive objects against the walls. The catapult and trebuchet are two of the best examples of energy systems offering a mechanical advantage beyond what human muscle could afford.

Simple catapults used the elastic properties of the arm material for propulsion, but as the need for more destructive force grew, energy was stored using a pulley system. Several men would use the mechanical advantage of a pulley system to store potential energy in twisted rope until a projectile was ready to be launched (remember balsa wood planes with their rubber band–powered propellers?). The trebuchet worked on a different energy storage principle—potential energy took the form of a large counterweight that provided force as it fell because of gravity. Again, it took the muscular energy of several men to raise the counterweight, but the effect was worth the effort because large trebuchets were able to launch 250-kg stones against fortified walls.

Humans continue to propel objects of war, but now they do so in competition. The shot put, javelin, and archery events seen in the Olympics are all descended from weapons of war. The mechanical design of the human body was the basis for larger weapons of war. The human body is constructed to take advantage of certain mechanical principles, including the building and use of machines. Think of all of the machines in a typical athletic or rehabilitation training facility. Think of the tools that you use on a daily basis. Imagine how much the invention of a machine called the *wheelchair* has changed lives. In addition to being able to construct machines, the musculoskeletal system is designed to gain different types of mechanical advantage across a variety of joints. Thus, anatomically, the human system itself can be said to be composed of many machine-like elements.

CONCEPTS

8.1 MUSCULOSKELETAL ANALOGY OF MACHINES

A **machine** is an apparatus or system that uses the combined action of several parts to apply mechanical force. Most people are comfortable with the idea that the human body uses machines to accomplish tasks. However, some are not comfortable with the idea that the human body is composed of machines, and they are even less comfortable with the notion of being the machine itself. The reality is that the human body is a system that uses the combined action of several parts to apply mechanical force. In fact, any given joint in the musculoskeletal system can technically be classified as an intrinsic machine. Even though these machines may not fit purist definitions, they do fit into our general definition of *machine*. So not only do humans use extrinsic and intrinsic machines, but also the human system itself can be classified as a natural machine. Even if there are no "true" machines within the human body, the analogy of machines is highly effective for understanding the mechanics of the human musculoskeletal system. Deeper knowledge of these machine-like musculoskeletal arrangements can at the very least help us to gain an appreciation of human motion in terms of the effects of applied force.

An understanding of machines and their uses brings together many concepts that we have covered in previous chapters (e.g., force and its application, torque, angular kinetics, and kinematics). Of course, according to our definition, many types of machines exist. Within the human body, three types of machines are used for various purposes: lever systems, wheel-and-axle systems, and pulley systems. Each machine is used to fulfill one or more of the following functions: (1) to *transmit a force* (e.g., tendons transmit muscle forces to bones), (2) to *increase the magnitude of a force* (i.e., less effort is required to move a given resistance), (3) to *increase the linear distance and velocity of a force* (i.e., a resistance is moved a greater distance or at a faster rate than the motive force), or (4) to *change the direction of a force* (i.e., a resistance is moved in a different direction than that of the motive force). This chapter describes the various machines used intrinsically and extrinsically by the human system and explains them in terms of their inherent abilities to fulfill the four functions listed here.

Machine An apparatus or system that uses the combined action of several parts to apply mechanical force.

8.2 LEVER SYSTEMS

A **lever system** consists of a rigid or semirigid object (the *lever*) that is capable of rotating about an axis called a **fulcrum** (Figure 8.1). The basic purpose of a lever system is to transmit energy (in the form of applied force) from one place to another. Because the fulcrum is an axis of rotation, the applied force must be off-axis (eccentric) to produce the torque necessary to rotate the lever. In fact, two types of torque operate in a lever system: motive and resistive. The **motive torque** comes in the form of an eccentrically applied force that attempts to rotate the lever in one direction about the fulcrum. The **resistive torque** is an eccentrically applied force that attempts to rotate the lever in the opposite direction.

Humans use extrinsic levers quite often and for a variety of purposes, often without realizing that the device they are using is a lever. But many arrangements within the body also serve as lever systems. For example, notice the parts that we need to form a lever system: a rigid lever, a fulcrum, and an eccentrically applied force can all be identified in the human body. A bone serves as the lever, a joint is the fulcrum about which the bone rotates, and a muscle provides the motive (and sometimes resistive) torque (Figure 8.2).

If the muscle is acting as a motive torque, the weight of the segment (and anything being held by the segment) provides the resistive torque.

Because the fulcrum is not always directly in the middle of the lever (especially in the case of the human body), the motive force or resistive force for a given lever has a mechanical advantage. The **mechanical advantage (MA)** of a lever system is the relationship of the motive force to a given resistive force—that is, the amount of one required to overcome the other. The mechanical advantage (or disadvantage) of any given lever system varies because as the location of the fulcrum is changed, the relative lengths of the moment arms for the motive and resistive forces vary. For example, if the fulcrum is in the middle, neither the motive nor resistive force has an inherent advantage because both of their moment arms are the same length. However, if the fulcrum is moved farther away from the motive force, then its moment arm (the **effort arm**) becomes longer, and it gains a mechanical advantage. If the fulcrum is closer to the motive force, then the moment arm for the resistive force (the **resistance arm**) becomes longer, and the resistive force has the advantage. The mechanical advantage of a lever system is often calculated numerically by using a ratio of the moment arm lengths:

$$MA = \frac{M_M}{M_R} \tag{8.1}$$

where

MA = mechanical advantage
M_M = length of the moment arm for the motive force (effort arm)
M_R = length of the moment arm for the resistive force (resistance arm)

The larger the number, the larger the advantage for the motive force, and the more leverage the given lever system provides. If the fulcrum is directly in the middle of the lever, the ratio is 1.00, and neither force has an advantage.

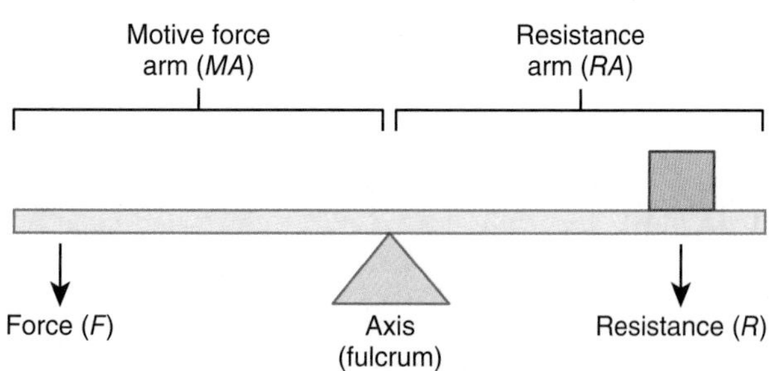

Figure 8.1 **Parts of a lever system.** In what situations do you most often use levers?

Lever system A machine consisting of a rigid or semirigid object that is capable of rotating about an axis.

Fulcrum The axis around which a lever rotates.

Motive torque An eccentrically applied force that attempts to rotate a lever in a given direction about a fulcrum.

Resistive torque An eccentrically applied force that attempts to rotate a lever in a direction opposite that of the motive torque.

Mechanical advantage (MA) The relationship of the motive force required to overcome a given resistive force with use of a lever system.

Effort arm The moment arm for the motive force.

Resistance arm The moment arm for the resistive force.

SAMPLE PROBLEM

For example, if a teeter-totter is 10 m in length, and the fulcrum is directly in the middle, the mechanical advantage is 5 m/5 m = 1.00, and neither user has an advantage. If a lever system provides a mechanical advantage greater than 1.00, it means the user does not have to generate a motive force equal to the resistive force. For example, if we slide the board on the teeter-totter so that it is 6 m from one end and 4 m from the other, the mechanical advantage is 6 m/4 m = 1.5. A mechanical advantage of 1.5 would mean that we could place an object weighing 50 N on the end that is 6 meters from the fulcrum and it would balance an object on the opposite end that weighs $50 \times 1.5 = 75$ N.

However, the calculation of mechanical advantage is somewhat misleading. The calculation assumes that the purpose of the lever is always to *create an advantage in force* (i.e., a given resistive force can be overcome with less motive force). But as you will learn, not all levers are meant to provide a mechanical advantage in the strictest sense of the term (i.e., motive moment arm advantage). As a matter of fact, a decreased mechanical advantage in terms of motive force is accompanied by advantages in other ways.

This chapter covers the various types of lever systems. Each lever system can be used to accomplish one or more of the different functions of a machine:

- to transmit a force,
- to increase the magnitude of a force,
- to increase the linear distance and velocity of a force, or
- to change the direction of a force.

The three types of lever systems are classified according to the positions of the motive and resistive forces relative to the fulcrum (Figure 8.3).

These relative positions, along with the mechanical advantage conferred by different moment arm lengths, affect the function of the lever system directly.

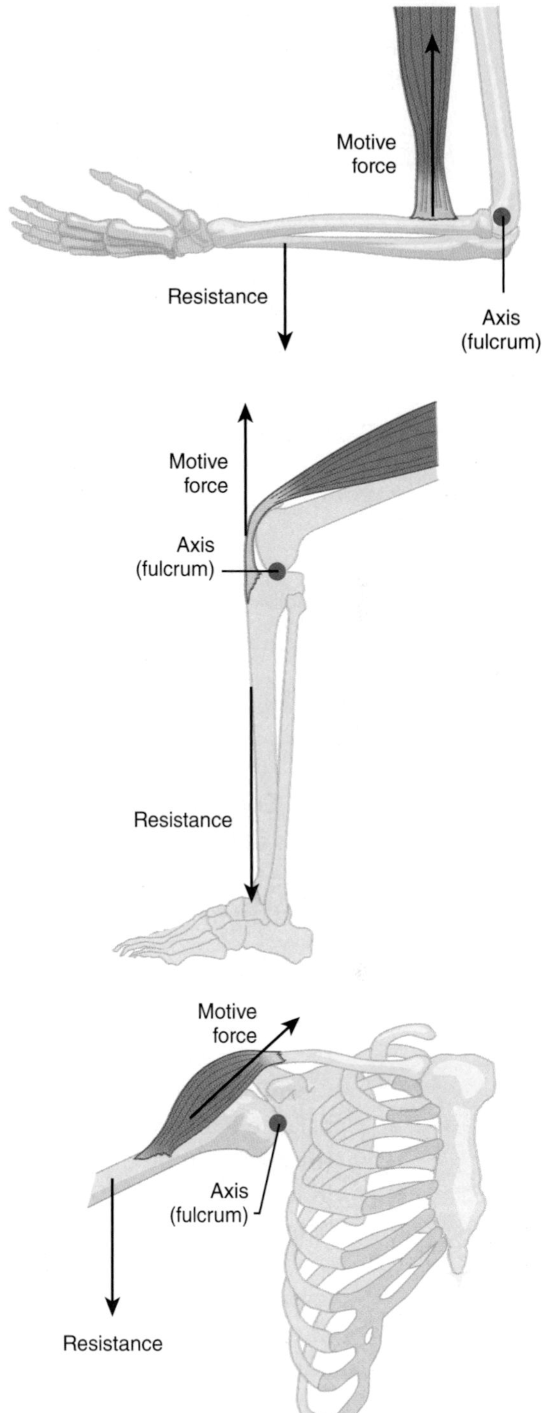

Figure 8.2 **Musculoskeletal lever system.** Can you think of other musculoskeletal levers?

First-Class Lever Systems

As mentioned in the previous section, lever systems are classified according to the relative positions of the fulcrum, the motive force, and the resistive force. In a **first-class lever system**, the fulcrum is between the motive and resistive forces (Figure 8.4). The first-class lever system is the most versatile of the three because it can fulfill any of the functions of a machine. Therefore, many uses arise for first-class levers in everyday life. Levers of the first class include household tools such a scissors and pliers. When using scissors and pliers, the fulcrum is in the middle, a motive force is applied by the hand, and a resistive force is being cut or held on the opposite side of the fulcrum.

First-class lever system A lever system in which the fulcrum is between the motive and resistive forces.

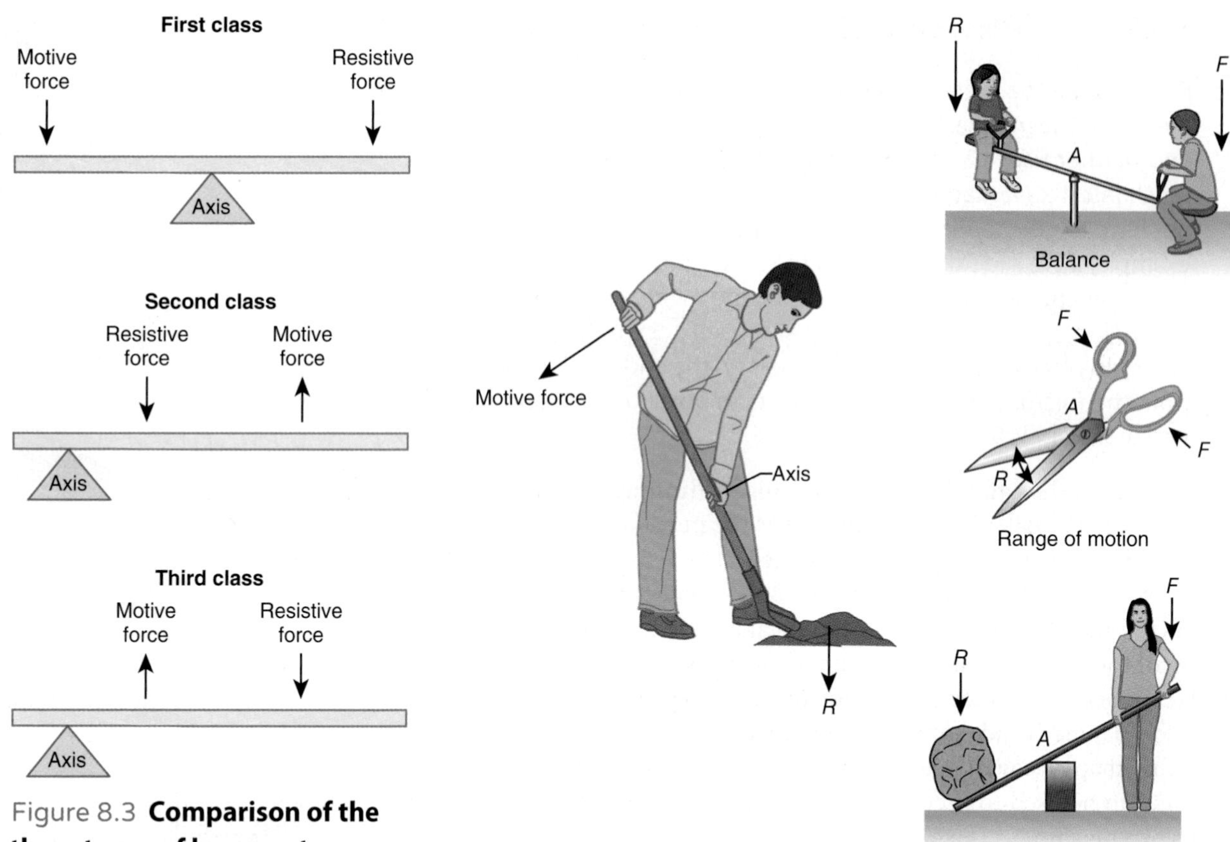

Figure 8.3 **Comparison of the three types of lever systems.** Which of the three types of levers do you most often use?

Figure 8.4 **Examples of first-class levers.** Can you think of other levers of the first class?

The specific use of the first-class lever in terms of the four functions of a machine depends on the exact location of the axis of rotation. For example, a teeter-totter is a first-class lever system in which the axis is located directly between the motive and resistive forces. Therefore, the moment arms for the forces are the same length (MA = 1.00), and in this case the lever is being used to transmit force to balance two forces. In contrast, the fulcrum could be moved farther from the applied motive force (increasing its moment arm length), and the first-class lever would be capable of creating an advantage in force (MA > 1.00). Tools used for prying (e.g., crowbars) are prime examples of first-class levers being used to gain an advantage in force.

Notice that when a first-class lever is used to gain a force advantage, the resistive force end of the lever is moved a miniscule distance relative to the motive force end. Recall from Chapter 6 that the linear distance traveled by a point on a rotating segment depends on the length of its radius of rotation ($l = \Delta\theta r$). In the case of a first-class lever being used to gain an advantage in force, the moment arm (which is actually the radius of rotation) for the resistive force is extremely short. So the linear distance traveled by a point on the resistive force end of the lever will be small. Note that mechanical advantage entails a tradeoff. A gain in mechanical advantage (in terms of force production) achieved by increasing the length of the effort arm will come at the expense of range of motion (linear distance traveled) because the resistance will have a relatively smaller displacement.

Of course we could move the fulcrum closer to the motive force, decreasing the length of the effort arm. In the strictest sense, this change would result in a mechanical disadvantage in terms of force production (MA < 1.00). However, the loss of length in moment arm for the motive force is gained by the resistive force. Therefore, even though more motive force will be required to move the resistance, the resistance will be moved a greater linear distance because its radius of rotation (resistance arm) will be longer.

For example, if an adult wants to play on a teeter-totter with a child, they will have to sit at different distances from the axis of rotation. The heavy adult will be balanced by a lighter child because of differences in moment arm length. As a result, the child will be moved a greater linear distance because of the additional length of the resistance arm.

In addition, all points on a rigid segment travel their respective distances in the same amount of time. Because the resistive end of the lever travels a greater linear distance in the same time, the resistive end of the lever in this case also has a greater linear velocity. So, in this situation, the first-class lever system is being used to gain an advantage in both linear range of motion and velocity. As long as the lever is rigid, those two qualities always occur together. This gain in range of motion at the expense of force is one reason why the mathematical calculation of mechanical advantage should be interpreted carefully. The equation assumes that one always seeks an advantage in force. If one wants a mechanical advantage in linear range of motion and velocity, then use the inverse of Equation 8.1 to calculate that advantage.

So we have seen that the first-class lever can be used to achieve an advantage in force, achieve an advantage in linear distance traveled (linear velocity), and balance two forces (Figure 8.5). The final function of a machine is to change the effective direction of an applied force. The first-class lever system also achieves this purpose. Because the fulcrum is between the applied forces, the resulting directions of the motive and resistive forces are opposite one another. In other words, one end of the teeter-totter goes up when the other end goes down.

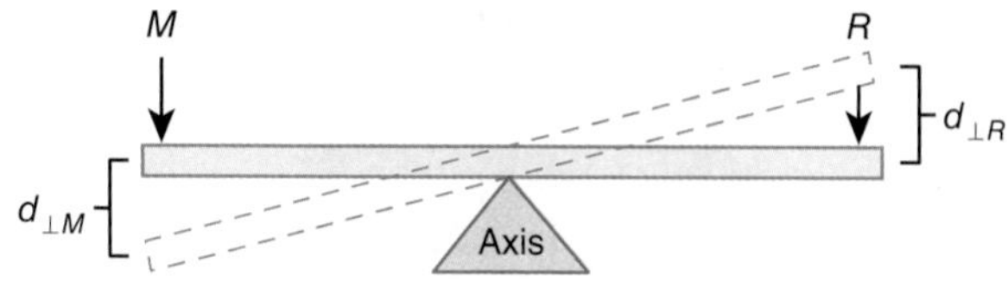

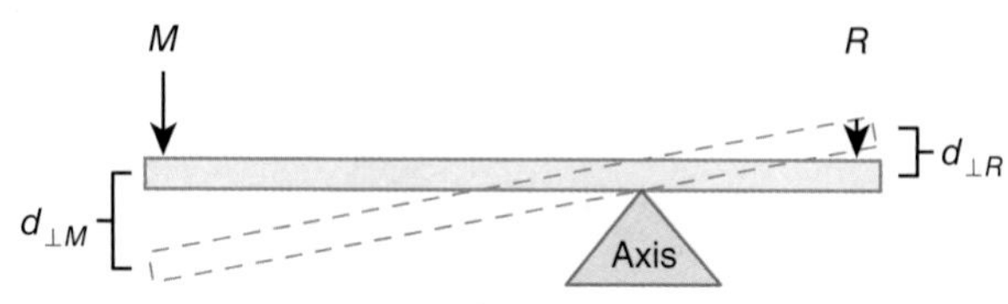

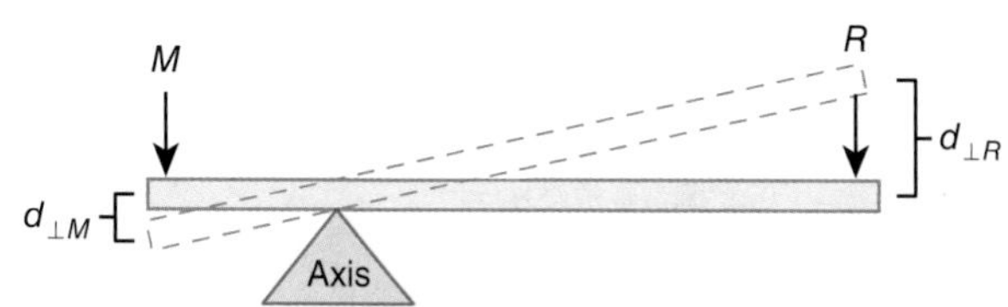

Figure 8.5 **Variations of first-class lever systems.** Have you ever constructed a first-class lever?

In the human body, intrinsic lever systems of the first class are rare. In our musculoskeletal system, the motive force is usually between the fulcrum and the resistive force. First-class levers found in the human musculoskeletal system tend to be arranged so that the fulcrum is near the motive force and an advantage in linear range of motion and velocity is gained. For example, extension of the elbow by the triceps muscle against a resistive force is a first-class lever system (Figure 8.6). The elbow is the fulcrum and is between the motive force of the triceps muscle and the resistive force applied by the weight of the forearm (and anything held by the hand).

Notice that the moment arm for the motive force is very small relative to the moment arm for the resistive force. Technically, this arrangement creates a mechanical disadvantage in force. We have seen before that muscle must generate large forces to compensate for short moment arms. The same is true if musculoskeletal arrangements function as first-class lever systems. But remember the advantage that is gained: greater linear range of motion and velocity. So in the case of extending the elbow, the triceps muscle causes a large linear displacement of the hand with only a small linear displacement of the insertion point of the triceps.

The musculoskeletal arrangement at the skull is an example of a first-class lever. The fulcrum is formed by the articulation of the base of the skull with the first cervical vertebra. The neck extensors in this situation provide the motive force to extend the neck; the resistive force is gravity acting through the center of gravity of the head. The skull is a spherical object, but it is a relatively rigid body that rocks on a fulcrum because of forces on opposite sides of that

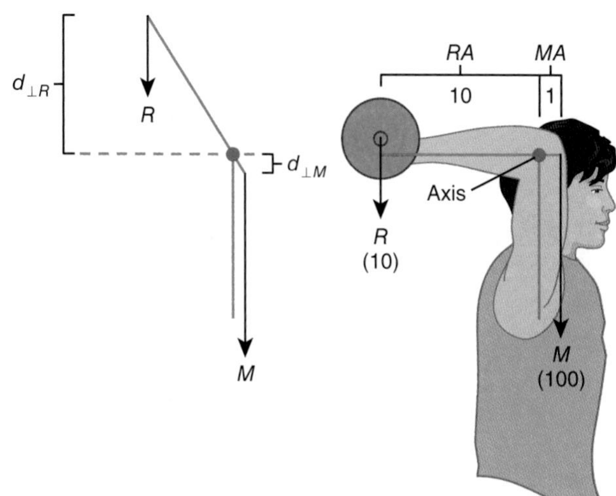

Figure 8.6 **Triceps extension of the elbow as a first-class lever system.** When is the elbow a lever of a different class?

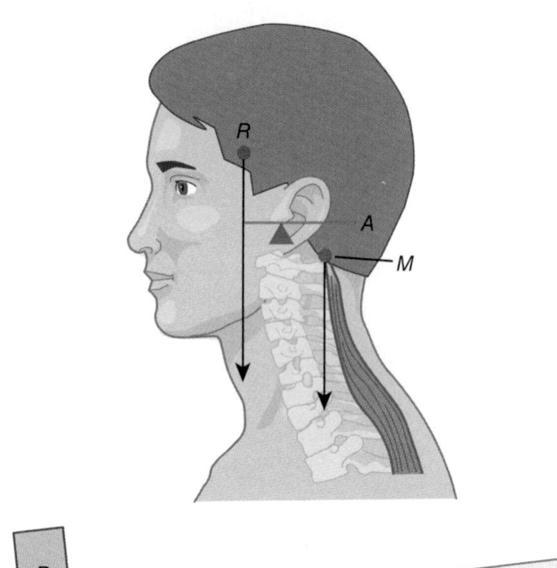

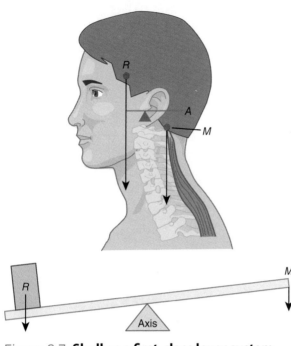

Figure 8.7 **Skull as a first-class lever system.** Could the skull be a different type of machine?

fulcrum. Therefore, this arrangement can be considered a first-class lever system (Figure 8.7).

Depending on the particular activity, another example of a musculoskeletal first-class lever system is the gastrocnemius and soleus muscles of the posterior calf contracting to plantar flex the ankle against resistance at the foot (i.e., pushing a resistive load away from the body). The ankle acts as the fulcrum and is between the motive force provided by the posterior calf muscles and the resistive force on the foot (Figure 8.8). Once again, notice that the fulcrum is closer to the motive force end of the lever, providing greater range of motion and velocity.

As we will see over and over again, the human system is highly adapted for increased range of motion and velocity. Another aspect of this particular musculoskeletal arrangement that one must be careful to clearly define is the parts of the lever system. Here we have defined the ankle as the axis of rotation. However, if the particular movement situation is changed, the ball of the foot may become the axis of rotation, and the type of lever system would not necessarily be of the first class. For example, plantar flexion exercises ("heel raises") are used to train the posterior calf muscles. The fulcrum during these exercises can be considered to be located at the ball of the foot, rather than at the ankle. The resistive force is provided by gravity, and the motive force is produced by the posterior calf muscles (Figure 8.9).

In this example, as long as the person's center of gravity is far enough forward, the lever is still of the first class, but the fulcrum is different and the resistance is now the body weight. So as you read about the other types of lever systems and the other machines, keep in mind that a particular musculoskeletal arrangement may form two levers of the same type, two levers of different types, or may even fit into more than one category of machine. Therefore, clearly defining the situation is absolutely necessary. This type of ambiguous categorization is one of the reasons for the opinion that the human body does not

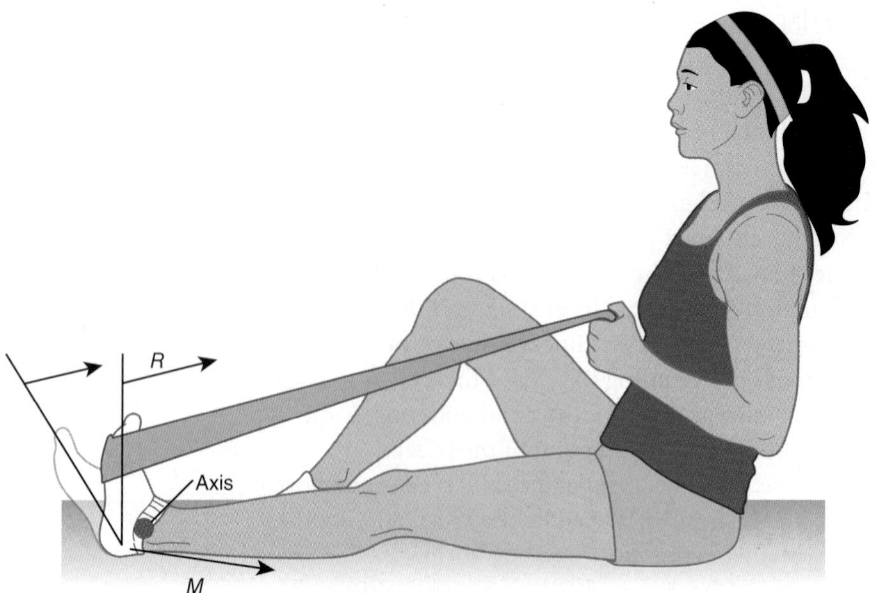

Figure 8.8 **Plantar flexion of the ankle as a first-class lever system.** Why is this lever of the first class?

contain any musculoskeletal machines. But again, the analogy of machines is helpful in understanding mechanics of the human system.

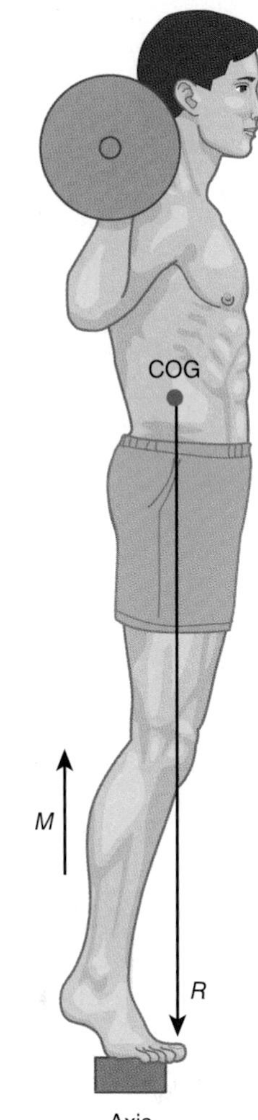

Figure 8.9 **Calf raise exercise as a first-class lever system.** Why is this lever of the first class?

Second-Class Lever Systems

In a **second-class lever system**, the resistive force is between the fulcrum and the motive force (Figure 8.10). The implication of this arrangement is that the moment arm for the motive force will always be greater than the moment arm for the resistive force (MA $>$ 1.00).

Therefore, levers of the second class are always used to gain an advantage in force production. Extrinsic second-class levers are tools such as wheelbarrows and nutcrackers (Figure 8.11).

Remember that this advantage in force comes at the price of linear range of motion and velocity. For example, a wheelbarrow can be used to lift quite heavy loads. However, notice that the resistance is only raised high enough to allow the wheel to roll. The resistive force has a short moment arm (radius of rotation), so its linear distance traveled is small relative to that of the motive end of the lever.

Therefore, second-class levers are not as versatile as first-class levers because they cannot be used to gain an advantage in range of motion. Also, the resistance is moved in the same direction as the applied force, so second-class lever systems cannot be used to change the effective direction of the applied force or to balance two forces.

Because the tendency in the human body is to opt for advantages in range of motion, the musculoskeletal system contains no exact analogues to the second-class lever. Instead, the human body tends to adopt positions or situations in which a second-class lever is formed. For example, a push-up is an exercise in which the entire body acts as a second-class lever (Figure 8.12). The fulcrum is formed by the tips of the toes, gravitational force acting through the center of gravity is the resistance, and the motive force is generated by the reaction force at the hands.

Previously, we used plantar flexion as an example of two different forms of the first-class lever system. If we slightly change our previous example, we can actually form a second-class lever. Recall that the fulcrum can be considered to be located at the ball of the foot (rather than at the ankle) during these exercises. The resistive force is provided by gravity, and the motive force is produced by the posterior calf muscles. The slight change in this situation is to make sure that the person is not leaning forward far enough for the line of gravity to fall on the side of the fulcrum opposite that of the motive force (in which case the lever would be of the first class; Figure 8.13).

Second-class lever systems are also formed if the situation is one in which the muscle provides the resistive force instead of the motive force (i.e., acts eccentrically; generates force while lengthening). An example of this situation would be the *lowering* phase of a bicep curl or other exercises. In this case, the elbow is the fulcrum, gravity is the motive force, and the muscle provides a resistive force to keep the weight from falling too rapidly (Figure 8.14).

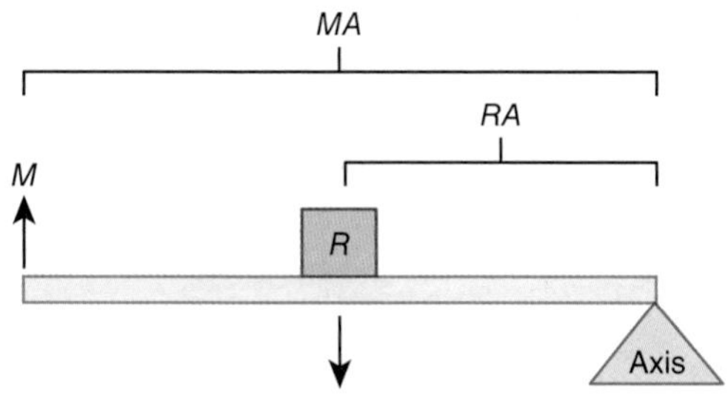

Figure 8.10 **A second-class lever system.** In what situations have you used a second-class lever?

Once again, notice that the situation must be clearly defined. We tend always to think of the muscles as providing motive force, but second-class lever situations show that this is not always the case. Also notice that whether the lever is first or second class, the muscle usually has the shorter moment arm.

Second-class lever system A lever system in which the resistive force is between the fulcrum and the motive force.

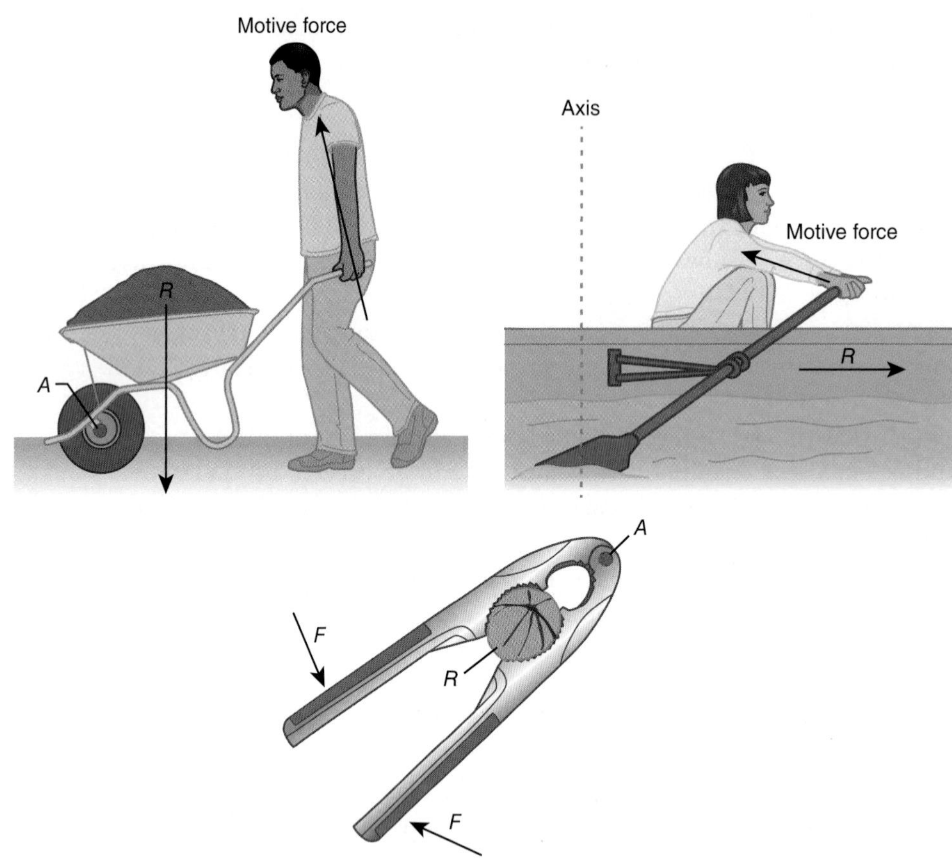

Figure 8.11 **Examples of second-class levers.** Can you think of other second-class levers?

Third-Class Lever Systems

In a **third-class lever system**, the motive force is between the fulcrum and the resistive force (Figure 8.15). The implication of this arrangement is that the moment arm for the motive force (effort arm) will always be less than the moment arm for the resistive force (MA $<$ 1.00).

Because the radius of rotation for the resistive force is larger than that for the motive force, its linear distance traveled is large relative to that of the motive end of the lever (Figure 8.16). Therefore, third-class levers are used to gain an advantage in linear range of motion and velocity (Figure 8.17).

Again, remember that this advantage in linear range of motion and velocity comes at the price of force. The motive force has a short moment arm and therefore a disproportionately large motive force must be produced to overcome the resistive force. Recall from previous chapters that we have calculated the muscle force necessary to maintain a given anatomical position. Each time, we found that a disproportionately large muscle force is required to overcome the resistive torque.

Third-class lever system A lever system in which the motive force is between the fulcrum and the resistive force.

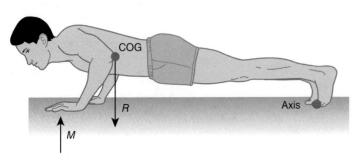

Figure 8.12 **Push-up exercise as a second-class lever system.** How could you change the mechanical advantage of this lever system?

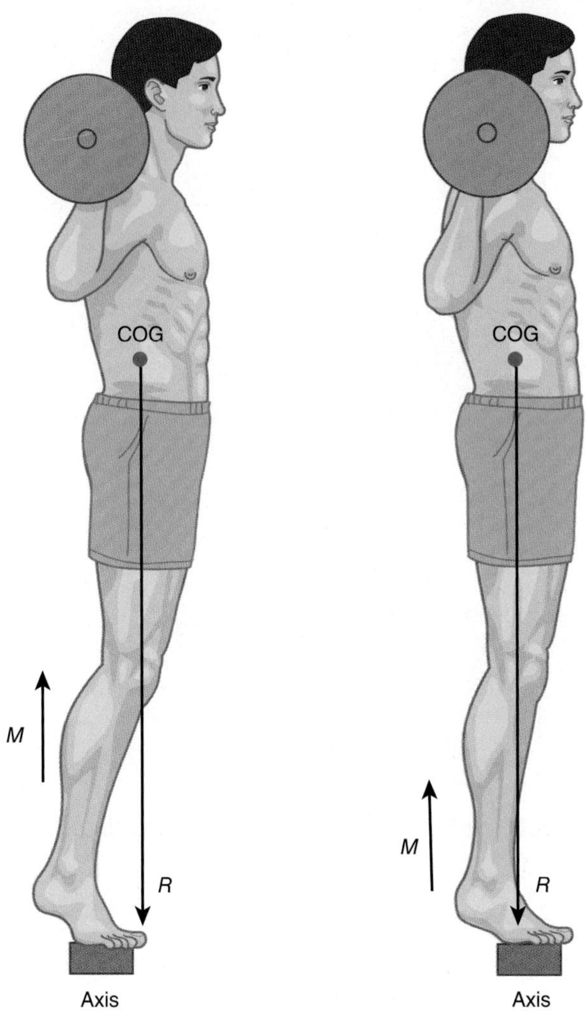

Figure 8.13 **Calf raise exercise as first-class and second-class lever.** Why does the class of lever change?

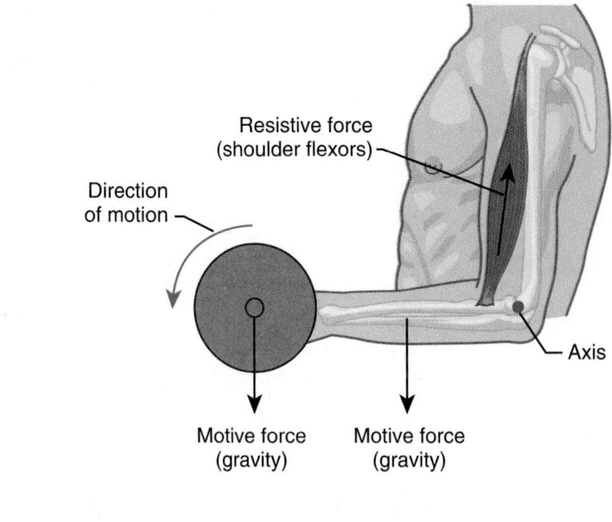

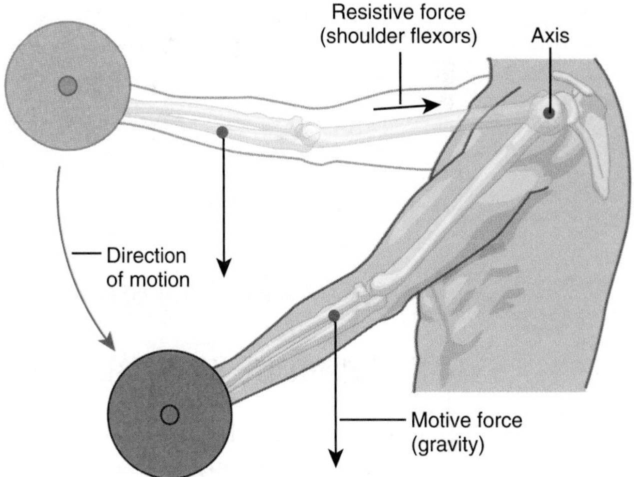

Figure 8.14 **Biceps curl and front raise exercises as second-class levers.** What could change these exercises to levers of a different class?

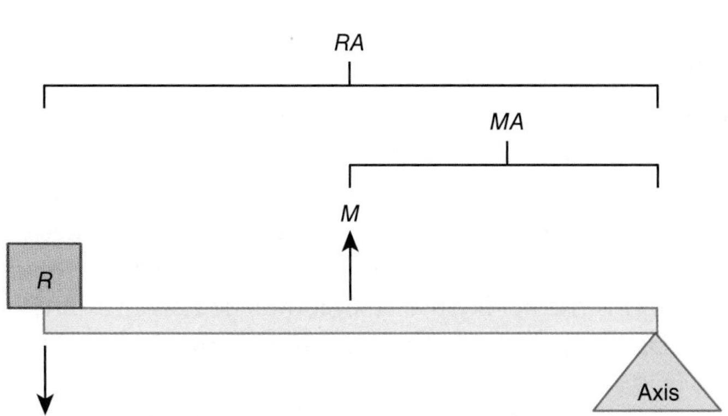

Figure 8.15 **A third-class lever.** In what situations do you use third-class levers?

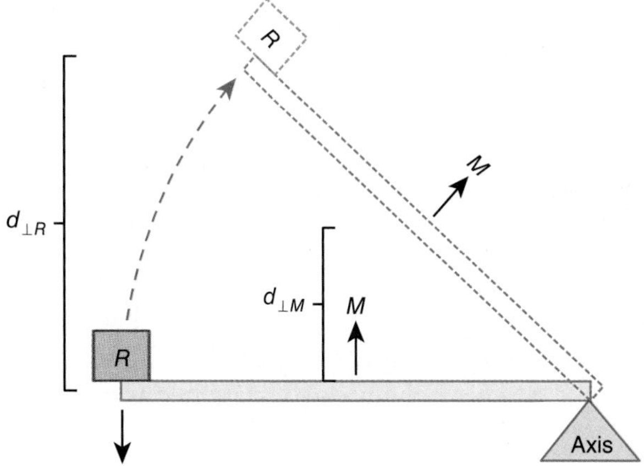

Figure 8.16 **Linear distance traveled by points on a third-class lever system.** What is the advantage of this type of lever system?

Figure 8.17 **Examples of third-class levers.** Can you think of other levers of the third class?

SAMPLE PROBLEM

For example, let's say that the moment arm for the muscle force (motive in this case) is 0.02 m and the moment arm for the resistive force is 0.40 m (Figure 8.18). That means that the ratio of motive moment arm length to resistive moment arm length is 1:20 (MA = 0.05).

If we also say that the resistive force is 10 N, then the torque caused by resistive force is 4.0 N · m (10 N times 0.40 m). In other words, the muscle (having a moment arm of only 0.02 m) will have to produce a force of 200 N (20 times that of the resistive force) to counteract the resistive torque and hold the arm in a static position!

Of course, we could regain some of the lost mechanical advantage by moving the insertion of the muscle farther from the fulcrum (increasing the moment arm for the motive force), but this solution would have huge implications in terms of range of motion (Figure 8.19).

Can you imagine your body shape if your biceps muscle were inserted on the wrist in order to gain an advantage in force? The implications for muscle insertion and their relationship to third-class levers and human shape will be discussed in a later chapter.

You have learned that the human musculoskeletal system is designed for increased linear range of motion and velocity. Therefore, it is no surprise that many musculoskeletal arrangements in the human body can be classified as third-class lever systems. This chapter has used the elbow flexors as an example of a third-class lever system (and of a second-class lever during eccentric actions). The flexors of the knee also function as a lever of the third class (Figure 8.20).

The flexors of the knee provide the motive force, the knee serves as a fulcrum, and the lower leg provides the resistive force. In addition, the deltoid muscle forms a third-class lever system when it abducts the shoulder (Figure 8.21).

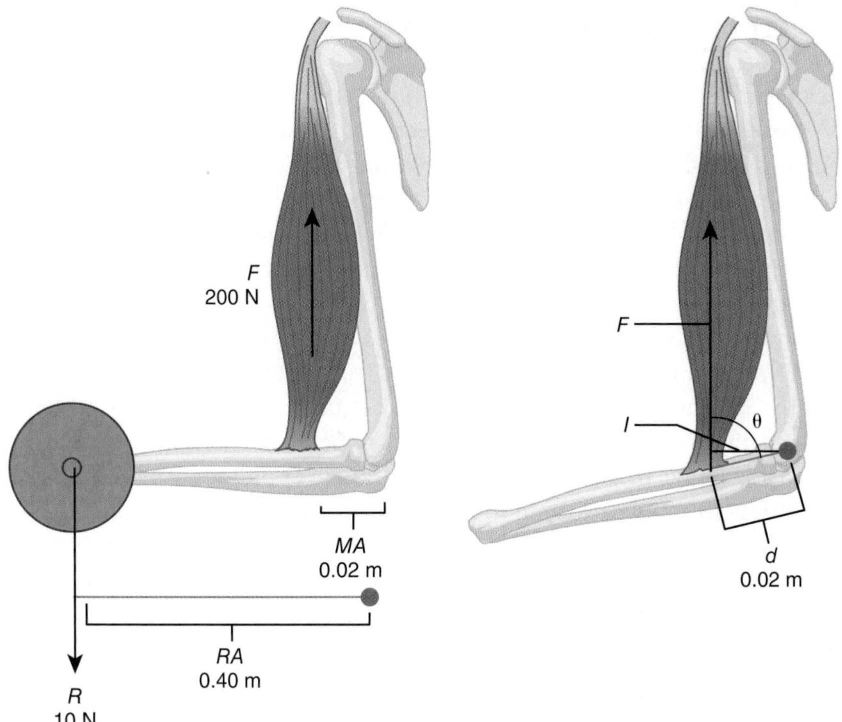

Figure 8.18 **Muscle contraction as a third-class lever system.** What is the disadvantage of this type of lever system?

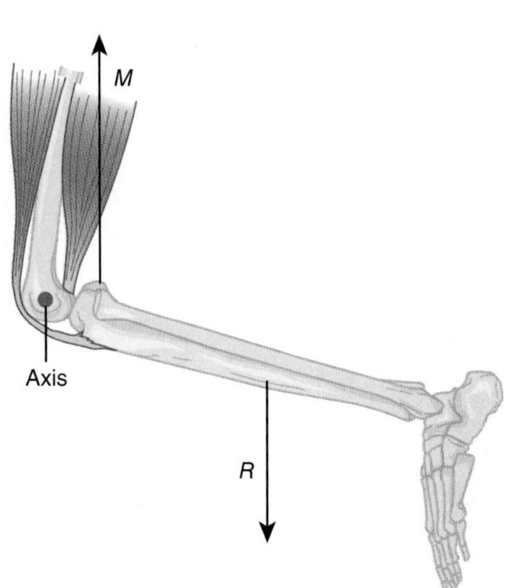

Figure 8.20 **Knee flexors as a third-class lever system.** Could this lever system be of a different class?

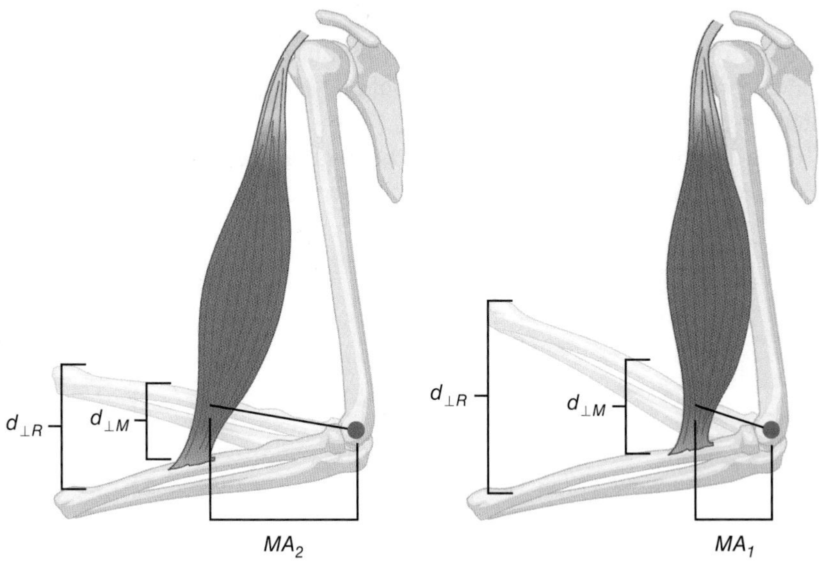

Figure 8.19 **Range of motion changes with biceps insertion point.** Why must muscles be inserted so close to the axis of rotation?

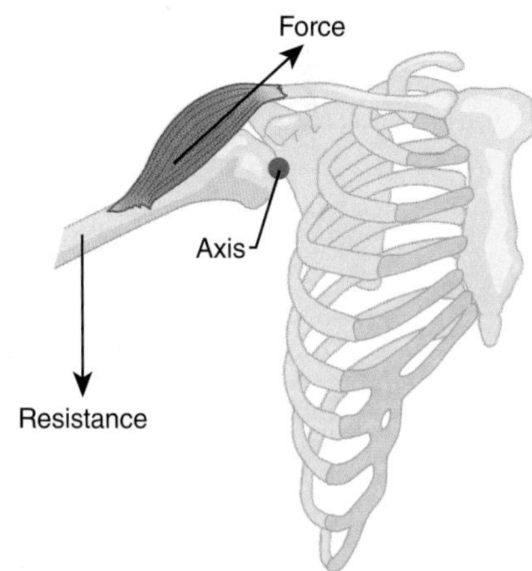

Figure 8.21 **Deltoid muscles acting as a third-class lever system.** Can this lever system be changed to the second class?

In this case, the deltoid muscle acts as a motive force while the glenohumeral joint serves as a fulcrum and the weight of the arms provides a resistive force.

Keep in mind that these same musculoskeletal arrangements can be considered second-class levers in situations of the muscle providing a resistive force (i.e., lowering the weight in a controlled manner).

Figure 8.22 **Machine pulley on lat pull exercise.** What other pulleys do you use?

8.3 PULLEY SYSTEMS

Another machine that can be found in the human musculoskeletal system is the pulley. A **pulley system** consists of an object that acts as a wheel around which a flexible cord or cable is pulled. Most people are familiar with machine pulleys, such as those found on cranes or exercise equipment.

One can see by observing exercise equipment that one major function of the pulley system is to change the effective direction of the applied force. A simple example is an exercise machine in which the motive force is applied in a downward direction and the resistance (weight stack) moves upward (Figure 8.22). But also notice that the cable of this exercise machine is transmitting force applied at the handle to the weight stack. In the human body, musculoskeletal pulley systems are used to change the effective direction of the applied force, transmit forces, and also gain an advantage in force by changing the angle of pull for the muscle. The **angle of pull** is the angle at which the muscle force acts relative to a given axis or lever. One common example is the pulley system formed by the interaction of the patella, the quadriceps muscles, and the patellar tendon (Figure 8.23).

The motive force of the quadriceps is focused in one direction while the resistance from the weight of the lower leg is moved in a different direction. Also, keep in mind that the tendon is transmitting muscular force to the tibia. But the directional change is not the only issue. Notice what would happen to the quadriceps muscles' angle of pull if the patella were removed (Figure 8.24).

Without the patella, the angle of pull would be much smaller. Remember that the equation for torque is: $T = F \times r \times \sin\theta$. The angle of pull for the muscle is the angle θ. Therefore, the larger the angle of pull (larger $\sin\theta$), the larger the resultant torque produced by the muscle. So without the pulley formed by the patella, the quadriceps would be at a large mechanical disadvantage in producing torque.

Other examples of musculoskeletal pulleys can be found by observing the shapes of bones. Notice that many bones, such as the femur and tibia,

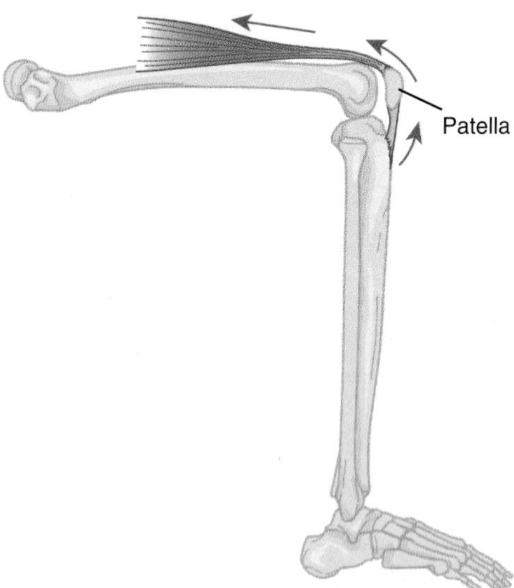

Figure 8.23 **The patella as a pulley.** What are all of the functions of the patella?

Pulley system A machine consisting of an object that acts as a wheel around which a flexible cord or cable is pulled.

Angle of pull The angle at which the muscle force acts relative to a given axis or lever.

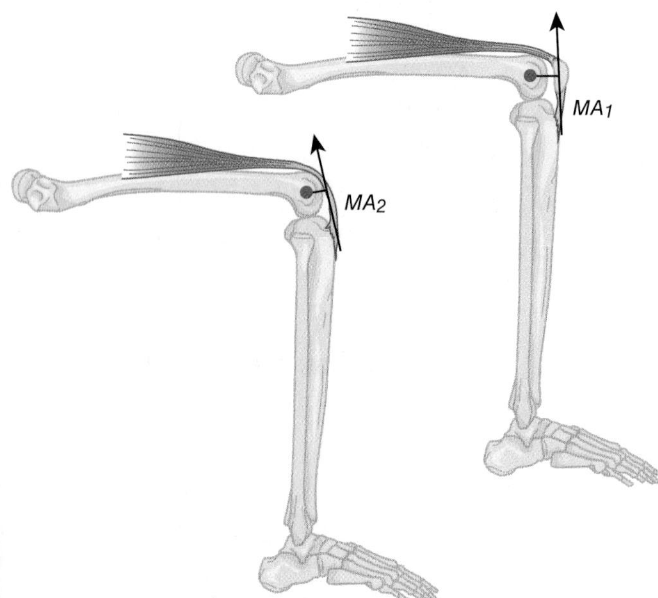

Figure 8.24 **The angle of pull changes when the patella is removed.**

are flared at the ends. The medial condyles of the femur and tibia serve as parts of a pulley system in which the gracilis muscle provides the motive force (Figure 8.25).

The gracilis muscle inserts into the tibia after passing around the condyles at the knee joint. Once again, this arrangement increases the angle of pull for the muscle and enhances its ability to produce torque. If the bones were not flared on the ends, the angle of pull would be much smaller, and therefore the torque from gracilis contraction would be minimized. A similar pulley arrangement exists at the ankle (Figure 8.26).

The peroneus longus muscle passes behind the lateral malleolus at the ankle and inserts into the undersurfaces of the cuneiform and first metatarsal bones. The peroneus longus is a strong evertor of the foot. But because its tendon passes behind the lateral malleolus, the effective direction of the applied muscle force is changed, and the peroneus longus also is able to assist in plantar flexion of the ankle (Floyd, 2018). If its tendon passed in front of the lateral malleolus, as some other muscles of the calf do, contracting the peroneus longus would result in eversion and dorsiflexion.

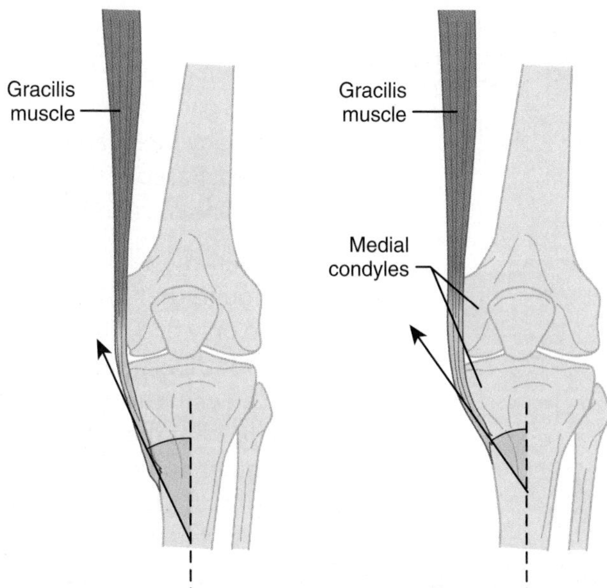

Figure 8.25 **The medial condyles and gracilis as a pulley system.**

So you can see that the pulley system has many implications for human motion. It changes the effective direction of applied muscle force, transmits force from one entity to another, and increases the angle of pull for the muscle. Before we move on, notice that some musculoskeletal arrangements that function as pulley systems could also be classified into one of the categories of lever systems. The applied motive force of the quadriceps is between the fulcrum at the knee and the resistive weight of the lower leg. This fact means that the knee could be classified as a third-class lever system. Once again, the line is blurred. However, when in doubt

FOCUS ON RESEARCH

As mentioned in the chapter, internal rotation of the shoulder with the elbow flexed forms a wheel-and-axle arrangement with the humerus serving as the axle and the distal segment forming a wheel. The muscle force is applied to the axle, and the wheel therefore travels a relatively large range of motion at sometimes extremely high velocity. In movements such as pitching in baseball, the large torque on the humerus along with the inertia of the forearm can place high stresses on the elbow. The medial ulnar collateral ligament (UCL) helps to stabilize the elbow as it is exposed to large valgus stress during throwing. With the repetitive stress that occurs over time in a pitcher's career, the UCL can gradually become stretched, develop microtears, or rupture. Historically, UCL injury was considered career ending—that is, until 1974 when surgeon Frank Jobe performed a UCL reconstruction (Jobe & Kvitne, 1991) on Tommy John, whose name is most frequently associated with the procedure. Since the development of the procedure, there has been some controversy over the effectiveness of the procedure in terms of player return rates and performance effectiveness after return.

Erickson et al. (2014) studied rate of return to pitching (RTP) and performance after Tommy John surgery in Major League Baseball (MLB) pitchers. Using MLB websites and Internet injury reports, the researchers identified MLB pitchers

(*continues*)

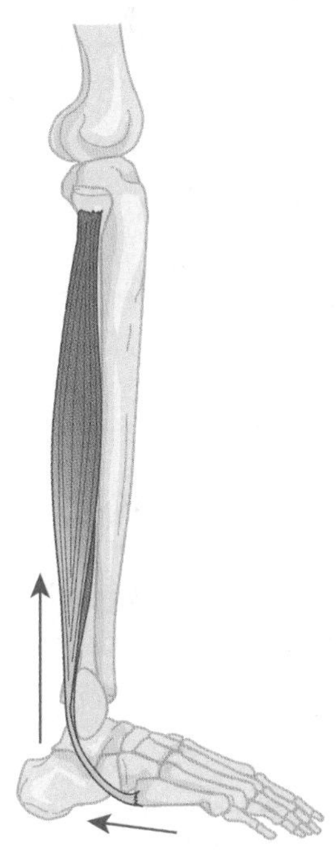

Figure 8.26 **The lateral malleolus as a pulley.** Why is this classified as a pulley?

(continued)

who underwent reconstruction after symptomatic UCL injury. Pitchers who had pitched at least one MLB game were included, and a pitcher was deemed to have returned to pitching if he pitched in any MLB game after surgery. Data on performance measures and demographics were collected. Based on the demographic data, a control group was selected for comparison to the cases. One hundred seventy-nine pitchers met inclusion criteria for analysis.

The researchers found that 148 of 179 pitchers (83%) were able to RTP in the major leagues. It was also found that 174 pitchers were able to RTP in MLB and minor league combined (97.2%), and only five pitchers (2.8%) were not able to RTP in either major or minor league baseball. After UCL reconstruction, pitchers returned to MLB at around 20.5 ± 9.72 months. Length of career in the majors after UCL reconstruction was 3.9 ± 2.84 years. But it was also noted that 56 of of the pitchers were still actively pitching in MLB at the start of the 2013 season. Another finding was that pitching performance declined significantly in the cases versus controls in the year before UCL reconstruction (reflected in number of innings pitched, games played, and wins and the winning percentage). In addition, pitchers showed significantly improved performance after surgery (fewer losses, lower losing percentage, lower earned run average (ERA), fewer walks, and fewer hits, runs, and home runs allowed). Finally, such cases had significantly fewer losses per season, a lower losing percentage, a significantly lower ERA, and allowed fewer walks and hits per inning pitched.

Based on the results, the researchers concluded that UCL reconstruction is actually associated with a predictably high, as well as successful, rate of return to Major League Baseball.

as to the classification of an arrangement, think of the function that a pulley can fulfill that a lever system cannot: increasing the angle of pull. For example, without the patella, the knee is still a third-class lever but has a smaller angle of pull. The third-class lever in this case acts to change the effective direction of applied force and gives an advantage in linear range of motion and velocity. Adding the patella to the arrangement changes the angle of pull without changing the parts of the third-class lever system. So an arrangement that is a lever system can also be considered a pulley when a bone or boney formation not only changes the effective direction of applied force but also changes the angle of pull. In other words, don't be afraid to classify a musculoskeletal arrangement into more than one category. Just remember that each machine provides a different advantage.

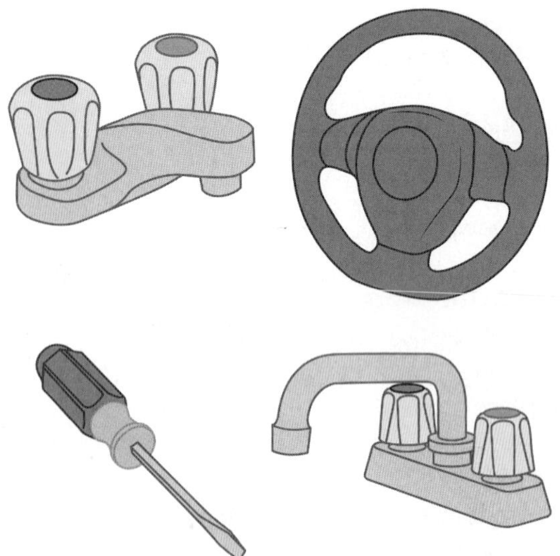

Figure 8.27 **Examples of wheel-and-axle systems.** When do you most often use wheel-and-axle systems?

8.4 WHEEL-AND-AXLE SYSTEMS

The final type of machine used by the human system is the wheel-and-axle arrangement. A **wheel-and-axle system** consists of an object acting as a wheel that is secured to a smaller wheel or shaft called the *axle*. A force is applied at a tangent to the circumference of the wheel or axle, and the other part of the system rotates whenever either the wheel or axle rotates (Figure 8.27). A wheel-and-axle arrangement is actually a type of lever, with a tangential force applied to a moment arm that is

Wheel-and-axle system A machine consisting of an object acting as a wheel that is capable of rotating about a shaft.

equal to the radius of the wheel or axle. The wheel-and-axle arrangement can transmit force. Force applied to the wheel is transmitted to a resistance at the axle and *vice versa*. The wheel-and-axle system can be used to gain an advantage in force or linear range of motion and velocity, depending on whether the motive force is applied to the wheel or the axle. For example, knobs on doors and faucets serve as wheel-and-axle systems.

In most wheel-and-axle systems, we tend to apply the motive force to the outside of the wheel. Because the force is eccentric, torque is produced. The larger the wheel, the more off axis the force is applied (greater radius of rotation). Therefore, a large wheel provides a large moment arm for the applied force (effort arm). Be careful not to confuse the effort arm of the wheel-and-axle system with the moment arm for the muscle that is applying the force (Figure 8.28).

Now let's turn our attention to the axle. Compared to the wheel, it has a small radius. This smaller moment arm means that torque production is compromised if the force is applied to the axle. Just imagine trying to turn on a faucet by turning the stem without the knob or using a screwdriver that does not have a handle. So applying motive force to the wheel instead of the axle provides an advantage in force. In addition, the larger the radius of the wheel compared to the radius of the axle, the greater the advantage in force. The advantage is greater because the motive force has a larger moment arm than the resistive force acting at the axle. As in a lever system, the ratio of the moment arm lengths can be used to numerically calculate the mechanical advantage of a wheel-and-axle system:

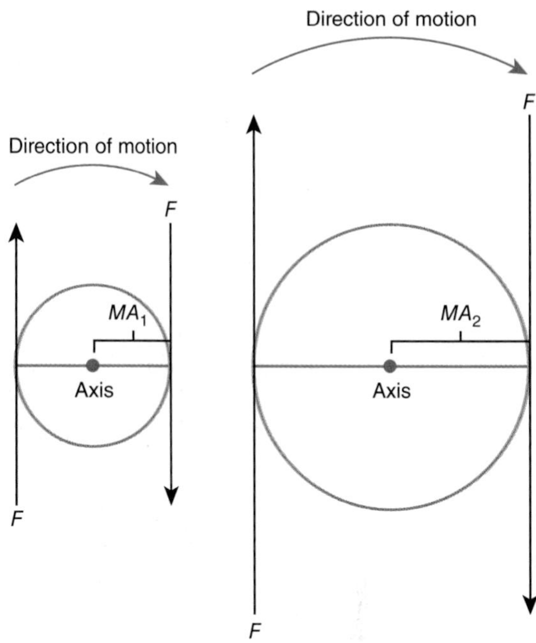

Figure 8.28 **The effort arm changes with diameter of the wheel.** How is a wheel-and-axle system similar to a lever system?

$$MA = \frac{r_w}{r_a} \tag{8.2}$$

where

MA = mechanical advantage
r_w = radius of the wheel (moment arm for the motive force)
r_a = radius of the axle (moment arm for the resistive force)

Therefore, the larger the wheel in relation to the axis, the greater the mechanical advantage in force. If the wheel and axle are the same diameter, then the ratio is 1.00, and no advantage exists.

SAMPLE PROBLEM

For example, if a screwdriver had no handle and a diameter of 0.75 cm, the mechanical advantage is 0.375 cm/0.375 cm = 1.00, and neither the user nor the screw has an advantage. If a wheel-and-axle system provides a mechanical advantage greater than 1.00, the user does not have to generate a motive force equal to the resistive force. For example, if we use a screwdriver that has a handle that is 4 cm in diameter and the shaft is still 0.75 cm, the mechanical advantage is 2.00 cm/0.375 m = 5.33! A mechanical advantage of 5.33 would mean that we could turn the handle by applying a force of 10 N and move a resistance that provides a force of 10 × 5.33 = 53.33 N. That's why screwdrivers have handles. This principle is also the reason that steering wheels in cars were huge before the age of power steering.

As usual, this advantage in force comes with compromised linear range of motion and *vice versa.* Because the axle's radius of rotation is small, the resistance is moved a small linear distance compared to distance traveled by the motive force. Therefore, if one wants an advantage in linear range of motion and velocity, one must apply the motive force to the axle. Doing so will require applying a greater force, but the linear distance traveled by the resistive force will be greater. For example, the torque provided by a car engine is applied to the axle, not the wheel. In this manner, the car travels a greater linear distance on its wheels for a given turn of the axle. Again, this process requires a large torque, but the gain is in linear distance traveled. Can you imagine the amount of torque necessary to rotate the axles of a "monster truck"?

Although few musculoskeletal arrangements are directly analogous to the wheel-and-axle system, humans do perform certain motions and adopt certain limb positions that can be considered this type of machine. First, the skull or torso rotating with the spine can be considered a wheel-and-axle arrangement, with the skull and torso being wheels that rotate with an axle formed by the spine. Also, rotations of bones around their longitudinal axes can be considered wheel-and-axle systems because the distal bone can serve as a wheel that rotates about a shaft formed by the proximal bone. For example, internal and external rotation at the shoulder joint with the elbow flexed forms a wheel-and-axle arrangement, with the proximal bone (humerus) serving as an axle, and the distal segments forming a wheel (Figure 8.29).

The muscle force in this case is applied to the axle. Once again, notice that the human opts for the advantage in linear range of motion and velocity by applying force to the axle. Though this option requires more torque, the most distal segment (the "wheel") travels a large linear distance relative to the axle. This is one reason to use implements; the implement increases the radius of the wheel (Figure 8.30). This increase translates into greater linear velocity of the implement's striking surface.

Not all implements serve this purpose because the proximal bone is not always being rotated around its axis during implement use. In some cases, implements create a lever system.

So, once again, the classification as one machine or another is sometimes ambiguous, but the analogy helps us understand human mechanics.

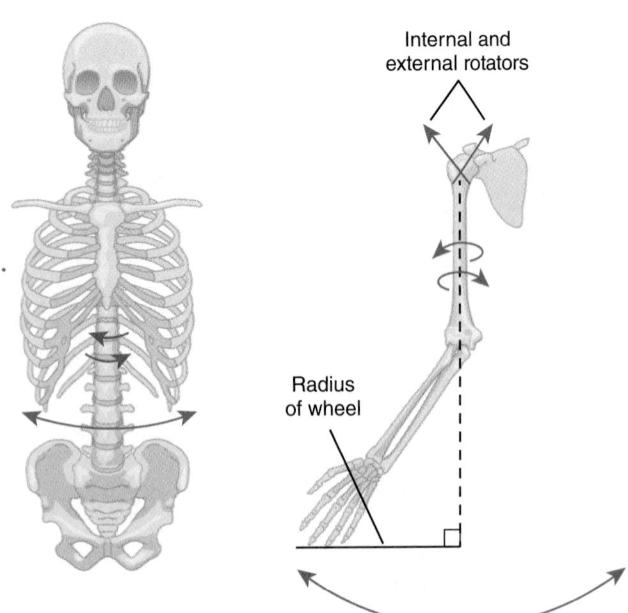

Figure 8.29 **Anatomical wheel-and-axle systems.** Can you think of other wheel-and-axle systems in the musculoskeletal system?

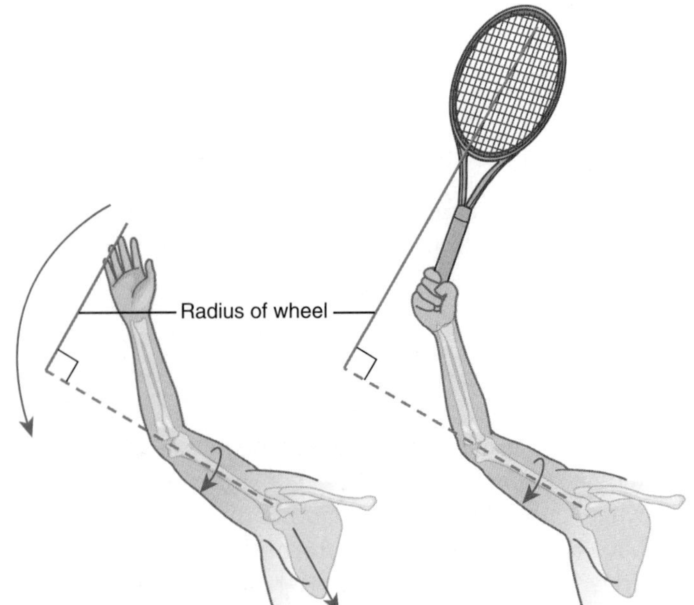

Figure 8.30 **Implements change the radius of the wheel.** In what situation is an implement classified as a different type of machine?

CONNECTIONS

8.5 HUMAN PERFORMANCE

Sport Science

The human body is a series of machine-like segments and joints that enable us to move ourselves within an environment and manipulate objects. Most of these structures are class-three levers that give us a large range of motion for a limited length of muscle contraction. This design gives us the ability to move the distal portion of our segments at a high velocity, ultimately at the expense of strength. Large range of motion lets us run fast and propel objects at fairly high velocities as long as their mass is small.

Primates are excellent climbers, which makes sense when you consider that trees offer protection from predators and provide food for their diet. Chimpanzees and apes can easily pull their entire body upward against gravity with one arm; think about how many one-armed pull-ups you can do. The reason for this enormous strength is not entirely in their muscle mass but in the design of their elbow joints. The attachment points of primate biceps and triceps are farther from the axis of rotation than in humans, giving primates a greater mechanical advantage in terms of strength. The tradeoff is that they lose the ability to move the lower arm at higher velocity, which seems less of a concern for them. There are unsubstantiated rumors that athletes of some communist nations in the 20th century had muscle attachment sites surgically relocated to provide advantages in either strength or velocity. Although this expedient is rather extreme, the theory is sound and based on simple physics. For example, Randy Johnson, a highly successful baseball pitcher now retired, had one of the highest velocity fastballs in Major League play. In a sport in which large players don't usually excel, Johnson stands out from the rest at 208 cm tall. This size would be a detriment to any other player on the field in terms of ability to move quickly toward a hit ball, but size gives an advantage to a pitcher. Beyond providing the intimidation of a person this tall standing on a raised mound less than 20 meters away, Johnson has arms that are longer than those of most pitchers. This length results in the ball being farther from the axis of rotation than it would be for an average size pitcher. If Johnson can develop the same angular velocity of the wheel and axle and in the levers of his shoulder and arm that a smaller pitcher can, physics tells us that the ball will leave his hand at a higher velocity.

Humans discovered that the use of implements could increase the body's mechanical advantage to an even greater degree. When the fabled David killed Goliath with a stone, he hurled it from a sling. The sling was an implement developed to give an advantage in velocity and distance when throwing objects. As long as the stone is not so massive that the arm can't accelerate it, the sling becomes an extension of the arm. With relatively small movements of the arm and shoulders, the stone is accelerated through several rotations to gain velocity that the arm itself could never attain. Add to this the length of the sling, which moves the stone farther from the axis of rotation, and you start to understand why this was a simple yet effective long-range weapon in its time. Most implements that we commonly use in sport today use the same principle as David's sling but are more rigid. Some are designed to strike an object, and some to throw and catch objects.

Lacrosse is a sport descended from a Native American game. There is a large field of play, several players on each team, and a goal into which a ball is thrown to score. The only way that the ball is propelled is by using a stick with a woven basket at the end. The ball is picked up from the ground or caught from the air in the basket and then passed to a teammate or shot on goal. The sticks used in lacrosse give players an advantage imparting velocity to the ball: At most times it acts as a class-three lever, with the axis in one hand and the motive force provided by the other (Figure 8.31).

Players also manipulate their hand position on the stick, depending on the situation. The stick's true axis of rotation is generally located within the nondominant hand (the one closest to the end of the stick). When a player passes the ball, it is usually over short distances, where maximal velocity

Figure 8.31 **Lacrosse.** Hands apart for passing (accuracy). Hands closer for shooting (velocity). Why change the hand position?

isn't required. In this instance, players move their hands apart on the stick, resulting in lower potential velocity. When players want to make a long pass or shoot on goal, they require high velocity and move the motive force (the dominant hand) closer to the axis of rotation. As lacrosse evolved, defenders were allowed to carry longer sticks. These sticks give them an increased range of motion for defending and also allow a greater mechanical advantage for the long passes they send downfield to offensive players.

Chapter 4's "Connection" section describes the ways that golf has been changed by new technology and design. Lighter-weight materials used in club heads and shafts have enabled manufacturers to increase the size of club heads without affecting a player's ability to accelerate them. Larger club heads result in larger "sweet spots" that make the clubs quite forgiving on off-center hits (less torque twisting the club). This trend is great news for amateur golfers who don't have enough time to practice for hours each day perfecting a swing. New materials such as titanium faces on drivers also have properties that increase the coefficient of restitution, providing a trampoline effect on contact with the ball. Because this chapter is about machines, let's look at the golf club as an extension of the human lever system.

A proper golf swing is a highly complex series of movements that deliver a club face to a stationary golf ball at velocities approaching 180 km/h and within a few degrees of perpendicular to the target line. At address and at impact, the club is a true extension of the lever created by the arms, with the spine regarded as the axis of rotation. However, at times during the backswing and through swing, it becomes a secondary lever system. When the club reaches a position parallel to the ground on the backswing, the wrists deviate radially, creating an "L" shape between the club and the forearms (Figure 8.32).

An axis of rotation for the body continues to be the spine, while the axis of rotation for the club changes to a point near the end of the club, closest to the body. This "L" is maintained through the rest of the backswing, which keeps the club closer to the body, and hence closer to the axis of rotation. Chapter 6 should provide a clue as to why this is a good position to accelerate on the through swing.

The "L" between the forearms and club is maintained even as the golfer's hands are passing the thigh of the back leg during the through swing. This more compact position allows the torso to rotate to a higher velocity before letting the club release farther from the axis of rotation and meet with the ball. When all of the body segments are sequenced in the correct order, an open kinetic chain is created. Power from the legs is transmitted to the torso, which adds more power through trunk rotation. That combined power is transmitted through the shoulders and arms, which add their own muscle forces. At impact, the club and arms are aligned to create a true one-lever extension system from the torso.

Figure 8.32 **A golfer:** "L" shape on the through swing (two levers) and at impact (one lever).

One of the more entertaining forms of golf has nothing to do with playing the game on an actual course, and it requires only one type of club. Long-driving competitions are becoming popular enough to be featured on ESPN because viewers are amazed to watch how far a golf ball can be hit. Although some competitors in these events are less than 180 cm tall, most are taller. What advantage does this give them? Not only do their bodies create longer levers, but also they can use longer implements more accurately. Shorter golfers in long driving events also use longer clubs than the average golfer, but they have to change their swing mechanics to compensate for the additional length. Tall golfers can maintain more appropriate mechanics while employing longer clubs. The club head moves farther from the axis of rotation, the lighter weight of new materials allows the club to be accelerated with little resistance to the body, and the end result is a club head meeting the ball with higher linear velocity. Chapter 9 helps explain how golf ball designers are also changing how the game is played.

Just as new materials have changed golf club design, they have also changed the nature of racquet sports. Racquets extend the natural levers of the body, allowing more range and greater velocity. Different racquet sports have benefited from new materials, and some have become much faster games through rule changes prompted by equipment design. The evolution of racquet sports from recreational entertainment to competitive events has driven these changes in design.

The game of tennis may be the most pure example of a racquet used as an extension of a body lever. In general, tennis swings are made with the arm in an extended position throughout the forehand swing. Until the 1970s, tennis racquets were made of wood. They were relatively heavy, and their mass hindered acceleration, but for the baseline rally game of their time they worked well. As the game evolved into a faster, more dynamic serve-and-volley form, manufacturers of tennis equipment began experimenting with ways to make racquets lighter and more powerful. Commercially successful metal racquets arrived in the 1970s, and Jimmy Connors made good use of their extra power and lighter weight to win major tournaments. Today's racquets are made from a variety of materials that combine light weight and great strength (Figure 8.33). These new "lever extensions" can be accelerated quickly and impart high velocity, but the mechanics of game skills remain unchanged.

Figure 8.33 **New and old racquets.** Has sport skill form changed along with equipment design?

Another trend in tennis evolution is the size of elite players. The average height of professional tennis players has been steadily increasing each decade, and taller players have several advantages on the court. The one that relates to this chapter is that taller players have longer natural levers (arms). Unlike long drive golfers, shorter players cannot get longer racquets to compensate for short stature because tennis has rules concerning maximum length of equipment. Tall players also tend to be stronger, which gives them the added bonus of higher motive force applied to their longer levers (assuming that long-limbed players have proportional strength to overcome the added inertia of their heavier limbs). Even more frustrating for short players is the fact that taller players can serve the ball with a steeper downward trajectory. In sum, tall players can generate higher ball velocity on the most important shot of any point—the serve. As power and velocity have increased, the game of tennis has evolved from a game that used to feature long baseline rallies to a serve-and-volley game with rare long rallies. These are two of the reasons that few professional tennis players are shorter than 180 cm. The women's game has also seen a similar shift toward taller players; the Williams sisters (Venus is 185 cm; Serena is 178 cm) from the United States and Maria Sharapova from Russia (184 cm) are examples.

Racquetball is a sport that evolved from a natural human lever activity called *handball*. Handball was hard on the hands, and this fact limited its appeal to new players, so short paddles and slightly different balls were introduced to handball courts. The paddles added length to the arm, increasing the length of the lever and providing added mechanical advantage. New materials have made current racquets lighter, but the primary reason for the increased speed of today's game is a rule change that allows longer racquets.

In an evolution similar to that in tennis, racquetball originally used wood racquets, which would later be replaced by metal and then composite materials. Current racquets are both lighter and longer than the original versions, so they can be accelerated more quickly and hit the ball at a farther distance from the axis of rotation. Remember from Chapter 6 that even if angular velocity remains constant, a point farther from the axis of rotation moves at a higher linear velocity. So new racquetball equipment achieves greater velocity than older equipment did in two ways: (1) the new racquets are lighter, so angular acceleration and velocity are increased, given the same motive force; and (2) the impact area is farther from the axis of rotation. Elite racquetball players can hit the ball in excess of 270 km/h.

A sport that truly adds to a natural body lever is jai alai, which is played on an indoor court similar to a racquetball court but much larger and with an open side. Proponents of the sport argue that it is the "fastest ball sport in the world," and they may have a point because the 125 g or pelota (ball) can travel in excess of 300 km/h! Compare this speed to the highest velocity of an object thrown by the natural levers of the human body: An elite baseball pitcher may occasionally throw close to 160 km/h. With knowledge gained from this chapter, you will realize that some type of implement had to be employed to generate such a high velocity. The implement is a wicker basket glove called a *cesta* that is approximately 63 to 70 cm long. Originally the cesta was held in the hand, but it was redesigned with an integral glove that essentially makes it a one-piece extension of the arm. As a child, you may have played in the backyard with a toy version of the cesta (Figure 8.34).

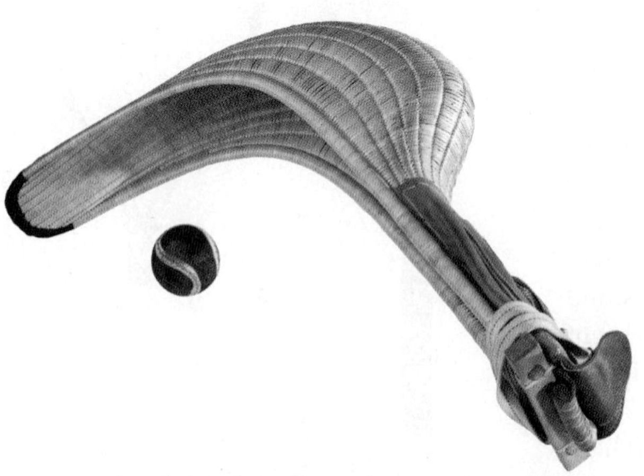

Figure 8.34 **Jai alai cesta.**

The toy version and the real jai alai glove are both designed with a curved shape. This shape makes the ball easier to catch and also imparts spin during a throw. The toy version is designed for fun and generates lots of spin. With a lighter ball, the spin results in entertaining amounts of curve as the ball travels between throwing partners. The cesta also imparts a high rate of spin, but its real advantage is increasing throwing velocity. Because the basket glove and ball combined are relatively light, athletes can accelerate this longer lever with little more effort than throwing a ball alone requires. With the axis of rotation at the shoulder or spine (depending on the type of throw), the lever becomes almost twice as long as the arm alone. If you look closely at the ball velocities of the baseball player and jai alai players, you'll notice almost the same ratio: 2 to 1. Coincidence? Or physics?

8.6 ERGONOMICS

Spinal Loading

Situations involving static equilibrium have been used several times throughout this text to solve for forces or torques necessary to maintain equilibrium. These same methods can be used to analyze situations in the field of ergonomics. One specific example is that of spinal loading during lifting tasks. When one lifts an object with the arms, the spine can be considered the fulcrum of a first-class lever system. The object being lifted applies a torque on one side of the fulcrum, while the extensor muscles of the spine apply an opposing torque (Figure 8.35).

The opposing torque of the spinal extensors must equal that of the torque caused by the object; if not, postural stability will be lost. The spinal extensors have a moment arm that is relatively small, maybe only 5 cm (Bridger, 2018). Keep in mind that this moment arm length is not highly variable. In contrast, the length of the moment arm for the object being lifted varies, depending on how far from the fulcrum the object is held. For example, if the object is held 50 cm from the fulcrum at the spine, then the spinal extensors are at a mechanical disadvantage of 10 to 1 (50:5). This disadvantage means that the spinal extensors must generate a force 10 times the magnitude of the weight of the object to maintain postural stability (Figure 8.36).

This mechanical disadvantage is further exaggerated if the weight is held farther from the body. At full extension of the arm, the object may be 100 cm from the fulcrum, and the disadvantage could reach 20:1. In terms of ergonomics, this situation is important, because compressive force on the spine could lead to injury. In this example, three compressive forces and associated reaction forces must be considered: (1) compressive force from the load, (2) compressive force from the weight of the upper body, and (3) compressive force from the spinal extensors (Bridger, 2018). With all of these compressive forces acting on the spine, one can no doubt see the importance of proper task design. This combination of forces is the reason for instructing people to keep objects close to the body while lifting. Keeping the object close

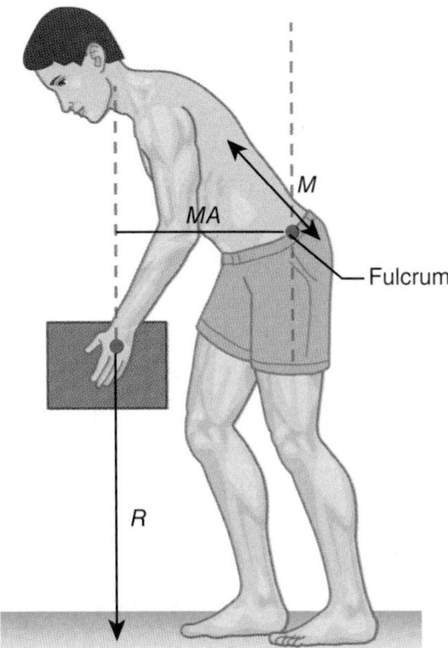

Figure 8.35 **The spine as a fulcrum in a first-class lever system.** Why is the spine a fulcrum?

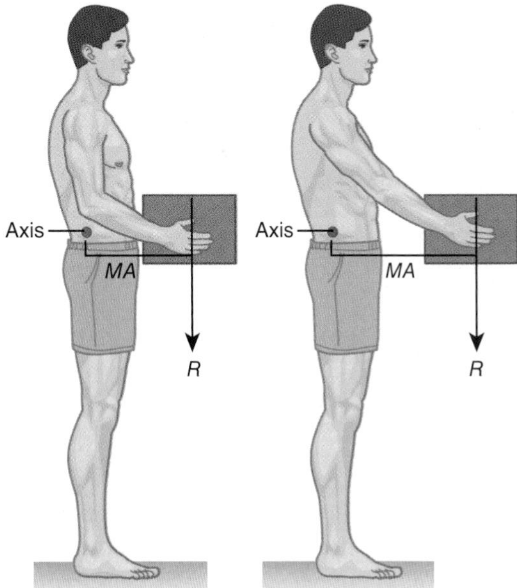

Figure 8.36 **Changing length of a moment arm.** What is the potential danger of holding heavy objects too far from the body?

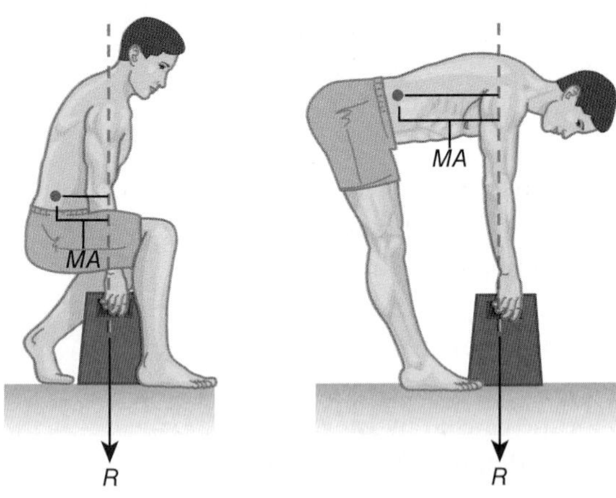

Figure 8.37 **Proper lifting form versus improper.** What are the differences between these situations?

to the fulcrum shortens its moment arm and therefore minimizes the compressive force exerted by the spinal extensors.

Of course this model is highly simplified, because it does not account for the acceleration of the object (i.e., it is a static analysis) or the degree of spinal flexion. If large accelerations are involved in the task, this type of analysis will underestimate the compressive force. Therefore, a slow lift is safer than one performed with greater changes in velocity. During spinal flexion, the trunk's center of gravity moves to a position in front of the lumbar spine. This shift means that the weight of the upper body now has a moment arm and becomes torque producing, which adds a shear component in addition to the compressive component (Bridger, 2018). Therefore, in addition to keeping the load close to the body and lifting slowly, one should also keep the spine as erect as possible (i.e., "back flat"; Figure 8.37).

© Jamie Roach/Shutterstock

© Martin Good/Shutterstock

Figure 8.38 **Racing wheelchair and handcycle.**

8.7 ADAPTED MOTION

Adapted Sport Science

People with lower extremity challenges have been racing wheelchairs competitively for decades. Early racers used regular wheelchairs and propelled them by grasping the edges of the wheels and pushing forward. One limitation of common wheelchairs is that above a certain velocity, the human body is not capable of generating enough arm velocity to apply force to the fast-moving wheel. Current racing chairs have evolved into serious purpose-built machines that address this limitation in two different ways. Before we discuss the mechanical advantage issue, note that racing chairs have received the same benefits in materials technology as other sports. Wheels, frames, seats, and other components are built using incredibly strong, lightweight materials, so the chairs are much lighter than traditional wheelchairs.

Wheelchair racing is now split into two separate divisions based on propulsion methods, and both divisions give athletes a distinct mechanical advantage over traditional chairs. Traditional racers still propel their chairs by applying force directly to the wheels. The difference in the new manual racing chairs is a smaller diameter ring attached to each wheel (Figure 8.38).

After reading about wheel-and-axle machines in this chapter, you will understand that a motive force applied closer to the axis of rotation results in the outer edge of the wheel moving at a higher velocity. Although it takes a larger motive force to initiate motion, once the chair is underway, shorter pushes on the smaller ring will equal higher speed at the outer diameter of the wheel. In essence, designers have changed the "gearing" of the machine. Notice also the friction device attached to the athlete's hands. The smaller wheel ring is made of a rubber-like material and the friction device is made of a similar material. Rather than grasping the smaller ring, the athlete need only make contact between the two rubber surfaces to impart motive force.

The second division of wheelchair racing doesn't even include the term *wheelchair,* and it is a much faster version of the sport. This newer event is called handcycle racing and combines the sports of chair and bicycle racing. The arms are still used as the motive force, but they work through a set of pedals similar to those on a bicycle and, more importantly, gain mechanical advantage through multiple gears identical to those used on multispeed bicycles.

The variety of gear ratios allows racers to adjust the level of mechanical advantage, depending on the situation. On flat surfaces and downhills, the larger top sprocket and smaller wheel sprocket let the athlete turn lower pedal revolutions per minute for higher speeds. At the start and when going uphill, the handcycles are shifted to a smaller top sprocket and larger wheel sprocket. In this configuration, the rider trades high speed for an advantage in power. Both forms of racing depend on two different machines for propulsion: the natural levers of the body, working through the wheel-and-axle systems of the chairs.

SUMMARY

A machine is an apparatus or system that uses the combined action of several parts to apply mechanical force. Most people are comfortable with the idea that the human body uses machines to accomplish tasks. However, some are not comfortable with the notion that the human body is composed of machines or is itself a machine. In fact, the human body is a system that uses the combined action of several parts to apply mechanical force, and therefore it can be correctly thought of as a machine. Within the human body, three types of machines are used for various purposes: lever systems, pulley systems, and wheel-and-axle systems. Each machine is used to fulfill one or more of the following functions:

1. to *transmit a force* (e.g., tendons transmit muscle forces to bones),
2. to *increase the magnitude of a force* (i.e., less effort is required to move a given resistance),
3. to *increase the linear distance and velocity of a force* (i.e., a resistance is moved a greater distance or at a faster rate than the motive force), and
4. to *change the direction of a force* (i.e., a resistance is moved in a different direction than that of the motive force).

A lever system consists of a rigid or semirigid object that rotates around an axis called a *fulcrum*. The basic purpose of a lever system is to transmit energy (in the form of applied force) from one place to another. Because there is an axis of rotation (fulcrum), the applied force must be eccentric to produce the torque necessary to rotate the lever. Two types of torque exist in a lever system: motive and resistive. The three types of lever systems are classified by the positions of the motive and resistive forces relative to the fulcrum. One major function of the pulley system is to change the effective direction of the applied force. The wheel-and-axle system can be used to gain an advantage in force or linear range of motion and velocity of movement. The advantage depends on whether the motive force is applied to the wheel or the axle.

REVIEW QUESTIONS

1. What are three types of musculoskeletal arrangements that serve as machines?
2. Of what does a lever system consist? What are the parts of the human body that correspond to parts of a lever system?
3. Explain a mechanical advantage equal to 1.00, less than 1.00, and greater than 1.00.
4. Give the arrangement of the axis of rotation and the motive and resistive forces for first-, second-, and third-class levers. Give two anatomical examples of each. Which lever system is most commonly found in the human body?

5. In a wheel-and-axle arrangement, what is the advantage if the motive force is applied to the wheel? What if it is applied to the axle? To what is the motive force most often applied in the musculoskeletal system? So for what activity are human beings best suited?
6. In what way do sports implements give an advantage?
7. What is the function of a pulley-like arrangement? Give two anatomical examples.
8. Name the functions of a machine that are fulfilled by the following items: (a) a pry bar, (b) a bicycle chain, (c) a tendon over a malleolus, (d) the hamstrings muscle acting to flex the knee, and (e) a trebuchet.
9. With golf club head velocities in excess of 150 km/h, what other factors may come into play that limit the ultimate velocity it can achieve?
10. Height may be an advantage for a tennis server, but can you think of any disadvantages that excess height may have in the same game?
11. Current racing wheelchairs weigh far less than older wheelchairs. Using Newton's laws, describe what effect this change will have during a race.

PRACTICE PROBLEMS

1. A teeter-totter 3 m in length is placed so that the fulcrum is 1.8 m from one end. Calculate the mechanical advantage of this lever system. If a child who weighs 178 N is placed on the 1.8-m end, calculate the weight of a child that could balance the teeter-totter.
2. The moment arm for a muscle force (motive in this case) is 0.025 m, and the moment arm for the resistive force is 0.50 m. Calculate the mechanical advantage.
3. If the resistive force in question 2 is 25 N, calculate the torque attributable to resistive force and the muscle force necessary to counteract the resistive torque and hold the arm in a static position.
4. A faucet has a handle with a diameter of 7 cm and a shaft (axle) with a diameter of 1 cm. Calculate the mechanical advantage. Calculate the amount of motive force necessary to overcome a resistance of 60 N.

EQUATIONS

Mechanical advantage of a lever system $$MA = \frac{M_M}{M_R} \quad (8.1)$$

Mechanical advantage of a wheel-and-axle system $$MA = \frac{r_w}{r_a} \quad (8.2)$$

REFERENCES AND SUGGESTED READINGS

Bridger, R. S. 2018. *Introduction to human factors and ergonomics*, 4th ed. Boca Raton, FL: Taylor & Francis.

Erickson, B. J., A. K. Gupta, J. D. Harris, C. Bush-Joseph, B. R. Bach, G. D. Abrams, A. M. San Juan, B. J. Cole, & A. A. Romeo. 2014. Rate of return to pitching and performance after Tommy John surgery in Major League Baseball pitchers. *American Journal of Sports Medicine*, 42(3): 536–543.

Floyd, R. T. 2018. *Manual of structural kinesiology*, 20th ed. New York, NY: McGraw-Hill.

Jobe, F. W., & R. S. Kvitne. (1991). Reconstruction of the ulnar collateral ligament of the elbow. *Techniques in Orthopedics*, 6(1): 39–42.

CHAPTER 9

SYSTEM MOTION IN A FLUID MEDIUM

© technotr/Getty Images

LEARNING OBJECTIVES

1. Describe buoyant fluid force.
2. Understand dynamic fluid force as it acts on the body and projectiles.

CONCEPTS

CONNECTIONS

If you're a fan of bicycle road racing, there is no better place to be in July than France. The names of Tour de France winners are well known, but have you ever noticed that they rarely ride at the front of the race? International long-distance riding has evolved into a team event because of the physics of fluid dynamics.

Each rider and bicycle are outfitted in ways that minimize aerodynamic drag—from the shape of bicycle frame tubes and helmets, to skin-tight suits, to rider position on the bike. All of these factors make the long days easier, but the primary benefit to the lead rider on each team is the drafting effect of following another rider. The contenders for the overall event win remain in the *peleton* (the large group of riders who remain tightly grouped), saving 30% to 40% in energy expenditures until the end of a stage (or a particularly tough climb). It is the job of the rest of their team to cut through the air to create the drafting effect for as long as possible to allow contenders to save energy and sprint for the win at the end of each stage or pull them up long climbs as far as possible. The rest of the team are called *domestiques* (servants); they do a tremendous amount of work for the good of the team with little reward or mention in the results.

Air and water are the primary fluids that affect motion of the human system and the sport implements used by humans. Boats, balls, discs, cyclists, and swimmers all move through fluids. So a multitude of fluid force applications exist, which means that comprehension of fluid forces is necessary not only for understanding human motion more deeply but also some interesting aspects of nature: Why does a curve ball curve? How do birds fly? Why is a raindrop shaped like a raindrop?

CONCEPTS

9.1 BUOYANT FLUID FORCE

In Chapter 4, we defined *buoyant force* as the vertical, upward-directed force acting on an object that is submerged or partially submerged in a fluid. Remember the two key concepts from the discussion of pressure that are needed to understand buoyant force: (1) Pascal's law, which states that pressure applied to a fluid is transmitted undiminished to every point of a fluid and to the walls of the container—therefore, all points of a submerged body, at a given depth, will experience the same pressure; and (2) the fact that pressure increases in large increments with relatively small changes in depth. If an object submerged in water is neither rising nor falling, the net force on the object must be zero. We can ignore the horizontal forces on a submerged object because they will cancel one another (Serway & Vuille, 2018). We know that downward force is exerted on the object by the water above. We also know that force acts on the bottom of the object. Because the bottom of the body is at a greater depth, the pressure is higher at the lower portion than at the top. This net force arising from the differences in fluid pressure is the *buoyant force*. But in this case, the object neither rises nor falls. The weight of the object itself is sufficient to balance the forces and cause them to sum to zero (preventing acceleration). The difference between the downward and upward forces (buoyant force) is equal to the weight of the water displaced by the object (Serway & Vuille, 2018). In this case, the weight of the object is equal to the weight of the displaced water. Therefore, there is no acceleration of the object upward or downward.

This idea can be stated as *Archimedes' principle:* A body submerged in a fluid will be buoyed up by a force that is equal in magnitude to the weight of the displaced water. If we replace the object with an empty container that fills the same amount of space (i.e., has volume equal to that of the displaced water), we can get a visual idea of Archimedes' principle. The water surrounding the empty container will behave the same as if the displaced object is still there. In other words, the water pressure at a given depth is a constant. So the empty container will experience the same downward force from the water pressure produced by the weight of the water above; and it will also experience the same upward force from the water pressure at the bottom of the container. Now let's be more specific about the magnitudes of these forces. First, one cubic meter (m^3) of water weighs approximately 9,810 N (9.81 N per liter). This means that the pressure exerted by water increases by approximately 9,810 N per square meter for every successive meter of depth that an object is submerged (Serway & Vuille, 2018).

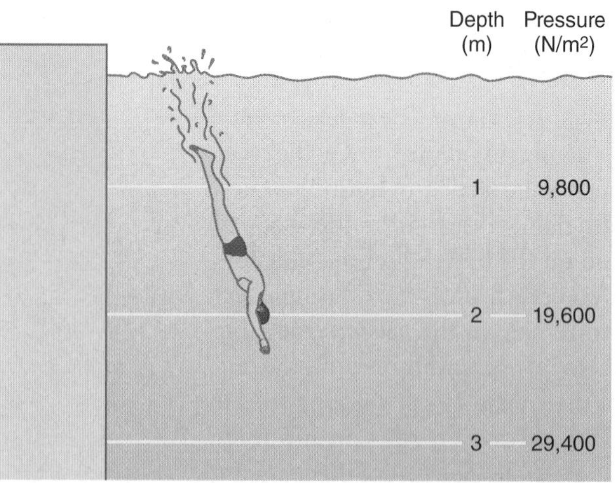

Figure 9.1 **Pressure of water at varying depth.** What causes ear pain in a pool?

If the empty container is submerged one meter below the surface of the water, it will experience a pressure of 9,810 N/m^2. If it is submerged to a depth of 2 m, it will experience a pressure of 19,620 N/m^2, and so on (Figure 9.1). Therefore, a submerged object will experience a greater amount of pressure at its bottom than at its top. You have learned that the net force attributable to the differences in fluid pressure is the *buoyant force*. As long as the submerged object weighs enough to compensate for the pressure difference, it will neither rise upward nor sink downward; no net force exists to cause acceleration. For example, let's say that the submerged object is water itself. In our introduction to buoyant force, we outlined a section of water in a swimming pool. If that section of water is one cubic meter and is submerged such that its top is just below the surface, the top of the cube will experience only a minimal downward force because only a small amount of water above is pushing down. However, the bottom of the cubic meter of water will experience 9,810 N because of the higher pressure at the greater depth of 1 m (net force = 9,810 N upward).

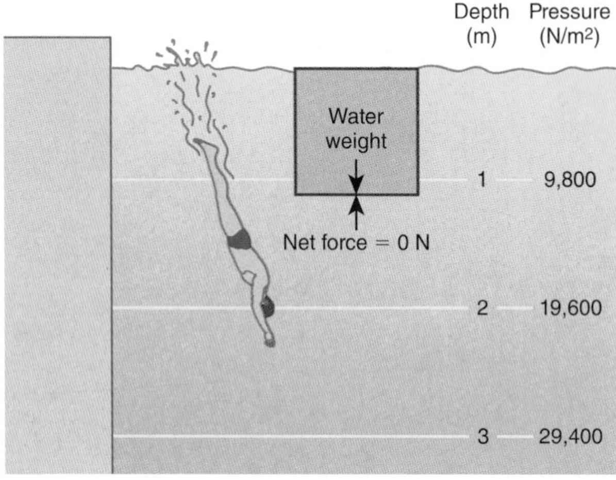

Figure 9.2 **Pressure on top and bottom of one cubic meter of water.**

But because one cubic meter of water weighs 9,810 N, it counteracts the difference between the downward force (0 N) and upward force (9,810 N), and the section of water is not accelerated (Figure 9.2). However, to demonstrate Archimedes' principle, imagine replacing the section of water with an empty container. The net force caused by the difference in water pressure is the same (9,810 N). The difference is that the container is empty and will not weigh enough to balance the difference between the upward and downward forces (i.e., it cannot balance the buoyant force). So, because a net force is now acting vertically upward, the container will be accelerated upward. The opposite situation will occur if the weight of the object is greater than the weight of the water it displaces. In other words, an object will be accelerated downward if its weight is greater than the net force caused by fluid pressure (buoyant force). This means that the net force will be downward, and the object will sink. The most important point to remember is that the buoyant force will be equal to the weight of the displaced water. In our example, we displaced 1 m^3 of water with an empty container. Because 1 m^3 of water weighs 9,810 N, the buoyant force experienced by the empty container is 9,810 N. If the object displaces 2 m^3 of water, then the buoyant force will be equal to 19,620 N. So for every cubic meter of water displaced by an object, the buoyant force increases by 9,810 N. That is the essence of Archimedes' principle: *A body submerged in a fluid will be buoyed up by a force that is equal in magnitude to the weight of the displaced water.*

Therefore, according to Archimedes' principle, for an object to float, it must weigh less than the weight of the displaced water.

When an object is floating, a net upward force buoys the object up, and the object is said to be **positively buoyant**. If the object weighs more than the displaced water, a net downward force exists, and the object is called **negatively buoyant**. In this case, the object will sink. Of course, the object could have the same weight as the displaced water (net force = 0), in which case it will have no tendency to either float or sink; such an object is called **neutrally buoyant**. Archimedes' principle explains why huge objects such as aircraft carriers are able to float; they have enormous weights, but they also displace huge amounts of water. So even though an aircraft carrier has an immense weight, it does not weigh as much as the water it displaces and remains positively buoyant.

A final important point to notice is that this principle applies at any depth. Remember that 1 m^3 of water weighs approximately 9,810 N (9.81 N per liter) and cannot be easily compressed. So the pressure exerted by water increases by approximately 9,810 N/m^2 for every successive meter of depth that an object is submerged. Therefore, if our 1 m^3 of water were submerged 1 m below the surface,

Positively buoyant Quality of an object that weighs less than the water it displaces, causing a net upward force that buoys the object up.

Negatively buoyant Quality of an object that weighs more than the displaced water and causes a net downward force that sinks the object.

Neutrally buoyant Quality of an object that has the same weight as the displaced water, in which case there is no net force and the object neither floats or sinks.

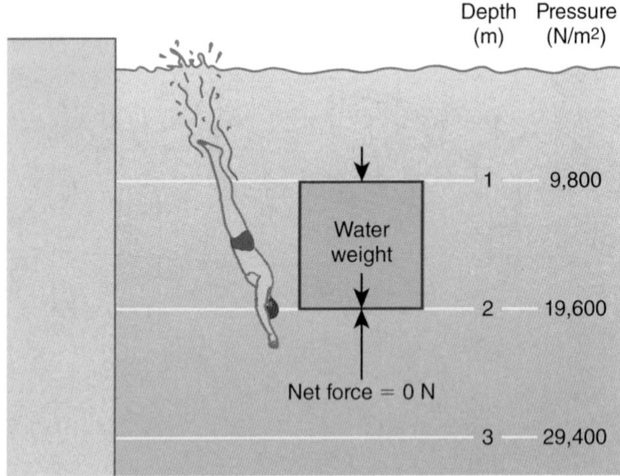

Figure 9.3 **Pressure on top and bottom of one cubic meter of water submerged to a depth of 1 meter.**

the top of the area of water would experience a force of 9,810 N downward from the pressure of the water above, and the bottom of the cubic meter of water will experience 19,620 N from the higher pressure at the greater depth of 2 m. The net force would still be 9,810 N upward, which is exactly equal to the weight of the cubic meter of water. Once again, the section of water neither sinks nor rises (Figure 9.3).

So even though the pressure exerted by water rises with depth, the magnitude of the buoyant force is always equal to the weight of the displaced water. If this is true, then why does a positively buoyant object released from a greater depth reach the surface of the water with a greater momentum than if it were released from just beneath the surface? We can answer this question with some information from a previous chapter: the *linear impulse* equation ($\Sigma F\,(\Delta t) = \Delta M$). The buoyant force for a given object is equal to the weight of the displaced water, no matter the depth. However, if the submerged object is released from a greater depth, buoyant force is applied for more time before it reaches the surface of the water. Therefore, the change in momentum will be greater and give the *appearance* of a greater buoyant force.

Finally, let's consider objects that are only partially submerged. So far, we examined only objects that are completely submerged. We began our discussion of Archimedes' principle with completely submerged objects because maximum buoyant force for a given object occurs only when it is completely submerged. However, one might notice that an object like a toy balloon seems to not be submerged at all. So, where is the buoyant force? Actually, if you look closely at a balloon floating on water, it is submerged to some extent. So, Archimedes' principle still applies: The balloon is buoyed up by a force equal to the weight of the displaced water. The tiny amount of water displacement by the balloon produces enough force to buoy it up because the balloon's weight is simply not enough to allow it to sink. Now consider a similar object that is slightly heavier, maybe a beach ball. On inspection, one would notice that the beach ball doesn't sink to a great degree either, but it sinks deeper than the toy balloon because it is slightly heavier and therefore requires more buoyant force to float. The additional buoyant force needed to hold up the beach ball requires that more water be displaced. Next, think of a basketball. The basketball sinks even deeper than the beach ball because it needs even more buoyant force to stay afloat. The point is that buoyant force increases as more of an object is submerged, reaching a maximum at the point where the object is completely submerged. The object will sink until it has displaced enough water for buoyant force to equal its weight; thus, different objects have varying floating depths. If water displacement never produces sufficient buoyant force to equal the weight of the object, then the object will sink completely. Objects that sink completely in water are often called *dense*. Therefore, density is our next topic of interest.

Density

From our discussion so far, we can determine that Archimedes' principle is based on two considerations: (1) the weight of the object and (2) the amount of water it displaces. The weight of the object, of course, is related to its mass. The amount of water the object displaces is related to the space occupied by its mass. **Volume** is the space occupied by an object's mass. Materials vary in their mass and volume relationships. For example, 5 kg of lead occupies less space (has lower volume) than 5 kg of Styrofoam (Figure 9.4). Their masses are the same (5 kg), but one material has a

Volume The space occupied by an object's mass.

larger volume for the same amount of mass. The relationship of mass to volume is called **density** and is expressed in the following ratio:

$$\rho = \frac{m}{V} \tag{9.1}$$

where

ρ = density of an object of uniform composition
m = mass of the object
V = volume or space occupied by the object's mass

Density has units of mass divided by units of volume. Because mass is measured in kilograms (kg) and volume is usually measured in cubic meters (m^3), density has Système International units of kilograms per cubic meter (kg/m^3). For example, if an object has a mass of 5 kg and occupies a volume of 3 cubic meters, then its density is equal to $5\ kg/3\ m^3 = 1.66\ kg/m^3$. Another object may have the same mass but has a volume of 6 cubic meters, in which case its density is $5\ kg/6\ m^3 = 0.83\ kg/m^3$. In our example of lead and Styrofoam, the lead has a greater density because it has the same amount of mass packed into a smaller space.

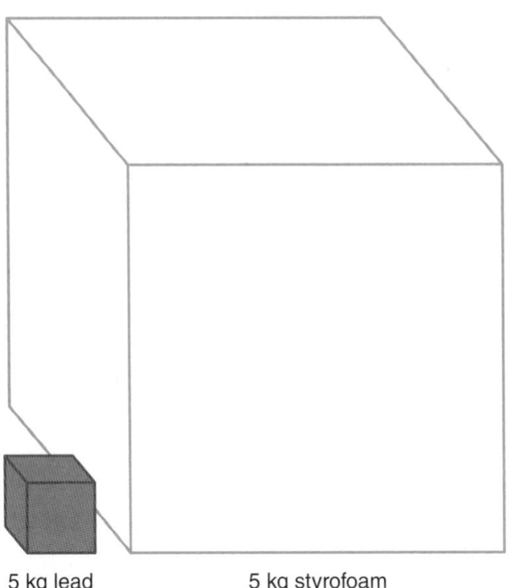

Figure 9.4 **Five kg of lead and 5 kg of Styrofoam.** Which is denser?

In contrast, two objects could have approximately the same volume but different amounts of mass. For example, we could blow up a toy balloon to the same volume as a bowling ball. Even though both objects are approximately the same volume, the bowling ball is more massive and therefore has a greater density.

Using these terms in a scientifically accurate way is important because they are often misunderstood or misused. A common misuse is to refer to objects of large volume as "massive." We all know what is meant by that statement. But from our toy balloon and bowling ball example, we know that volume is not necessarily a great indicator of mass. Another misstatement related to density is to say that an object of greater density "weighs more" than another object. For example, many people say that "muscle weighs more than fat." Muscle is denser than fat, but we cannot say that it weighs more: 1 kg of muscle weighs the same as 1kg of fat. Muscle has a lower volume for the same amount of mass and therefore has a greater density, but we cannot say that it "weighs" more. As we will see, the densities of various materials (especially biological materials) have many implications, both biomechanically and physiologically (Table 9.1; Figure 9.5).

Specific Gravity

Specific gravity is a concept related to and often confused with density. **Specific gravity** is the ratio of the density of a given substance to the density of water ($1.0 \times 10^3\ kg/m^3$) (Serway & Vuille, 2018). So if the specific gravity of a substance is 2.0, we know that its density is two times that of water ($2.0 \times 10^3\ kg/m^3$). Because specific gravity is a comparison to the density of water, it is sometimes defined as the ratio of the weight of an object to the weight of an equal volume of water. This definition is actually another way of making the same comparison. If the volumes are held constant, the comparison is really one of mass or weight. Either way that one wants to think about it, specific gravity is still a dimensionless ratio that compares the weight of a substance to that of water.

The importance of specific gravity for our purposes goes back to Archimedes' principle. Remember that a body in water will be buoyed up by a force equal to the weight of the displaced water. If the body weighs more than the displaced water, it will sink. If the body weighs less than

Density The relationship of an object's mass relative to the space that it occupies.

Specific gravity The ratio of the density of a given substance to the density of water.

Table 9.1
Densities of Various Materials

Substance	ρ(kg/m³)[a]	Substance	ρ(kg/m³)[a]
Air	1.29	Ice	0.917×10^3
Aluminum	2.70×10^3	Iron	7.85×10^3
Benzene	0.879×10^3	Lead	11.3×10^3
Bone	1.28×10^3	Mercury	13.6×10^3
Copper	8.92×10^3	Muscle	1.06×10^3
Ethyl alcohol	0.806×10^3	Oxygen	1.43
Fat	0.9×10^3	Platinum	21.4×10^3
Glycerin	1.26×10^3	Silver	10.5×10^3
Gold	19.3×10^3	Styrofoam	0.10×10^3
Helium	1.79×10^{-1}	Uranium	18.7×10^3
Hydrogen	8.99×10^{-2}	Water	1.00×10^3

[a]All values are at standard atmospheric temperature and pressure (STP), defined as 0˚C (273 K) and 1 atm (1.013×10^5 Pa). To correct to gramubic centimeter, multiply by 10^{-3}

the displaced water, it is positively buoyant and will float. Another way to look at Archimedes' principle is to realize that an object with a specific gravity greater than 1.0 (density greater than 1.0×10^3 kg/m³) will be negatively buoyant and sink, but an object with a specific gravity less than 1.0 (density less than 1.0×10^3 kg/m³) will float. We can observe this fact from the point of view of both of our definitions of specific gravity. In reference to our first definition, specific gravity greater than 1.0 means that the substance has a density greater than that of water. In other words, the substance has more mass for a given volume than water. So when the object is submerged, its mass or weight is greater than that of the displaced water, and it sinks (it is negatively buoyant). This concept automatically fits with our second definition of specific gravity. If specific gravity is greater than 1.0, the weight of the object is greater than that of an equal volume of water. In this case, the comparison volume of water is the amount displaced by the submerged object. Once again, because the object weighs more than the displaced water, it is negatively buoyant and sinks.

Let's go back to our previous examples to demonstrate this point. First, let's revisit our lead and Styrofoam example. Imagine 5 kg of lead and 5 kg of Styrofoam. They have the same mass, but the Styrofoam occupies much more space (has a greater volume) than the lead. Therefore, the Styrofoam is less dense than the lead. In addition, the specific gravity of the Styrofoam is less than 1.0 (its density is less than that of water). But let's also think of this situation in terms of Archimedes' principle. If we submerge both the lead and the Styrofoam under water, we will see that the Styrofoam displaces a large amount of water because of its volume (Figure 9.6).

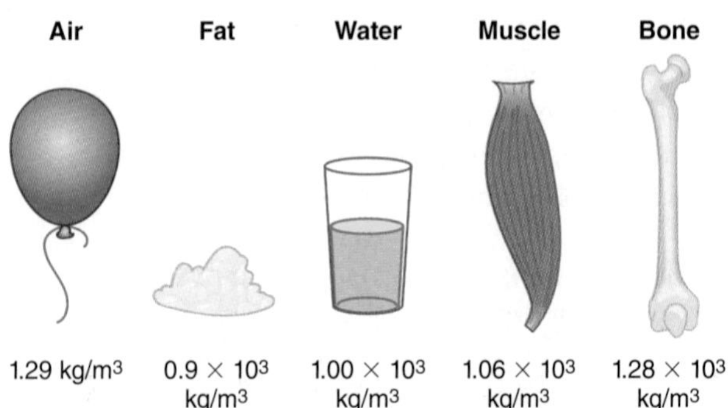

Figure 9.5 **Substances possessing different densities.** Which will sink in water?

The lead object will displace relatively little water in comparison. The buoyant force is equal to the weight of the displaced water. The Styrofoam, because it displaces more water, will be subjected to a greater buoyant force than the lead. Both objects have the same mass and therefore the same ability to apply a downward force because of their weight. However, the weight of the Styrofoam will be acting against a greater buoyant force. The weight of the Styrofoam will be unable to balance the upward force caused by fluid pressure and, as a result, will be accelerated upward. The lead, on the other hand, will be subjected to a smaller buoyant force relative to its weight. In this case, the forces will be unbalanced, but in the opposite direction. The lead will weigh more than the displaced water and will be accelerated downward. This is an example of two objects of the same mass being subjected to two different magnitudes of buoyant force.

The balloon and bowling ball provide an example of two objects subjected to the same buoyant force but having two different masses. As long as we inflate the toy balloon to the same size as the bowling ball, it will occupy approximately the same space (same volume). However, the toy balloon is not a solid object and is composed of less mass than the bowling ball. Therefore, the toy balloon has less mass per unit volume (is of a lower density). Also, the density of the balloon will certainly be less than that of water, so its specific gravity will be less than 1.0. According to Archimedes' principle, as long as these objects have the same volume, they will displace the same amount of water.

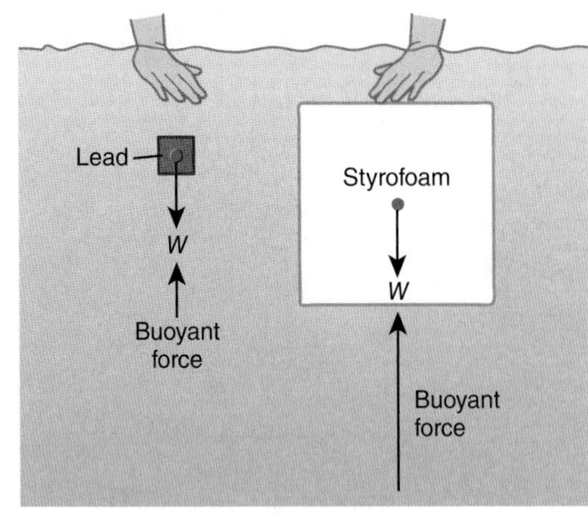

Figure 9.6 **Forces acting on 5 kg of lead and 5 kg of Styrofoam submerged under water.**

This equivalent displacement means that they are exposed to the same buoyant force. However, the balloon does not weigh enough to cancel the affect of the upward buoyant force and will be accelerated upward. A bowling ball having a mass greater than 6 kg weighs more than the water it displaces and will be accelerated downward to a floating position lower than that of a balloon.

Archimedes' principle as it applies to both density and specific gravity is especially important. As you will see, these concepts are tied to floating positions of the human body. Later, in the "Connections" section, we will see that the concepts are directly related to the measurement of body composition.

Floating Position

In previous chapters, you have read about the center of mass and the center of gravity. The cardinal planes pass through the **center of mass**. Gravitational pull is concentrated at the center of mass. So in the vertical axis, the center of mass can be considered to be the same as the **center of gravity** (the point at which the force of gravity seems to be concentrated). In other words, the center of mass (or center of gravity) of the system is at the intersection of the three cardinal planes. According to this definition, the center of mass of an object that possesses uniform mass and mass distribution (i.e., a symmetrical object made of the same material throughout its structure) is directly in the middle of the object. Another definition of the center of gravity is the point around which all of the torques (or moments) sum to zero.

Recall that because humans are not of uniform mass or mass distribution, our center of gravity is not measured directly between the top of the head and the bottom of the feet. For example, the human torso does not consist of as much mass as the lower body. The head is relatively solid and located above the torso. The lower body tapers so that more mass is located near the hips. In the context of our current topic, these various portions of the human body have different densities. The torso contains cavities such as the lungs and stomach that contain air and gases. Air and gas are not as dense as the bone and muscle that make up the lower body. The final result of this complicated mass distribution is that the center of gravity varies between approximately 53% and 59% of standing height measured from the soles of the feet, depending on gender (Braune & Fischer, 1889; Dempster, 1955; Hay, 1973; Heinrichs, 1990; Plagenhoef Evans, & Abdelnour, 1983; Winter, 2009). Females tend to have a center of gravity that varies between 53% and 55% of standing height, whereas males have a range of 54% to 59%.

Center of mass The point at which all of a system's mass is concentrated.

Center of gravity The point at which the force of gravity seems to be concentrated.

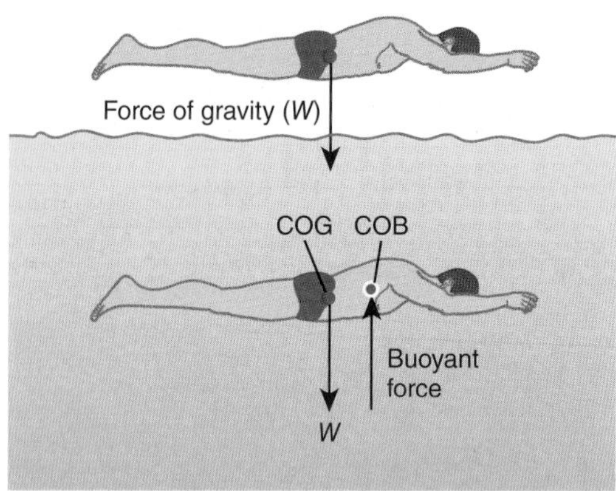

Figure 9.7 **Center of gravity (COG) and center of buoyancy (COB) of a submerged person.** Why are the COG and COB not in the same location?

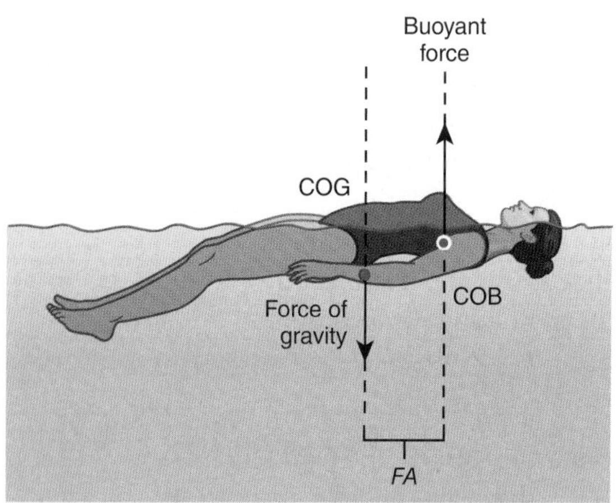

Figure 9.8 **Lines of action of the force of gravity and buoyant force on a submerged person.**

We have mentioned the displacement of water many times. Imagine removing that displaced water and observing it. Think about the shape of that displaced water. Isn't it the exact shape of the submerged object? So a submerged ball displaces a ball or sphere-shaped volume of water, and a submerged human displaces a human body–shaped volume of water. That displaced volume of water also has a center of gravity. The center of buoyancy is the underwater analog of the center of gravity. The **center of buoyancy** is the center of gravity of the displaced volume of water, equal in shape and volume to the submerged object that displaced it, or the point at which the force of buoyancy seems to be concentrated. Because the displaced water is of consistent density, the center of buoyancy can also be defined as the center of volume of the object displacing the water. Just as gravitational force acts through the center of gravity of an object, the buoyant force acts through the center of buoyancy of the submerged object.

Where will the center of buoyancy be located compared to the center of gravity? It depends. For instance, if an object is of relatively uniform density, then its center of gravity and center of buoyancy will have the same location. For example, the center of gravity of a golf ball would be directly in the center of the ball. If we then submerge the ball, it will displace a golf ball–shaped volume of water. The center of buoyancy will be directly in the center of that displaced volume of water.

This distinction may not seem important, but what about the relative locations of the human body's center of gravity and center of buoyancy? You have just learned that the density of the human body is not consistent and that the center of gravity is approximately 56% of the height from the soles of the feet. But remember that the water displaced by a submerged human *does* have a consistent density (i.e., it doesn't have a torso full of cavities). Therefore, if the body is oriented vertically with the head pointing toward the surface of the water, the center of buoyancy in a human will be located superior to the center of gravity (Figure 9.7).

Again, the center of gravity is the point around which all of the torques sum to zero. Compared to the actual human body, the displaced volume of water has relatively more mass located superiorly. Therefore, the balance point (center of buoyancy) of the displaced water is superior to the balance point (center of gravity) of the human body.

The implication for the difference in location of the center of gravity and the center of buoyancy is actually a torque-related issue. The line of action of gravitational force through the center of gravity will be vertically downward while the line of action of buoyant force through the center of buoyancy will be vertically upward (Figure 9.8).

Because these forces are noncolinear (eccentric), a moment arm is present, and therefore so is torque. The torque caused by these forces causes the human body to rotate around an axis (fulcrum) at the center of buoyancy and assume a more vertical position in the water (Figure 9.9).

Center of buoyancy The center of gravity of the volume of water displaced by an object; the point at which the force of buoyancy seems to be concentrated.

The body will stop rotating when the moment arm is no longer present. For this to occur, the lines of action for the gravitational and buoyant forces must become colinear. Colinearity is achieved once the center of gravity has rotated to a position directly below the center of buoyancy. In this position, a moment arm no longer exists; therefore, no torque is present and rotation stops (i.e., the body comes to equilibrium).

The magnitude of rotation, and therefore the ultimate floating position, depends on the relative locations of the person's center of gravity and center of buoyancy. In other words, the farther apart the forces are, the longer the moment arm. A longer moment arm will create more torque for a given amount of force, and the person will rotate more. This fact means that the final floating position will be relatively vertical. On the other hand, if the center of buoyancy and center of gravity are relatively close together, the moment arm is short. Therefore, the center of gravity rotates a shorter distance before falling underneath the center of buoyancy. In this case, the person assumes a more horizontal floating position.

The floating position can also be affected by repositioning body segments. Remember that any change in body position redistributes a portion of the mass. The center of gravity shifts in the same direction as the redistributed mass. The center of buoyancy is not affected to as great a degree as the center of gravity. Therefore, simply repositioning body segments can change the distance between the center of gravity and the center of buoyancy (Figure 9.10).

So a person who normally floats in a relatively vertical position can float more horizontally by shifting the center of gravity toward the center of buoyancy (e.g., bend at the knees and raise the arms behind the head.

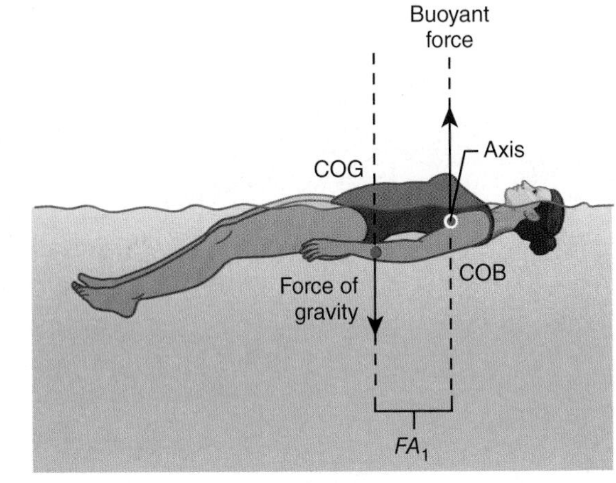

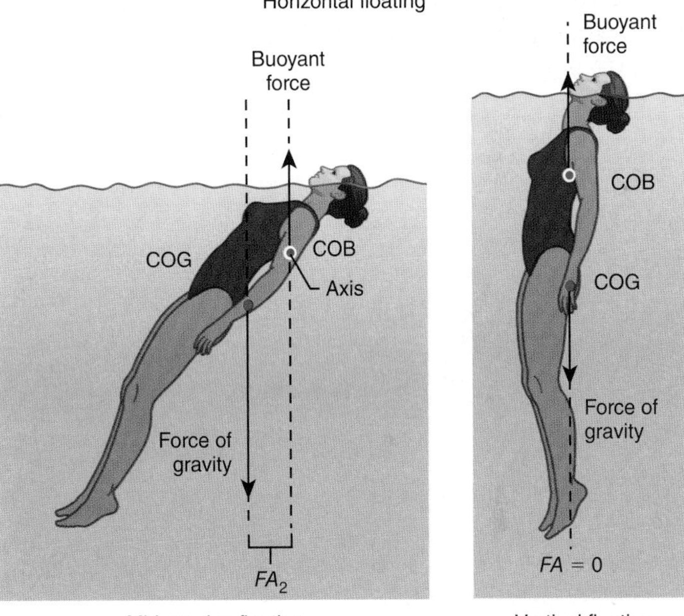

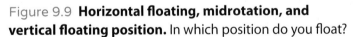

Figure 9.9 **Horizontal floating, midrotation, and vertical floating position.** In which position do you float?

A water survival technique called *drownproofing* (Lanoue, 1963) takes advantage of the natural buoyancy of the body, using proper positioning of the body and its segments to help people stay afloat for hours. Most people are buoyant to some degree simply because air in the lungs creates an area of low density (i.e., low mass in the chest cavity in comparison to its volume). However, their bodies may not provide enough buoyancy to float with the head completely out of the water. The drownproofing technique involves floating in an upright posture (rather than on the back),with the back of the head just breaking the surface of the water (Figure 9.10). The elbows are bent and the arms are allowed to float in a position in which the hands are in front of the shoulders. When another breath of air is needed, the person brings the face out of the water only far enough to inhale. Any farther out of the water, and the person will sink too far as they sink back into position.

Notice the interaction of various concepts. There is a relationship between torque and the center of gravity. There is also a relationship between the concepts of center of gravity and center of buoyancy as well as to torque. Buoyancy is related to Archimedes' principle. And ultimate floating position (position of equilibrium in the water) is related to all of these concepts. As you will see, the interactions do not stop here.

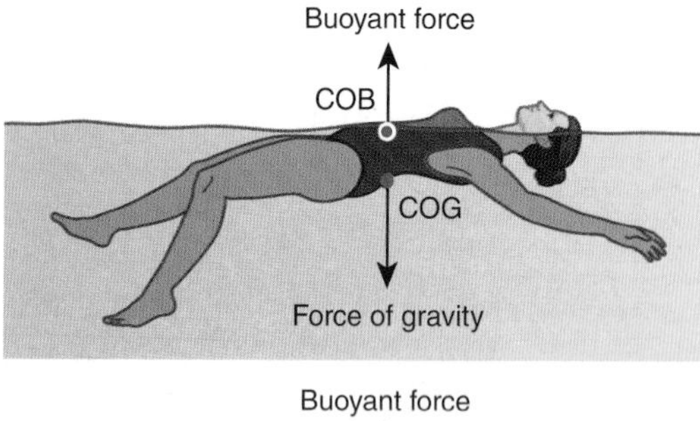

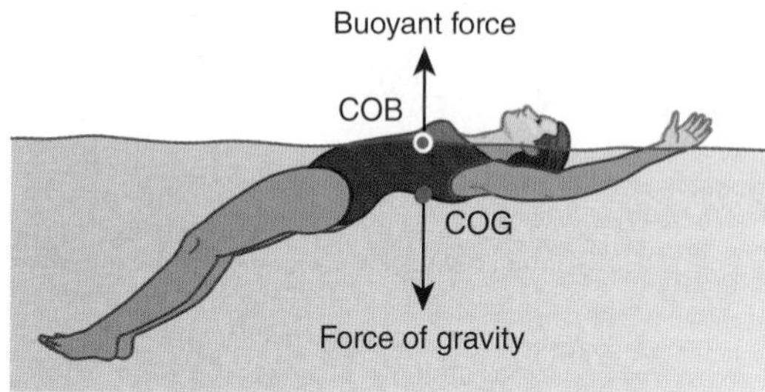

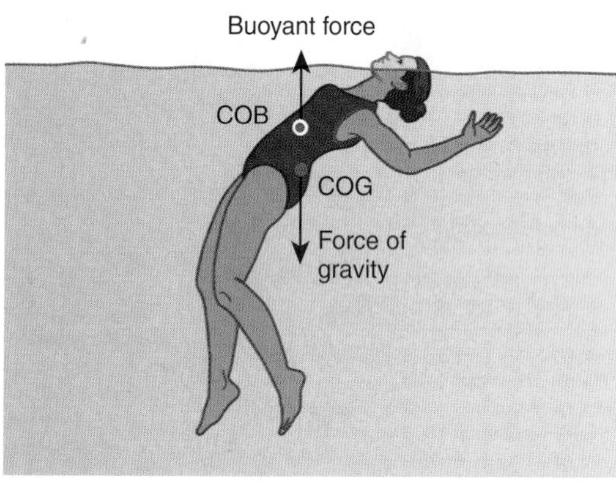

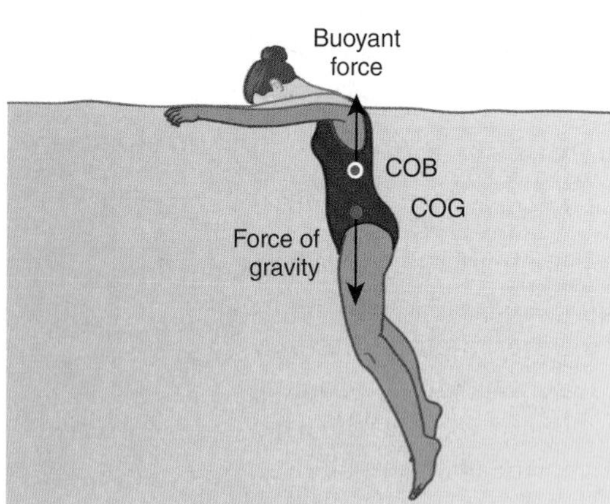

Figure 9.10 **Floating position can change because of the repositioning of segments.** Which position would be best to prevent drowning?

9.2 DYNAMIC FLUID FORCE

As mentioned in the previous section, buoyant force acts on any object that is submerged. Recall that **dynamic fluid force** acts on a system that is moving through a fluid. When moving through a fluid, the system (whether an airplane, a bird, a car, a dolphin, or a human) causes particles of the fluid to be deflected so the system can move forward. In other words, the system applies a force to the fluid particles that moves them along the outside of the system. If the system is tilted (such as the wing of an airplane or bird), the deflection is even greater.

According to Newton's third law of motion, an equal and oppositely directed force must be applied to the system by the fluid particles. Dynamic fluid force is the equal and oppositely directed force of the fluid particles in reaction to the applied force of the system moving through the fluid. Similar to other forces, the dynamic fluid force vector can be resolved into two components: a parallel component called *drag* and a perpendicular component called *lift*.

To fully elucidate these concepts, recall the idea of relative motion and introduce the concept of viscosity.

Relative Motion

Chapter 2 defined **relative motion** as the motion of one object with respect to a reference object. Recall the example of the two hockey players skating next to each other at 5 kilometers per hour (km/hr). They are both moving at 5 km/hr relative to the rink, but neither is moving relative to the other (0 km/hr). If another hockey player skates past them in the same direction at 15 km/hr, this third player is moving at 15 km/hr relative to the rink and 10 km/hr relative to the other two players (the numbers are subtracted if both objects are moving in the same direction).

On the other hand, one player could be skating at 15 km/hr toward another player moving at 10 km/hr. In this case, their relative motion or **closing velocity**, which is the sum of the absolute values of the players' velocities, is 25 km/hr. To truly understand the concept, think of the *appearance* of the situation. If you are one of the hockey players and maintain eye contact with the player skating next to you at the same velocity, he will not *appear* to be moving.

Another example would be driving in a car at the same velocity as the car next to you. If both cars are traveling 50 km/hr, one can glance over and see that the other driver is neither passing nor falling back (relative velocity = 0 km/hr). If the other driver then speeds up to 55 km/hr, she can be observed to be passing, but not at a rapid rate. In this case, the relative velocity is only 5 km/hr. If another car is traveling in the opposite direction at 50 km/hr, it will appear that it is moving more rapidly because the relative velocity is 100 km/hr!

The concept of relative motion helps us understand why a ceiling fan appears to be moving more slowly if a person visually focuses on one blade and follows it around its path. Because the eyes are tracking the blade at the same velocity that it rotates, the relative motion is less and rotation appears to be slower. However, if a person observes the fan without following the blades with the eyes, the relative motion is greater and the fan appears to be moving faster.

If you imagine swimming *against* rather than *with* the current of a river, you can begin to understand that the concept of relative motion is important in reference to dynamic fluid force.

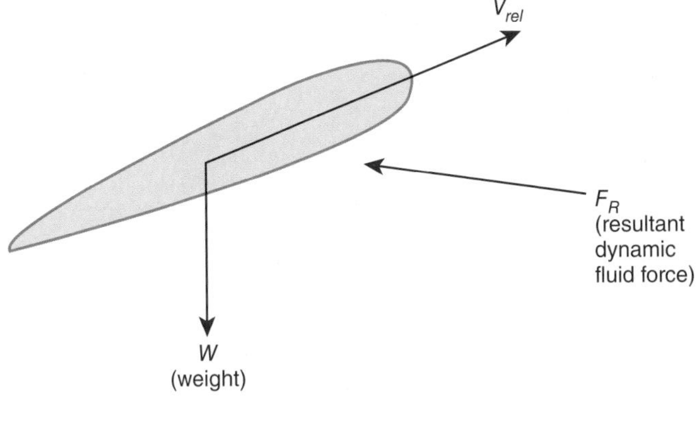

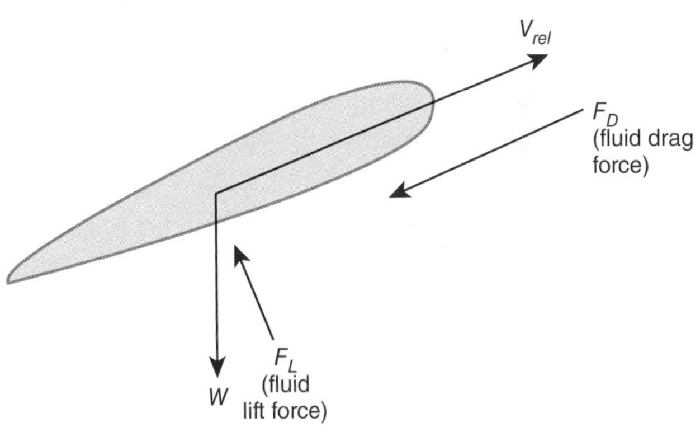

Figure 9.11 **The drag component of the dynamic fluid force vector.**

Viscosity

Another important concept related to fluids is viscosity. **Viscosity** is a measure of the resistance of two adjacent layers of fluid molecules moving relative to each other. So viscosity is actually a specific type of friction and is often defined as *internal friction of a fluid*, or *viscous force*. A more viscous fluid flows more slowly, and a less viscous fluid flows more quickly. Imagine the flow velocity of water poured from a glass compared with that of molasses poured from the same glass. In terms of energy transfer, viscosity causes some of the kinetic energy of a flowing fluid to be converted into internal energy (Serway & Jewett, 2019). So although the concept of viscosity refers specifically to the flow of fluids, it has ramifications for a system moving through the fluid. Similar to the difficulty encountered while attempting to slide on a rough surface, more *work* is required to move through a viscous fluid than to move through a less viscous fluid. More work is required because *the work performed by externally applied forces other than gravity changes the energy of the object acted upon*. In this case, viscous force performs work that transfers kinetic energy to internal energy. As a consequence, the system must perform more work to maintain its kinetic energy. Imagine swimming in molasses instead of water. It's an extreme example, but you get the point. In terms of human motion, the fluids involved are usually air and water. Air and water are both fluids and follow fluid mechanical principles. However, air and water have different densities and viscosities. This distinction leads us to our first major dynamic fluid force concept: drag.

Drag

You read in Chapter 4 that the parallel component of dynamic fluid force acts in the opposite direction of system motion with respect to the fluid and is called **drag force** (Figure 9.11).

Drag force tends to resist motion of the system through the fluid (in other words, drag force decelerates the object) and is therefore also called *fluid resistance*. The force of drag can easily be experienced by holding your hand out the window while riding in a car. If your palm is forward, you will experience the force of drag attempting to push your hand in the direction opposite that of the motion of the car.

Dynamic fluid force The equal and oppositely directed force of the fluid particles in reaction to the applied force of the system moving through the fluid.

Relative motion The motion of one object with respect to a reference object.

Closing velocity The sum of the absolute values of velocities.

Viscosity A measure of the resistance of two adjacent layers of fluid molecules moving relative to each other.

Drag force The parallel component of dynamic fluid force that acts in the opposite direction of system motion with respect to the fluid; tends to resist motion of the system through the fluid.

As you will see, many factors affect the force caused by drag. The mathematical relationship of the factors affecting drag force is expressed as follows:

$$F_D = ½C_D\rho A_p v^2 \quad (9.2)$$

where

F_D = drag force
C_D = coefficient of drag
ρ = "rho" is the density of the fluid through which the object is moving
A_p = projected area of the object (area of the object oriented perpendicular to fluid flow)
v^2 = relative velocity of the object with respect to fluid flow

Drag is a force and therefore is measured in Newtons.

SAMPLE PROBLEM

Let's use a parachute as an example. We will say that an open parachute has a coefficient of drag equal to 1.2, the density of air is 1.225 kg/m³, the projected area of the parachute is 16 m², and the velocity of the falling skydiver is 5 m/sec after the parachute is open. Drag force is calculated as follows: $F_D = (0.5 \times 1.2) \times 1.225\ kg/m^3 \times 16\ m^2 \times (5\ m/sec)^2 = 294\ kg \cdot m/sec^2 = 294\ N$.

The **coefficient of drag** is a unitless quantity that expresses the capability of a particular object to create drag. The quantity is laboratory calculated and depends on the physical characteristics of the object (e.g., its shape and texture), as well as the object's presentation (orientation) relative to the fluid flow. An object with a high coefficient of drag has a greater capability to generate drag force (Table 9.2). For example, a spherical body moving through air has a coefficient of drag of approximately 0.50, and the Earth-facing side of an open parachute has a coefficient of drag of approximately 1.20.

The effect of fluid density is a little more obvious. The denser the fluid, the greater the drag force. For example, drag force is greater in water than in air. If a greater cross-sectional area is presented perpendicular to the fluid flow, drag increases proportionally. Recall the example of holding your hand out the window while riding in a car. You will notice more drag if the palm is facing forward relative to the flow than if the palm is

Coefficient of drag A unitless quantity that expresses the capability of a particular object to create drag.

Table 9.2
Drag Coefficients of Various Objects

Object	C_D	Object	C_D
Human (upright position)	1.0–1.3	Penguin	0.0044
Human running	0.5	Falcon	0.24
Racing cyclist	0.4	Aircraft	0.02–0.09
Ski jumper	1.2–1.3	Smooth sphere	0.1–0.5
Skier	1.0–1.1	Golf ball	0.25–0.3
Parachutist	1.0–1.4	Car	0.35–0.5
Human swimming	0.035	Sports car	0.3–0.4
Trout	0.015	Motorcycle	0.5–1.00
Dolphin	0.0036	Truck	0.6–1.00
Seal	0.004	Tractor-trailer	0.6–1.20

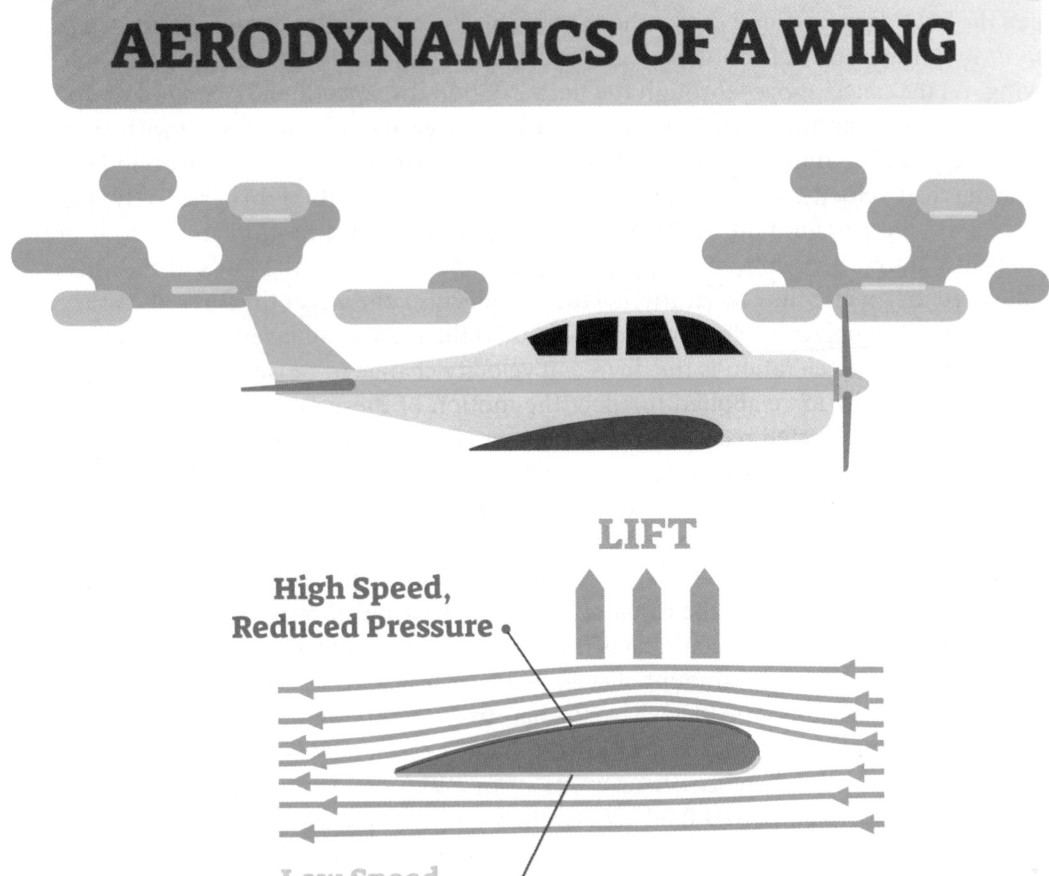

Figure 9.12 **Laminar flow.** Notice that the flow lines do not cross.

down. This change in orientation affects the coefficient of drag because the shape presentation has changed; with the palm forward, however, more area is also facing the flow direction. Finally, notice that relative flow velocity has an exponential effect on generating drag. So, if velocity is doubled, drag force is quadrupled. Also, don't forget that this is *relative* velocity, as discussed earlier. Relative velocity and therefore drag of an object moving at 10 m/sec is greater when it is moving through an oncoming fluid than when it is moving at the same rate through a fluid that is flowing in the same direction.

Let's turn our attention to the two sources of drag force: *surface drag* and *form drag*.

Surface Drag

As a system moves through a fluid, the layer of fluid in immediate contact with the system is called the **boundary layer**. If the molecules of fluid in the boundary layer follow the same smooth, unbroken path completely around the object, the fluid flow is called **laminar flow** (Figure 9.12).

The path followed by the molecules of a layer of fluid is called a **streamline**. In laminar flow, the streamlines do not cross each other; therefore, laminar flow is

Boundary layer The layer of fluid in immediate contact with the system in motion.

Laminar flow Fluid flow in which the molecules of fluid in the boundary layer follow the same smooth unbroken path completely around the object.

Streamline The path followed by the molecules of a layer of fluid

often called *streamlined flow.* Even during this smooth laminar flow, drag exists because of friction between the boundary layer and the surface of the object. **Surface drag** (also called *skin friction* or *viscous drag*) is the result of friction between the surface of the body and the fluid through which it is moving. As the object moves through the fluid, friction disrupts the flow of fluid molecules and slows the velocity of the boundary layer. The boundary layer of fluid is in contact with another layer of fluid molecules. Friction disrupts the flow of fluid molecules adjacent to the boundary layer, in turn slowing their velocity. This disruption of flow (change in velocity) is progressively smaller with each successive layer of fluid molecules until ultimately no relative motion (friction) is present in the outermost layers. Remember that F = ma; therefore, the greater the mass of the affected fluid molecules and the greater the acceleration of those molecules, the larger the force necessary to affect the motion of those molecules. In our case, the mass is the number of layers of fluid molecules that are affected, and the acceleration is the degree of velocity change in those molecules. According to Newton's third law, the force applied to affect the motion of these molecules will be met with an equal and oppositely directed reaction force. This reaction force is the surface drag force.

Surface drag is, of course, affected by the coefficient of drag, density of the fluid, projected area, and relative velocity (Equation 9.2). Recall that the coefficient of drag is affected by the object's physical characteristics and its presentation to the fluid. Shape and presentation are fairly obvious issues. In surface drag, the texture of an object's surface is an important determinant in the coefficient of drag. Because surface drag is essentially the result of friction, any change in surface texture has an effect on drag. An object with a rough surface creates friction between the object's surface and the boundary layer and therefore increases the amount of surface drag. This fact is one reason why aquatic athletes wear supersmooth swimsuits and often shave the hair from their bodies.

In addition, remember that *viscosity* is a measure of the resistance of two adjacent layers of fluid molecules to moving relative to each other. Therefore, more force is required to change the velocity of adjacent layers of molecules in a more viscous fluid because the layers are more resistant to relative motion. Once again, imagine swimming through molasses instead of water. Because the viscosity (internal friction) of a fluid affects the force necessary to move through it, surface drag is often called *viscous drag.* Less obvious is that the viscosity of the fluid affects the coefficient of drag. If the fluid is viscous, then the boundary layer actually clings to the surface of the object, changing its shape. This change in shape because of a "sticky" boundary layer of fluid changes the coefficient of drag.

Surface drag is the primary source of drag during situations of laminar flow, which tend to occur at slow relative flow velocities. However, at high relative flow velocities, an additional source of drag called *form drag* predominates.

Form Drag

You have read that laminar flow exists when the molecules of fluid in the boundary layer follow the same smooth, unbroken path completely around the object, and the streamlines do not cross each other. In situations of laminar flow, surface drag is the major source of drag. However, if an object has a high relative flow velocity, abrupt changes in surface features, or both, flow of the boundary layer of fluid may not remain smooth. This situation of irregular fluid flow in which the streamlines may be broken is referred to as **turbulent flow**. This turbulent flow is a second source of drag called **form drag**, sometimes called *eddy drag, pressure drag,* or *shape drag.* In the case of turbulent flow, the molecules of the boundary layer do not follow the same smooth, unbroken path completely around

Surface drag The result of friction between the surface of the body and the fluid through which it is moving.

Turbulent flow A situation of irregular fluid flow in which the streamlines may be broken; the molecules of the boundary layer do not follow the same smooth, unbroken path completely around the object.

Form drag Drag due to a relatively higher pressure on the leading edge of the object compared to its trailing edge.

the object (Figure 9.13). More specifically, the trailing edge of the object is left without contact with the boundary layer.

Recall that the pressure exerted by a fluid is equal in all directions, so fluid particles apply a force to the leading edge of an object that is opposite in direction to that of the object's motion. Particles on the trailing edge apply a force that is in the same direction as object motion. If no fluid molecules contact the trailing surface, less pressure is exerted against the object in this area. Therefore, pressure on the leading edge of the object is higher than the pressure on its trailing edge. This pressure gradient (pressure differential) creates a suction force (or vacuum) that is directed opposite that of the direction of flow. This suction force is the cause of form drag. Because it is caused by a pressure differential, it is also called *pressure drag*. The low pressure zone behind the trailing edge of the object creates a region of turbulence in which the streamlines do not remain parallel but cross each other. One can observe this area of turbulence by noticing eddies in the wake of a boat.

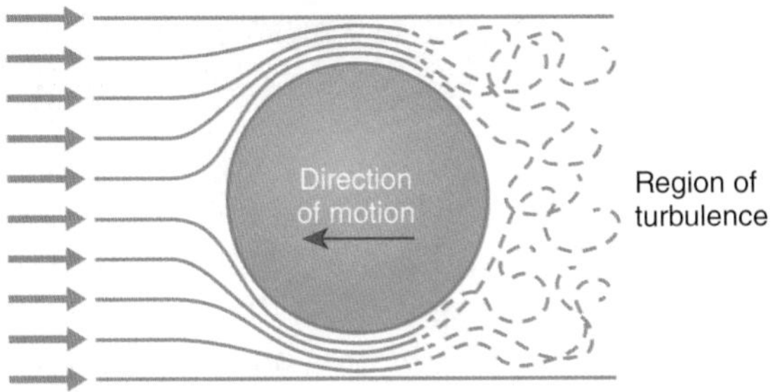

Figure 9.13 **Turbulent flow.** What is the cause of the turbulence?

You may have noticed just how strong this suction force is if you have ever accidentally gotten your car too near the trailing edge of a large truck. You may have felt that your car was being pulled toward the truck—and it was! Your car was being pulled into the low pressure zone created by the form drag of the truck.

To gain a deeper understanding, picture a raindrop. Raindrops are not round as they fall. The bottom edge remains essentially round, but while descending, the leading edge pushes air molecules aside, which creates turbulence on the trailing edge. The net effect is that when the laminar flow is disrupted, a vacuum is created that is strong enough to draw some of the water backward, which creates the shape we know as the "teardrop." The round leading edge and tapering trailing edge combine to maximize laminar flow and minimize disturbance of the fluid through which the drop is traveling (air). Nature provides an excellent variety of examples that speak to the effectiveness of this shape in maximizing velocity through a fluid. Most birds have bodies that are pointed or rounded near the head and taper toward the tail. Fish, whales, and other mammals that travel through water also have bodies that are designed the same way: they taper toward the tail. Is this a coincidence? Not at all—this shape moves efficiently through fluids and therefore is a shape that engineers strive to achieve to maximize fluid dynamics in sport.

Form drag, like surface drag, is affected by the coefficient of drag, density of the fluid, projected area, and relative velocity (Equation 9.2). In the case of form drag, the coefficient of drag is greatly affected by the profile or shape of the object (thus, the terms *profile drag* and *shape drag*). More specifically, form drag will be increased greatly if the object has a shape with contours that vary widely or abruptly from front to back. These types of contours are difficult for the streamlines to follow, leading boundary layers to separate and create a low pressure zone behind the moving object. Think about the trailing edge of the truck in our previous example. The truck is shaped basically like a box, and therefore its trailing edge is flat. So an abrupt change (90°) occurs in contour as the streamlines flow around the sides and trailing edge.

This abrupt change creates a large low pressure zone behind large trucks. One way to reduce form drag is **streamlining** the object so that changes in contour are gradual from front to back. This shape allows the molecules in the streamlines to follow the same smooth, unbroken path completely around the object, decreasing the size of the low pressure zone and reducing form drag.

Streamlining Designing the shape of an object such that changes in contour are gradual from front to back, which allows the molecules in the streamlines to the follow the same smooth, unbroken path, thereby reducing drag.

FOCUS ON RESEARCH

With more exposure comes more participation in sports for people with disabilities. The Paralympic games are now getting worldwide coverage on television and online, and the level of competition is rising as more participants learn about and train for high-level sports. As the competitiveness increases, more records are falling and more research is being conducted to find small gains in performance that might make the difference between a podium finish and fourth place.

Tandem racing (two riders on one bicycle) for competitors with visual impairments is a fairly recent development in Paralympic cycling and takes advantage of the same aerodynamic advantages that solo racers use. However, tandem riding has unique airflow characteristics that solo riders do not share. In Paralympic tandem racing, a visually impaired rider is seated in the rear position and called the *stoker*. The front rider is sighted and called the *pilot*.

Research on bicycle racing has been ongoing for decades, but little has been conducted on tandem riding. With 70% to 90% of cycling power working to overcome aerodynamic drag, research on solo riding has focused on ways to reduce it. From the bicycle to the rider, every aspect has been studied to reduce the amount of energy it takes to move through air. Bicycle frames and wheels are streamlined, cables are routed inside the frame, riders wear skin-tight racing clothing, and have specially designed helmets. Even the rubber tires are constructed to minimize drag.

The study of body position effect on aerodynamics for solo riding is extensive, but until recently there was little related to the position of tandem riders. Mannion and colleagues (2018) set out to close this gap by studying drag of tandem cyclists in different riding positions. Two body positions most prominent in bicycle racing are *crouched* (also called *dropped*) and time trial (TT). The dropped position is used for road racing and refers to using the lowest section of a set of conventional curved handlebars. The TT position is considered the fastest because its use of a specialized handlebar allows a rider to rest her forearms on pads that are inside the plane of the torso, with the forearms parallel to the ground.

Mannion's research team used a wind tunnel and computational fluid dynamics to test the drag on tandem riders in three different positions: upright, dropped, and TT. As expected, both dropped and TT positions were significantly more aerodynamic than the upright position. Because of the arm position of the TT setup there was less drag than in the dropped position, and a further reduction was noted for a modified position known as the frame clench (FC) in which the stoker grasps the frame near the center of the bicycle rather than the handlebars.

It was found that both the stoker and the pilot experienced slightly lower drag in the FC position. Although the results may seem insignificant between the two TT positions, it is estimated that the frame clench position would translate to an eight-second gain over the course of a 10-k time trial. To put this in perspective, over a 30-k TT course in the 2016 Paralympics, second through fourth places in the tandem women's race were separated by only 12 seconds. As the level of competition in Paralympic events progresses, athletes will strive to find even the smallest of advantages.

Because streamlining requires a gradual change in shape, the trailing end of the object is often extended into the low pressure zone (Figure 9.14).

If the low pressure zone is filled, then the pressure gradient is reduced. These streamlined shapes are often observed in situations in which air resistance is a major issue. For example, the helmets worn by competitive cyclists are tapered to reduce form drag (Figure 9.15).

The coefficient of drag can also be affected by the texture of the object surface. A rough surface can cause the boundary layer to separate from the object, leading to turbulence. Therefore, in most cases, a smooth surface helps to decrease form drag.

Of course, relative flow velocity has the largest effect on form drag. If the relative flow velocity is large, the fluid molecules cannot apply force to the trailing edge of the object before it has passed. If they move at different velocities, two objects of the same shape can experience different amounts of drag. Don't forget that relative velocity is a squared factor in the drag force equation, so it has the largest effect. One can see the combined affects of object shape and relative velocity while observing a slow moving canoe and a speed boat moving at a high velocity (Figure 9.16).

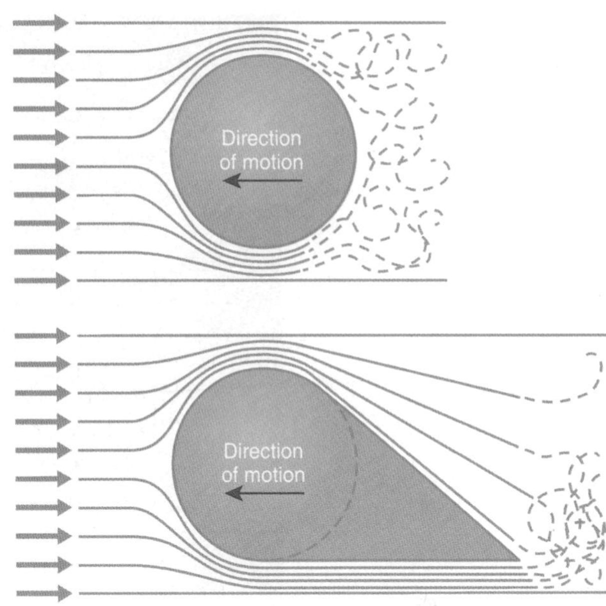

Figure 9.14 **A streamlined versus a nonstreamlined object.**

The canoe has a streamlined shape and moves at a slower relative velocity, therefore creating only a small amount of turbulence. The speedboat has a flat trailing edge and moves at a higher velocity, so it creates a greater amount of turbulence because of its large lower pressure zone.

The projected area of the object can also have an effect on form drag. The greater the area of the object facing the oncoming fluid, the larger will be the number of particles that come in contact with the leading edge. Remember that these particles apply an oppositely directed force. This force is the reason that the leading edge, as well as the trailing edges, of a streamlined object is often tapered.

Effects of Drag on Falling Objects

If the acceleration caused by gravity is 9.81 m/sec^2, why do all objects not fall at the same rate? The answer is drag force. You read in Chapter 7 that a system in *dynamic equilibrium* is in motion, but it is experiencing no change in velocity or direction. Drag is the source of the dynamic equilibrium of falling objects. An object that is falling experiences a downward vertical force from gravity that is equal to its weight. This gravitational force will initially cause the object to be accelerated at a rate of 9.81 m/sec^2. However, as the velocity of the object increases, so does the drag force acting in the opposite direction. Drag force will continue to increase and eventually equal the weight of the object. Acceleration requires a net force. When drag force is equal to the weight of the falling object, a net force no longer exists (i.e., the negative vertical force supplied by gravity is balanced by an equal and oppositely directed force caused by drag). At this point, a state of dynamic equilibrium is attained, and the object will no longer accelerate. Of course

Figure 9.15 **Aerodynamic helmet.** What is the advantage of this helmet?

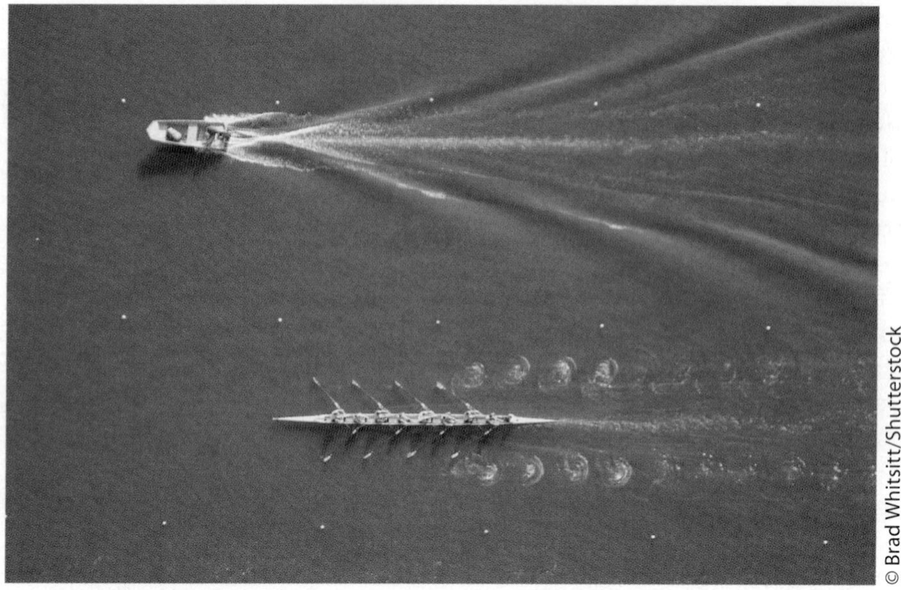

Figure 9.16 **A racing scull and a motor boat.** Which is designed for less drag?

Figure 9.17 **A skydiver.** Can he still fall faster?

this does *not* mean that the object stops falling, it simply falls with a constant velocity (its **terminal velocity**; Figure 9.17).

Acceleration (in this case, deceleration) is proportional to the applied net force and inversely proportional to the mass of the object. Therefore, two objects that possess the same shape and volume but have different masses will have different terminal velocities. The objects will experience the same drag force, but the more massive object will have a greater resistance to being decelerated. Therefore, the drag force has to be applied to the more massive object for a longer time before terminal velocity is achieved. This example refers to objects of the same shape and volume. But remember that drag force is also affected by object shape and projected area. Therefore, even though gravity

Terminal velocity Velocity at which drag force is equal to the weight of the falling object, and the object will no longer accelerate.

Table 9.3
Terminal Velocities of Various Objects

Object	Mass (kg)	Cross-Sectioned Area (m²)	VT(m/s)
Sky diver	75	0.70	60
Baseball (radius 3.7 cm)	0.145	4.2×10^{-3}	43
Golf ball (radius 2.1 cm)	0.046	1.4×10^{-3}	44
Hailstone (radius 0.50 cm)	4.8×10^{-4}	7.9×10^{-5}	14
Raindrop (radius 0.20 cm)	3.4×10^{-5}	1.3×10^{-5}	9.0

SOURCE: Serway, R. A., and J. W. Jewett Jr. 2019. *Physics for Scientists and Engineers*, 6th ed. Belmont, CA: Brooks/Cole-Thomson Learning.

can be considered constant, drag force is highly variable, so objects attain terminal velocity depending on their individual characteristics (Table 9.3).

For example, a skydiver falling at terminal velocity experiences a change in drag force when her parachute opens. This event causes a brief net force directed opposite to that of gravity, which causes a large deceleration. A new, much smaller terminal velocity will be reached once a state of dynamic equilibrium is again reached (i.e., when a net force no longer exists).

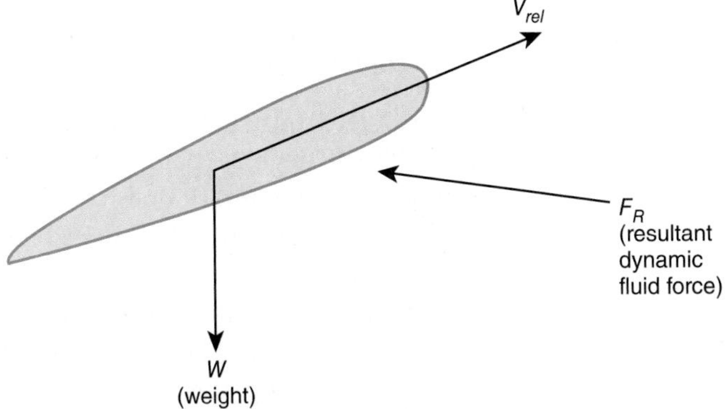

Lift

You have read that drag force is the parallel component of dynamic fluid force and acts in the opposite direction of system motion with respect to the fluid. The perpendicular component of dynamic fluid force can act in any direction that is perpendicular to system motion with respect to the fluid, even though it is termed **lift force** (which implies upward motion). Therefore, lift tends to change the direction of system motion (Figure 9.18). For example, air particles tend to be deflected downward if a wing is tilted with the leading edge up. This effect means that the reaction force applied by the particles to the wing will cause upward motion (Anderson & Eberhardt, 2010; Craig, 2002; Weltner, 1987). You can also experience this effect with your hand as it moves through oncoming air. Simply change the position of your hand from completely palm forward to a slanted position, with the leading edge up. You will experience lift force (along with the drag force) that causes the hand to rise.

So far, you have encountered lift only in general terms. Now we must gain a deeper understanding because lift applies to more situations than may be readily apparent. For instance, when we think of lift, we often think of airplanes and birds. But lift force is also experienced by

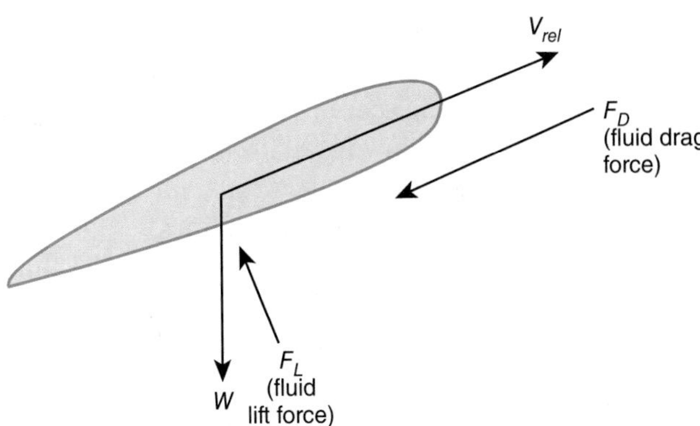

Figure 9.18 **The lift component of dynamic fluid force vector.**

Lift force The perpendicular component of dynamic fluid force that can act in any direction that is perpendicular to system motion with respect to the fluid; tends to change the direction of system motion through the fluid.

spinning objects. So in the realm of sports, lift applies to any situation in which an object is in flight or is rotating (or both). First let's examine lift in terms of the most obvious question: How do airplanes fly?

Lift and Newton's Laws of Motion

A complete understanding of airplane design is not necessary to gain insight into object flight. One must simply have an understanding of Newton's laws of motion and a little imagination (Anderson & Eberhardt, 2010). First, let's review the laws of motion:

1. *Law of Inertia:* Every body perseveres in its state of rest, or of uniform motion in a straight line, unless it is compelled to change that state by forces impressed thereon.
2. *Fundamental Law of Dynamics or Law of Acceleration:* The alteration of motion is ever proportional to the motive force impressed; and is made in the direction of the right line in which that force is impressed.
3. *Law of Reciprocal Actions or Law of Action-Reaction:* To every action there is always opposed an equal reaction; or the mutual actions of two bodies upon each other are always equal, and directed to contrary parts.

Next, let's construct a visual image of exactly what happens as a wing travels through air. Tilt your hand 30 to 40 degrees (palm-side down) with the thumb as the leading edge. Now imagine scraping your hand across sand (it doesn't have to be sand; it could be water, snow, etc.). Notice that sand collects under your hand and is pushed forward.

Also notice that the hand moves along, it leaves a trench behind it. A wing does essentially the same thing that your hand is doing. Fluid particles are piled up in front of and below the wing. And in the case of a wing (unlike your hand in sand), some fluid particles are also deflected downward. Let's start with the pileup; this concentration of fluid particles causes an area of high pressure underneath and in front of the wing. Now let's move to the trench; fluid molecules have essentially been scooped out of the area, causing an area of low pressure. Notice that a pressure differential is the basic source of drag and lift. So what does that have to do with Newton's laws of motion?

First, remember that the underside of the wing deflects some fluid particles vertically downward. But in addition, notice that we have a low pressure zone on top of the wing. Atmospheric air above the "trench" is pulled downward by the pressure differential. It's really very simple: air must be pulled into the space ("trench") behind the wing, or airplanes would leave voids in the atmosphere. Pulling the air downward lowers the pressure above the wing (notice the similarity to form drag). But the key to lift is that *wings cause air to be moved vertically downward.* In other words, wings bend or deflect the paths of the streamlines.

In terms of Newton's laws, the fluid particles have been accelerated. According to Newton's first law, no acceleration occurs without an applied force. Newton's second law ($F = ma$) helps us understand that the greater the total amount of fluid particles accelerated (mass) and the greater the deflection of those particles (acceleration), the greater the applied force must be. And finally, according to Newton's third law, there will be a force equal in magnitude and opposite in direction to this applied force. This equal and opposite force is the lift force (i.e., the "action" is the wing's deflection of fluid particles, and the "reaction" is the lift force). You can easily experience the deflection of air particles by wings at home by simply standing under a ceiling fan. First, notice that the blades are tilted and pay attention to the direction of rotation. If the fan is rotating so that the leading edge is tilted upward, the fan will divert air downward toward you. If the direction of rotation is changed so that the leading edge is downward, the fan will divert air toward the ceiling. So your ceiling fan is creating lift. Once again, notice that lift doesn't have to be upward. Lift is simply perpendicular to the direction of object motion through the fluid.

Now that we understand the source of lift, the factors that affect it will be easier to understand. Because lift is a component of the same dynamic fluid force as drag, it is affected by the same factors. The mathematical relationship of the factors affecting lift force is expressed as follows:

$$F_L = ½C_L\rho A_p v^2 \qquad (9.3)$$

where

F_L = lift force
C_L = coefficient of lift
ρ = "rho" is the density of the fluid through which the object is moving
A_p = projected area of the object (area of the object oriented perpendicular to fluid flow)
v^2 = relative velocity of the object with respect to fluid flow

Like drag, lift is a force and therefore is measured in Newtons.

SAMPLE PROBLEM

Let's use an airplane as an example. We will say that the airplane at a given angle of attack has a coefficient of lift equal to 0.65, the density of air is 1.225 kg/m^3, the projected area of the airplane is 15 m^2, and the velocity of flight is 65 m/sec. Lift force is calculated as follows: $F_L = (0.5 \times 0.65) \times 1.225\ \text{kg/m}^3 \times 15\ \text{m}^2 \times (65\ \text{m/sec})^2 = 25{,}231.17\ \text{kg} \cdot \text{m/sec}^2 = 25{,}231.17\ \text{N}$.

The **coefficient of lift** is a unitless quantity that expresses the capability of a particular object to create lift. The quantity is laboratory calculated and depends on the physical characteristics of the object (e.g., shape and texture), as well as the object's presentation (orientation) relative to the fluid flow. An object with a high coefficient of lift has a greater capability to generate lift force. For example, it is obvious that an airplane wing would have a higher coefficient of lift than a sphere. But don't get the idea that a wing is the only shape that can generate lift. The Wright brothers got off the ground, and toy planes made of balsa wood or paper will fly. A **foil** is any object that can generate lift while moving through a fluid. The key component to deflection is actually the presentation (orientation) of the object relative to the fluid flow, which is called the **angle of attack**. Remember that the ceiling fan can deflect air downward or upward, depending on its angle of attack. The generation of lift force increases with attack angle until the angle is too steep for fluid particles to flow around. But remember that lift and drag are components of the same force, so changing one will change the other. So if the angle of attack is increased in an attempt to maximize lift, be aware that the area exposed to the oncoming fluid flow increases as well, creating greater form drag (Figure 9.19).

So the optimal angle of attack not only maximizes lift but also minimizes drag. This relationship is expressed in the **lift–drag ratio**. At angles of attack that are too steep, drag can become greater than lift, and flight is no longer possible. This situation is sometimes referred to as the **stall angle**.

The effect of fluid density may not be as obvious. The more dense the fluid, the greater the lift force. Once again, lift is generated by deflecting (pulling) fluid particles downward. The more particles accelerated, the greater

Coefficient of lift A unitless quantity that expresses the capability of a particular object to create lift.

Foil Any object that can generate lift while moving through a fluid.

Angle of attack The presentation (orientation) of the object relative to the fluid flow.

Lift–drag ratio Ratio that expresses the relationship of the amount of lift to the amount of drag at a given angle of attack.

Stall angle The angle of attack at which drag can become greater than lift and flight is no longer possible.

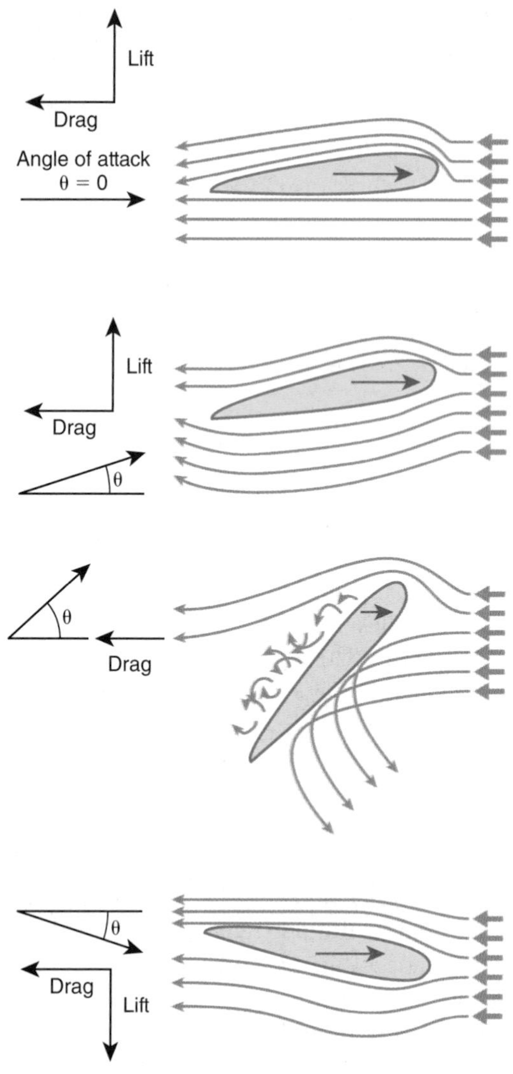

Figure 9.19 **Drag and lift at different angles of attack.** How does changing the angle of attack change the magnitude of the drag and lift components?

the lift. For example, air is less dense at higher altitudes. So airplanes must adjust the angle of attack to compensate for fewer particles of air being deflected at a given velocity. If a greater cross-sectional area is presented perpendicular to the fluid flow, lift increases proportionally. Again, this increase occurs because of the amount of fluid deflected. The greater the area presented to the oncoming fluid, the greater the amount of fluid molecules that will be deflected. Finally, notice that relative flow velocity has an exponential effect on generation of lift. So, if velocity is doubled, then lift force is quadrupled. Again, don't forget that this is relative velocity, as discussed previously. For example, lift can be generated more easily by taking off into the wind because of increased relative velocity. As the relative velocity increases, the amount of fluid diverted increases proportionally.

If you are ever confused about how lift is generated, remember that foils deflect (pull or accelerate) fluid particles. This deflection is met with an equal and oppositely directed lift force. Also, don't assume that lift is always upward. For example, spoilers on race cars produce downward lift that pushes the tires into the racetrack (Figure 9.20).

This action increases the friction between the tires and the track, helping to prevent sliding or even flight of the car itself. A lesser known aspect of lift is that it can be caused by rotation.

Lift Caused by Rotation

As mentioned previously, lift is usually associated with planes and birds. However, lift can be generated by less obvious objects if they are rotating. For example, a baseball doesn't have the appearance of an object that could produce a large amount of lift. But if you throw the ball with spin, it will curve or "break." Also, we can kick a soccer ball off-center, and it will curve (a "banana kick"). In these situations we use the word *curve*, but what is occurring is lift. Remember that lift can act in any direction perpendicular to object motion through the fluid. This lift force caused by rotation is called **Magnus force**, and the phenomenon is referred to as the **Magnus effect**. To understand the Magnus effect, we must first become familiar with **Bernoulli's principle**, which states that a fluid moving with high relative flow velocity exerts less pressure than a fluid moving with a low relative flow velocity. At first glance, this concept may seem difficult to understand, but it is actually relatively simple. Imagine yourself and several other people as particles of a fluid. You are all in the same hallway, but not necessarily facing the same directions. Now imagine that I give the cue for everyone to begin running, but in no particular direction. No doubt there will be many collisions between you, the other people, and the walls. Also, the resultant velocity of the total fluid will probably be tiny. If the fluid pressure is represented by the collisions, you can see that a lot of pressure (collisions) occurs at a lower velocity. Now imagine that we have all of the people facing the same direction and give the cue to run as fast as possible. Because everyone is now moving in the same direction, collisions will be fewer and impact will be less intense when they do occur. Also, notice that the resultant velocity of the total fluid will be high. Therefore, faster-moving fluids exert lower lateral pressures. You will notice the

Magnus force Lift force due to rotation.

Magnus effect Phenomenon of a rotating object experiencing a curved path because of the Magnus force.

Bernoulli's principle A fluid moving with high relative flow velocity exerts less pressure than a fluid moving with a low relative flow velocity.

same phenomenon the next time your shower curtain billows in at you. The fluid on one side of the curtain is moving with a relatively higher velocity (creating low pressure) than the fluid on the other side (which creates high pressure). This pressure gradient pushes the curtain toward you.

We can use Bernoulli's principle as it applies to rotating objects to understand the Magnus effect. We will use the example of a curveball in baseball, but you can apply the same explanation to any situation in which a rotating object curves (e.g., a golf ball or soccer ball). First notice that when an object rotates (whether with backspin, sidespin, or topspin) as it moves through a fluid, one side of the object rotates in the opposite direction of the streamlines of the oncoming fluid. The other side of the object is rotating in the same direction as the streamlines of the fluid (Figure 9.21).

Figure 9.20 **Formula 1 race car.** Why does it have so many wings?

As fluid particles in the streamlines are met with the surface of the ball that is spinning in the opposite direction to their flow, they are slowed down. In accordance with Bernoulli's principle, these slower-moving particles will exert a relatively higher pressure. Also, the opposing motion of the ball's surface makes it difficult for the fluid particles to cling to the surface of the ball. This has the effect of early separation of the boundary layer on that side of the ball—that is, the particles are deflected. (Remember how a wing causes lift? By deflecting air particles.) In contrast, the fluid particles on the opposite side of the ball are not slowed down as much because the outer surface of the ball on that side is spinning in the same direction as they are flowing. Bernoulli's principle explains that this higher flow velocity will result in a lower pressure on that side of the ball. In addition, the fluid particles cling to a surface more readily if its motion is in the same direction as their flow. In this case, the fluid particles are not deflected, and the boundary layer clings to the surface on that side of the object for a longer time (Figure 9.22).

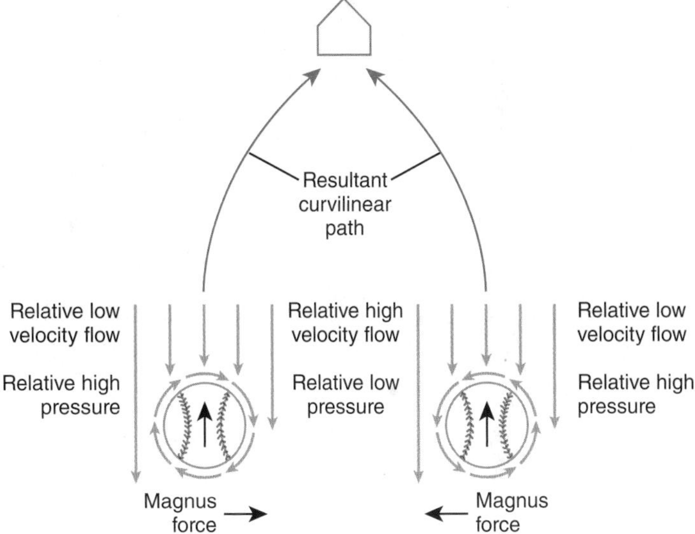

Figure 9.21 **The curvilinear path of a spinning object.** What causes the curve?

So spin actually causes two phenomena. First, a pressure gradient is created from one side of the spinning ball to the other in accordance with Bernoulli's principle. Second, because the boundary layer clings to one side longer than to the other, fluid particles have been deflected (or accelerated). The result is an equal and opposite reaction force gradually curving the path of object motion.

In the sport of baseball, many types of spin are used by pitchers to confuse batters. The typical curveball is the most obvious example. But topspin also can be applied to a ball to make it drop faster than the batter is expecting it to. Backspin causes the ball to not drop as quickly. Notice that we are saying that it does not drop as quickly, not that it will break upward. Some players claim that pitchers can throw an upward breaking ball. However, a human cannot throw a baseball with enough velocity to generate a Magnus force of sufficient magnitude to overcome the weight of the ball. The upward breaking baseball is probably an optical illusion caused by the batter's expectations about how fast the ball should drop. In

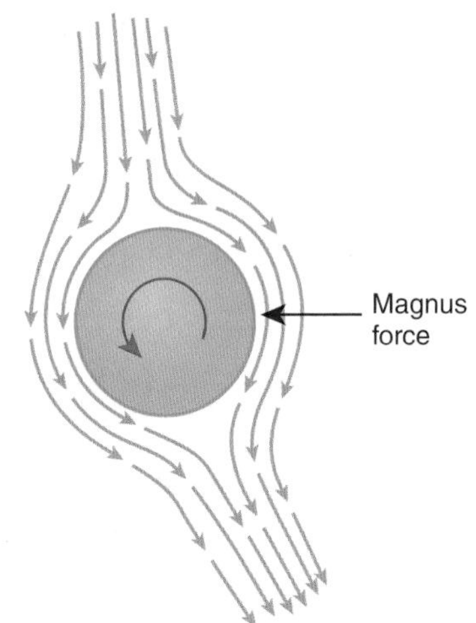

Figure 9.22 **The boundary layers around a spinning object.**

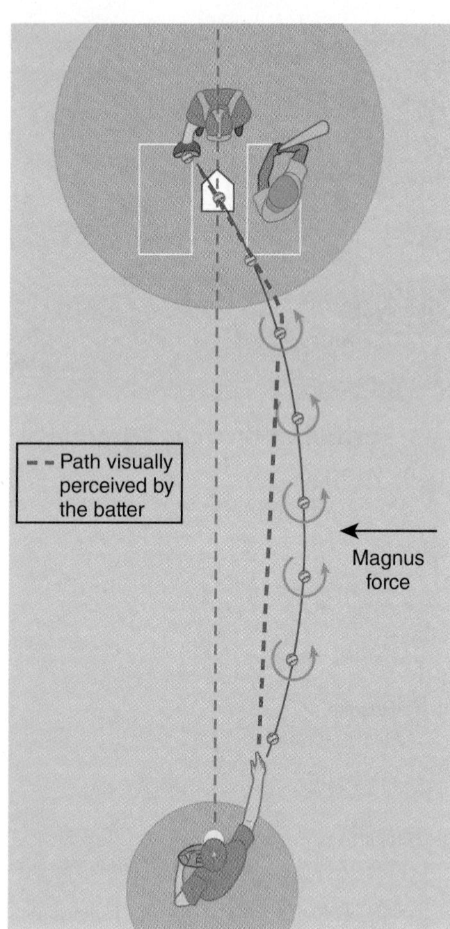

Figure 9.23 **The trajectory of a curveball.**

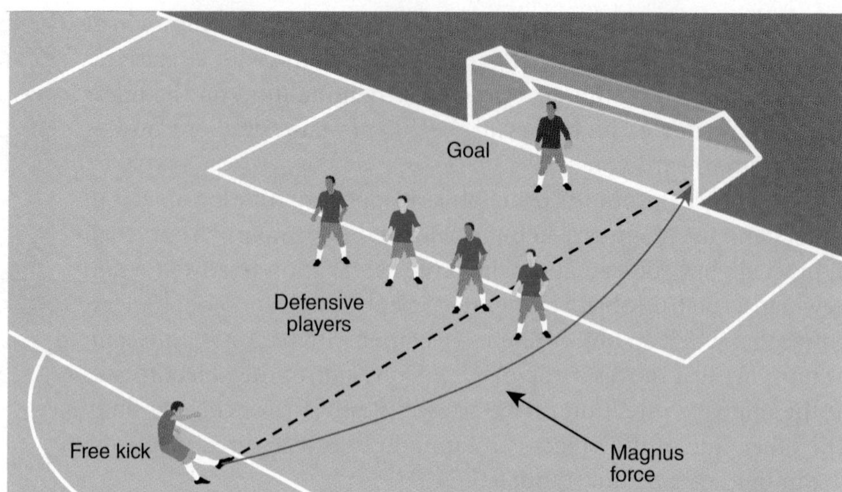

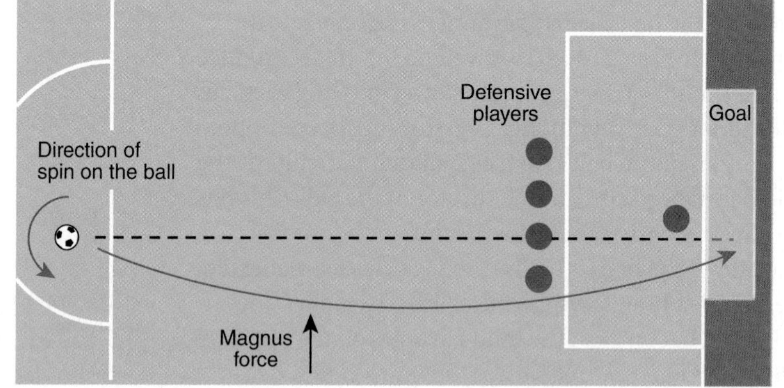

Figure 9.24 **Trajectory of a banana kick in soccer.** In what situations is a banana kick advantageous?

addition, the word *break* is misleading because it implies that the Magnus effect occurs suddenly. In reality, a ball that curves because of the Magnus effect actually follows a regular arc-shaped trajectory (Figure 9.23). Once again, the batter is confused by an optical illusion.

In the sport of golf, the club does impart enough force to the golf ball to cause a trajectory with an upward arc. Chapter 8 describes the advantage that implements provide, and of course the golf ball weighs less than a baseball. In soccer, the Magnus effect can be used to arc a ball around a defender when taking a shot on goal (Figure 9.24).

There is no end to the advantages of being able to place spin on a ball. We use the Magnus effect in billiards, volleyball, and all racquet sports.

CONNECTIONS

9.3 HUMAN PERFORMANCE

Exercise Physiology

You know that specific gravity is the ratio of the density of a given substance to the density of water (1.0×10^3 kg/m^3). So if the specific gravity of a substance is 2.0, then we know that its density is two times that of water (2.0×10^3 kg/m^3). Because specific gravity is a comparison to water, it is sometimes defined as the ratio of the weight of an object to the weight of an equal volume of water. Recall that an object with a specific gravity greater than 1.0 (density greater than 1.0×10^3 kg/m^3) will be negatively buoyant and sink, whereas an object with a specific gravity less than 1.0 (density less than 1.0×10^3 kg/m^3) will float. These concepts are used in the field of exercise physiology to determine body fat. In the process of **hydrostatic weighing**, a person's weight in the water is compared to her weight outside of water as a means of estimating her body volume.

Once body volume is estimated, density can be calculated as mass per unit volume. A person with more body fat will be buoyed up, or "lose weight" in water, because fat is not as dense as water (i.e., the person will weigh less than the water displaced). Remember the Styrofoam example? In contrast, a person consisting of proportionately more lean tissue will tend to sink in water because lean tissue (muscle, bone, etc.) has a higher density than water (just like lead in our earlier example). The percentage of total body weight attributable to fat can be calculated based on the change in weight during submersion in water.

Sport Science

Air and fluid dynamics influence all sport objects that pass through either medium—including the human body. The equipment used in many sports has evolved to take advantage of the physics of fluids. Track events, golf, swimming, cycling, running, and more have seen changes that either lessen the effect of fluid dynamics or use it for an advantage, depending on the circumstances.

Swimming

For decades, swimmers have altered variables to decrease the amount of drag as they pass through water. There are two ways to decrease drag: (1) through body position (pressure drag) and (2) through strategies that reduce the overall surface drag of the body. Swimmers who keep their body position as close to parallel with the surface of the water as possible get the most efficient shape for higher speed and lower energy expenditure. The farther from parallel a swimmer gets in the line of travel, the more pressure drag will occur (Figure 9.25). Beginning swimmers tend to keep their heads high, which makes their legs drop. These beginners will expend more energy as well as move more slowly than an experienced swimmer.

You can get an idea of this effect by holding your open hand outside the window of a moving car. If you keep your fingers straight and pointing forward and your palm parallel to the line of travel, it takes little energy to hold it in position. If you tip your hand 10 degrees upward at the fingers, the energy required to hold your hand in place increases dramatically.

Elite swimmers look for any advantage in reducing drag in the pool. When 10ths or even 100ths of a second separate first and third place, any strategy to increase speed or reduce energy expenditure becomes relevant. Traditional ways to reduce surface drag include wearing tight-fitting rubber swim caps and removing any exposed body hair. Shaving arms, legs, and torso may seem extreme and insignificant, but some research has indicated that "there is indeed a physical benefit to shaving down (most likely a reduction in body drag) and that the benefits are not solely due to a psychological response" (Sharp et al., 1988). Reduced drag

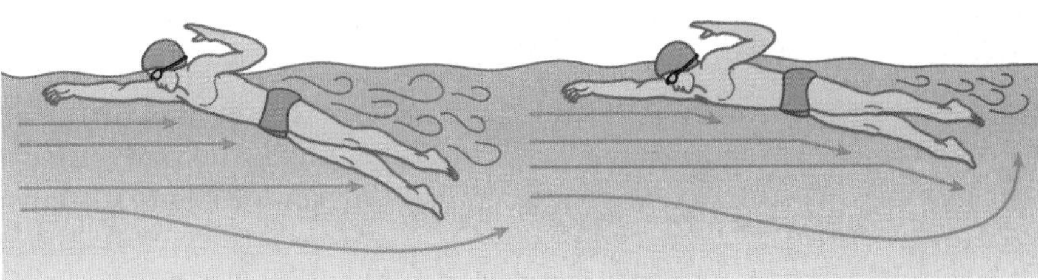

Figure 9.25 **Different body positions and drag of swimmers.**

Hydrostatic weighing A process in which a person's weight in the water is compared to their weight outside of water in order to estimate their body volume.

is also part of the controversy that led to a ban on certain types of swimsuits after the 2008 Olympic Games. In the 2000s, several companies released new swimsuits designed to reduce surface drag. Using specially designed raised patterns—some look like tiny raised V's and some like golf ball dimples—engineers succeeded in decreasing the amount of surface drag below that of shaved human skin. The new high-tech suits covered the body from shoulders to ankles, and many were custom designed for an individual swimmer based on that swimmer's body characteristics to enhance flow characteristics in the water. The Speedo Fastskin® suit followed a design found in nature—sharkskin. Sharks have dermal denticles or placoid scales that are microscopic raised areas covering their skin that disrupt the flow of water and decrease surface drag by breaking up the boundary layer of water and creating tiny vortices. Speedo's LZR suit went even further by trapping air between the suit and skin to aid buoyancy. After 23 of 25 world records were broken by athletes in LZR suits, full-body suits were banned in 2009.

Cycling

Bicycle racers are also keen about aerodynamics. Decades ago, they learned that the human body accounts for 80 percent of the drag while riding at 30 km/hour, and they found the most aerodynamic body position by conducting thorough wind-tunnel tests. The equation for drag explains why improving aerodynamics is important to cyclists and other athletes who travel at high velocities. Remember that relative flow velocity has an exponential effect on drag. A cyclist who wants to double his velocity will have to expend four times the energy! The human body can develop only a finite amount of energy, so the only way to gain velocity or decrease energy expenditure is to reduce drag. The current time-trial bicycles and rider equipment used during the Tour de France gives some indication of the lengths to which engineers can improve a sport by applying sound physics principles (See Figure 9.26).

Traditional bicycle wheels contain 32–36 spokes that are round in shape. Each round spoke creates an area of turbulence that results in drag. New wheels come in two varieties: the solid disc and three- or four-spoke aerodynamic wheels, whose spokes have teardrop-shaped cross-sections. All of these new wheels result in less turbulence and therefore less drag. Frames have evolved from round material to oval and then to teardrop shape for the same reason. The riders wear form-fitting suits to minimize drag on their bodies and wear specially designed helmets during time trials that are rounded in the front and tapered in the back! These helmets have been shown to reduce drag even when compared to a shaven head. Why do you think this is?

Even with all of the new technology to reduce drag, an area of turbulence will always be present behind each rider. Long-distance cyclists count on this area of turbulence and use it to their advantage in a tactic called *drafting* (Figure 9.26). This approach is part of the reason why teams of riders dominate most long-distance bicycle races.

Figure 9.26 **Bicycle racers drafting.** What is the advantage of the pack?

The lead rider pushes air out of the way, but because the airflow doesn't close cleanly behind the leader, it creates a vacuum effect, or draft. Another rider following closely enough behind the lead rider expends less energy to maintain the same velocity. But there is still an area of low pressure behind the last rider, so what is the benefit of drafting if the low-pressure zone still exists? The low-pressure zone *is* still the same size. However, during drafting the low-pressure zone is being resisted by two riders (or more) instead of just one (Figure 9.27).

Teams have a designated rider they hope will win the race, and the rest of the bicylists are along to help the lead rider by towing her along in the draft for most of each race leg's distance. Because the designated rider has expended less energy for most of the day, that rider has more energy to sprint at the end. It is rare to see winners like Chris Froome leading the way for any appreciable amount of time

Figure 9.27 **The low-pressure zone behind cyclists.** Which rider benefits most?

in a long race. This same principle is used among drivers in car racing events such as NASCAR. In the case of cars, the horsepower of two engines resists the low-pressure zone, helping the speed and fuel economy of both cars.

SCUBA Diving

SCUBA diving is a great example of Archimedes' principle and applied buoyant force. First, a question: Why does a human body float when the lungs are full of air and sink when the lungs are empty? (Body composition may affect this dynamic, but not for the reasons you might think; that's explained in the preceding section about hydrostatic weighing.) If you answered that the air causes you to float, you're correct, although the explanation *why* may surprise you. The air itself has no effect other than giving your body more volume with little increase in mass. Remember Archimedes' principle, buoyant force is equal to the weight of the water that an object displaces.

The human body on average has a density nearly identical to that of water, which makes sense when you consider the fact that humans are made up of a high percentage of water. Fat has a specific gravity of about 0.9, and the figure is around 1.1 for muscle. In essence, the average body is almost neutrally buoyant. Based on Archimedes' principle, the only way that we can increase buoyant force is by increasing our volume. That's exactly what taking a deep breath does. The thoracic cavity expands, we displace more weight in water, and the body rises because it has more buoyant force.

The difference in overall body volume between lungs empty and lungs full may seem trivial, but it makes sense when you understand two important pieces of information: (1) The body is close to neutrally buoyant, and (2) water is heavy. Even the small change in the volume of the body from taking a deep breath will displace a substantial amount of water weight. Even a 4.45-N (1-lb.) buoyant force is enough to cause the body to move toward the surface.

By now you may be asking yourself why, if this is the case, does the part of the body with the air stay near the surface while floating, while the legs eventually point toward the bottom of the pool? Density is the answer. When you took that deep breath before performing a survival float, your thoracic cavity expanded with air and became less dense. The less dense area of any object floats at a higher level than a more dense section. Density also explains how an object filled with 80 cubic feet of air can sink like a stone. A SCUBA tank can contain that volume of air, but it's held in a small volume of space under extremely high pressure. If you make the mistake of dropping a full SCUBA tank in the water, you'll end up diving to retrieve it.

SCUBA diving is based on the concept of neutral buoyancy and involves a complex balance of factors that affect volume. A diver can counter the effects of buoyancy (positive and negative) by expending energy in the form of contracting leg muscles to drive swim fins. Swimming wastes energy and air, both of which are limited commodities, so divers manipulate other variables to achieve neutral buoyancy.

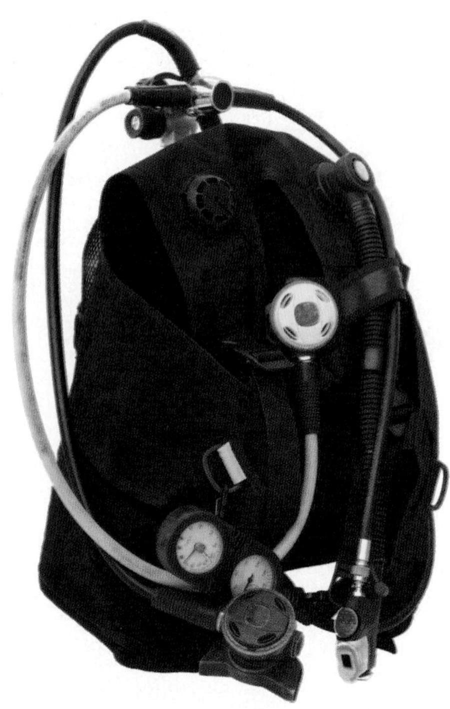

Figure 9.28 **A buoyancy compensator.**

A wetsuit adds buoyancy for the same reason that taking a deep breath does—it adds volume with a less dense material. Divers also wear *buoyancy compensators* (BCs), which are worn like vests and can be inflated and deflated while diving (Figure 9.28).

The BC and air tank add more volume to the diver, thereby increasing buoyancy to the point where most divers need to wear lead weights to overcome it. Finding the proper amount of lead weight is one factor in manipulating buoyancy. The weight is correct when a diver descends slowly with no air in the BC.

As divers descend from the surface, they come under increasing pressure from the surrounding water. This pressure causes the neoprene wetsuit material to compress, and the diver displaces less volume. Because the mass of the diver has not changed, the decrease in volume creates an increase in density, which causes negative buoyancy. With your knowledge of Archimedes' principle, you'll realize that there is only one way for the diver to regain neutral buoyancy. Assuming the diver would like to remain at this depth, what would you advise?

If you answered that increasing the volume of the diver in some way would solve the problem, you're right, and that's where the BC gets its name. With increasing depth and water pressure, the wetsuit and BC compress and become denser. To compensate, the diver presses a button to inflate the BC with air from the tank, the BC expands just like the lungs did when floating on the surface, more water is displaced, and the diver regains neutral buoyancy. Again, it's not the air itself that creates the buoyancy; it's the air that creates more volume. As divers change depths, they constantly adjust the volume of the BC to maintain neutral buoyancy.

Golf

Golf balls have changed dramatically since the origins of the game in Scotland more than 600 years ago. New materials have played a significant role in the evolution of the game. Chapter 4 examines the effects of elastic energy and coefficient of restitution, which can explain much of the reason that balls fly farther than ever. Golf balls are now made of materials that lose less energy at impact, resulting in higher velocity off the club face. A second variable comes into play that affects distance and also plays a significant role in accuracy—the dimple.

In 1848, a close predecessor of the current ball was introduced to the game. It was made of a rubber called *gutta percha*, which could be easily molded when boiled and cooled into a solid form. The first "gutties," as they came to be known, were smooth skinned, but an interesting phenomenon resulted in a change that still drives golf ball design today. Players noticed that the more nicks and scratches the balls accumulated, the better they flew. Gutties had a reputation for falling apart rather quickly, so golf ball makers began scribing lines into brand new balls. Now they flew better from the first shot.

Using fluid dynamics, we can explain the reason why the gutties flew better with the nicks, scratches, and scribed lines. Any object passing through a fluid encounters pressure drag and surface drag. Surface drag creates friction that results in the ball losing energy, but this is only a small fraction of the energy the ball loses because of pressure drag. A smooth sphere traveling through fluid creates an area of turbulence in its wake, with laminar flow exiting tangentially from the top and bottom of the ball. This tangential flow results in a turbulence area as large as the cross section of the ball at its largest point. The rough surface of the scribed balls creates turbulence in the boundary layer along the surface of the ball, which keeps the airflow on the ball for a longer time. The area of resultant turbulence behind the ball is reduced, leading to a decrease in pressure drag. Golf ball dimple design relies on these principles, and every major manufacturer has engineers on staff who strive to design the perfect dimple shape and size that will minimize pressure drag (Figure 9.29).

Designers wouldn't have such a difficult time coming up with the perfect design if there were not another factor that affects ball flight. When a golf ball is struck properly, backspin is imparted by the club face. This spin creates a differential in airflow between the top and bottom of the ball, which is explained in the concepts section of this chapter as the Magnus effect. With backspin, the airflow across the top of the ball moves at a higher velocity than the flow on the bottom of the ball. According to Bernoulli's principle, an area of lower pressure exists at the top of the ball, and an area of higher pressure on the bottom. This differential creates an imbalance, and a Magnus force is created at right angles to the airflow; the result is that the golf ball rises like the wing of an airplane (Figure 9.30).

Current research agrees that the friction from the surface of a rotating sphere only affects airflow motion in a thin layer near its surface called the *boundary layer*. The separation of the boundary layer is delayed on the top of the ball and occurs early on the bottom. This difference disrupts the laminar flow around the boundary layer of a spinning sphere and is responsible for the Magnus force and resultant lift. The conundrum for golf ball designers is deciding which is more important when it comes to disrupting the boundary layer: decrease pressure drag or increase lift? Will less pressure drag increase velocity enough to compensate for a loss in lift? Or will added lift, and therefore longer time of flight, compensate for decreased velocity? The debate continues.

One debate that ended long ago concerned the reason golf balls hooked and sliced. In the absence of other factors such as a side wind, a well-hit golf ball will travel straight and produce enough lift to rise against gravity. After reading about why a baseball curves and a golf balls rises, you should be able to understand the phenomenon of a golf ball curving right or left (baseballs can be seen doing the same thing when viewed from a catcher's perspective). The obvious answer is that sidespin was imparted on the ball at impact. Airflow velocity is higher on one side than on the other, and the ball is deflected in the direction of higher relative airflow velocity. Most amateur golfers impart slice spin but don't understand why their shots slice dramatically with the driver, less so with middle irons, and possibly not at all with short irons and wedges. The answer lies in the proportion of sidespin to backspin generated by different clubs. A driver puts little backspin on a ball, so the sidespin imparted is the predominant spin. Conversely, a shot hit with a wedge has tremendous backspin that will keep the ball traveling straight against the relatively weak sidespin that can be imparted with that club.

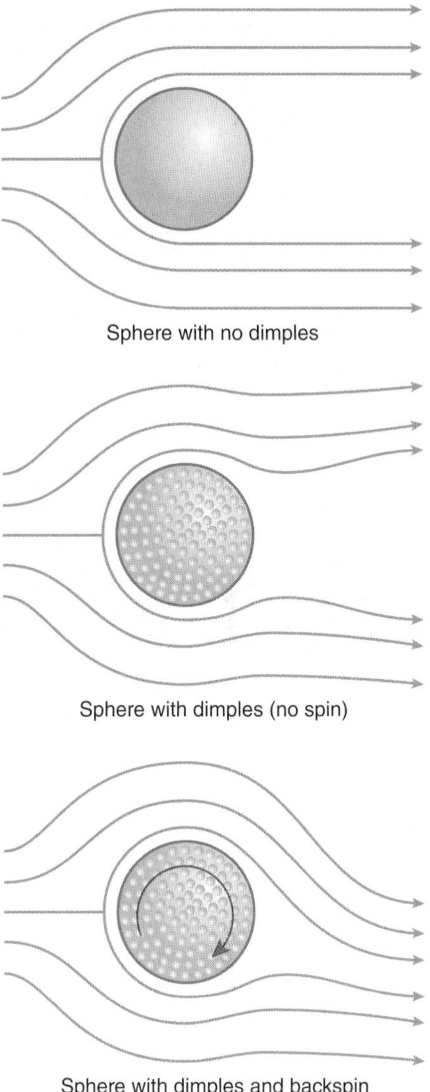

Figure 9.29 **Golf balls.** How do dimples help?

Baseball

Now that you understand what makes a curveball curve, a sinker drop, and a golf ball rise and turn right or left, you might enjoy learning why it's extremely difficult to hit the slowest pitch in baseball. In an earlier chapter we learned that rotation provides stability. The rotation of a baseball might make it curve, sink, or "rise," but it will move in a relatively predictable fashion. The same can't be said for the knuckleball—a pitch that is thrown with little or no rotation. Because the pitcher's grip on the ball is limited, it can only be thrown at two-thirds the velocity of a fastball. Then why is it so hard to hit? The answer lies in the ball's unpredictability. Without rotation, the ball is unstable in flight. It becomes even more unpredictable and effective when it has a slight rotation. Baseballs have asymmetrical raised seams where the two halves of the cover are

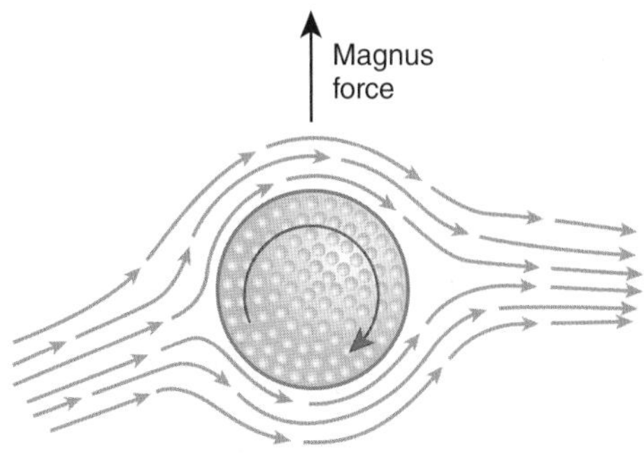

Figure 9.30 **Magnus force acting on a golf ball.**

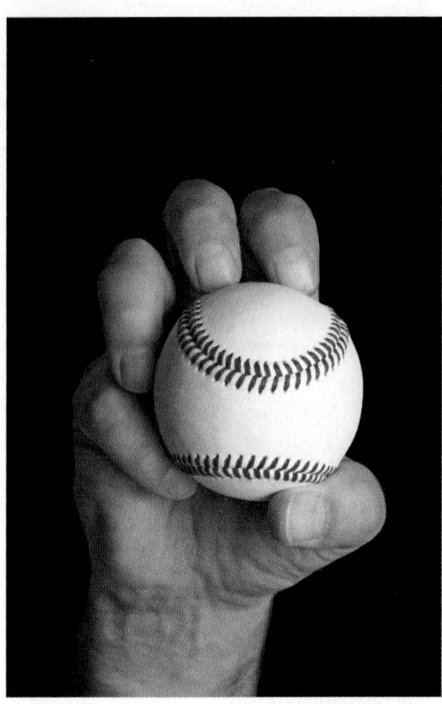

Figure 9.31 **A knuckleball pitch.** What makes it so hard to hit?

stitched together, and good knuckleball pitchers have learned that by changing the angle of the seams with each pitch, the airflow past the seams affects the ball differently. As the ball moves toward the batter, it presents a constantly changing profile to the approaching airflow, and it may rise, fall, or turn right or left at any time as the seams move in and out of the flow (Figure 9.31).

The batter has no idea which way the ball will be moving as it arrives in the hitting area, and this author would argue that the pitcher couldn't predict it either! It may be fun for fans to watch a professional batter look helpless while swatting at a slow-moving ball, but any catcher will tell you that his job is a nightmare when a knuckleball pitcher is on the mound.

The same situation is faced by the receiving team when a volleyball player uses a "float" serve. The float serve is struck with less power and velocity than a top-spin serve but is contacted directly in the center of the ball so there is no eccentric component. The ball leaves the server's hand with no rotation and flies over the net on an inconsistent trajectory that is constantly changing. This makes it extremely difficult for a passer to handle when it arrives.

Track and Field

In high school physics you learned that if you wanted to propel an object for maximum distance, the optimum launch angle is 45°. Given no air resistance, this would hold true for all objects; but our atmosphere is full of air, and we actually use this fact to increase the distance we can propel some objects. Imparting backspin on a golf ball creates lift that results in longer flight time, even with less than 45° of launch angle. Two field events at track meets also use lift created by airflow to enhance maximal throwing distance: the javelin and discus. In fact, the rules of the javelin design were altered in 1986 when athletes began throwing them too far and jeopardized the safety of spectators and other athletes!

Javelin design is guided by strict rules concerning length, weight, center of mass, and surface preparation. Optimum launch angle is 30°, but the javelin itself has a relative angle to the ground of 37°, creating an angle of attack that is designed to maximize airflow to create lift (Figure 9.32).

A change in a few degrees of either launch angle or angle of attack dramatically alters the distance a javelin will travel. If the angle of attack is less than optimal, lift will not be maximized; and if it is too

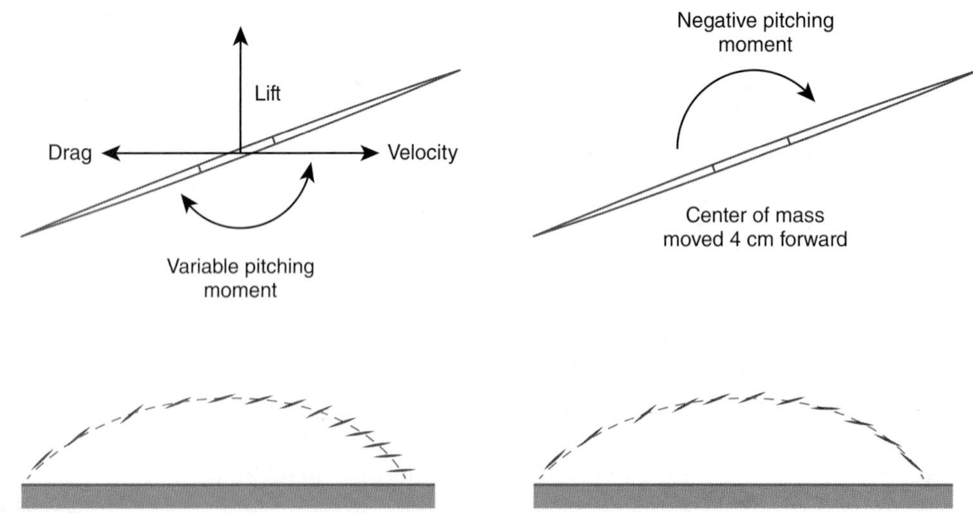

Figure 9.32 **The angle of attack of a javelin.**

high, the javelin will stall in midflight (because more drag is produced than lift). Athletes who found the perfect angles under good conditions were able to throw a javelin more than 100 meters in the early 1980s.

This ability became a safety issue in smaller stadiums, and one relatively minor change was made to javelin weighting that reduced Olympic throws by 15 meters between the 1980 and 1988 games. When an old-rules javelin (pre-1986) was launched, the center of pressure was ahead of the center of mass. The center of pressure is the point where aerodynamic drag and lift forces act on the javelin. As the javelin travels through an arc during flight, the angle of attack increases as the javelin rotates around its center of mass. This rotation allows the javelin to remain in an aerodynamically advantageous position (creating lift) for a longer time. New-rules javelins have the center of mass moved forward 4 cm. This change has moved the center of pressure behind the center of mass. As the new-rules javelin travels through a flight arc, the rotation around the center of mass decreases the angle of attack, thereby reducing total lift during flight. It also encourages a tip-down attitude, making the new javelins easier to throw. The previous discussion is a simplified overview of factors involved in javelin flight, but it should give you a better idea of how fluid dynamics can alter sporting events.

The discus is also thrown at less than 45°, and it creates lift in a fashion similar to that of an airplane wing. A cross section of a discus reveals that the upper and lower surfaces are identical in shape, but when it is thrown with a slightly upward tilt (angle of attack), air is deflected (accelerated). This creates a pressure differential that results in lift. Studies have shown that a discus thrown into a moderate headwind at the proper trajectory will actually travel farther than in still air. The additional lift provided by the higher relative airflow in a headwind results in longer flight time and can increase distances by up to 5 m. As with the javelin and airplane wing, at a particular angle of attack, lift is lost to excessive separation of airflow. For the discus, this stall angle is around 25 degrees (Figure 9.33).

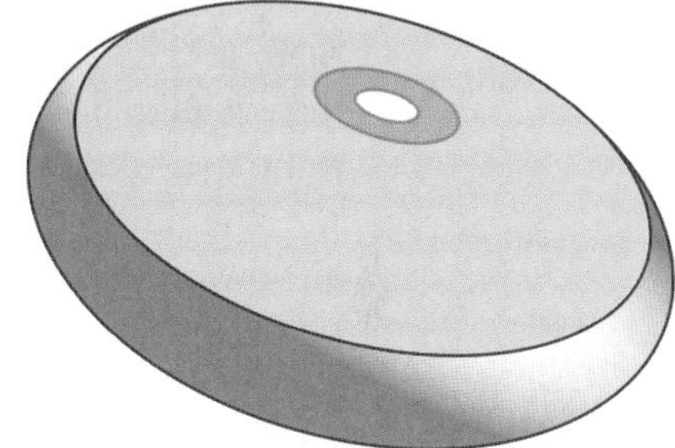

Figure 9.33 **The angle of attack of a discus.**

Flying Discs

The Frisbee® traces its origin back to empty pie tins. The pies were sold to colleges in the Northeast by the Frisbie Pie Company and embossed with the words "Frisbie Pies." When students finished eating the pies, they discovered that throwing and catching the pie tins provided hours of enjoyment. The secret to the fun was that the tins would fly relatively long distances, a fact that can be attributed to the shape of the pie tin when turned upside-down. Think for a moment about what that shape would look like from the side. Flat on the bottom, with a sloping leading edge—basically a modified airfoil.

The Frisbee as we know it today is a better version of the Frisbie Pie tin. The shape is more aerodynamic, and small ridges have been added to increase stability (Figure 9.34).

When thrown parallel to the ground, the Frisbee produces enough lift to counteract the force of gravity and remain traveling on a path parallel to the ground. As with the discus, stability in flight is maintained through angular momentum (spin).

Several sports and games have been developed around the flying disc, and one in particular deserves mention in this chapter. Disc golf has evolved during the last 30 years from a game in which participants threw a regular Frisbee at trees and lampposts into a sport with highly specialized discs that use aerodynamic principles to enhance certain flight qualities. Like a golfer who carries 14 clubs in a bag, disc golfers now carry a variety of discs around a course. If you look at the original Frisbee patent drawings, you'll notice that it's fairly "thick" from top to bottom. This thickness becomes a

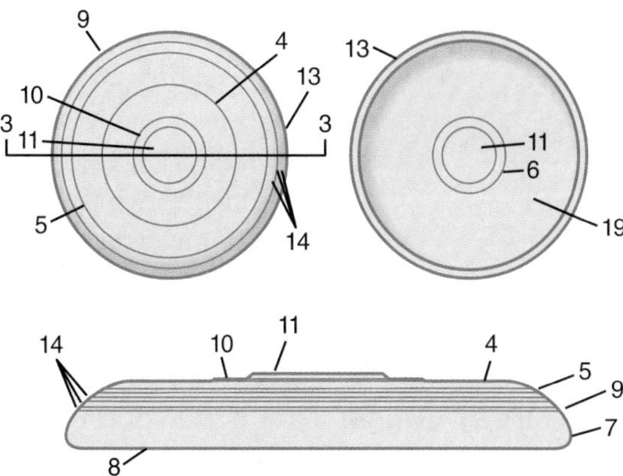

Figure 9.34 **The original patent drawings of the Frisbee®.**

Figure 9.35 **Ski jumper.** How does this position maximize distance?

limiting factor in how far the disc will travel because it causes aerodynamic drag. Specialized golf discs are much thinner and have sharply angled edges. These angles are designed to produce lift, and the thin profile reduces drag better than a regular Frisbee can. The long-distance driver is especially suited for maximum distance with an ultrathin profile, and some have been thrown more than 250 meters by professionals. At the extreme of flying disc records is the Aerobie®, a thin ring that took years of work to design and create based on aerodynamic principles. The world record for this disc is more than 400 meters. With your understanding of fluid dynamics, can you explain how the Aerobie can fly so much farther than an original Frisbee?

Ski Jumping

At least one sport has adapted new techniques that use drag to increase performance. The sport of ski jumping was changed dramatically in 1989 with no change in equipment. The essence of this sport is to descend down a ramp gaining maximal velocity and then leave the ramp surface to fly through space before landing downhill more than 100 meters later. Jumpers lean forward over their skis, creating a modified airfoil shape that produces an angle of attack similar to that of a javelin or discus in flight. This shape doesn't create enough force to lift the skiers' mass, but it keeps them in the air for a longer period of time by slowing their rate of descent. The first part of the ski jumping equation has not changed at all: Old-style and new-style jumpers maintain a tuck position down the ramp to decrease their drag against airflow, and they will eventually reach nearly 100 km/h at takeoff.

The newer style of jumping differs once the skier leaves the ramp and is affected by gravity and airflow only. Traditional ski jumpers aligned their skis parallel to each other directly in front of their bodies to minimize drag. Then, in 1989, a jumper revolutionized the sport by spreading his skis in a V-shape with the opening forward (Figure 9.35).

This shape presented a much larger area of drag that enabled the jumper to stay in the air even longer and created more lift effect (slower descent) at the end of a jump. This is just one example of increasing drag to increase performance. Can you think of any others?

9.4 VETERINARY MEDICINE

High-rise syndrome is a term used in veterinary medicine to refer to traumatic injuries sustained by animals as a result of falling from substantial height (usually two or more stories). Whitney and Mehlhaff (1987) examined 132 cases of high-rise syndrome in cats. The distance that the cats fell ranged from two to 32 stories, with a mean fall of 5.5 stories. Overall, 90% of the treated cats survived! To give you some perspective, almost 100% of human falls from heights greater than six stories are fatal. In this study, the injury rate increased approximately linearly with distance fallen up to approximately seven stories. The most interesting finding of the study was that the injury rate did not continue to increase for falls greater than seven stories, and the fracture rate actually decreased at heights greater than seven stories. Of the cats that fell farther than seven stories, only one of 22 died. Of the 13 cats that fell farther than nine stories, only one suffered a fracture.

High-rise syndrome A term used in veterinary medicine in reference to animals sustaining traumatic injuries as a result of falling from substantial height (usually two or more stories).

Figure 9.36 **A cat falling.** How do they survive long falls?

One cat actually fell 32 stories onto concrete and was released after 48 hours of observation! It had suffered a mild pneumothorax and a chipped tooth. How is this possible? There are several biomechanical concepts related to these extraordinary survival statistics. First, you have read that cats have a great ability to twist while falling and position themselves with the feet down. Second, an average sized cat achieves a terminal velocity of approximately 96.54 km/hr (60 mph) after falling five stories. As would be expected, the injury rate was correlated with the distance fallen up to seven stories (a distance just higher than that required to reach terminal velocity). But why did the fracture rate actually decline in falls that were from a distance greater than seven stories? There are two possible explanations, and both probably play a role. First, it is speculated by the researchers that the vestibular apparatus is stimulated as long as the cat is still accelerating (before terminal velocity). This effect causes the cat to stiffen its forelimbs reflexively. However, at terminal velocity, the vestibular system is no longer stimulated, and the cat may actually relax the forelimbs in the absence of acceleration. This relaxation may lead to a more horizontal positioning of the limbs (like a flying squirrel), which maximizes drag (Figure 9.36).

But on landing, this position would mean that the impact would be more evenly distributed along the ventral surface of the body (remember pressure?). Another contributor may also be related to the relaxation that accompanies terminal velocity. What would you do to make your landing more comfortable? Would you land stiffly, not allowing your hips or knees to bend on impact? Intuitively, most people would say the opposite: to "give" on impact. We have asked these questions before while discussing the linear impulse equation ($F \times t = \Delta M$). If the cat is relaxed, its body is more likely to "give" on impact and absorb the impact force over a longer period.

SUMMARY

Buoyant force is the vertical, upward-directed force acting on an object that is submerged in a fluid. Buoyant force is explained by Archimedes' principle: *A body submerged in a fluid will be buoyed up by a force that is equal in magnitude to the weight of the displaced water*. Density is the relationship of the mass of an object to its volume, and specific gravity is the ratio of the density of a given substance to the density of water. The center of buoyancy is the center of gravity of the displaced volume of water, and it is equal in shape and volume to the submerged object that displaced it. Floating position depends on the relative locations of the center of gravity and the center of buoyancy of the object.

Dynamic fluid force is the equal and oppositely directed force of the fluid particles in reaction to the applied force of the system moving through the fluid. Similar to other forces, the dynamic fluid force vector can be resolved into two components: a parallel component called *drag* and a perpendicular component called *lift*. Drag force tends to resist motion of the system through the fluid (or decelerate the object) and is therefore called *fluid resistance*. There are two sources of drag: surface drag and form drag. The perpendicular component of dynamic fluid force can act in any direction that is perpendicular to system motion with respect to the fluid, even though it is termed *lift force*. Lift can also be caused by rotation of an object, a phenomenon called the *Magnus effect*. Because drag and lift are components of the same dynamic fluid force, they are affected by the same factors: the drag or lift coefficient, the density of the fluid through which the object is moving, the projected area of the object, and the relative velocity of the object with respect to fluid flow.

REVIEW QUESTIONS

1. Give the equation for drag force and explain each part in your own words.
2. Explain what causes skin friction drag (surface drag).
3. Compare laminar flow and turbulent flow.
4. Why is profile drag also called *pressure drag*?
5. How can we reduce the effects of profile drag? Explain the technical details.
6. I am driving with my hand out the window and my palm facing the oncoming wind. If I place my palm down, which factor(s) in the drag force equation change(s)?
7. What would have a greater effect—the hand-position change mentioned previously or increasing the velocity of the car? Why?
8. If mass doesn't affect drag force, why does a golf ball hit the ground before a ping-pong ball if they are both dropped from 50 feet? Use the linear acceleration equation to explain your answer.
9. Give the equation for lift force and explain each part in your own words.
10. Using Newton's laws of motion, explain how lift force causes flight.
11. What is the Magnus effect? Use it to explain a curve ball.
12. Explain Bernoulli's principle.
13. How is Bernoulli's principle related to the Magnus effect?
14. What happens at the stall angle?
15. What would be the advantage of a plane taking off into the wind?
16. Why would a cyclist with a cleanly shaven head be at a disadvantage to a rider with a well-designed aerodynamic helmet?
17. What happens when a Frisbee is subjected to a sudden headwind? Why does this happen?
18. Does a heavy ball curve because of the same principles that cause a baseball to curve? Can you think of a sport where a larger ball curves dramatically?

PRACTICE PROBLEMS

1. One substance has a mass of 1.06 g and occupies a volume of 1 ml. A second substance has a mass of 0.9 g and occupies the same volume as the first substance. Calculate the density and specific gravity of both substances. What are two biological substances that fit the above profiles?
2. Two cars are traveling next to each other at 113 km/hr. Calculate their relative velocity. Calculate the relative velocity of another car that passes them in the same direction at 115 km/hr. Calculate the relative velocity of another car that drives toward the other three at 100 km/hr.
3. A car has a coefficient of drag equal to 0.6, the density of air is 1.225 kg/m^3, and the projected area of the car is 2.5 m^2. Calculate the drag force at a velocity of 40 m/sec and 45 m/sec.
4. An airplane at a given angle of attack has a coefficient of lift equal 0.55, the density of air is 1.225 kg/m^3, and the projected area of the airplane is 15 m^2. Calculate the lift force at 65 m/sec and 70 m/sec.

EQUATIONS

Density $$\rho = \frac{m}{V} \tag{9.1}$$

Drag Force $$F_D = \tfrac{1}{2} C_D \rho A_p v^2 \tag{9.2}$$

Lift force $$F_L = \tfrac{1}{2} C_L \rho A_p v^2 \tag{9.3}$$

REFERENCES AND SUGGESTED READINGS

Anderson, D. F., & S. Eberhardt. 2010. *Understanding flight*, 2nd ed. New York: McGraw-Hill.

Braune, W., & O. Fischer. 1889. *The center of gravity of the human body as related to the German infantryman*. Leipzig. Available as report ATI 138 452, National Technical Information Service, Springfield, VA.

Craig, G. M. 2002. *Introduction to aerodynamics*. Anderson, IN: Regenerative Press.

Dempster, W. T. 1955. *Space requirements of the seated operator: Geometrical, kinematic and mechanical aspects of the body with special reference to the limbs*. Technical Report WADC-TR-55-159, Wright-Patterson Air Force Base, OH: Wright Air Development Center.

Hay, J. G. 1973. The center of gravity of the human body. *Kinesiology*, III: 20–44.

Heinrichs, R. N. 1990. Adjustments to the center of mass proportions of Clauser et al. (1969). *Journal of Biomechanics*, 23: 949–951.

Lanoue, F. 1963. *Drownproofing: A new technique for water safety*. Englewood Cliffs, NJ: Prentice-Hall.

Mannion, P., Y. Toparlar, B. Blocken, E. Clifford, T. Andrianne, & M. Hadjukiewicz. 2018. Aerodynamic drag in competitive tandem para-cycling: Road race versus time trial positions. *Journal of Wind Engineering and Industrial Aerodynamics*, 179: 92–101.

Plagenhoef, S., F. G. Evans, & T. Abdelnour. 1983. Anatomical data for analyzing human motion. *Research Quarterly for Exercise and Sport*, 54(2): 169–178.

Serway, R. A., & J. W. Jewett Jr. 2019. *Physics for scientists and engineers with modern physics*, 10th ed. Boston: Cengage Learning.

Serway, R. A., & C. Vuille. 2018. *College physics*, 11th ed. Boston: Cengage Learning.

Sharp, R. L., A. C. Hackney, S. M. Cain, & R. J. Ness.1988. The effect of shaving body hair on the physiological cost of freestyle swimming. *Journal of Swimming Research*, 4: 9–13.

Weltner, K. A. 1987. A comparison of explanations of the aerodynamic lifting force. *American Journal of Physics*, 55: 50–54.

Whitney, W. O., & C. J. Mehlhaff. 1987. High-rise syndrome in cats. *Journal of the American Veterinary Medical Association*, 191: 1399–1403.

Winter, D. A. 2009. *Biomechanics and motor control of human movement*, 4th ed. Hoboken, NJ: John Wiley & Sons.

CHAPTER 10

THE SYSTEM AS A PROJECTILE

LEARNING OBJECTIVES

1. Understand the force factors related to projectile motion.
2. Describe factors that influence trajectory of a projectile.
3. Describe forces that influence the horizontal and vertical distance a projectile travels.

CONCEPTS

CONNECTIONS

Folklore describes the story of William Tell shooting an arrow though an apple perched on his son's head. Either he was extremely close to his son when he let the arrow fly or he understood Newtonian principles. To hit a specific target with a projected object at a distance, you have to take into account how far it will drop over time and distance because of gravity. Although other factors such as fluid dynamics may affect the flight of a projectile, even a bullet shot from a high-powered rifle will be affected by the pull of gravity. Hunters and target shooters understand this effect and spend large amounts of time sighting their guns to account for the drop of the projectile at different distances. To confound the effect even further, the height of release will also affect how a projectile travels. Competitive shooters need to understand that there is a big difference between shooting from a prone position and shooting from a standing position.

Target archers have many options to help them accurately hit targets at various distances. Because the targets are set at standard distances, they have sighting systems for each distance. A three-pin system is common, with three different colored pins coded for specific distances. Once set for a particular bow, a shooter needs only to align the correct colored pin with the second sighting pin to account for the vertical drop the arrow will experience from gravity over that distance. As you'll see in this chapter, many variables could negatively affect even a perfectly sighted arrow.

Projectiles are subject to linear, rotary, and fluid forces. Therefore, their flight can only be fully understood at this point in the text. In this chapter, you will gain an understanding of the mechanics of projecting an object for vertical and horizontal distance as well as for accuracy. It begins with a review of previously covered concepts related to projectiles and then elucidates the mechanisms by which these factors affect projection. Finally, it connects these concepts to several disciplines such as motor development in which structural and physiological milestones lead to changes in projection ability.

CONCEPTS

10.1 FORCE FACTORS RELATED TO PROJECTILE MOTION

A **projectile** is a body whose motion is subject only to the forces of gravity and fluid resistance. Almost every sport involves the projection of an object, whether the object is thrown, kicked, or struck with an implement. The human system itself can be a projectile. Projectiles are subject to the influence of gravity and fluid resistance and can exhibit linear, rotary, or general motion. Therefore, let's review some previous concepts that are important to understanding projectile motion.

Kinetic Chain Concept

Even though an object can be projected in various ways (e.g., a bow, a gun, or a catapult), most projectile motion is the result of force applied by the distal segment of a kinetic chain. Therefore, let's review the kinetic chain concept. Recall that a **kinetic chain** is simply a system of linked rigid bodies subject to force application. Also remember that a kinetic chain can be classified according to its complexity and the mobility of its distal segment. A **simple kinetic chain** (Figures 10.1a and 10.1b) is one in which each segment participates in no more than two linkages (i.e., the hand is linked to the forearm; the forearm is linked to the hand and upper arm; and the upper arm is linked to the forearm and the shoulder).

A **complex kinetic chain** is one in which a segment is linked to more than two other segments (Figures 10.1c and 10.1d). For example, the human torso (if modeled as

Projectile A body whose motion is subject only to the forces of gravity and fluid resistance.

Kinetic chain System of linked rigid bodies subject to force application.

Simple kinetic chain Kinetic chain in which each segment participates in no more than two linkages; also called a *serial kinetic chain*.

Complex kinetic chain Kinetic chain in which a segment is linked to more than two other segments.

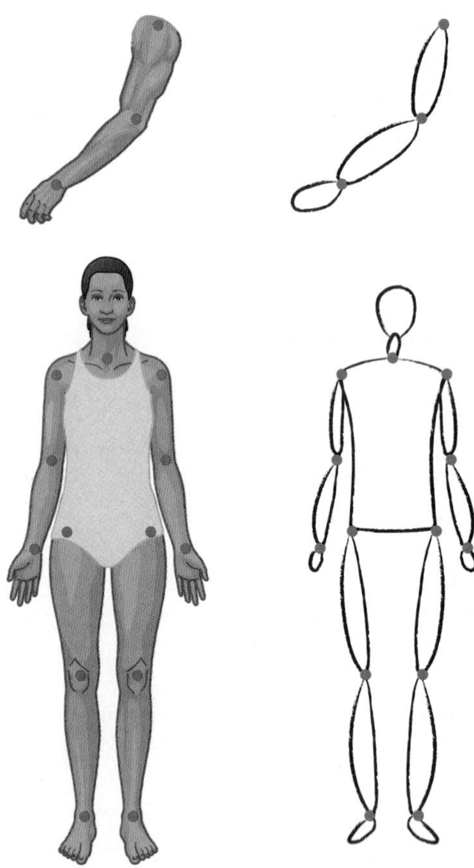

Figure 10.1 **A simple kinetic chain and a complex kinetic chain.**

one segment) has two links at the hips, one at the neck, and two at the shoulders. In a **closed kinetic chain** or *closed kinematic chain*) (Figure 10.2a), the distal segment is stationary ("closed"), and therefore the total chain has less mobility (fewer degrees of freedom). In other words, the distal segment is in contact with the reference frame or portion of the reference frame (e.g., Earth or another object) that provides enough resistance to prohibit free motion. An **open kinetic chain** (Figure 10.2b) is one in which the most distal (or terminal) segment is free ("open") to move. Another classification is that motion can occur at one link (joint) in the chain without cooperative motion at other links. Therefore, open kinetic chains have greater mobility (more degrees of freedom) than closed ones do (Zatsiorsky, 1998).They have a more proximal segment (or proximal end of the same segment) at which the force is usually applied to initiate motion of the chain. The momentum of this proximal segment can then be transferred throughout other links in the chain to ultimately apply force to project an object.

In accordance with the previous definitions, you can see that projection of an object by the human body often involves an open complex kinetic chain, because the entire body is usually involved, and the distal segment can move freely. For example, proficient throwing involves generating and transferring force from the base segments (legs) to the hips, then the torso, upper arm, and forearm, and finally the hand (the distal segment). So the forward momentum of one segment in the chain (caused by applying force or torque) is transferred to the next segment, and so on. Notice that in throwing, properly timing segmental participation produces "lag." In other words, the hips rotate forward while the upper torso is still rotating in the opposite direction ("lagging behind"). Then as the upper torso rotates forward, the humerus lags, and so on. The final result is a motion of the entire kinetic chain that is similar to that of a whip.

Because of the knowledge you have gained, you can now understand that this "lag" of some of the segments is caused by backward momentum imparted to them during the backswing. Therefore, as force is imparted to the most proximal segment to cause forward momentum, the remaining segments still have inertia in the opposite direction that must be overcome.

This kinetic chain concept applies even if an implement is used to cause the projection. The implement becomes the most distal segment of the chain, but it has a larger radius of rotation and therefore a greater range of motion and linear velocity.

Of course the kinetic chain is used to apply force to the object to project it. Once the object has been released from the distal segment of the chain, it is officially a projectile and subject only to the forces of gravity and fluid resistance.

Closed kinetic chain Kinetic chain in which motion at one link is possible only with cooperative movement at other links; also called a *closed kinematic chain*.

Open kinetic chain Kinetic chain in which motion can occur at one link in the chain without cooperative motion at other links; also called an *open kinematic chain*.

Influence of Gravity on Projectile Motion

Newton's law of universal gravitation states, *Every body in the universe attracts every other body with a force directed along the line of centers for the two objects that is directly proportional to the product of their masses and inversely proportional to the square of the separation between the two objects.* The acceleration of an object caused by Earth's gravitational force (g) is measured experimentally, and for most situations it can be assumed to be a constant value of 9.81 m/sec^2

Figure 10.2 **A closed kinetic chain and an open chain.**

(or 9.81 m/sec/sec) vertically downward. This constant means that if no other forces act on a projectile, it will fall at a rate of 9.81 m per second for the first second, and then it will fall an additional 9.81 m per second faster for each additional second (i.e., it will be falling at 19.62 m/sec after two seconds, etc.). The acceleration caused by gravity is a constant, no matter what the object's weight. Therefore, in the absence of any other force than gravity, the center of gravity of any object projected with some horizontal component will follow a predictable parabolic path. A **parabola** is a curved symmetrical shape (Figure 10.3). The height of the parabolic path can vary and depends upon the angle at which the object is projected and its vertical velocity.

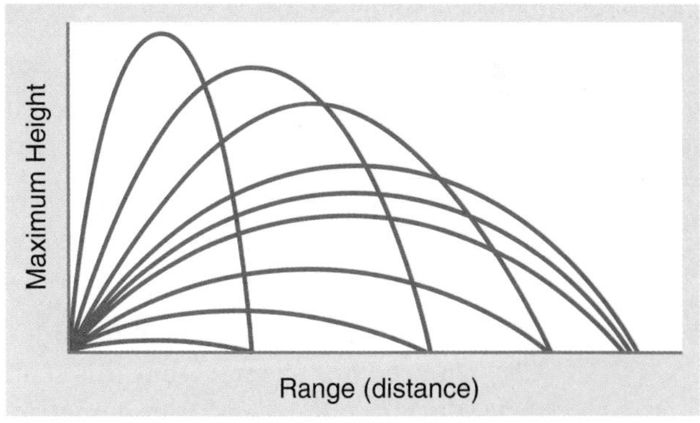

Figure 10.3 **Parabolic paths caused by gravity.**

Even if an object is projected perfectly horizontal to the ground, it will follow a parabolic path, but only one half of the parabola is observed by the viewer (Figure 10.4).

This perfectly symmetrical parabolic path will, of course, occur only in the absence of forces other than gravity. We know that the projectile is also affected by another force.

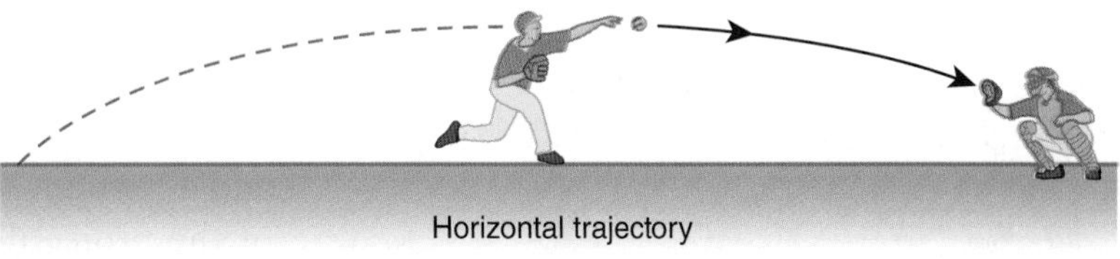

Figure 10.4 **Parabolic path of object projected perfectly horizontally.**

Parabola A curved symmetrical shape.

Influence of the Fluid Medium on Projectile Motion

The other force acting on a projectile is that applied by the fluid medium through which it is moving. Recall that *dynamic fluid force* acts on a system that is moving through a fluid. When moving through a fluid, the system (a projectile in this case), causes particles of the fluid to be deflected so that it can move forward. In other words, the system applies a force to the fluid particles that moves them along the outside of the system. According to Newton's third law of motion, an equal and oppositely directed force must be applied to the system by the fluid particles. Dynamic fluid force is the equal and oppositely directed force of the fluid particles in reaction to the applied force of the system moving through the fluid. As with other forces, the dynamic fluid force vector can be resolved into two components: a parallel component called *drag* and a perpendicular component called *lift*. The parallel component of dynamic fluid force acts in the direction opposite that of the system's motion through the fluid and is called *drag force.* Drag force tends to resist the system's motion through the fluid (that is, it decelerates the object). The perpendicular component of dynamic fluid force can act in any direction that is perpendicular to the system's motion through the fluid, even though it is termed *lift force* (which implies upward motion). Therefore, lift tends to change the direction of system motion. Because drag and lift are components of the same dynamic fluid force, they are affected by the same factors: the drag or lift coefficient, the density of the fluid through which the object is moving, the projected area of the object (area of the object oriented perpendicular to fluid flow), and the relative velocity of the object with respect to fluid flow. Also recall lift occurs because of spin, a phenomenon called the *Magnus effect*. Chapter 9 explains Magnus force in accordance with *Bernoulli's principle*.

Because this dynamic fluid force is applied to the projectile, the symmetrical shape of the parabolic path may be altered (Figure 10.5).

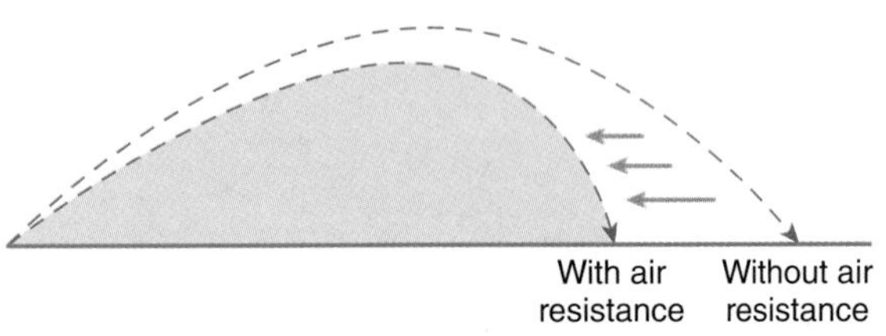

Figure 10.5 **Parabolic path altered by air resistance.**

Dynamic fluid force is sometimes ignored, depending on the situation. For example, if the object is relatively massive, it is less susceptible to alterations in the parabolic path. However, less massive objects (or spinning objects) can be subject to large distortions caused by drag or lift (e.g., a golf ball). Also, if the length of the projection path is short, dynamic fluid force is applied for a shorter time. So in cases of short projection distance (e.g., shot put), fluid force is often ignored because its effect is minimal.

Vector Resolution

Vector analysis is critical to understanding the behavior of a projectile. Vector resolution is used in previous chapters, but this section reviews the basic concepts because of its great importance. Both graphical and trigonometric methods are used throughout this text, and both methods are used in this chapter: graphical because it provides a better visual image, and trigonometric because of its accuracy.

To project an object, force is applied to cause acceleration. The applied force causes the system to travel with a given velocity. Velocity is a vector quantity, and as such it has a magnitude and direction. Because of its known direction, the velocity vector of the projectile can be resolved into component vectors. The velocity vector of a projectile possesses two perpendicular components: the *vertical component* (also called *perpendicular*) and the *horizontal component* (also called *parallel*). Graphically, we can use the parallelogram method to find the magnitudes of the component velocity vectors (Figure 10.6).

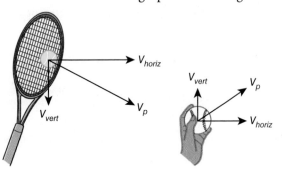

Figure 10.6 **Vector resolutions of two different projectiles.**

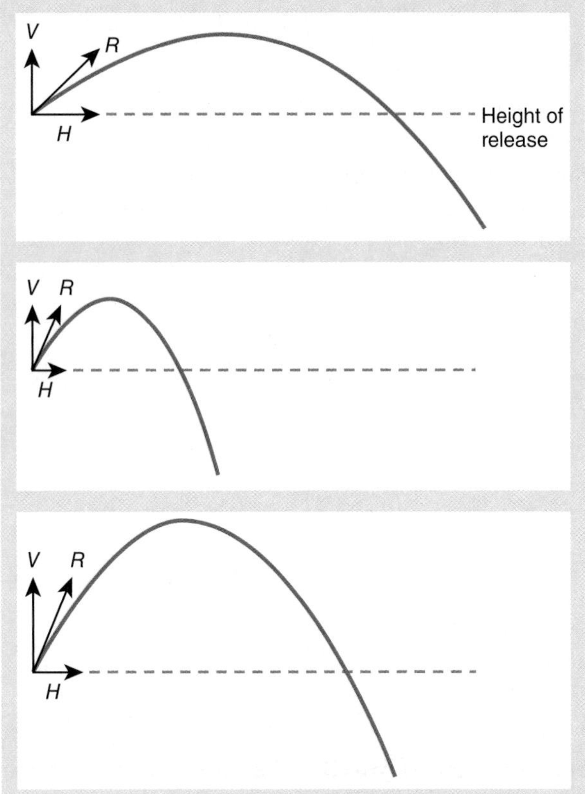

Figure 10.7 **Vector resolutions of three different parabolic paths.**

Visually, the relative sizes of the component vectors help us predict and understand the path of the projectile (Figure 10.7).

We can also use graphical vector analysis to understand the effect of gravity on the projectile (Figure 10.8).

Notice that the relative sizes of the resultant and vertical components change throughout the flight of the projectile because of the acceleration caused by gravitational force. The horizontal component has been kept constant to isolate the effect of gravity. But remember that fluid force varies with the square of velocity. Therefore, as the projectile slows down, drag force can decrease considerably. This change in drag force, as well as other environmental conditions such as wind, affects the horizontal component of the velocity vector.

Although graphical techniques of vector analysis provide good visual pictures of projectile motion, trigonometric methods provide more accurate information. Recall that vectors are drawn within the coordinate system so that the tail is at the origin (0, 0), and the tip has x and y coordinates. The magnitude is represented by r, and the orientation of the vector is given by θ relative to the positive x-axis. Drawing a line that connects the tip of the vector to the x-axis forms a triangle. Trigonometric vector resolution is simply the process of finding the x and y coordinates (components). In this case, the x and y coordinates represent the location of the tips of two component vectors (Figure 10.9).

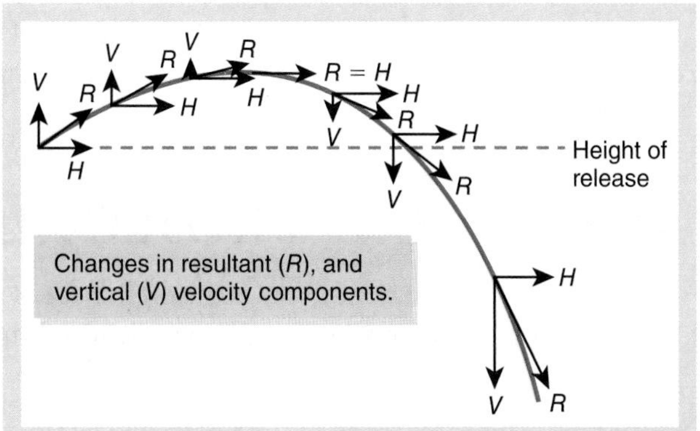

Figure 10.8 **Vector resolution of the parabolic path caused by gravity.**

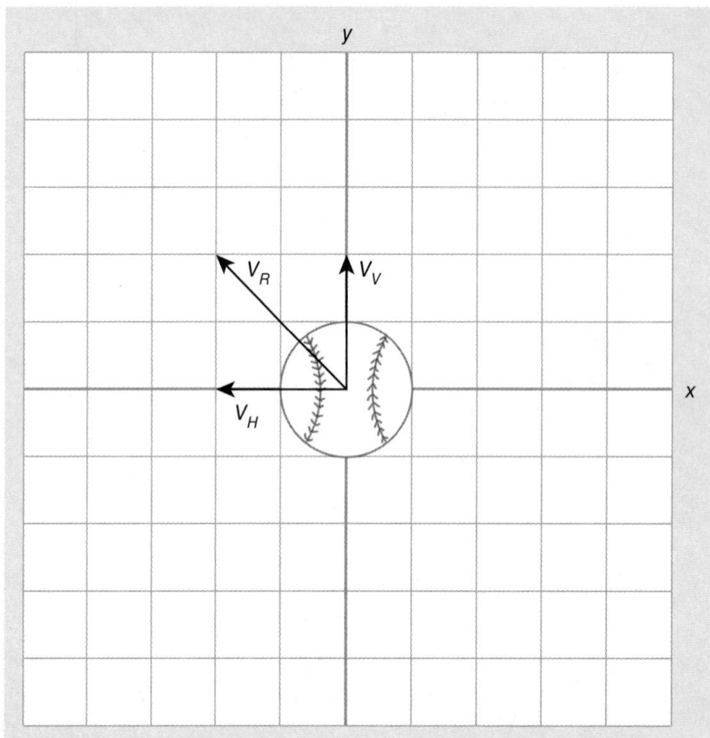

Figure 10.9 **Location of the center of gravity of a ball in a coordinate reference system.**

SAMPLE PROBLEM

For example, let's say that on release a projectile has a velocity of 10 m/sec and is released at an angle of 50° relative to the positive *x*-axis. Recall that Equations 3.1 and 3.2 can be used to transform the polar coordinates representative of our given velocity vector (10 m/sec, 50°) to rectangular coordinates:

$$y = r \sin \theta = (10 \text{ m/sec}) \sin 50° = \mathbf{+7.66 \text{ m/sec}}$$

$$x = r \cos \theta = (10 \text{ m/sec}) \cos 50° = \mathbf{+6.43 \text{ m/sec}}$$

So the initial velocity vector of 10 m/sec has a vertically directed component of 7.66 m/sec and a horizontally directed component of 6.43 m/sec. The plus signs indicate the positive *x* and *y* directions. As you can see, the use of trigonometry easily provides exact values for the magnitudes of the vertical and horizontal components and is relatively easy to use.

10.2 PROJECTILE TRAJECTORY

The **trajectory** is the flight path of a projectile; more specifically, the flight path of the center of gravity of a projectile. You have learned that once the projectile is released, this trajectory is affected by both gravity and fluid resistance. Gravitational and fluid forces are not under the performer's control. However, three biomechanical factors affect trajectory: (1) projection angle; (2) projection velocity; and (3) relative projection height (Figure 10.10).

Trajectory The term for the flight path of a projectile; more specifically, the flight path of the center of gravity of a projectile.

Of course, these three factors exert their effects before the projectile is released. Once it is released, its trajectory is influenced only by gravity and fluid resistance.

Projection Angle

You have learned that in the absence of any force other than gravity, the center of gravity of any object projected horizontally follows a predictable parabolic trajectory. However, because dynamic fluid force is applied to the projectile, the symmetrical shape of the parabolic trajectory may be altered. So fluid resistance is an uncontrollable factor that dictates the shape of projectile trajectory. The shape of the parabolic path can also vary, depending upon the angle at which the object is projected, which is called the **projection angle** or *angle of attack*. We can observe through graphical vector analysis that this change in the shape of the trajectory occurs because the relative magnitudes of the vertical and horizontal components vary with the angle of the projectile velocity vector (Figure 10.11).

To gain a deeper understanding, one can also resolve the velocity vector into its vertical and horizontal components through trigonometric methods. Let's begin with one extreme situation: a projectile that is released perfectly horizontally (i.e., a projection angle of 0° relative to the positive x-axis) possessing a vector magnitude of 20 m/sec (Figure 10.12).

Equations 3.1 and 3.2 can be used to transform the polar coordinates representative of our given velocity vector (20 m/sec, 0°) to rectangular coordinates:

$$y = r \sin \theta = (20 \text{ m/sec}) \sin 0° = \mathbf{+0.00 \text{ m/sec}}$$

$$x = r \cos \theta = (20 \text{ m/sec}) \cos 0° = \mathbf{+20.00 \text{ m/sec}}$$

So the initial velocity vector of 20 m/sec has a vertically directed component of 0.00 m/sec and a horizontally directed component of 20.00 m/sec. This result simply means that the projectile will not travel at all in the positive vertical direction ($+y$). However, it does *not* mean that the projectile will have a trajectory that is a perfectly straight line (rectilinear). Remember that gravity begins to act vertically downward on the projectile as soon as it is free of support. As we can see, this force will create a trajectory that is half of a parabola.

Let's now analyze a situation that illustrates the opposite extreme, a projectile that is released perfectly vertically (i.e., a projection angle of 90° relative to the positive x-axis) and possesses a vector magnitude of 20 m/sec. With graphical vector analysis, this situation would involve colinear vectors.

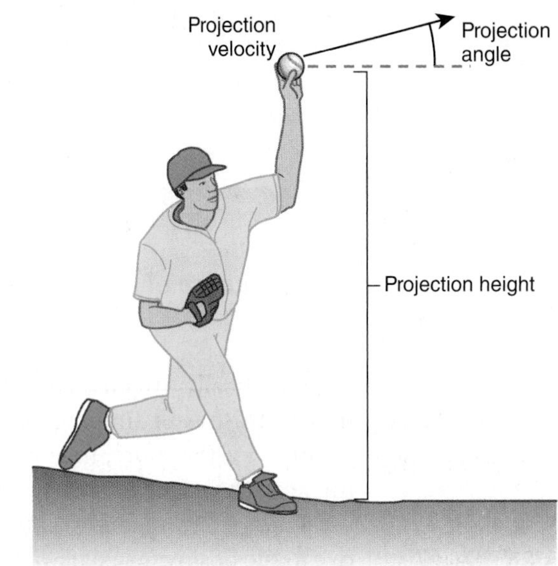

Figure 10.10 **Three factors affecting trajectory.**

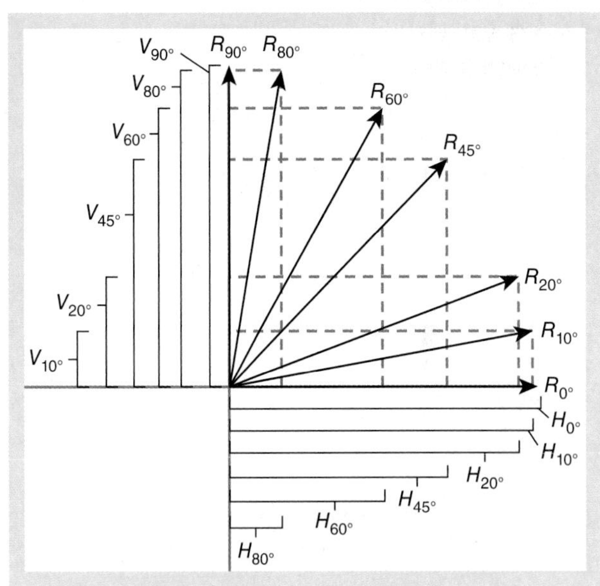

Figure 10.11 **Relative component magnitudes depending on angle of projection.**

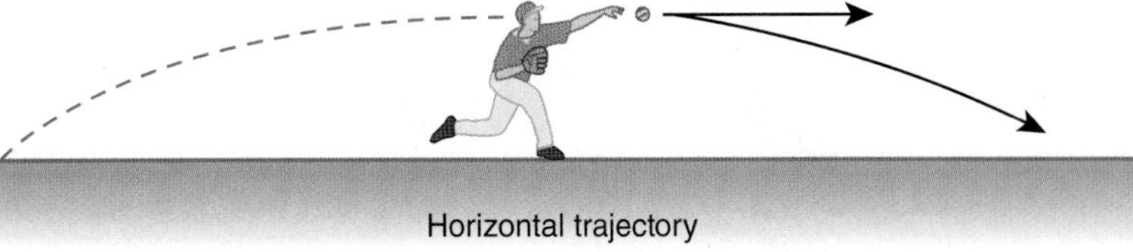

Figure 10.12 **Parabolic path of a projectile released perfectly horizontally.**

Projection angle The angle at which the object is projected; also called *angle of attack*.

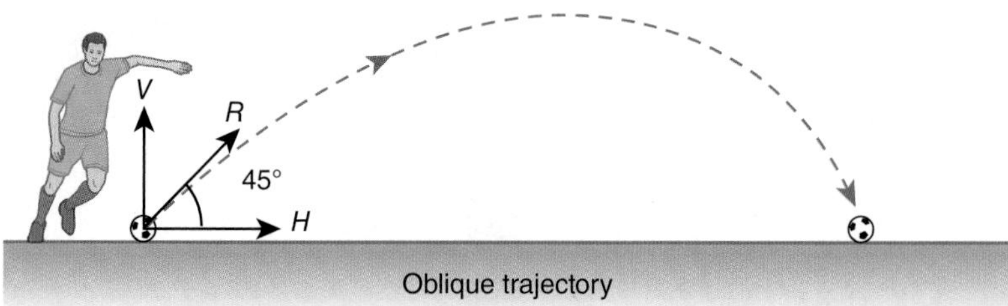

Figure 10.13 **Vector resolution of a projectile released at 45°.**

We again use Equations 3.1 and 3.2 to transform the polar coordinates representative of our given velocity vector (20 m/sec, 90°) to rectangular coordinates:

$$y = r \sin \theta = (30 \text{ m/sec}) \sin 90° = \mathbf{+20.00 \text{ m/sec}}$$

$$x = r \cos \theta = (30 \text{ m/sec}) \cos 90° = \mathbf{+0.00 \text{ m/sec}}$$

So the initial velocity vector of 20 m/sec has a vertically directed component of 20.00 m/sec and a horizontally directed component of 0.00 m/sec. This result simply means that the projectile will not travel at all in the horizontal direction (+ or $-x$). Therefore, the projectile will have a rectilinear path vertically upward and then a rectilinear path vertically downward caused by the force of gravity.

Finally, if we analyze a situation of an oblique projection angle (any angle between 0° and 90° relative to the positive x-axis), we find that it has both a horizontal and a vertical component. Let's analyze a projectile released at 45° relative to the positive x-axis, which possesses a velocity vector magnitude of 20 m/sec (Figure 10.13).

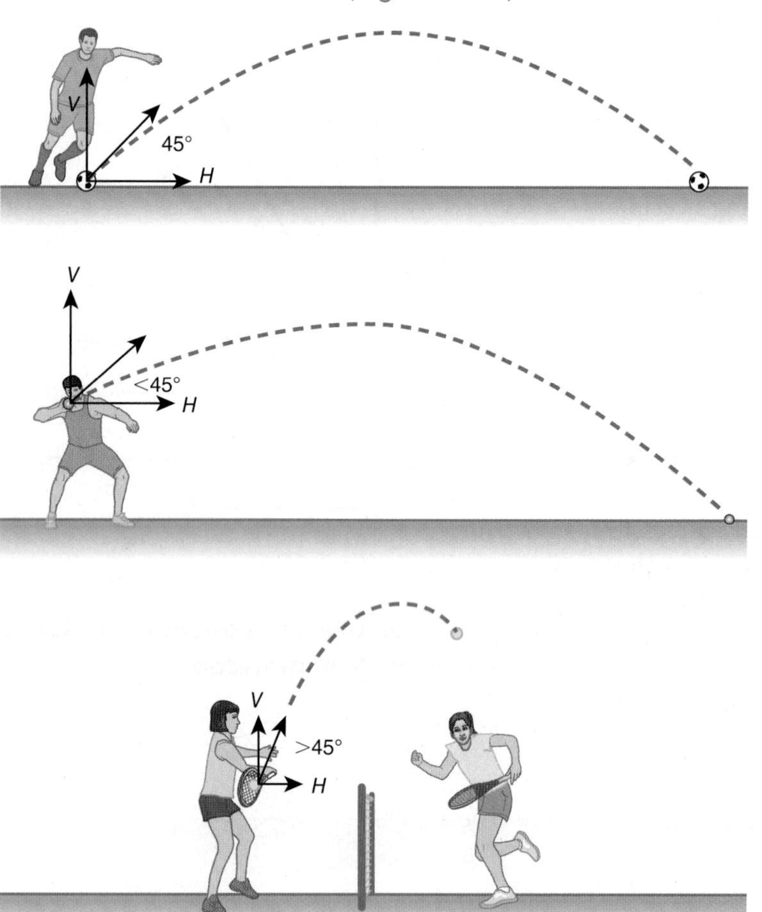

Figure 10.14 **Parabolic shape of trajectory of three oblique projection angles.**

We again use Equations 3.1 and 3.2 to transform the polar coordinates representative of our given velocity vector (20 m/sec, 45°) to rectangular coordinates:

$$y = r \sin \theta = (20 \text{ m/sec}) \sin 45° = \mathbf{+14.14 \text{ m/sec}}$$

$$x = r \cos \theta = (20 \text{ m/sec}) \cos 45° = \mathbf{+14.14 \text{ m/sec}}$$

So the initial velocity vector of 20 m/sec has a vertically directed component of 14.14 m/sec and a horizontally directed component of 14.14 m/sec. This result means that the projectile will travel in both the x direction and y direction and describe a parabolic trajectory because of the force of gravity. Equal vertical and horizontal components occur only at a projection angle of 45°. Also consider that fluid resistance may somewhat distort the parabolic shape of the trajectory, depending on the mass of the projectile and the flight time.

Any other oblique projectile angle (angles between 0° and 90° relative to the positive x-axis) also produces a parabolic trajectory, but the shape of the parabola (specifically its steepness) depends on the relative sizes of the vertical and horizontal components. And, of course, those relative sizes depend on the projection angle (Figure 10.14).

Projection Velocity

Whereas projection angle and fluid resistance influence the shape of the projectile's trajectory, a variable that affects the height and length (or size) of the trajectory is the **projection velocity** (initial velocity or velocity at release). Note that projection *velocity* is a vector quantity, possessing both magnitude and direction. The *magnitude* is projection speed, and the *direction* is projection angle. So both characteristics are represented by the initial velocity vector. Once the projection velocity is resolved into its horizontal and vertical components, each component possesses a given *speed* in its respective direction.

Let's demonstrate the concept of projection velocity using the same three situations used in the section on projection angle: perfect horizontal projection, perfect vertical projection, and oblique projection. In this case, though, we will calculate two different projection velocities while holding the projection angle constant.

We begin with projectiles that are released perfectly horizontally (i.e., a projection angle of 0° relative to the positive *x*-axis; Figure 10.15).

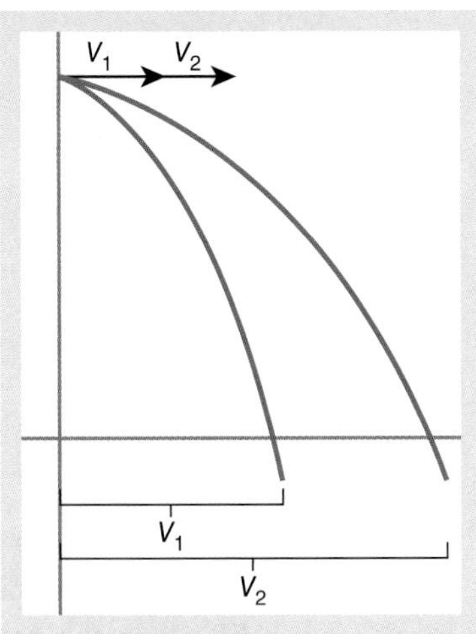

Figure 10.15 **Projectiles released perfectly horizontally with two different velocities.**

Projection velocity The initial velocity or velocity at release of a projectile.

SAMPLE PROBLEM

We found previously that a projectile released at 0° with a projection velocity of 20 m/sec had the following rectangular coordinates:

$$y = r \sin \theta = (20 \text{ m/sec}) \sin 0° = \mathbf{+0.00 \text{ m/sec}}$$

$$x = r \cos \theta = (20 \text{ m/sec}) \cos 0° = \mathbf{+20.00 \text{ m/sec}}$$

The initial velocity vector of 20 m/sec has a vertically directed component of 0.00 m/sec and a horizontally directed component of 20.00 m/sec. This result means that the projectile will not travel at all in the positive vertical direction (+*y*), and this absence of vertical motion will create a trajectory that is half of a parabola.

If we maintain the projection angle at 0° but change the projection velocity to 30 m/sec, we find the following rectangular coordinates:

$$y = r \sin \theta = (20 \text{ m/sec}) \sin 0° = \mathbf{+0.00 \text{ m/sec}}$$

$$x = r \cos \theta = (20 \text{ m/sec}) \cos 0° = \mathbf{+30.00 \text{ m/sec}}$$

In this case, the initial velocity vector of 30 m/sec has a vertically directed component of 0.00 m/sec and a horizontally directed component of 30.00 m/sec. This result is simple in mathematical terms, but it helps to illustrate our point. If projection angle and flight time are held constant, projection velocity will determine the *length* of the half-parabola trajectory.

Let's now analyze the opposite extreme, a projectile that is released perfectly vertically (i.e., a projection angle of 90° relative to the positive *x*-axis) and possesses a vector magnitude of 20 m/sec (Figure 10.16).

(continues)

(continued)

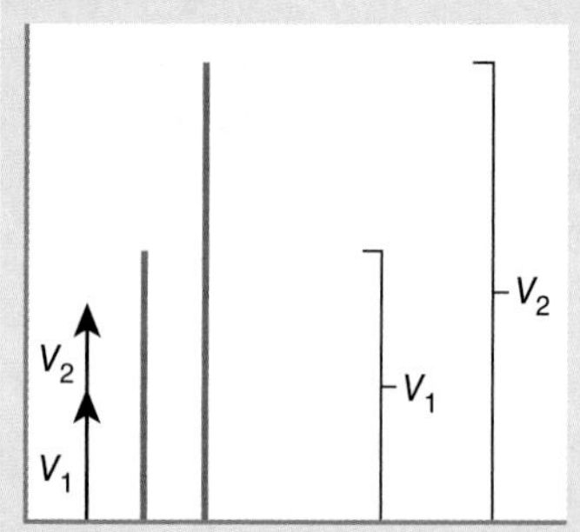

Figure 10.16 **Projectiles released perfectly vertically with two different projection velocities.**

We found earlier that the transformed polar coordinates representative of our given velocity vector (20 m/sec, 90°) have the following rectangular coordinates:

$$y = r \sin \theta = (30 \text{ m/sec}) \sin 90° = \mathbf{+20.00 \text{ m/sec}}$$

$$x = r \cos \theta = (30 \text{ m/sec}) \cos 90° = \mathbf{+0.00 \text{ m/sec}}$$

So the initial velocity vector of 20 m/sec has a vertically directed component of 20.00 m/sec and a horizontally directed component of 0.00 m/sec. Therefore, the projectile will have a rectilinear path vertically upward and then a rectilinear path vertically downward because of the force of gravity.

If we maintain the projection angle at 90° but change the projection velocity to 30 m/sec, we find the following rectangular coordinates:

$$y = r \sin \theta = (30 \text{ m/sec}) \sin 90° = \mathbf{+30.00 \text{ m/sec}}$$

$$x = r \cos \theta = (30 \text{ m/sec}) \cos 90° = \mathbf{+0.00 \text{ m/sec}}$$

So the initial velocity vector of 30 m/sec has a vertically directed component of 30.00 m/sec and a horizontally directed component of 0.00 m/sec. Again, this result is simple in mathematical terms, but it helps illustrate our second point. If projection angle and flight time are held constant, projection velocity will determine the *height* or *apex* of the trajectory.

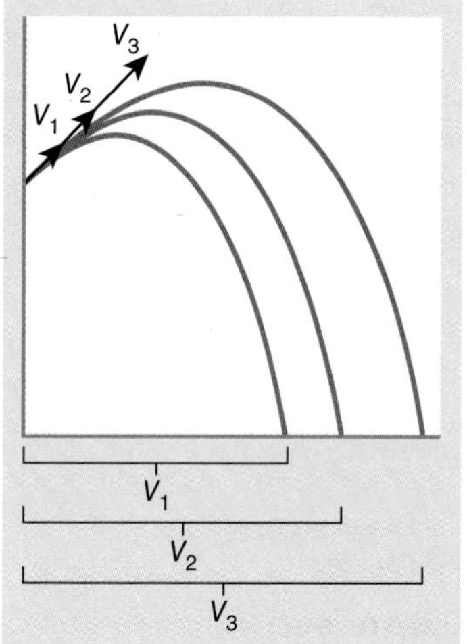

Figure 10.17 **Three projectiles released at 45° with different projection velocities.**

Finally, if we analyze a situation of an oblique projection angle (any angle between 0° and 90° relative to the positive *x*-axis), we find that both a horizontal and a vertical component exist. Let's analyze projectiles released at 45° relative to the positive *x*-axis (Figure 10.17).

Previously we used Equations 3.1 and 3.2 to transform the polar coordinates representative of our given velocity vector (20 m/sec, 45°) to rectangular coordinates:

$$y = r \sin \theta = (20 \text{ m/sec}) \sin 45° = \mathbf{+14.14 \text{ m/sec}}$$

$$x = r \cos \theta = (20 \text{ m/sec}) \cos 45° = \mathbf{+14.14 \text{ m/sec}}$$

So the initial velocity vector of 20 m/sec has a vertically directed component of 14.14 m/sec and a horizontally directed component of 14.14 m/sec and describes a parabolic trajectory caused by the force of gravity.

If we maintain the projection angle at 45° but change the projection velocity to 40 m/sec and then 60 m/sec, we find the following rectangular coordinates:

$$y = r \sin \theta = (40 \text{ m/sec}) \sin 45° = \mathbf{+28.28 \text{ m/sec}}$$

$$x = r \cos \theta = (40 \text{ m/sec}) \cos 45° = \mathbf{+28.28 \text{ m/sec}}$$

$$y = r \sin \theta = (60 \text{ m/sec}) \sin 45° = \mathbf{+42.43 \text{ m/sec}}$$

$$x = r \cos \theta = (60 \text{ m/sec}) \cos 45° = \mathbf{+42.43 \text{ m/sec}}$$

Again, we notice that a projection angle of 45° produces equal vertical and horizontal components. Also, the trajectory describes a parabola because the projection angle is oblique. But our example also shows that when projection angle is held constant, the height and length (or size) of the parabolic trajectory is determined by the projection velocity. So these examples illustrate three parabolic trajectories, each with a different height of apex and length.

Projection Height

Relative projection height is the final factor that can influence the trajectory of a projectile. **Relative projection height** (Figure 10.18) is the mathematical difference between the height of projectile release (**projection height**) and the height of projectile impact or landing (**impact height**).

Relative projection height is zero if a ball is kicked from the playing surface and then lands on a part of the surface with the same height. Relative projection height of a baseball pitch is a positive value, because the projection

Relative projection height The mathematical difference between the height of projectile release (projection height) and the height of projectile impact or landing (impact height).

Projection height The height of projectile release.

Impact height The height of projectile impact or landing.

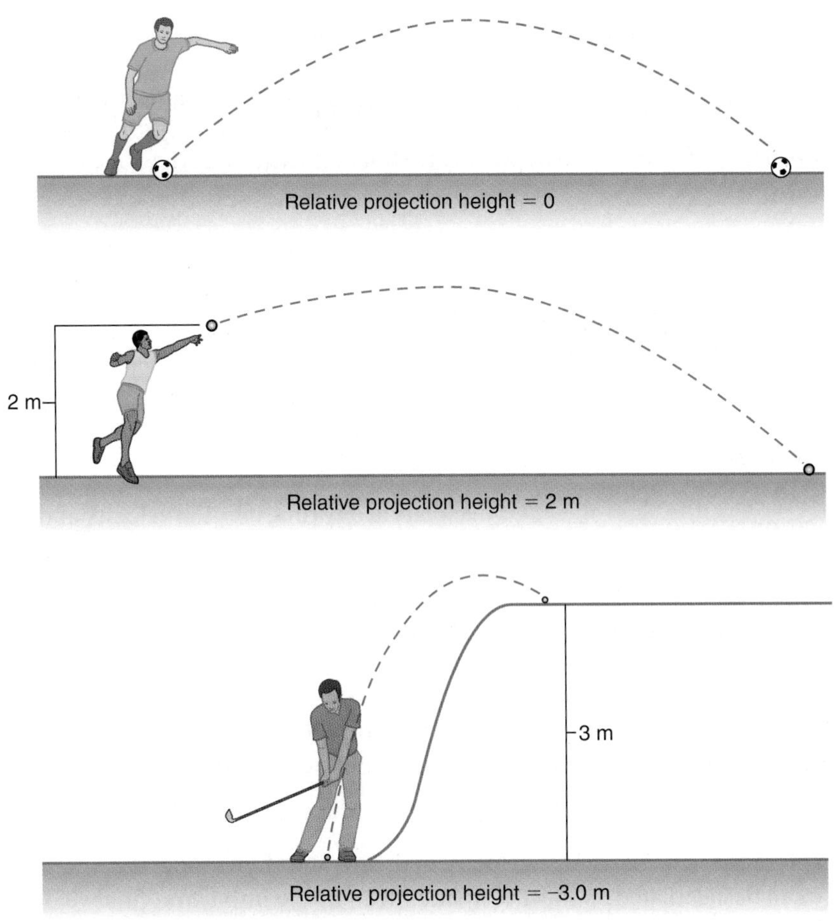

Figure 10.18 **Comparison of different relative projection heights.**

height is a greater value than the impact height (assuming the ball is caught by the catcher). Relative projection height is a negative number if one chips a golf ball from inside of a bunker up to the green. It's negative because the impact height is greater than the projection height.

With all other factors held constant, the greater the relative projection height, the longer the flight time of the projectile. Flight time is extended because a positive relative projection height produces a parabolic trajectory that has a relatively high apex, whereas a negative projection height truncates the parabolic path. However, this rule does not mean that greater relative projection height is always associated with optimum projection. As we shall see, optimal relationship between projection angle, projection velocity, and projection height is delicate, depending on the ultimate goal of the projection.

10.3 LAWS OF UNIFORMLY ACCELERATED MOTION

One final concept that we must understand before analyzing specific forms of projection is the **laws of uniformly accelerated motion**, which was originally derived by Galileo Galilei (1564–1642). Through experimentation, Galileo made several observations that relate acceleration, displacement, time, and velocity in the condition of constant gravitational acceleration. These ideas apply directly to projectile motion. Let's begin with an equation with which we are already familiar, that for linear acceleration:

$$a = \frac{\Delta v}{\Delta t} \tag{5.3}$$

where

a = the acceleration, or rate of change in linear velocity
Δv = $\text{velocity}_{\text{final}} - \text{velocity}_{\text{initial}}$ (change in velocity)
Δt = $\text{time}_{\text{final}} - \text{time}_{\text{initial}}$ (the change in time, or time interval)

If we simply rearrange Equation 5.3 by solving for final velocity, we derive the **first law of uniformly accelerated motion**:

$$v_f = v_i + at \tag{10.1}$$

where

v_f = final velocity
v_i = initial velocity
a = the acceleration, or rate of change in linear velocity
t = time

This equation enables us to calculate a projectile's velocity at any time if initial velocity and a constant acceleration are known (Serway & Jewett, 2019).

Laws of uniformly accelerated motion Several observations made by Galileo that relate acceleration, displacement, time, and velocity in the condition of constant gravitational acceleration.

First law of uniformly accelerated motion A projectile's final velocity is related to its initial velocity and the value of constant acceleration.

SAMPLE PROBLEM

For example, if a ball rebounds off a wall at 5 m/sec and is struck by an implement (moving in the same direction) with enough force to accelerate the ball at a rate of 10 m/sec^2, we can calculate final velocity if we know that the implement and the ball made contact for 0.5 sec: v_f = 5 m/sec + (10 m/sec^2 × 0.5 sec) = 10 m/sec.

Next let's explore the idea of average velocity in greater depth. As long as acceleration is constant, average velocity can be calculated as the arithmetic mean of the initial and final velocities:

$$v_a = \frac{v_i + v_f}{2} \tag{10.2}$$

where

v_a = average velocity
v_i = initial velocity
v_f = final velocity

Next, recall that linear displacement (d) = $\text{position}_{\text{final}} - \text{position}_{\text{initial}}$ and the equation for velocity:

$$v = \frac{d}{\Delta t} \quad (5.2)$$

where

v = the velocity, or rate of motion in a specific direction
$d = \text{position}_{\text{final}} - \text{position}_{\text{initial}}$ (displacement, or change in position)
$\Delta t = \text{time}_{\text{final}} - \text{time}_{\text{initial}}$ (the change in time, or time interval)

If we use Equation 10.2 along with the equations for displacement and velocity and solve for final position, we arrive at the following equation:

$$p_{final} = p_{initial} + \frac{(v_i + v_f)t}{2} \quad (10.3)$$

where

p_{final} = final position
$p_{initial}$ = initial position
v_i = initial velocity
v_f = final velocity
t = time

This equation enables us to calculate final position of a projectile with respect to initial and final velocities (Serway & Jewett, 2019).

Second law of uniformly accelerated motion The final position of a projectile is related to its initial velocity and acceleration.

SAMPLE PROBLEM

For example, an implement has an initial velocity of 20 m/sec and a final velocity of 5 m/sec. The flight time for the implement is 3.5 seconds. We can calculate its final position as: $p_{final} = 0 +$ [(20 m/sec + 5 m/sec) 3.5 sec]/2 = 43.75 m.

We can substitute Equation 10.1 into Equation 10.3 to derive the **second law of uniformly accelerated motion**:

$$p_{final} = p_{initial} + v_i t + \frac{at^2}{2} \quad (10.4)$$

where

p_{final} = final position
$p_{initial}$ = initial position
v_i = initial velocity
t = time
a = acceleration

This equation enables us to calculate the final position of a projectile with respect to initial velocity and acceleration (Serway & Jewett, 2019).

SAMPLE PROBLEM

For example, suppose that a person on top of a building throws a ball perfectly downward in the negative *y direction* with an initial velocity of 7 m/sec. We can calculate its position 3 seconds later as: $p_{final} = 0 + (7\text{ m/sec} \times 3\text{ sec}) + [(9.81\text{ m/sec}^2)(3\text{ sec})^2]/2 = 61.145\text{ m}$.

Finally, we can substitute the value of *t* from Equation 10.1 into Equation 10.3 to arrive at the **third law of uniformly accelerated motion**:

$$v_f^2 = v_i^2 + 2a(d) \tag{10.5}$$

where

v_f^2 = final velocity squared
v_i^2 = initial velocity squared
a = the acceleration, or rate of change in linear velocity
d = position$_{\text{final}}$ − position$_{\text{initial}}$ (displacement, or change in position)

This equation enables us to calculate the final velocity of a projectile with respect to acceleration and displacement (Serway & Jewett, 2019).

SAMPLE PROBLEM

Consider the previous example. The person on top of the building throws a ball perfectly downward in the negative *y direction* with an initial velocity of 7 m/sec. We can calculate final velocity of the ball after it has traveled the 61.145 m that we calculated: $v_f = \sqrt{[(7\text{ m/sec})^2 + (2 \times 9.81\text{ m/sec}^2 \times 61.145\text{ m})]} = 35.34\text{ m/sec}$. Note that these are laws of constantly accelerated motion, so this value reflects the absence of drag force.

At first glance, the equations representing the laws of uniformly accelerated motion may seem far removed from the everyday projection situations to which we are accustomed. In other words, math is usually not the first thought that comes to mind when one is throwing a football. However, in the following sections we will see that it is directly applicable.

10.4 PROJECTION FOR VERTICAL DISTANCE

In sports, few instances of perfectly vertical projection exist. However, in some situations, the vertical component of projection must be maximized (e.g., high jumping and pole vault). When analyzing these events, the same factors we have just examined must be considered: (1) gravity, (2) air resistance, (3) relative projection angle, (4) projection velocity, and (5) relative projection height. Therefore, an analysis of vertical projection must include these factors as well as the laws of uniformly accelerated motion. Previously in this chapter, you read about the potential effects of gravity and dynamic fluid forces. This section focuses on the factors that are under the performer's control.

The third law of uniformly accelerated motion can give us a clearer understanding of vertical projectile motion:

$$v_f^2 = v_i^2 + 2a(d) \tag{10.5}$$

By making a few assumptions, we can simplify Equation 10.5. First, let's assume that vertical velocity of the projectile at the apex of its trajectory is equal to zero (i.e., $v_f^2 = 0$). Also remember that the acceleration in this case is caused by gravity (g) and is therefore equal to 9.81 m/sec^2.

Third law of uniformly accelerated motion The final velocity of a projectile is related to its acceleration and displacement.

In addition, gravity is acting vertically downward (in the negative *y* direction) and therefore has a negative value. Finally, the projectile velocity vector may not be perfectly vertical. With these assumptions in mind, Equation 10.5 takes on a more simplified form:

$$d = \frac{v_i^2}{2g} \qquad \text{OR} \qquad d = \frac{(v_i \sin\theta)^2}{2g} \tag{10.6}$$

where

d = $\text{position}_{\text{final}} - \text{position}_{\text{initial}}$ (vertical displacement, or change in position)
v_i = initial vertical velocity squared
g = acceleration caused by gravity (9.81 m/sec^2)

We can use this equation to calculate the vertical displacement of an object projected with a given initial velocity. Note that this displacement is vertical from the point of projection. Also notice that if projection is perfectly vertical, either version of the equation can be used, because sin 90° is 1.00. If, however, projection is not perfectly vertical, we must account for the fact that some horizontal displacement will occur.

SAMPLE PROBLEM

For example, if a projectile is released perfectly vertically (90° relative to the horizontal) and has an initial velocity of 25 m/sec, it will travel a vertical distance of 31.86 m. If, however, the object has the same initial velocity (25 m/sec) but an angle of projection of 75°, it will travel a vertical distance of 29.72 m. The loss in vertical height occurs because the resultant velocity vector now has a horizontal component. In the extreme case of a perfectly horizontal projection angle (0° relative to the horizontal), positive vertical displacement is zero.

Now let's return to our list of factors that affect projectile trajectory: (1) projection angle, (2) projection velocity, and (3) relative projection height. Using Equation 10.6, we can definitely understand in our examples that the size of the vertical component of the velocity vector varies with projection angle. The closer the angle of projection to 90°, the greater the size of the vertical component.

Also, we can clearly see that the projection velocity is a major factor in vertical displacement of a projectile. Therefore, we must optimize use of the kinetic chain to impart as much velocity as possible to the projectile before release or takeoff. Sometimes we use devices to maximize projection velocity. For example, springboards in diving help add projection velocity because of their elastic recoil.

The effect of relative projection height may not be as readily apparent when using Equation 10.6. Recall that with all other factors held constant, the greater the relative projection height, the longer the flight time of the projectile. Equation 10.6 is used to calculate vertical displacement from the point of projection. Although relative projection height is not a factor in the equation itself, it does affect the height of the apex of the trajectory.

SAMPLE PROBLEM

For example, we calculated that if a projectile is released perfectly vertically (90° relative to the horizontal) and has an initial velocity of 25 m/sec, it will travel a vertical distance of 31.86 m. In this example, no relative projection height is specified. Let's now compare two relative projection heights for the same projectile. If the object is released 1 m from the ground, the apex of its trajectory will be 32.86 m *relative to the ground* (1 m + a vertical displacement of 31.86 m from the point of projection). Just to emphasize, the apex will be 31.86 m relative to the projection point, but it will be 32.86 m relative to the ground. If the same object is released 2 m from the ground, the apex of its trajectory will be 33.86 m *relative to the ground*.

Figure 10.19 **Straddle versus back layout high jump styles.**

Relative projection height can be especially important in situations in which the human body is the projectile—for example, in high jumping. Recall that the trajectory is the flight path of the *center of gravity* of a projectile. You have learned that the human body's center of gravity can be shifted by relocating mass—that is, the center of gravity will be relocated closer to the more massive area. Therefore, during a high jump (while still on the ground), one can raise the center of gravity of the body by flexing the arms about the shoulders and one of the thighs about the hip before takeoff. Raising the center of gravity increases the relative projection height and therefore leads to greater vertical displacement (or apex) relative to the ground (Figure 10.19).

Of course, many factors must be maximized to be successful in the high jump event, but maximizing relative projection height must not be ignored.

As you can see from this discussion of vertical projection, some degree of horizontal displacement is almost always present. So let's examine horizontal projection.

10.5 PROJECTION FOR HORIZONTAL DISTANCE

In most situations of projection, at least some degree of horizontal displacement occurs. The total horizontal displacement of a projectile is called the **range**. One important fact to remember is that the horizontal component of the velocity vector is perpendicular to the vector representing the force of gravity. Therefore, unlike the vertical component of the velocity vector, the horizontal component remains relatively unchanged throughout the trajectory of the projectile. Of course, fluid resistance is present as described previously, but we are assuming its effect to be negligible so we can focus on using the equations for uniformly accelerated motion. With this assumption in mind, we can simplify Equation 10.4 in the case of horizontal projection to arrive at the following variation of the *second law of uniformly accelerated motion*:

$$d = v_i t \quad (10.4)$$

Range The total horizontal displacement of a projectile.

where

d = position$_{final}$ − position$_{initial}$ (horizontal displacement, or change in position)
v_i = initial *horizontal* velocity
t = time of flight

This calculation is relatively straightforward as long as initial velocity and flight time are known and the projectile is released perfectly horizontal to the ground. For example, if an object is projected perfectly horizontally (0° relative to the positive x-axis) with an initial velocity of 20 m/sec and a flight time of 2 sec, the object projectile should travel 40 m horizontally. What if the projectile is released at some other angle? We know from the previous section that the angle of projection affects the magnitude of the vertical component of the initial velocity vector. This relationship means that changing the projection angle changes the vertical displacement of the projectile, which in turn affects the total time of flight. Therefore, the range of a projectile is affected by both the projection velocity and the projection angle.

In situations of projection with angles other than 0°, we must return to the *first law of uniformly accelerated motion*:

$$v_f = v_i + at \tag{10.1}$$

If we recall that final vertical velocity is at the apex of the trajectory and is equal to zero at that point, we can use Equation 10.1 to solve for flight time.

SAMPLE PROBLEM

Let's say that a projectile is released with a projection velocity of 20 m/sec and a projection angle of 10°. In this case, the equation will appear as follows:

$$0 = (20 \text{ m/sec} \times \sin 10°) + (-9.81 \text{ m/sec}^2)t$$

Remember that acceleration is caused by gravity and acts vertically downward; thus, the value is negative. We can then solve for time and find a *time to apex* of 0.354 sec. Notice that this value is the *time to apex*. If we assume that the path is parabolic, then it should take another 0.354 sec for the projectile to travel from the apex of the trajectory to the landing site; that is, if the relative projection height is equal to zero. Therefore, total flight time of the projectile is 0.708 sec (2 × 0.354 sec). Now that flight time is known, we can use Equation 10.4 to solve for the range:

$$d = (20 \text{ m/sec} \times \cos 10°)(0.708 \text{ sec}) = 13.945 \text{ m}$$

Note that another equation is sometimes used to calculate the range of a projectile when flight time is unknown. This equation is derived by substituting the value of t from Equation 10.1 into Equation 10.4 using the same assumptions as in Equations 10.6. Remember, however, that the value of t is the time to apex. The full parabolic path requires two times this value:

$$d = \frac{(v_i^2 + \sin 2\theta)}{g} \tag{10.7}$$

where

d = position$_{final}$ − position$_{initial}$ (horizontal displacement, or change in position)
v_i = initial velocity squared
g = acceleration caused by gravity (9.81 m/sec²)

If the values from the previous example (projection velocity of 20 m/sec and a projection angle of 10°) are substituted into Equation 10.7, we find the same answer of 13.945 m.

Once again, we return to the factors that affect projectile trajectory: (1) projection angle, (2) projection velocity, and (3) relative projection height. From the preceding example, you can see that projection for horizontal distance is affected by both the projection angle and projection velocity (Table 10.1).

More specifically, the range of a projectile depends on initial velocity and flight time. The flight time depends on the projection angle and relative projection height. Therefore, horizontal range depends on both the horizontal and vertical components of the initial velocity vector. Also recall from our previous sections that the relative sizes of the vertical and horizontal components depend on the angle of projection: The closer the projection angle is to 90°, the larger the vertical component; the

Table 10.1
Range as a Function of Velocity and Projection Angle

Projection Angle (°)	Projection Velocity (m/s)	Range (m)
10	10	3.49
20	10	6.55
30	10	8.83
40	10	10.04
45	10	10.19
50	10	10.04
60	10	8.83
70	10	6.55
80	10	3.49
10	20	13.94
20	20	26.21
30	20	35.31
40	20	40.15
45	20	40.77
50	20	40.15
60	20	35.31
70	20	26.21
80	20	13.94
10	30	31.38
20	30	58.97
30	30	79.45
40	30	90.35
45	30	91.74
50	30	90.35
60	30	79.45
70	30	58.94
80	30	31.38

closer to 0°, the larger the horizontal component. However, we must now think of this statement in terms of maximizing range (horizontal displacement). We now know that a smaller projection angle creates a horizontal component of greater magnitude but a smaller total flight time. A greater projection angle creates a smaller horizontal component, but a greater total flight time (because of the larger vertical component). Therefore, if the relative projection height is equal to zero, we tend to find that the optimal angle for horizontal projection at a given projection velocity is 45° because this angle maximizes both the vertical and horizontal components (Figure 10.20 and Table 10.2).

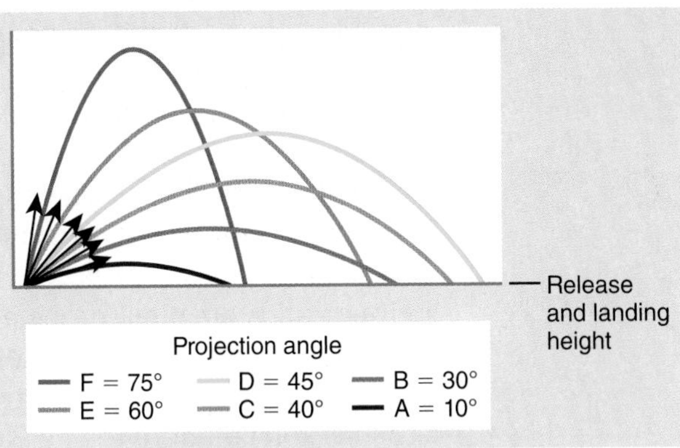

Figure 10.20 **Range for a given projection velocity with varying projection angle.**

Of course, this optimal angle is for projecting maximum horizontal *displacement*. In addition, many sport-related projectiles have a relative projection height that is not equal to zero. Also, notice no mention of projection horizontally for *maximum velocity*. In other words, in some activities the range is not the major issue. Instead, it is velocity when the projectile arrives at the target that is most important. Pitching in baseball is an example of an activity in which the range is a constant, and maximal velocity is desired. In this case, the projection angle can be minimized to maximize the horizontal component of the velocity vector. Also, neither of the previous sections consider the accuracy of the projectile. The next section covers these issues.

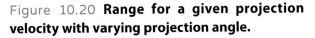

Table 10.2
Range as a Function of Velocity and Projection Angle

Projection Angle (°)	Projection Velocity (m/s)	Range (m)
15	10	5.1
15	20	20.4
15	30	45.9
30	10	8.8
30	20	35.3
30	30	79.5
45	5	2.5
45	10	10.2
45	15	22.9
45	20	40.8
45	25	63.7
45	30	91.7
60	10	8.8
60	20	35.3
60	30	79.5
75	10	5.1
75	20	20.4
75	30	45.9

10.6 PROJECTION FOR ACCURACY

In many situations, simply projecting for maximum vertical or horizontal displacement is not the goal. Some sports require maximum velocity, accuracy, or both instead of maximum displacement. For example, in baseball pitching or shooting a foul shot in basketball, the distance is a constant. Therefore, accuracy is a key factor in both skills. Also, as you have read, in baseball pitching, horizontal velocity of the projectile is a major component. In addition, relative projection height for a given skill is often out of the performer's control. Therefore, when analyzing any skill, one must decide the appropriate balance of projection angle and projection velocity.

Recall that a smaller projection angle creates a horizontal component of greater magnitude but a smaller total flight time. A greater projection angle creates a smaller horizontal component but a greater total flight time (because of the larger vertical component). With this information in mind, we tend to find that the optimal angle for horizontal projection at a given projection velocity is 45° because this projection angle maximizes both the vertical and horizontal components. Also, the height and length (or range) of the trajectory are affected by projection velocity. The greater the projection velocity, the greater the magnitudes of both the vertical and horizontal components of the initial velocity vector.

With this information in mind, some skills are easier to analyze than others. For instance, when projecting for maximum vertical distance, one must maximize projection velocity and increase projection angle as much as possible up to 90°. If projection for horizontal distance is necessary, projection velocity must once again be maximized, and the optimal angle of projection is 45°. However, this optimal projection angle of 45° holds only if the relative projection height is zero (i.e., no difference exists between the projection height and the landing height). This is the point at which the analysis becomes more complicated, and a delicate balance must be struck between projection height, projection angle, and projection velocity.

Projection and landing heights are often dictated by the event and therefore are out of the performer's control. Also, projection velocity is chosen according to the need for maximal distance. So in terms of projecting for accuracy, the projection angle becomes the most important factor. You have learned that for a given projection velocity, a projection angle of 45° is optimal for horizontal displacement as long as the projection height is equal to zero (e.g., kicking a soccer ball from the ground). However, when velocity is held constant, relative projection height determines the optimal projection angle. In general, as relative projection height increases, the optimal projection angle decreases; and as relative projection height decreases (becomes more negative), the optimal projection angle increases (Figures 10.21 and 10.22).

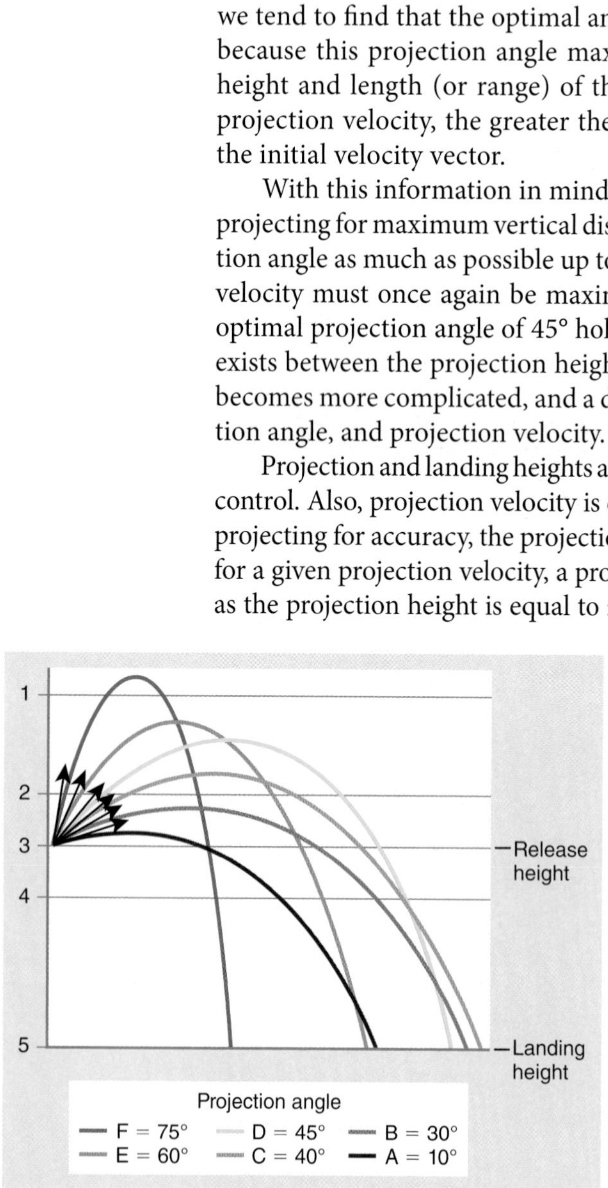

Figure 10.21 **Influence of relative projection height on optimal projection angle.**

Take the example of pitching a baseball. The range of the pitch is determined by the rules of the game and is therefore a constant. Relative projection height of a baseball pitch is a positive value, because the projection height is a greater value than the impact height. With the demands of the skill, dictated range, and relative projection height in mind, we know that the optimal projection angle must be less than 45°. What are the reasons for this difference in optimal projection angle? First we must understand the demands of the skill. Successful baseball pitching requires accuracy and maximal horizontal velocity. If projection velocity is maximized, and the projection angle is too large, the ball will overshoot the target (that is, accuracy will be poor). We cannot simply lower the projection velocity to maintain the theoretical optimal 45° projection angle because doing

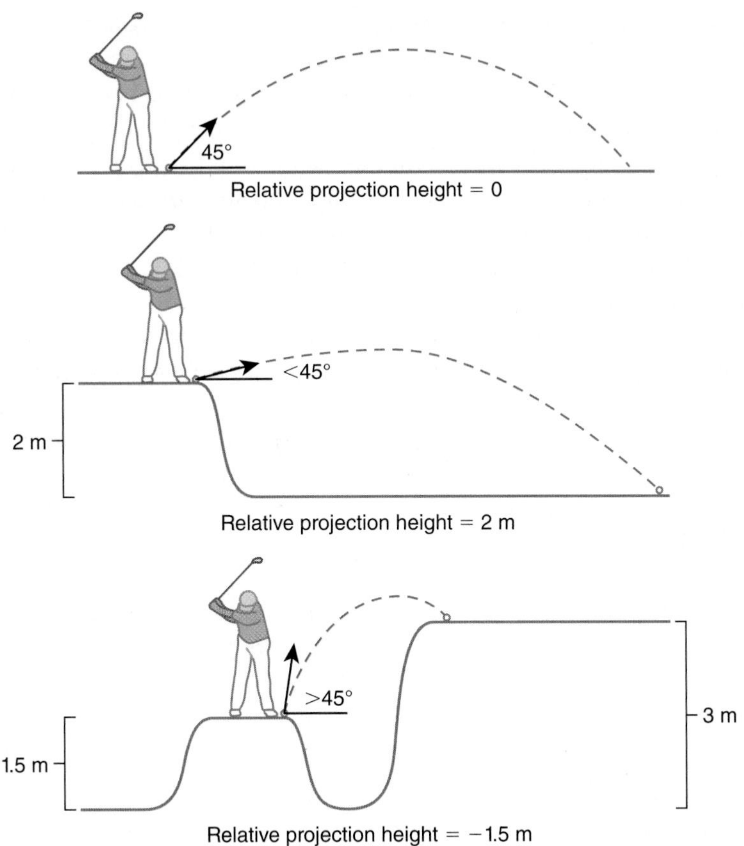

Figure 10.22 **Influence of relative projection height on optimal projection angle.**

so would allow the batter a longer look at the ball. Therefore, the projection angle must be lowered to hit the target while maximizing the horizontal component of the velocity vector. On the opposite end of the spectrum, the ball will hit the ground before reaching the target if the pitcher has chosen a projection angle that is too small. Doing so maximizes the horizontal component of the velocity vector, but the vertical component is too small for the ball to travel the entire distance to the plate. So, as you can see, accuracy is a delicate balance of these projection-related factors, especially when projection velocity must be maximized.

Even though accuracy is paramount in both skills, shooting in basketball is actually the opposite situation from baseball pitching in terms of constraints. First, relative projection height is a negative number if one is shooting a basketball free throw (or jump shot for most people). The number is negative because the impact height is greater than the projection height. Second, the velocity on arrival is actually minimized in basketball shooting because maximizing velocity will likely lead to the ball bouncing off the hoop or backboard. In terms of accuracy, the margin for error is miniscule for most shots in basketball because the hoop is only twice the diameter of the ball. One can experience this small margin for error by shooting free throws at a carnival or arcade. Both places tend to have basketball hoops that are almost imperceptibly smaller than regulation hoops. With the constraints of this skill in mind, it is easy to see that basketball shooting is a fine balance of projection height, projection angle, and projection velocity.

First, relative projection height varies in basketball shooting, depending on whether the shot is a free throw, jump shot, and so on. However, the relative projection angle is a negative value. This fact means that the optimal projection angle will likely be greater than 45°. Again, let's analyze the reason for this variation in optimal projection angle. First, accuracy is of utmost importance because the margin of error is relatively small. Excessively small projection angles are usually observed in novice players. If the projection angle is too small, the margin for error is also tiny for a couple of reasons. First, a small projection angle produces a trajectory that is relatively flat and long. This flat trajectory is likely to result in the ball bouncing off the front of the hoop because the trajectory is too flat (Figure 10.23) or even hitting the back of the hoop and bouncing straight back out because the trajectory is long (i.e., the horizontal component is too large).

Of course the shot is not impossible; it's just that the projection velocity must be perfect for the ball to make it over the front edge of the hoop and then fall in without hitting the back of the hoop with too much velocity. The second problem is simply that a ball shot with a small projection angle is easier for opponents to block. Therefore, a larger projection angle must be used to maximize the margin for error (Figure 10.24).

A larger angle has several advantages. First, a larger projection angle produces a steeper parabolic trajectory. This trajectory minimizes the likelihood of hitting the front of the hoop. In addition, increasing the projection angle reduces the horizontal component of the initial velocity vector. Therefore, the ball will rebound less on striking the backboard or back of the hoop (don't forget what you already know about kinetic energy, momentum, and elastic recoil). A reduced horizontal component, along with some backspin, can help the ball simply hit the back of the hoop or backboard and roll backward into the hoop. In addition, a ball projected at a larger angle is more difficult to block. So with these issues

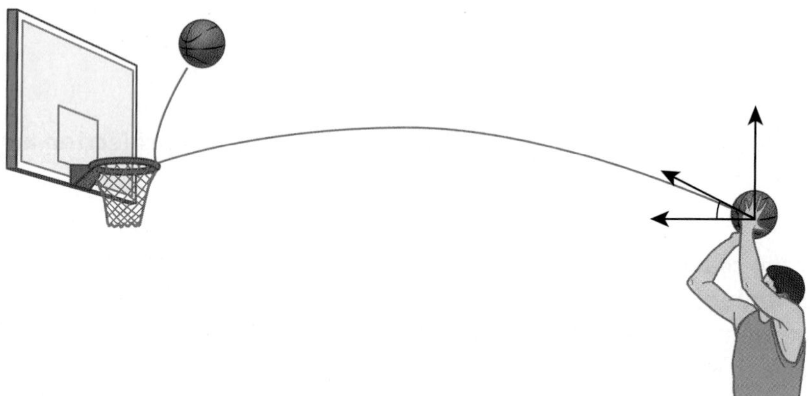

Figure 10.23 **Basketball shot with errant projection angle.**

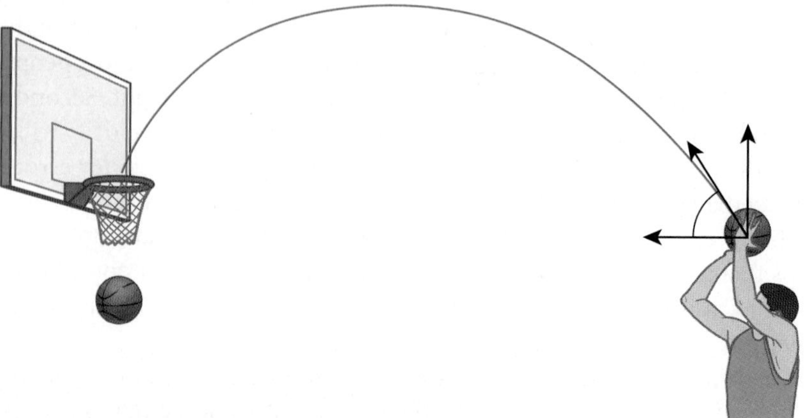

Figure 10.24 **Basketball shot with appropriate projection angle.**

in mind, should a player just shoot the basketball with the highest projection angle possible? If she did, she would encounter different potential problems. First, the horizontal component is progressively reduced as the projection angle increases. This situation increases the odds that the ball will fall short of the hoop unless the projection velocity is greatly increased. In addition, because the trajectory is so steep in the case of a large projection angle, the ball will arrive at the target possessing a large vertical component. Therefore, if accuracy is not near perfect, the ball is likely to bounce straight upward on striking the hoop. In this case, the bouncing ball may or may not fall into the hoop.

In addition to these situations, remember that shooting in basketball is not always from the same relative projection height or distance from the target. For example, a jump shot increases the relative projection height (i.e., it's no longer as negative). As you've learned, a greater projection height means that the projection angle need not be as large. Also, unless the shot is a free throw, the range of the shot is highly unpredictable in basketball. As the range changes, the optimal projection angle also changes because of the need to increase horizontal displacement. Remember that increased range can be accomplished by increasing projection velocity, decreasing the projection angle (i.e., increasing the magnitude of the horizontal component), or both.

We have used only two examples of projection for accuracy. But accuracy must be considered in many situations (archery, darts, golf, horseshoe pitching, etc.). In any of these situations, one must first analyze the demands and constraints of the task. Then the optimal combination of projection angle, projection velocity, and relative projection height must be chosen. The balance chosen for a given situation may not always be appropriate, even within a given sport, because any of the projection factors can vary as necessitated by the situation. A change in one projection factor must be compensated for by a change in another. Projection angle is highly variable, depending on the sport and individual circumstances within a sport. Margin for error varies with any change in projection angle, but the change may be necessitated. For example, when shooting from a short distance from the target in archery, one can use a flatter trajectory (smaller projection angle). But the farther from the target, one must increase the vertical component by using a larger projection angle so that the arrow doesn't drop excessively and fall short of the target. So in this case, the dictated range necessitates changes in other projection factors. In some sports, accuracy is not enough. One may also have to interfere with the other opponents' projectiles (e.g., pitching horseshoes or bocce). If the goal is to land a projectile on a target with maximum accuracy, then a larger vertical component is needed (larger projection angle). However, to move an opponent's projectile, one must use a smaller projection angle. This smaller angle is required because the horizontal component of the velocity vector is larger, providing a greater capacity for the projectile to perform work on the opponent's projectile. Finally, in some sports such as golf, both the range and relative projection height can change on every single shot. This variability means that projection angle and projection velocity must also be changed with each new situation.

CONNECTIONS

10.7 HUMAN PERFORMANCE

Sport Science

Several sport and recreational activities require the human body to become a projectile. It's easy to envision the highly athletic jumps performed to dunk basketballs and spike volleyballs as examples, but even simple children's activities like jumping rope and leapfrog require strength and timing to counter the force of gravity. Two track-and-field events attempt to maximize human potential as a projectile, one for vertical height and one for linear displacement (distance). The next two sections describe factors that lead to optimal performance in the high jump and long jump, and they may help you understand why humans can propel their center of gravity only one-third as far vertically as they can horizontally.

Projection of the Body for Height

Projectile trajectories are measured through an object's center of gravity. Solid inanimate objects maintain a consistent center of gravity, but the human body has the ability to change shape and therefore change the position of its center of gravity relative to segments of the body. Shape change is one critical feature in clearing a high jump bar, a horizontal bar that is often set beyond the height of the jumpers in high-level competition. The predominant jumping style used in world high jump competition today is called the *Fosbury Flop*, named for the first person to use it in competition. The performer's center of gravity just clears the bar, while the body continuously changes shape by reorienting segments. The inverted "U" body position attained at the peak of this jumping style moves the center of gravity outside the body and results from back and neck extension combined with knee flexion. Note that some parts of the body are *below* the bar at all times during a jump, and even though the center of gravity may remain below the bar, the whole body can clear it by changing shape (Figure 10.25).

Figure 10.25 **High jumper using the Fosbury Flop.**

High jumping is a highly complex skill that requires a precise approach and takeoff to propel the center of gravity to the maximal height at exactly the right moment. The world record in this event is 2.45 m, but it's quite possible that jumpers have exceeded this height, just not at the right time to clear a bar. Humans produce enough force to move horizontally at 10.5 m/sec, but because of the force of gravity can project vertically at only 3.5 m/sec (Linthorne, Guzman, & Bridgett, 2005). This limitation helps account for the fact that we can jump approximately three times farther horizontally than vertically: The world record long jump is 8.95 m. Two factors help attain maximal height at the peak of a jump: projection velocity and projection height.

To attain maximal vertical projection velocity, high jumpers use a few strategies to generate optimal force. They start with a multistep approach that generates momentum in a horizontal direction. The intent is not to reach maximal velocity but to accelerate consistently. The final step of the approach is slightly shorter than the rest in preparation for changing the direction of the horizontal inertia to a vertical orientation. The jumper doesn't direct the force into a purely vertical vector because he needs to maintain some horizontal velocity to travel over the bar. The last step is called the *plant,* and the plant leg flexes at the knee for a couple of reasons: (1) to absorb the force of the inertia built up during the approach and (2) to generate vertical extension force over a longer time. With a little help from stored strain energy in the ligaments and tendons, and also from the muscular stretch reflex (see Chapter 11), the body accelerates upward.

Once the plant foot leaves the ground, only gravity influences the trajectory of the body's center of gravity. However, given the same upward force, if the center of gravity starts at a higher place, it can attain a higher peak. This is the second factor in optimizing high jump potential: maximizing projection height. Elite jumpers raise their center of gravity as high as possible while still in contact with the ground. While the plant foot is still on the ground, the jumper raises the other leg and both arms forcefully. The upward motion of the arms and one leg accomplishes two things: moving the body's center of gravity higher, and increasing inertia in a vertical orientation. Raising the center of gravity results in a greater projection height, and therefore increases the ultimate height of the center of gravity *relative to the ground* at the apex. If the jumper hits the takeoff mark precisely and changes shape in the proper sequence during flight, the chances are optimized that the center of gravity will clear the bar.

Projection of the Body for Distance

From the discussion in the concepts section of this chapter, you know that the optimal projection angle to maximize horizontal displacement is 45°. Given a consistent projection velocity and height, this rule holds true. The discussion of the high jump in the previous section notes that humans can project vertically at only one-third the velocity with which they can move horizontally. This fact might help you understand the reason why the optimal projection angle of world-class long jumpers is approximately 22° (Linthorne et al., 2005). If a long jumper could maintain the same horizontal velocity independent of projection angle, the 45° rule would hold true, but projection velocity decreases significantly when the jumper transfers some of the horizontal force into vertically directed force. For optimal long jump performance, projection velocity is a more important factor than attaining a theoretical optimum launch angle, and the human body is far more capable of producing horizontal velocity than vertical (Figure 10.26).

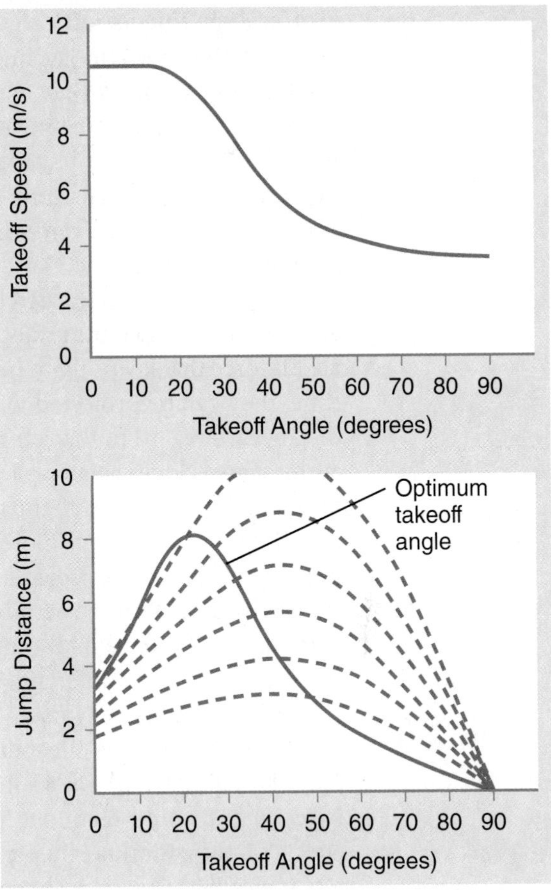

Figure 10.26 **Jump distance as a function of takeoff velocity and takeoff angle.**

Just as in the high jump, projection height also plays a part in explaining the lower optimal projection angle for long jumpers. Projection angle for optimal horizontal distance changes slightly depending on relative projection height. As the relative projection height is increased, the optimal angle is decreased (Linthorne et al., 2005). In the case of the long jump, even though the surface of the takeoff area and the sand landing area are at the same level, the height of the center of gravity is significantly higher on takeoff than at landing. At takeoff, the jumper forcefully extends the leg in contact with the ground and plantar flexes the foot. He also extends at least one arm upward, raising the center of gravity even farther. During flight, the jumper changes the shape of his body by raising and extending his legs, which allows him to drop farther before contacting the sand. The center of gravity is much lower at landing than it was at takeoff (Figure 10.27).

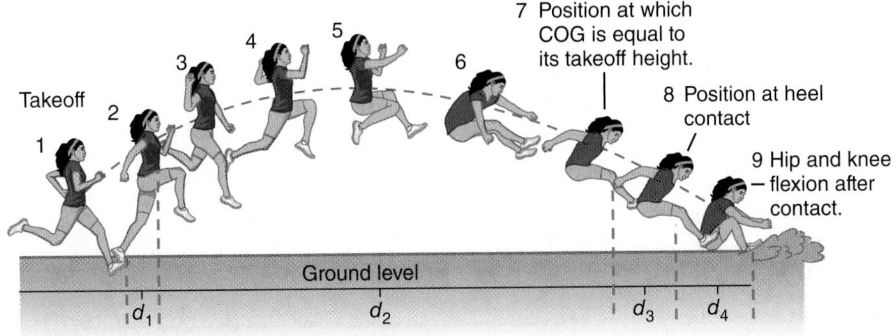

Figure 10.27 **A long jumper.**

Projecting Objects with the Body for Distance

What do the shot put, javelin, and discus have in common other than functioning as projectiles in the Olympic Games? They all descend from tools of warfare. Humans have employed projectiles for thousand of years, usually with the intent to cause injury to animals or other humans. From early hand-launched projectiles to the use of mechanical implements (e.g., catapult, trebuchet, and cannon), humans have gained a strong understanding of weapons ballistics. We learned that maximum range does not always result from some theoretically optimum launch angle, especially for objects propelled by human muscle force.

In Chapter 9, you learned that aerodynamics play a role in the optimal projection angle of a discus and javelin. Between the projection angle and the angle of attack relative to airflow, these projectiles create lift force that keeps them in the air for a longer time. If the objects can remain in the air for a longer time, they can be projected with a larger horizontal velocity and a smaller angle to create maximum distance. In regard to throwing objects, the human factor also comes into play; humans cannot throw at the same velocity at all projection angles. The human musculoskeletal system is designed better for horizontal acceleration than for vertical. When you try to increase the vertical component, the horizontal vector suffers. One more variable that influences optimum projection angles for these field-event implements is relative projection height. Each projectile is released approximately 2 m higher than the impact height. With the aerodynamic effects on javelin and discus flight, this variable does not make a dramatic difference in optimal projection angle, which may decrease a couple of degrees. However, the shot put is quite different (Figure 10.28).

The shot put is known as a *free-flight* projectile because it is massive enough that forces other than gravity have negligible effect during flight. The free-flight nature and relatively short range of the shot put is barely affected by a positive 2-m projection height. Because aerodynamics plays little part in the trajectory, it would seem natural that the optimal projection angle would be close to 45°. When the effects of projection height are taken into account, this angle decreases slightly to 42°. Elite shot putters typically throw their longest distances at angles between 30° and 40°. Why the discrepancy? The answer is the same as that for the long jump: The human body expends significantly more energy to produce vertical force against gravity, so less energy is available for horizontal projection. A men's shot put has a mass of 7.3 kg, and as projection angle increases, more energy is required to throw it upward against the force of gravity. Elite putters sacrifice projection height for projection velocity, because no inherent force exists to inhibit horizontal velocity, as is present in the case of gravity influencing vertical velocity. The 10° variation in an elite thrower's optimal projection angle results from the different rates of velocity loss with increased projection angle (Figure 10.29). Every athlete is different in this respect, so individual coaching and analysis are important.

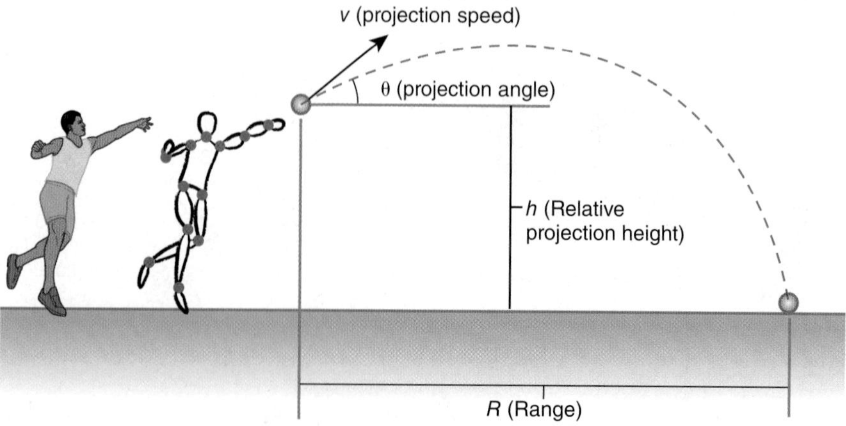

Figure 10.28 **Free-body diagram of shot putting.**

Projecting Objects with the Body for Accuracy

At times, maximal range is not the goal of sending a projectile away from the body. In these activities, the intention is to have the projectile hit a particular location. Target games such as horseshoes and darts come to mind when describing throwing for accuracy, and each of these activities can be played successfully using two different methods of delivering the projectile to the target. Some dart and horseshoe players throw with less velocity but a higher projection angle, and some throw with high velocity but smaller projection angle. In both games, the variables remain constant, including the weight of the projectiles and the distance from the target.

In the game of darts, greater velocity does not share the same consequences as horseshoes when the intended location is missed by a small distance. An errant dart thrown with high velocity will still hit the dartboard and possibly even score points. A horseshoe that misses the stake at high velocity, however, will travel away from the target and rarely ends up in a position to score points. Conversely, a horseshoe thrown in a high arc with a slightly slower velocity and a higher projection angle will land at a particular distance and remain relatively close to that position. If the distance is correct, and the shoe doesn't hit the stake, it will more likely remain in a position to score points.

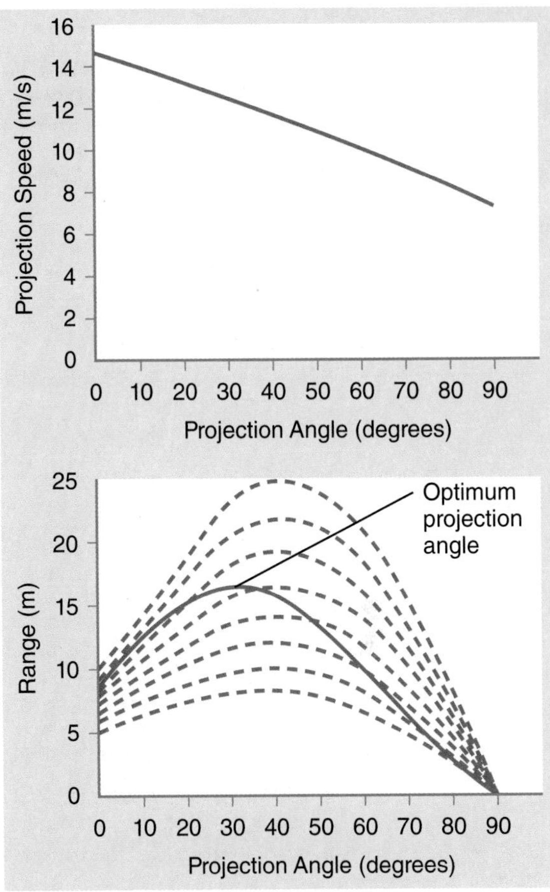

Figure 10.29 **Decrease in projection velocity with increasing projection angle.**

Dance

In some situations, maximum jump height is actually not the desired outcome. In the field of dance, choreographers must plan jumps according to the tempo of the music (Laws, 1984, 2008). Also, the dancer performing the piece must be fully aware of the consequences of errant jump height. These constraints make for a delicate balance between performing the choreographed movements and still maintaining the tempo. The dancer must jump high enough to perform the desired body movement before landing. However, if the jump is too high, the performer may not land fast enough to keep the tempo. In other words, the dancer can control how high she jumps but can fall only so fast (Figure 10.30).

So jump height is constrained by the time allotted between movements (Table 10.3). For example, a dancer who jumps a vertical height of 15 cm will have a flight time of approximately 0.35 sec. The given flight time means that this dancer can perform 14 jumps in 5 seconds. Another dancer who jumps a vertical distance of 30 cm will have a flight time of approximately 0.50 seconds, which restricts that dancer to 10 jumps in 5 seconds. The choreographer must remember that jump height (and therefore flight time) varies, depending on the size and physical capabilities of the dancer. A more novice dancer may not have the expertise to control jump height to the same degree a more experienced dancer can. The more experienced dancer may have the physical ability to jump higher, which in turn necessitates restraint to keep the given tempo.

Figure 10.30 **A dancer jumping.**

Table 10.3
Vertical Jump Height and Flight Time

Takeoff Velocity (m/sec)	Vertical Jump Height	Flight Time
0.981	4.9	0.2
1.226	7.7	0.25
1.472	11.0	0.3
1.717	15.0	0.35
1.962	19.6	0.4
2.207	24.8	0.45
2.453	30.7	0.5
2.698	37.1	0.55
2.943	44.1	0.6
3.188	51.8	0.65
3.434	60.1	0.7
3.679	69.0	0.75
3.924	78.5	0.8
4.169	88.6	0.85
4.415	99.3	0.9
4.659	110.6	0.95

FOCUS ON RESEARCH

Projection for accuracy can be tremendously difficult under the best conditions. Using an implement to launch the projectile can be even more complicated. At least in most sports, the playing venue, including the target of the projectile, is a known and constant. However, in the sport of golf, a player must use an implement to launch a projectile accurately toward the target in a playing venue that is constantly changing. One variable that changes on a golf course is slope.

Blenkinsop et al. (2018) investigated the effect of uphill and downhill slopes on weight transfer, alignment, and shot outcome in golf. A computer-assisted rehabilitation environment system was used to create 5° slopes. Force plates along with motion capture were used to collect 3-D kinematics and kinetics of the swings. Twelve elite male golfers hit 30 shots, with 10 shots being from each slope condition (flat, uphill, downhill). Golfers were asked to hit straight shots using their own six iron, golf shoes, and glove using a well-known brand of golf ball. The investigators found there was a shift in the center of pressure throughout the swing when performed on a slope, with the average position moving approximately 9% closer to the lower foot. The researchers stated that the golfers were attempting to remain perpendicular to the slope (based on coaching advice), resulting in a weight transfer towards the lower foot. All golfers achieved this at the hips, but a complete adjustment was not found at the shoulders. However, at ball contact there was a significant difference in shoulder alignment for uphill versus downhill slopes, with the golfers tending to be more "open" (left of target) on the downhill slope. The golfers also adopted a wider stance on both slopes, with the ball being played farther forward in the stance for uphill compared to downhill slopes. It was also found that there were no consistent adjustments made to alignment at address and azimuth to counter the additional sidespin created by playing from a slope. The authors state that this means the change in final shot dispersion resulted from the ball's lateral spin. Although launch angle and ball spin were significantly affected by the slope, ball speed was not. Golfers were also found more likely to hit shots to the left from an uphill slope and to the right for a downhill slope, which the authors state is consistent with coaching literature.

In this research study, the only variable that changed was slope. Imagine how difficult projection of a golf ball for accuracy can be considering that golf includes changes in slope, elevation, distance, and grass texture, along with trees, water, and sand traps.

Another consideration is that when a dancer becomes a projectile, the projectile must land. Repeatedly jumping over the course of a career could result in injury caused by multiple landing impacts. Indeed, injuries to the lower extremities, many of which are chronic, are common in ballet, jazz, and modern dancers (Hardaker, Margello, & Goldner, 1985; Silver, 1985). Researchers have provided evidence that landing at greater jump distances is associated with increased joint reaction and muscle axial forces, increased total joint axial forces, and high rates of axial force application at the knee and ankle joints (Simpson & Kanter, 1997), as well as greater shear force loading in some dancers (Simpson & Pettit, 1997).

10.8 MOTOR BEHAVIOR

Motor Development

Basic motor skills are learned activities, and throwing is one of the most important for participating successfully in a variety of sport and recreational activities. The intent of throwing is to propel an object for distance, accuracy, or a combination of both. During the learning process, humans progress through different stages that are defined by characteristic movements that the body performs. As we progress through the stages of throwing from initial attempts to proficient performance, it's easy to identify changes in the neuromuscular system. Through each progressive stage of throwing skill, more parts of the body are used to provide force (i.e., added to the kinetic chain), and they are recruited with increased synchronicity. The progression from initial to mature stages of throwing helps to maximize projection velocity while the child learns to control projection angle and trajectory in different situations. Monica Wild conducted seminal work in describing throwing pattern development in children (Wild, 1938). She examined 32 boys and girls between ages 2 and 12 and described four distinct stages that they progressed through in acquiring a mature throwing pattern (Table 10.4).

The first stage describes a throwing motion that is arm dominated, with the thrower facing the intended target (Figure 10.31a). The only preparatory motion is bringing the arm upward, either with forward or sideward direction. No trunk rotation is evident, and the feet remain stationary during the entire throwing act. With only the arm muscles responsible for propelling the ball, little chance exists to develop much velocity.

The second stage describes the body moving horizontally rather than relying on the anterior–posterior arm-only motion of stage one. The throwing arm moves in a high plane above the shoulder,

Table 10.4
Wild's Four Developmental Stages of Throwing

Stage-I (2–3 Years)	Stage-II (3½–5 Years)	Stage-III (5–6 Years)	Stage-IV (6½–12 Years)
Throw is arm dominated. Preparatory arm movements involve bringing the arm sideways–upward or forward–upward. Thrower faces the direction of intended throw at all times. No rotation of trunk and hips is evident. Feet remain stationary during the entire throwing act.	Body moves in horizontal plane instead of anterior–posterior plane. Throwing arm moves in a high oblique plane or horizontal plane above the shoulder. Throwing is initiated predominately by arm and elbow extension. Feet remain stationary, but rotary movement of the trunk is observable.	Forward step is unilateral to the throwing arm. Arm is prepared by swinging it obliquely upward over the shoulder with a large degree of elbow flexion. Arm follows through forward and downward and is companied by forward flexion of the trunk.	Forward step is taken with the contralateral leg. Trunk rotation is clearly evident. Arm is horizontally adducted in the forward swing.

Source: Data from Payne and Isaacs, 1987, as interpreted from Wild, 1938

and the throwing hand reaches back past the ear (Figure 10.31b). From this preparatory position, a child has the potential to apply force to the object for a longer time, and although the feet still remain stationary in this stage, the trunk begins to rotate, which further increases the potential time that force can be applied. The throwing motion is still initiated by arm and elbow extension, but with increased degrees of freedom around more joints, the child gains potential to project the object with greater velocity.

By the third stage, the neuromuscular system is maturing and allowing the thrower to add more joints and increase range of motion. Arm preparation includes swinging it upward over the shoulder with a large amount of elbow flexion (Figure 10.31c). There still isn't a high degree of trunk rotation, but the child steps forward on the leg that's on the same side as the throwing arm. The step forward adds velocity before the arm extends forward, and the additional flexion of the elbow allows increased time to apply force; this opportunity results in increased velocity. With the step forward and aggressive forward motion as the arm extends, the thrower must dissipate energy with a follow-through with the arm moving downward and the trunk flexing forward.

Stage four is a mature throwing motion that displays the optimal body motions to create maximal projection velocity. The nonthrowing side is oriented toward the target to increase preparatory backward rotation. The arm is brought down, then back and up behind the shoulder. All of the weight of the thrower is placed on the foot away from the target at the end of the preparatory phase, which allows the nondominant leg to take a large step forward to initiate the throw (Figure 10.31d). When sequenced properly, the throwing motion is a complex kinetic chain that begins with a leg drive as the dominant-side knee extends. The force is transmitted and added to the hip and trunk rotation forces and then transferred through the shoulder, arm, and eventually to the hand. When a professional baseball player synchronizes all the muscles in the proper sequential order, he is able to propel a ball at nearly 160 km/h. The drawback of projecting an object at that velocity is the strain it puts on the shoulder joint, which can internally rotate at almost 7,000° per second, one of the fastest movements the human body is capable of producing (Fleisig et. al., 2006).

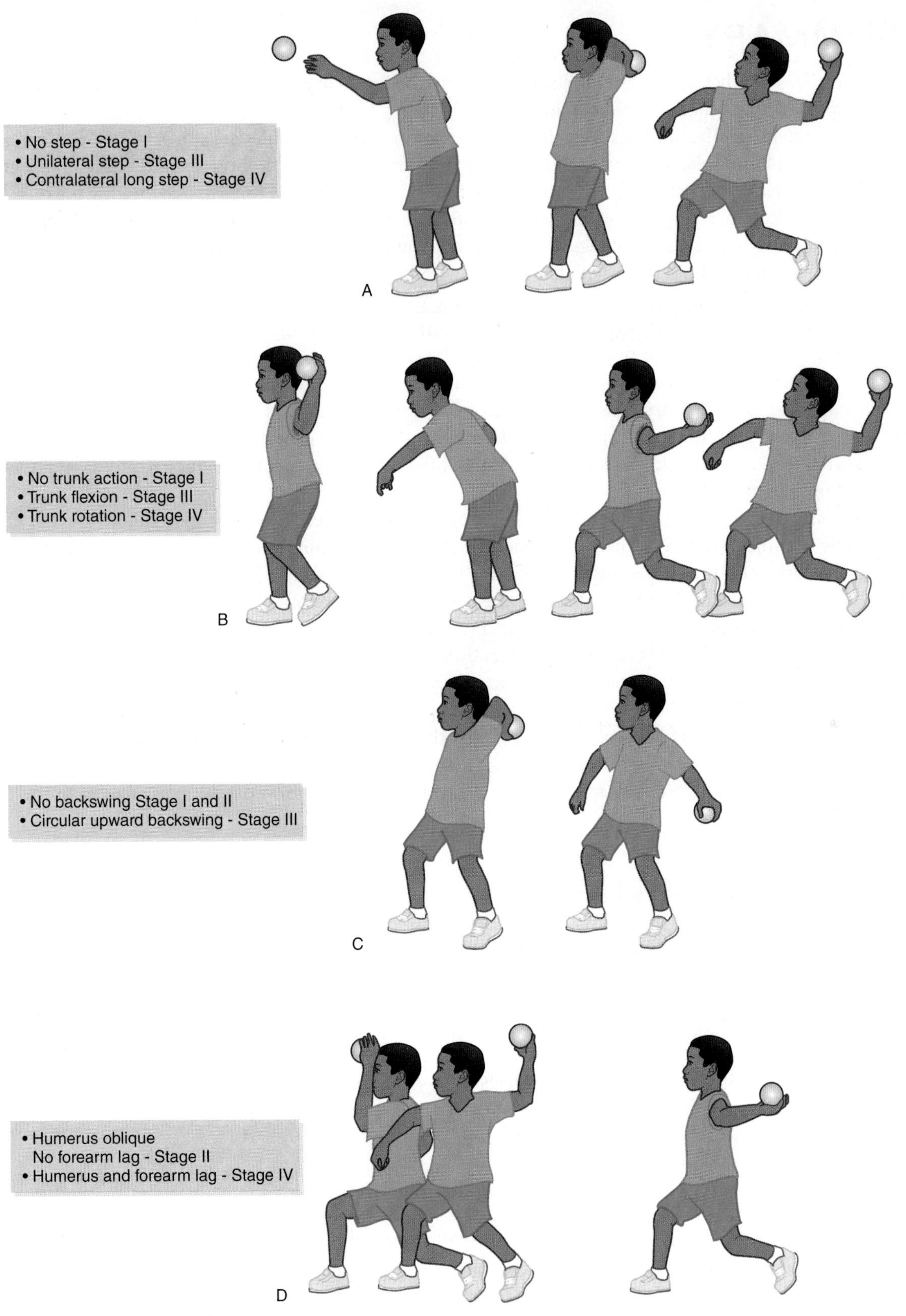

Figure 10.31 **Developmental stages of throwing.**

SUMMARY

A projectile is a body that has a motion subject only to the forces of gravity and fluid resistance. Almost every sport involves the projection of an object, whether it is caused by throwing, kicking, or striking with an implement. The human system itself can be a projectile. Projectiles are subject to the influence of gravity and fluid resistance and can exhibit linear, rotary, or general motion. Even though an object can be projected in various ways, most projectile motion is the result of force application by the distal segment of a kinetic chain. In the absence of any force other than gravity, the center of gravity of any object projected horizontally follows a predictable parabolic path. Along that path, the projectile is subjected to dynamic fluid force, which is the equal and oppositely directed force of the fluid particles in reaction to the applied force of the system moving through the fluid.

Velocity is a vector quantity and as such has magnitude and direction. Because of its known direction, the velocity vector of the projectile can be resolved into component vectors. The velocity vector of a projectile possesses two perpendicular components: the vertical component (also called *perpendicular*) and the horizontal component (also called *parallel*). *Trajectory* is the term for the flight path of a projectile; more specifically, the flight path of the projectile's center of gravity. Three biomechanical factors affect trajectory: (1) projection angle, (2) projection velocity, and (3) relative projection height.

The size of the vertical component of the velocity vector varies with projection angle. The closer the angle of projection to 90°, the greater the size of the vertical component. In the extreme case of a perfectly horizontal projection angle (0° relative to the horizontal), positive vertical displacement is zero. Some degree of horizontal displacement is almost always present. The total horizontal displacement of a projectile is called the *range*. We tend to find an optimal angle of 45° for maximizing range at a given projection velocity because this projection angle maximizes both the vertical and horizontal components. In any situation necessitating accuracy, one must first analyze the demands and constraints of the task. Then the optimal combination of projection angle, projection velocity, and relative projection height must be chosen. Remember that the balance chosen for a given situation may not always be appropriate, even within a given sport. It varies because any of the projection factors can vary as necessitated by the situation. A change in one projection factor must be compensated for by a change in another.

REVIEW QUESTIONS

1. Explain all factors that can affect the trajectory of a projectile.
2. Name some sports in which dynamic fluid forces play a large role in projection.
3. Explain why the path of a projectile is a parabola.
4. Discuss the factors related to projectile trajectory.
5. Discuss the meaning of a negative relative projection height and its implications for the parabolic path of a projectile.
6. What are the most important factors to consider in projection for vertical distance?
7. What are the most important factors to consider in projection for horizontal distance?
8. What are the major considerations when projecting for accuracy?

PRACTICE PROBLEMS

1. A projectile has an initial velocity of 15 m/sec. Calculate the horizontal and vertical components of the initial velocity vector if the projectile is released at 30° relative to the positive *x*-axis versus 60°.
2. A projectile is released at an angle of 25° relative to the positive *x*-axis. Calculate the horizontal and vertical components of the initial velocity vector if the projectile is released at 5 m/sec versus 8 m/sec.
3. A ball traveling 25 m/sec is struck by an implement (moving in the opposite direction) with enough force to accelerate the ball at a rate of 130 m/sec^2. Calculate final velocity if the implement and the ball made contact for 0.5 sec.
4. A ball is kicked with an initial velocity that has a horizontal component of 15 m/sec. The ball's flight time is 2.5 sec, and its final velocity is 5 m/sec. Calculate the distance traveled by the ball.
5. A person on top of a building throws a ball perfectly upward in the positive *y* direction with an initial velocity of 10 m/sec. Calculate its position 3 seconds later. Calculate the final velocity of the ball after it has traveled the distance that you calculated.
6. You are attempting a long jump. You leave the takeoff board with an initial velocity of 10 m/sec at an angle 20° relative to the horizontal. Calculate your: (a) vertical velocity, (b) horizontal velocity, (c) flight time, and (d) range.

EQUATIONS

Linear acceleration
$$a = \frac{\Delta v}{\Delta t} \tag{5.3}$$

First law of uniformly accelerated motion
$$v_f = v_i + at \tag{10.1}$$

Average velocity
$$v_a = \frac{v_i + v_f}{2} \tag{10.2}$$

Linear velocity
$$v = \frac{d}{\Delta t} \tag{5.2}$$

Final position of a projectile
$$p_{final} = \frac{p_{initial} + (v_i + v_f)t}{2} \tag{10.3}$$

Second law of uniformly accelerated motion
$$p_{final} = p_{initial} + v_i^t + \frac{at^2}{2} \text{ OR } d = v_i^t \tag{10.4}$$

Third law of uniformly accelerated motion
$$v_f^2 = v_i^2 + 2a(d) \tag{10.5}$$

Vertical displacement of a projectile
$$d = \frac{v_i^2}{2g} \quad \text{OR} \quad d = \frac{(v_i \sin\theta)^2}{2g} \tag{10.6}$$

The full parabolic path of a projectile
$$d = \frac{(v_i^2 + \sin 2\theta)}{g} \tag{10.7}$$

REFERENCES AND SUGGESTED READINGS

Blenkinsop, G. M., Y. Liang, N. J. Gallimore, & M. J. Hiley. (2018). The effect of uphill and downhill slopes on weight transfer, alignment and shot outcome in golf. *Journal of Applied Biomechanics*, 13: 1–25.

Fleisig, G. S., D. S. Kingsley, J. W. Loftice, K. P. Dinnen, R. Ranganathan, S. Dun, R. F. Escamilla, & J. R. Andrews. 2006. Kinetic comparison among the fastball, curveball, change-up, and slider in collegiate baseball pitchers. *American Journal of Sports Medicine*, 34(3): 423–430.

Hardaker, W. J., S. Margello, & J. L. Goldner. 1985. Foot and ankle injuries in theatrical dancers. *Foot and Ankle*, 6: 59–69.

Laws, K. L. 1984. *The physics of dance*. New York: Schirmer Books.

Laws, K. L. 2008. *Physics and the art of dance: Understanding movement*, 2nd ed. New York: Oxford University Press.

Linthorne, N. P., M. S. Guzman, & L. A. Bridgett. 2005. Optimum take-off angle in the long jump. *Journal of Sports Sciences*, 23(7): 703–712.

Payne, V. Gregory, and Larry D. Isaacs. 1987. *Human motor development: a lifespan approach*. Mountain View, Calif: Mayfield Pub. Co.

Serway, R. A., & J. W. Jewett Jr. 2019. *Physics for scientists and engineers with modern physics,* 10th ed. Boston: Cengage Learning.

Silver, D. M. 1985. Knee problems and solutions in dancers. *Kinesiology for Dance*, 8: 9–10.

Simpson, K. J., and Kanter, L. 1997. Jump distance of dance landings influencing internal joint force: I: Axial forces. *Medicine & Science in Sports & Exercise*, 29(7): 916-927.

Simpson, K. J., and Pettit, M. 1997. Jump distance of dance landings influencing internal joint force: II: Shear forces. *Medicine & Science in Sports & Exercise*, 29(7): 928-936.

Wild, M. 1938. The behavior pattern of throwing and some observations concerning its course of development in children. *Research Quarterly*, 9, 20–24.

Zatsiorsky, V. M. 1998. *Kinematics of human motion*. Champaign, IL: Human Kinetics.

CHAPTER 11

BIOMECHANICS OF THE MUSCULOSKELETAL SYSTEM

© technotr/Getty Images

LEARNING OBJECTIVES

1. Describe muscle structure and function as it relates to human movement.
2. Describe biomechanical factors that relate to muscle location, origin, and insertion.
3. Understand the effects of muscle architecture on biomechanics.

CONCEPTS

CONNECTIONS

A power lifter takes his position in preparation for a bench press. He unracks the 800-lb barbell and gradually lowers it toward his chest. As the bar touches his chest, he suddenly contracts his muscles with enough force to drive the barbell explosively back toward the rack.

The bench press and other exercises are seemingly simple motions, but the accomplishment of an 800-lb bench press is highly impressive when one considers that muscle is a relatively soft tissue in its resting state. Skeletal muscle comprises microscopic cellular components whose individual contributions to gross motor movements of the human body are relatively minor. But when the individual subcellular components act together, the whole muscle is capable of incredible feats of strength.

This chapter helps to explain how the components of skeletal muscle act, together and in conjunction, with the skeletal system to produce effective and efficient movement. It covers the biomechanical implications of muscle location, shape, and design, as well as the way that muscles work together to produce movement and reduce injury. This chapter also discusses how the structure and function of the musculoskeletal system are related to other concepts discussed in this book. For example, it uses concepts from all of the previous chapters to explain biomechanical reasons for our physical shape and resulting motion. Concepts from this chapter are important to all movement-related disciplines because muscle itself is the basis for motion.

CONCEPTS

11.1 REVIEW OF MUSCLE PHYSIOLOGY

At the most basic level, motion of the human system is the result of muscles applying forces to bones through the process of tension development. Because the study of biomechanics is essentially the physics (mechanics) of a system's motion, no text in this field would be complete without an explanation of the contractile process that produces that motion. However, the purpose of this chapter is not to cover every detail of muscle physiology and action. Its goal is to review basic muscle physiology concepts and focus on the aspects of muscle function that are most important to our understanding of biomechanics and the human system. For a more in-depth understanding of muscle physiology, see one of the many books that deal with this topic in greater detail (Brooks, Fahey, & Baldwin, 2005; Enoka, 2015; Hochachka, 1994; Lieber, 2010; McArdle et al., 2006; McIntosh, Gardiner, & McComas, 2006; Nigg, MacIntosh, & Mester, 2000; Sherwood, 2016). Many aspects of muscle physiology can be understood only after learning basic biomechanical concepts. Therefore, this chapter is placed near the end of the book and refers back to many of the previously covered concepts.

Muscle Structure and Function

As you have learned, the function of skeletal muscle is to apply forces to the bones. This function is intimately related to muscle structure. Therefore, let's begin with a brief review of muscle structure and then proceed to a discussion of how this structure is used to produce human motion. The process is detailed because it involves the cooperation of three major systems of the human body: (1) the neurological system, (2) the muscular system, and (3) the skeletal system. A detailed explanation of this process is beyond the scope of this text. Therefore, this section reviews only the major concepts.

Muscle Structure

To understand the function of muscle, one must first understand its structure. Muscle is composed of approximately 75% water, 20% protein, and 5% inorganic salts; other substances such as enzymes; the high-energy phosphates adenosine triphosphate (ATP) and phosphocreatine; ions such as Na^+, K^+, and Cl^-; minerals such as calcium; and particles of lipid and carbohydrate (McArdle, 2015). This section examines structure first macroscopically and then at the microscopic level. As you read, notice this repeated theme: The basic structure of a muscle is a bundle of bundles (Figure 11.1).

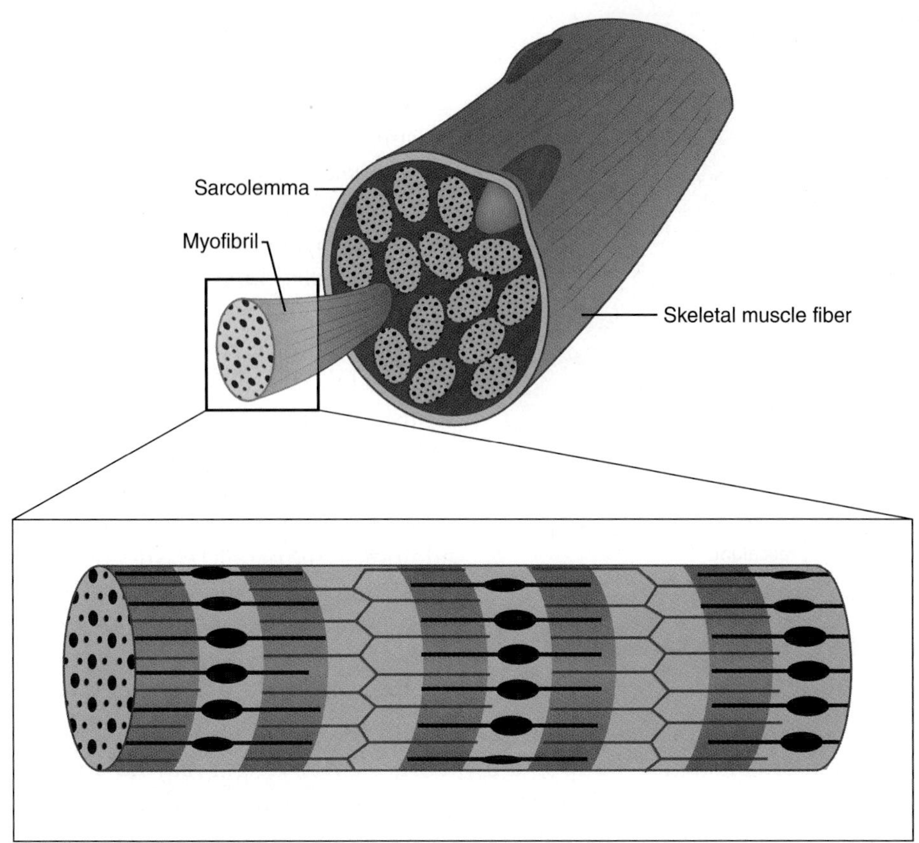

Figure 11.3 **Muscle fiber structure showing sarcolemma.**

At the next smaller level, a muscle fiber is a bundle of specialized threadlike structures called **myofibrils**. Each myofibril is approximately 1 μm in diameter and extends the entire length of the fiber. The number of myofibrils per muscle fiber depends upon the diameter of the muscle fiber. A muscle fiber approximately 100 μm in diameter may contain as many as 8,000 myofibrils.

Although myofibrils constitute approximately 80% of the volume of the muscle fiber, each fiber also contains cellular proteins, enzymes, particles of lipid and glycogen, organelles such as mitochondria, and two structures specialized to aid in the excitation–contraction coupling process (described in the next section of this chapter): **sarcoplasmic reticulum** and **transverse tubules**. The sarcoplasmic reticulum is a system of channels that surround and run parallel to the myofibrils, serving as a storage site for calcium. Enlarged portions of the sarcoplasmic reticulum that run perpendicular to the myofibrils are called **terminal cisternae**. Transverse tubules (T-tubules) are actually a continuation of the sarcolemma into the interior of the fiber. They are channels that run perpendicular to the myofibrils and connect the sarcoplasmic reticulum to pores in the sarcolemma (Figure 11.4).

Myofibrils A bundle of specialized threadlike structures making up the muscle fiber.

Sarcoplasmic reticulum A system of channels that surround and run parallel to the myofibrils and serve as a storage site for calcium.

Transverse tubules A continuation of the sarcolemma into the interior of the fiber; channels that run perpendicular to the myofibrils, connecting the sarcoplasmic reticulum to pores in the sarcolemma; also known as *T-tubules*.

Terminal cisternae Enlarged portions of the sarcoplasmic reticulum running perpendicular to the myofibrils.

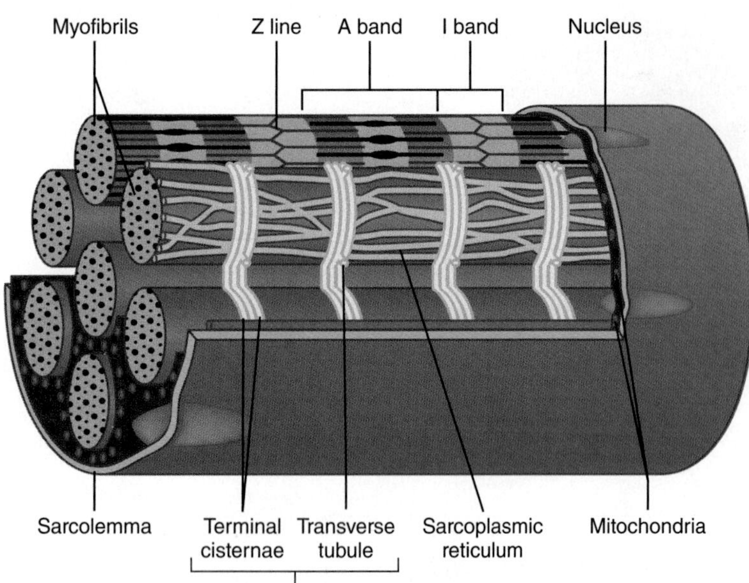

Figure 11.4 **Muscle fiber structure showing sarcolemma, sarcoplasmic reticulum, and T-tubules.**

One T-tubule passes between two terminal cisternae; this combination is often called the **triad of the reticulum**. As you will read later, the sarcoplasmic reticulum and T-tubules play a major role in muscle contraction.

Much like all of the muscle structures described so far, the myofibrils are subdivided into a series of segments called **sarcomeres** that are strung end to end (Figure 11.5). The sarcomere is the functional unit of the muscle. Each sarcomere is connected to the next by a structural protein called a **Z-disk**. The Z-disks are often referred to as *Z-lines* because of their appearance under the microscope. The Z-disks not only define the ends of the sarcomere but also contribute a noncontractile element of elasticity in addition to the parallel elastic component. Because the Z-disks are perpendicular to the muscle fiber and create a *series* of sarcomeres, the noncontractile elasticity contributed by the Z-disks is another component of the *series-elastic component.*

The number of sarcomeres per myofibril varies with the length of the myofibril. The length of one sarcomere is approximately 2.5 μm, so a myofibril 1 cm in length may have more than 4,000 sarcomeres joined in series. The sarcomere is called the *functional unit* of the muscle because it is the smallest unit that can perform the function of the entire organ (the whole muscle). So essentially all of the action takes place within sarcomeres. The important function of the sarcomere is attributable to the contractile proteins that it contains. Although several different proteins are found within the sarcomere, two filamentous structures are most crucial to muscle contraction: **thick filaments** (which have a diameter of approximately 15 nm and a length of approximately 1.5 μm) and **thin filaments** (with a diameter of approximately 7 nm and a length of approximately 1.0 μm) (Sherwood, 2016). The thick filaments are large polymers (assemblies) of the protein **myosin**, whereas the thin filaments are polymers of the protein **actin**. A sarcomere 1 μm in diameter contains approximately 450 thick filaments and 1,800 thin filaments in a highly organized arrangement. The thin filaments are arranged in a hexagonal array around the thick filaments (Figure 11.6).

The thin filaments do not traverse the entire length of the sarcomere; half are attached to the Z-disk at one end of the sarcomere, and half are attached to the Z-disk that defines the other end of the sarcomere (Figure 11.7).

The thick filaments are centered within the sarcomere, with three surrounding each thin filament. The arrangement of the thick and thin filaments relative to each other is highly organized, but the protein structure of the individual filaments is also impressively precise.

As you have read, the thick filaments are polymers of several hundred molecules of the protein myosin. Its shape is usually described as resembling one of various sports implements: a golf club, a hockey stick, or an oar, all of which have in common a long tail with a protruding

Triad of the reticulum One T-tubule passing between two terminal cisternae.

Sarcomeres Segments of the myofibrils that are strung end to end and serve as the functional unit of the muscle.

Z-disk Structural protein connecting each sarcomere to the next.

Thick filaments Large polymers (assemblies) of the protein myosin.

Thin filaments Polymers (assemblies) of the protein actin.

Myosin A contractile protein that serves as the major structural component of the thick filaments in muscle fibers.

Actin A contractile protein that serves as the major structural component of the thin filaments in muscle fibers.

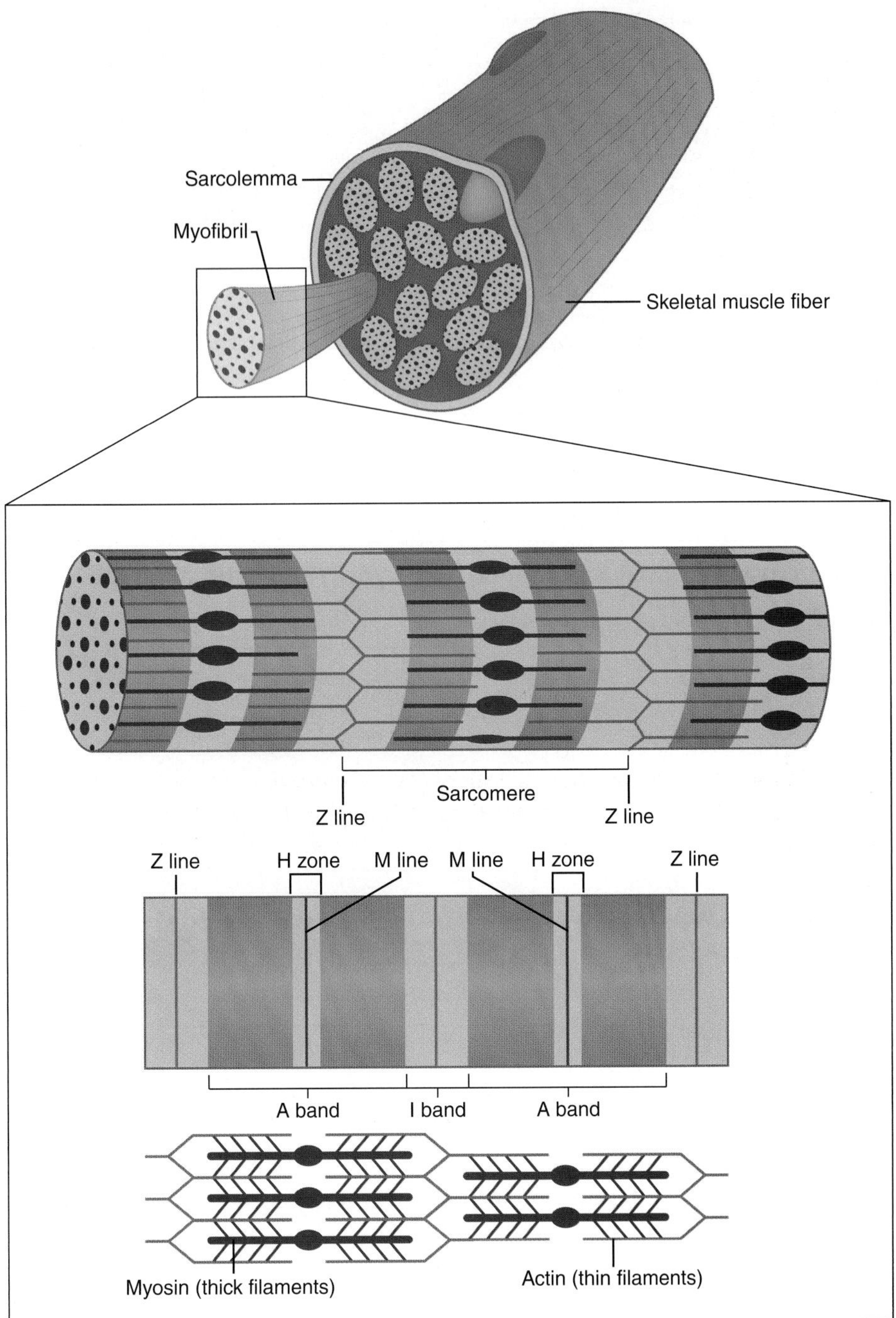

Figure 11.5 **A muscle fiber, myofibrils, sarcomeres, and Z-lines.**

globular head called a *crossbridge* (Figure 11.8). (Some sources do not refer to the myosin head and neck as a crossbridge when it is in an unbound state.)

More specifically, each myosin molecule is composed of two identical subunits. Each subunit has a long tail and a globular head. The tails of the subunits are intertwined with each other, and the heads protrude outward. The globular head actually consists of two subfragments (S-1 and S-2), one of which is designated the *head* (S-1), and the other of which is designated the *neck* (S-2). The head has two sites necessary for muscle contraction: (1) a site for binding to actin, and (2) a site for splitting ATP. The thick filament itself is composed of two bundles of these myosin molecules arranged in a staggered array

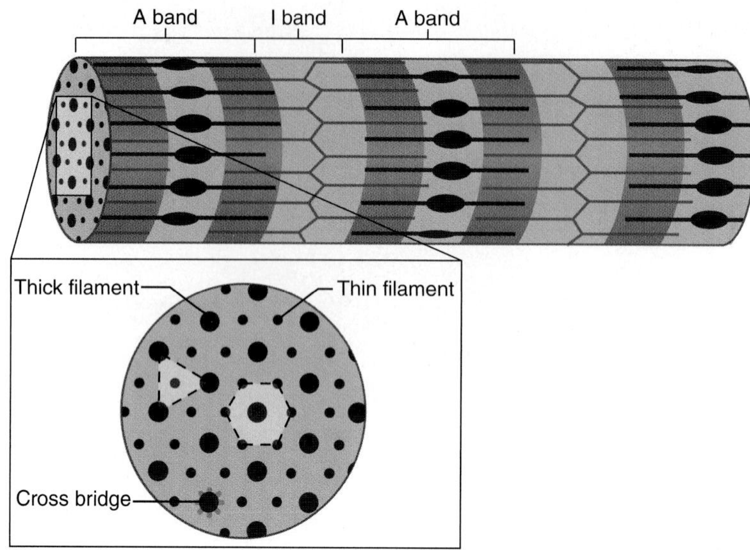

Figure 11.6 **The cross-sectional arrangement of myosin and actin filaments.**

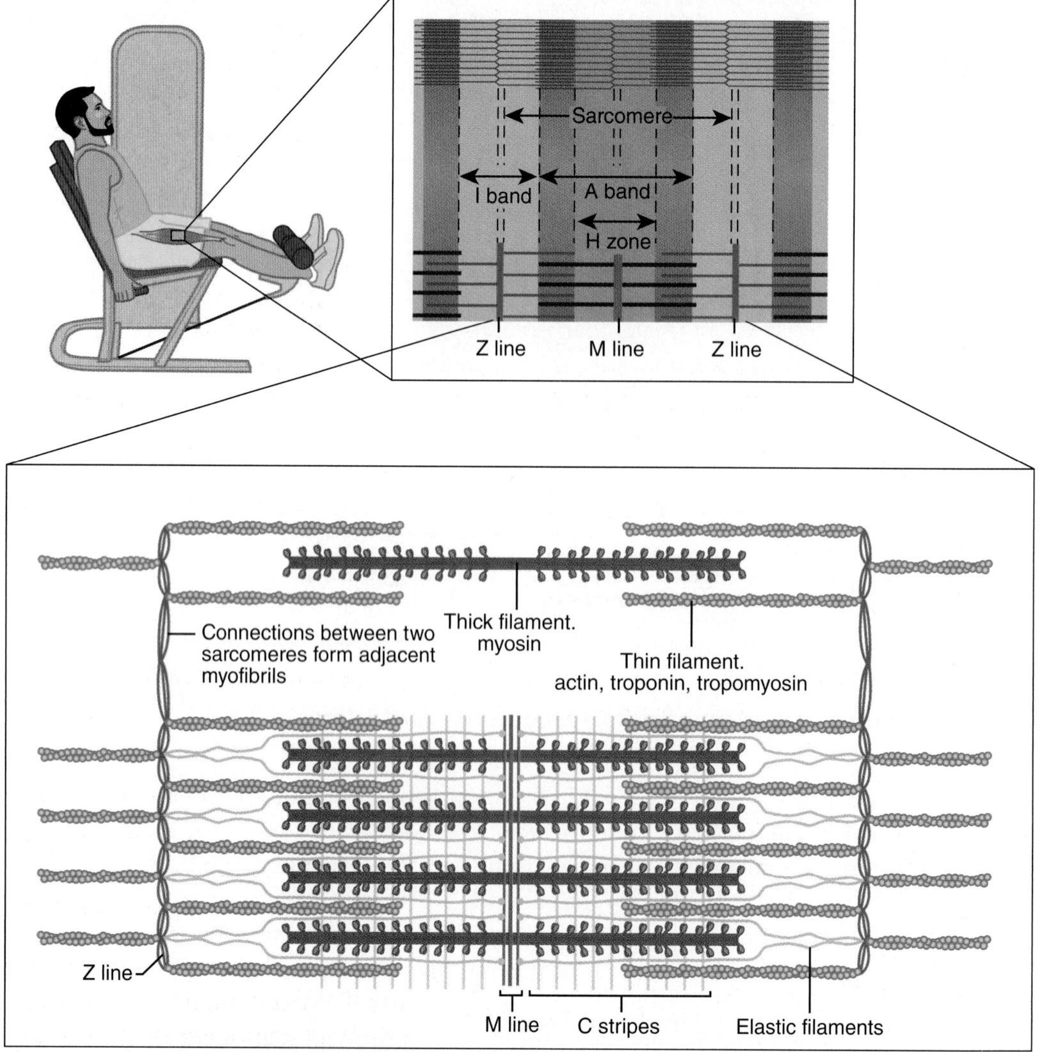

Figure 11.7 **The arrangement of myosin and actin filaments within the sarcomere.**

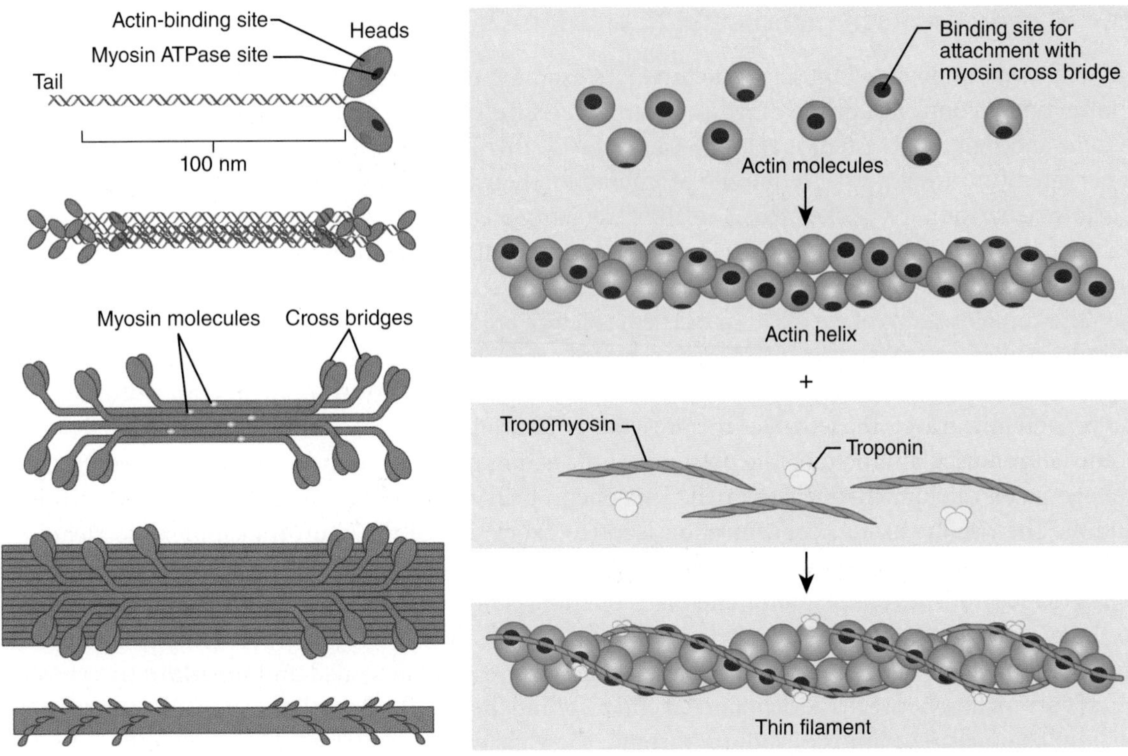

Figure 11.8 **The structure of a myosin molecule.**

Figure 11.9 **The structure of an actin molecule.**

with the tails oriented toward the center (imagine two bundles of golf clubs joined at the handle ends). Therefore, the final structure of the thick filament is a polymer of several hundred myosin molecules with globular heads that protrude outward toward the surrounding thin filaments.

The thin filament is composed mainly of the protein actin, but it also includes the proteins **tropomyosin** and **troponin** (the *troponin–tropomyosin complex*). The overall structure of the thin filament is often described as resembling two strands of pearls twisted together. The actin molecules themselves are spherical in shape and possess a binding site for the myosin head (crossbridge). The actin molecules are arranged in chains. Two chains of actin molecules are intertwined in a helical shape to form the primary structure of the thin filament (Figure 11.9).

Although a binding site for the myosin head is present on each actin molecule, binding is prevented in the resting state by the other two protein structures of the thin filament (tropomyosin and troponin). The tropomyosin molecule is made up of long, thin, threadlike proteins that spiral around the chains of actin molecules and block the myosin binding sites. This arrangement prevents "strong" binding of the myosin heads to actin until contraction is needed (i.e., actin and myosin are believed to be always bound but with a bond of varying strength). Troponin is a protein complex spaced at intervals along the thin filament. Troponin has three subunits, each with a different binding site: (1) an actin-binding site, (2) a calcium-binding site, and (3) a tropomyosin-binding site. In the relaxed state, troponin acts as a clamp that maintains tropomyosin in a position that blocks the myosin binding sites of actin (maintaining the "weak" bond). As you will see, the calcium-binding site is used during the contractile process.

Tropomyosin A molecule made up of long, thin threadlike proteins that spiral around the chains of actin molecules, blocking the myosin binding sites.

Troponin A protein complex spaced at intervals along the thin filament, that in the relaxed state, acts to maintain tropomyosin in a position to block the myosin binding sites of actin.

The Neuromuscular Junction and Excitation-Contraction Coupling

Now that you understand muscle structure, let's examine the process of muscle contraction. Muscle contraction is controlled by the nervous system; therefore, we begin at the synapse, where the nervous system and muscular system meet. The **neuromuscular junction** is the site where muscle fiber is innervated (Figure 11.10). At this site of interaction between the motor neuron and muscle fiber, a shallow depression, or pocket, exists in the sarcolemma called the **motor end plate**. The terminal button (the end of the motor neuron) is not in physical contact with the motor end-plate; instead, they are separated by a small gap or cleft called the **neuromuscular cleft**.

The complex series of events in muscle contraction are set in motion by an electrical motor-nerve impulse (called an **action potential**). The series of events in which the action potential makes its way to the muscle cell and initiates the contractile process is called **excitation–contraction coupling**. The action potential travels the length of the presynaptic neuron until it reaches the axon's terminal button at the neuromuscular junction. The presence of the action potential in the terminal button stimulates voltage-gated calcium channels to open. This opening allows an influx of calcium into the terminal button. The rise in calcium concentration leads to exocytosis of the neurotransmitter acetylcholine from synaptic vesicles into the neuromuscular cleft. Acetylcholine then diffuses across the cleft and binds to receptor sites on the motor end plate, opening cation channels. If enough acetylcholine binds to the receptor sites, then a relatively large influx of sodium into the muscle cell compared to movement of potassium out of the cell depolarizes the motor end plate (called an **end-plate potential**). Local current flow between the depolarized end plate and the adjacent sarcolemma opens voltage-gated sodium channels. The influx of sodium through these channels initiates an action potential, which is then propagated across the sarcolemma. Remember that the T-tubules are continuous with pores in the sarcolemma. Therefore, the T-tubules are able to transmit the newly formed action potential to the network of sarcoplasmic reticulum in the interior of the fiber.

Recall that the sarcoplasmic reticulum is a system of storage channels for calcium. The arrival of the action potential from the T-tubules triggers the release of calcium from the terminal cisternae of the sarcoplasmic reticulum into the sarcoplasm. The calcium is then free to bind to sites on troponin. This binding is the event that initiates the process of muscle contraction.

Neuromuscular junction The site of interdigitation (physical interaction) of the motor neuron and the muscle fiber.

Motor end plate A shallow depression or pocket in the sarcolemma located at the neuromuscular junction.

Neuromuscular cleft A small gap or cleft between the terminal button and the motor end plate.

Action potential An electrical motor nerve impulse.

Excitation–contraction coupling The series of events in which the action potential makes its way to the muscle cell and initiates muscle contraction.

End-plate potential Depolarization of the motor end plate caused by acetylcholine binding to receptor sites.

Power stroke The tilt of the myosin head toward the center of the sarcomere.

Muscle Contraction

Troponin is the protein complex that actually initiates the contraction process. In the resting state, troponin keeps tropomyosin positioned so that the myosin binding sites on actin are blocked. However, binding of calcium to troponin initiates a conformational change in the protein (Figure 11.11). This change in shape is believed to allow tropomyosin to move into the groove formed by the helical structure of the actin chains and off the myosin binding sites. Once tropomyosin has moved, myosin is free to form a strong bond with the actin sites. Once uncovered, the myosin binding sites on actin are often called *active-actin sites.*

At the moment that strong binding to actin is achieved, the myosin crossbridge (myosin's globular head and neck) is believed to undergo a conformational change: The head of the crossbridge experiences a strong intermolecular attraction to the neck, causing it to tilt in that direction. This tilt of the myosin head toward the center of the sarcomere—the **power stroke**—is believed to coincide with the release of inorganic phosphate (P_i) from the ATP molecule that has

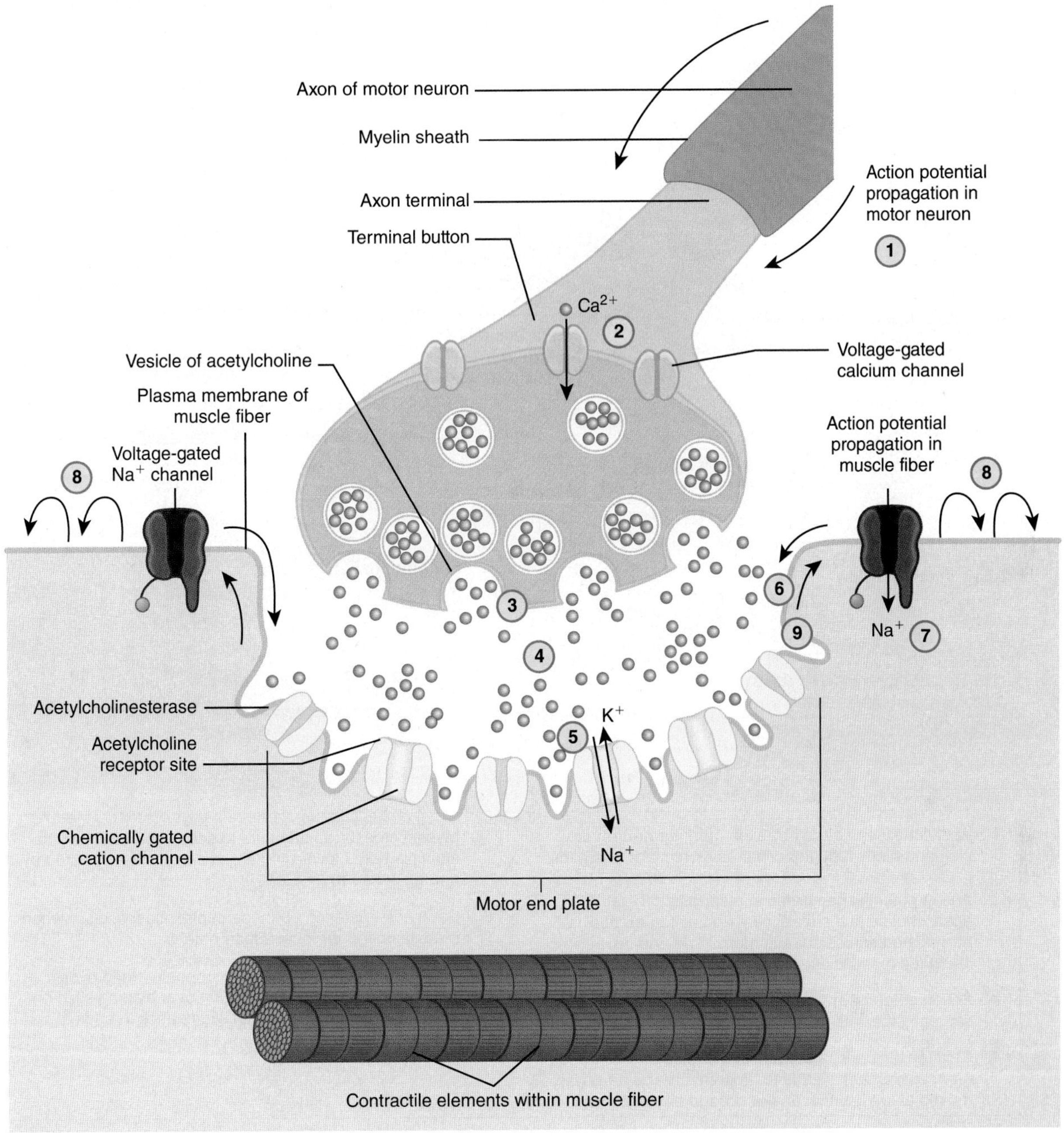

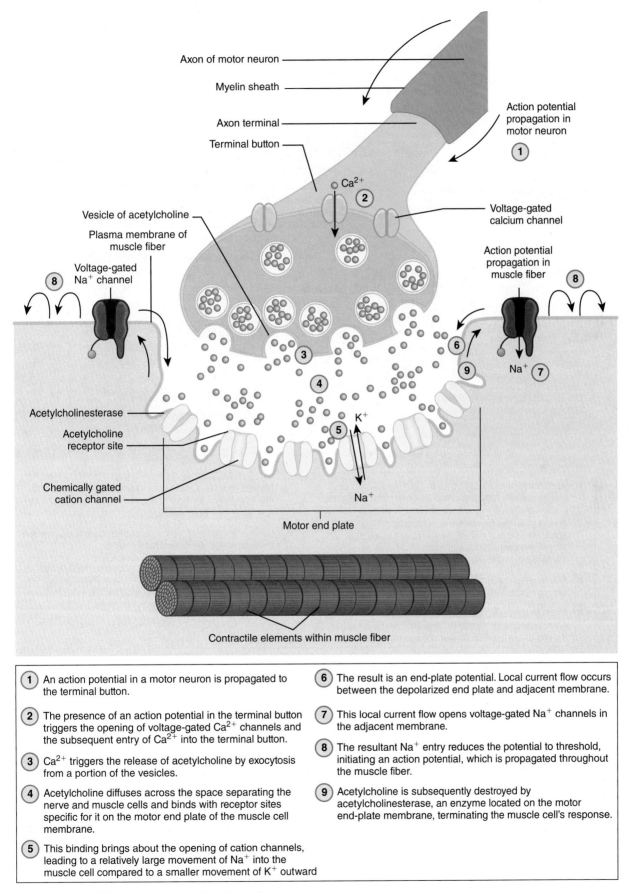

1. An action potential in a motor neuron is propagated to the terminal button.
2. The presence of an action potential in the terminal button triggers the opening of voltage-gated Ca^{2+} channels and the subsequent entry of Ca^{2+} into the terminal button.
3. Ca^{2+} triggers the release of acetylcholine by exocytosis from a portion of the vesicles.
4. Acetylcholine diffuses across the space separating the nerve and muscle cells and binds with receptor sites specific for it on the motor end plate of the muscle cell membrane.
5. This binding brings about the opening of cation channels, leading to a relatively large movement of Na^+ into the muscle cell compared to a smaller movement of K^+ outward
6. The result is an end-plate potential. Local current flow occurs between the depolarized end plate and adjacent membrane.
7. This local current flow opens voltage-gated Na^+ channels in the adjacent membrane.
8. The resultant Na^+ entry reduces the potential to threshold, initiating an action potential, which is propagated throughout the muscle fiber.
9. Acetylcholine is subsequently destroyed by acetylcholinesterase, an enzyme located on the motor end-plate membrane, terminating the muscle cell's response.

Figure 11.10 **The neuromuscular junction.**

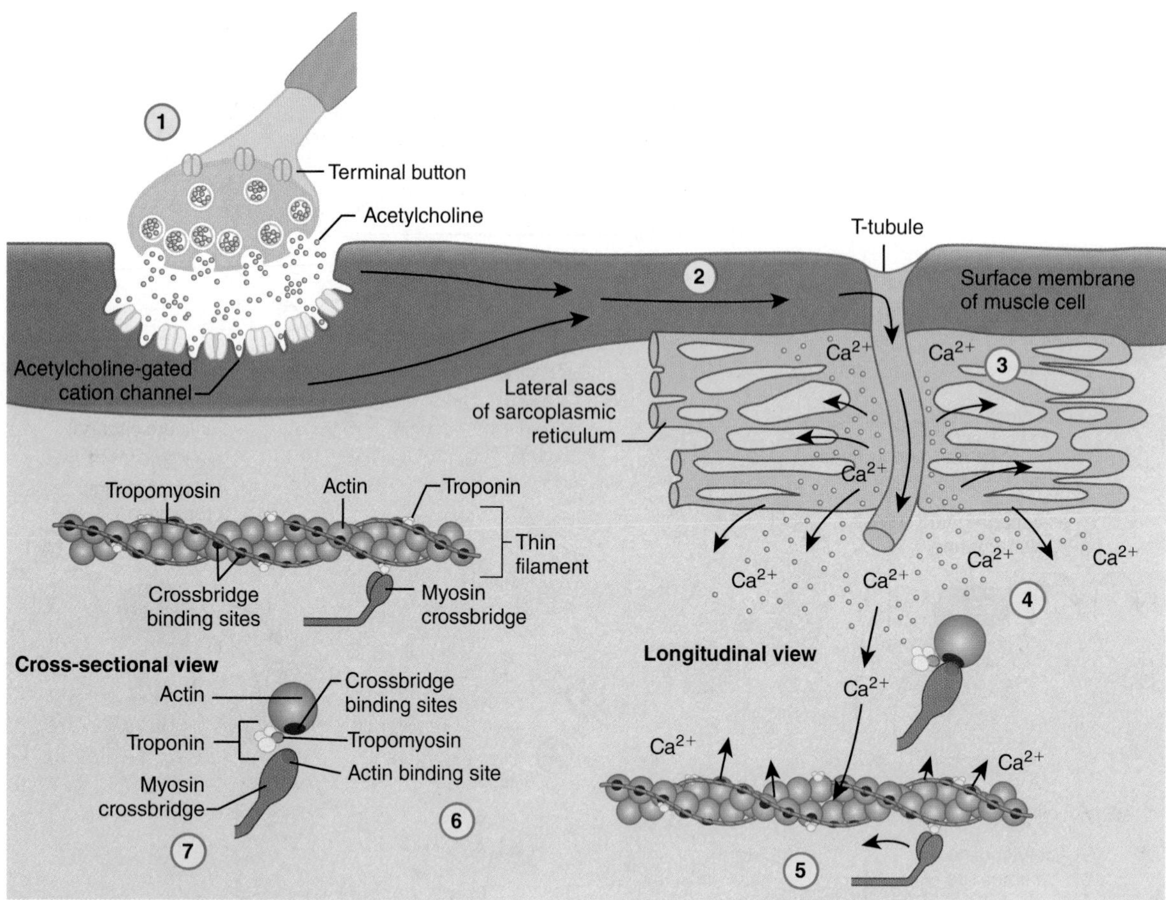

1. Acetylcholine released by axon of motor neuron crosses cleft and binds receptors/channels on motor end plate.
2. Action potential generated in response to binding of acetylcholine and subsequent end plate potential is propagated across surface membrane and down T-tubules of muscle cell.
3. Action potential in T-tubule triggers Ca^{2+} release from sarcoplasmic reticulum.
4. Calcium ions released from lateral sacs bind to troponin on actin filaments; leads to tropomyosin being physically moved aside to uncover crossbridge binding sites on actin.
5. Myosin crossbridges attach to actin and bend, pulling actin filaments toward center of sarcomere; powered by energy provided by ATP.
6. Ca^{2+} actively taken up by sarcoplasmic reticulum when there is no longer local action potential.
7. With Ca^{2+} no longer bound to troponin, tropomyosin slips back to its blocking position over binding sites on actin; contraction ends; actin slides back to original resting position.

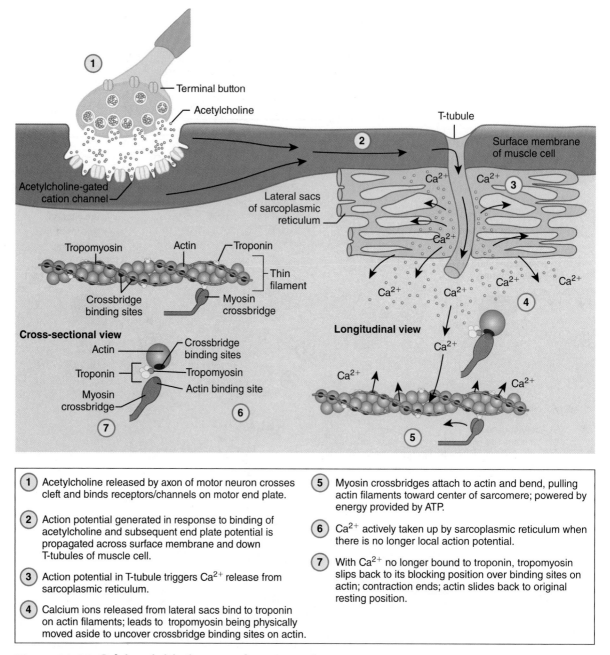

Figure 11.11 **Calcium initiating muscle contraction.**

provided the energy (McLester, 1997). The crossbridge power stroke pulls the six surrounding thin filaments toward the center of the sarcomere simultaneously (Figure 11.12).

As you can imagine, one power stroke pulls the thin filament only a tiny distance because the displacement is only the linear distance traveled by one myosin head. Therefore, full shortening of the muscle is achieved by repeated cycles of detachment, binding, and tilting. During these repeated cycles of crossbridge interaction, the thin filaments are sliding across the thick filaments. This sliding of the thick and thin filaments relative to one another is often called the **sliding filament theory of muscle contraction**.

Sliding filament theory of muscle contraction Theory of muscle force generation in which tension is generated by the interaction of myosin with actin leading to myofibrillar translation.

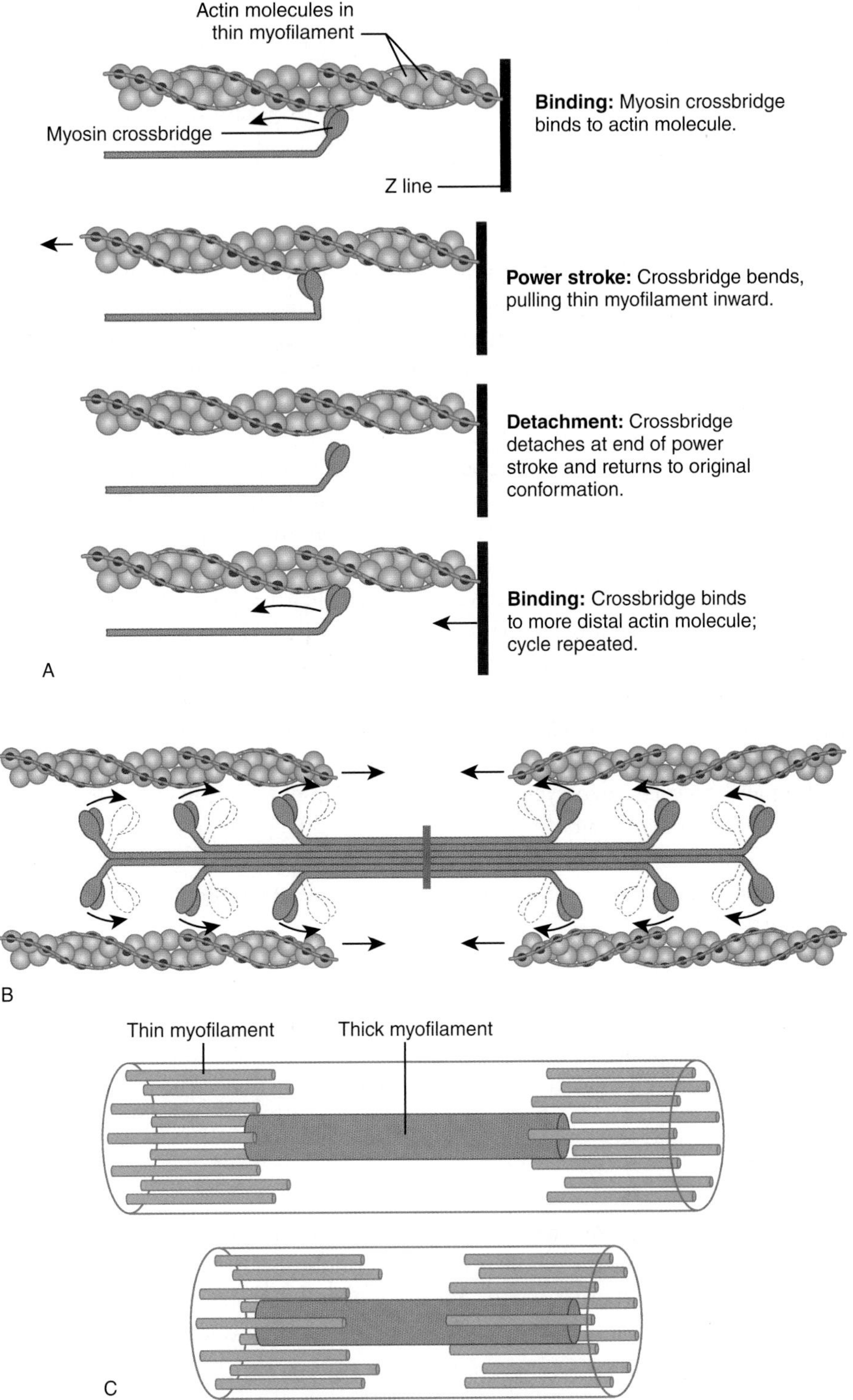

Figure 11.12 **The crossbridge power stroke.**

So what keeps the thin filaments from sliding back into place during detachment of the myosin heads? First, remember that each myosin molecule has two heads. These two heads work independently, with only one attached to an actin molecule at any given time. In addition, not all of the myosin heads of a thick filament are attached at the same time. Therefore, as some are undergoing their power-stroke phase, others are detaching and preparing to reattach to the next actin site, which is now closer. Think of how you would go about climbing a rope. Would you pull yourself up with both hands and then release them both at the same time to grasp higher on the rope? Probably not! You would probably choose to climb hand over hand, holding yourself on the rope with one hand while reaching higher with the next. In just this way, some of the myosin crossbridges pull the thin filament toward the center of the sarcomere while others prevent it from slipping back into the resting position.

As you have read, the myosin head has a site for splitting ATP. ATP is split as soon as it is bound to the site on myosin. However, the ADP and P_i remain bound to the myosin until released during the crossbridge cycle (Figure 11.13). The energy released during the breakdown of the ATP molecule is stored within the myosin head until ready for the power stroke (similar to an arrow that has been drawn back in a bow but not yet released). Recall that the power stroke occurs simultaneously with the release of P_i from the myosin head.

Detachment of the crossbridge is also coupled with chemical steps associated with ATP usage. After the power stroke is complete, ADP is released. With both P_i and ADP removed from the myosin head, it is now ready for detachment. However, the myosin head will not detach from the actin-binding site until a new ATP is bound (**rigor mortis** is muscle stiffness caused by inability to detach crossbridges because ATP is not available after death.). Once ATP is bound to the myosin head, it detaches and returns to its original conformation. Once in its original position, it is prepared to attach to a new active-actin site that has now been pulled closer. This cycle of binding, power stroke, detachment, and reattachment continues as long as calcium concentration in the sarcoplasm remains at sufficient levels to inhibit the troponin–tropomyosin complex (and as long as ATP is available).

Each binding and power-stroke cycle pulls the thin filament slightly closer to the middle of the sarcomere. To truly understand muscle contraction, always keep the entire structure of the muscle in mind. The actin protein chains are anchored to the Z-disks of the sarcomere; therefore, the pull of myosin forces the entire sarcomere to shorten (Figure 11.14). Just imagine a strong man pulling two large objects toward him.

Because myofibrils are made up of sarcomeres, the myofibrils shorten. Muscle fibers are made up of myofibrils, so they shorten in turn. Muscle fibers make up fasciculi, and therefore the entire muscle shortens. The final result is that the entire muscle belly shortens. This shortening applies force to the tendons and, in turn, to the bones to which they are attached. So where does everything in the muscle go as it shortens? It is all forced toward the middle of the muscle; that's why muscles bulge when they are contracted.

Muscle Relaxation

Recall that crossbridge cycling continues as long as calcium levels are sufficient in quantity to bind to troponin. Now let's examine how the whole process is brought to a halt. It's really a simple concept: It has to be stopped at the point where it starts—the arrival of an action potential. When action potentials stop arriving at the terminal button, acetylcholine is no longer released into the neuromuscular cleft. The acetylcholine that was previously released is broken down by the enzyme acetylcholinesterase. Once acetylcholine is removed, end-plate potentials are no longer produced. Without the action potentials, calcium release from the terminal cisternae of the sarcoplasmic reticulum stops. Calcium previously released into the sarcoplasm is actively transported by calcium pumps back into the sarcoplasmic reticulum. As calcium is removed from troponin, tropomyosin moves back into its original position. Once binding sites are blocked by tropomyosin, myosin can no longer achieve a strong binding state with actin. Crossbridge action is prevented, and relaxation of the muscle is achieved passively.

Rigor mortis Muscle stiffness caused by inability to detach myosin crossbridges because of the lack of available ATP after death.

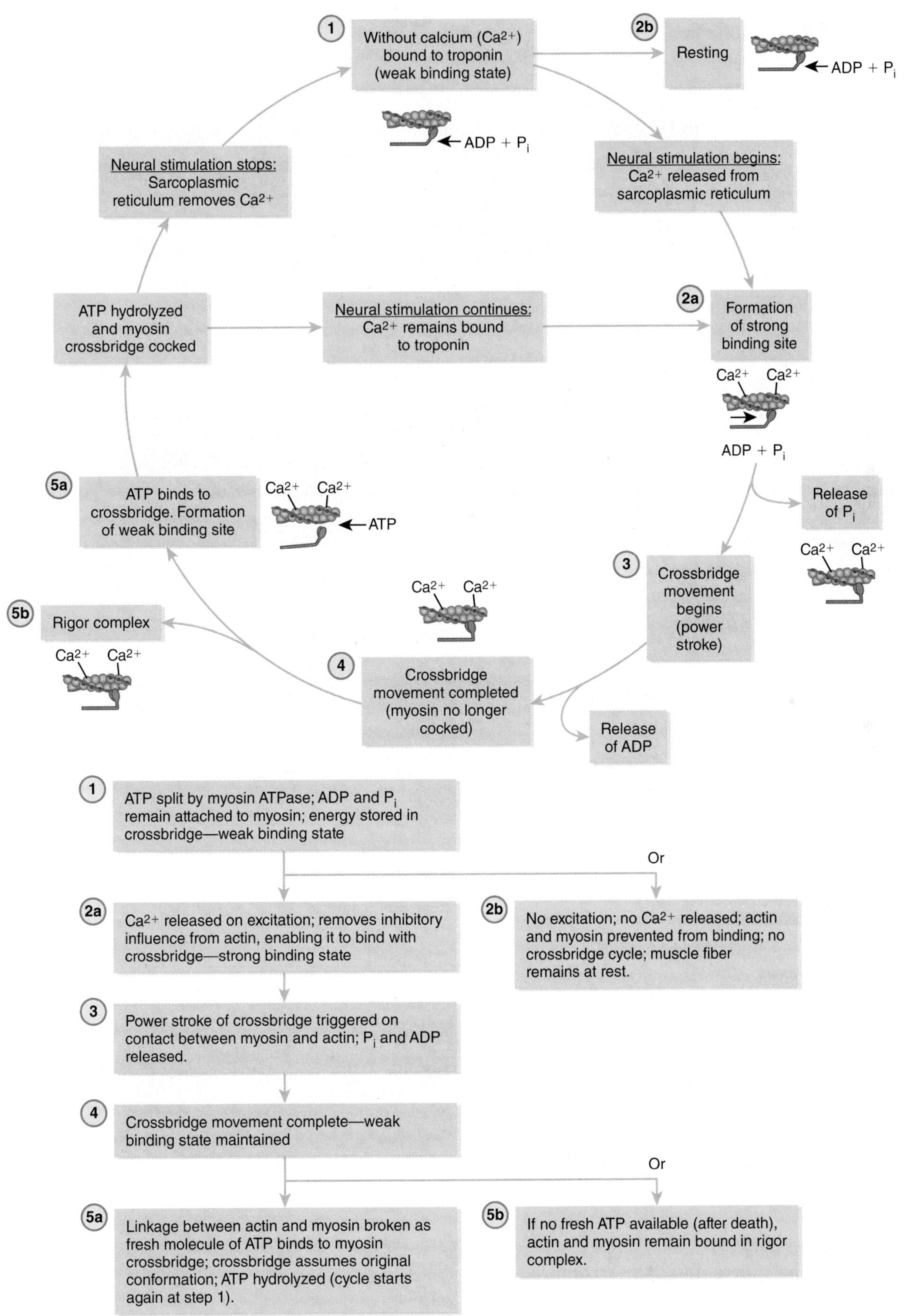

Figure 11.13 **ADP and Pi relationship to contractile process.**

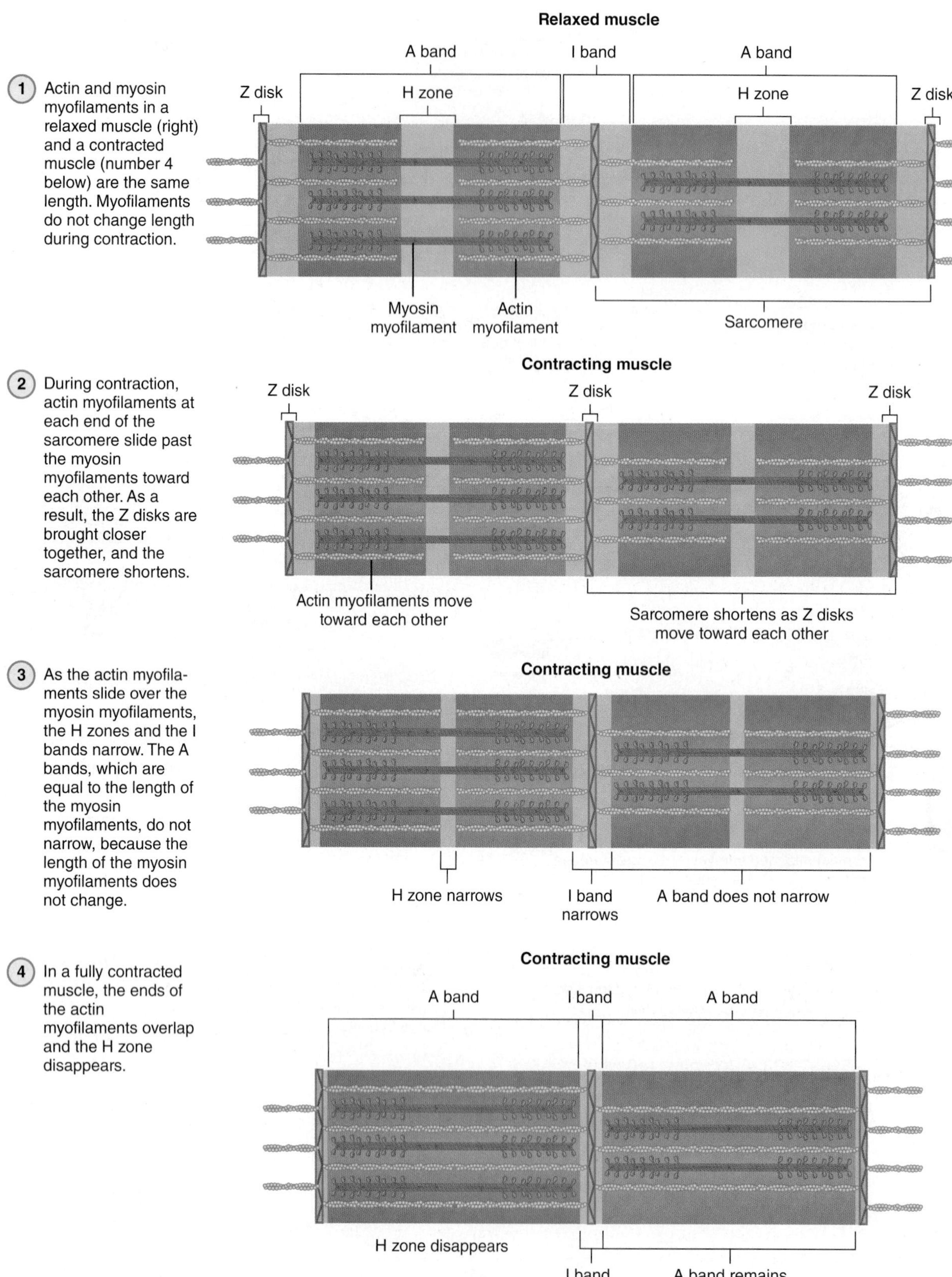

Relaxed muscle
A band
I band
A band
1 Actin and myosin myofilaments in a relaxed muscle (right) and a contracted muscle (number 4 below) are the same length. Myofilaments do not change length during contraction.
Z disk
H zone
H zone
Z disk
Myosin myofilament
Actin myofilament
Sarcomere
Contracting muscle
2 During contraction, actin myofilaments at each end of the sarcomere slide past the myosin myofilaments toward each other. As a result, the Z disks are brought closer together, and the sarcomere shortens.
Z disk
Z disk
Z disk
Actin myofilaments move toward each other
Sarcomere shortens as Z disks move toward each other
Contracting muscle
3 As the actin myofilaments slide over the myosin myofilaments, the H zones and the I bands narrow. The A bands, which are equal to the length of the myosin myofilaments, do not narrow, because the length of the myosin myofilaments does not change.
H zone narrows
I band narrows
A band does not narrow
Contracting muscle
4 In a fully contracted muscle, the ends of the actin myofilaments overlap and the H zone disappears.
A band
I band
A band
H zone disappears
I band narrows further
A band remains unchanged

Figure 11.14 **The shortening of a sarcomere.**

Types of Muscle Actions

The term *muscle contraction* implies that every time the events just described occur, the muscle shortens. But you know that this is not the case. For example, what if you apply force to an object that is too heavy for you to move? In this case (load torque ≥ muscle torque generated) tension is generated within the muscle but the bone to which the muscle force is applied does not move. This situation is called a **static muscle action**. If the body segment moves during the generation of muscle tension, the situation is referred to as a **dynamic muscle action**. However, two types of dynamic muscle actions occur: concentric and eccentric. During a **concentric muscle action**, the muscle torque generated is greater than the load torque, and the muscle is therefore allowed to shorten. This muscle action is called *concentric* because the ends of the muscle move toward the center as the muscle shortens. Alternatively, a person may lower a weight that has been lifted (e.g., the lowering phase of a bicep curl exercise). In this case, the muscle generates tension, but the resultant torque is less than the load torque, and the muscle lengthens. This kind of situation in which the muscle lengthens while generating force is called an **eccentric muscle action** because the ends of the muscle move away from the center.

Factors Related to Contractile Force

As you already know, the neuromuscular system can produce many different degrees of force. The same person who can perform the fine motor movements necessary to use a scalpel with precision can also train his body to bench press more than 150 kg. Now that we understand the contractile process that produces force, we can gain a deeper knowledge of the mechanisms by which this force is controlled. In general, two major factors affect whole muscle tension: (1) the number of fibers contracting, and (2) the tension developed by each individual fiber involved in the contraction. The next section describes the individual mechanisms by which these two factors are associated with such varied degrees of force.

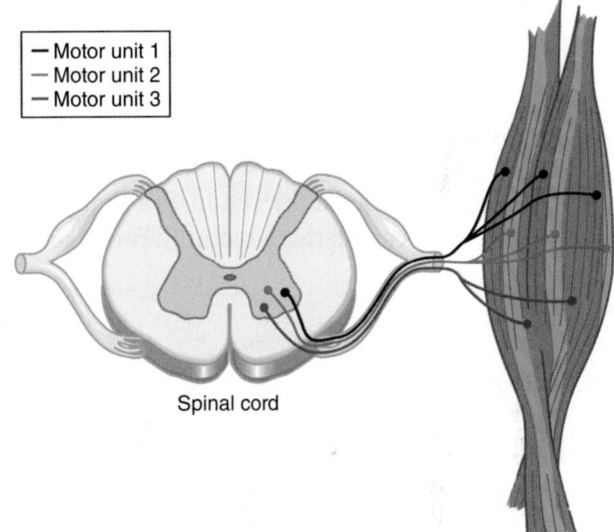

Figure 11.15 **Motor units of varying innervation ratios.**

The Motor Unit

It makes sense that the number of muscle fibers that contract affects the tension generated by a whole muscle. But how are the number of fibers that contract actually regulated? Each whole muscle in the body is innervated by a different number of motor neurons. Each motor neuron innervates a different number of muscle fibers. However, each muscle fiber is only innervated by one motor neuron (Figure 11.15).

An individual motor neuron and the group of fibers that it innervates are collectively called a **motor unit**. When a given motor neuron is activated (recruited) by the nervous system, all of the fibers it innervates are stimulated to act

Static muscle action Tension is generated within the muscle, but the bone to which the muscle force is applied does not move.

Dynamic muscle action Movement of the body segment occurs during generation of muscle tension.

Concentric muscle action The muscle torque generated is greater than the load torque, and the muscle is therefore allowed to shorten.

Eccentric muscle action The resultant muscle torque is less than the load torque and the muscle lengthens during generation of tension.

Motor unit A given motor neuron and the group of muscle fibers that it innervates.

simultaneously. The muscle fibers of the motor unit are distributed throughout a whole muscle (i.e., they are not all lying parallel to each other in one area). In this way, their tension is not localized but is evenly dispersed throughout the muscle. The size of the motor unit (number of fibers innervated by the motor neuron) determines how much force is generated by its stimulation. The number of fibers in a given motor unit can vary widely, depending upon the location of the muscle. For example, motor units of muscles used mainly for precise or delicate motions (e.g., eye muscles or muscles of the hand) may have as few as 10 to 12 fibers. The motions are more delicate because so few fibers contract in response to activation. In contrast, muscles of the legs may be divided into motor units containing as many as 2,000 fibers. Of course, motions of the legs are used mainly for gross motor activities that may be more powerful but less precise.

The importance of motor units in controlling muscle force should not be underemphasized. Because muscles have varying numbers of motor units as well as motor units of different sizes, many varying degrees of force can be achieved, depending on the situation. The number of combinations of recruitment is quite impressive. If the movement is a precise one and doesn't require large amounts of force, then a small number of motor units, or motor units of small size, can be recruited. For example, a surgeon using a scalpel needs a highly precise amount of muscular force. In this case, one would not want to recruit large motor units (or many of them) because all of the fibers within that unit will contract simultaneously. Therefore, a small motor unit is recruited. If the surgeon then chooses a slightly heavier instrument, the previous small unit plus other small units can be recruited until the desired degree of force is attained. In contrast, large numbers of motor units and units of larger size would be recruited to perform a gross motor activity such as a squat exercise.

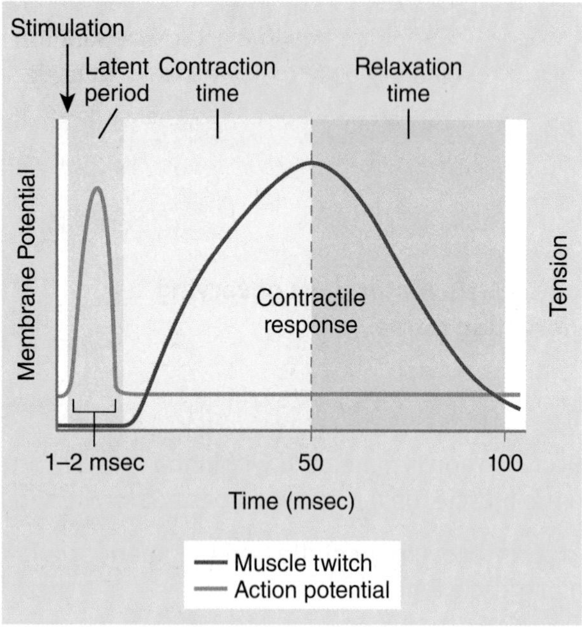

Figure 11.16 **Time course of a muscle twitch.**

Twitch A weak, extremely brief action of the fiber caused by the arrival of a single action potential.

Latent period A period of a few milliseconds following the occurrence of the action potential, during which the fiber has not yet begun to generate tension.

Neural Stimulation Frequency

Dividing the whole muscle into motor units provides a large number of possibilities for generating force. However, we can also control the tension developed by the fibers of a motor unit once it is recruited. In other words, they all contract simultaneously but they do not necessarily contract maximally. The arrival of a single action potential in the muscle fiber produces a weak, extremely brief action of the fiber called a **twitch**. A muscle twitch is not useful for most activities because the contraction is weak. You may have experienced a twitch in one of your muscles. It can be observed as movement (many times involuntary) of one tiny localized area of a muscle. You probably also noticed that this twitch did not result in complete contraction of the muscle; it is simply too weak. However, understanding a single muscle twitch can give you a better understanding of how muscle force is controlled by the frequency of neural stimulation. First, we must understand the time course of a single twitch. The single action potential that stimulates the muscle fiber to twitch lasts approximately 1 to 2 msec. A **latent period** of a few milliseconds occurs following the action potential when the fiber has not yet begun to generate tension. This latent period occurs because the amount of time necessary for all of the steps in the excitation–contraction coupling process to take place is finite (Figure 11.16).

Following the latent period is the actual **tension generation period**. Depending on the type of muscle fiber, approximately 50 msec are required for the fiber to reach peak tension (note, however, that the time to peak tension varies widely). Remember that the contraction will last until the calcium has been actively transported back into the sarcoplasmic reticulum. Therefore, a **relaxation period** occurs, which also lasts approximately 50 msec. The important point to note is that the action potential lasts only 1 to 2 msec, but the resulting muscle fiber action may last 100 msec or more. This time difference is critically important to controlling muscle contraction force.

The importance of this time differential is that more than one action potential can be sent to the muscle fiber before it has had time to completely relax. So how does this overlap help? A single action potential stimulates the release of enough calcium to interact with all of the troponin within the muscle fiber. This abundance of calcium means that all of the crossbridges are free to interact. However, active transport of calcium into the sarcoplasmic reticulum is constant. Only during the action potential does release of calcium from the sarcoplasmic reticulum exceed uptake. As soon as the action potential is over, the net active transport of calcium is back into the sarcoplasmic reticulum. As calcium is removed from the sarcoplasm, fewer and fewer crossbridges are free to participate because tropomyosin is moving back into the position in which it blocks actin-binding sites. With fewer crossbridge interactions, the resultant force gradually subsides. So the force produced by a single twitch is not large because it is essentially coming to an end as soon as it gets started.

So how do we solve this problem and achieve a greater amount of force? We send more action potentials. But we can't just send more action potentials; the key is to send more before the fiber has had time to relax from the first. If we send another action potential before the fiber has completely relaxed from the first, we will simply get another twitch (Figure 11.17).

Tension generation period Period of time following the latent period when the muscle fiber actually generates tension.

Relaxation period Period following tension generation in which the muscle fiber returns to the resting state.

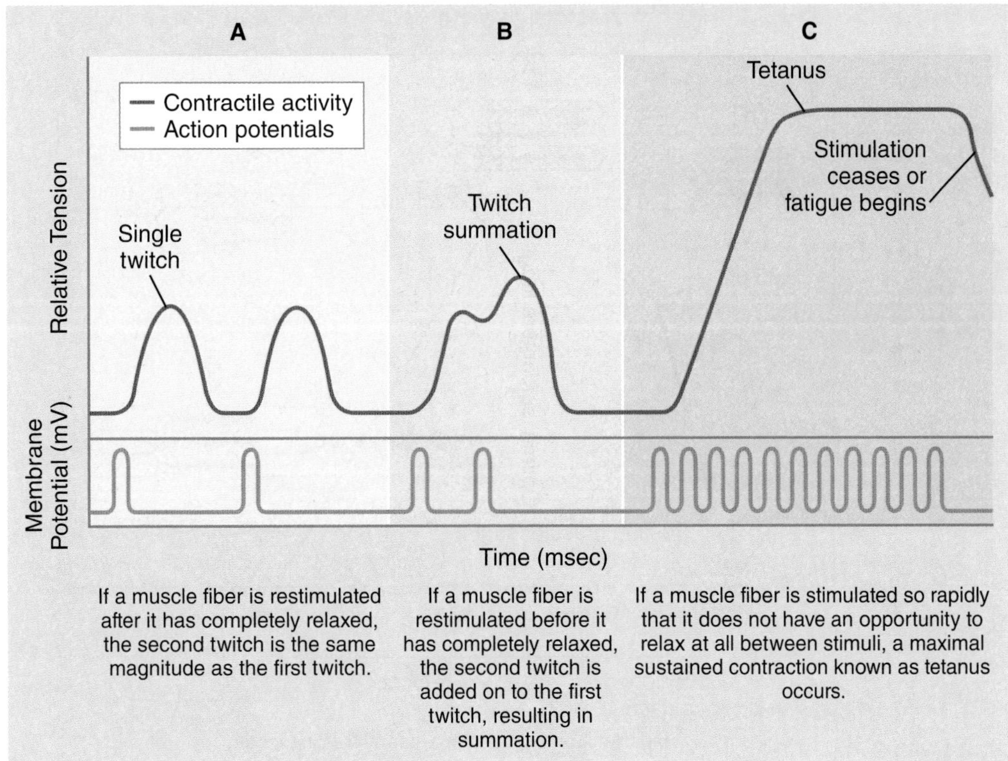

Figure 11.17 **Summation versus tetanus.**

If we can send another action potential before complete relaxation, a second release of calcium will be added to the calcium remaining from the first action potential. The longer the time that calcium is available, the more crossbridge cycling can take place. As a result, the force of the second twitch is *summed* with the force remaining from the first twitch (**twitch summation**). With each additional action potential sent before relaxation, the force will increase accordingly because of the increased calcium available. However, there is a point at which the maximum number of crossbridges are participating in the muscle action. At this point, called **tetanus**, the muscle fiber reaches peak tension development. This tetanic contraction is smooth and sustained and therefore more useful for normal body motion.

Length–Tension Relationship

Another factor that can affect the tension developed by an individual muscle fiber is its length before it is stimulated (Podolsky & Shoenberg, 1983). This relationship between length before stimulation and subsequent tetanic tension development is called the **length–tension relationship** (Figure 11.18). The length–tension relationship is understandable as long as one understands the sliding filament theory of muscle contraction. For each muscle, an optimal length exists for developing tetanic tension. This optimal length

Twitch summation The force of a second twitch is *summed* with the force remaining from the first twitch.

Tetanus The point at which the maximum number of crossbridges are participating in the muscle action and the muscle fiber reaches peak tension development.

Length–tension relationship Relationship between muscle fiber length prior to stimulation and subsequent tetanic tension development.

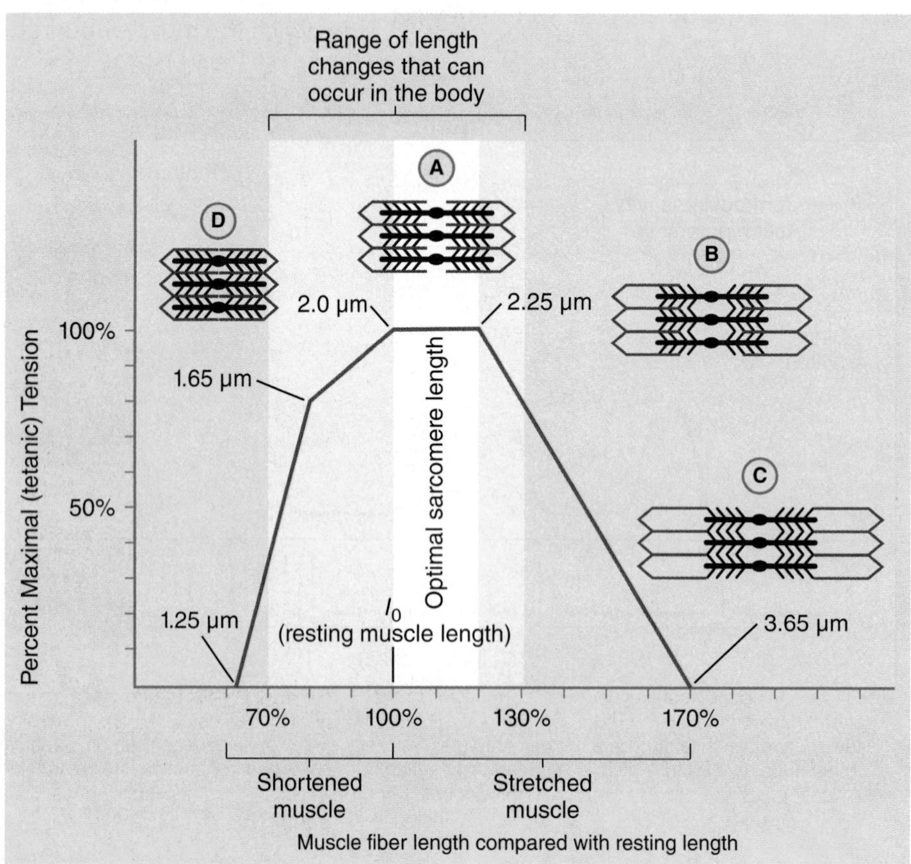

Figure 11.18 **The length–tension relationship.**

corresponds with maximum overlap of the thick and thin filaments (Gordon, Huxley, & Julian, 1966). In other words, the development of tension is optimized because the maximal number of active-actin sites is available for attachment of myosin crossbridges.

Force output steadily declines as the sarcomere is stretched farther from the optimal length. This decline occurs because the thin filaments are pulled away from the thick filaments, preventing myosin heads from binding to the active-actin sites. Therefore, as more and more sites become unavailable for binding, force decreases accordingly. In an intact human, the theoretical optimal length roughly corresponds to a muscle's resting length. However, a muscle's ability to produce force is actually greatest at approximately 120% to 130% of resting length because of the elastic force contributed by the parallel and series-elastic components. The force of the elastic recoil is enough to more than compensate for loss of force from nonparticipating crossbridges (Figure 11.19).

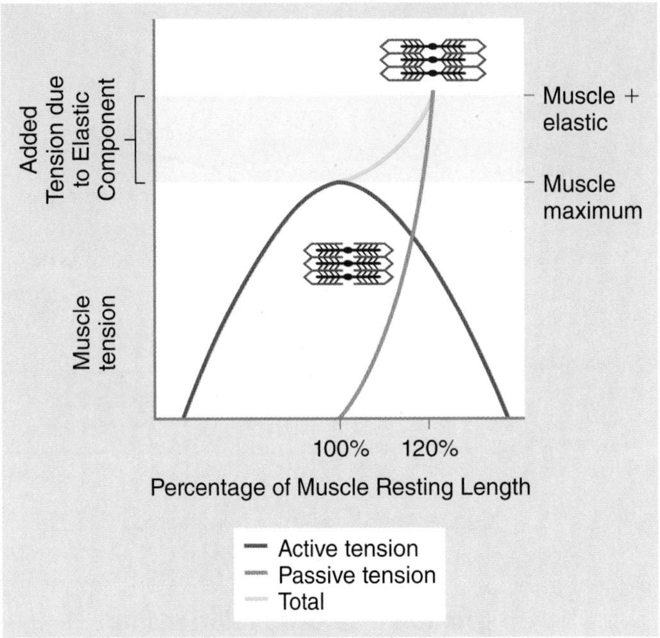

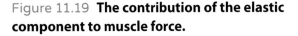

Figure 11.19 **The contribution of the elastic component to muscle force.**

However, major reductions in force tend to occur at lengths greater than 120% to 130% of resting length (i.e., when the fiber has been stretched 30% beyond resting length). If a fiber is stretched approximately 70% beyond its resting length, the myosin crossbridges cannot reach any of the active-actin sites and contraction is no longer possible. However, this limitation is found experimentally by removing fibers from the skeleton and stretching them before artificially stimulating them. Because of the constraints imposed by the intact musculoskeletal system, a muscle cannot be stretched to 70% of its optimal length. In fact, these constraints usually prevent a stretch beyond 30% of optimal (Sherwood, 2016). So the skeletal system actually acts as a protective mechanism, preventing excessive stretch and the related decline in force production.

Force-production capability also declines as a fiber is shortened. The sliding filament theory also helps explain this phenomenon. Several structural reasons exist for the decline in force that occurs with fiber shortening. First, as the thin filaments are pulled toward the center of the sarcomere, they overlap. Because of this overlap, some of the active-actin sites on one thin filament are blocked by another advancing thin filament. So this overlap physically prevents myosin crossbridges from binding. In addition, remember that the thick filament is composed of two bundles of myosin molecules arranged with the tails oriented toward the center. Therefore, no myosin heads are present in the center region of the thick filament. So as the thin filaments are pulled toward the center of the thick filaments, fewer crossbridges are available to interact with the active-actin sites. Finally, the construction of the sarcomere itself can provide an obstruction. Remember that the ends of the sarcomere are defined by Z-disks. So as the fiber is shortened, the ends of the thick and thin filaments eventually are forced into the Z-disks. This collision physically prevents further shortening. Excessive shortening is also thought to inhibit the binding of calcium to troponin. This inhibition would translate into fewer available active-actin sites. Major reductions in contractile force usually begin to occur at approximately 70% to 80% of optimal length (i.e., when the muscle has been shortened by approximately 30%). But once again, because of the construction of the musculoskeletal system, a fiber cannot be shortened by more than 30% of optimal length. So even though contractile force varies with the length of the fiber, the human body usually maintains a length that is nearly optimal.

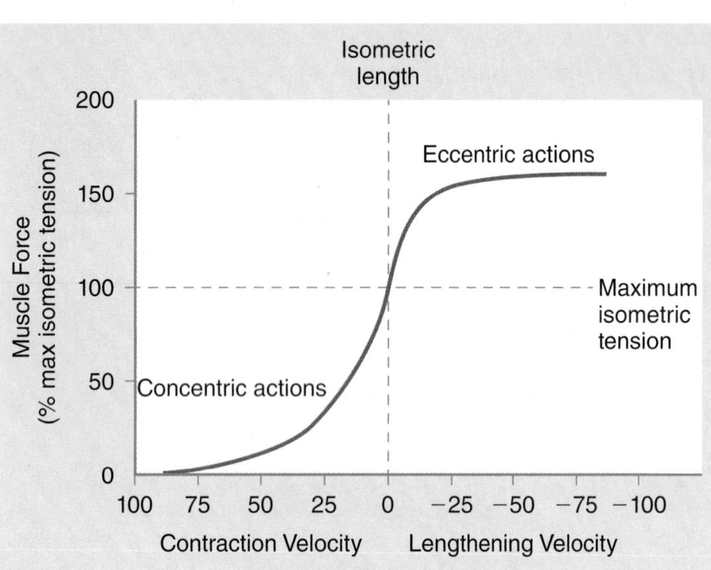

Figure 11.20 **The force-velocity relationship.**

Force–Velocity Relationship

Velocity of contraction is the final factor that can affect force produced by an individual muscle fiber. In general, as the velocity of a concentric muscle action increases, the force produced during the action decreases. Another way of stating this **force–velocity relationship** is: The greater the load against which a muscle must contract, the lower the velocity of that contraction (Figure 11.20). You have experienced this firsthand if you have ever noticed that you cannot move a heavy weight with the same velocity at which you can move a lighter weight. In fact, the force of a muscle action is maximal at zero velocity (static muscle action).

As the concentric action velocity increases, force-production capability decreases. This relationship means that maximal velocity of contraction occurs when no load is opposed against that contraction. Once again, we must look to the sliding filament theory for an explanation. It has probably become obvious by this point that the greater the number of myosin crossbridges attached to actin, the greater the force produced. With this in mind, we must also realize that it takes a finite amount of time for a crossbridge to attach and detach. Looking at the situation from one perspective, one could simply speculate that if the load is increased, it takes longer for the power stroke to occur. But that's really the same as saying that velocity decreases as the load increases. It doesn't really provide an explanation. So let's look at an example. Imagine that you are working on an assembly line, and the conveyer belt is stopped (velocity = 0). No time constraints exist, so making contact with objects on that belt is easy. Now imagine that the conveyer belt starts moving (the velocity increases), but you are able to move your arms at a given rate. You will no doubt make contact with many of the objects, but you will also miss some because you simply do not have time to make contact with all of them. If the velocity of the belt progressively increases, you will likely miss more and more of the objects. Myosin crossbridges encounter a similar situation. At zero velocity (static action), myosin crossbridges can easily attach to the active-actin sites because they have no time constraints. Therefore, we have maximal crossbridge interaction and maximal force output. However, as the velocity of contraction increases, the thin filament moves past the thick filament at a faster and faster rate. Under these conditions, some myosin crossbridges simply do not have time to bind to active-actin sites because they attach and detach at a given rate. Therefore, force is diminished because fewer crossbridges are connected. The faster the contraction, the fewer crossbridge interactions, and force production drops precipitously.

The force–velocity relationship for eccentric actions is not the same as that for concentric conditions. Actually, the maximal voluntary contractile force during eccentric actions is relatively independent of lengthening velocity (Enoka, 1996; Lieber, 2010). This phenomenon is quite interesting and still the subject of speculation (Enoka, 1996). First, we must understand that greater force can be generated during a maximal eccentric action than during a concentric action at a lower energy cost (Stauber, 1989) and with a lower level of nervous system activation (Bigland & Lippold, 1954). How is this possible? The lower energy cost is probably because fewer crossbridges are cycling and fewer fibers are being recruited during an eccentric action (Stauber, 1989), and some evidence shows that higher-threshold fibers are

Force–velocity relationship The greater the load against which a muscle must contract, the lower the velocity of that contraction.

selectively activated (Nardone, Romanò, & Schieppati, 1989). The increased force is probably the result of some of the crossbridges not cycling during an eccentric action but remaining bound. Therefore, as new crossbridges are formed, the total amount of crossbridge interaction may exceed that of a concentric action (Stauber, 1989). In addition, actomyosin crossbridge bonds are probably mechanically disrupted during eccentric actions as opposed to the ATP-dependent detachment that occurs during concentric actions (Flitney & Hirst, 1978). Therefore, eccentric actions are less metabolically costly. In addition, this mechanical detachment probably places high stresses on the crossbridges, likely being a source of the tissue damage that occurs during eccentric actions (Enoka, 1996). In summary, it seems that the contractile filaments perform differently during eccentric and concentric actions, with force being maintained through additional crossbridge attachment while the muscle acts eccentrically. In addition, the force is maintained during an eccentric action even at high velocity because the series elastic component is stretched during the action, adding passive tension to the active contractile elements (Åstrand et al., 2003). This observation seems to mean that skeletal muscle is resistant to stretch. This resistance to stretch is a good thing because it could protect against unexpected lapses in muscle tension in response to an externally applied force.

The Compromise for Power

The force–velocity relationship has ramifications for sports performance and many other areas of human movement because most sports have some power component. Recall that power is the product of force and velocity. To increase power production at any given instant, we cannot simply maximize either value. According to the force–velocity relationship, if we maximize velocity, the force-production capability will decrease. In turn, maximal force production occurs at zero velocity. So to achieve maximal power, a compromise must be reached. Maximal power output occurs at approximately 30% of maximal contraction velocity (Hill, 1970). The contractile force is approximately 30% of maximum at this velocity (Komi et al., 2000). *Velocity* is one factor in the *power* equation, so power generated by a muscle increases as velocity of movement increases up to a peak value at approximately 200–300 deg/sec (Perrine & Edgerton, 1978).

At this point (~30% of maximal velocity), power production decreases with increases in velocity. The decrease occurs because force production is inhibited at higher velocities, and *force* is the other factor in the *power* equation (Figure 11.21). In terms of training for power sports, we can interpret this observation to mean that training programs must include both a force component and a velocity component.

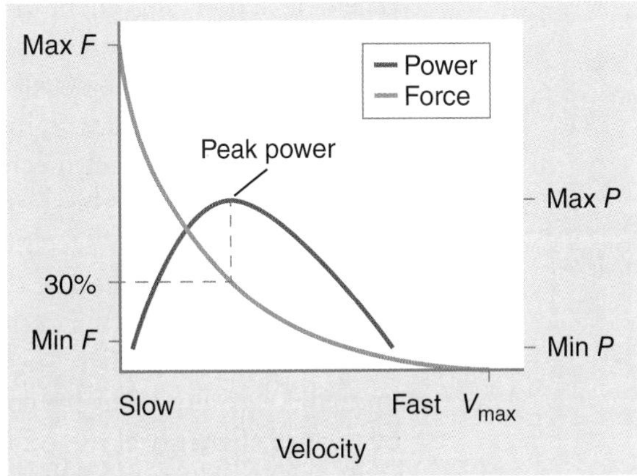

Figure 11.21 **The force, velocity, power relationship.**

The Stretch–Shorten Cycle

A topic of great interest to biomechanists and athletes alike is the behavior of a muscle that has been subjected to an eccentric action immediately followed by a concentric action. Extensive research (and anecdotal experience) has found that muscle force production is enhanced if an active muscle is stretched to generate eccentric tension *immediately* before the concentric action (Buchanan, 1997; Doan et al., 2002; Finni et al., 2003; Ishikawa & Komi, 2004; Komi & Gullhofer, 1997; Kubo et al., 2000a; Kubo et al., 2000b; Trimble, Kulkulka, & Thomas, 2000). This phenomenon of enhanced force through use of an eccentric–concentric sequence is called the **stretch–shorten cycle**. Examples of stretch–shorten cycles can be seen in activities such jumping, running, and

Stretch–shorten cycle The phenomenon of enhanced force by generating eccentric tension immediately prior to the concentric action.

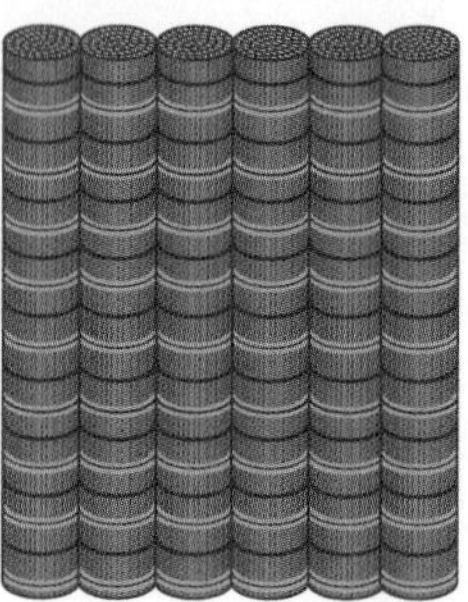
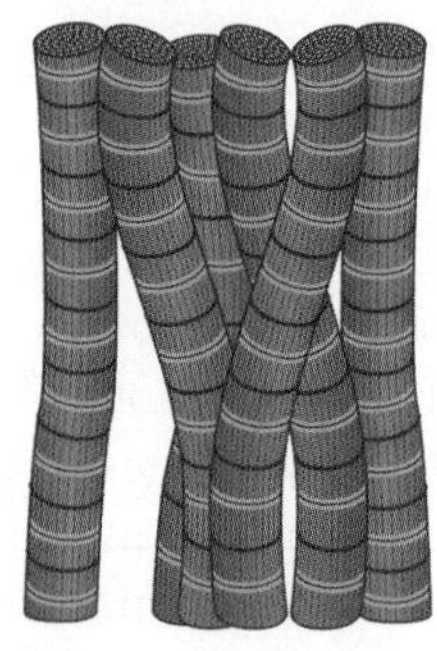

Figure 11.22 **Theoretically "slack" fibers.**

throwing. Often a countermovement is used in the preparation phase of a vertical jump. Also, during each stride of a running pattern, the plantar flexors generate eccentric tension and then are immediately contracted. In sports that involve throwing or the use of implements, often a backswing precedes the actual throw or strike.

Although the exact mechanisms involved in the stretch–shorten cycle are not completely understood, many contributing factors have been proposed and studied (Buchanan, 1997; Doan et al., 2002; Finni et al., 2003; Ishikawa & Komi, 2004; Komi & Gullhofer, 1997; Kubo et al., 2000a; Kubo et al., 2000b; Trimble et al., 2000). This section covers the basics of some of those research findings. First is the simple storage and subsequent use of elastic energy contributed by the parallel and series elastic components. You read earlier that this elastic recoil is a factor in the length–tension relationship. However, simple elastic recoil cannot account for all of the enhanced performance accompanying the stretch–shorten cycle. Otherwise, an athlete could pause between the eccentric and concentric phases and get the same performance effect. However, that approach doesn't work; the concentric action must be performed immediately after the eccentric action to reap the benefits of the stretch–shorten cycle. For example, to benefit from the stretch–shorten cycle, the concentric phase of a vertical jump must be performed immediately after the countermovement with no pause. Another proposed mechanism is the stimulation of the muscle spindle reflex that occurs because of the sudden stretch. Muscle spindles are proprioceptors that are embedded within the muscle belly, and they monitor the velocity and degree of stretch. Remember that the muscle spindles detect sudden or exaggerated stretches and respond by initiating a muscle contraction. Stimulating this reflex may increase stiffness of the muscle during the eccentric phase. The increased stiffness may allow the elastic components of the muscle to store more energy for subsequent use during the concentric action. A third contributing factor may be related to the length–tension relationship. Not all of the fibers in a resting muscle are stretched to the same degree; some are relatively slack (Figure 11.22). It's important to note that the fibers aren't literally slack, but not every fiber in the muscle is at the same *optimal* length at rest.

A slack fiber is not at its theoretically optimal resting length. By prestretching the fiber during the eccentric action, alignment of crossbridges in the previously slack fibers could be enhanced. A final contributor may be more chemical in nature and related to the fiber type. Slow-twitch muscle fibers require a longer time to develop full tension because of different rates of myosin ATPase activity. The eccentric action provides an increased time for tension to develop within the slow-twitch fibers.

No matter the exact mechanisms, the stretch–shorten cycle must be considered during sports training. For example, the stretch–shorten cycle is the basis for plyometric training because plyometric exercises are simply eccentric actions followed immediately by concentric actions. One must also consider the contribution of the stretch–shorten cycle to activities in terms of metabolic cost. With use of the stretch–shorten cycle, more mechanical work can be performed for a given metabolic cost. For example, in an activity such as running, the ground contact phase causes eccentric loading that is immediately followed by the concentric action of a toe-off. So the metabolic cost of running is lowered by the use of energy at toe-off that was stored at ground contact. Think of the many eccentric actions performed by the human body everyday and you can begin to grasp the importance of the stretch–shorten cycle in minimizing metabolic cost.

11.2 BIOMECHANICS OF MUSCLE LOCATION, ORIGIN, AND INSERTION

Nearly the entire skeletal framework of the body is covered with muscles, each with a known and predictable location across the entire human population. Because of its specific location, each individual muscle is capable of carrying out a different movement function. In general, a muscle has two points of attachment: a proximal attachment (sometimes called an **origin**) that tends to be on a relatively immoveable location, and a distal attachment on a relatively more moveable segment (or **insertion**). The origin and insertion points of the muscle (and therefore the two segments to which it is attached) are brought closer together during muscle contraction. Therefore, the origin and insertion points chiefly determine the function of a particular muscle. However, some aspects related to muscle location, origin, and insertion are not so obvious.

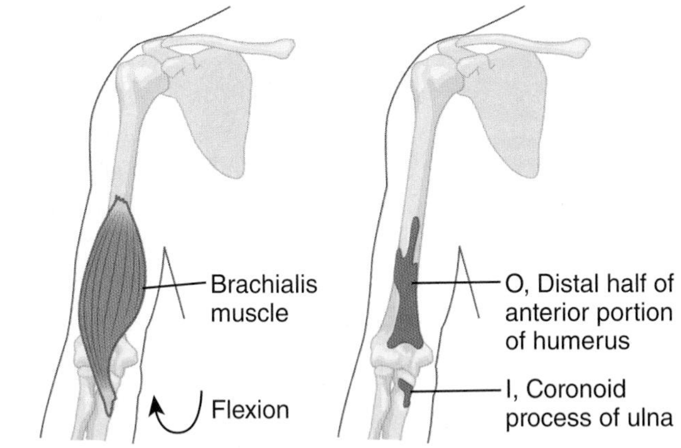

Figure 11.23 **Brachialis origin (O) and insertion (I).**

Uniarticular Versus Multiarticular Muscles

In addition to origin and insertion, a major factor related to the function of an individual muscle is the number of joints (articulations) that it crosses. A muscle can affect the actions of any joint that it crosses. Of course, many muscles only cross a single articulation and are therefore called **uniarticular (single-joint) muscles**. The brachialis of the elbow is an example of a uniarticular muscle. It has its origin on the humerus and insertion on the ulna and therefore can cause only elbow flexion (Figure 11.23).

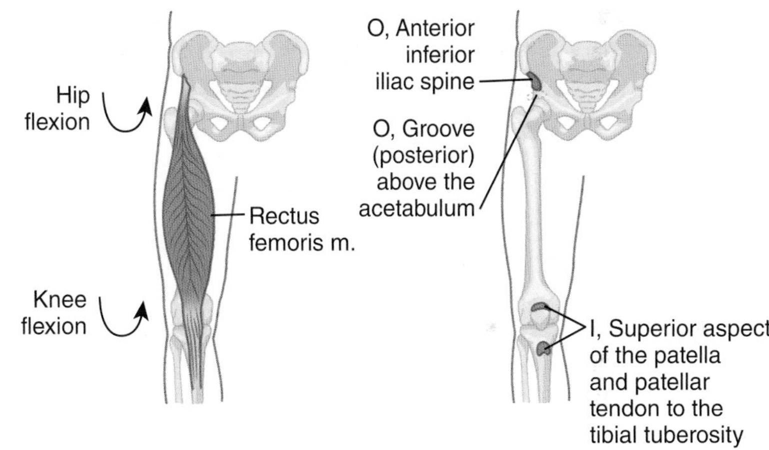

Figure 11.24 **Rectus femoris origin (O) and insertion (I).**

Many muscles of the body are **multiarticular compound** or **multicompound-joint**), crossing two or more articulations. During action, these multiarticular muscles can cause motion at one of those articulations or all of them simultaneously. The rectus femoris is a multiarticular muscle and can therefore cause hip flexion and knee extension (Figure 11.24).

One advantage to having multiarticular muscles is efficiency because more than one joint is moved with a single muscle contraction. Another advantage is that multiarticular muscles are capable of maintaining a relatively constant length, enabling force to be produced more consistently. Constant length is maintained by motion at one joint compensating for motion at another. For example, in performing a leg press exercise, the hip and knee begin in a flexed position.

During the actual execution phase of a leg press, the knee and hip extend simultaneously. Knee extension alone would shorten the rectus femoris muscle, and the force produced would progressively decline. However, the relative degree of shortening of the rectus femoris is counteracted to some degree by extension at the hip.

Active and Passive Insufficiency

As always, exceptions to the rule exist. Although multiarticular muscles provide some advantages, in some

Origin The point of muscle attachment that tends to be on a relatively immovable or proximal location.

Insertion The point of muscle attachment that tends to be on a relatively more movable segment (or distal location).

Uniarticular (single-joint) muscle A muscle that crosses only a single articulation.

Multiarticular (multi- or compound-joint) muscle A muscle that crosses two or more articulations.

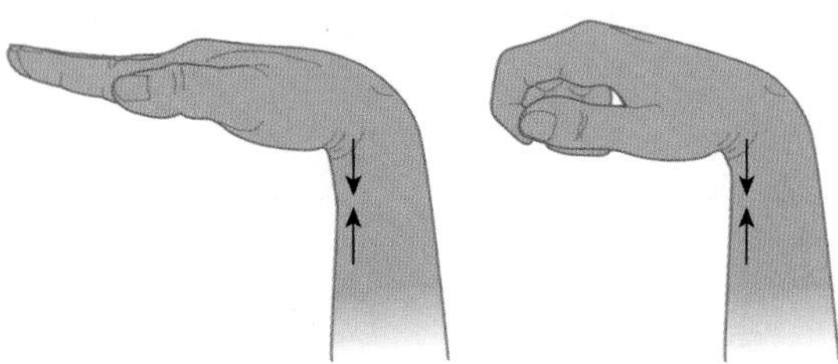

Figure 11.25 **Wrist and finger flexors in active insufficiency.**

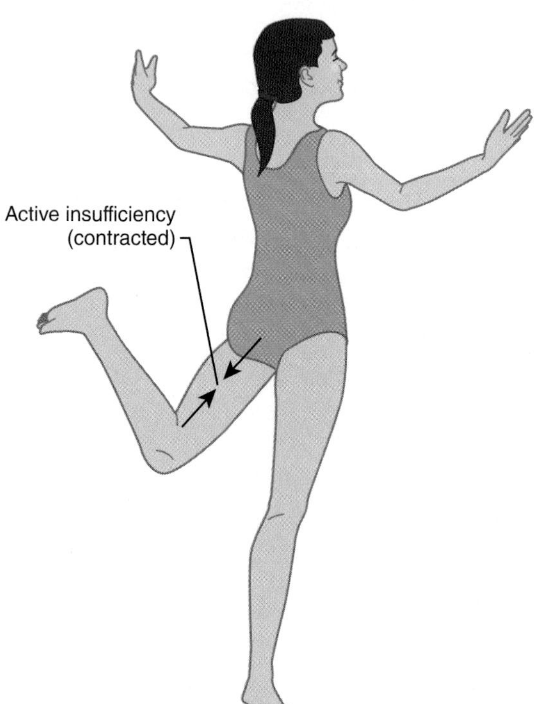

Figure 11.26 **Hamstrings in active insufficiency.**

situations these muscles display inadequacies. For example, although a multiarticular muscle can cause actions at all of the joints it crosses at the same time, it cannot *shorten* adequately enough to move all of those joints through a full range of motion simultaneously. This inability to generate sufficient force throughout all degrees of freedom is called **active insufficiency**. One example of active insufficiency is simultaneous flexion of the wrist and finger joints (Figure 11.25). With the wrist in a neutral position, fully flexing the joints of the fingers presents no difficulty.

However, fully flexing the fingers if the wrist is placed in flexion first is extremely difficult. A similar situation exists if we attempt to fully flex the knee while the hip is in hyperextension (Figure 11.26).

The hamstring muscles are capable of causing either of these actions, but when they are performed at the same time, range of motion is lost.

A multiarticular muscle is also incapable of *lengthening* to a degree that will allow simultaneous full range of motion at all of the joints it crosses. This incapacity is called **passive insufficiency**. For example, normally the hamstring muscles are not capable of stretching to an extent that would allow full hip flexion and knee extension. Passive insufficiency often causes people in the initial stages of a flexibility routine to flex the knees when attempting to touch the toes (hip flexion) (Figure 11.27).

Degrees of Freedom

You already know that the many joints of the body vary in their degrees of freedom (DOF). Therefore, a wide range of muscles and muscle attachments is needed to move the segments of our skeletal system through all of the available degrees of freedom. Recall from Chapter 2 that DOF is the number of independent ways in which a system can move or the number of values required to completely describe system motion relative to the established coordinate reference frame. Now let's review this concept in the context of uniarticular and multiarticular muscles. In some cases, increasing the available DOF at a segmental link is accomplished by having many muscles of both a uniarticular and multiarticular nature (e.g., the shoulder area has many muscles). Having one multiarticular muscle perform more than one function can also provide added DOF (e.g., the gastrocnemius is a prime mover in plantar flexion but also aids in knee flexion). In addition, a multiarticular muscle may have a specific insertion point that adds additional capabilities. For example, the biceps brachii can cause elbow flexion and shoulder flexion. But because its insertion is on the radius rather than on the ulna, it is also a powerful supinator of the forearm (Figure 11.28).

Therefore, when analyzing the full capabilities of a muscle, one must consider location and number of articulations in addition to specific origin and insertion.

Active insufficiency Inability of a multiarticular muscle to generate sufficient force throughout all degrees of freedom.

Passive insufficiency Inability of a multiarticular muscle to lengthen to a degree that allows full range of motion at all of the joints it crosses simultaneously.

Agonists Versus Antagonists

Depending upon their relative locations, extremely important relationships called *agonist–antagonist relationships* exist between opposing muscles. **Agonists** are the muscles that are responsible for causing a particular motion. **Prime movers** are the agonists that are most directly involved in causing the motion. **Synergists** are the agonists that are indirectly involved and play a more assistive role in bringing about the motion (i.e., stabilizers or neutralizers). The synergists are often ignored when analyzing motions, but their roles are hugely important. A synergist may assist by performing an action that is redundant to that of the prime mover, or it may serve in a less obvious role that helps the movement succeed. An example of redundancy of the prime mover and the synergist would be shoulder adduction during a "lat pull" exercise. The prime mover for shoulder adduction is the latissimus dorsi, and one of the synergists is the teres major, which performs the same action. The rhomboid muscles act as synergists in this same situation. However, the role of the rhomboids is not redundant with that of the prime mover. Instead, the rhomboids act to stabilize the scapula. Without this stabilization, the scapula would move toward the humerus because of tension from the teres major. In fact, the action of a synergist is often scapular stabilization if the movement is at the shoulder. Synergists are also especially important if the agonist prime mover is a multiarticular muscle. Recall that a multiarticular muscle has multiple actions. For example, the rectus femoris crosses the hip and knee joints and is therefore capable of hip flexion and knee extension. During the performance of a squat exercise, the rectus femoris acts as a prime mover in extending the knee. However, a person would like to rise from the squatting position without excessive trunk and hip flexion. Therefore, the hip extensors (such as the gluteus maximus) act as synergists by counteracting the potential hip flexion that the rectus femoris would cause.

The **antagonist** in the relationship is the muscle that performs the joint motion opposite that of the agonist. The antagonist is, therefore, usually located on the opposite side of the joint from the agonist. The relationship between the agonist and the antagonist is a fine balance. For the agonist to carry out the desired motion, the antagonist must relax to some degree to allow that motion. However, this relaxation does not mean that the antagonist is always inactive during agonist activity. In fact, the antagonist may act eccentrically to stabilize the joint and decelerate the limb at the end of the motion. For example, during running, a critically important agonist–antagonist relationship exists between the hip flexors and hip extensors. The hip flexors act as agonists to

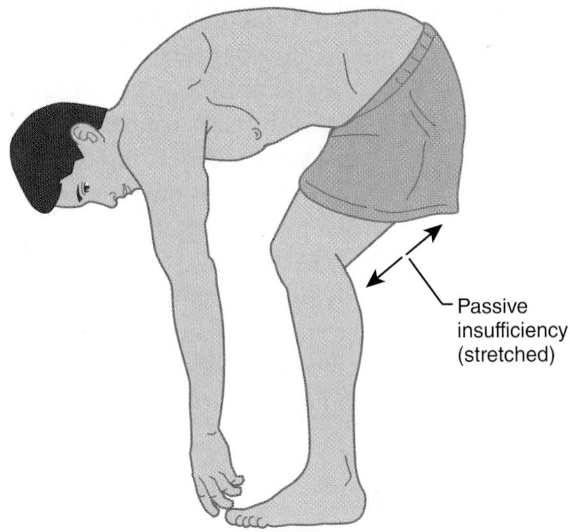

Figure 11.27 **Passive insufficiency of the hamstrings while touching the toes.**

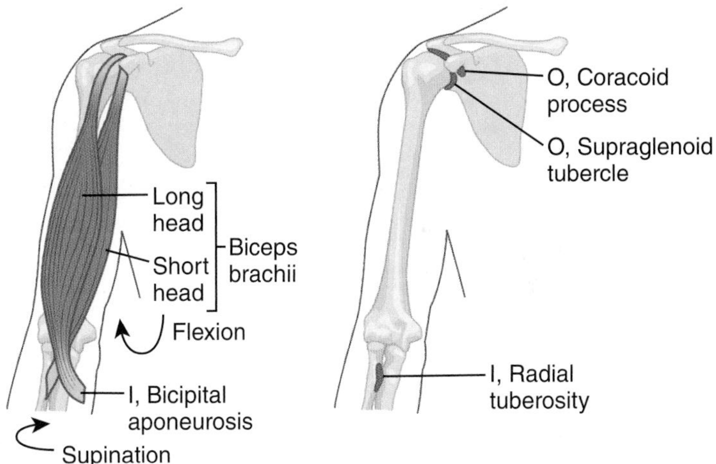

Figure 11.28 **Biceps brachii origin (O) and insertion (I).**

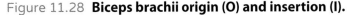

Agonists The muscles that are responsible for causing a particular motion.

Prime movers The agonist muscles that are most directly involved in causing the motion.

Synergists The agonist muscles that are indirectly involved and play a more assistive role in bringing about the motion (i.e., stabilizers or neutralizers).

Antagonist The muscle that performs the joint motion opposite that of the agonist.

accelerate the femurs forward during the hip-flexion phase of the running stride. Simultaneously, the hip extensors play an antagonistic role by acting eccentrically to safely decelerate the limbs at the end of the hip-flexion phase. In this way, the antagonists act to prevent the body from injuring itself. In relationships such as this one, the antagonist often gets injured in conditions of muscle imbalance. For example, because the hamstrings decelerate the leg after it has been accelerated by the quadriceps during running, a person may be at risk for a hamstring injury if those muscles are disproportionately weak compared to the quadriceps.

As you have learned, the antagonist does have to relax to some degree for the agonist to perform the primary motion. Otherwise, the two muscles would simply fight each other, and the result would be either a static action or a highly inefficient and jerky motion. Therefore, the nervous system helps prevent this undesirable situation by inhibiting the motor units of the antagonist while stimulating the motor units in the agonist.

This neural inhibition of the antagonistic motor units is called **reciprocal inhibition** (Figure 11.29) or **reciprocal innervation**. Reciprocal inhibition is the basis for the popular stretching technique called proprioceptive neuromuscular facilitation. For example, it is easier to stretch the hamstrings if the quadriceps muscles are contracted.

Reciprocal inhibition/innervation Neural inhibition of the antagonistic motor units during actions of the agonist muscle.

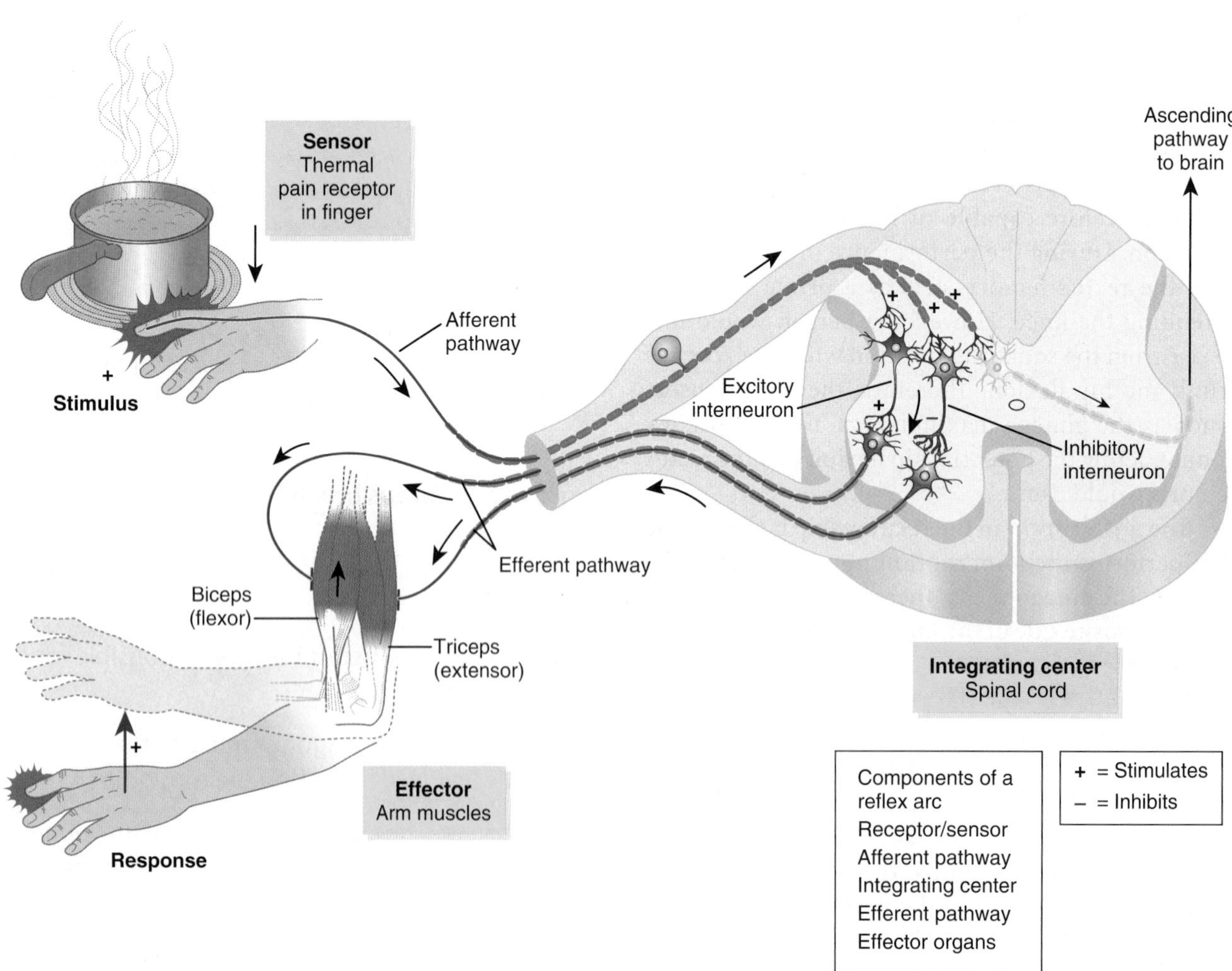

Figure 11.29 **Reciprocal inhibition.**

Angle of Muscle Pull

The muscle's *angle of pull* is the angle formed by the muscle force vector relative to the bone on which the muscle is attached. Distinguishing between the joint angle and the angle of pull is especially important. The joint angle is that formed by the bones of the articulation; the angle of pull is formed by the muscle and the bone of attachment. The angles are not necessarily the same, so do not make that assumption (Figure 11.30).

The importance of the angle of pull is that it is related to many factors covered in previous chapters. Specifically, the angle of pull affects the relative magnitudes of the perpendicular and parallel components of the resultant force vector, as well as the size of the moment arm for the resultant muscle torque.

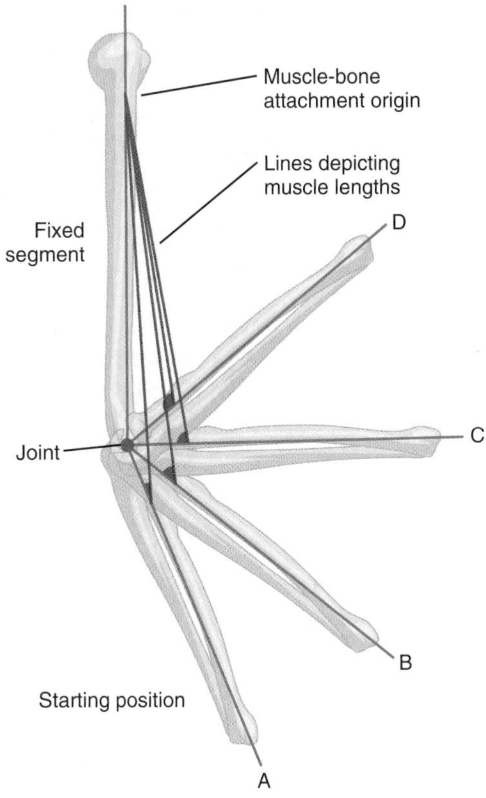

Figure 11.30 **Variation in angle of muscle pull with joint angle.**

Rotary and Stabilizing Components

You learned in Chapter 3 that when resolving muscle force vectors, we define the vertical or perpendicular component as the *rotary component*, because if it were allowed to act alone, it would cause joint rotation. Also, if the horizontal (parallel) component acted alone, it would cause the bones of the articulation to be either pushed tightly together or pulled apart. So, we define the horizontal (parallel) component of a muscle force vector as either the *stabilizing component* or the *destabilizing component* (Figures 11.31 and 11.32).

Whether the parallel component causes stabilization or destabilization depends on the joint angle. For example, if the elbow angle is changed so that it is less than 90° and causes the biceps tendon angle to exceed 90°,

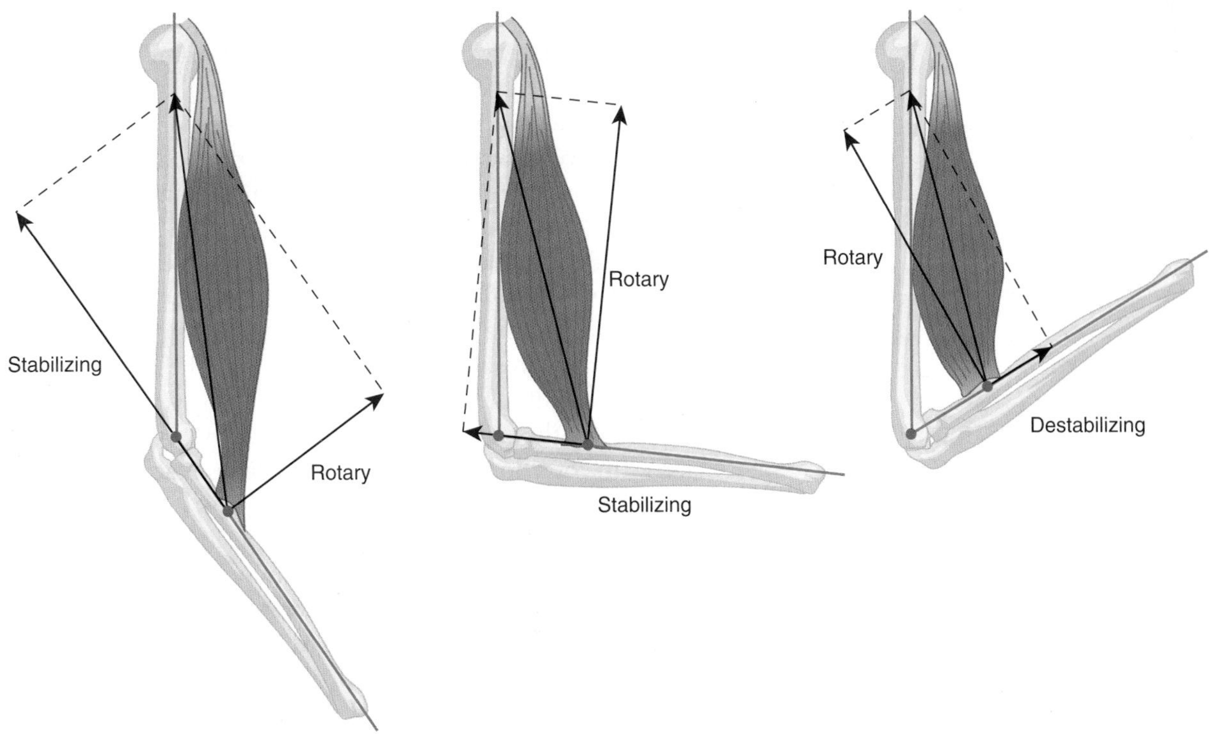

Figure 11.31 **Vector resolution of biceps brachii muscle force.**

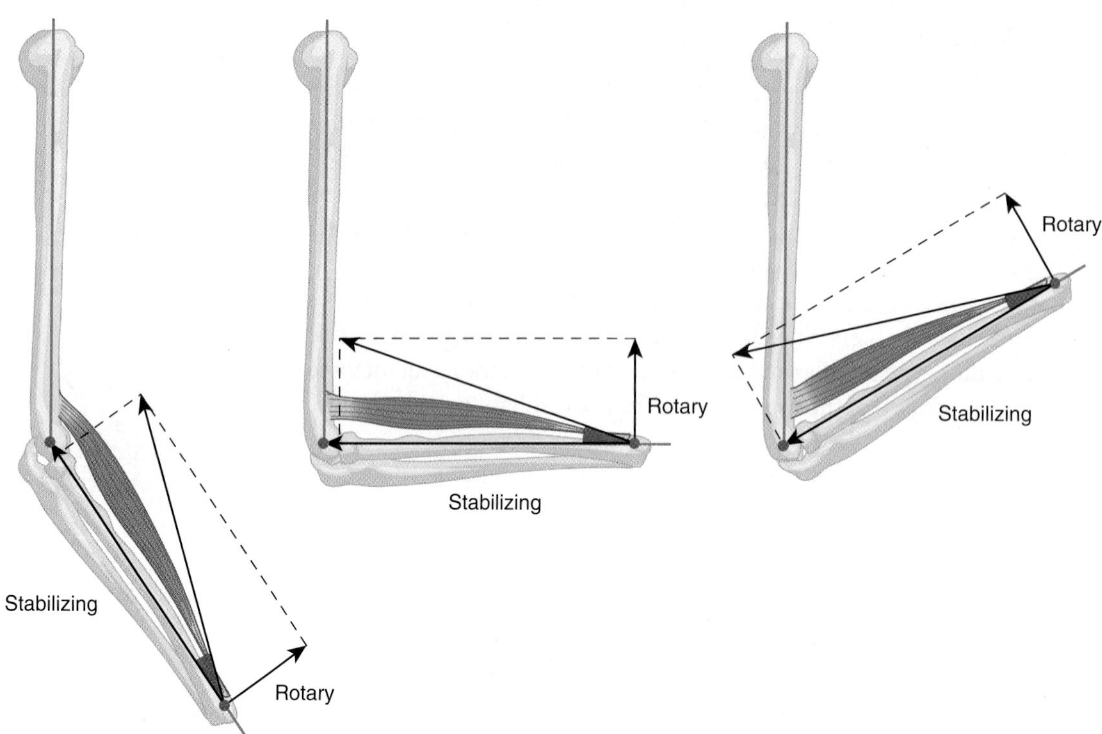

Figure 11.32 **Vector resolution of brachioradialis muscle force.**

then the vector representing the parallel component points away from the elbow joint (Figures 11.31 and 11.34). So, in this case, the horizontal (parallel) component, acting alone, would actually pull the forearm away from the humerus—thus, it is said to be *destabilizing*.

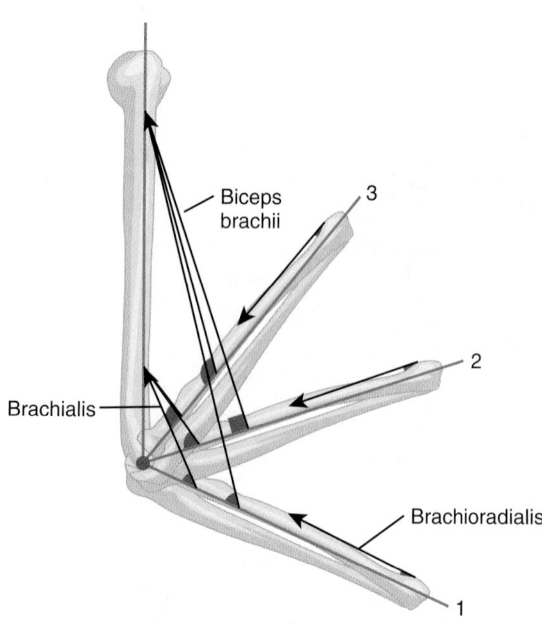

Figure 11.33 **The angle of pull for each of the three elbow flexors at different joint angles.**

With these concepts in mind, let's return to the analysis of muscle force vectors. However, we revisit this topic with new concepts to apply. First, consider the sizes of the perpendicular and parallel components. The relative magnitudes of the muscle force vector components vary with the joint angle, as well as with the origin and insertion of the individual muscle. As the joint angle changes, so does the angle of muscle pull. For example, the flexors of the elbow are the biceps brachii, brachialis, and brachioradialis. Because of their varying origins and insertions, they all have a different angle of pull at a given elbow joint angle (Figure 11.33).

When comparing the relative magnitudes of the perpendicular and parallel components, we are concerned with the angle of pull. Just remember that although the angle of pull changes with the joint angle, they are not the same. Let's begin the analysis with the elbow joint in its resting position and start with the biceps brachii because it is familiar. We can then superimpose other muscles. If we resolve the biceps brachii muscle force vector in the resting position (a tiny angle of pull), we find that the parallel (stabilizing) component is rather large in comparison to the perpendicular (rotary) component (Figure 11.34).

This finding actually meshes quite well with the length–tension relationship. At resting length, the angle of pull is acute, meaning that much of the resultant force will be directed into the joint itself rather than causing rotation. Therefore, it makes sense that our muscles tend to be strongest at resting length. If we then resolve the biceps brachii force vector with the elbow in slight flexion (a slightly greater angle of pull but still acute), we notice that the relative sizes of the components have changed. Because the angle of pull is larger, more of the force is directed perpendicularly to the bone. Therefore, the rotary component is larger and the stabilizing component has been reduced in magnitude. If the elbow is flexed so that the angle of pull is exactly 90°, we find that 100% of the resultant force is directed perpendicular to the bone (sin 90° = 1.00; cos 90° = 0.00). So, once again, as the angle of pull becomes larger, the rotary component increases in magnitude while the stabilizing component decreases in magnitude (all the way to 0). At angles of pull greater than 90°, the pattern reverses and a portion of the resultant force is once again directed parallel to the bone. However, at obtuse angles of pull, the parallel component is directed away from the articulation and is therefore *destabilizing*. The rotary component is reduced in magnitude at obtuse angles. At an even more obtuse angle, the rotary component again becomes smaller in magnitude while the destabilizing component becomes larger. This finding again works especially well with the length–tension relationship. Notice that as a large destabilizing component is developed, the biceps brachii becomes shorter and shorter. The more shortened the muscle, the less force it can produce. This relationship is advantageous because most of the force at this point would be used to pull the bones of the articulation apart. So, in this situation, the muscular system is protecting the skeletal system.

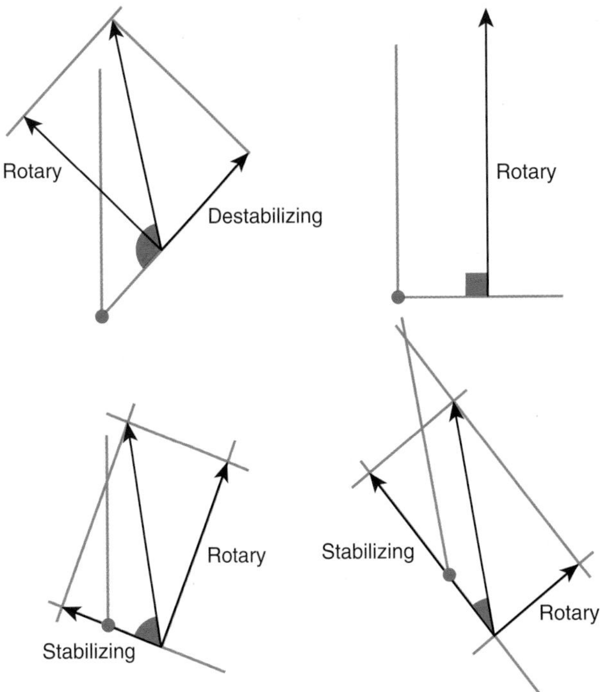

Figure 11.34 **Vector resolution of biceps brachii muscle force.**

Do not get the idea that all muscles are capable of the progression described above with a small rotary component and a large stabilizing component at acute angles; 100% of the resultant muscle force rotary at a 90° angle of pull; and finally a destabilizing component developing at obtuse angles of pull while the rotary component becomes smaller. Because of the constraints of origin and insertion, some muscles simply cannot achieve every angle of pull. For example, the brachioradialis muscle is an elbow flexor. However, its origin is on the distal humerus, and its insertion is on the distal radius. This configuration means that no matter the elbow joint angle, the angle of pull of the brachioradialis will always be acute (Figures 11.32 and 11.35).

This is not to say that this muscle's angle of pull will not change, but only that the muscle can never achieve a 90° angle of pull. As a matter of fact, it will never achieve even a 45° angle of pull. This fact has implications for this muscle's function as an elbow flexor. If the angle of pull is always less than 45°, the stabilizing component of this muscle's force vector will always be larger than its rotary component. So it can contribute to elbow flexion, but its primary action is to stabilize the joint. On the other hand, the biceps brachii can develop rather large rotary and stabilizing components, which help to maintain joint integrity if a heavy object is being held in the hands while the elbow is fully extended, or if the body is hanging by the hands (i.e., *brachiating*).

The final elbow flexor, the brachialis, actually does much of the work. The brachialis can achieve an angle of pull equal to 90°. However, because of its origin and insertion, its angle of pull at any given joint angle varies from that of the bicep (Figure 11.36).

Therefore, the relative magnitudes (in terms of percentage of resultant force) of its parallel and perpendicular components will not necessarily match those of the biceps brachii (or the brachioradialis for that matter) at a given elbow position. Having three elbow flexors, all varying from each other in angle of pull, ensures smooth joint action and maximum efficiency of individual muscle contributions throughout the range of motion.

You have learned that muscle redundancy is used to enhance degrees of freedom. Now you can see that having multiple muscles perform similar actions is also complementary in terms of parallel and

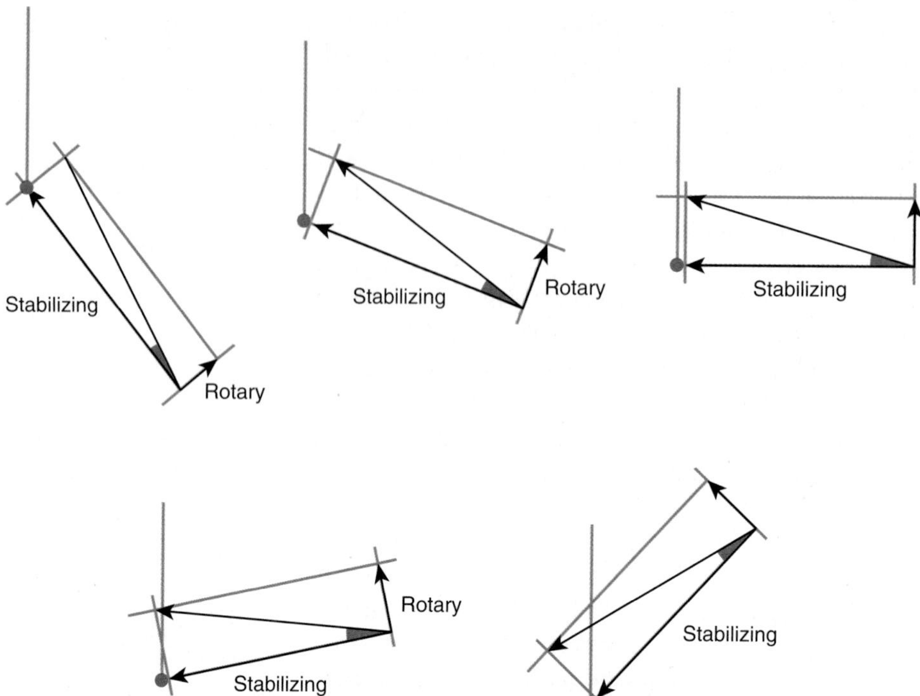

Figure 11.35 **Vector resolution of brachioradialis muscle force.**

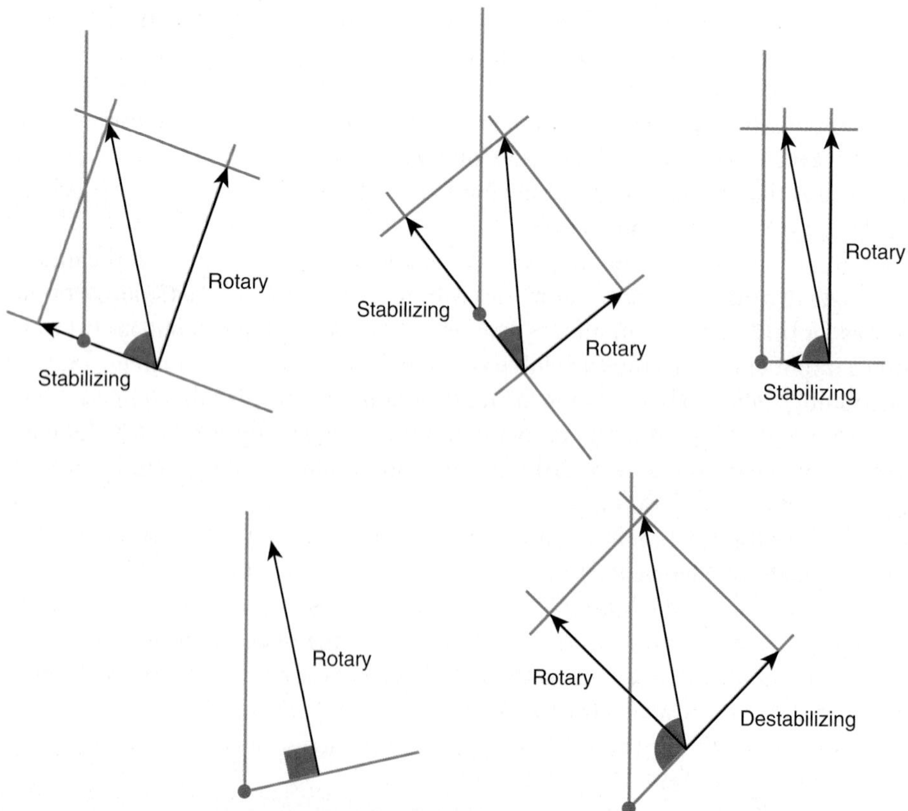

Figure 11.36 **Vector resolution of brachialis muscle force.**

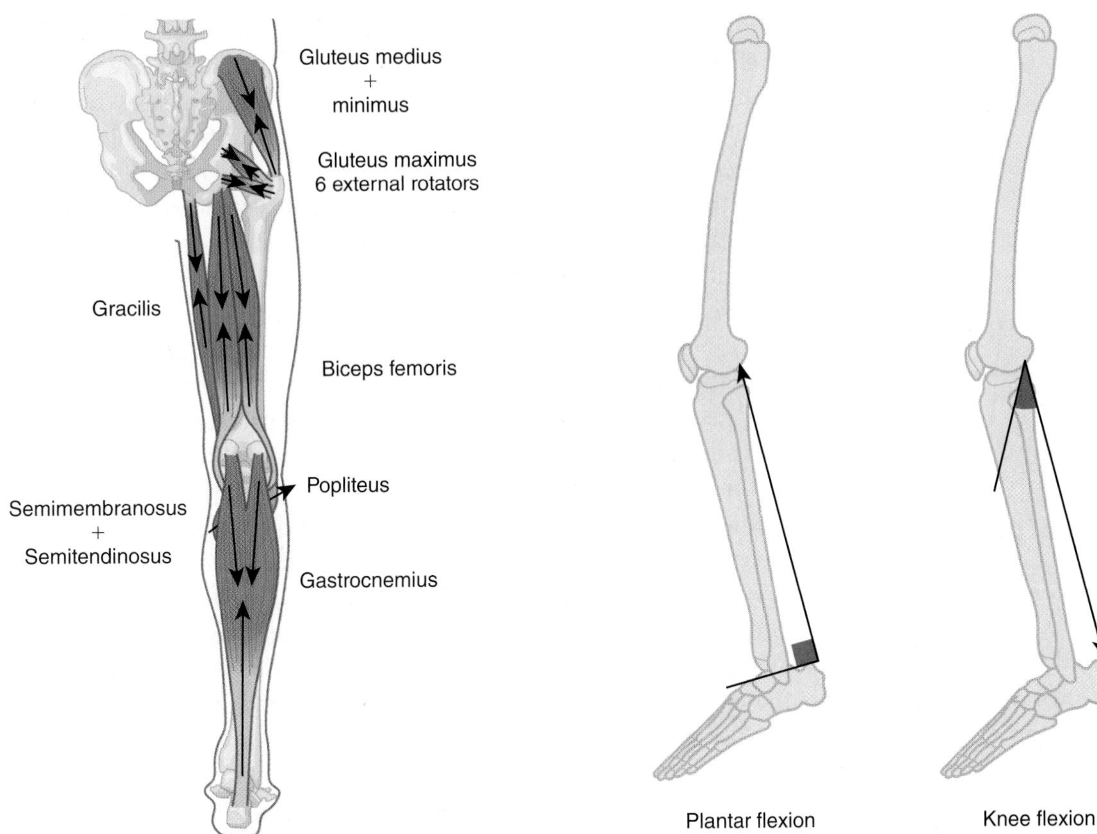

Figure 11.37 **Arrangement of hamstrings, popliteus, and gastrocnemius muscles.**

Figure 11.38 **A comparison of gastrocnemius angle of pull during plantar flexion versus knee flexion.**

perpendicular components. This type of arrangement in which muscles contribute to the same action while varying in function (rotation or stability) is a repeated theme throughout the human system. For example, instead of the elbow joint, it could be the knee (Figure 11.37).

The hamstrings play a role similar to that of the biceps brachii, possessing the capability to achieve various angles of pull that are greater than 90°. The gastrocnemius is also active during knee flexion and is positioned to act similarly to the brachioradialis. In other words, the gastrocnemius has a large stabilizing component throughout knee flexion range of motion. Even the little-known popliteus muscle has a position similar to that of the brachialis of the elbow.

A final point about perpendicular and parallel components: A muscle that acts primarily as a stabilizer during one action may act as the primary rotator in a different action. The gastrocnemius is an example of a muscle that has a dual role because it is a multiarticular muscle. The origin of the gastrocnemius is on the distal femur, and the insertion is on the calcaneus. As a prime mover in plantar flexion, the gastrocnemius can achieve multiple angles of pull (including 90°), and therefore it develops a large rotary component. However, the bone of interest is the tibia in the case of knee flexion. So, although the gastrocnemius can achieve a 90° angle of pull in relation to the calcaneus for action at the ankle, it cannot achieve angles that large in relation to the tibia for action at the knee. Therefore, the gastrocnemius serves as a stabilizer during knee flexion, extremely similar to the action of the brachioradialis during elbow flexion (Figure 11.38).

One final example of muscles acting primarily as stabilizers because of their angle of pull is the group of muscles collectively referred to as the *rotator cuff* muscles. The **rotator cuff** muscle group comprises the supraspinatus, infraspinatus, teres minor, and subscapularis. Although

Rotator cuff The muscle group comprising the supraspinatus, infraspinatus, teres minor, and subscapularis.

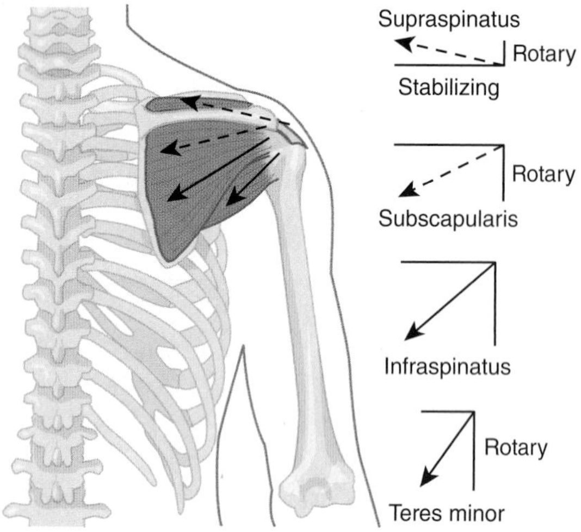

Supraspinatus
Rotary
Stabilizing
Rotary
Subscapularis
Infraspinatus
Rotary
Teres minor

Figure 11.39 **Angles of pull of the rotator cuff muscles.**

this group of muscles is known for rotation (both internal and external), all of these muscles have tiny angles of pull (Figure 11.39).

This conformation means that the rotator cuff muscles can *help* with rotation at the shoulder, but their primary function is stabilization. More specifically, their primary role is to stabilize the head of the humerus in the especially shallow glenoid fossa. This stabilization is necessary because unlike the hip joint, the shoulder is not a deep ball-and-socket joint. Therefore, dislocation is more likely without multiple muscles that have large stabilizing components.

Now let's analyze the perpendicular and parallel components in relation to the concept of torque. Remember that an *off-axis* force is necessary if a change in angular motion is desired. *Torque* is the tendency of an off-axis force to rotate an object around an axis. Torque is also called moment of force, or simply *moment.* Not everyone agrees on the specific differences and usages of the term *torque* versus *moment*, but there are some consistencies. Torque is sometimes defined as the rate of change of angular momentum of an object (a movement force). In other words, usage of the term *torque* would imply that the angular velocity or the moment of inertia of an object (or both) is changing. In contrast, *moment* is the term often used for the tendency of one or more applied forces to rotate an object around an axis but not necessarily changing the angular momentum of the object (a static force). In the case of rotary motion at an articulation, the joint becomes the *axis of rotation* (the point about which a body rotates). So torque depends on two factors: (1) the magnitude of externally applied force and (2) the distance from the axis of rotation at which the force is applied. The perpendicular distance between the axis of rotation and the line of action of the applied force is called the *force arm, moment arm,* or *lever arm.* Recall that the *line of action* is an imaginary line extending infinitely along the vector through the point of application and in the direction of the applied force. Also remember that the line of action is sometimes called the *line of pull* when referring to muscles.

Mathematically, torque is calculated as follows:

$$T = F \times r \times \sin \theta \qquad (6.1)$$

where

T = torque (or moment of force)
F = magnitude of the externally applied force
r = distance between the axis and the point of force application (or radius of rotation of the point of force application)
θ = angle at which the force acts relative to the horizontal (positive x-axis)

The *perpendicular* distance between the axis of rotation and the line of action of the applied force is given by $r \times \sin \theta$. So if the moment arm is measured with a ruler as the perpendicular distance ($\theta = 90°$) between the line of action of the force and the axis of rotation, then $\sin \theta = 1.00$ and Equation 6.1 can be restated as:

$$T = F \times d \qquad (6.1)$$

where

T = torque (or moment of force)
F = magnitude of the externally applied force
d = moment arm (*perpendicular* distance between the axis of rotation and the line of action of the applied force)

Please keep in mind that the quantity d (or moment arm) is $r \times \sin \theta$ and not simply the distance between the axis and the point of force application.

In terms of rotary and stabilizing components, notice their individual vectors in relation to the axis of rotation (Figure 11.40). The line of action (pull) of the perpendicular component never passes through the axis of rotation; therefore, a moment arm always exists for this component. So the perpendicular component is always torque producing and is therefore called the *rotary* component. Next, notice that the line of pull of the parallel component always passes through the axis of rotation. This constant means that no moment arm exists for the parallel component because there is no distance between the axis of rotation and the line of action. Therefore, the parallel component cannot produce torque ($T = F \times 0 = 0 \text{ N} \cdot \text{m}$). This is the reason that the parallel component of a muscle force vector either causes stabilization or destabilization but not rotation. It can cause only *linear* motion: linear motion of the bones *toward* each other (stabilization) or *away* from each other (destabilization).

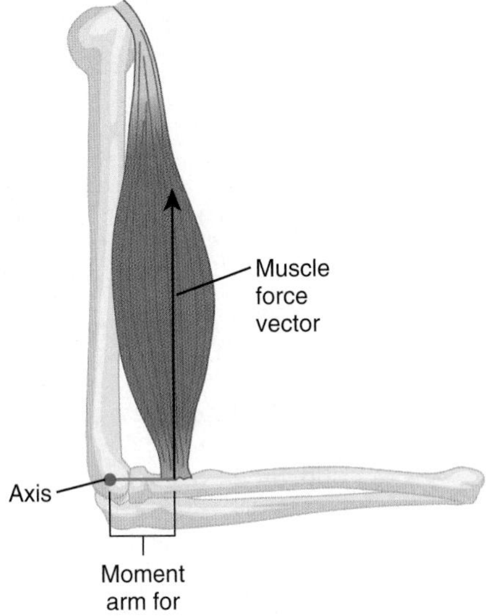

Figure 11.40 **Muscle force vector, moment arm, and axis of rotation.**

Moment Arm Length

Let's also look at the size of the moment arm as the joint angle changes. Actually, two moment arms are of interest: (1) the moment arm for the muscle and (2) the moment arm for the resistance. You already know that the angle of pull changes as the joint angle changes. This varying angle of pull must be considered because the resultant torque produced by a muscle changes throughout its range of motion because of changes in that angle. For example, as the elbow is flexed from the anatomical position, the angle of pull of the biceps brachii becomes larger. As the angle of pull becomes larger, so does $\sin \theta$ (Figure 11.41).

This relationship continues until the angle of pull reaches 90°, at which point $\sin \theta = 1.00$. Therefore, the size of the moment arm for the biceps brachii increases until peaking at the point where the angle of pull reaches 90°. As the angle of pull increases beyond 90°, $\sin \theta$ gradually decreases. So at angles of pull larger than 90°, the moment arm of the biceps brachii decreases. This change in moment arm length can also be observed with use of the other form of the torque equation ($T = F \times d$). Simply stated, as the angle of pull approaches 90°, the perpendicular distance between the axis of rotation and the line of pull gets progressively larger. As the angle of pull then becomes greater than 90°, the perpendicular distance between the axis of rotation and the line of pull gets progressively smaller (Figure 11.42).

So no matter the method of calculation, the mechanical advantage changes throughout the entire range of joint motion but peaks at angles of pull equal to 90°. In addition, one must keep in mind the opposing perspective on moment arm length. The moment arm for the *resistive* force also changes throughout the range of joint motion, peaking at 90° and decreasing at progressively more acute or obtuse angles (Figure 11.43).

Along with angle of pull are the length–tension relationship and the relative size of the rotary component. The biceps brachii can generate the most force at its resting length, but its moment arm and rotary component are smaller because of the small angle of pull. At an angle of pull that is exactly

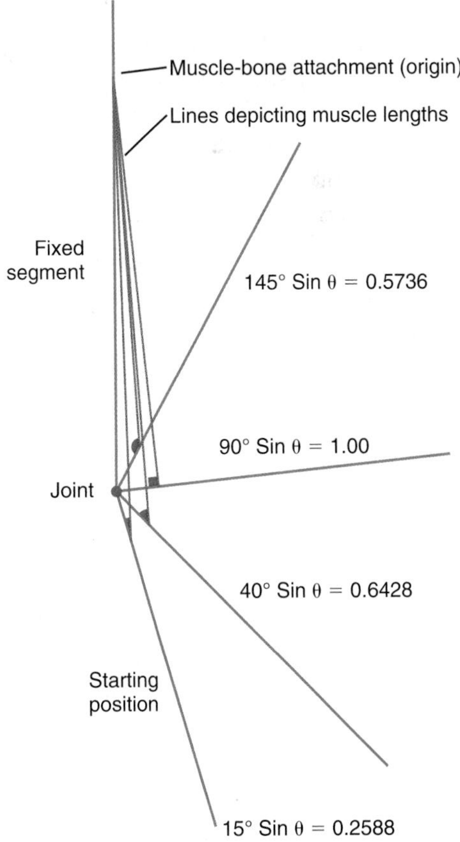

Figure 11.41 **Change in muscle angle of pull and length of moment arm.**

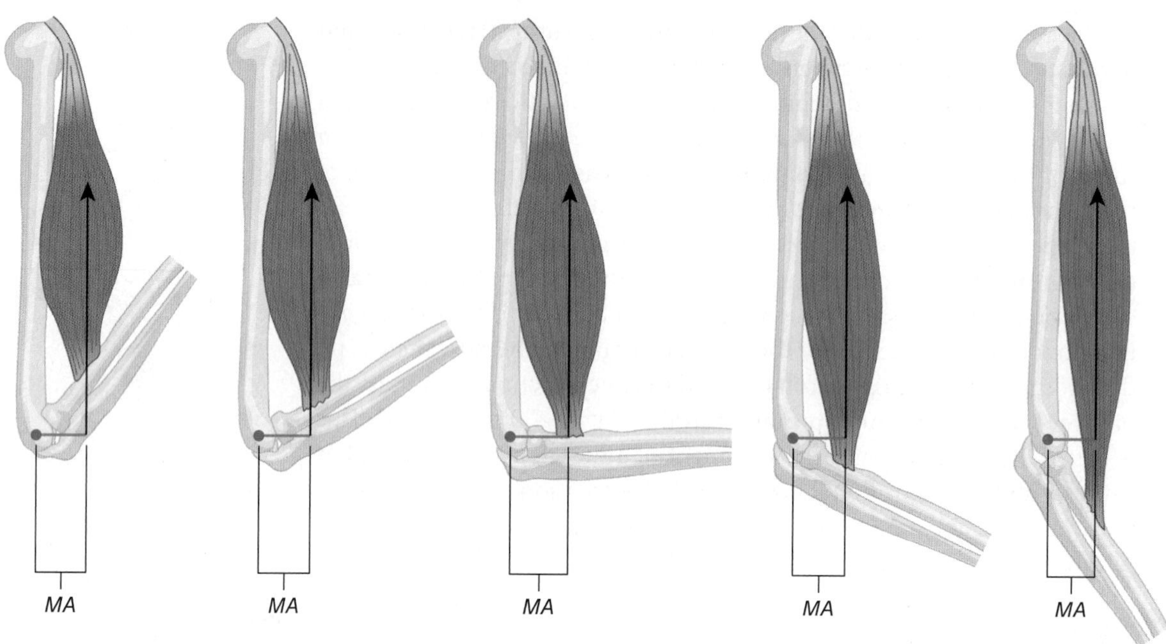

Figure 11.42 **Change in muscle angle of pull and length of moment arm.**

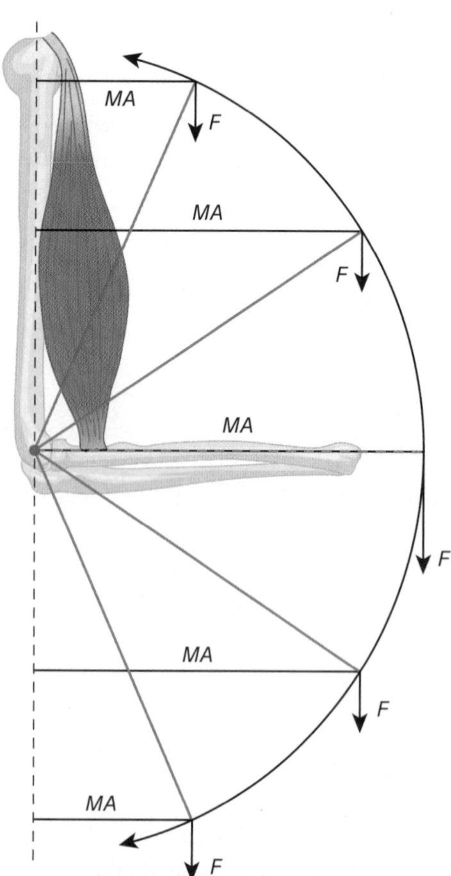

Figure 11.43 **Moment arm for the resistive force.**

90°, the moment arm is largest but the muscle is shorter. However, at 90° all of the resultant force is rotary. So even though muscle force production is diminished, all of that muscle force is being used to rotate the joint. This exclusive use helps compensate for the fact that muscle force-production capability is diminished at the point where the moment arm for the resistive force is largest. At angles of pull larger than 90°, the moment arm and muscle force-production capability are further diminished. This phenomenon is not exactly terrible because much of the muscle resultant force at angles of pull greater than 90° is acting to destabilize the joint.

So as you can see, the resultant muscle torque is a complex interplay of muscle force, moment arms, and force-vector components. This interaction is the result of changing muscle length and angle of pull. Changes in muscle length can be affected by the number of articulations they cross, which depends on the location of its proximal and distal attachment sites. The angles of pull that any given muscle is capable of achieving are also determined by its unique origin and insertion.

Anatomical Force Couples

In Chapter 6, you learned that the structure of the human body is such that it is sometimes necessary to use concurrent torques to actually prevent linear displacement. For example, the scapula is not directly attached to the thorax, so no bony structures are present to inhibit unwanted motion. Because of this lack of inhibition, the muscles of the posterior thorax must stabilize the scapula. Upward rotation of the scapula is achieved by contraction of both the upper and lower trapezius muscles (Figure 11.44). The torques produced by these muscles are equal in magnitude, opposite in direction, and noncolinear.

Two muscles co-contracted in opposite directions in an attempt to accomplish one action or prevent an unwanted motion (or both) are called an *anatomical force couple* (Oatis, 2017; Robertson et al., 2014). So, in this case, because the trapezius muscles contract concurrently in this manner, there is no linear displacement of the scapula; only rotation occurs. For example, if the upper trapezius acted alone, the scapula would exhibit rotation (because the force is eccentric), but some degree of superior linear displacement would also occur. Let's review this concept in light of this new information. In this example, notice that both muscles act on the same bone, but they have opposite actions because their origins and insertions differ. Also notice that the result of their co-contraction is stabilization. Recall that stabilization is often the role of a synergist. In this example, the anatomical force couple is made up of two synergists that act to stabilize the scapula. This stabilization is often necessary to prevent unwanted motion while larger prime movers such as the latissimus dorsi and pectoralis major cause the desired motion.

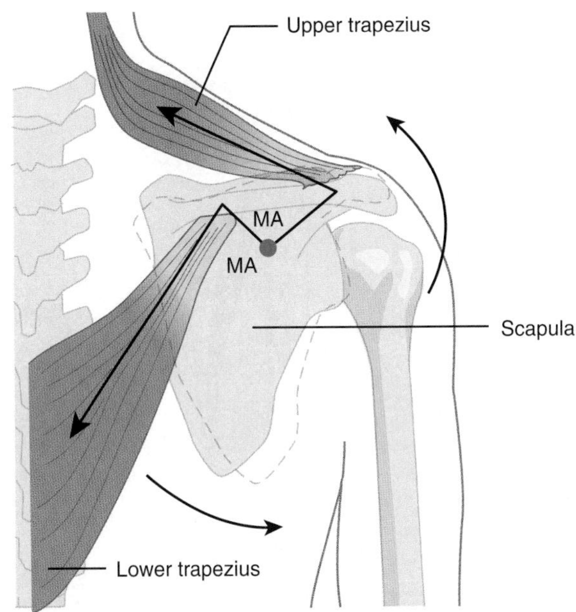

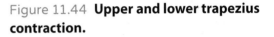

Figure 11.44 **Upper and lower trapezius contraction.**

11.3 BIOMECHANICS OF MUSCLE ARCHITECTURE

Muscles are classified in many different ways: action (flexor, extensor, levator), direction (oblique, transverse), location in relation to anatomical directions (anterior, posterior, medialis, lateralis), location on the skeleton (brachii, femoris, gluteus, pectoralis, scapulae), origin and insertion (brachioradialis, sternocleidomastoid), number of heads (biceps, triceps), shape (gracilis, rhomboid, teres, trapezius), and size (maximus, medialis, minimus). Most names for muscles are derived from any number of combinations of these classifications (for example, biceps femoris, gluteus maximus, levator scapulae, pectoralis major, serratus anterior). Learning the names of muscles is important from a purely anatomical perspective. However, in biomechanics we are more interested in muscle function arising from architecture.

Whole muscles vary widely in their maximal force-production capability. The reported values of stress exerted by a maximally contracting muscle usually range from approximately 10 N/cm^2 to 50 N/cm^2 of active cross-sectional area (the cross-sectional area of the fibers actively producing tension) (Brand, Beach, & Thompson, 1981; Buchanan, 1995; Davies, Thomas, & White, 1986; Fukunaga et al., 1996; Haxton, 1944; Lieber, 2010; Maganaris et al., 2001; Narici, Landoni, & Minneti, 1992; Oatis, 2017; Powell et al., 1984; Zajac, 1992). The **anatomical cross-sectional area** of the muscle is measured perpendicular to the longitudinal axis of the whole muscle at its widest point (Figure 11.45).

It only makes sense that the force produced by a muscle would increase with cross-sectional area; greater area means more myofibrils with potential for contractile function. However, predicting force output cannot be as simple as measuring anatomical cross-sectional area. Many factors can affect muscle force production. A biomechanical factor that may contribute to the large variation previously noted is the architecture of the muscle in terms of fiber arrangement and length. In other words, just looking at a cross-section perpendicular to the largest portion of the muscle may not reveal the entire picture because not all of the fibers in different muscles are arranged in the same way.

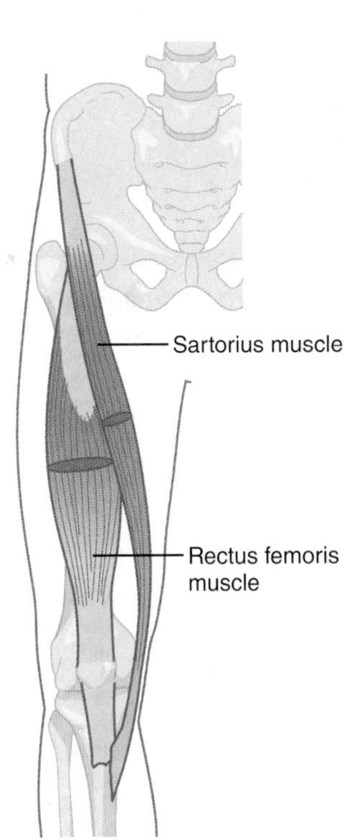

Figure 11.45 **Anatomical cross-sectional area.**

Muscle Fiber Arrangement and Length

The previous section listed many ways in which muscle can be classified. A classification of a more biomechanical

Anatomical cross-sectional area The cross-sectional area of a muscle measured perpendicular to the longitudinal axis of the whole muscle at its widest point.

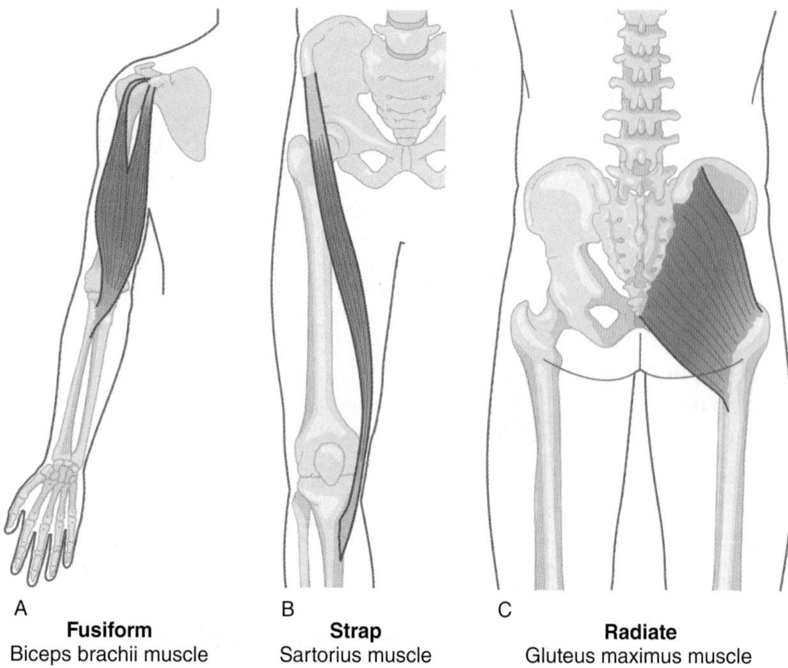

Figure 11.46 **Examples of longitudinal muscles (biceps brachii, sartorius, and gluteus maximus).**

nature is according to fiber arrangement. Even within this system, many muscle classifications exist. But overall, two basic categories are used based on fiber arrangement: (1) longitudinal (or fusiform) and (2) pennate ("feather-like"). **Longitudinal** or **fusiform fibers** run somewhat parallel to the muscle's longitudinal axis (Figure 11.46a). These fibers are often called *fusiform* because they taper at the tendons on each end of the muscle. However, some muscles have less prominent tendons and therefore taper less (e.g., the sartorius muscle; Figure 11.46b). Muscles such as the sartorius that are not as spindle shaped are still considered longitudinal but are sometimes classified into a subcategory called **strap muscles**. Also, some longitudinal muscles have an exaggerated taper to the tendon and are called **radiate muscles**. An example of a radiate muscle would be the gluteus maximus (Figure 11.46c).

Longitudinal fibers of a given muscle are not necessarily the same length. In other words, no fiber runs the entire length of the muscle. However, fibers classified as longitudinal share the characteristic of being oriented at an angle of approximately 0° relative to the longitudinal axis of the muscle.

In contrast to the alignment of longitudinal fibers, **pennate fibers** are arranged obliquely to a tendon that runs along the longitudinal axis of the muscle (much like the appearance of a feather). The pennate fiber arrangement also has subdivisions: unipennate, bipennate, and multipennate.

In a **unipennate fiber** formation (Figure 11.47a), only one set of fibers is present, and they are parallel to each other but oblique to the tendon. Examples of unipennate arrangements would be the tibialis posterior and wrist and finger flexors. An arrangement with **bipennate fibers** uses two groups of pennate fibers (Figure 11.47b). The fibers within a group are parallel to each other but are oblique to the other group and to the tendon. The gastrocnemius and rectus femoris are examples of bipennate arrangements. In a **multipennate** arrangement, two or more bipennate muscles converge on one tendon to form a single muscle (Figure 11.47c). The deltoid muscle is an example of a multipennate arrangement.

The angle at which fibers in a pennate arrangement are oriented relative to the longitudinal axis (**pennation angle**) varies from muscle to muscle. Pennation angle can be as much as 30° but is rarely larger

Longitudinal/fusiform fibers Muscle fibers that run somewhat parallel to the muscle's longitudinal axis.

Strap muscles Longitudinal muscles that have less prominent tendons and therefore taper less.

Radiate muscles A longitudinal muscle with an exaggerated taper to the tendon.

Pennate fibers Fibers that are arranged obliquely to a tendon that runs along the longitudinal axis of the muscle.

Unipennate fibers One set of pennate fibers that are parallel to each other but oblique to the tendon.

Bipennate fibers Two groups of pennate fibers, with the fibers in one group parallel to each other but oblique to the other group and to the tendon.

Multipennate Arrangement in which two or more bipennate muscles converge on one tendon to form a single muscle.

Pennation angle The angle at which fibers in a pennate arrangement are oriented relative to the longitudinal axis.

than 15°. For example, the fibers of the rectus femoris and vastus lateralis have pennation angles of 5°, whereas the soleus fibers have a pennation angle of 25° (Lieber, 2010; Lieber, Fazeli, & Botte, 1990; Lieber et al., 1992; Wickiewicz et al., 1983). Remember that pennation angle changes substantially from rest to maximal voluntary contraction. The degree of pennation and size of the pennation angle are not small factors and are actually directly related to the maximal force production and total range of motion of the muscle (Figure 11.48). To fully understand this concept, vector analysis is necessary. Let's begin with a longitudinal fiber arrangement in which the pennation angle is near zero. The pennation angle is measured relative to the longitudinal axis of the muscle. The longitudinal axis of the muscle is usually designated as the *x*-axis when analyzing individual muscle fiber vectors; therefore, we will use the cosine function to resolve the vector. For example, in a fiber arrangement with a 0° pennation angle, all of the force will be directed along the longitudinal axis of the fiber because cos 0° = 1.00 (sin 0° = 0.00). Thus, 100% of the force is in the *x* direction (along the longitudinal axis).

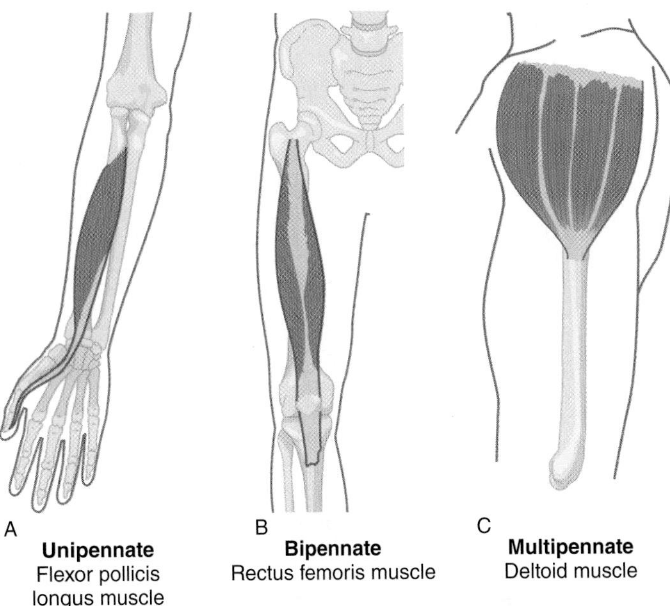

Figure 11.47 **Examples of pennate muscles (finger flexors, rectus femoris, and deltoid).**

If the pennation angle is equal to 5°, 99.62% of the force is directed along the long axis of the muscle (cos 5° = 0.9962). Soleus fibers have a pennation angle equal to 25°, so 90.63% of the force is directed along the longitudinal axis of the whole muscle. Notice that as the pennation angle increases, less force is directed along the longitudinal axis. So why have pennated muscle fiber arrangements? Well, pennation angle is not the whole story. The advantage to pennation is that it allows more fibers to be packed into a given cross-sectional area. The overall force-production capacity of a muscle is the vector sum of all of the individual fibers contracting. As you have read, an anatomical cross-sectional area is measured perpendicular to the longitudinal axis of the whole muscle at its widest point. If a simple cross-section is taken of a pennated muscle such as the rectus femoris, however, some of the fibers contributing to whole muscle tension are missed.

Therefore, *anatomical cross-sectional area* underestimates the total number of fibers in a pennated muscle. A better estimation of the force capacity of a muscle is **physiological cross-sectional area (PCSA)** (Figure 11.49), which is taken perpendicular to all of the fibers in the muscle and is therefore approximately equal to the sum of all of the cross-sectional areas of the individual fibers in that muscle. So, PCSA takes into account the angle of pennation of the fibers in the muscle. PCSA is calculated as follows (Lieber, 2010):

$$PCSA = \frac{m \times \cos\theta}{\rho \times l} \tag{11.1}$$

where

$PCSA$ = Physiological cross-sectional area in cm^2

m = muscle mass in grams

θ = angle of pennation

ρ = density of the muscle in g/cm^3 (= 1.056 g/cm^3 in mammalian muscle)

l = fiber length in cm

Physiological cross-sectional area (PCSA) A cross section taken perpendicular to all of the fibers in the muscle and therefore approximately equal to the sum of all of the cross-sectional areas of the individual fibers in that muscle.

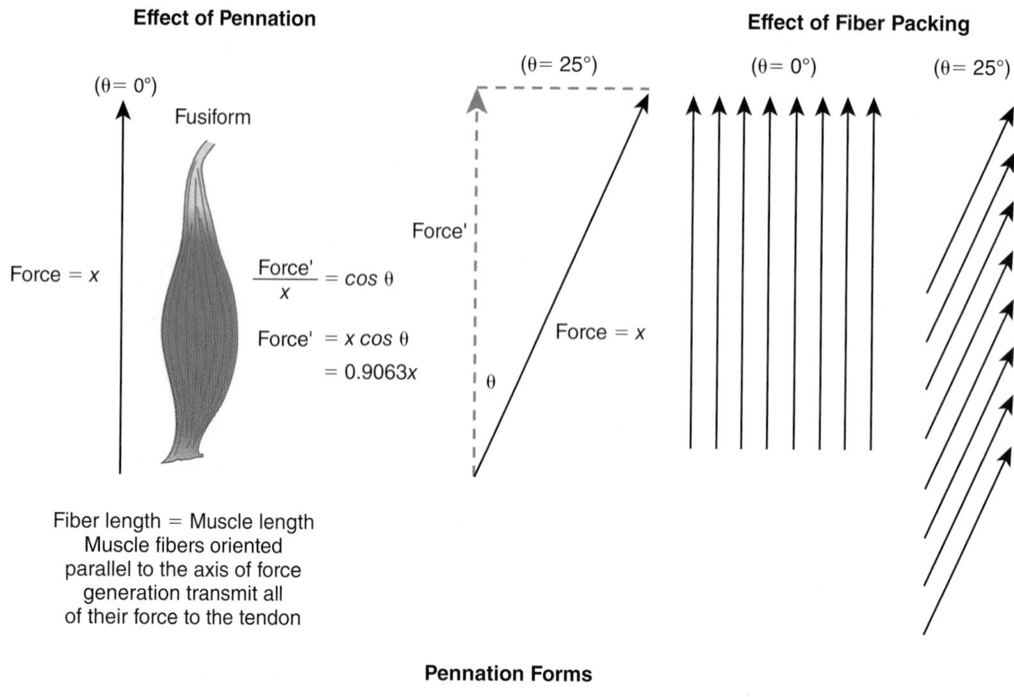

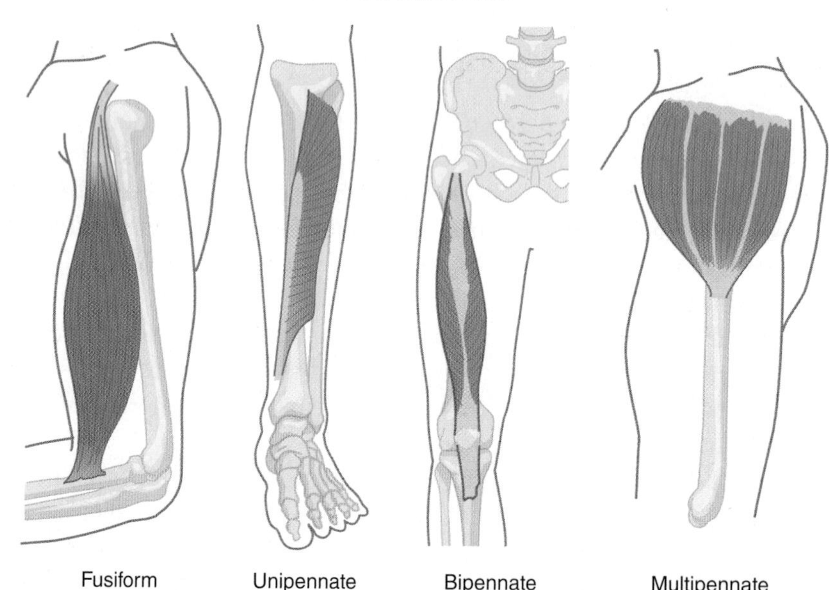

Figure 11.48 **Force along the longitudinal axis as pennation angle changes.**

Equation 11.1 can be restated in a simplified version (Lieber, 2010):

$$PCSA = CSA \times \cos\theta \tag{11.2}$$

where

$PCSA$ = Physiological cross-sectional area in cm^2

CSA = cross-sectional area (cm^2) of a cylinder with equal length to that of the fibers

θ = angle of pennation

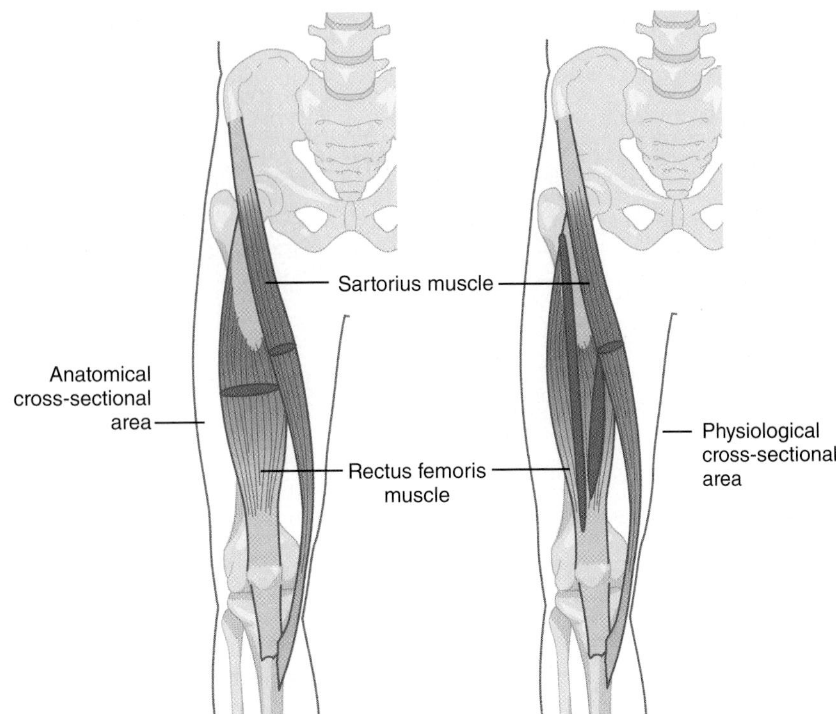

Figure 11.49 **Cross sections of the rectus femoris.**

In either equation, the key point is that pennation is taken into account; and the greater the angle of pennation, the greater the physiological cross-sectional area (with all other factors held constant). In a longitudinal fiber arrangement, the anatomical cross-sectional area is approximately equal to the physiological cross-sectional area. However, in a pennated fiber such as the rectus femoris, the physiological cross-sectional area is significantly larger than the anatomical cross-sectional area. A large cross-sectional area means that more sarcomeres are contributing to contractile force. You have learned that soleus fibers have a pennation angle equal to 25°, so 90.63% of the force is directed along the longitudinal axis of the whole muscle. This percentage means that approximately 9.37% of force-generation capacity is lost because of the pennation. However, the increased physiological cross-sectional area and the accompanying increase in sarcomeres that contribute to force generation more than compensate for this loss. Therefore, by using pennation, the body can save space by packing more fibers into a smaller area.

If pennation provides such an advantage, why do we have some muscles that have pennation angles close to zero? There is a tradeoff for the gain in force. Recall that a sarcomere can shorten only to a point where the myosin filaments come in contact with the Z-disks. In other words, a sarcomere can shorten only to a certain degree (i.e., the length of the myosin filaments). Also recall that the sarcomeres are arranged in a series to make up the myofibrils. Because of this arrangement, the degree to which a muscle fiber can shorten is dictated by the number of sarcomeres in series. For example, the length of one sarcomere is approximately 2.5 μm, and the length of the thick filaments is approximately 1.5 μm. Let's assume that the sarcomere can shorten only to the length of the thick filament because it is obstructed by the Z-disks. This shortening distance is equal to approximately 50% of original length. If a myofibril is made up of 5 sarcomeres (total length of 12.5 μm), it has the capacity to shorten by 6.25 μm (0.50 × 12.5 μm). Similarly, a myofibril with 20 sarcomeres in series (total length of 50.0 μm) can shorten a total distance of 25.0 μm. Therefore, the more sarcomeres in series, the greater the total amount of fiber shortening that is possible (Figure 11.50).

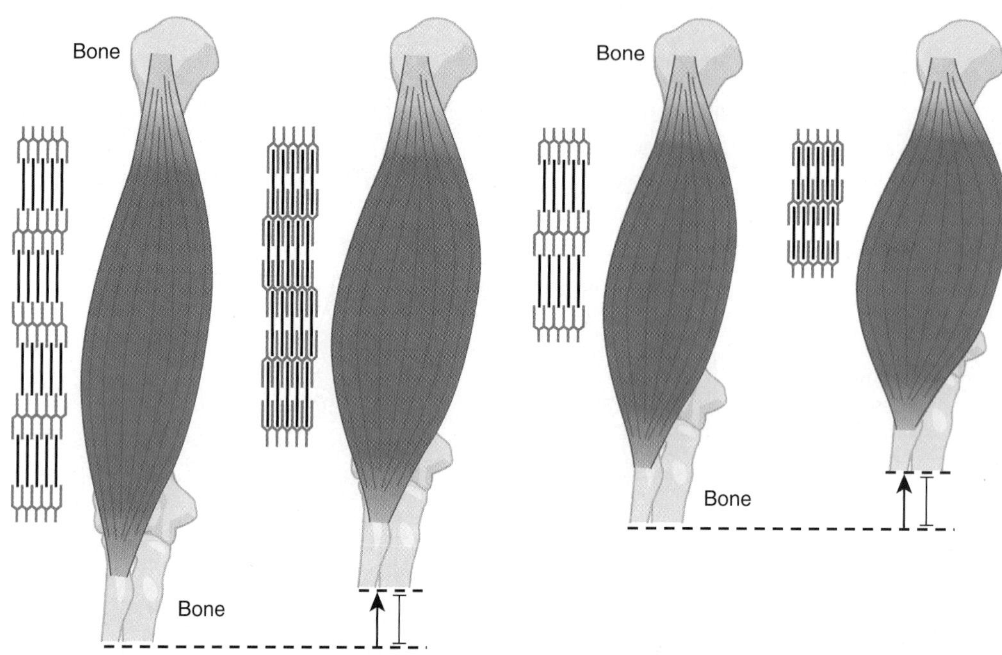

Figure 11.50 **Variation in shortening capability depends on number of sarcomeres.**

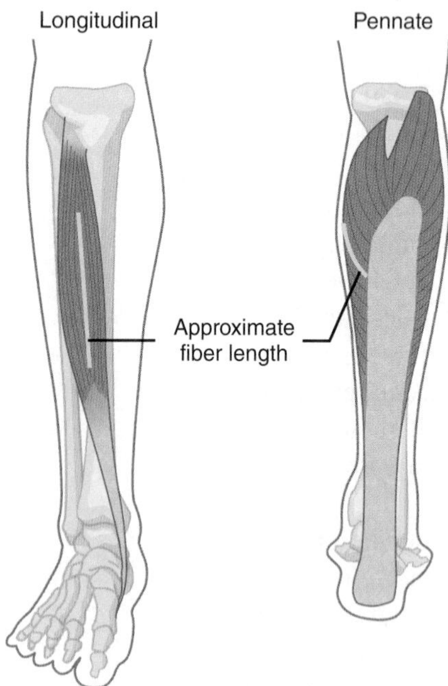

Figure 11.51 **Length of longitudinal fiber compared to pennate.**

This potential is critically important in relation to pennation angle. Although fibers do not run the entire length of the muscle in longitudinal fiber arrangements, fibers are longer than those found in pennate arrangements (Figure 11.51).

The fibers of a pennate muscle are arranged obliquely to a tendon that runs along the longitudinal axis of the muscle; as a result, longitudinal muscle fibers have more sarcomeres *in series* than pennate fibers and are therefore capable of a greater degree of total shortening. The pennate fibers have more sarcomeres *per unit of cross-sectional area* and therefore produce more force. However, this additional force comes at the expense of total shortening capacity. Therefore, longitudinal fiber arrangements tend to be found in muscles designed for producing greater range of motion at a joint or greater **joint excursion** (for example the hamstrings muscle group). In contrast, pennate formations tend to be located in areas requiring high force production but not necessarily large joint excursion (e.g., plantar flexors such as the gastrocnemius and soleus). A final characteristic affected by pennation is the velocity of muscle shortening. Recall that all of the fibers in a motor unit contract simultaneously. Thus, the sarcomeres of a myofibril are not shortening one at a time but all at once. Because total fiber shortening is greater in longitudinal arrangements in any given period, the velocity of muscle shortening is also proportional to the number of sarcomeres in series. Therefore, longitudinal muscle arrangements have the advantage of greater range of motion and velocity of contraction, and pennated muscle has the advantage of force production. So in answer to our question about pennation angle, we need both types of fiber arrangements, depending upon the needs of the particular joint.

Joint excursion Total range of motion at a joint.

FOCUS ON RESEARCH

As is stated in this chapter, the pennation angle is the angle at which the fibers in a pennate arrangement are oriented relative to the longitudinal axis. The degree of pennation and size of pennation angle are not small factors, and they are actually directly related to the maximal force production and total range of motion of the muscle. It was long believed that the pennation angle of fibers was uniform throughout a given muscle. However, enhanced technology in the field of muscle physiology is allowing for a better understanding of muscle structure and function.

Infantolino and Challis (2014) used magnetic resonance imaging (MRI) to study the variability in pennation angle throughout the first dorsal interosseous (FDI) muscle, which is a bipennate muscle of the hand with two heads. The authors explain that MRI produces images in the three anatomical planes throughout the muscle, and in two of those planes (frontal and sagittal) pennation angles can be measured. Therefore, the variability of pennation angle throughout a muscle can be quantified by measuring the pennation angle in both planes for all MRI images. The researchers used the FDI muscle of two cadavers and found that along the medial–lateral axis the first cadaver muscle was 16.8 mm long, resulting in 448 analyzed slices. The second muscle was 10.8 mm long (287 slices) in the medial–lateral axis. In the anterior–posterior axis, the first muscle was 11.4 mm long (305 slices), with the second cadaver muscle being 8.9 mm long (237 slices). It was observed that there was a change in pennation angle for both axes of measurement in both muscles with changing depth of analysis plane. In the FDI for cadaver 1, pennation angle in planes along the medial–lateral axis ranged from 2.2° to 24.4°, whereas the anterior–posterior axis ranged from 3.3° to 14.0°. For cadaver 2, pennation angle in planes along the medial–lateral axis ranged from 3.2° to 22.6°, whereas the anterior–posterior axis ranged from 3.1° to 24.5°. Based on their findings, the investigators conclude that such non-normal distributions of pennation angles is suggestive of a much more complex distribution of fascicles. In addition, they state that use of a single pennation angle to represent an entire muscle may not be accurate.

This study is important for a couple of reasons. First, it demonstrates that muscle is much more complex in structure and function than it was already believed to be. Second, the study demonstrates that care should always be taken when making broad assumptions from small samples. It was previously believed that one pennation angle is representative of an entire muscle, but this study shows that is no longer a safe assumption.

11.4 THE SYSTEM AS A HUMAN

This text covers many biomechanical concepts and it also attempts to show the connections between these concepts. The concepts are not only related to each other but are also highly integrated with the human form. At this point, let's review several concepts from previous chapters and their relationship to the human musculoskeletal system. Remember that all of the concepts in this text are in some way related to human motion. The next section reviews the topics that dictate human form.

Human Skeletal Link System

The human skeletal system is a kinetic chain, a system of linked rigid bodies subject to force application. The links are in the form of various types of joints, and the bones are semirigid bodies. Force is applied to the segments by muscle contractions, causing motion at the joints. A lever system consists of a rigid

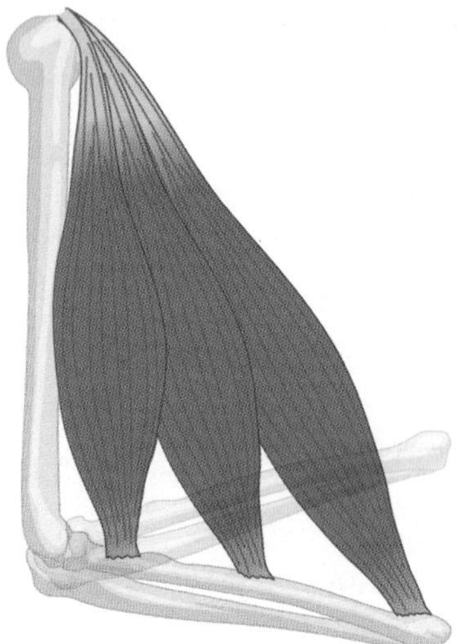

Figure 11.52 **Muscle attached normally and then distally.**

or semirigid object (lever) that is capable of rotating about an axis called a *fulcrum*. Therefore, the human skeleton is analogous to a group of connected lever systems. As you learned in Chapter 8, three types of lever systems exist: first, second, and third class. Many musculoskeletal arrangements in the human body can be classified as third-class lever systems. In a third-class lever system, the motive force (muscle) is between the fulcrum (joint) and the resistive force (segment). Because the radius of rotation for the resistive force is larger than that for the motive force, its linear distance traveled is large relative to that of the motive end of the lever. Therefore, third-class levers are used to gain an advantage in linear range of motion and velocity of movement. This advantage in linear range of motion and velocity of movement comes at the price of force because the motive force has a short moment arm, and therefore a disproportionately large motive force must be produced to overcome the resistive force. Of course, we could regain some of the lost mechanical advantage by moving the insertion of the muscle farther from the fulcrum (increasing the moment arm for the motive force), but this change would have huge implications in terms of range of motion. Also, think of the implications for human shape.

By moving the muscle insertion to the distal end of the segment (Figure 11.52), humans could be extremely strong. However, the range of motion of the joint would be minimized and the shape of the body would be dramatically altered.

The wheel-and-axle system can be used to gain an advantage in force or linear range of motion and velocity of movement. The advantage depends on whether the motive force is applied to the wheel or the axle. Although no musculoskeletal arrangements are directly analogous to the wheel-and-axle system, humans do adopt certain limb positions that can be considered to be this type of machine. Rotations of bones about their longitudinal axes can be considered wheel-and-axle systems (Figure 11.53).

For example, when throwing an object, the proximal bone serves as an axle while the distal segments form a wheel. The muscle force in this case is applied to the axle. However, the axle has a smaller radius compared to the wheel. This smaller moment arm means that torque production is compromised if the force is applied to the axle. Doing so will require application of a greater force, but the linear distance traveled by the resistive force will be greater.

In both cases, notice that the human system has opted for increased range of motion and velocity of movement rather than an advantage in torque production. This choice provides humans with an

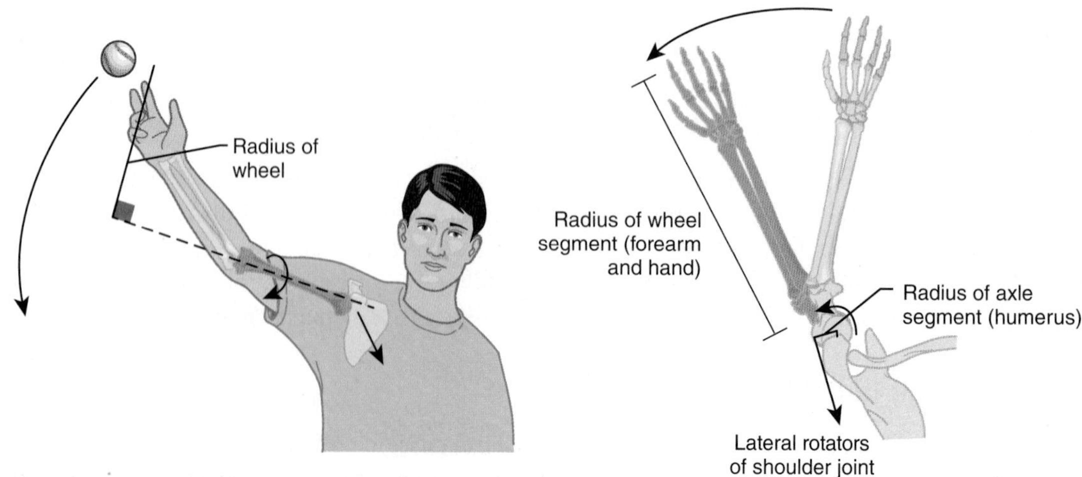

Figure 11.53 **Throwing as a wheel-and-axle system.**

enhanced capability to manipulate the environment. Enhanced manipulative skills provide humans with the opportunity to create tools to compensate for the disadvantage in torque. For example, humans can create machines that mimic second- and third-class lever systems as well as wheel-and-axle systems in which we can apply force to the wheel. In addition, opting for enhanced range of motion has direct implications for human form because these choices create constraints for the muscular system, which is superimposed on the skeletal link system.

Human Skeletal and Muscular System Interaction

Due to the design of the skeletal system, the muscular system has certain constraints. Because an axis of rotation exists at each segmental link, the resolved muscle force vector will have components with specifically defined roles. You know that when resolving muscle force vectors you define the perpendicular component as the rotary component because if it were allowed to act *alone* it would cause joint rotation. The rotary component does not pass through the axis of rotation, so it always has a moment arm and is therefore torque producing. Also, if the horizontal (parallel) component acted *alone*, it would cause the bones of the articulation to be either pushed tightly together or pulled apart. So, the parallel component of a muscle force vector is defined as either the stabilizing component or the destabilizing component. Because the parallel component always passes through the axis of rotation, it never has a moment arm and is therefore not capable of producing torque. The primary consideration related to these components is that the angle of muscle pull changes throughout the range of motion at the joint. Because of the changing angle of pull, the relative sizes of the components vary throughout joint excursion. This variation wouldn't be a huge problem, except that the parallel component becomes destabilizing at angles of pull greater than 90°.

This potential problem forces the issue of muscle redundancy. You know that having multiple muscles contributing to the same action increases degrees of freedom. Also, having multiple muscles perform similar actions is necessary in terms of parallel and perpendicular components. Because of the constraints of origin and insertion, some muscles simply cannot achieve large angles of pull. Thus, no matter the joint angle, some muscles always have a large stabilizing component (Figure 11.54).

Having a muscle with a constant stabilizing effect protects the integrity of the joint from destabilization by other muscles. Therefore, we need multiple muscles contributing to a joint action.

Varying the angle of pull also places constraints on internal muscle physiology. At resting length, the angle of pull is acute, so much of the resultant force is directed into the joint itself rather than causing rotation. Also, at this angle of pull, the moment arm is at its shortest. Therefore, it works out well that our muscles tend to generate the greatest amount of force at resting length. According to the length–tension relationship, the muscles are strongest at their optimal length (slightly stretched beyond rest) because this position provides maximum interdigitation of myosin and actin filaments. On the opposite end of the spectrum, as a large destabilizing component is developed, the muscle tends to be greatly shortened. The more shortened the muscle, the less force it can produce. This is a good thing because most of the force at this point would be used to pull the bones of the articulation apart. So the length–tension relationship of the muscle works well with the varying conditions produced by the angle of pull. Let's also not forget that the skeleton prevents the muscle from being stretched to a length that would interfere with tension development. Therefore, the muscle works well with

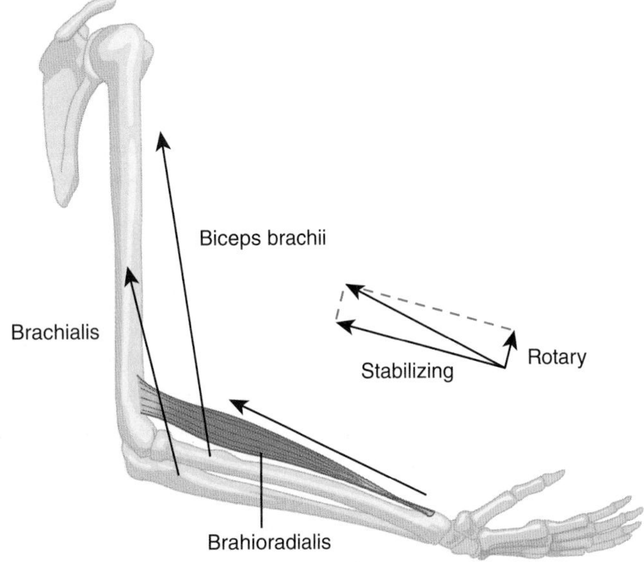

Figure 11.54 **Brachioradialis muscle force components.**

the constraints produced by the skeletal system, and the skeletal system helps to prevent force loss by maintaining muscle length within optimal limits.

So we need multiple muscles contracting to produce the same motion. The skeletal link system is designed for maximum range of motion. That motion is the result of muscle contraction. The angle of pull varies throughout the muscle contraction. Different angles of pull result in varying magnitudes of the perpendicular and parallel components, as well as changes in length of the moment arm. Development of a destabilizing component caused by one muscle necessitates the presence of a stabilizing component from another muscle. Redundant muscles are necessary to compensate for the constraints provided by the skeletal system. However, redundancy also has implications for human form.

Human Limb Shape

Because of the construction of the skeletal system, we need multiple muscles. However, we cannot just place them indiscriminately. Remember, human torque production has a mechanical disadvantage because it uses a third-class lever system (i.e., small moment arms). This disadvantage presents a potential problem in angularly accelerating the limb. Remember that angular acceleration is directly proportional to the sum of the torques and inversely proportional to rotational inertia:

$$\alpha_a = \frac{\Sigma T_a}{I_a} \quad (6.12)$$

where

α_a = angular acceleration about a given axis, or change in angular motion of the system
ΣT_a = the sum of the torques (net torque) about a given axis
I_a = moment of inertia about a given axis (rotational inertia)

According to Equation 6.12, if humans are at a disadvantage in producing torque, we must be highly creative with rotational inertia of the limb we wish to accelerate. Recall that the moment of inertia is related to the mass and distribution of the mass:

$$I_a = \Sigma m_i r_i^2 \quad (6.10)$$

where

I_a = moment of inertia about a given axis
m_i = mass of a given particle i
r_i^2 = radius of rotation of the particle i to the axis of rotation

According to Equation 6.10, we can reduce a segment's moment of inertia by reducing the mass, by keeping the mass close to the axis of rotation, or both. The human body has managed to accomplish both to compensate for the disadvantage in torque production.

Limb Mass

First, let's discuss the mass, looking at both skeletal and muscular solutions. In the skeletal system, the bones are not solid all the way to the core. This structure keeps the mass of the segment minimal. Notice that with disuse, muscles atrophy. It makes sense that disuse atrophy is also accompanied by bone loss. Weaker muscles have difficulty accelerating a more massive segment. So as the muscle becomes weaker, the mass of the skeletal system is reduced, thus lowering rotational inertia. Also notice that if a cast is placed on a broken limb, the muscles atrophy. Although we tend to think of atrophy as a bad thing, it could be the human system's solution to extra mass being added to a segment (especially if the segment is more distal). Although the body is adapted to minimize mass of the limbs, one of its most creative strategies is in dealing with the mass distribution of the limbs.

Limb Mass Distribution

Now let's consider limb mass distribution, which has direct implications for the radius of rotation of the limb. This point may seem obvious, but the muscles that move one segment are located on the more proximal segment. Only the distal tendon is attached to the segment of interest. For example, the muscle bellies of the strongest finger flexor and extensor muscle are located on the forearm (Figure 11.55a). The bellies of the strongest elbow flexors and extensors are located on the humerus (Figure 11.55b), and the prime movers of the shoulder are located on the torso (Figure 11.55c).

Notice the effect this design strategy has on human form and rotational inertia. The total mass of the hand is smaller than that of the forearm, and the total mass of the forearm is smaller than that of the upper arm. Two points should be noticed. First, each successively proximal segment must have more muscle mass because more torque is needed to accelerate each progressively more massive segment. In other words, the muscles that move the upper arm must be large because the upper arm contains the muscles that move the forearm, which in turn contains the muscles that move the hand. This situation is exacerbated further because proximal segments are usually needed to drive the inertias of all of the distal segments. Second, proximal bones are more massive than distal bones. Together these variations in muscle and bone mass have direct implications for human shape in terms of rotational inertia. Basically, because each successively distal limb is progressively less massive, a tapering effect is created (Figure 11.56).

This tapering tends to apply not only to the overall limb but also to each individual segment. For example, the more distal portion of the forearm is less massive than the more proximal end. The distal portion of the thigh is less massive than the more proximal portion and so on. So each segment is tapered and then superimposed on this shape; the entire limb is tapered. This fact may seem obvious visually, but it is highly ingenious. Overall, most of the muscle and bone mass of a limb is located near the axis of rotation. This conformation lowers the rotational inertia of the total limb by decreasing the radius of rotation. Notice that the radius of rotation is squared in the rotational inertia equation, so the effect of tapering is large. Because humans are at a mechanical disadvantage in torque production, lower rotational inertia means that less torque is necessary to angularly accelerate the limb. Can you imagine if it was the other way around, and most of the mass was distal to the axis? Not only would our shape be dramatically altered (Figure 11.57), but also it would be extremely difficult to accelerate a limb segment.

Muscle shape and fiber arrangement also play a role as described in this chapter. Pennated fiber arrangements save space by packing more fibers into a smaller cross-sectional area. More force production per cross-sectional area means that the muscle can produce more force without large amounts of muscle spanning the entire length of the segment (i.e., most of the mass can be kept close to the axis). Notice that many pennated muscle arrangements occur in the more distal segments. For example, the muscles located on the lower legs and forearms tend to be highly pennated. (Of course, this is not always the case. The deltoids are multipennate, but you get the point.) Pennation helps minimize the total amount of mass located distally on the limb segment, which decreases rotational inertia. If that limb happens to be distal, keeping the mass low is especially important because otherwise radius of rotation of the entire limb (which is the larger factor in the equation) would be affected. Could this be the reason for difficulty in developing huge calf and forearm muscles? Notice that many great sprinters have well developed thighs and gluteus muscles, but their calves may not necessarily be proportionate.

The next time you observe the human form, consider these points. The human system needs a specific range of motion and velocity of movement. This setup requires a body made up of many third-class lever systems. Because of this arrangement, the body has a disadvantage in torque production capability, and many muscles are required to move and protect the joints. However, mass must be added carefully in both location and amount. The human system has dealt with this dilemma by tapering both the individual segments and the overall limb. This tapering reduces the radius of rotation and, in turn, the rotational inertia of the limb. In addition, mass is kept to a minimum by using a "hollow" bone

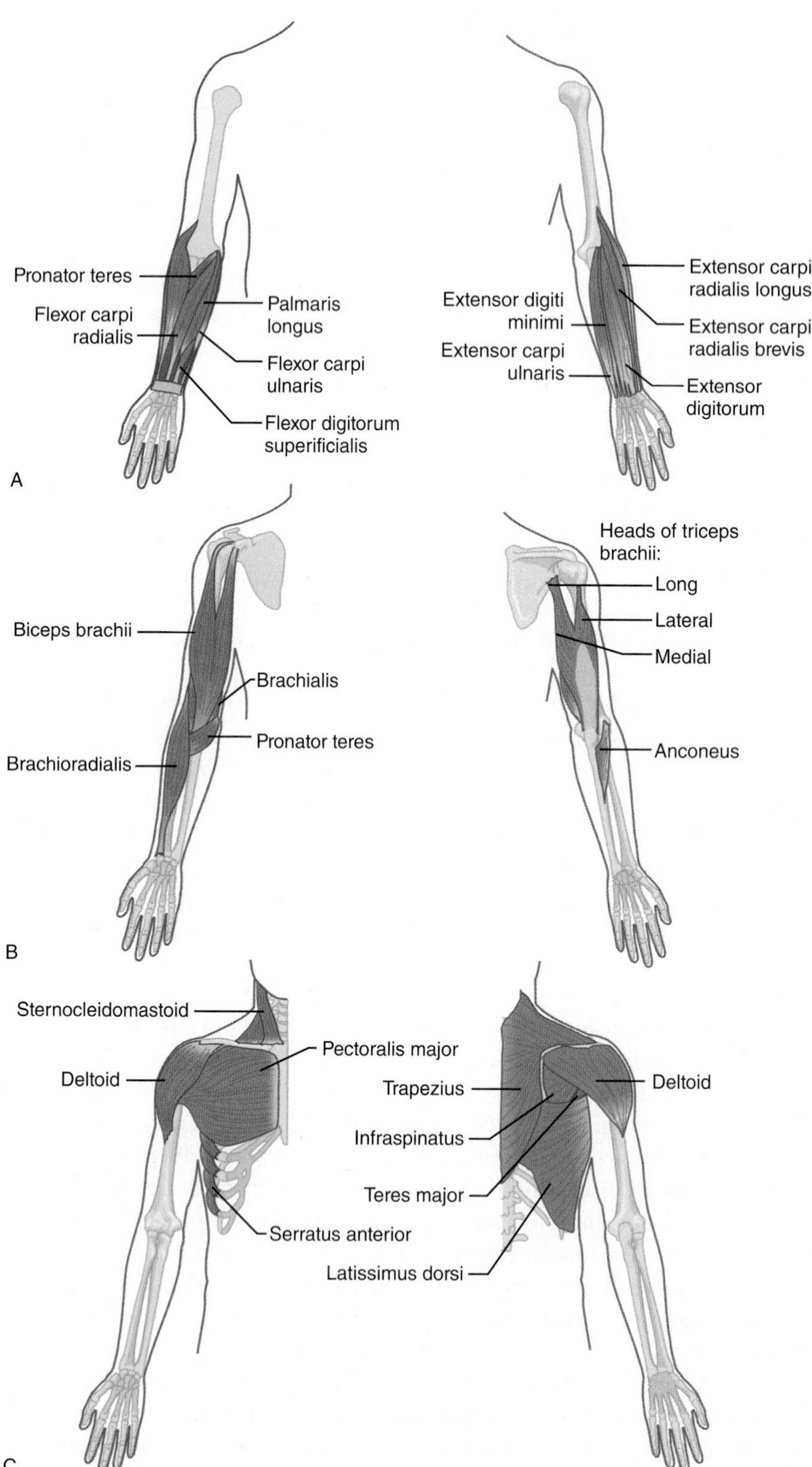

Figure 11.55 **Relative locations of the muscles of fingers, muscles of elbow, and muscles of the shoulder.**

structure and pennated muscle fiber arrangements. These aspects of structure lower the mass of the individual limb segment and help reduce the radius of rotation of the overall limb. Overall, the rotational inertia is lower for the limb, which helps compensate for the choice of the third-class lever system. So the human form itself may give rise to demands to which the human system adapts. These adaptations further amplify the form. Therefore, the final human form is a set of structures that allow for a multitude of complex motions given the constraints of its environmental situation.

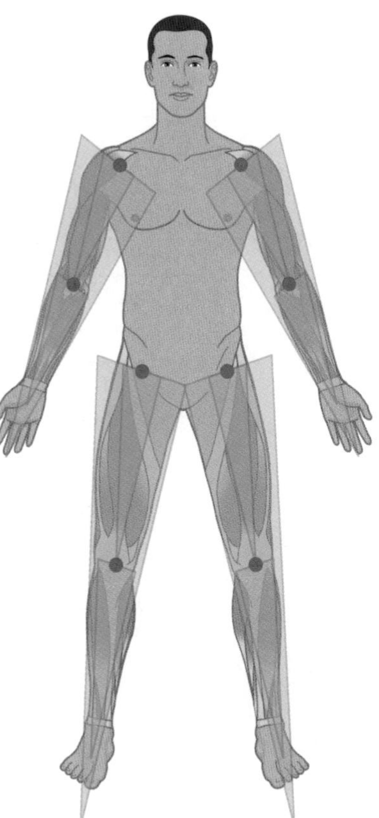

Figure 11.56 **Tapering of the human form.**

CONNECTIONS

11.5 EXERCISE SCIENCE

This chapter describes how skeletal muscle varies in its capacity to produce torque depending on length, velocity, and joint angle. Exercise machines use various apparatus (e.g., cables, pulleys, etc.) to control the amount of resistance as well as its direction and pattern. However, machines that use cams are making an attempt to work with the natural structure of the human body to maximize exercise effectiveness (Figure 11.58).

The idea behind the cam-based machine (sometimes called a *variable resistance machine*) is to try to match resistive torque to muscle torque throughout the range of motion of the joint. The way that matching is accomplished is by using a cam of variable radius. Because the radius of the cam varies throughout the range of motion, so does the length of moment arm through which the resistive force acts. In theory, this variation would provide (1) more resistive torque at joint angles in the range of motion where muscle torque production is high and, (2) less resistive torque at joint angles where muscle torque is low. These types of machines have advantages such as safety and form. However, they have some difficulty in matching the torque capability patterns of humans (Harman, 1983; Johnson, Colodny, & Jackson, 1990). This fact should not be a surprise because humans are extremely complex organisms and are not all built exactly alike.

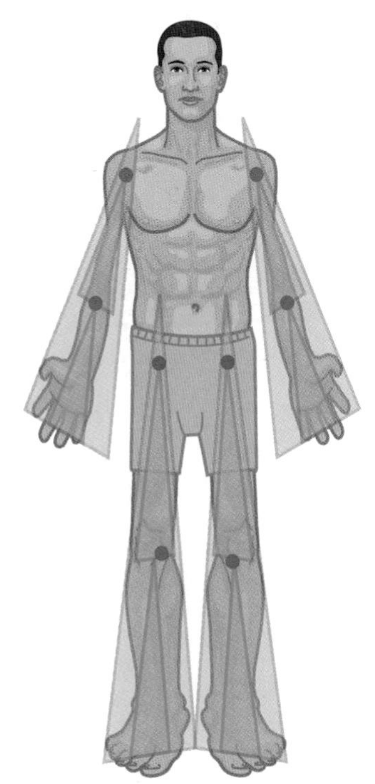

Figure 11.57 **Tapering opposite that of the natural human form.**

11.6 PHYSICAL EDUCATION

In this chapter, you have read about the properties of muscle and have also learned something about the inertial characteristics of the human limbs attributable to shape. A couple of issues relate muscle to health through the human form. Most practitioners in physical education and related fields are currently aware of the risk of osteoporotic fractures with age, and many are interested in ways of treating osteoporosis. However, most practitioners would agree that preventing osteoporosis is a better way to approach the problem than treating the condition once it has begun. In addition, everyone across all movement-related disciplines is acutely aware of the growing obesity problem among our youth. So let's discuss these two health issues in the context of strength training.

Keeping prevention of osteoporosis in mind, a study by Conroy et al. (1993) investigated the relationship of bone mineral density (BMD) to muscular strength in 25 elite junior weightlifters with an average age of 17.4 years. The average training experience for the group was 2.7 years. The BMD in the lumbar spine and proximal femur (femoral neck) of the weightlifters was compared to that of 11 age-matched controls. The BMD for the lifters was found to be significantly greater at all sites when

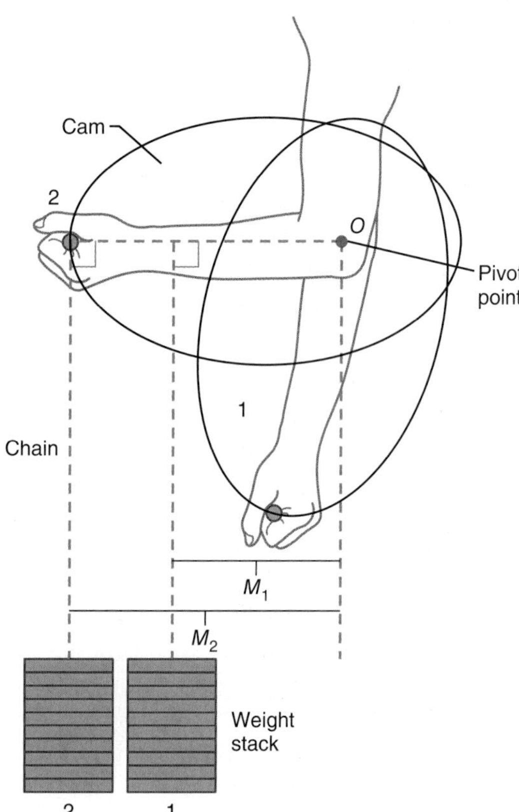

Figure 11.58 **A cam-based exercise machine.**

compared to the control group. In addition, the lifters also had higher BMD when compared to adult reference data (20- to 39-year-old men). These are critically important findings, especially when considering that optimization and maintenance of bone mass during youth could reduce the risk of fractures from osteoporosis later in life (Dalsky, 1987).

From an inertial point of view, as the muscles get stronger, they are capable of accelerating a limb of greater mass to the same degree. So a bone of higher density is less likely to be fractured but poses no disadvantage in terms of human motion because the density was gained using exercise that increases the muscle's ability to produce torque.

A special population in which strength training could potentially have enormous benefit is that of the overweight teenager. This consideration is critically important in light of the fact that the number of overweight teens in the United States continues to rise. It is imperative at this time that we encourage our youth to engage in regular physical activity. To some, the obvious solution is to encourage overweight teens to engage in regular aerobic exercise with the noble goal of burning calories. However, aerobic activities set up a situation in which most overweight teens are uncomfortable (both physically and psychosocially), which also means they are prone to failure. Some of the physical discomfort is, of course, attributable to the training state (or lack thereof) of the teenager. In terms of body shape, a limb that is large because of obesity is difficult to accelerate. In other words, the limb has a lot of inertia but does not have the advantage of a strong muscle to accelerate it. An alternative choice to aerobic exercise, at least in the initial stages of an exercise routine for an obese teen, is resistance training. Weight training is an activity in which overweight teens are more likely to succeed (they are often the strongest kids in the class) and therefore feel good about their performance, which in theory may enhance their level of participation (Faigenbaum, 2002). Not only does success in a physical activity improve psychosocial well-being, but also exercise improves weight control, bone mineral density, and cardiovascular risk profile (Faigenbaum & Westcott, 2000). In addition, a strengthened muscular system allows the overweight teen to perform activities of daily living with greater ease and more energy (Faigenbaum, 2002). The activities are easier because the inertial properties of the limbs are likely to change as they lose mass at the same time muscular strength is being developed. These improvements in physical and mental well-being could encourage the overweight teen to engage in activities of a more aerobic nature (and thereby improve health to an even greater degree) or at the very least establish a habit of regular physical activity.

11.7 PHYSICAL THERAPY

Physical therapists use machines to clinically assess strength to track patient progress. Therefore, physical therapists must be aware of the affect of joint position on torque production. Remember that joint position changes the length of the muscle fibers as well as the joint angle. We know that length of the muscle fiber at onset of contraction affects the amount of tension that it can develop. Also, a change in joint angle affects the length of the moment arm for the muscle force. In addition, machines used to test torque production can be set for different velocities of motion. Again, velocity affects muscle torque production.

No matter the angle or velocity chosen by the therapist, the important consideration is that these variables be kept the same from test to test (Croce, Miller, & St Pierre, 2000). Otherwise, the therapist will have difficulty distinguishing between differences in torque production attributable to intervention and differences stemming from variation in muscle length, moment arm, and velocity of action (Oatis, 2017). By the way, this same advice applies to anyone who is testing torque production as a means of progress.

SUMMARY

At the most basic level, motion of the human system is the result of muscles applying forces to bones through the process of contraction. The sarcomere is called the functional unit of the muscle because it is the smallest unit that can perform the function of the entire organ (the whole muscle). The function of the sarcomere is possible because of the contractile proteins it contains. Though several different proteins are found within the sarcomere, two filamentous structures are most crucial to muscle contraction: thick filaments and thin filaments. The thick filaments are large polymers (assemblies) of the protein myosin, and the thin filaments are polymers of the protein actin. Once binding sites on actin are exposed, myosin is free to form a strong bond with the actin sites. At the moment that strong binding to actin is achieved, the myosin crossbridge (globular head and neck) is believed to undergo a conformational change: The head of the crossbridge experiences a strong intermolecular attraction to the neck, causing it to tilt in that direction. This tilt of the myosin head toward the center of the sarcomere is called the *power stroke*. Each binding and power stroke cycle pulls the thin filament slightly closer to the middle of the sarcomere. Because myofibrils are made of sarcomeres, the myofibrils shorten. Muscle fibers are made up of myofibrils, so they shorten. Muscle fibers make up fasciculi, and therefore the entire muscle shortens. The final result is shortening of the entire muscle belly. This shortening applies force to the tendons and, in turn, the bones to which they are attached.

In general, two major factors affect whole muscle tension: (1) the number of fibers contracting and (2) the tension developed by each individual fiber involved in the contraction. The number of fibers contracting depends on the size of the stimulated motor unit and the number of motor units recruited. Tension developed by an individual muscle fiber is affected by the frequency of stimulation, the length of the fiber at onset of stimulation, the velocity of contraction, and elasticity of the tissues.

Because of its specific location, each individual muscle is capable of carrying out a different movement function. The function is primarily dictated by its origin and insertion. The origin and insertion determine the number of articulations of the muscle and its angle of pull. A biomechanical factor that may be a contributor to the large variation in whole muscle tension development is the architecture of the muscle in terms of fiber arrangement and length. Overall, two basic categories exist based on fiber arrangement: (1) longitudinal (or fusiform) and (2) pennate ("feather-like"). Longitudinal muscle arrangements have the advantage of greater range of motion and velocity of contraction, and pennated muscle has the advantage of force production.

The skeletal system of the human body is made of many third-class lever arrangements. This type of lever system provides enhanced range of motion but limits torque output. The human system has dealt with this dilemma by tapering both the individual segments and the overall limb to reduce the radius of rotation and, in turn, the rotational inertia of the limb. In addition, mass is kept to a minimum by using a hollow bone structure and pennated muscle fiber arrangements. These aspects of bone and muscle structure lower the mass of the individual limb segment and aid in reducing the radius of rotation of the overall limb. Overall, the rotational inertia is lower for the limb, which helps compensate for the choice of the third-class lever system.

REVIEW QUESTIONS

1. What prevents muscle contraction from occurring involuntarily?
2. Name all of the proteins involved in the sliding filament theory and their respective roles.
3. Explain how muscle excitation and contraction are coupled.

4. Compare the types of muscle actions.
5. Discuss the ways in which whole muscle tension is controlled.
6. Discuss the force–velocity relationship and give mechanisms.
7. Discuss the length–tension relationship and give mechanisms.
8. Discuss the similarities between the force–velocity relationship and the length–tension relationship in terms of mechanisms.
9. Why does jump height increase when the concentric action is immediately preceded by an eccentric action? Give at least three possible mechanisms.
10. How is optimal power output achieved? What are the mechanisms?
11. What happens to the force arm as the angle of tendon insertion approaches 90°? What about beyond that point?
12. What happens to the rotary and stabilizing components as the angle of tendon insertion approaches 90°? What about at exactly 90°? What about greater than 90°?
13. At angles of pull greater than 90°, what does the stabilizing component become?
14. Name some muscles whose insertion angle never attains 90°. Why is the presence of these muscles a distinct advantage?
15. Give the arrangement of the axis of rotation and the motive and resistive forces for first-, second-, and third-class levers. Which of the lever systems is most commonly found in the human body? What are the parts of the human body that correspond to the parts of this lever system?
16. In a wheel-and-axle arrangement, what is the advantage if the motive force is applied to the wheel? What if it is applied to the axle? In the human body, is the force most often applied to wheel or axle?
17. Based on your answers to the previous two questions, are humans best suited for torque production or range of motion? Explain your answer.
18. Human limbs tend to taper distally. Using both of the following equations and knowledge of moment arm length, fully explain why this shape is a distinct advantage.

$$\mathrm{I} = \mathrm{mr}^2 \qquad \alpha = \frac{\mathrm{T}}{\mathrm{I}}$$

19. Explain the advantage of muscle pennation in terms of rotational inertia and angular acceleration.

EQUATIONS

Torque	$T = F \times r \times \sin\theta$	(6.1)
Physiological cross-sectional area	$PCSA = \dfrac{m \times \cos\theta}{\rho \times l}$	(11.1)
	or	
	$PCSA = CSA \times \cos\theta$	(11.2)
Angular acceleration around a given axis	$\alpha_a = \dfrac{\Sigma T_a}{I_a}$	(6.12)
Moment of inertia around a given axis	$I_a = \Sigma m_i r_i^2$	(6.10)

REFERENCES AND SUGGESTED READINGS

Åstrand, P., K. Rodahl, H. A. Dahl, & S. B. Strømme. 2003. *Textbook of work physiology: Physiological bases of exercise*, 4th ed. Champaign, IL: Human Kinetics.

Bigland, B., & O. C. J. Lippold. 1954. The relation between force, velocity and integrated electrical activity in human muscles. *Journal of Physiology* (London), 123: 214–224.

Brand, P. W., R. B. Beach, & D. E. Thompson. 1981. Relative tension and potential excursion of muscles in the forearm and hand. *Journal of Hand Surgery* [Am], 6(3): 209–219.

Brooks, G. A., T. D. Fahey, & K. M. Baldwin. 2005. *Exercise physiology: Human bioenergetics and its applications*, 4th ed. New York: McGraw-Hill.

Buchanan, T. S. 1995. Evidence that maximum muscle stress is not a constant: Differences in specific tension in elbow flexors and extensors. *Medical Engineering & Physics*, 17(7): 529–536.

Buchanan, T. S (ed.). 1997. Mechanics and energetics of the stretch-shortening cycle—Special issue. *Journal of Applied Biomechanics*, 13(4).

Conroy, B. P., W. J. Kraemer, C. M. Maresh, S. J. Fleck, M. H. Stone, A. C. Fry, P. D. Miller, & G. P. Dalsky. 1993. Bone mineral density in elite junior Olympic weightlifters. *Medicine & Science in Sports & Exercise*, 25(10): 1103–1109.

Croce, R. V., J. Miller, & P. E. St Pierre. 2000. Effect of ankle position fixation on peak torque and electromyographic activity of the knee flexors and extensors. *Electromyography and Clinical Neurophysiology*, 40: 365–373.

Dalsky, G. P. 1987. Exercise: Its effect on bone mineral content. *Clinical Gynecology*, 30: 820–831.

Davies, C. T. M., D. O. Thomas, & M. J. White. 1986. Mechanical properties of young and elderly human muscle. *Acta Medica Scandinavica*, 711: 219–226.

Doan, B. K., R. U. Newton, J. L. Marsit, N. T. Triplett-McBride, L. P. Koziris, A. C. Fry, & W. J. Kraemer. 2002. Effects of increased eccentric loading on bench press 1 RM. *Journal of Strength and Conditioning Research*, 16(1): 9–13.

Enoka, R. M. 1996. Eccentric contractions require unique activation strategies by the nervous system. *Journal of Applied Physiology*, 81(6): 2339–2346.

Enoka, R. M. 2015. *Neuromechanics of human movement*, 5th ed. Champaign, IL: Human Kinetics.

Faigenbaum, A. D. 2002. Strength training for overweight teenagers. *Strength and Conditioning Journal*, 24(5): 67–68.

Faigenbaum, A., & W. Westcott. 2000. *Strength and power for young athletes*. Champaign, IL: Human Kinetics Publishers.

Finni, T., S. Ikegawa, V. Lepola, & P. V. Komi. 2003. Comparison of force-velocity relationship of vastus lateralis muscle in isokinetic and in stretch-shortening cycle exercises. *Acta Physiologica Scandinavica*, 177: 483–491.

Flitney, F. W., & D. G. Hirst. 1978. Cross-bridge detachment and sarcomere "give" during stretch of active frog's muscle. *Journal of Physiology* (London), 276: 449–465.

Fukunaga, T., R. R. Roy, F. G. Shellock, J. A. Hodgson, & V. R. Edgerton. 1996. Specific tension of human plantar flexors and dorsiflexors. *Journal of Applied Physiology*, 80(1): 158–165.

Gordon, A. M., A. F. Huxley, & F. J. Julian. 1966. The variation in isometric tension with sarcomere length in vertebrate muscle fibres. *Journal of Physiology (London)* 184: 170–192.

Harman, E. 1983. Resistive torque analysis of 5 Nautilus exercise machines. *Medicine & Science in Sports & Exercise*, 15(2): 113.

Haxton, H. A. 1944. Absolute muscle force in the ankle flexors of man. *Journal of Physiology* (London), 103(3): 267–273.

Hill, A. V. 1970. *First and last experiments in muscle mechanics*. New York: Cambridge University Press.

Hochachka, P. W. 1994. *Muscles as molecular and metabolic machines*. Boca Raton, FL: CRC Press, 1994.

Infantolino, B. W., & J. H. Challis. 2014. Short communication: pennation angle variability in human muscle. *Journal of Applied Biomechanics*, 30(5): 663–667.

Ishikawa, M., & P. V. Komi. 2004. Effects of different dropping intensities on fascicle and tendinous tissue behavior during stretch-shortening cycle exercise. *Journal of Applied Physiology*, 96(3): 848–852.

Johnson, J. H., S. Colodny, & D. Jackson. 1990. Human torque capability versus machine resistive torque for four Eagle resistance machines. *Journal of Applied Sport Science Research*, 4(3): 83–87.

Komi, P. V., & A. Gullhofer. 1997. Stretch reflexes can have an important role in force enhancement during SSC exercise. *Journal of Applied Biomechanics*, 13(4): 451–459.

Komi, P. V., V. Linnamo, P. Silventoinen, & M. Sillanpää. 2000. Force and EMG power spectrum during eccentric and concentric actions. *Medicine & Science in Sports & Exercise*, 32 (10): 1757–1762.

Kubo, K., H. Kanehisa, Y. Kawakami, Fukunaga, & T. Fukunaga. 2000a. Elasticity of tendon structures of the lower limbs in sprinters. *Acta Physiologica Scandinavica*, 168(2): 327–336.

Kubo, K., H. Kanehisa, D. Takeshita, Y. Kawakami, S. Fukashiro, & T. Fukunaga. 2000b. In vivo dynamics of human medial gastronemius muscle-tendon complex during stretch-shortening cycle exercise. *Acta Physiologica Scandinavica*, 170(2): 127–136.

Lieber, R. L. 2010. *Skeletal muscle structure, function, and plasticity: The physiological basis of rehabilitation*, 3rd ed. Baltimore: Lippincott Williams & Wilkins.

Lieber, R. L., B. M. Fazeli, & M. J. Botte. 1990. Architecture of selected wrist flexor and extensor muscles. *Journal of Hand Surgery*, 15A: 244–250.

Lieber, R. L., M. D. Jacobson, B. M Fazeli, R. A. Abrams, & M. J. Botte. 1992. Architecture of selected muscle of the arm and forearm: anatomy and implications for tendon transfer. *Journal of Hand Surgery*, 17A: 787–798.

Maganaris, C. N., V. Baltzopoulos, D. Ball, & A. J. Sargeant. 2001. In vivo specific tension of human skeletal muscle. *Journal of Applied Physiology*, 90(3), 865–872.

McArdle, W. D., F. I. Katch, & V. L. Katch. 2015. *Exercise physiology: Nutrition, energy, and human performance*, 8th ed. Baltimore: Williams & Wilkins.

McIntosh, B. R., P. F. Gardiner, & A. J. McComas. 2006. *Skeletal muscle: Form and function*, 2nd ed. Champaign, IL: Human Kinetics.

McLester, J. R. 1997. Muscle contraction and fatigue: The role of adenosine 5′-diphosphate and inorganic phosphate. *Sports Medicine*, 23(5): 287–305.

Nardone, A., C. Romanò, & M. Schieppati. 1989. Selective recruitment of high-threshold human motor units during voluntary isotonic lengthening of active muscles. *Journal of Physiology* (London), 409: 451–471.

Narici, M. V., L. Landoni, & A. E. Minneti. 1992. Assessment of human knee extensor muscle stress from in vivo physiological cross-sectional area and strength measurements. *European Journal of Applied Physiology*, 65(5): 438–444.

Nigg, B. M., B. R. MacIntosh, & J. Mester (eds). 2000. *Biomechanics and Biology of Movement*. Champaign, IL: Human Kinetics.

Oatis, C. A. 2017. *Kinesiology: The mechanics & pathomechanics of human movement*. Baltimore: Wolters Kluwer.

Perrine, J. J., & V. R. Edgerton. 1978. Muscle force-velocity and power-velocity relationships under isokinetic loading. *Medicine & Science in Sports & Exercise*, 10: 159–166.

Podolsky, R. J., & M. Shoenberg. 1983. Force generation and shortening in skeletal muscle. Pp. 173–187 in *Handbook of Physiology*. Baltimore: American Physiological Society.

Powell, P. L., R. R. Roy, P. Kanim, M. A. Beloo, & V. R. Edgerton. 1984. Predictability of skeletal muscle tension from architectural determinations in guinea pig hindlimbs. *Journal of Applied Physiology*, 57(6): 1715–1721.

Robertson, D. G. E., G. E. Caldwell, J. Hamill, G. Kamen, & S. N. Whittlesey. 2014. *Research methods in biomechanics*, 2nd ed. Champaign, IL: Human Kinetics.

Sherwood, L. 2016. *Human physiology: From cells to systems*, 9th ed. Boston: Cengage Learning.

Stauber, W. T. 1989. Eccentric action of muscles: Physiology, injury, and adaptation. *Exercise and Sports Science Reviews*, 17: 157–186.

Trimble, M. H., C. G. Kulkulka, & R. S. Thomas. 2000. Reflex facilitation during the stretch-shortening cycle. *Journal of Electromyography & Kinesiology*, 10(3): 179–187.

Wickiewicz, T. L., R. R. Roy, P. L. Powell, & V. R. Edgerton. 1983. Muscle architecture of the human lower limb. *Clinical Orthopaedics and Related Research*, 179: 275–283.

Zajac, F. E. 1992. How musculotendon architecture and joint geometry affect the capacity of muscles to move and exert force on objects: a review with application to arm and forearm tendon transfer design. *Journal of Hand Surgery* [Am], 17(5): 799–804.

CHAPTER 12

CONNECTION BY APPLICATION

LEARNING OBJECTIVES

1. Understand how the concepts of previous chapters can be applied to a single sport.
2. Understand the connections that support the discipline of biomechanics in analyzing human motion in sport and related fields.

THE FINAL CONNECTIONS

- 12.1 Chapter 1: Biomechanics and Related Movement Disciplines
- 12.2 Chapter 2: Describing the System and Its Motion
- 12.3 Chapter 3: Paradigms for Studying Motion of the System
- 12.4 Chapter 4: Interaction of Forces and the System
- 12.5 Chapter 5: Linear Motion of the System
- 12.6 Chapter 6: Angular Motion of the System
- 12.7 Chapter 7: System Balance and Stability
- 12.8 Chapter 8: The System as a Machine
- 12.9 Chapter 9: System Motion in a Fluid Medium
- 12.10 Chapter 10: The System as a Projectile
- 12.11 Chapter 11: Biomechanics of the Musculoskeletal System

This chapter provides a more detailed look at two popular sports using concepts from throughout this text. It serves as a review of concepts presented in previous chapters and shows the integrated nature of the field of biomechanics. Rather than just targeting individual concepts as in previous chapters, this chapter gives an actual example of sport analysis in a more comprehensive way.

THE FINAL CONNECTIONS

Throughout previous chapters you've learned specific biomechanical concepts related to human performance. Included in these chapters have been relevant, real-life connections exploring how movement professionals use the physics, equations, and techniques of measurement and evaluation in biomechanics. This final chapter uses two popular sports to exemplify many of the ideas presented in previous chapters. We use golf and soccer as examples for several reasons: (1) Both are played around the world, (2) both are watched by millions of fans on a weekly basis, and (3) they encompasses a wide variety of the biomechanical concepts explained in previous chapters.

Both sports use a round ball that can be manipulated in many ways that affect its movement. Soccer skills are generally open and dynamic in nature but performed within a standard boundary. Golf skills are typically closed, but the conditions from one shot to any other are rarely the same. Golf also adds the requirement of implements that make direct contact with the ball rather than the body. As you read through this chapter try to make similar connections to any sport or activity that interests you. A detailed examination of golf is presented for all 11 chapters in the book and is followed by a discussion of research-based biomechanical concepts related to soccer.

GOLF

12.1 CHAPTER 1: BIOMECHANICS AND RELATED MOVEMENT DISCIPLINES

By this point, you should understand that sport activities and skills fall into a variety of movement-related sciences. *Motor development* can help determine when people are most ready to learn and produce skilled movements. Learning is enhanced through good teaching, and sport psychologists and sociologists help us understand, attain, and maintain consistent high-quality performance. *Anatomy* describes structure and function of the muscular and skeletal systems, and *physiology* describes how the muscles contract to produce torque for motion. *Functional anatomy* combines the knowledge of anatomical structure with an understanding of function. *Biomechanics* takes functional anatomy a step further, by helping students understand the physical principles related to the function of systems. These principles are based on laws of physics that have been proven to hold true in all situations so far in human history. A meaningful understanding of these principles and how they affect systems has led to amazing advances in sports performance. Advances based on biomechanical principles in the sport of golf include new materials, new designs, and better training. New materials used in manufacturing golf balls and clubs increase the *coefficient of restitution* and conserve more energy from impact. The end result is that the average driving distance of professionals during the past decade alone has increased by more than 5 m. A better indicator of how technology has improved distance can be seen in older professionals on the Champions Tour (for players 50 years of age and older). Many of the top players on the Champions Tour are driving the ball 10 m to 25 m farther today than they did in their prime 20 to 30 years ago! Club design has improved play for the average golfer by making the effective hitting area of the clubs (the sweet spot) larger and by moving more mass to the perimeter of the clubface to reduce twisting on off-center hits. Ball design is a secretive and ongoing project at each company that produces them. There are balls with round or hexagonal dimples, variations in dimple size, patterned dimples, and dimples of different depths. There are balls with softer or harder materials in their outer shells, and softer and harder inner layers. Some balls are designed to maximize distance, and others will sacrifice distance to maximize spin and control. The combinations are endless, but the purpose of continuing research is the same: to provide consumers with a product that improves their game.

Figure 12.1a **Dustin Johnson.** Taller equals longer levers.

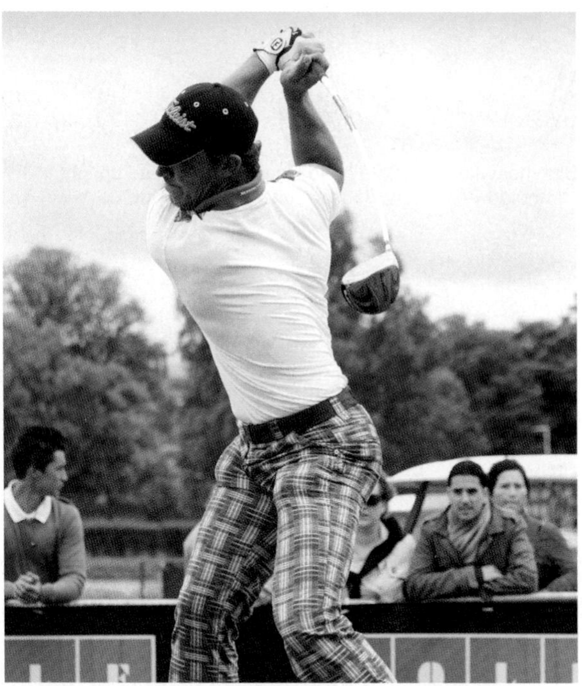

Figure 12.1b **Extreme flexibility in a long drive athlete.**

Some of the distance gains in professional golf are the result of improving the human system to produce more clubhead velocity at impact. Taller players have an inherent biomechanical advantage that derives from their longer limbs. For a given angular velocity, a point farther away from the axis of rotation will have a higher linear velocity. One player who exemplifies this is Dustin Johnson, who stands 193 cm tall (Figure 12.1a). He is strong and flexible but swings smoothly with seemingly little effort. The ball is naturally farther from the axis of rotation in taller players. Players of shorter stature can change other variables to increase clubhead velocity.

Two variables that are easily changed are muscular strength and range of motion. The sport of long driving is quite popular, and many of the athletes in this discipline train for both strength and flexibility (Figure 12.1b). Most golf instructors teach their students to take the club back to a position parallel to the ground during the backswing. Notice that the long-drive competitor continues the turn past parallel, which allows him to apply force for a longer amount of time, resulting in higher velocity at impact. For a thorough overview of how biomechanics can affect distance and accuracy of golf shots, Hume, Keogh, and Reid (2005) provide a comprehensive literature review.

12.2 CHAPTER 2: DESCRIBING THE SYSTEM AND ITS MOTION

A golf shot includes two major systems of interest: the golfer and the ball. Each body segment of the golfer can be broken down into its own subsystem. With sound knowledge of anatomy and the knowledge you've gleaned from the chapters in this textbook, you could easily do so. However, this task is beyond the scope of this chapter or book. Instead, we focus on other concepts in Chapter 2.

An adage among golf instructors states, "There is not one perfect golf swing, but there is a perfect swing for each person." This statement is based mainly on the field of *anthropometrics*. Humans come

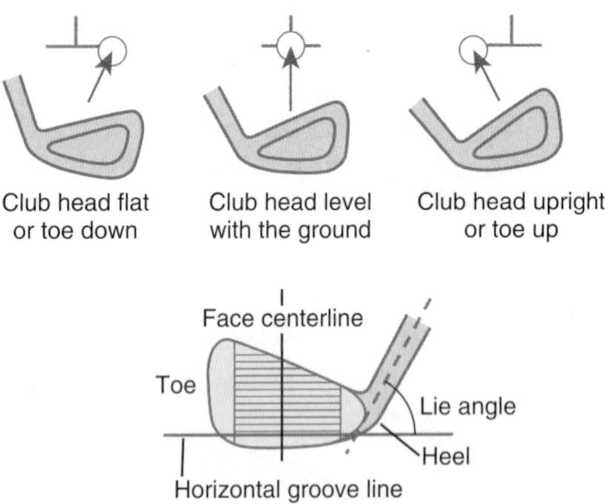

Figure 12.2 **Lie angle.**

in a wide variety of shapes and sizes, and many develop movement limitations for various reasons that can affect the golf swing. Arthritis can limit joint mobility, back pain may limit bending at the waist, tall players typically use average-sized clubs rather than have custom clubs made, and overweight people often modify their swings to enable their hands to clear their stomachs. Veteran golf instructors have learned to deal with each of these situations and depend upon physical measurements of their students to improve their games in the best way. Information such as height, weight, range of motion, previous injury, and other limitations are taken into account to find the most efficient swing for a particular golfer. Note that although a perfect swing exists for each golfer, some of these swings require compensations that deviate from fundamental movements of a theoretically ideal swing.

One of the more recent developments based on anthropometrics is the process of club fitting. Pro shops and larger golf-equipment suppliers can match clubs to a player's size and swing. The idea is to find the proper angle between the shaft and clubhead at impact, which is measured using a value called *lie angle* (Figure 12.2). Golfers who are appreciably taller or shorter than average need different length shafts to maintain proper posture, and the angle between the shaft and clubhead will change because of these length differences. A quality golf fitting will include shaft length, shaft flexibility, lie angle, and even grip size.

Many books and papers describe the specific *anatomical positions* of golfers during the three phases of the motion: backswing, through swing, and follow-through. Any library or Internet search including the terms *biomechanics* and *golf* produces an enormous variety of information in both scope and depth. In essence, the golf club shaft should ideally scribe a diagonal *plane*, with the spine as the *axis of rotation*. Maintaining the position of the club along this plane is a crucial component in striking the ball consistently, and an amazing variety of training aids have been designed to help golfers in this regard. There are specially designed training clubs with hinges built into the shaft that help train a golfer to maintain the proper swing plane and tempo—if he doesn't get both factors correct, the hinges collapse and cause the shaft to fall out of alignment. The Explanar (Figure 12.3) allows a golfer to "feel" proper club and body position, especially when the club travels out of view behind the body.

The muscle actions and movements that individual body segments produce to keep the club traveling along this plane are complex, but one element is critical: maintaining the position of the spine in space. The body should not shift laterally away from, or toward, the target during a swing; it should only rotate around the spine. As mass shifts during different phases of the swing, the *center of gravity* will shift slightly as body segments move the club around this axis. At address, the body is in a symmetrical position and mass is balanced equally between the two points of contact with the ground (feet). As the arms move the club back during the takeaway phase, the center of gravity shifts toward the foot farthest from the target although the body itself should not shift along a *frontal plane*.

This shift of mass is important in generating maximum force during the swing. Remember that the golf swing is an *open kinetic chain activity*, and the first link in the chain that develops the velocity of the clubhead is the lower body. The backswing sets the club and body into the proper position to initiate the through swing. The through swing is the movement that guides the clubhead from the end of the backswing to impact with the ball. The initiation of the through swing begins when muscle forces in the leg push the rear foot into the ground as the hips begin to rotate toward the target. Because more of the body's mass is over the rear foot at this point in the swing, this motion provides a more solid and stabilizing foundation. Strong players who generate tremendous force in the lower body have a difficult time preventing the rear foot from sliding away from the target,

Figure 12.3 **The Explanar golf training device.**
© Guang Niu/Getty Images AsiaPac/Getty Images

and this is the primary reason that golf shoes have spikes protruding from their soles. The spikes provide additional *friction*, allowing the forces generated by the lower body to be transferred to the club rather than wasted in slippage.

At impact, the body shifts back to a relatively symmetrical position for an extremely short time; then the center of gravity shifts toward the target as the mass of the arms and club move in that direction. Although the ball is long gone from the clubface milliseconds after impact, this shift in the center of gravity is again important for the golfer. At the point of impact, the body has produced enough force that the clubhead is traveling nearly 44 m/sec (up to 55 m/sec for professionals), and this energy has to be controlled and dissipated if the golfer intends to remain on his feet. The foundation for harnessing this energy is now the forward foot, which bears more than 90% of the body's mass by the end of the swing. Imagine what controlling this energy would be like without those spikes on the bottoms of golf shoes.

Golf can be considered an *open discrete* or *closed discrete* skill, depending upon the situation. By the rules, a golf ball cannot be moving when struck, so what could be changing that could classify it as an open skill? The answer is the constantly varying environment. The best example of a closed skill in golf is a tee shot with a driver in which the intent is to hit the ball as far as possible along an intended path. On any respectable course, the teeing areas are level, and the grass is mown to a consistent height. The player places a tee into the ground to support the ball at a certain height, the driver is the same length each time, and the swing should be identical each time. You may be asking, "What if there is a cross wind? Don't I need to vary the shot for that context?" The answer is yes, but the skill itself does not change. The golfer may align her body differently to account for the wind, but the swing should be exactly the same.

Putting is another skill that could be considered closed but on further examination is not. Contrary to teeing areas, putting greens are rarely level and flat. To increase the challenge of the game, course designers plan each green on a course to be different in shape, size, and contour. At times, you may be putting uphill or downhill, in which case a 2-meter putt would require a significantly different force to reach each hole. At times a golfer also needs to consider that a ball may break, or curve, to the right or left as it is rolling. At the highest levels of golf, players even take into account which way the blades of grass are growing, which is known as the *grain*. Taken together, these variables change the act of putting into an open skill. In addition, any shot made between the teeing area and the green is an open skill. Consider the variables that a golfer could face: angle of the ground where the ball lies, depth of grass, hitting from nongrass areas such as pine needles or sand, length of each shot, wind, and obstacles such as trees. If this variability did not exist, few people would join golf courses and choose to play the same course every day.

12.3 CHAPTER 3: PARADIGMS FOR STUDYING MOTION OF THE SYSTEM

Two general categories of golfers help describe the concepts in this section: "feel" golfers and "analytical" golfers. Feel golfers play shots based on what their bodies feel like during the swing and are usually better able to adapt to unusual situations by slightly altering the plane of the swing. They generally have a good understanding of the physics involved in the game, and they develop the ability to make the ball fly at will left to right, right to left, high, and low. Analytical golfers generally are more conscious about exactly what their bodies are doing during the swing, and they practice to produce exactly the same swing each time. Each type of golfer can be highly effective, but teaching the two types requires drastically different techniques.

The feel golfer is better served by *qualitative analysis* and by teaching methods that encourage kinesthetic concepts. Several computer-based software packages are available that work in tandem with digital video to analyze golf swings. Teachers take digital video of students swinging and import the video to a software package. The accuracy of the software allows teachers to measure angles, distances, and velocities and even compare a student's swing side by side to that of a professional with a similar body type. Although this type of analysis may help feel golfers, the exact measurements are of little concern to them. Instead of measurements, these golfers would rather have a teacher who can help them understand what they're doing wrong by how it feels.

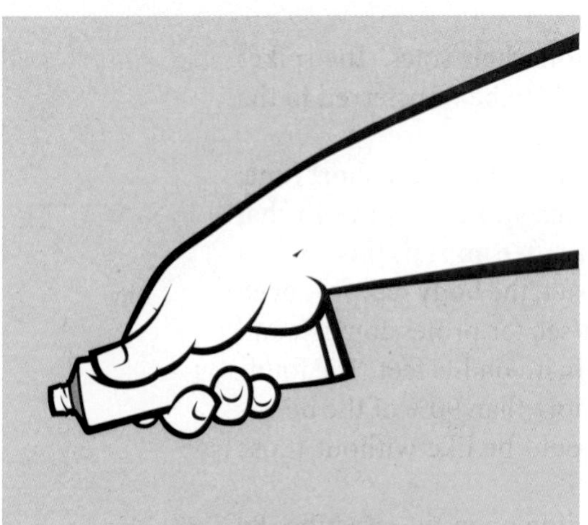

Figure 12.4 **Tube of toothpaste.**

One interesting method for working with feel students is to use metaphors (St. Pierre, 2002). Metaphors give students a mental image to which they can relate in a kinesthetic way. Rather than tell a student the exact pressure required to hold a club, a teacher may say, "Hold it like a tube of toothpaste with the cap off—don't squeeze any out." (Figure 12.4) To prevent a student from making any lateral movement during a swing, the teacher might say, "Pretend your upper body is turning inside a barrel." Instead of measuring the angle of the dominant hand at the end of the backswing, a student might be asked to "imagine holding a drink tray" to keep the palm parallel to the ground. The skill of chipping could be equated to the smooth action of a rocking chair, the proper back alignment in preparation to swing is "sitting on the edge of a bar stool," and the notion of accelerating toward the ball becomes "sweeping dirt from a crack." These metaphors give feel golfers a learning tool that is more relevant for them than quantitative measures.

Analytical golfers, on the other hand, get exactly what they need from the video and software packages. You can pick out these golfers at almost any course or practice facility: they are the ones who stop their swings at various points, then turn their heads to see exactly where the club is and how it is aligned. Some get so obsessed with the proper angles and planes that they sometimes lose the ability to swing consistently—a condition called "paralysis by analysis." At the highest levels of golf, professional players rely on the precise measurements afforded by new technology. A one-degree variation in a clubface to the right or left away from the intended target line might seem insignificant, but when you consider that a ball can travel 300 m, one degree can be the difference between playing the next shot from the fairway or from the woods.

12.4 CHAPTER 4: INTERACTION OF FORCES AND THE SYSTEM

Forces possess the capability to change the motion of a system. During the golf swing, muscles produce force to move body segments in a certain sequence that will generate high velocity at the end of a club. The impact between the club and the ball imparts force to transform the ball from a stationary object to a high-velocity projectile—in essence, a change in motion. Remember that force does not cause motion, but it has the ability to change the motion of a system. Slow-motion analysis of a golf club and ball at impact (Figure 12.5) shows that the ball actually changes *shape.* The force imparted by the golf club creates changes in motion within the molecular structure of the ball and causes certain areas of the ball to move from their original position. After leaving the clubface, these molecules return to their original positions relative to the others because of the elastic properties of the material—a stored force within the ball that creates motion and returns the ball to its original shape.

From the instant of initial contact with the club at impact, the golf ball is subject only to the laws of physics. Some golfers plead with their golf ball as it flies, but the only person a ball listens to and respects is *Sir Isaac Newton. Newton's Laws of Motion* can be used to calculate every detail about the golf ball in motion, whether in the air or rolling along a putting green. Golf balls in flight can gain altitude against gravity, turn right and left, fly high or low, and occasionally be legally propelled without ever touching a clubface. Newtonian laws of motion govern all of these situations.

Figure 12.5 **Deformation of a golf ball on impact.**

Newton's First Law, the *Law of Inertia,* states that a system at rest will remain at rest, and a system in motion will remain in motion in a straight line until acted on by an external force. The system of most interest in golf is the ball. By rule, a certified golf ball in the United States must have a mass of approximately 46 grams, and hence a weight of 0.44 N. This mass provides a certain amount of *inertia* (resistance to change in motion). A golf ball is not a massive object, so overcoming its inertia and setting it into motion with an external force is quite easy.

The external force—the clubhead—is a more massive object, traveling at close to 44 m/sec. Research in golf club design has determined that the average golfer will achieve the highest ball velocity with a clubhead mass of approximately 200 grams (weight of 2.06 N). The force of a golfer's hands against the club and the mass of the club shaft also affect the energy imparted to the ball to a small degree, but for the sake of this example we'll focus on the masses of the clubhead and the ball. The collision applies a force to the ball and, during the few instants after contact, the shape of the ball changes as it resists the force. Remember, this observed change in shape is actually motion of the molecules to different positions within the ball. Because the club is more massive, the ball can resist moving as a system only for a short time. Robert, Jones, and Rothberg (2001) report that impact duration between a ball and club can vary, depending on the type of ball and clubhead velocity, but generally it is around half a *millisecond.* After the ball leaves the clubface, Newton's first law states that it will remain in motion in a straight line unless acted on by an external force. As we know, the ball eventually returns to a resting state, so other forces must be acting on it; these forces will be described later.

Newton's Second Law is the *Fundamental Law of Dynamics*, or *Law of Acceleration.* This law states that a system's acceleration (change in state of motion) is directly proportional to the applied force. Let's assume that a golfer produces a clubhead velocity of 44 m/sec. If the weight of the clubhead is 2.06 N, the ball will accelerate at a certain rate. The true calculation of this ball velocity is quite complicated given the variables involved, including coefficients of restitution for different clubs and balls. To give you some idea of this value in real life, the figures given here would result in a ball velocity of approximately 66 m/sec. Newton's second law reveals that if the same player swung a clubhead with twice the weight (4.12 N) at the same velocity, the ball would accelerate to twice the velocity, or 132 m/sec. Achieving such velocity would seem to be a great idea for increasing maximal driving distance, but we would have to assume that the golfer could accelerate a heavier club to the same velocity as a lighter one, and this is not the case. In reality, humans can accelerate light clubheads to approximately 55 m/sec and heavier clubs to about 44 m/sec. Although the lighter club is traveling at a greater velocity, it maintains an amount of kinetic energy similar to that of the heavy club swung at a slower speed.

Newton's Third Law is the *Law of Reciprocal Actions* or *Law of Action–Reaction.* Forces exist in pairs, and one extremely common way of stating this law is "for every action there is an equal and opposite reaction." This definition can be confusing if you assume that the observable reactions are identical. For example, let's take a look at the moment of impact between a ball and clubhead, which is known as a *contact force.* The clubhead imparts a force on the ball (an *action*), causing a change in motion as the ball is accelerated to 66 m/sec. If we took Newton's third law and applied the common definition, an equal and opposite *reaction* would have the clubhead traveling backward away from the ball. We know that doesn't happen, so obviously the "equal and opposite" statement is overly simplistic. An equal and opposite pair of forces does exist, but because the mass of the clubhead is five times the mass of the ball, Newton's second law helps us understand why the clubhead maintains it original direction of travel, and 90% of its original velocity, after impact.

Newton's most well known law is the *Law of Universal Gravitation*, also known as the *Law of Gravitation* or *Law of Attraction.* In general, every particle (body) in the universe is attracted to every other particle (body). The force of that attraction increases proportionally with the masses of those bodies (i.e., the more massive the bodies, the greater the force of attraction). These attractive forces are known as *field forces.*

When struck properly, a golf ball gains altitude away from Earth because of the angle of projection and the lift created by backward rotation of the ball. Earth is a massive body and exerts an attractive force on the ball known as *gravity.* On Earth, this force accelerates the center of mass of the ball toward the center of mass of Earth at a rate of 9.81 m/sec^2. The moon is a less massive object than Earth. Therefore, the force of gravity is much lower on the moon, approximately one-sixth the attractive force, so the ball would return to the surface of the moon more slowly. During an early Apollo mission, Alan Shepherd "played golf" on the moon. Even with a one-handed swing and impaired mobility from wearing a space suit, he was able to propel a golf ball between 200 and 400 m. The ball was able to stay in flight longer and therefore was subject to the horizontal component of the force vector for a longer time. The lack of atmosphere also played a part: The ball was subject to one less external force that causes deceleration on Earth: *drag* caused by air molecules.

Because air is present in Earth's atmosphere, a golfer can create lift force by imparting backward spin to the ball. Backward spin is the result of *friction* between the clubface and the ball. Early golf clubs had smooth faces that would provide some level of friction if the ball and club were dry but little friction if the conditions were damp or grass was between the clubface and ball at impact. The first golf balls were made of wood and later of stuffed leather that had the consistency of wood. These balls provided little in the way of potential for friction. The next ball that revolutionized the game was made of a type of natural rubber, and this material offered players an advantage by providing backspin to the ball because of its higher *coefficient of friction*. Today's golf club design adds to a player's ability to impart backspin through the use of horizontal grooves on the clubface. As the ball compresses into the clubface, small areas of the ball surface are pressed into the grooves and create an extremely high friction force that makes the balls roll up the clubface rather than sliding. This rolling effect generates high spin rates.

One factor that limits ball flight distance is friction between the ball and the air that it passes through (Figure 12.6). Energy of the ball is used to move air particles aside on the front of the ball, which creates a turbulent boundary layer around the ball. Friction occurs between this boundary layer and layers of air farther from the ball. This principle is the same as riding a bicycle against a headwind; your body has to waste energy pushing air molecules aside, and the energy to move air becomes exponentially higher with higher velocities. The friction that slows the golf ball down during flight can also lead to longer flight time and hence distance because of the lift created by backward rotation. This concept is explained in greater detail in a later section.

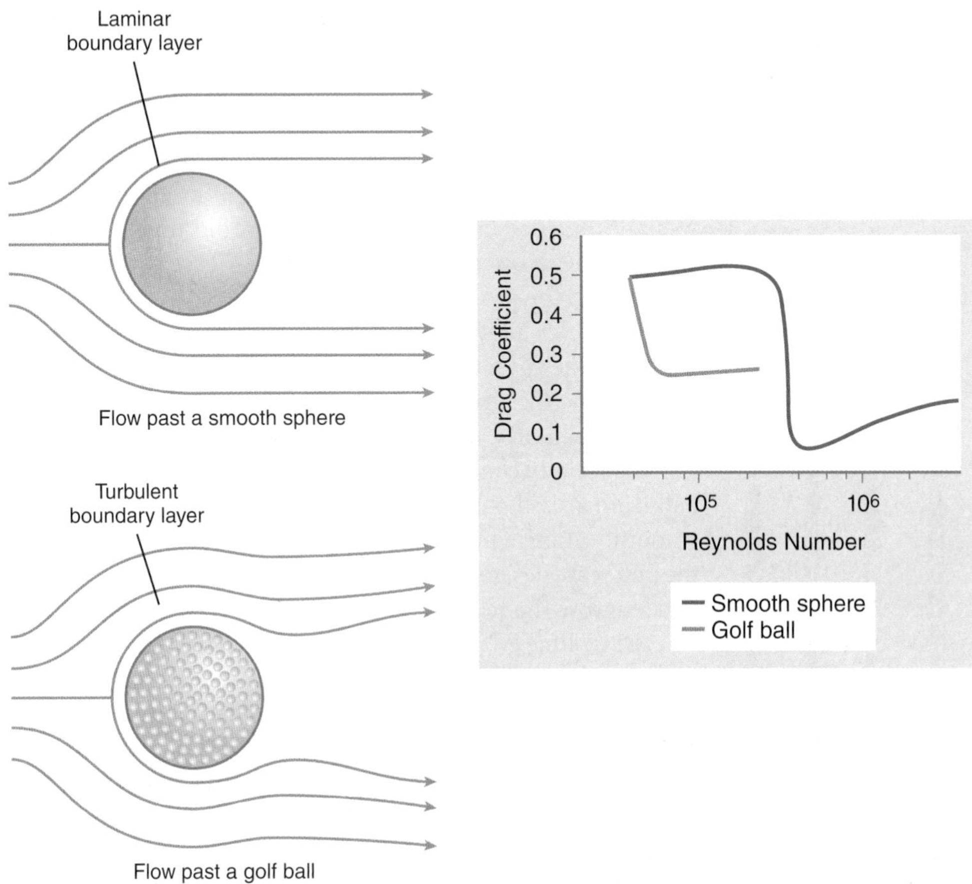

Figure 12.6 **Laminar flow of air around a golf ball.**

Once the ball comes to rest on a green, gravitational force is still acting on it. This force creates *pressure* between the ball and the grass on the putting green and results in *static friction.* When a player makes a putting stroke to propel the ball toward the hole, the ball becomes subject to *rolling friction,* which is a specialized type of static friction. The ball must push aside individual blades of grass that create friction along its relatively hard surface. The two materials don't stick together and pull apart like rubber tires against a road surface, but the friction is enough to slow the ball as it rolls. Putting greens can be prepared by maintenance personnel to be fast, slow, or some speed in between. Although other variables can influence green speeds, the length of the blades of grass is the primary factor. Special mowers can vary the height of particular grasses; depending on the type of grass and required speed, grass can be as short as 3 mm. Shorter grass equals less friction on the surface of the ball, so the ball loses energy at a slower rate (rolls farther for a given impulse force). At the Masters Tournament, the greens are prepared to be extremely fast and were designed with severe changes in elevation. Golf balls have been known to be struck uphill toward a hole, miss the hole, and begin accelerating back toward and past the golfer because of gravity and the lack of rolling friction. The greens at most courses that cater to average golfers leave the grass longer on putting greens to increase friction and decrease frustration.

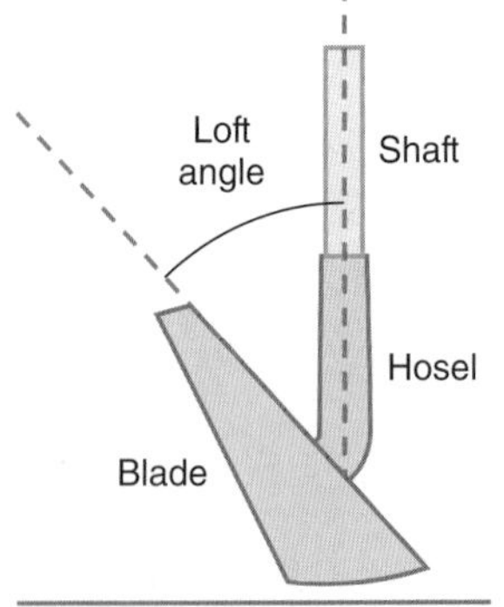

Figure 12.7a **Loft Angle.**

12.5 CHAPTER 5: LINEAR MOTION OF THE SYSTEM

Golf is a sport of *acceleration.* A player creates rotary motion around the axis of the spine to accelerate the clubhead from a standstill at the beginning of the through swing to 44 m/sec at impact with the ball. The club travels along an angular path but propels the ball into a linear state of motion. The angle of the clubface relative to vertical, also known as the *loft angle* of a club (Figure 12.7), also causes the ball to rotate backward as it travels up the clubface and is influenced by horizontal grooves. A more lofted club will result in a higher rate of backward rotation. As the ball flies, it travels along a curvilinear path and has a rotary component that influences the direction of the curve. In the absence of other environmental influences and forces, the ball would travel infinitely along a straight line. However, on Earth, various forces eventually bring the ball back to the ground.

Figure 12.7b **Clubs with varying degrees of loft.**

Golf rules allow a player to carry as many as 14 clubs during a round. Some specialty clubs are used only in certain situations. One such club is the driver, which is only used when a player is free to place the ball on a raised tee for the first shot on most longer holes. Other specialty clubs are the sand wedge, which is used to extricate a ball from a sand bunker; and the putter, which is used to roll the ball smoothly along a putting green. The clubs in between the driver and the putter are designed to present a continuum of progressive loft angles and lengths. The reason for this variety of clubs is to make golf easier.

An 18-hole golf course usually doesn't have two holes that are the same in terms of length, shape, or obstructions (e.g., water, sand, or trees). Even if a player hits the ball exactly the same distance with a driver on every hole, she is often left with different distances, or *linear displacement*, to reach the putting green. The theory behind carrying a variety of clubs with different lofts is that a golfer can use the same swing to propel the ball different distances by changing the loft angle of the club used. Early players of the game found that it was easier to use the same swing and change the nature of the club rather than use the same club and change the nature of the swing for each shot. At this point, a comparison of *linear displacement* and *linear distance* can help. Linear displacement is the length of a straight line between the point at which the ball begins its flight and that at which it stops. Linear distance is the actual distance the ball travels during its flight along a curvilinear path (Figure 12.8).

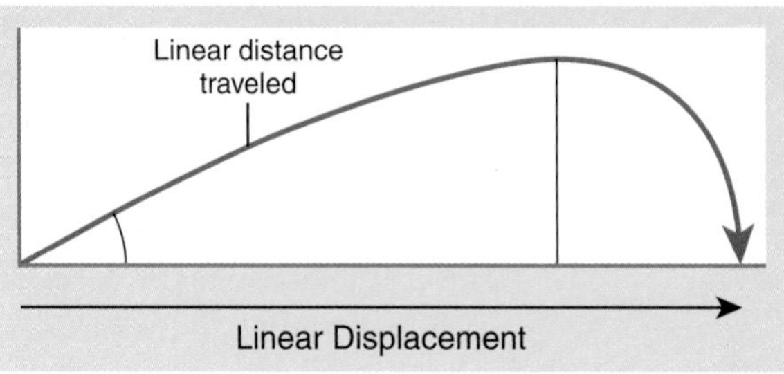

Figure 12.8 **Curvilinear path of golf ball flight.**

Most golfers in the United States attempt to land the ball near the hole on a putting green, whereas in golf's land of origin (Scotland), players strike the ball with a low trajectory so that it rolls a significant distance on its way to the hole. Longer shots require the use of clubs with less loft (a more upright clubface angle) that send the ball along a lower trajectory. The drawback of hitting a long shot onto a putting green is that it still maintains a fair amount of energy in the *x* direction and tends to roll a long distance after hitting the ground because of its *momentum.* As a player gets closer to a green, he makes the same swing and imparts virtually the same amount of energy to the ball; in choosing a club with a higher degree of loft, however, he attempts to accelerate the ball in a more vertical orientation. It's not uncommon for a ball to attain an altitude of 25 m or more on shots that are 70 m from the hole. The negative acceleration forces of both *fluid drag* and gravity dissipate energy as the ball travels along a modified parabolic arc, and the ball returns to the ground with less velocity in the horizontal direction. The ball stops close to where it hits the ground. The ball that lands with less linear displacement may actually travel a longer distance than a ball hit toward a green from farther away (Figure 12.9).

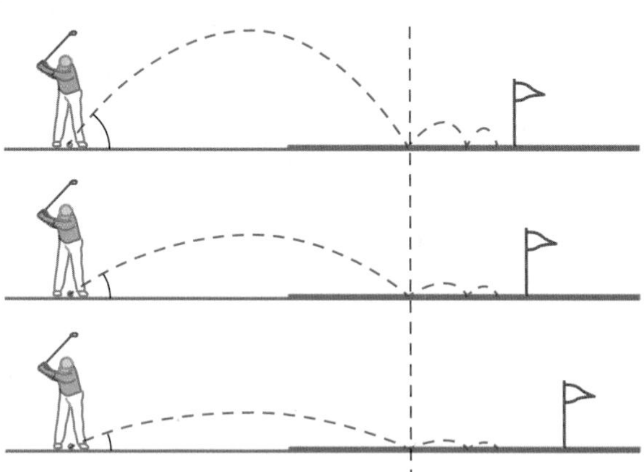

Figure 12.9 **Parabolic path of a golf ball depending on club loft.**

One of the more memorable shots in golf history illustrates the difference between displacement and distance along a more horizontal plane. During the final round of the 2005 Masters Tournament, Tiger Woods hit his tee shot over the 16th green and left himself a difficult shot about 12 m in length—if it were measured in terms of absolute linear displacement. What made the shot difficult was that the surface of the green was tilted significantly from left to right; if he chose to strike the ball directly toward the hole, gravity was going to influence the ball as it rolled and take it far to the right of the hole. After a few moments of thought, Tiger hit the ball to a point 8–9 m left of the hole, then watched as gravity returned the ball down the hill and into the hole. In terms of linear displacement, the ball traveled 12 m, but in terms of linear distance, it traveled significantly farther (Figure 12.10).

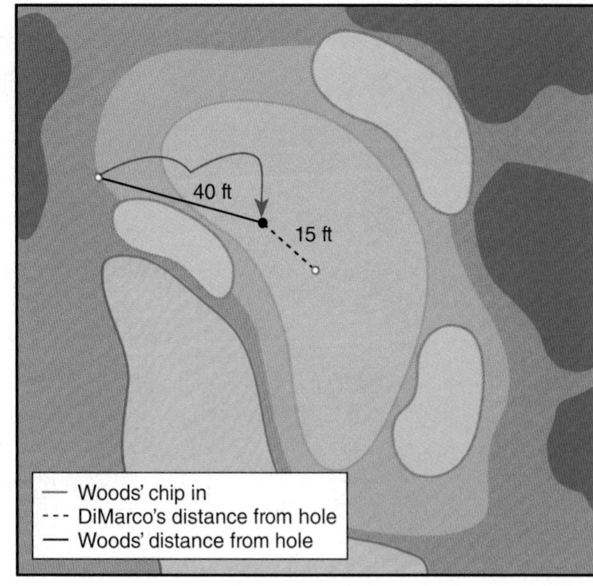

Figure 12.10 **Tiger Woods's iconic chip in at the 2005 Masters Tournament.**

Human muscular force is transferred through a club to accelerate a golf ball from 0 to 66 m/sec in half a millisecond, an especially short period of *positive acceleration.* From the instant the ball leaves the clubface, it is subject to forces that result in *negative acceleration*, which is more commonly known as *deceleration.* Remember that acceleration is a change in a system's state of motion, which can be positive or negative; deceleration is a change in the state of motion in the direction opposite the line of travel. The ball is still traveling toward the target, but the acceleration is negative. By rules, a ball cannot be in motion when struck, so every shot in the game of golf requires accelerating it from a state of *equilibrium*. Once the ball leaves the club it begins decelerating, so a golf ball is always in a state of acceleration while in motion. If the ball is propelled into the air, it decelerates in the positive x direction and the positive y direction until it reaches its maximum altitude; then it accelerates in the negative y direction. These accelerations (or decelerations) caused by air resistance and gravity result in the curvilinear path that a golf ball travels.

Elastic strain energy is a major factor in the high launch velocity of a golf ball off a clubface. The ball deforms dramatically at the moment of impact as it resists the force applied by the club, and *potential energy* is stored in the materials of the ball. This potential energy is released as the ball reverts back to its original shape (Figure 12.11).

Like a ball storing and releasing energy, the human body uses the same principles. During the backswing, a golfer's body rotates with motion through several joints. Across each joint are muscles, tendons, and ligaments that have the ability to store energy. A metaphor used to help students learn the golf swing is to "coil like a spring" and then "uncoil" during the through swing. This metaphor is valid because, just like a real spring, the tissues of the body can store *elastic strain energy* as they are stretched (deform), a form of *potential energy* that can be used to accelerate the club to a higher velocity as the tissues return to their original shape and position.

As mentioned previously in this chapter, a special situation exists in golf in which the ball is legally propelled toward the hole without ever being touched by a club. This situation occurs when a ball comes to rest in a sand bunker, or on rare occasions, in shallow water. Although the club does not directly contact the ball in either case, it does provide the energy to move the ball indirectly. The objective of a sand shot is to swing the clubhead under the ball with a layer of sand between the two objects (Figure 12.12a). The club moves the sand, and the sand exerts force on the ball. The depth of this sand layer varies, depending on the nature of the sand, which can be soft or packed, coarse or fine, dry or wet. This layer of sand results in an inelastic collision, so the golfer must swing fairly hard to move the ball a short distance. This technique is known as an *explosion shot*, and often some of the sand flies farther than the ball (Figure 12.12b).

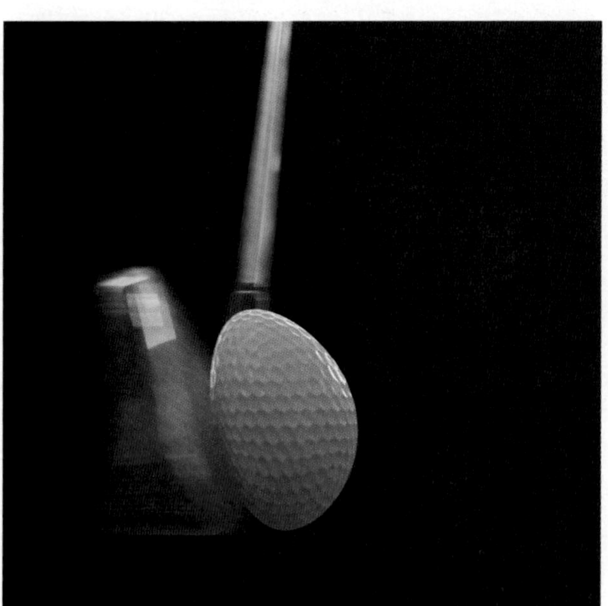

Figure 12.11 **A golf ball at impact with an iron.**
© Boogaloo/Shutterstock

Energy cannot be created or destroyed; it can only be transferred from one form to another. Changes in the sum of the *kinetic*, *potential*, and thermal forms of energy produced by applying force externally equals the *mechanical work* performed. As an example of the *work–energy relationship*, let's analyze the energy change during a golf drive. Before being struck by the club, the golf ball has a tiny amount of potential energy because of its position on the tee. The ball has zero kinetic energy because it has zero velocity. However, the club does have kinetic energy as it approaches the ball. The kinetic energy of the club gives the clubhead the ability to perform work on the ball. At contact, the energy of the club is used to deform the ball and propel it forward and upward. The work performed on the ball (the energy change) can be observed in the changes in both its potential and kinetic energy. The potential energy of the ball increases as it gains height, and its kinetic energy decreases because of its change in velocity. So again, the total work performed on the ball is equal to the change in the sum of its kinetic and potential energy.

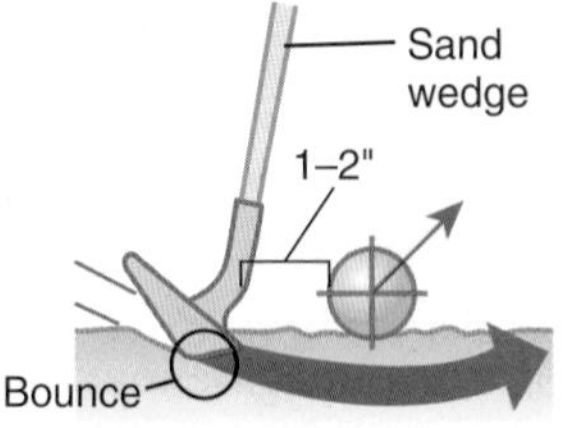

Figure 12.12a **Sand Shot.**

Figure 12.12b **Sand shot.**

12.6 CHAPTER 6: ANGULAR MOTION OF THE SYSTEM

The collision at impact sends a golf ball along a linear path, but the forces involved in getting the clubhead to the ball are almost purely rotary. The general body motion is rotation around the axis of the spine, and most joints that participate in this process are subject to *angular motion* within each joint. Muscle forces act across joints to create rotation around the joint axis. For example, during the through swing, the triceps contract to extend the elbow, and the forearm rotates about the elbow joint. The ball is also subject to rotation as it travels along its curvilinear path, and in this section we examine not only backward rotation of the ball but also sidespin that results from off-center contact with the clubface.

The design of a golf club produces backward rotation of the ball. Between the loft angle and the horizontal grooves of the clubface, a golfer has the potential to generate enough backspin to create lift while the ball is in flight and also to back the ball up on the putting green on approach shots given the right conditions. Golf clubs are designed to contact the ball off-center (*eccentrically*) in a transverse plane; that is, they should hit the ball below the equator. Remember from the previous section that if the force applied to an object is *concentric*, it will result in linear motion. In a perfect world, the golf club would contact the center of the ball relative to a sagittal plane to avoid imparting sidespin, but few players have the ability to strike the ball consistently with the clubface perfectly square to the intended target line (*concentric force* in the sagittal plane). When the club strikes the ball *off-center* to the right or left, it imparts both backspin and a degree of sidespin; the ball actually spins on a diagonal plane.

Off-center force produces *torque* that results in rotation of a system around its axis. It's possible to contact a ball so that the line of action passes different distances from the center of mass. This distance is called the *moment arm*, and for a given force, the longer the moment arm, the more rotation is generated. This relationship means that the more off-center a ball is struck in a sagittal plane, the farther it will deviate from the intended line of travel as it flies. We'll examine this phenomenon in more detail later. So far we've mentioned only off-center collisions in which the object is struck, but the club is also subject to the same physics as the ball. Because of the collision force between the clubface and ball, it is also possible that the clubhead will be subject to rotation if the ball contacts the clubface away from its center of mass.

Let's assume for the following discussion that each shot strikes the ball centrically—that is, through the ball's center of mass. The first shot is performed with good form, and the center of mass of the club exerts force directly through the center of mass of the ball. This shot will be subject to normal backward rotation, have no sidespin, and will not create any torque within the club itself. Now imagine that the ball contacts the clubface 15 mm toward the toe of the clubhead, the portion farther from the golfer's body. The impact with the ball will create a moment arm 15 mm long in the clubhead, resulting in torque that will turn the clubface open while the ball is still in contact. The muscles of the hands and arms resist this motion to a certain degree, but the force of the collision is of a magnitude great enough to twist the shaft of the club. The degree to which the clubface rotates depends on how far off-center the ball is struck, the stiffness of the club shaft, and the friction force between the hands and the grip. Quite possibly it could twist more than 10 degrees and the ball will be propelled in a different direction than intended. The opposite effect occurs when the ball contacts the clubface toward the heel, the portion closer to the body. In this case, the clubhead twists closed, and the ball travels left of the intended target line of a right-handed golfer.

When you consider the fact that few golfers have the ability to match the club's center of mass with the ball's center of mass consistently, the combination of eccentric forces on both club and ball can cause an amazing variation of ball flight paths. The two worst flight paths are commonly known as the *push slice* and the *pull hook*, and both result from the additive effects of eccentric forces applied to the ball and to the clubface simultaneously at impact. The slice occurs when the ball contacts the toe of the club, and the club strikes the ball eccentrically on the side closest to the golfer. The golf ball then starts traveling away from the intended target line and, because of fluid forces and rotation, deviates even farther as it curves away with sideways lift (like a curveball in baseball). The hook is exactly the opposite of the slice. The ball contacts the clubhead closer to the heel and is struck on the side away from the golfer. The clubface closes, the ball starts traveling along a different path than intended and then turns away even farther because of sidespin. It is entirely possible to contact the face of the club toward the toe and the ball on the side away from the body. This impact would result in a ball flight that starts by traveling along a line away from the intended target because of the open clubface angle but then during flight curves back toward the target (Figure 12.13).

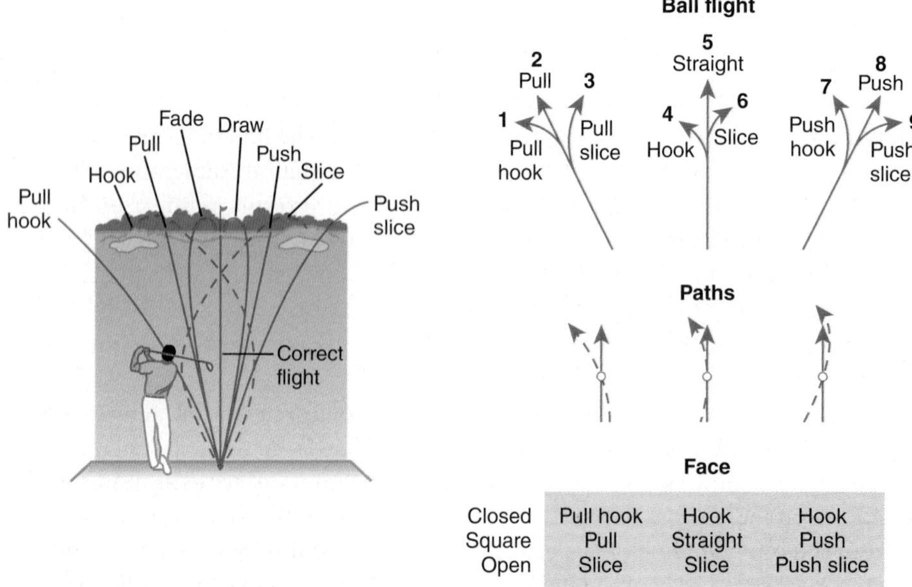

Figure 12.13 **Golf ball path (straight, hook, and slice).**

The good news for golfers is that club designers understand physics. The more loft a club has, the more potential backspin it can create. As the rate of backspin increases, the effects of sidespin are lessened. If you're a golfer, this interaction might explain the mystery of why your pitching wedge shots don't seem to curve (the high rate of backspin counters any sidespin), and your drives veer wildly off the intended target line. The clubs that impart less backspin are called *woods*, even though they aren't made of wood anymore. Club designers learned years ago that they could counteract some of the sidespin created by eccentric contact by shaping the clubface a certain way, a feature known as *bulge*. If you look at the clubhead of a driver from the top, you'll notice that the center of the face bulges out in relation to the toe and heel sections. When a ball contacts the clubface of a wood toward the toe, the radius of the bulge generates "draw" sidespin that brings the ball back toward the intended target (Figure 12.14).

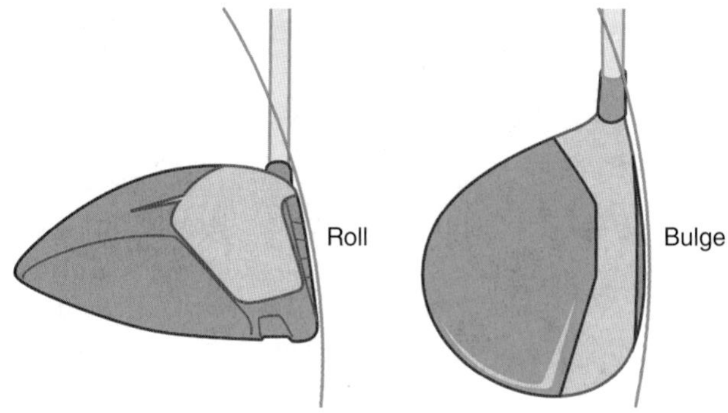

Figure 12.14 **Bulge and roll of driver face.**

Remember *inertia* from Chapter 4? Club designers have found ways to use it to a golfer's advantage. Inertia is resistance to change in motion, and golf clubs designed in the past 30 years have taken advantage of newer construction techniques to change the inertia within a clubface. The new clubs are called "game improvement clubs," and they are designed to help beginning and midlevel golfers who don't strike the ball consistently through the center of mass of the clubhead. For centuries, the irons in a golf set were made of a forged metal, which means they took their shape under the pressure of hammers or forges. The technique of casting, or pouring metal into a mold to create a desired shape, has revolutionized the game for less-skilled players. Casting allowed club makers to move some of the weight of the clubhead from the center to the perimeter of the clubface. With the weight placed toward the perimeter, twisting the club on eccentric hits becomes harder because the *moment of inertia* has been shifted away from the center of mass. The toe and heel contact that caused significant twisting in forged clubs is dramatically reduced by the perimeter weighting of cast clubs. Because the clubs twist less, the shots stay closer to their intended target line. New drivers use the same theory by allowing players to adjust the perimeter weighting to fit their swings. By changing the position of removable weights, the center of gravity and moment of inertia vary and can be custom tailored to each golfer.

Golfers and figure skaters share an interesting physical concept as they perform. When a skater wants to increase his angular velocity and spin faster, he moves as much mass toward the axis of rotation as possible. He pulls his arms tightly against his torso and locks his legs together as closely as possible. When he wishes to slow the rate of spin, he extends body parts away from the rotational axis. Golfers attempt to maximize *angular rotation* and *acceleration* in the same way. During the backswing, a golfer flexes the dominant arm at the elbow and hinges the wrists radially (toward the thumbs). This posture moves the mass of the arms and club closer to the axis of rotation in preparation for the through swing. Muscle forces produced by the body can accelerate the club more quickly in the early part of the through swing, resulting in higher *angular velocity*. Just as for skaters, golfers' performance can suffer dramatically if body parts are extended too early, slowing rotation. A skater could underrotate during a jump and land with the blade oriented in the wrong direction. This mishap is the cause of most falls in figure skating. A golfer can extend the arms and wrists early, slowing her angular velocity and creating problems in maintaining the club along the desired plane. This mistake is known as *casting*, and it requires golfers to compensate during the swing to get the club back on plane to strike the ball. These golfers not only lose angular and linear velocity but also causes them to pull the club through the ball and create slice sidespin because of the necessary compensation. The end result is lower initial ball velocity and more lost distance when energy is used as the ball curves rather than traveling along a straight path.

12.7 CHAPTER 7: SYSTEM BALANCE AND STABILITY

Stability and *balance* are essential to perform well in the game of golf, although there are few times during a round of golf when the body or golf ball is in a state of *equilibrium*. Equilibrium occurs when the resultant force or torque acting on an object is equal to zero. This doesn't mean that the object is not in motion; it could be moving or rotating at a constant velocity as a result of zero net force being applied. In a theoretical sense, the latter scenario can occur, but because of Earth's atmosphere, a moving golf ball is always subject to external forces and will never be in a state of *dynamic equilibrium* while in motion. This rule holds true for both *linear* and *rotational equilibrium*. The only time a golf ball assumes a state of equilibrium is when it is at rest, a state known as *static equilibrium*.

Stability is the resistance of an object to having its equilibrium disturbed. *Stable equilibrium* is necessary for a golfer to counteract the forces generated during the swing and maintain the axis of rotation in a consistent position relative to the ball. When a player is ready to strike a ball, the position is known as *address*. At address, a player wants to be as stable as possible on two planes: frontal and sagittal. Remember that a proper golf swing is essentially rotation around the spine, and pure rotation allows the clubhead to return to the exact same location it occupied at address. A common mistake that many golfers make is to let their bodies shift laterally away from the ball during the backswing and then shift toward the ball during the through swing. The odds of returning to the exact position of address in this situation are slim. To attain maximal stability on the frontal plane and be able to resist disruption of equilibrium, a player will assume a wider stance, thereby creating a larger *base of support*. This strategy is used commonly in sport when stability is required. When forces must be resisted along a sagittal plane, the stance is widened anteroposteriorly such as in wrestling or a rugby scrum. Because much of the force production in a golf swing occurs along a frontal plane, the stance is widened laterally.

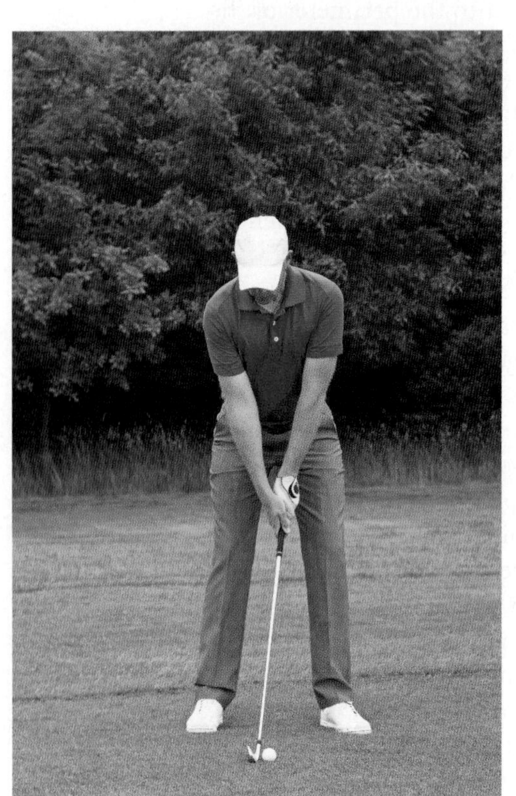

Figure 12.15 **Golfer balanced at address.**

Attaining maximum stability in a sagittal plane is a little more difficult, but a combination of two factors increases the force necessary to disturb equilibrium in this plane. The first is to balance the weight of the body evenly on the soles of the feet. The second is to position segments of the body in a way that maintains the center of mass through the base of support. At address, a golfer should have his knees flexed and his torso flexed at the waist (Figure 12.15). This waist flexion puts the spine into the proper angle for rotation, and the knee flexion allows the joints in the legs to turn more freely. The overall effect lowers and centralizes the center of mass over the base of support. If body weight is not centered anteroposteriorly, the body may rock back onto the heels or fall forward onto the toes. Neither situation is conducive to striking the ball consistently.

Balance is the ability to control the current state of equilibrium, and it implies conscious effort and coordination of the neuromuscular system. To strike the ball consistently, a golfer must maintain stability by controlling muscle actions accurately (Figure 12.16). Coordinated muscle action both produces and counteracts the forces involved in a golf swing, and knowing which muscles to contract and relax in the proper sequence requires a process of learning. The golf swing is a complicated task that involves every major joint in the body, and the forces required

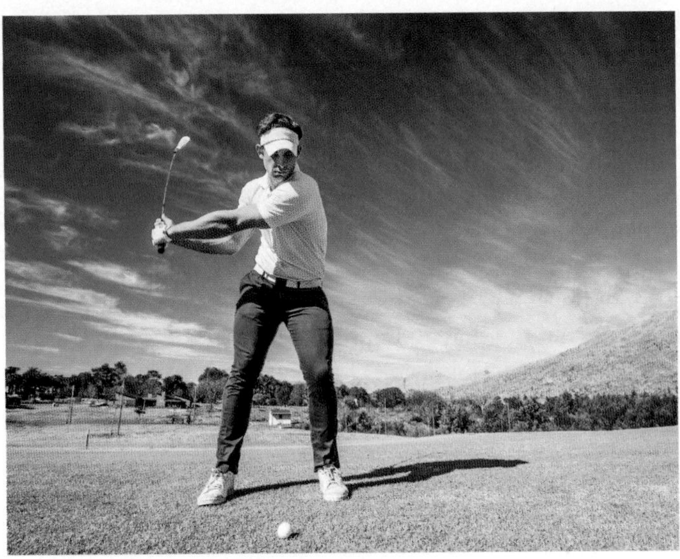

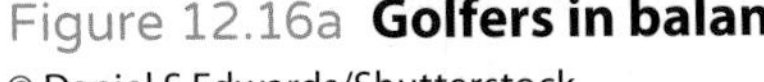

Figure 12.16a **Golfers in balance.**

© Daniel S Edwards/Shutterstock

Figure 12.16b **Balanced follow-through.**

© Joaquin F. Ruiz/Shutterstock

can disrupt equilibrium in two planes and in both directions along these planes. Losing balance in the frontal plane usually occurs when golfers attempt to swing with great amounts of force and fall away from the target (onto the foot away from the target). The center of mass shifts rearward, and the arms can no longer extend the clubhead to its position at address. The result is that the contact point on the clubface is moved to a lower and more toe-oriented position, and it strikes the upper portion of the ball. If the ball gets airborne, it takes off with a lower trajectory and tends to be projected off the intended target line as the clubhead twists open. Many shots hit with the body weight shifted to the rear foot are propelled downward into the ground. The same results occur when a golfer loses balance away from the ball along the sagittal plane; the arms cannot extend the clubface far enough to contact the ball in its center. With quality teaching and practice, golfers can learn the proper muscular sequences necessary to maintain balance and stability (Figure 12.16).

An earlier section in this chapter described how significantly different forces might be necessary to strike a 2-m putt. If the putt is uphill, more force is required to reach the hole than for a downhill putt. On a flat green, the ball is as close to neutral equilibrium as possible. Although it is still losing velocity because of rolling friction, the ball's center of mass remains directly over the point of contact. Therefore, gravity acts through the center of mass, and no torque is produced. When the surface of the green is tilted in any direction, the line of gravity falls outside the base of support of the ball and creates a disruptive torque (Figure 12.17). The ball falls downhill relative to the direction of the tilt, which on a putting green may be straight downhill, to the left, or to the right. On a purely uphill putt, the surface tilts toward the golfer, and the disruptive torque decreases linear velocity more quickly. Every golf course green is different in size, shape, and contour. Putts that are purely uphill and downhill are rare, so golfers not only have to judge for more or less velocity but also consider how far the ball may curve right or left at the same time.

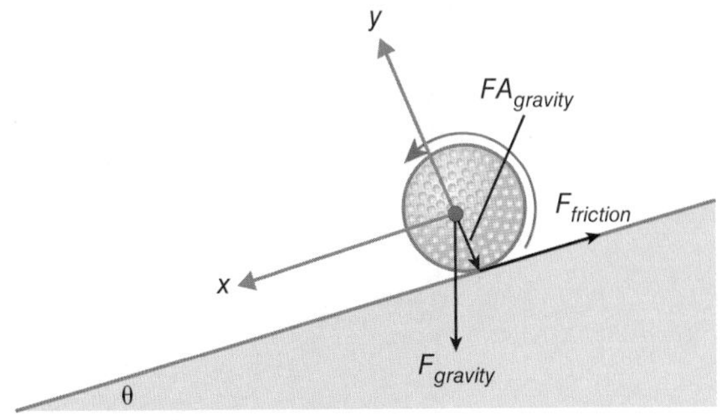

Figure 12.17 **Gravity acting eccentrically creates torque.**

Putting could be considered an art form given all the variables involved to propel a 43-mm ball into a 106-mm hole. Players must account for the firmness of the surface, the type and length of the grass, the contour between the ball and the hole, moisture, and possibly even the time of day. Grass blades grow toward the sun, which changes position from morning to evening, and the direction of the grass blades affects the amount of rolling friction. Two general categories of putting style exist, and two of the greatest names in golf serve as examples. Jack Nicklaus was a lag putter: he preferred to propel the ball with only enough force to get it to the hole. The negative consequence of this style of putting was that he had to choose a perfect line that accounted for the speed and contour that would influence the roll of the ball. Arnold Palmer was an aggressive putter, choosing to strike the ball with more force than Nicklaus did. The intent was to increase the *stability* of the ball through increased velocity so that the ball would not turn right or left to the same degree if the contour tilted. This approach is called "taking the break out" of a putt. The increased momentum of the ball makes it harder to disrupt its course side to side. For this reason, professional golfers would much rather have an uphill putt; they can strike the ball with more force, which keeps it on line better because inertia is greater. One negative consequence of aggressive putting is that if the ball misses the hole, it travels well past and leaves a long putt coming back.

12.8 CHAPTER 8: THE SYSTEM AS A MACHINE

The "Connections" section in Chapter 4 describes how golf has been changed by new technology and design. Lighter weight materials used in clubheads and shafts have allowed manufacturers to increase the size of clubheads with no change in a player's ability to accelerate them. Larger clubheads result in larger sweet spots, which make the clubs especially forgiving on off-center hits (because less torque is twisting the club). This advance is great news for amateur golfers who can't practice for hours each day to perfect a swing. New materials such as titanium faces on drivers also have properties that increase the coefficient of restitution, providing a trampoline effect on contact with the ball. This chapter is about machines, and now we'll look at the golf club as an extension of the human lever system.

A proper golf swing is a complex series of movements that deliver a clubface to a stationary golf ball at velocities approaching 180 km/h and within a few degrees of perpendicular to the target line. The club is a true extension of the lever created by the arms at address and at impact, with the spine as the axis of rotation. However, at times during the backswing and through swing, it becomes a secondary lever system. When the club reaches a position parallel to the ground on the backswing, the wrists hinge radially, creating an "L" shape between the club and the forearms (Figure 12.18).

Figure 12.18 **Through swing: two levers into a single lever at impact.**

© arsenik/E+/Getty Images

The spine continues to be an axis of rotation for the body, while that axis of rotation for the club changes to a point near the end of the club closest to the body. This "L" is maintained through the rest of the backswing, which keeps the club closer to the body and therefore closer to the axis of rotation. Chapter 6 explains why this is a good position to accelerate on the through swing.

The "L" between the forearms and club is maintained even as the golfer's hands are passing the thigh of his back leg during the through swing. This more compact position allows the torso to rotate to a higher velocity before letting the club release farther from the axis of rotation and meet with the ball. When all of the body segments are sequenced in the correct order, a kinetic chain is created. Power from the legs is transmitted to the torso, which adds more power through trunk rotation. Then that sum is transmitted through the shoulders and arms, which add their own muscle forces. At impact, the club and arms are aligned to create a true one-lever extension system from the torso (Figure 12.18).

12.9 CHAPTER 9: SYSTEM MOTION IN A FLUID MEDIUM

A golf ball in flight is subject to *dynamic fluid forces*, and these forces have the capacity to both decrease and increase the distance of a golf shot. *Dynamic fluid force* is the equal and oppositely directed force of the fluid particles in reaction to the applied force of the system moving through the fluid. Similar to other forces, the dynamic fluid force vector can be resolved into two components: a parallel component called *drag* and a perpendicular component called *lift*.

The *relative motion* of the two systems of interest (the ball and the atmosphere) is subject to variation in several ways. First, golf clubs are designed to propel the ball at different trajectories. A ball leaving the face of a lower-lofted club will have a larger horizontal vector parallel to drag that is more directly influenced by the fluid forces acting against it. Higher-lofted clubs propel the ball with decreased velocity parallel to drag and increased velocity vertically against gravity. Second, as the ball loses velocity during flight, the fluid forces acting on it decrease exponentially. Finally, as a player navigates a course, the fluid forces may change in magnitude because of wind. A golf shot directed into a headwind will be affected differently than a shot hit downwind, and side winds create additional problems for golfers.

Although it's not something most golfers think about, the *density* of the atmosphere also affects ball flight. A weekend golfer probably won't notice a change in distance between a warm and a cold day, or between a dry and a humid day. However, professionals take these conditions into account because they have the ability to consistently hit the ball with little variation in distance. A weekend golfer may be happy to hit anywhere on a putting green, and a high-level amateur is content with hitting a smaller section of the green near where the hole is located. To make money as a professional, a player must place the ball near the hole each time, so a change of 2–3 m in distance caused by humidity or temperature must be considered. A density change that all golfers must consider is the effect of altitude. Air molecules are closer together at sea level because of the pressure of the columns of air above them, but pressure decreases as altitude increases, and fewer air molecules are present per unit volume. Because air molecules create the fluid force against a golf ball, the ball encounters less resistance in the thinner air at higher altitudes and flies farther. This is a great bonus when driving for maximum distance but creates problems for golfers who are used to playing at lower altitudes. As an example, an average male golfer hits a seven iron approximately 145 m at sea level, but he may hit the same club 160 m at 2,000 m of elevation. When golf professionals play tournaments at high altitudes, they arrive early and spend a great deal of time at the practice range to recalculate the distances that each club sends the ball.

At altitude, the drag force against a ball is lower because air is less dense, but the same change in density also affects lift. Aerodynamic lift is created by pressure differentials as an object moves through a fluid. Two phenomena are responsible for this lift effect of a golf ball: the *Magnus effect* and *Bernoulli's principle*. A properly struck golf ball has some degree of backward rotation. In this situation, the bottom of the ball is traveling against the direction of airflow, which slows the flow velocity and creates an area

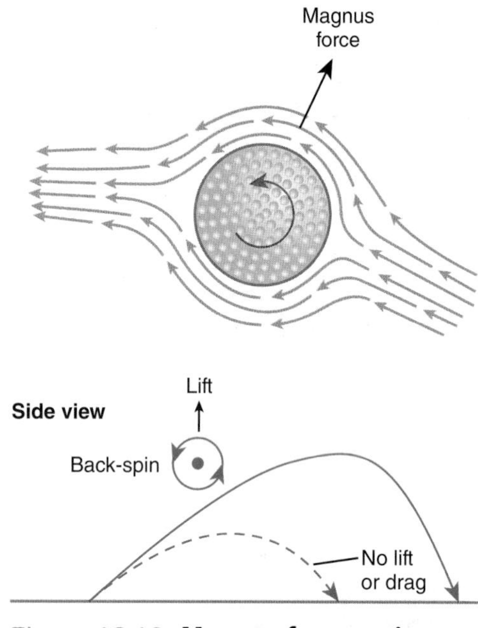

Figure 12.19 **Magnus force acting on golf ball.**

of higher pressure on that side of the ball. Conversely, the top of the ball is traveling in the direction of airflow, so the air molecules are moving faster and create an area of lower pressure on that side of the ball. This imbalance results in a Magnus force perpendicular to the direction of travel and parallel to the direction of rotation. Greater rotational velocity results in increased lift, and enough lift force can be created to counteract the force of gravity and gain altitude beyond that expected from the launch angle of the clubface. This lift keeps the ball in flight for a longer time, allowing it to travel farther before returning to the ground. The Magnus force resulting from a properly struck ball maximizes distance, but most golfers see the negative effects of Magnus more often (Figure 12.19). As you'll see, "lift" does not necessarily mean upward against gravity.

There's an amusing adage in golf: "If it goes right it's a slice, if it goes left it's a hook, and if it goes straight it's a miracle." Few players, including professionals, are able to hit the ball perfectly straight, with no sidespin, on a consistent basis. Anyone who has visited a golf practice facility has watched shots that curve dramatically to the left or right. By now you probably understand the phenomenon of a golf ball curving right or left (baseballs, viewed from a catcher's perspective, can be seen doing the same thing). The obvious explanation is that sidespin was imparted on the ball, and Bernoulli and Magnus took over. Take another look at Figure 12.19 and imagine that you are looking down from the top rather than from the side. Airflow velocity is higher on one *side* than the other, and the ball is deflected in the direction of higher relative airflow velocity. Even though it's directed left or right, the perpendicular force is called *lift*. Can you guess why a shot hit with sidespin deviates significantly more left or right than a ball with the same vertical rotation rises? Vertically directed lift is working not only against fluid forces but also against gravity. Although a ball hit with significant sidespin is also affected by gravity, the right- or left-directed lift forces have to work only against fluid forces.

Most amateur golfers impart slice spin because of poor swing mechanics but don't understand why their shots slice dramatically off course with the driver, less so with middle irons, and possibly not at all with short irons and wedges. The answer lies in the proportion of sidespin to backspin generated by different clubs. The less-lofted clubs are designed to send a ball a long way and impart much less backspin and potential for a substantial degree of sidespin if not struck properly, so the ball tends to curve off line to a much greater degree. Not enough backspin occurs to counteract the sidespin. Conversely, a shot hit with a high-lofted wedge has tremendous backspin that will keep the ball traveling straight against the relatively weak sidespin that might be imparted with that club (Figure 12.20).

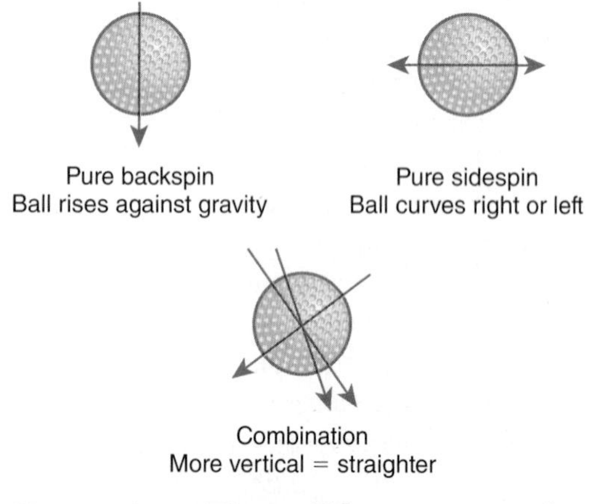

Figure 12.20 **Three different types of spin on a golf ball.**

Beyond the ball, fluid dynamics can also play a part in other equipment design, although the effects are far less important than marketing departments would like consumers to believe. Current-generation drivers already have heads with a modified teardrop shape, so pressure drag is low when you consider the enormous clubface presented to the oncoming airflow. However, some club designers offer interesting new concepts that include flat shafts that cut through the air more easily than traditional round shafts and clubheads that "accelerate themselves" because of aerodynamic forces similar to those of an airplane wing. After reading this textbook, we hope that you will laugh heartily at the last statement. A club cannot accelerate beyond the forces applied to it, and the only forces acting to accelerate a club are muscular forces produced by the human body. The limiting factor in generating extra clubhead velocity is not the negligible effects of enhanced aerodynamic equipment design. A sound understanding of physics will help you reach the best methods for increasing velocity: (1) increase the distance of the clubhead from the axis of rotation or (2) apply more force or the same force over a longer time.

12.10 CHAPTER 10: THE SYSTEM AS A PROJECTILE

Through the point of contact, the biomechanical forces produced by the body during a golf swing serve one purpose: to impart force on a stationary object that results in the object becoming a *projectile.* Muscles throughout the body make up an open complex kinetic chain to produce maximum velocity at the point of contact. The lower body initiates the swing, and that force is transferred through the torso, shoulders, arms, and hands. The hands impart this additive force to the final lever, the golf club itself. A close look at a slow-motion photo as the body nears impact shows the "lag" that produces the whiplike snap at the end of a swing. The hands are close to their position of contact while the clubhead lags a significant angular distance behind (Figure 12.21). In less than 0.0005 seconds (the duration of impact), the ball is transformed from a stationary object in equilibrium to a projectile that is affected only by gravity and fluid forces.

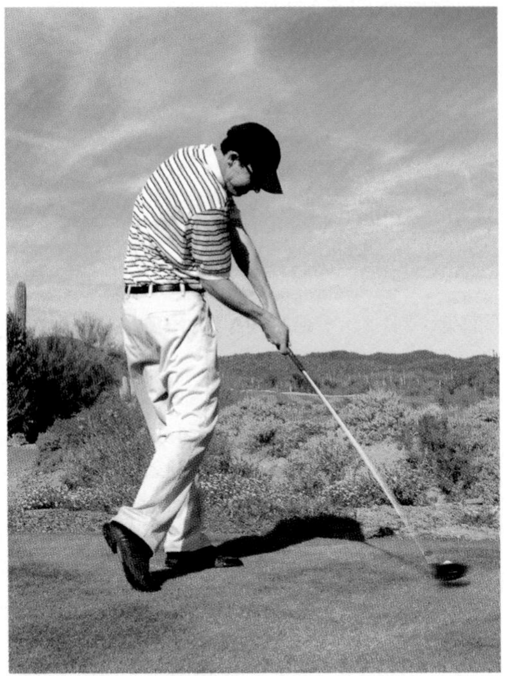

Figure 12.21 **Shaft lag at impact creating a whip effect.**

Assuming that the ball is hit well, it will begin traveling along a *parabolic arc.* In the absence of any external forces, it would travel forever at the same velocity and same initial direction that it traveled when it left the club. The force that creates the parabolic shape is gravity, and the shape of the parabola will change, depending on the launch angle and initial velocity (Figure 12.22). A golf ball rarely scribes a perfect parabolic arc, but it comes close during extremely short chip and pitch shots that don't have enough velocity or time to be affected significantly by air molecules. Most golf shots during a round are affected by air and travel along paths that can vary dramatically. A shot made with an iron may start out along a straight path and then rise against gravity because of the Magnus force of rotation before beginning its descent to the ground as gravitational force exceeds the Magnus force. A shot hit with less backspin will begin scribing a parabola, but as it encounters air resistance it loses velocity and drops more steeply than the angle with which it leaves the clubface.

The *trajectory* of full-swing (normal clubhead velocity) golf shots is primarily affected by the *projection angle,* but it can be affected to a lesser degree by *projection velocity* and can be modified to adjust for *projection height.* With 14 clubs to choose from that vary in loft angle, projection angle is the primary factor in trajectory. Although golfers try to swing each club with the same angular velocity, club sets are designed to progressively decrease in length as the loft angle increases. When the length of the lever is shortened, the clubhead will not reach the same velocity as longer clubs do, and the ball will have progressively less projection velocity.

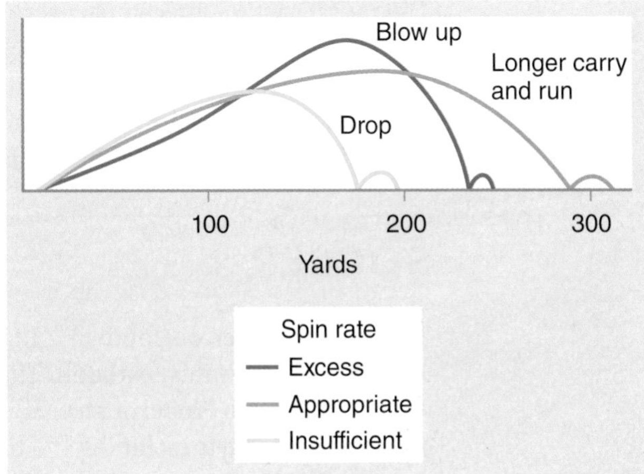

Figure 12.22 **Relationship of spin rate and trajectory.**

Although projection height differs little between a ball on the ground and a ball sitting 2 inches up on a tee, projection height can come into play on a golf course. At times, the contour of a course requires a player to aim for a target well above or below the starting position of the ball. In some cases, this

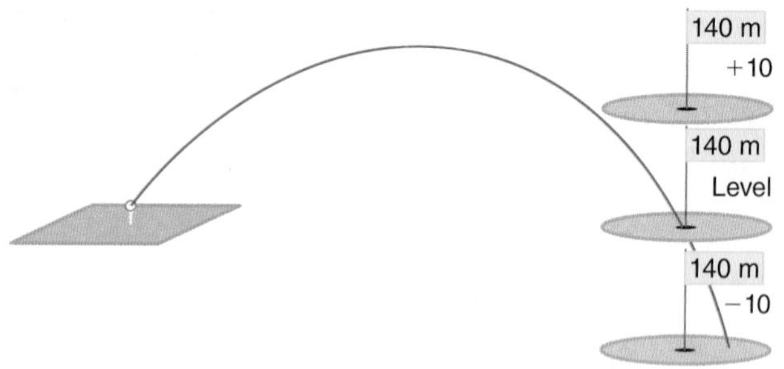

Figure 12.23 **Relative projection height and trajectory.**

change in elevation can be dramatic and require forethought in selecting the proper club.

Let's assume that a player comes across three different par-3 holes during a round. Each hole is 140 m in length and should be easily hit with a good tee shot. The first hole is level from the tee to the green, so the projection height is the same as the impact height. Our golfer chooses a seven iron and hits the middle of the green. The next hole is uphill, with a 10-m change in elevation. Now the projection height is below the impact height, so the *relative projection height* is a negative value. After success on the previous hole, our golfer again takes out his seven iron and strikes it well, but this time the ball lands short of the green. Slightly perplexed, our golfer heads to the last hole, which has a 10-m drop in elevation from the tee to the green, or a *positive relative projection height*. He takes out the seven iron again after coming up short on the previous hole and sends this shot over the back of the green. As you can see in Figure 12.23, physics is behind the results of our golfer's three tee shots. Our golfer knew that his seven iron was good for the 140-m shot but forgot to take into account that with negative relative projection height, the ball would be in the air for less time and therefore would not be able to travel the full distance of a shot on level ground. For the shot with positive relative projection height, the ball was in the air for a longer time, giving it more time to gain linear distance.

At times, a golfer can use different trajectories to his advantage. A shot hit for maximum distance with a tailwind should be projected higher to let the wind push the ball for a longer time. Conversely, a shot hit for maximum distance into a headwind should be kept low, to minimize the destabilizing effects of higher relative velocity. Remember that higher velocity increases lift, and lift forces to the left and right (slices and hooks) will be exacerbated by a headwind.

12.11 CHAPTER 11: BIOMECHANICS OF THE MUSCULOSKELETAL SYSTEM

Finally, let's not forget the system that provides the force to swing a golf club: the musculoskeletal system. Muscle contraction provides the force necessary to both swing the club and control it. The previous section mentions that muscular force developed through the point of impact is used to turn a stationary golf ball into a projectile. The muscles also function well beyond the point of impact, but this muscle action has nothing to with the projectile, which left the clubface milliseconds after impact. The muscle action after impact serves to dissipate the force created during the through swing (i.e., it decelerates the club). The body generates a tremendous amount of torque through the point of impact, and the antagonist muscles work almost as hard to slow the rotation down and keep the body in a balanced position at the finish of the swing. For a thorough look at the muscles responsible for a golf swing, McHardy and Pollard (2005) provide a comprehensive review of literature on this topic.

SOCCER

The sport of soccer, or "football" outside the United States, can also be examined through a lens related to each chapter in this textbook. The human body generates forces that allow movement around the field of play and to propel or receive an object in a variety of ways including kicks, throws, and rebounds off other body parts including the head. Interactions between the body, the ball, and the ground can be influenced by multiple forces, including friction, fluid, pressure, and more. This section is designed to present a selection of information related to biomechanical principles that have been studied in the sport of soccer, including those related to skills, equipment, and injury.

SKILLS

Figure 12.24 **Soccer power kick.**

One of the most studied aspects of soccer is the act of kicking, which makes sense because it is the primary means of moving the ball around the field of play. Of particular interest to biomechanists has been the *power kick* used during goal kicks, corner and free kicks, and penalty shots. The objective is to impart the maximum amount of force into the ball to attain high velocity.

To understand maximal force, several biomechanical concepts come into play. The first is an approach to the ball that develops energy in the form of *inertia*. Muscular forces within the body combine to move the body toward the ball and also move the kicking leg into a position to develop maximal velocity at the kicking foot. The knee of the kicking leg is flexed, which moves the lower leg closer to the *axis of rotation* of two pivot points—the hip and knee (Figure 12.24).

The non-kicking foot (called the plant foot) lands just beside the ball and provides solid point of *friction* from which to start a *kinetic chain*. The momentum of the forward body motion is translated into rotary force as the hips rotate around the plant foot pivot point as the kicking leg lags slightly behind. Once the hips have rotated toward the target the kicking leg acts like a whip, first by flexing at the hip to pull the entire leg forward, then extending the knee to propel the foot forward with maximum velocity.

When this kinetic chain is performed properly an experienced male player will contact the ball for approximately10 milliseconds and generate velocities between 20–30 m per second. During the short time of contact the ball first deforms as it *resists a change in motion* (similar to a golf ball but more elastic), then reforms into its original shape as it leaves the foot. For maximal velocity the contact point should be directly through the center of the ball to impart motion in a linear direction. Any off center contact will result in force applied to a *moment arm* and create rotation of the ball.

There are times when rotation of the ball is desired, and experienced players learn to control this technique for a tactical advantage. The banana kick—so called for the shape of its flight path—inspired a movie called *Bend It Like Beckham*. David Beckham is a highly popular player from England who mastered the art of kicks that curve to the right and left. This technique can be used during corner kicks, penalty kicks, and free kicks to guide a ball around a goalie or wall of defenders who are blocking the goal. In this kick a player intentionally contacts the ball *eccentrically* to generate side spin along with a fairly powerful forward velocity.

As the ball rotates and moves through air it creates a *Magnus effect*. The side of the ball moving against the flow of air creates a higher pressure area and the side moving with airflow creates an area of lower pressure. The ball veers toward the side with lower pressure. This technique is also used to gain a tactical advantage in volleyball, tennis, table tennis, and pitchers in cricket and baseball. The intent in each sport is to make a ball move in a particular direction to make it more difficult for an opponent to play. Experienced players in all of these sports also learn how to read the spin on a ball to defend against it more easily because the curve is relatively consistent. There is another special power kick, similar to a knuckleball in baseball that makes defending almost impossible.

The knuckleball in both baseball and soccer, along with the float serve in volleyball, depend on a lack of rotation to be effective. When thrown or struck directly through the center, each of these balls will move through the air with little or no rotation. While each ball moves through the air the seams and panels on the balls will react to the passing airflow in different ways that are entirely unpredictable, even to the players that set them into motion. Even though knuckleballs are generally moving at submaximal velocities they are often the most difficult for opponents to stop or return because they could move in any direction at any time.

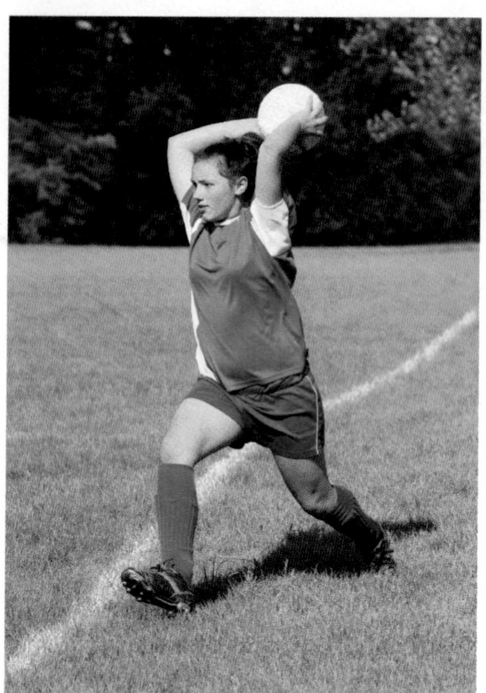

Figure 12.25 **Soccer throw-in.**

Although used at a much lower rate, the skill of throwing can be an offensive weapon in soccer. Experienced players have the ability and strength to throw a ball far enough to put it near a goal from the sideline. One method to increase throwing distance is a specialized technique called a *flip throw* in which a player makes a run-up toward the sideline, then bends forward to touch the ball to the ground. *Momentum* carries the player's body over the ball in a forward flipping motion and the rotation brings the player back to their feet. The body's spin rate is quite fast, and when timed correctly the arms of the player lag behind, pulling the ball off the ground and propelling it overhead into play. This is the same principle that is used in the discus throw where a horizontal spin motion is used to apply force over a longer time. This skill is rarely used because of the critical timing necessary to perform it well.

The throw-in method used most often includes a short run toward the sideline, a hop just before setting both feet on the ground, and force applied by both hands to the ball from a position behind the head. You probably already realize that the run-up creates *inertia*, and strong shoulders and long arms can increase the force applied to the ball, but there are two other factors that have been studied relating to propelling a soccer ball for maximum distance. The first is *projection angle,* or the angle at which the ball is released. Given a consistent velocity, the projection angle that produces maximum distance is 45 degrees. However, Linthorne and Everett (2006) found that soccer players produced maximum distance at projections angles between 30° and 35°. It turns out that humans are not able to apply the same force to a soccer ball at all projection angles because of the constraints of the skill, particularly the rule stating that the ball must be propelled by both hands over the head (Figure 12.25). They found that experienced players can apply significantly more force at 30 degrees, enough so that the ball will travel farthest at this angle.

In a later study, Linthorne and Thomas (2016) examined spin rate on a soccer ball to see if it had any effect on throw-in distance. Similar to a backward-spinning golf ball, a soccer ball should produce some amount of lift—more likely a slower rate of drop—with backward rotation. They found that applying spin did not affect the player's ability to produce high projection velocity and that higher spin rates resulted in a 7% increase in distance for a given projection velocity.

EQUIPMENT AND INJURY

With the rising interest and incidence of brain injuries in American football, there has been an increase in research into the possible negative effects of heading a soccer ball. Chronic traumatic encephalopathy (CTE) has been found in participants of contact sports such as American football and hockey, where concussive forces to the head are quite high. Even with the use of padded helmets to spread force loads, impacts with other hard objects (other helmets, sticks, high-speed pucks, goal frames, and boards) result in high rates of *acceleration* and *deceleration* of the brain within the skull. The brain is cushioned to a degree by fluid, but the intense accelerations of the head during collisions result in the brain contacting the skull. Over time, these collisions cause negative changes in brain structure and function and potentially CTE.

Even though the energy transfer to the head during the act of heading is lower than head impacts in American football and hockey, there is evidence that repeated subconcussive impacts may have long-term deleterious effects on brain function (Spiotta, Bartsch, & Benzel, 2012). However, a review of studies on the effects of heading a soccer ball on brain function reports that evidence of damage is inconclusive (Rodrigues, Lasmar, & Caramelli, 2016). One reason why heading a soccer ball results in lower impact forces is the cushioning effect of the ball itself (Figure 12.26). The materials that balls are made from are relatively soft, and the interior is filled with air to maintain a round shape. Both the material and air are *elastic* in nature and change shape dramatically when a ball comes into contact with the head. This spreads the force over time and area and results in a collision without the intense forces of helmet-to-helmet (almost inelastic) collisions of other sports.

Figure 12.26 **Soccer ball heading.**

Changes in equipment have made a difference in the impact forces to the head, primarily because of the material changes during the past 30 years. Traditional balls were made of leather and could increase in mass nearly 20% because of water absorption on wet fields. Additional mass at similar velocities results in higher impact forces and the potential for more damage to the brain when heading a ball. Newer ball materials are resistant to absorbing water so mass remains consistent even during rainy matches. However, it must be noted that a newer ball that doesn't gain mass from moisture can be kicked at higher velocities, so the overall change in impact force to the head may be insignificant.

Although a review of literature on brain trauma from heading a soccer ball reports inconclusive results, the evidence from many studies does suggest that soccer players are subject to forces that result in negative effects such as decreased cognitive function. One problem in making a strong correlation between heading and brain damage is that soccer players often sustain head contact from objects that are not balls and that are not as forgiving. These include impact with an opponent's head, knee, elbow, or foot; impact with the ground; and occasionally with the goal itself. Without being able to separate these collisions, it is impossible to prove which events cause neurological damage.

Because the act of heading a soccer ball repeatedly cannot be ruled out as a cause of changes in the brain, research is being conducted that may help to reduce the impact forces of this skill. In a "better safe than sorry" strategy in the United States, children under the age of 11 are not allowed to practice or use the skill of heading. To reduce impact forces, it is possible to change the ball and the area of head impact. One way to change the ball is to make it more elastic. Shewchenko and colleagues (2005) found that reductions from standard ball pressure resulted in reduced linear and angular head accelerations, while higher than standard pressure resulted in a modest increase in both measures. A second method of reducing force is to use a lighter ball, a common practice since the advent of smaller balls designed for different age groups of children. In recent years, the development of headgear for soccer players has increased dramatically. The intent of headgear is to spread impact forces from both balls and opponents' heads and reduce forces to a subconcussive level. Research about the effectiveness of headgear is also inconclusive at this time.

Another injury that keeps players off the field for long periods of time is an anterior cruciate ligament (ACL) tear in the knee. This injury can be caused by contact with another player, but it is just as often experienced during the normal course of play from the forces generated while quickly changing speed and direction (cutting, stopping, and accelerating). Research indicates that females experience ACL injuries at a significantly higher rate than males, with soccer players being particularly susceptible (Agel, Arendt, & Bershadsky, 2005).

There have been a variety of reasons presented to explain the higher rates of ACL injuries in female soccer players, and most are physics related. The hamstring muscle group is considered a protector of the ACL, yet is typically trained less and is less strong than the antagonist quadriceps group whose forces the ACL counteracts. If the hamstring group is not as strong initially or fatigues more quickly during a game, then it loses the ability to provide *stabilization* forces within the knee joint.

A second indicator of ACL injury in female players is joint angles because of differences in body shape relative to males. After puberty, females tend to have wider hips than males, which results in varying degrees of varus in the upper leg and valgus (angle away from the midline) in the lower leg.

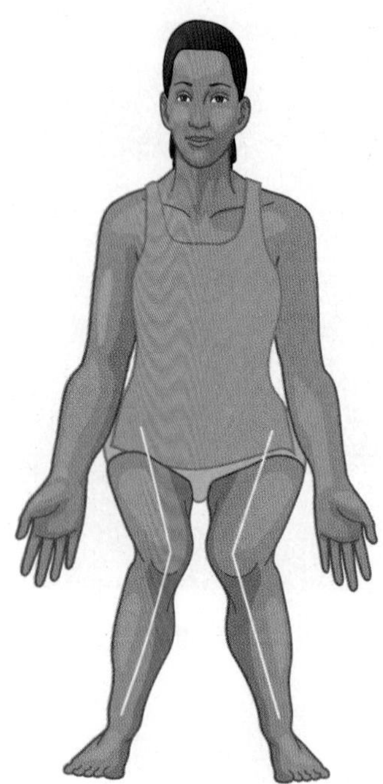

Figure 12.27 **Knee valgus.**

Males tend to have straighter lines between the hip and ankle. When muscle or impact forces (landing or cutting) are generated through the knee, any alignment issues produce *eccentric* forces that put additional stress on the structures within the joint. If the musculature and ligaments cannot maintain joint stability, then something has to give—and it is often the ACL. In a study of knee kinetics between male and female athletes, Schilaty and colleagues (2018) found that females demonstrated significantly higher knee abduction moments during landing movements (Figure 12.27). Alentorn-Geli and colleagues (2009) provide a more comprehensive examination of ACL injury mechanisms in a review of literature and a second review of prevention programs.

A quick search of the Internet will show a wide variety of results for connections between soccer and biomechanics. For a game that has not changed appreciably during the past 50 years in terms of skills, rules, and equipment, there is an expanding foundation of research. From maximizing ball velocity to decreasing the risk of injury, scientists continue to study the forces produced by, and acting on, the human system.

SUMMARY

In some ways, this chapter is a summary. It shows the relationship of soccer and golf skills to the various topics covered in this text. But the chapter should not be seen as simply a soccer or golf analysis. Think of any sport skill. How many connections can you make to the chapters in this text? More than likely you can connect your skill to every chapter. So this chapter is about more than soccer and golf: It is about the human body and its ability to produce forces that allow us to interact with objects and the environment, to play the games that we enjoy, and to move about to accomplish everyday activities that enrich our lives.

REFERENCES AND SUGGESTED READINGS

Agel, J., E. A. Arendt, & B. Bershadsky (2005). Anterior cruciate ligament injury in National Collegiate Athletic Association Basketball and Soccer: A 13-year review. *American Journal of Sports Medicine*, 33(4): 524–531.

Alentorn-Geli, E., G. D. Meyer, H. J. Silvers, G. Samitier, D. Romero, C. Lazaro-Haro, & R. Cugat (2009). Prevention of non-contact anterior cruciate ligament injuries in soccer players. Part 1: Mechanisms of injury and underlying risk factors. *Knee Surgery, Sports Traumatology, Arthroscopy*, 17(7): 705–729.

Hume, P. A., J. Keogh, & D. Reid. 2005. The role of biomechanics in maximizing distance and accuracy of golf shots. *Sports Medicine*, 35(5): 429–449.

Linthorne, N. P., & D. A. Evereett (2006). Release angle for attaining maximum distance in soccer throw-in. *Sport Biomechanics*, 5: 243–260.

Linthorne, N. P., & J. M. Thomas (2016). The effect of ball spin rate on distance achieved in a long soccer throw-in. *Procedia Engineering*, 147: 677–682.

McHardy, A., & H. Pollard. 2005. Muscle activity during the golf swing. *British Journal of Sport Medicine*, 39: 799–804.

Robert, J. R., R. Jones, & S. J. Rothberg. 2001. Measurement of contact time in short duration sports ball impacts: An experimental method and correlation with the perceptions of elite golfers. *Sport Engineering*, 4: 191–203.

Rodrigues, A. C., R. P. Lasmar, & P. Caramelli (2016) Effects of soccer heading on brain structure and function. *Frontiers in Neurology* (online), 7:38.

Schilaty, N. D., N. A. Bates, C. Nagelli, A. J. Krych, & T. E. Hewett (2018). Sex-based differences in knee kinetics with anterior cruciate ligament strain on cadaveric impact simulations. *Orthopedic Journal of Sports Medicine*, 6(3).

Shewchenko, N., C. Withnall, M. Keown, R. Gittens, & J. Dvorak. 2005. Heading in football. Part 3: Effect of ball properties on head response. *British Journal of Sports Medicine*, 39(suppl 1): i33–39.

Spiotta, A. M., A. J. Bartsch, & E. C. Benzel. 2012. Heading in soccer: Dangerous play? *Neurosurgery*, 70: 1–11.

St. Pierre, P. E. 2002. The role of metaphor in expert golf instruction. *Proceedings of the 4th Annual World Scientific Congress of Golf*. St. Andrews, Scotland.

APPENDIX A
KEY EQUATIONS

CHAPTER 2

Body mass index	$BMI = \frac{kg}{m^2}$
Ponderal index	$PI = \frac{kg}{m^3}$
Crural index	$CI = \frac{\text{length of the tibia}}{\text{length of the femur}} \times 100$

CHAPTER 3

Sine $\sin\theta = \frac{\text{side opposite } \theta}{\text{hypotenuse}} = \frac{y}{r}$ or $y = r\sin\theta$ or $r = \frac{y}{\sin\theta}$

Cosine $\cos\theta = \frac{\text{side adjacent to } \theta}{\text{hypotenuse}} = \frac{x}{r}$ or $x = r\cos\theta$ or $r = \frac{x}{\cos\theta}$

Tangent $\tan\theta = \frac{\text{side opposite } \theta}{\text{side adjacent to } \theta} = \frac{y}{x}$

Pythagorean Theorem	$r^2 = x^2 + y^2$ or $r = \sqrt{x^2 + y^2}$
Commutative law of addition	$A + B = B + A$
Associative law of addition	$(A + B) + C = A + (B + C)$
Negative of a vector	$A + (-A) = 0$
Vector subtraction	$A - B = A + (-B)$
Vector cross product	$A \times B = C$
	$A \times B \times (\sin\theta) = C$

CHAPTER 4

Coefficient of static friction	$\mu_s \le \frac{f_s}{F_n}$
Static friction force	$f_s \le \mu_s F_n$
Coefficient of kinetic friction	$\mu_k = \frac{f_k}{F_n}$
Kinetic friction force	$f_k = \mu_k F_n$
Pressure	$P = \frac{F}{A}$
Stress	$\sigma = \frac{F}{A}$
Strain	$\varepsilon = \frac{\Delta l}{l_i}$
Poisson's ratio	$\nu = \frac{\varepsilon_t}{\varepsilon_a}$
Elastic modulus	$E = \frac{\sigma}{\varepsilon}$
Young's modulus	$Y = \frac{F/A}{\Delta l / l_i}$
Coefficient of restitution	$e = \sqrt{h_{rebound}/h_{drop}}$ or $e = \frac{v_{af} - v_{bf}}{v_{bi} - v_{ai}}$ or $-e = \frac{v_{af} - v_{bf}}{v_{ai} - v_{bi}}$

CHAPTER 5

Speed	$s = \frac{l}{\Delta t}$
Linear velocity	$v = \frac{d}{\Delta t}$
Linear acceleration	$a = \frac{\Delta v}{\Delta t}$
Relationship of acceleration and force	$a = \frac{\Sigma F}{m}$ $\Sigma F = ma$

Gravitational force	$F = \frac{Gm_1m_2}{r^2}$
Acceleration caused by gravity	$g = \frac{Gm_2}{r^2}$
Gravitational force	$F = mg$
Linear momentum	$M = mv$
Linear momentum in an *elastic* collision	$(m_1v_{1i} + m_2v_{2i}) = (m_1v_{1f} + m_2v_{2f})$
Coefficient of restitution	$e = \frac{v_{af} - v_{bf}}{v_{bi} - v_{ai}}$
Linear momentum in an *inelastic* collision	$(m_1v_{1i} + m_2v_{2i}) = (m_1 + m_2)v_f$
Final velocity after an inelastic collision	$v_f = \frac{m_1v_{1i} + m_2v_{2i}}{m_1 + m_2}$
Derivation of linear impulse	$\Sigma F = \frac{m\Delta v}{\Delta t}$
	$\Sigma F\,(\Delta t) = m\Delta v$
	or
	$\Sigma F\,(\Delta t) = m(\text{velocity}_{\text{final}} - \text{velocity}_{\text{initial}})$
Linear impulse	$\Sigma F\,(\Delta t) = \Delta M$
	or
	$\Sigma F\,(\Delta t) = \Delta m$
Work	$W = F(d)$
Power	$P = \frac{W}{\Delta t}$
	$P = \frac{F(d)}{\Delta t}$
	$P = Fv$
Gravitational potential energy	$GPE = wh$
	$GPE = mgh$
Strain energy	$SE = ½k\Delta x^2$
Kinetic energy	$KE = ½mv^2$
Conservation of mechanical energy	$(KE + PE)_i = (KE + PE)_f$
Relationship of work and energy	$W = \Delta E = \Delta KE + \Delta PE + \Delta TE$

CHAPTER 6

Torque	$T = F \times r \times \sin\theta$
Theta in radians	$\theta = \frac{l}{r}$
Linear displacement of a point on a rotating segment	$l = \Delta\theta r$
Angular speed	$\delta = \frac{\varphi}{\Delta t}$
Angular velocity	$\omega = \frac{\Delta\theta}{\Delta t}$
Tangential linear velocity	$v_T = \frac{\Delta\theta r}{\Delta t} = \omega r$
Angular acceleration	$\alpha = \frac{\Delta\omega}{\Delta t}$
Tangential linear acceleration	$a_T = \frac{\Delta\omega r}{\Delta t} = \alpha r$
Centripetal acceleration	$a_c = \frac{v_T^2}{r}$
Moment of inertia	$I_a = \Sigma m_i r_i^2$
	$I_a = mk_a^2$
Relationship of acceleration and force	$a = \frac{\Sigma F}{m}$
	$\Sigma F = ma$
Relationship of angular acceleration and torque	$\alpha_a = \frac{\Sigma T_a}{I_a}$
	$\Sigma T_a = I_a\alpha_a$
Relationship of Newton's second law and centripetal force	$\Sigma F = ma_c = m\frac{v_T^2}{r}$
Linear momentum	$M = mv$
Angular momentum	$L_a = I_a\omega_a$
Linear impulse	$\Sigma F(\Delta t) = \Delta M$
Angular impulse	$\Sigma T_a(\Delta t) = \Delta L_a$
Work	$W = F(d)$

Work performed about an axis $W_a = F_T(r\Delta\theta) = T_a(\Delta\theta)$

Power $P = Fv$

Power about an axis $P_a = F_T \dfrac{(r\Delta\theta)}{\Delta t}$

or

$$P_a = T_a \frac{(\Delta\theta)}{\Delta t}$$

or

$$P_a = T_a\omega$$

Kinetic energy $KE = \frac{1}{2}mv^2$

Rotational kinetic energy $KE_R = \frac{1}{2} I_a\omega_a^2$

Conservation of mechanical energy $(KE + PE)_i = (KE + PE)_f$

$$(KE_L + KE_R + PE)_i = (KE_L + KE_R + PE)_f$$

Relationship of work and energy $W = \Delta E = \Delta KE + \Delta PE = \Delta TE$

$$W = \Delta E = \Delta KE_L + \Delta KE_R = \Delta PE + \Delta TE$$

CHAPTER 7

Equations satisfying the condition of static equilibrium in 3-D space:

$$\Sigma F_x = 0 \qquad \Sigma F_y = 0 \qquad \Sigma F_z = 0$$

$$\Sigma T_x = 0 \qquad \Sigma T_y = 0 \qquad \Sigma T_z = 0$$

Equations satisfying the condition of static equilibrium in 2-D space:

$$\Sigma F_x = 0 \qquad \Sigma F_y = 0 \qquad \Sigma T_a = 0$$

Equations satisfying the condition of dynamic equilibrium in 3-D space:

$$\Sigma F_x - ma_x = 0 \qquad \Sigma F_y - ma_y = 0 \qquad \Sigma F_z - ma_z = 0$$

$$\Sigma T_x - I_x\alpha_x = 0 \qquad \Sigma T_y - I_y\alpha_y = 0 \qquad \Sigma T_z - I_z\alpha_z = 0$$

Equations satisfying the condition of dynamic equilibrium in 2-D space:

$$\Sigma F_x - ma_x = 0 \qquad \Sigma F_y - ma_y = 0 \qquad \Sigma T_a - I_a\alpha_a = 0$$

Moment of inertia about a given axis

$$I_a = mk_a^2$$

Angular acceleration about a given axis

$$\alpha_a = \frac{\Sigma T_a}{I_a}$$

CHAPTER 8

Mechanical advantage of a lever system $$MA = \frac{M_M}{M_R}$$

Mechanical advantage of a wheel-and-axle system $$MA = \frac{r_w}{r_a}$$

CHAPTER 9

Density $$\rho = \frac{m}{V}$$

Drag Force $$F_D = ½C_D\rho A_p v^2$$

Lift force $$F_L = ½C_L\rho A_p v^2$$

CHAPTER 10

Linear acceleration $$a = \frac{\Delta v}{\Delta t}$$

First law of uniformly accelerated motion $$v_f = v_i + at$$

Average velocity $$v_a = \frac{v_i + v_f}{2}$$

Linear velocity $$v = \frac{d}{\Delta t}$$

Final position of a projectile $$p_{final} = \frac{p_{initial} + (v_i + v_f)t}{2}$$

Second law of uniformly accelerated motion $$p_{final} = p_{initial} + v_i^t + \frac{at^2}{2} \text{ OR } d = v_i^t$$

Third law of uniformly accelerated motion $$v_f^2 = v_i^2 + 2a(d)$$

Vertical displacement of a projectile $$d = \frac{v_i^2}{2g} \text{ OR } d = \frac{(v_i \sin\theta)^2}{2g}$$

The full parabolic path of a projectile $$d = \frac{(v_i^2 + \sin 2\theta)}{g}$$

CHAPTER 11

Torque $$T = F \times r \times \sin\theta$$

Physiological cross-sectional area $$PCSA = \frac{m \times \cos\theta}{\rho \times l}$$

or

$$PCSA = CSA \times \cos\theta$$

Angular acceleration around a given axis $\alpha_a = \frac{\Sigma T_a}{I_a}$

Moment of inertia around a given axis $I_a = \Sigma m_i r_i^2$

NUMBERED EQUATIONS

$$BMI = \frac{kg}{m^2} \tag{2.1}$$

$$PI = \frac{kg}{m^3} \tag{2.2}$$

$$CI = \frac{\text{length of the tibia}}{\text{length of the femur}} \times 100 \tag{2.3}$$

$$\sin\theta = \frac{\text{side opposite } \theta}{\text{hypotenuse}} = \frac{y}{r} \quad \text{or} \quad y = r\sin\theta \quad \text{or} \quad r = \frac{y}{\sin\theta} \tag{3.1}$$

$$\cos\theta = \frac{\text{side adjacent to } \theta}{\text{hypotenuse}} = \frac{x}{r} \quad \text{or} \quad x = r\cos\theta \quad \text{or} \quad r = \frac{x}{\cos\theta} \tag{3.2}$$

$$\tan\theta = \frac{\text{side opposite } \theta}{\text{side adjacent to } \theta} = \frac{y}{x} \tag{3.3}$$

$$r^2 = x^2 + y^2 \tag{3.4}$$

$$A + B = B + A \tag{3.5}$$

$$(A + B) + C = A + (B + C) \tag{3.6}$$

$$A + (-A) = 0 \tag{3.7}$$

$$A - B = A + (-B) \tag{3.8}$$

$$A \times B = C \tag{3.9}$$

$$A \times B \times (\sin\theta) = C \tag{3.10}$$

$$\mu_s \leq \frac{f_s}{F_n} \tag{4.1}$$

$$f_s \leq \mu_s F_n \tag{4.2}$$

$$\mu_k = \frac{f_k}{F_n} \tag{4.3}$$

$$f_k = \mu_k F_n \tag{4.4}$$

$$P = \frac{F}{A} \tag{4.5}$$

$$\sigma = \frac{F}{A} \tag{4.6}$$

$$\varepsilon = \frac{\Delta l}{l_i} \tag{4.7}$$

$$\nu = \frac{\varepsilon_t}{\varepsilon_a} \tag{4.8}$$

$$E = \frac{\sigma}{\varepsilon} \tag{4.9}$$

$$Y = \frac{F/A}{\Delta l/l_i} \tag{4.10}$$

$$e = \sqrt{h_{rebound}/h_{drop}} \tag{4.11}$$

$$e = \frac{\nu_{af} - \nu_{bf}}{\nu_{bi} - \nu_{ai}} \tag{4.12}$$

$$-e = \frac{\nu_{af} - \nu_{bf}}{\nu_{ai} - \nu_{bi}} \tag{4.13}$$

$$s = \frac{l}{\Delta t} \tag{5.1}$$

$$v = \frac{d}{\Delta t} \tag{5.2}$$

$$a = \frac{\Delta v}{\Delta t} \tag{5.3}$$

$$a = \frac{\Sigma F}{m} \tag{5.4}$$

$$\Sigma F = ma \tag{5.5}$$

$$F = \frac{Gm_1 m_2}{r^2} \tag{5.6}$$

$$g = \frac{Gm_2}{r^2} \tag{5.7}$$

$$F = mg \tag{5.8}$$

$$M = mv \tag{5.9}$$

$$(m_1 v_{1i} + m_2 v_{2i}) = (m_1 v_{1f} + m_2 v_{2f}) \tag{5.10}$$

$$e = \frac{v_{af} - v_{bf}}{v_{bi} - v_{ai}} \tag{4.12}$$

$$(m_1 v_{1i} + m_2 v_{2i}) = (m_1 + m_2) v_f \tag{5.11}$$

$$v_f = \frac{m_1 v_{1i} + m_2 v_{2i}}{m_1 + m_2} \tag{5.12}$$

$$\Sigma F = \frac{m \Delta v}{\Delta t} \tag{5.13}$$

$$\Sigma F\,(\Delta t) = m\Delta v \tag{5.14}$$

or

$$\Sigma F\,(\Delta t) = m(\text{velocity}_{\text{final}} - \text{velocity}_{\text{initial}})$$

$$\Sigma F\,(\Delta t) = \Delta M \tag{5.15}$$

or

$$\Sigma F\,(\Delta t) = \Delta m$$

$$W = F(d) \tag{5.16}$$

$$P = \frac{W}{\Delta t} \tag{5.17}$$

$$P = \frac{F(d)}{\Delta t} \tag{5.18}$$

$$P = Fv \tag{5.19}$$

$$GPE = wh \tag{5.20}$$

$$GPE = mgh \tag{5.21}$$

$$SE = ½k\Delta x^2 \tag{5.22}$$

$$KE = ½mv^2 \tag{5.23}$$

$$(KE + PE)_i = (KE + PE)_f \tag{5.24}$$

$$W = \Delta E = \Delta KE + \Delta PE + \Delta TE \tag{5.25}$$

$$T = F \times r \times \sin\theta \tag{6.1}$$

$$\theta = \frac{l}{r} \tag{6.2}$$

$$l = \Delta\theta r \tag{6.3}$$

$$\delta = \frac{\varphi}{\Delta t} \tag{6.4}$$

$$\omega = \frac{\Delta\theta}{\Delta t} \tag{6.5}$$

$$v_T = \frac{\Delta\theta r}{\Delta t} = \omega r \tag{6.6}$$

$$\alpha = \frac{\Delta\omega}{\Delta t} \tag{6.7}$$

$$a_T = \frac{\Delta\omega r}{\Delta t} = \alpha r \tag{6.8}$$

$$a_c = \frac{v_T^2}{r} \tag{6.9}$$

$$I_a = \Sigma m_i r_i^2 \tag{6.10}$$

$$I_a = mk_a^2 \quad (6.11)$$

$$a = \frac{\Sigma F}{m} \quad (5.4)$$

$$\Sigma F = ma \quad (5.5)$$

$$\alpha_a = \frac{\Sigma T_a}{I_a} \quad (6.12)$$

$$\Sigma T_a = I_a \alpha_a \quad (6.13)$$

$$\Sigma F = ma_c = m\frac{v_T^2}{r} \quad (6.14)$$

$$M = mv \quad (5.9)$$

$$L_a = I_a \omega_a \quad (6.15)$$

$$\Sigma F(\Delta t) = \Delta M \quad (5.15)$$

or

$$\Sigma F(\Delta t) = \Delta m$$

$$\Sigma T_a(\Delta t) = \Delta L_a \quad (6.16)$$

$$W = F(d) \quad (5.16)$$

$$W_a = F_T(r\Delta\theta) = T_a(\Delta\theta) \quad (6.17)$$

$$P = Fv \quad (5.19)$$

$$P_a = F_T \frac{(r\Delta\theta)}{\Delta t} \quad (6.18)$$

$$P_a = T_a \frac{(\Delta\theta)}{\Delta t} \quad (6.19)$$

$$P_a = T_a \omega \quad (6.20)$$

$$KE = \tfrac{1}{2}mv^2 \quad (5.23)$$

$$KE_R = \tfrac{1}{2} I_a \omega_a^2 \quad (6.21)$$

$$(KE + PE)_i = (KE + PE)_f \quad (5.24)$$

$$(KE_L + KE_R + PE)_i = (KE_L + KE_R + PE)_f \quad (6.22)$$

$$W = \Delta E = \Delta KE + \Delta PE = \Delta TE \quad (5.25)$$

$$W = \Delta E = \Delta KE_L + \Delta KE_R = \Delta PE + \Delta TE \quad (6.23)$$

$$I_a = mk_a^2 \quad (6.11)$$

$$\alpha_a = \frac{\Sigma T_a}{I_a} \quad (6.12)$$

$$MA = \frac{M_M}{M_R} \quad (8.1)$$

$$MA = \frac{r_w}{r_a} \quad (8.2)$$

$$\rho = \frac{m}{V} \quad (9.1)$$

$$F_D = ½C_D\rho A_p v^2 \quad (9.2)$$

$$F_L = ½C_L\rho A_p v^2 \quad (9.3)$$

$$a = \frac{\Delta v}{\Delta t} \quad (5.3)$$

$$v_f = v_i + at \quad (10.1)$$

$$v_a = \frac{v_i + v_f}{2} \quad (10.2)$$

$$v = \frac{d}{\Delta t} \quad (5.2)$$

$$p_{final} = p_{initial} + \frac{(v_i + v_f)t}{2} \quad (10.3)$$

$$p_{final} = p_{initial} + v_i^t + \frac{at^2}{2} \text{ OR } d = v_i^t \quad (10.4)$$

$$v_f^2 = v_i^2 + 2a(d) \quad (10.5)$$

$$d = \frac{v_i^2}{2g} \quad \text{OR} \quad d = \frac{(v_i \sin\theta)^2}{2g} \quad (10.6)$$

$$d = \frac{(v_i^2 + \sin 2\theta)}{g} \quad (10.7)$$

$$T = F \times r \times \sin\theta \quad (6.1)$$

$$PCSA = \frac{m \times \cos\theta}{\rho \times l} \quad (11.1)$$

$$PCSA = CSA \times \cos\theta \quad (11.2)$$

$$\alpha_a = \frac{\Sigma T_a}{I_a} \quad (6.12)$$

$$I_a = \Sigma m_i r_i^2 \quad (6.10)$$

UNNUMBERED EQUATIONS

$\Sigma F_x = 0$ $\quad \Sigma F_y = 0$ $\quad \Sigma F_z = 0$

$\Sigma T_x = 0$ $\quad \Sigma T_y = 0$ $\quad \Sigma T_z = 0$

$\Sigma F_x = 0$ $\quad \Sigma F_y = 0$ $\quad \Sigma T_a = 0$

$\Sigma F_x - ma_x = 0$ $\quad \Sigma F_y - ma_y = 0$ $\quad \Sigma F_z - ma_z = 0$

$\Sigma T_x - I_x\alpha_x = 0$ $\quad \Sigma T_y - I_y\alpha_y = 0$ $\quad \Sigma T_z - I_z\alpha_z = 0$

$\Sigma F_x - ma_x = 0$ $\quad \Sigma F_y - ma_y = 0$ $\quad \Sigma T_a - I_a\alpha_a = 0$

APPENDIX B
CONVERSION FACTORS

	Length					
	m	cm	km	in.	ft	mi
1 meter	1	10^2	10^{-3}	39.37	3.281	6.214×10^{-4}
1 centimeter	10^{-2}	1	10^{-5}	0.3937	3.281×10^{-2}	6.214×10^{-6}
1 kilometer	10^3	10^5	1	3.937×10^4	3.281×10^3	0.6214
1 inch	2.540×10^{-2}	2.540	2.540×10^{-5}	1	8.333×10^{-2}	1.578×10^{-5}
1 foot	0.3048	30.48	3.048×10^{-4}	12	1	1.894×10^{-4}
1 mile	1 609	1.609×10^5	1.609	6.336×10^4	5 280	1

	Mass			
	kg	g	slug	u
1 kilogram	1	10^3	6.852×10^{-2}	6.024×10^{26}
1 gram	10^{-3}	1	6.852×10^{-5}	6.024×10^{23}
1 slug	14.59	1.459×10^4	1	8.789×10^{27}
1 atomic mass unit	1.660×10^{-27}	1.660×10^{-24}	1.137×10^{-28}	1

Note: 1 metric ton = 1000 kg.

	Time				
	s	min	h	day	yr
1 second	1	1.667×10^{-2}	2.778×10^{-4}	1.157×10^{-5}	3.169×10^{-8}
1 minute	60	1	1.667×10^{-2}	6.997×10^{-4}	1.901×10^{-6}
1 hour	3 600	60	1	4.167×10^{-2}	1.141×10^{-4}
1 day	8.640×10^4	1 440	24	1	2.738×10^{-5}
1 year	3.156×10^7	5.259×10^5	8.766×10^3	365.2	1

	Speed			
	m/s	cm/s	ft/s	mi/h
1 meter per second	1	10^2	3.281	2.237
1 centimeter per second	10^{-2}	1	3.281×10^{-2}	2.237×10^{-2}
1 foot per second	0.3048	30.48	1	0.681 8
1 mile per hour	0.4470	44.70	1.467	1

Note: 1 mi/min = 60 mi/h = 88 ft/s.

	Force	
	N	lb
1 newton	1	0.224 8
1 pound	4.448	1

	Work, Energy, Heat		
	J	ft · lb	eV
1 joule	1	0.737 6	6.242×10^{18}
1 foot-pound	1.356	1	8.464×10^{18}
1 electron volt	1.602×10^{-19}	1.182×10^{-19}	1
1 calorie	4.186	3.087	2.613×10^{19}
1 British thermal unit	1.055×10^{3}	7.779×10^{2}	6.585×10^{21}
1 kilowatt hour	3.600×10^{6}	2.655×10^{6}	2.247×10^{25}
	cal	Btu	kWh
1 joule	0.2389	9.481×10^{-4}	2.778×10^{-7}
1 foot-pound	0.3239	1.285×10^{-3}	3.766×10^{-7}
1 electron volt	3.827×10^{-20}	1.519×10^{-22}	4.450×10^{-26}
1 calorie	1	3.968×10^{-3}	1.163×10^{-6}
1 British thermal unit	2.520×10^{2}	1	2.930×10^{-4}
1 kilowatt hour	8.601×10^{5}	3.413×10^{2}	1

	Pressure	
	Pa	atm
1 pascal[a]	1	9.869×10^{-6}
1 atmosphere	1.013×10^{5}	1
1 centimeter mercury[2]	1.333×10^{3}	1.316×10^{-2}
1 pound per square inch	6.895×10^{3}	6.805×10^{-2}
1 pound per square foot	47.88	4.725×10^{-4}

	cm Hg	lb/in.^{2}	lb/ft^2
1 pascal[a]	7.501×10^{-4}	1.450×10^{-4}	2.089×10^{-2}
1 atmosphere	76	14.70	2.116×10^{3}
1 centimeter mercury[2]	1	0.1943	27.85
1 pound per square inch	5.171	1	144
1 pound per square foot	3.591×10^{-2}	6.944×10^{-3}	1

[a]At 0 °C and at a location where the free-fall acceleration has its "standard" value, 9.80665 m/s^2.

APPENDIX C

ANSWERS TO ODD-NUMBERED PRACTICE PROBLEMS

CHAPTER 3

1. **a.** Stabilizing component = 150 N; rotary component = 70 N
   b. Stabilizing component = 100 N; rotary component = 180 N
   c. Stabilizing component = 0 N; rotary component = 150 N
   d. Destabilizing component = 100 N; rotary component = 130 N

As the elbow flexes, the stabilizing component becomes smaller and smaller, eventually equaling 0, and then it becomes a destabilizing component. Simultaneously, the rotary component increases in value, ultimately becoming 100% of the muscle force, and then it decreases in size as a destabilizing component develops.

3. **a.** Vertical component = 424.26 N; horizontal component = 424.26 N
   b. Vertical component = 597.72 N; horizontal component = 52.29 N
   c. Vertical component = 253.57 N; horizontal component = 543.78 N

When force is applied at 45°, equal amounts of the initial force are imparted in the vertical and horizontal directions. If force is applied at 85°, the majority of the force is directed vertically rather than horizontally. When force is directed at 25°, most of the force is directed horizontally rather than applied vertically.

CHAPTER 4

1. 160 N
3. 52.22 N/mm^2
5. **a.** 2.43 m
   b. 1.08 m
   c. 0.27 m

CHAPTER 5

1. 10 km/hr/sec
3. 500 kg · m/sec
5. 5 seconds

7. **a.** 900 N · m
   b. 900 N · m/sec
   c. 1,800 N · m/sec
9. Jack: M = 292.5 kg · m/sec KE = 658.125 J
   Jill: M = 292.5 kg · m/sec KE = 570.375 J

Jack will jump higher. He has greater kinetic energy to cause board deformation but the same resistance to board rebound.

CHAPTER 6

1. **a.** −84.2 N · m Clockwise rotation
   b. 0 N · m No rotation will occur
3. Angular velocity = 6 rad/sec Linear velocity = 6 m/sec
5. **a.** 1,000 kg · m^2
   b. 2,000 kg · m^2
   c. 4,000 kg · m^2
7. **a.** 1,666.67 N · m
   b. 3,333.33 N · m
   c. 6,666.67 N · m
9. 13,000 N · m
11. **c.** 50,000 J

CHAPTER 7

3. −180 N
5. −24.79° or +335.21°

CHAPTER 8

1. **a.** 0.6666 or 1.5
   b. 267 N
3. **a.** 12.5 N · m
   b. 500 N

CHAPTER 9

1. **a.** Density = 1.06 g/ml and 0.9 g/ml
   b. Specific gravity = 1.06 and 0.9
   c. Muscle and fat
3. 1,470 N and 1,860.47 N

CHAPTER 10

1. Horizontal component: 30° = 12.99 m/sec 60° = 7.5 m/sec
   Vertical component: 30° = 7.5 m/sec 60° = 12.99 m/sec
3. 40 m/sec
5. **a.** 14.13 m below release
   b. 16.65 m/sec

GLOSSARY

A

Abduction Motion in a frontal plane and around an anteroposterior axis that moves the segment away from the anatomical position.

Acceleration A change in the state of motion of the system caused by an applied force; a change in magnitude or direction of the velocity vector with respect to time.

Actin A contractile protein that serves as the major structural component of the thin filaments in muscle fibers.

Action force The initially applied force.

Action potential An electrical motor nerve impulse.

Active insufficiency Inability of a multiarticular muscle to generate sufficient force throughout all degrees of freedom.

Adapted movement Movement patterns that emerge because of compensation for changes to the physical body.

Adapted Physical Education The process of modifying equipment or the environment or both to successfully teach movement activities to all populations.

Adduction Frontal plane motion that returns the segment to the anatomical position.

Agonists The muscles that are responsible for causing a particular motion.

Anatomical cross-sectional area The cross-sectional area of a muscle measured perpendicular to the longitudinal axis of the whole muscle at its widest point.

Anatomical force couple Two muscles co-contracted in opposite directions in an attempt to accomplish one action, prevent an unwanted motion, or both.

Anatomical position Reference position defined by standing erect with all joints extended, feet parallel, palms facing forward, and fingers together.

Angle of attack The presentation (orientation) of the object relative to the fluid flow.

Angle of pull The angle at which the muscle force acts relative to a given axis or lever.

Angular acceleration A change in magnitude or direction (or both) of the angular velocity vector with respect to time.

Angular displacement The change in angular position of a segment or any point on the rotating segment.

Angular impulse The interval of torque application.

Angular momentum A system's quantity of angular motion.

Angular position The location of the given point, defined by its distance (radius) *r* from the origin, and the angle (the Greek letter *theta*) between the chosen reference axis and the line formed by connecting the given point to the origin.

Angular speed The scalar rate of angular motion.

Angular velocity The vector rate of angular motion, or rate of angular motion in a specific direction.

Antagonist The muscle that performs the joint motion opposite that of the agonist.

Anteroposterior axis Axis that runs horizontally from front to back and is perpendicular to the frontal plane of motion.

Anthropometry The discipline that studies measurements of the body and body segments in terms of height, weight, volume, length, breadth, proportion, inertia, and other properties related to shape, mass, and mass distribution.

Archimedes' principle A body submerged in a fluid will be buoyed up by a force that is equal in magnitude to the weight of the displaced water.

Associative law of addition The sum of three or more vectors is independent of the grouping of the vectors for addition: $(A + B) + C = A + (B + C)$.

Average acceleration The rate of change in velocity divided by the entire interval over which it changed.

Average angular acceleration The change in angular velocity divided by the entire interval over which it changed.

Average speed or average velocity The average rate of motion.

Axis of rotation The point around which a body rotates.

The ability to control the current state of equi- , implies conscious effort and coordination.

of support The entire area bounded by a perimeter ned by the points of contact with a resistive surface ıat provides a counterforce.

Bending Occurs when two off-axis forces are applied such that tension stress is caused on one side of the system and compression stress occurs on the other side.

Bernoulli's principle A fluid moving with high relative flow velocity exerts less pressure than a fluid moving with a low relative flow velocity.

Biomechanics Physics (mechanics) of motion exhibited or produced by biological systems.

Bipennate fibers Two groups of pennate fibers, with the fibers in one group parallel to each other but oblique to the other group and to the tendon.

Body mass index (BMI) Ratio of body mass to height used to describe stature.

Boundary layer The layer of fluid in immediate contact with the system in motion.

Brittle (or stiff) The quality of a material that can withstand high stress but fails with relatively low strain.

Buoyant force The vertical, upward-directed force acting on an object that is submerged in a fluid.

C

Cardinal plane Plane that passes directly through the midline of the body.

Cartesian or rectangular coordinate system A frame of reference defined by an origin and two or three orthogonal axes, each passing through the origin and defining one spatial dimension.

Cat rotation or cat twist A maneuver in which upper- and lower-body moments of inertia are manipulated to rotate while airborne.

Center of buoyancy The center of gravity of the volume of water displaced by an object; the point at which the force of buoyancy seems to be concentrated.

Center of gravity The point at which the force of gravity seems to be concentrated.

Center of mass The point at which all of a system's mass is concentrated.

Center of mass The point that represents the average location of a system's mass.

Center of percussion Spot on an implement that, when struck, will cause the implement to experience pure rotation around the suspension point with no vibration.

Center of pressure The point at which the sum of the weight distribution forces passes through the body.

Centric force A force having a line of action that passes through the center of mass of an object.

Centripetal acceleration An acceleration caused by change in *direction* of the velocity vector; a center-seeking acceleration.

Centripetal force Any force that causes a system to exhibit circular motion.

Closed kinetic chain Kinetic chain in which motion at one link is possible only with cooperative movement at other links.

Closed skill Skill performed under standardized or predictable environmental conditions.

Closed-loop Movements that can change during performance because of sensory feedback.

Closing velocity The sum of the absolute values of velocities.

Coaching Study of principles and methods of instructing athletes.

Coefficient of drag A unitless quantity that expresses the capability of a particular object to create drag.

Coefficient of friction An experimentally measured dimensionless value representing the proportion of friction force resisting sliding motion of the object to the normal force holding the objects together.

Coefficient of lift A unitless quantity that expresses the capability of a particular object to create lift.

Coefficient of restitution (or coefficient of elasticity) A parameter observed after reformation that indicates the ability of an object to return to its original shape after deformation.

Colinear vectors Vectors that have the same line of action.

Commutative law of addition The sum of vectors added together is independent of the order of addition: $A + B = B + A$.

Compensatory movements Adaptations at normal kinetic chain links as a result of abnormal motion at another link.

Complex kinetic chain Kinetic chain in which a segment is linked to more than two other segments.

Component approach Qualitative analysis approach that views the body in component sections, with each section progressing through more refined steps toward mature movement patterns.

Component vectors The individual vectors that are representative of each of the multiple effects that one vector represents.

Composite approach Qualitative analysis approach that views the whole body as a system that progresses through stages or phases as it refines movement patterns.

Compression stress The result of two forces being applied to the system in opposite directions toward each other.

Concentric muscle action The muscle torque generated is greater than the load torque, and the muscle is therefore allowed to shorten.

Contact forces Forces that are the result of physical contact between two bodies.

Continuous skill Cycles of motion performed repeatedly with no well-defined beginning or end points.

Creep The term for the property of experiencing increasing strain (continued deformation) under a constant stress.

Crural index Ratio of the length of the tibia to the length of the femur.

D

Degrees of freedom (DOF) The number of independent ways in which a system can move.

Density The relationship of an object's mass relative to the space it occupies.

Destabilizing component The horizontal (parallel) component of a muscle force vector directed away from the joint, representing the amount of force that would tend to destabilize the joint.

Deviation Frontal plane motion around an anteroposterior axis at the wrist.

Direction Sense or way in which a force is applied; represented by the tip of a vector.

Directional (or planar) stability A state in which a body is more stable in the direction that the base is widened.

Discrete skill Motion that has a definite beginning and end point.

Disruptive force A force that acts to move an object away from its original position.

Drag force The parallel component of dynamic fluid force that acts in the opposite direction of system motion with respect to the fluid; tends to resist motion of the system through the fluid.

Ductile (or pliant) The quality of a material that fails at low stress but can withstand a large strain.

Ductility The force per unit area required to deform a material and is represented by the steepness (slope) of the stress–strain curve.

Dynamic equilibrium A situation in which a system is in motion, but it is experiencing no change in velocity or direction.

Dynamic fluid force The equal and oppositely directed force of the fluid particles in reaction to the applied force of the system moving through the fluid.

Dynamic muscle action Movement of the body segment that occurs during generation of muscle tension.

Dynamics Branch of mechanics concerned with objects in a state of accelerated or changing motion.

E

Eccentric force A force having a line of action that does not pass through the center of mass of an object.

Eccentric muscle action The resultant muscle torque is less than the load torque and the muscle lengthens during generation of tension.

Ectomorphic Somatotype described as being linear and relatively thin for height.

Effort arm The moment arm for the motive force.

Elastic collision A collision in which two objects collide and bounce off one another.

Elastic modulus An expression of the relationship of stress and strain for a given material and type of deformation.

Elastic potential energy (or strain energy) The potential energy stored in a deformed object.

Elastic region The linear portion of any given stress–strain curve; a material will return to its original shape if the tensile stress is removed within this range.

Electromagnetic force Force that occurs between electric charges.

Endomorphic Somatotype described as being rounder and relatively heavy for height.

Endomysium The layer of connective tissue surrounding each muscle fiber.

End-plate potential Depolarization of the motor end plate caused by acetylcholine binding to receptor sites.

Epimysium A deep layer of connective tissue that surrounds the entire muscle; sometimes called *deep fascia.*

Equilibrium State in which the system is either at rest or moving with a constant velocity.

Ergonomics Discipline concerned with human–machine interaction.

Eversion Frontal plane motion around an anteroposterior axis such that the sole of the foot rotates outward or laterally.

Excitation–contraction coupling The series of events in which the action potential makes its way to the muscle cell and initiates muscle contraction.

Exercise physiology The study of physiology under conditions in which physical work has caused disrupted homeostasis.

Extension Returns a segment to the anatomical position in a sagittal plane around a mediolateral axis and is described as increasing the angle at the joint.

External forces Forces that interact with the system from the outside.

F

Failure strength The stress at which a material actually breaks or ruptures.

Fascia The layer of connective tissue that holds the muscle belly in position and separates it from other muscles.

Fasciculus or fascicle A bundle of the muscle fibers held together by perimysium.

Field force A force that acts at a distance without making contact with the object that it affects.

Fine movement Motion that is precise and generally controlled by small muscle groups.

First-class lever system A lever system in which the fulcrum is between the motive and resistive forces.

First law of uniformly accelerated motion A projectile's final velocity is related to its initial velocity and the value of constant acceleration.

Flexion Segmental motion in a sagittal plane, around a mediolateral axis and away from the anatomical position.

Foil Any object that can generate lift while moving through a fluid.

Force arm (moment arm or lever arm) The perpendicular distance between the axis of rotation and the line of action of the applied force.

Force couple Two forces that are not colinear but have parallel lines of action and act in opposite directions.

Force Something that possesses the capability to cause a change in motion or shape of the system.

Force–velocity relationship The greater the load against which a muscle must contract, the lower the velocity of that contraction.

Form drag Drag resulting from a relatively higher pressure on the leading edge of the object compared to its trailing edge.

Free-body diagram A simplified representation of a system free of the movement environment.

Friction The force that resists the sliding of two objects in contact.

Frontal plane Vertical plane dividing the body into anterior and posterior halves.

Fulcrum The axis around which a lever rotates.

Functional anatomy Study of the specific functions of individual structures that make up an organism.

Functional kinetic chain A complex kinetic chain in which some links are involved in open chain motion and others are engaged in closed chain motion.

G

Glide Sliding or pure translation in which a point on one surface glides or skids over many points of an opposing surface.

Global reference frame A fixed frame of reference that allows the location of any point to be specified with respect to a defined origin.

Gravitational force Force that exists between bodies of mass.

Gravitational potential energy The potential energy that an object has based on its position relative to a reference surface (often Earth).

Gross movement Motion that is the result of large muscle group activity and requires little precision.

Ground reaction force An equal and oppositely directed normal force from Earth.

Gyroscopic stability Stability of an object caused by its rotation.

H

Hemiparesis Partial paralysis of one side of the body.

High-rise syndrome A term used in veterinary medicine in reference to animals sustaining traumatic injuries as a result of falling from substantial height (usually two or more stories).

Hoop stress Stress caused by compressive forces applied to intervertebral discs.

Hydrostatic weighing A process in which a person's weight in the water is compared to his or her weight outside of water in order to estimate body volume.

I

Impact height The height of projectile impact or landing.

Impending motion The moment immediately before an object begins to slide because of the application of a force.

Inelastic collision A collision in which two objects collide and stick together.

Inertia A body's resistance to having its state of motion changed by application of a force.

Insertion The point of muscle attachment that tends to be on a relatively more movable segment (or distal location).

Instantaneous acceleration The rate of change in velocity at one specific instant in time.

Instantaneous angular acceleration The change in angular velocity at one specific instant in time.

Instantaneous angular speed (velocity) The rate of angular motion at one given instant in time.

Instantaneous speed Speed at one given instant in time.

Instantaneous velocity The rate of motion at one given instant in time.

Internal forces Forces that act within the defined system.

Inversion Frontal plane motion around an anteroposterior axis such that the sole of the foot rotates inward or medially.

J

Joint excursion Total range of motion at a joint.

K

Kinematics Study or description of the spatial and temporal characteristics of motion without regard to the causative forces.

Kinesiology Multidisciplinary study of human motion, including the anatomical, biomechanical, cultural, motor, pedagogical, physiological, psychological, and sociological aspects of motion.

Kinetic chain System of linked rigid bodies subject to force application.

Kinetic energy The energy (potential to perform work) associated with motion.

Kinetic friction Friction that exists when two surfaces are already sliding relative to each other.

Kinetics Study of forces that inhibit, cause, facilitate, or modify motion of a body.

L

Laminar flow Fluid flow in which the molecules of fluid in the boundary layer follow the same smooth unbroken path completely around the object.

Latent period A period of a few milliseconds following the occurrence of the action potential during which the fiber has not yet begun to generate tension.

Law of conservation of mechanical energy In the absence of externally applied forces other than gravity, the total mechanical energy of a system remains constant.

Laws of uniformly accelerated motion Several observations made by Galileo that relate acceleration, displacement, time, and velocity in the condition of constant gravitational acceleration.

Length–tension relationship Relationship between muscle fiber length before stimulation and subsequent tetanic tension development.

Lever system A machine consisting of a rigid or semirigid object that is capable of rotating around an axis.

Lift force The perpendicular component of dynamic fluid force that can act in any direction that is perpendicular to system motion with respect to the fluid; tends to change the direction of system motion through the fluid.

Lift–drag ratio Ratio that expresses the relationship of the amount of lift to the amount of drag at a given angle of attack.

Line of action An imaginary line extending infinitely along a vector through both the tip and tail, representing the path along which the vector would travel if moved forward or backward.

Line of gravity A vertical line representing gravity that passes though a system's center of mass.

Line of pull Resultant line of action of a muscle force vector.

Linear displacement The change in linear position of the system in a straight line.

Linear distance traveled The total length of the path traveled by the system of interest.

Linear impulse The interval during which force is applied.

Linear or translational equilibrium A state achieved when the net of the external forces acting on a system is equal to zero ($\Sigma F = 0$).

Linear stability The resistance of a body to having its linear equilibrium disrupted.

Local reference frame A frame of reference attached to and moving with the system of interest.

Longitudinal or fusiform fibers Muscle fibers that run somewhat parallel to the muscle's longitudinal axis.

Lordosis A posture in which there is exaggerated anterior lumbar curvature accompanied by excessive anterior tilt of the pelvis.

M

Machine An apparatus or system that uses the combined action of several parts in order to apply mechanical force.

Magnitude Size or amount of an applied force; represented by the length of a vector.

Magnus effect Phenomenon of a rotating object experiencing a curved path because of the Magnus force.

Magnus force Lift force resulting from rotation.

Mass (m) The quantity of matter of which a body is composed.

Maximal oxygen uptake (VO_2 max) Maximal capability to utilize oxygen in metabolic processes to make adenosine triphosphate (ATP).

Mechanical advantage (MA) The relationship of the motive force required to overcome a given resistive force with use of a lever system.

Mechanics Branch of physics concerned with the effect of forces and energy on the motion of bodies.

Mediolateral axis Axis that passes horizontally side to side and is perpendicular to the sagittal plane.

Mesomorphic Somatotype described as being muscular, strong, and possessing weight relatively proportional to height.

Mobility The total degrees of freedom of a kinetic chain.

Moment The tendency of one or more applied forces to rotate an object around an axis but not necessarily change the angular momentum of the object (a static force).

Moment of inertia The mathematical relationship of mass, mass distribution, and rotational inertia.

Momentum A system's quantity of motion.

Motion A change in position with respect to both spatial and temporal frames of reference.

Motive force A force that tends to cause a change in motion in the form of increased velocity or change in direction of the system.

Motive torque An eccentrically applied force that attempts to rotate a lever in a given direction around a fulcrum.

Motor control Mechanisms used by the nervous system to control and coordinate the movements of the musculoskeletal system.

Motor development Progression of motor control throughout the lifespan because of maturation.

Motor end plate A shallow depression or pocket in the sarcolemma located at the neuromuscular junction.

Motor learning Relatively permanent changes in proficiency of motor control through experience or practice.

Motor unit A given motor neuron and the group of muscle fibers that it innervates.

Multiarticular (multi- or compound-joint) muscle A muscle that crosses two or more articulations.

Multipennate Arrangement in which two or more bipennate muscles converge on one tendon to form a single muscle.

Muscle belly The whole muscle between the tendons.

Myofibrils A bundle of specialized threadlike structures making up the muscle fiber.

Myosin A contractile protein that serves as the major structural component of the thick filaments in muscle fibers.

N

Negative of a vector Another vector that, when added to the first, gives a sum equal to zero.

Negatively buoyant Quality of an object that weighs more than the displaced water, causing a net downward force that sinks the object.

Neuromuscular cleft A small gap or cleft between the terminal button and the motor end plate.

Neuromuscular junction The site of interdigitation (physical interaction) of the motor neuron and the muscle fiber.

Neutral equilibrium A state in which there is no tendency to fall in one direction or the other.

Neutrally buoyant Quality of an object that has the same weight as the displaced water, in which case there is no net force and the object neither floats, or sinks.

Normal force Force that acts downward on one surface and upward on another.

O

Occupational biomechanics Specialized area of biomechanics focused on human mechanics in work environments.

Occupational therapy Field focused on helping people to improve their ability to carry out activities of daily living and self-care tasks (i.e., "occupations") after an injury, disability, or other health condition.

Open kinetic chain Kinetic chain in which motion can occur at one link in the chain without cooperative motion at other links.

Open skill Skill performed in a changing or unpredictable environment.

Open-loop Movements occurring too rapidly to be modified by sensory feedback.

Orientation The alignment or inclination of the vector in relation to the cardinal directions.

Origin (O) A stationary point in the environment from which all measurements are made.

Origin The point of muscle attachment that tends to be on a relatively immovable or proximal location.

P

Parabola A curved symmetrical shape.

Pascal's law Pressure applied to a fluid is transmitted undiminished to every point of a fluid and to the walls of the container.

Passive insufficiency Inability of a multiarticular muscle to lengthen to a degree that allows full range of motion at all of the joints it crosses simultaneously.

Peak rate of motion Maximum rate of motion achieved.

Pedagogy Study of principles and methods of instruction.

Pennate fibers Fibers that are arranged obliquely to a tendon that runs along the longitudinal axis of the muscle.

Pennation angle The angle at which fibers in a pennate arrangement are oriented relative to the longitudinal axis.

Perimysium The layer of connective tissue surrounding each fasciculus.

Periosteum The outermost layer of the bone into which the tendons are inserted.

Physical therapy Field dedicated to evaluating and treating movement abnormalities.

Physiological cross-sectional area (PCSA) A cross section taken perpendicular to all of the fibers in the muscle and therefore approximately equal to the sum of all of the cross-sectional areas of the individual fibers in that muscle.

Pitch Rotation of a system around the z-axis.

Plane polar coordinates Coordinates (r, θ) representing the location of a point within a polar coordinate system.

Plastic region The nonlinear response of the material after the yield point; some degree of deformation will persist after removal of the stress.

Point of application The point or location at which a system receives an applied force; usually defined by the tail of the vector.

Poisson's ratio An expression of the tendency of a material to exhibit transverse (lateral) strain simultaneously with axial (longitudinal) strain.

Polar coordinate system A coordinate system in which the location of the given point is defined by its distance (radius) r from the origin, and by the angle θ between the chosen reference axis and the line formed by connecting the given point to the origin.

Ponderal index Ratio used to describe stature.

Positively buoyant Quality of an object that weighs less than the water it displaces, causing a net upward force that buoys the object up.

Postural sway Oscillations of the center of mass within the base of support during normal erect standing posture because of variations in muscle tension.

Potential energy (or stored energy) The capacity of an object to perform work based on its position, deformation, or configuration.

Power stroke The tilt of the myosin head toward the center of the sarcomere.

Power The amount of mechanical work performed in a given time interval.

Pressure The magnitude of applied force acting over a given area.

Prime movers The agonist muscles that are most directly involved in causing the motion.

Principle of conservation of linear momentum In the absence of a net externally applied force, the total momentum of a system that comprises multiple bodies remains constant in time.

Principal moment of inertia Term used when referring to the moment of inertia with respect to one of the principle axes of rotation.

Principle of conservation of angular momentum In the absence of a net externally applied torque, the total angular momentum of a system comprised of multiple bodies remains constant in time.

Principle of work and energy The work performed by externally applied forces other than gravity causes a change in energy of the object acted on.

Projectile A body whose motion is subject only to the forces of gravity and fluid resistance.

Projection angle The angle at which the object is projected; also called *angle of attack.*

Projection height The height of projectile release.

Projection velocity The initial velocity or velocity at release of a projectile.

Pronation Rotation at the radioulnar joint around a superoinferior axis that causes the palm to turn toward the body (medially and posteriorly).

Proprioceptors Responsible for providing information concerning joint position, velocity, muscle tension and rate of change in length, and other important information related to maintaining equilibrium.

Pulley system A machine consisting of an object that acts as a wheel around which a flexible cord or cable is pulled.

Pythagorean theorem An expression of the relationships (ratios) between the lengths of the sides of a right triangle: $r^2 = x^2 + y^2$ or $r = \sqrt{x^2 + y^2}$.

Q

Q-angle (quadriceps angle) The angle formed by the longitudinal axes of the femur and tibia, which approximates the resultant line of action of the quadriceps muscles.

R

Radial expansion The bulging of an intervertebral disc in accordance with Poisson's ratio.

Radian The ratio of the length of a circular arc to the length of the circle's radius; 1 rad 57.3°.

Radiate muscles A longitudinal muscle with an exaggerated taper to the tendon.

Radius of gyration The distance that represents how far the mass of a rigid body would be from an axis of rotation if its mass were concentrated at one point.

Radius of rotation The distance from the axis of rotation to a given point on the rotating segment.

Range The total horizontal displacement of a projectile.

Reaction force The simultaneous equal counterforce acting in the opposite direction to the action force.

Reciprocal inhibition or innervation Neural inhibition of the antagonistic motor units during actions of the agonist muscle.

Relative motion The motion of one object with respect to a reference object.

Relative projection height The mathematical difference between the height of projectile release (projection height) and the height of projectile impact or landing (impact height).

Relaxation period Period following tension generation in which the muscle fiber returns to the resting state.

Repeated discrete Cycles of motion that are seemingly continuous but require a recovery phase between propulsive actions.

Resistance arm The moment arm for the resistive force.

Resistive force A force that tends to prevent changes in motion by other external forces or decrease the velocity of a system that is already in motion.

Resistive torque An eccentrically applied force that attempts to rotate a lever in a direction opposite that of the motive torque.

Restorative force A force that acts to return an object to its original position.

Resultant A single vector representative of the sum of multiple vectors.

Rigor mortis Muscle stiffness caused by inability to detach myosin crossbridges because of the lack of available adenosine triphosphate (ATP) after death.

Roll A combination of rotation and translation in which each point on a surface contacts a unique location on the other surface.

Roll Rotation of a system around the *x*-axis.

Rolling friction Friction that exists whenever one surface is rolling over another but not sliding across it.

Rotary component The vertical or perpendicular component of a muscle force vector representing the amount of force that would tend to cause joint rotation.

Rotation Motion around a fixed axis and therefore in a circular path; angular motion.

Rotational equilibrium A state achieved when the net of the external torques (moments) acting upon a system is equal to zero ($\Sigma T = 0$).

Rotational inertia (angular inertia) The resistance of an object to having its state of angular motion changed.

Rotational kinetic energy The energy (potential to perform work) associated with angular motion.

Rotational power The amount of angular mechanical work performed during a given interval.

Rotational stability The resistance of a body to having its rotational or angular equilibrium disrupted.

Rotational work Angular displacement of an object around an axis caused by the application of a torque; a transfer of energy.

Rotator cuff The muscle group comprising the supraspinatus, infraspinatus, teres minor, and subscapularis.

S

Sagittal plane Vertical plane dividing the body into right and left halves.

Sarcolemma The cell membrane of a muscle fiber.

Sarcomeres Segments of the myofibrils that are strung end to end and serve as the functional unit of the muscle.

Sarcoplasm The cytoplasm inside of a muscle fiber.

Sarcoplasmic reticulum A system of channels that surround and run parallel to the myofibrils and serve as a storage site for calcium.

Scalar quantity A quantity that can be fully specified simply with a single numerical magnitude of appropriate units.

Second-class lever system A lever system in which the resistive force is between the fulcrum and the motive force.

Second law of uniformly accelerated motion The final position of a projectile is related to its initial velocity and acceleration.

Serial or simple kinetic chain Kinetic chain in which each segment participates in no more than two linkages.

Serial skill Movement that comprises a series of discrete motions.

Shear stress Occurs from application of two parallel forces that tend to simultaneously displace one part of a system in a direction opposite another part of the system.

Sliding filament theory of muscle contraction Theory of muscle force generation in which tension is generated by the interaction of myosin with actin leading to myofibrillar translation.

Somatotyping System of body-type description based on weight and muscularity relative to height.

Spatial Relating to, or with respect to, the three-dimensional world.

Specific gravity The ratio of the density of a given substance to the density of water.

Speed The scalar rate of motion.

Spin Occurs if all points on one articulating surface come into contact with one point on another articulating surface.

Sports medicine Field dedicated to the prevention, immediate treatment, and rehabilitation of injuries that occur during sports participation.

Stability The resistance of an object to having its equilibrium disturbed.

Stabilizing component The horizontal (parallel) component of a muscle force vector directed toward the joint; represents the amount of force that would tend to stabilize the joint.

Stable equilibrium A state in which a large force or torque is required to disrupt a body's current position.

Stall angle The angle of attack at which drag can become greater than lift and flight is no longer possible.

Static equilibrium A situation in which the system is in linear and rotational equilibrium ($\Sigma F = 0$ and $\Sigma T = 0$) *and* possesses zero linear or rotational velocity.

Static friction Friction that exists when two contacting surfaces are not currently sliding relative to each other but do possess the *potential* for movement.

Static muscle action Tension is generated within the muscle, but the bone to which the muscle force is applied does not move.

Statically indeterminate A situation in which there is more than one unknown value in an equation, causing there to be an infinite number of solutions.

Statics Branch of mechanics concerned with objects in a state of equilibrium (at rest or in a constant state of motion).

Strain The resulting magnitude of deformation as a result of an applied stress.

Strap muscles Longitudinal muscles that have less prominent tendons and therefore taper less.

Streamline The path followed by the molecules of a layer of fluid.

Streamlining Designing the shape of an object such that changes in contour are gradual from front to back, which allows the molecules in the streamlines to the follow the same smooth, unbroken path, thereby reducing drag.

Strength The maximum stress or strain that a material can withstand without permanent deformation.

Stress relaxation The eventual decrease in stress that will occur as fluid is no longer exuded.

Stress The external force acting to deform a material.

Stretch–shorten cycle The phenomenon of enhanced force by generating eccentric tension immediately before the concentric action.

Strong nuclear force Force that occurs between subatomic particles, preventing the nucleus of an atom from exploding due to protons producing a repulsive electric force.

Superoinferior axis Axis that passes up and down and is perpendicular to the transverse plane.

Supination Transverse plane motion that returns the radioulnar joint toward the anatomical position (palm moves anteriorly).

Surface drag The result of friction between the surface of the body and the fluid through which it is moving.

Synergists The agonist muscles that are indirectly involved and play a more assistive role in bringing about the motion (i.e., stabilizers or neutralizers).

System Any structure or organization of related structures whose state of motion is of analytical interest.

T

Tangential linear acceleration The linear acceleration of a point on a rotating segment.

Tangential linear velocity The linear velocity of a point on a rotating segment.

Teaching cues Single words or short phrases that identify critical elements of a skill.

Temporal Relating to, or with, respect to time.

Tendon An extremely strong connective tissue formed by the convergence of epimysium, perimysium, and endomysium.

Tensile strength The maximum stretch that a material can withstand without rupture.

Tension generation period Period of time following the latent period during which the muscle fiber actually generates tension.

Tension stress Occurs when two forces are applied to a system in opposite directions away from each other.

Terminal cisternae Enlarged portions of the sarcoplasmic reticulum running perpendicular to the myofibrils.

Terminal velocity Velocity at which drag force is equal to the weight of the falling object and the object will no longer accelerate.

Tetanus The point at which the maximum number of crossbridges are participating in the muscle action and the muscle fiber reaches peak tension development.

Thick filaments Large polymers (assemblies) of the protein myosin.

Thin filaments Polymers (assemblies) of the protein actin.

Third-class lever system A lever system in which the motive force is between the fulcrum and the resistive force.

Third law of uniformly accelerated motion The final velocity of a projectile is related to its acceleration and displacement.

Torque (moment of force or moment) The tendency of an eccentric force to rotate an object around an axis.

Torsion Caused by two forces being applied in such a way that part of the system is rotated around its longitudinal axis in a direction opposite to the rotation of another part of the system.

Toughness The total energy required to cause material failure.

Trajectory The term for the flight path of a projectile; more specifically, the flight path of the center of gravity of a projectile.

Translation Motion along one of the *x*-, *y*-, or *z*-axes; linear motion.

Transverse plane Horizontal plane dividing the body into superior and inferior halves.

Transverse tubules (T-tubules) A continuation of the sarcolemma into the interior of the fiber; channels that run perpendicular to the myofibrils, connecting the sarcoplasmic reticulum to pores in the sarcolemma.

Triad of the reticulum One T-tubule passing between two terminal cisternae.

Tropomyosin A molecule is made up of long thin threadlike proteins that spiral around the chains of actin molecules, blocking the myosin binding sites.

Troponin A protein complex spaced at intervals along the thin filament that in the relaxed state acts to maintain tropomyosin in a position to block the myosin binding sites of actin.

Turbulent flow A situation of irregular fluid flow in which the streamlines may be broken; the molecules of the boundary layer do not follow the same smooth, unbroken path completely around the object.

Twitch A weak, brief action of the fiber caused by the arrival of a single action potential.

Twitch summation The force of a second twitch is *summed* with the force remaining from the first twitch.

U

Ultimate strength The maximal stress that a material can withstand before failing.

Uniarticular (single-joint) muscle A muscle that crosses only a single articulation.

Unipennate fibers One set of pennate fibers parallel to each other but oblique to the tendon.

Unstable equilibrium A state in which little force or torque is required to disrupt a body's current position.

V

Valgus Alignment of a joint in the frontal plane such that the angle formed by the proximal and distal segments opens laterally; the distal segment is displaced laterally.

Varus Alignment of a joint in the frontal plane such that the angle formed by the proximal and distal segments opens medially; the distal segment is displaced medially.

Vector chain A form of vector analysis in which vectors are arranged tip to tail.

Vector composition Process by which two or more vectors are summed to determine a single resultant vector.

Vector equality Property that two vectors are considered equal ($A = B$) as long as they possess the same magnitude and orientation.

Vector quantity A quantity that can only be fully specified with a magnitude of appropriate units and a precise direction.

Vector resolution The process by which individual directional component vectors of a single vector are determined.

Velocity The vector rate of motion, or rate of motion in a specific direction.

Viscoelastic The quality of a material whose deformation is affected by both the rate of loading and the length of time it is subjected to a constant load.

Viscosity A measure of the resistance of two adjacent layers of fluid molecules moving relative to each other.

Volume The space occupied by an object's mass.

W

Waist-to-hip ratio Ratio of waist circumference to hip circumference often associated with disease risk.

Weak nuclear force Force that is a product of some radioactive decay processes.

Weight Measure of the force with which gravity pulls on an object's mass.

Wheel-and-axle system A machine consisting of an object acting as a wheel that is capable of rotating around a shaft.

Work Displacement of an object caused by the application of a force; a transfer of energy.

Y

Yaw Rotation of a system around the y-axis.

Yield point The point at which an applied stress can lead to permanent deformation.

Yield strength Stress at the yield point of a material.

Young's modulus Term for the elastic modulus; specifically refers to a condition of tension stress.

Z

Z-disk Structural protein connecting each sarcomere to the next.

INDEX

Note: Page numbers followed by *f*, or *t*, indicate material in figures, or tables, respectively

A

B

C

D

G

H

I

J

K

M

S

U

V

CHAPTER 1

DISCIPLINES

NAME

SECTION DATE

1.1 List some other courses within your major to which you think biomechanics might apply. In what way do you think they are related?

1.2 List three occupations in which you are potentially interested and how biomechanics might be used in that occupation.

CHAPTER 2

THE SYSTEM

NAME

SECTION DATE

2.1 Categorize yourself as an endomorph, mesomorph, ectomorph, or a combination of two of the classifications. What physical traits do you possess that you feel cause you to fit into that category?

2.2 Even if you have never played a sport, list two sports for which you think you are best physically suited and why.

2.3 For the next few days, observe other people walking. Pay attention to their individual body segments as they move (arm swing, hip and leg movements, foot strike, etc.). Describe the most interesting movement pattern that you observed. How is that person similar and different from you in the way they walk?

CHAPTER 3

VECTOR ANALYSIS

NAME

SECTION DATE

3.1 Estimate your own Q-angle: Using a marker, draw a dot on your anterior superior iliac spine of the pelvis, the midpoint of the patella, and your tibial tuberosity. Stand with your knees extended and quadriceps muscle relaxed and the medial borders of your feet touching. Now have someone take a photo of you in the described position with those three points visible. Print the photo and then draw a line from the ASIS to the midpoint of the patella and a second line from the tibial tuberosity through the midpoint of the patella. The angle formed by the intersection of the two lines is your Q-angle. Using a protractor, measure your Q-angle and then compare it to normal values.

3.2 Resolve each of the following muscle force vectors (1 cm = 100 N). One series represents the biceps brachii while the second series represents the brachioradialis. Based on your resolutions, describe the primary functions of the two muscles.

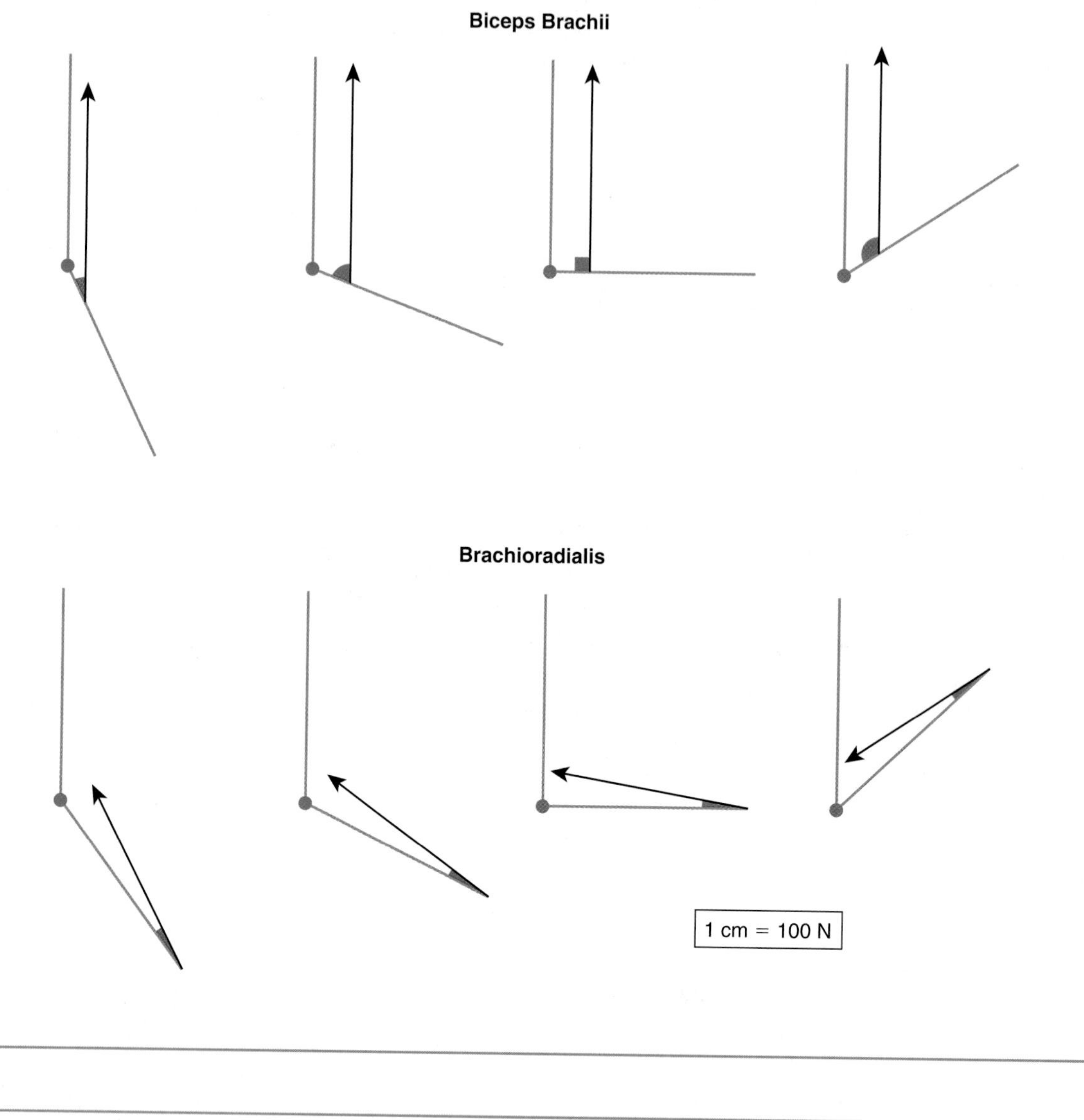

3.3 Compose the following vectors to find the resultant (1 cm = 25 N).

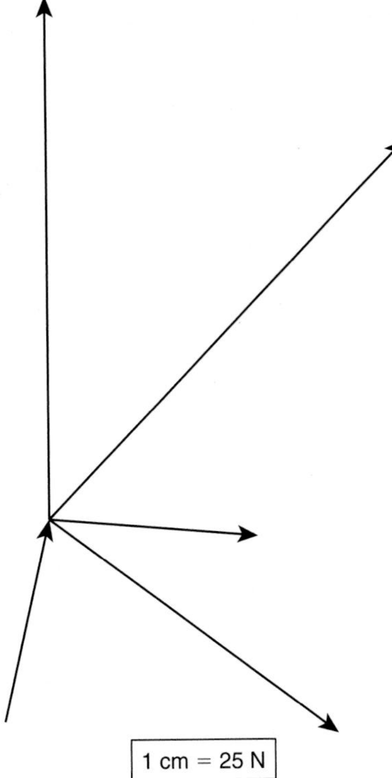

3.4 Compose the following vectors to find the resultant (1 cm = 30 N).

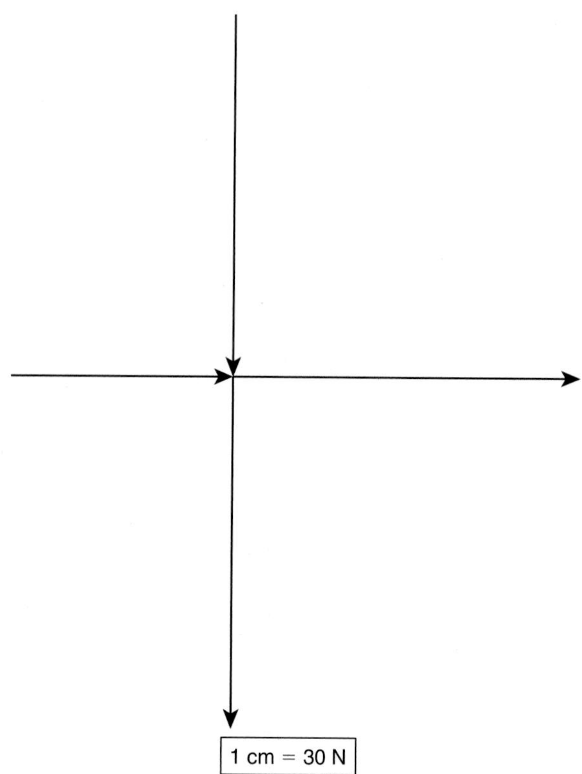

CHAPTER 4

FORCES

NAME

SECTION DATE

4.1 Give the equation for determining how much centripetal force is needed to constrain an object in a circular path. According to the equation, what would be the most effective way for a person riding a bike on a wet surface to keep from slipping? Why? What would be the next best thing? Why?

What is the cause of the "centrifugal force" in the preceding example? What is the cause of the centripetal force?

What would happen in the above example if "centrifugal force" were greater than centripetal force needed?

4.2 You are driving your car and suddenly must activate the brakes to avoid a collision. Which of Newton's laws best explains why you feel your body and head continue to move forward as your car stops? Why?

If you apply the same force to two objects of different mass, which object will be affected the most? Which of Newton's laws explains your answer?

4.3 A wooden box with a student trapped in it weighs 1,100 N and is on a wooden floor. The coefficient of static friction for wood on wood is 0.50. How much applied force is required to slide the box? The coefficient of kinetic friction for wood on wood is 0.20. How much force will be required once the box is sliding?

Explain why the coefficient of kinetic friction is lower than the coefficient of static friction.

4.4 A person steps on a rock with a downward force of 800 N applied over a cross-sectional area of 2 cm^2. How much pressure was experienced?

4.5 In your own words, Archimedes principle basically explains:

If a person sinks in water, what was the cause and what is it called?

Buoyant force is most directly related to what?

A person's ability to float is most directly related to what?

4.6 What value represents how much a material will strain with an applied force? Give the equation. Explain why this value would be different for tendon, ligament, and bone.

4.7 A viscoelastic material is one that possesses what properties?

4.8 The basilisk (Jesus Christ) lizard is known for its ability to run on top of water. This ability would be to the product of what two concepts from the chapter?

Explain why barefoot water skiing must be performed at a much higher speed than with the use of skis.

CHAPTER 5

LINEAR MOTION

NAME

SECTION

DATE

5.1 What is the momentum of a 60-kg cheetah running at 10 m/s?

5.2 A 150-kg object is traveling 2 m/sec. What is the magnitude of force that would have to be applied to stop the object in 3 seconds?

5.3 A 100-kg object is traveling 3 m/sec. How long will it take to stop the object by applying a force of 50 N?

5.4 Using an equation, explain how air bags and crumple zones in cars protect passengers.

5.5 How much mechanical work is done by a person who pushes a 200-N object 4 m across the floor?

5.6 A person having a mass of 70 kg standing at the edge of a diving board 5 m above the water possesses how much gravitational potential energy?

5.7 How much kinetic energy is present in a 5-kg ball dropping with a velocity of 10 m/s?

5.8 Player A (mass of 80 kg and velocity of 5 m/sec) collides with player B (mass of 100 kg and velocity of 0 m/sec). What is their final velocity after the collision?

5.9 If velocity is constant, then what is the acceleration?

5.10 Explain terminal velocity and how it is achieved.

5.11 A barbell weighing 500 N is displaced 2 meters. How much work was performed? If the displacement takes place in 0.5 sec, what is the power?

5.12 A runner moving at a velocity of 6 m/s increases the velocity in a 2 sec time interval to a rate of 10 m/s. What is the rate of acceleration?

5.13 Jack has a mass of 100 kg and hits a mini-trampoline at 4.0 m/sec. Jill has a mass of 64 kg and hits the diving board at 5.0 m/sec.

a. Which has the greater resistance to having their motion changed? Why?

b. Which has the greater ability to perform work on the minitrampoline? Why?

c. Which will jump higher? Why?

5.14 How much energy will be stored in the sole of a shoe that has a spring constant of 10^7 N/m and deforms 2.7 mm on contact?

CHAPTER 6

ANGULAR MOTION

NAME

SECTION DATE

6.1 When running, we bend our knee before swinging our leg forward. Why does bending the knee make running more economical? Explain using two equations.

6.2 What happens to angular velocity when a diver moves from a layout position to a tuck position? Explain using at least two equations.

6.3 When a gymnast lands from a vault ,what are the sources of difficulty and danger?

6.4 List at least four everyday items that work according to the right-hand–thumb rule.

6.5 For the following system, calculate the individual torque magnitudes and the net torque and indicate the direction of system motion (1 cm = 10 N).

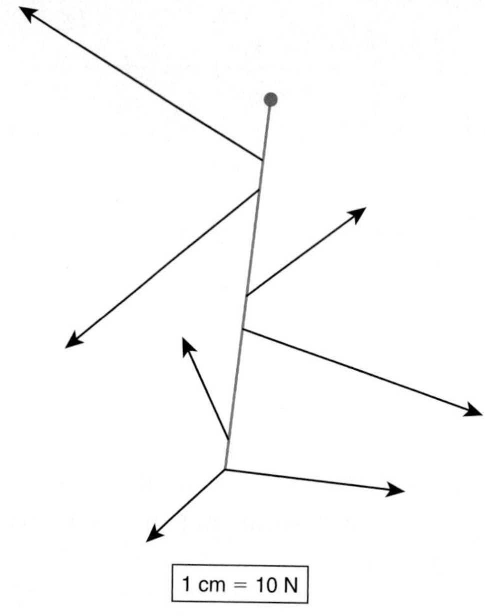

6.6 For the following system, calculate the individual torque magnitudes and the net torque and indicate the direction of system motion (1 cm = 10 N).

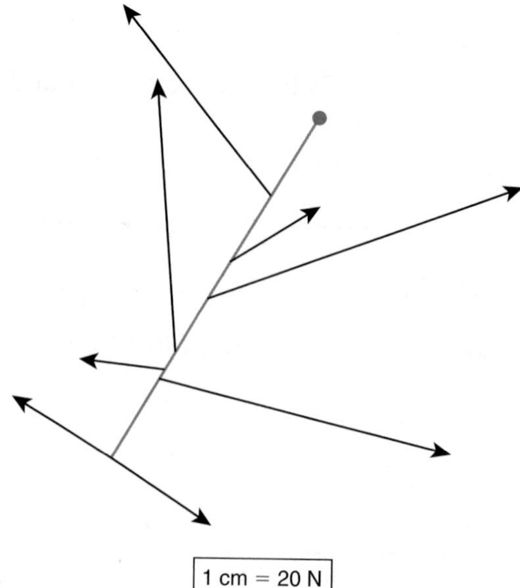

6.7 For the following system, calculate the muscle force necessary to hold the dumbbell in a static position. Show all work (1 cm = 12 N).

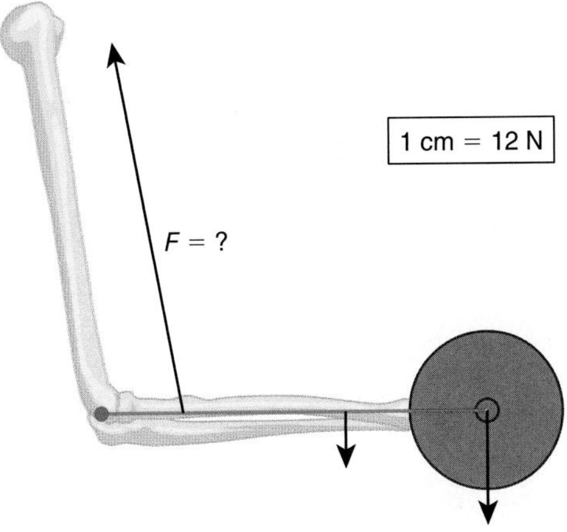

CHAPTER 7

BALANCE AND STABILITY

NAME

SECTION DATE

7.1 Explain the use of long poles by tightrope walkers.

7.2 If you stand with your heels and back against a wall, why is it difficult to bend over while keeping your heels in place?

7.3 If you are trying to push a large box across the floor, why do we often push down and forward rather than squat down and push straight forward?

7.4 Explain the usefulness of the defensive sprawl used by wrestlers.

7.5 Explain why a ball placed on a sloped surface will roll.

____

____

____

7.6 If you place a domino and a die on a board and slowly begin to raise the incline of the board, which will fall over first? Why?

____

____

____

7.7 While lifting heavy objects it is recommended to keep your back flat and bend at the knees. Why not just bend over at the waist?

____

____

____

7.8 Describe who would be the best anchor in a tug-of-war and what stance would be most effective. Why?

____

____

____

7.9 Describe five stances in sports and the reason for using those particular stances.

____

____

____

CHAPTER 8

MACHINES

NAME

SECTION DATE

8.1 List five machines that you use almost every day. What type are they and what advantage do they provide?

8.2 What advantage is gained by having a screwdriver with a large handle?

8.3 To which class of levers do a catapult, baseball bat, leaf rake, and door belong? Draw free-body diagrams to explain your answers.

8.4 What is the purpose of a cam in a variable-resistance exercise machine?

8.5 The moment arm for the biceps muscle is 0.02 m, and the moment arm for the resistive force held in the hand is 0.45 m. Calculate the mechanical advantage.

8.6 What advantage(s) does a person with longer limbs have in throwing and striking movements? How might this be a disadvantage in resistance-training movements?

8.7 In what situations are you most likely to use a pulley in your daily life?

CHAPTER 9

FLUID MOTION

NAME

SECTION DATE

9.1 If an object is submerged in water, why is the net force in the horizontal direction equal to zero?

9.2 When your ceiling fan rotates clockwise, which way does air move? Why? Which way does air move when the fan is turning counterclockwise? Why?

9.3 Explain the flight path of a boomerang.

9.4 How would you determine the density of an irregularly shaped object?

9.5 Once a ski jumper takes flight, why does he lean forward and keep his arms at his sides?

9.6 How can ants survive falls from great heights relative to their size?

9.7 Explain the advantages to a swimmer having a long torso relative to her leg length?

9.8 What is the purpose of drafting in racing?

9.9 Why do golf balls have dimples?

9.10 How can seams affect the flight path of a baseball?

9.11 Will a boat float higher in seawater or freshwater? Why?

9.12 Explain the purpose of a spoiler on a race car.